CONSTITUTIVE MODELS FOR RUBBER XIII

Constitutive Models for Rubber XIII is a comprehensive compilation of the oral and poster contributions to the XIII European Conference on Constitutive Models for Rubbers (İstanbul, Türkiye, 26-28 June 2024). The XIII edition again brought together researchers from the industry and the academia working in the field of elastomer technology and science to discuss the most recent advancement in the following topics:

- Constitutive models
- Micro-structural investigations
- Experimental methods and characterization
- Numerical methods
- Fatigue and fracture
- Aging
- Industrial applications
- Smart elastomer materials: applications and modelling
- Recyclable elastomer systems design and modelling

Including 53 contributions from authors from around the world, this book aims at professionals and academics interested in recent advances in elastomer technology and science.

PROCEEDINGS OF THE 13TH EUROPEAN CONFERENCE ON CONSTITUTIVE MODELS FOR RUBBER (ECCMR 2024), JUNE 26–28, 2024, İSTANBUL, TÜRKIYE

Constitutive Models for Rubber XIII

Editor

Hüsnü Dal

Middle East Technical University, Ankara, Türkiye

CRC Press
Taylor & Francis Group
Boca Raton London New York

CRC Press is an imprint of the
Taylor & Francis Group, an **informa** business
A BALKEMA BOOK

Designed cover image: Hüsnü Dal

First published 2025
by CRC Press/Balkema
4 Park Square, Milton Park, Abingdon, Oxon, OX14 4RN

and by CRC Press/Balkema
2385 NW Executive Center Drive, Suite 320, Boca Raton FL 33431

CRC Press/Balkema is an imprint of the Taylor & Francis Group, an informa business

Typeset by Integra Software Services Pvt. Ltd., Pondicherry, India

British Library Cataloguing-in-Publication Data
A catalogue record for this book is available from the British Library

Library of Congress Cataloging-in-Publication Data
A catalog record has been requested for this book

ISBN: 978-1-032-85138-9 (hbk)
ISBN: 978-1-032-85164-8 (pbk)
ISBN: 978-1-003-51688-0 (ebk)
DOI: 10.1201/9781003516880

Constitutive Models for Rubber XIII – Dal (Ed.)
© 2025 The Editor, ISBN 978-1-032-85138-9

Table of contents

Aging

Industrial applications

Smart elastomer materials applications and modelling

Recyclable elastomer systems design and modelling

© 2025 The Editor, ISBN 978-1-032-85138-9

Foreword

This book collects papers from invited talks, oral presentations, and poster sessions presented at the 13th European Conference on Constitutive Models for Rubber (ECCMR 2024). The biennial conference, in its 13th edition, was held in 2024 in Istanbul, at a venue situated along the Bosphorus, where Europe meets Asia. Since its inauguration in 1999, ECCMR has become the major international forum where, international delegates from around the world, representing diverse disciplines from both academia and industry, had the opportunity to come together and exchange ideas and discuss the latest trends and developments on elastomer mechanics.

The current volume covers a wide range of topics, including constitutive models, microstructural investigations, experimental methods and characterization, numerical methods, fatigue and fracture, aging, rheology and processing, industrial applications, smart elastomer materials, and recyclable elastomer systems.

The editor would like to thank the local organization team members Mustafa Öksüz, Kemal Açıkgöz, Alp Kağan Açan, Selçuk Başdemir, Bülent Efe Tanış, Recep Durna and Yasin Gargı. Special thanks go to Claude Foubert from VERTCOM for his technical support. I would like to express my sincere gratitude to the members of the international scientific committee — Claudia Marano, James Busfield, Michael Kaliske, Leif Kari, William Mars, Yann Marco, and Michael Johlitz — for their long-term efforts in maintaining this conference series as the central hub of the rubber research community Last but not least, I would like to thank the plenary speakers — Laura de Lorenzis, Michael Kaliske, and Mokarram Hossain — for their invaluable contributions to the success of the event.

Hüsnü Dal
September 2024

Constitutive Models for Rubber XIII – Dal (Ed.)
© 2025 The Editor, ISBN 978-1-032-85138-9

Foreword (Volume 1)

The extraordinary stress–strain behaviour of rubber has presented an opportunity for inventive engineers and a challenge for scientists since the mid-nineteenth century, and continues to do so today. Major branches of theory, such as the statistical theory of rubber elasticity and finite strain elasticity theory, have been spawned by the properties of rubber. Until recently, however, the theoretical framework for large deformations found little application among rubber engineers because the mathematics rapidly becomes intractable for all but the simplest components. The advent of affordable and powerful computers has changed all this, and brought the challenge of rubber to new sets of people – software engineers and desktop, as opposed to empirical, designers.

The development of the statistical theory of rubber elasticity in the 1940s, of finite strain elasticity theory in the 1950s, and of convenient forms for the strain energy function in the 1970s, all focused on modelling the elastic characteristics of rubber. Although much literature has appeared in recent years following this theme, the Physics of Rubber Elasticity by L.R.G. Treloar (3rd Edition, Clarendon Press, Oxford, 1975) and the proceedings of a Discussion on Rubber Elasticity (Proc.Roy.Soc.London, 1976, A351, No. 1666, 295–406) remain very valuable reviews.

The treatment of rubber as a 'hyperelastic' material – that is, a material modelled by a strain-energy function for finite strains – was implemented into finite strain element analysis in the 1980s and is now widely available in commercial software packages.

However, only a few engineering elastomers – such as unfilled natural rubber and some grades of polyurethane – really conform to the "hyperelastic" ideal. Most other engineering elastomers incorporate "reinforcing" fillers, needed to confer adequate strength properties and also to improve processing characteristics and to enable adjustment of hardness over a wide range. The stress–strain characteristics of such filled elastomers depart significantly from elasticity. While ways of thinking about these departures – such as "dynamic-to-static ratio" of rubber springs – may have satisfied a previous generation of design engineers, there is now an opportunity to apply more sophisticated models.

One major current challenge is thus to model those aspects of the inelastic behaviour that are relevant to engineers, and to do this in such a way that the models are implementable in finite element analysis.

Although potentially the involvement of representatives of several disciplines should facilitate progress, this is only the case if they talk to each other. In practice, software engineers might rely on the literature and on desktop designers as sources of information about rubber, and fail to achieve as good a balance of understanding as they could if they listened also to experimental rubber scientists and empirical designers. Applied mathematicians might develop phenomenological models which address issues of secondary interest to designers, or which misrepresent important aspects of the experimentally observed behaviour. Experimentalists might develop models without reference to the existing framework of continuum mechanics, resulting in internal inconsistencies and difficulty in implementation in software packages. The First European Conference on Constitutive Models for Rubber sprang from the idea of providing a forum for multi-disciplinary discussion, seeking to bring the fragmented strands of recent research together.

Within the UK a start has been made in this direction – through a workshop on Deformation Modelling for Solid Polymers (Oxford University, 1997) and a seminar on Finite

Element Analysis of Elastomers (Institution of Mechanical Engineers, London, 1997). The proceedings of the latter are available as a publication of the same name (Professional Engineering Publications, London, 1999). Similarly, in Germany a workshop of Finite Element Analysis – Basics and Future Trends was organised by the Deutsche Institute für Kautschuk Technology (Hannover, 1998). The interest in these essentially national meetings suggested that further cross-fertilisation should be stimulated by providing a European forum for discussion.

The contributions to this Proceedings cover a wide range of subjects. Consistent with the analysis given above, relatively few authors chose to present hyperelastic models for rubber; however, readers interested in this topic will find ample references to earlier work. Several contributions address inelastic effects associated with filled elastomers – such as Mullins' effect and quasi-static hysteresis. For others – most obviously in processing uncured rubber – the interest is in modelling viscoelasticity. In addition to stress–strain behaviour, work is presented on frictional contact and on mechanical failure. Looking at the applications side, computational techniques are addressed and applied to a diverse range of components, including tyres, earthquake isolation bearings and intervertebral discs. Overall, the authors have achieved progress in a wide range of areas – including experimental results, theory and practical utility. They raise many questions as well, as one might expect from the first forum of this kind.

We would like to thank our colleagues on the Scientific Committee (R.W. Ogden, Chairman; D. Besdo, R.de Borst, K.N.G. Fuller, H.A. Mang, H. Menderez, G. Meschke and H. Rothert) and all the authors who have worked with us to produce this book.

Al Dorfmann
Alan H. Muhr
Vienna/Hertford, June 1999

Constitutive Models for Rubber XIII – Dal (Ed.)
© 2025 The Editor, ISBN 978-1-032-85138-9

Sponsors

Organized by

ORTA DOĞU TEKNİK ÜNİVERSİTESİ
MIDDLE EAST TECHNICAL UNIVERSITY

ODTÜ BİLTİR Merkezi
METU CAD/CAM Robotics Center

Executive sponsor

MELOS®
Active Zinc Oxide

Platinium sponsors

COESFELD MATERIALTEST

metravib material testing

petlas

Endurica Get Durability Right®

bayrak®

numesys NUMERICAL SYSTEMS

Gold sponsors

SUMITOMO RUBBER INDUSTRIES

Hankook driving emotion

Keynote lectures

Constitutive Models for Rubber XIII – Dal (Ed.)
© 2025 The Author(s), ISBN 978-1-032-85138-9

Automated discovery of hyperelastic material models

L. De Lorenzis
Department of Mechanical and Process Engineering, ETH Zürich, Switzerland

ABSTRACT: This paper reviews the recent research on automated discovery of hyperelastic material models carried out by the group of the author and collaborators. Overall, this research proposes a shift in paradigm, moving away from the classical *calibration*of the unknown parameters in an a priori chosen material model, towards the new task of model *discovery*, i.e. simultaneous selection of the most appropriate model and calibration of its unknown parameters, by leveraging tools such as sparse, Bayesian or symbolic regression. Detailed aspects including choice of the modeling space, type of data, objective function, optimization method and physics constraints are discussed in a modular fashion. Avenues for future research are highlighted.

1 INTRODUCTION

Material models are an essential component of any engineering simulation aiming at the prediction of mechanical phenomena. Thus, their accurate determination is of great relevance across a wide range of engineering applications. The task of identifying a material model is described as follows by Lemaitre and Chaboche's classical textbook (Lemaitre and Chaboche 1994) (italic added by the author):

"Identification (of material models) can be defined as all the work which consists in *specifying the functions which appear in the model*and in *finding the numerical values of the coefficients*which define the functions for each material. This represents a difficult task which does not follow any rigorous rules and in which experience and 'art of model construction' play a major role in steering between theory and experiment".

This sentence highlights the fundamental separation of the material model identification task in two sub-tasks: *model selection*, i.e. specifying the functional form of the mathematical expression defining the material model, and *parameter identification*, i.e., determining the values of the unknown material constants. In conventional material modeling, the first sub-task is carried out by the modeler based on experience, and possibly on the availability of models from the literature known to be accurate for the material at hand. The second sub-task, for a given material model, is carried out automatically through an optimization procedure which minimizes some objective function. This conventional way of proceeding relies on a trial-and-error iterative loop. If, upon manual selection of a model, the subsequent optimization does not deliver any parameter values for which the selected model well fits the data, the modeler needs to retry with a different expression, and the procedure continues until a good combination of model and parameters is found. The tedious and time-consuming nature of this "manual" exercise, which Lemaitre and Chaboche recognize not to "follow any rigorous rules", sharply contrasts with the high level of automation reached by laboratory testing procedures.

Recently, the author with her group and collaborators proposed a shift in paradigm, moving away from the conventional procedure outlined above, towards the new task of fully automated model *discovery*, i.e. automated and simultaneous selection of the most appropriate model and calibration of its unknown parameters, by leveraging tools such as sparse, Bayesian or symbolic regression, based on experimental data. The objectives of this paper are to provide a review of the related research and to point out open issues and promising avenues for further work, in order to ultimately stimulate advancements in this exciting area. Following the first paper (Flaschel et al. 2021), relevant work has been carried out by other research groups. However, due to space limitations, this paper reviews only the contributions by the group of the author and collaborators. The review is intentionally modular, in the sense that different aspects are looked at separately since they can possibly be combined in different ways than demonstrated so far in the overviewed papers. While here the focus lies on *hyperelastic isotropic*materials, the vast majority of the concepts and methods can be extended to other material types, see e.g. (Flaschel et al. 2022, Flaschel et al. 2023, Marino et al. 2023).

DOI: 10.1201/9781003516880-1

2 MODELING SPACE

The first choice to be made concerns the modeling space in which the target model is to be searched for (which we also refer to as ansatz for the material model). This choice is partially related with the decision on the strategy to find the "best" model (based on data and criteria to be defined later). In this section, we review a few different assumptions for the modeling space, which will later correspond to recently investigated strategies for model selection, generation, or encoding. We note preliminarily that, since the material is assumed to be hyperelastic, our final goal is to determine its underlying strain energy density $W(\boldsymbol{F})$, where $\boldsymbol{F}$ is the deformation gradient, from which the first Piola-Kirchhoff stress tensor can be derived as $\boldsymbol{P} = \partial W / \partial \boldsymbol{F}$.

2.1 *Modeling space based on a library*

In the earliest proposed approach, the modeling space consists of a potentially very large library of material models, from which the best model has to be selected (Flaschel et al. 2021, Flaschel et al. 2023, Joshi et al. 2022). The construction of the library, performed upfront by the modeler, should consider three main aspects: the size, the type of dependency on the unknown material parameters (linear or non-linear), and the physics constraints to be satisfied a priori by the model (see Section 5).

The library can be arbitrarily large, in principle including all models known from the literature for the material at hand, and/or a desired number of terms from a functional series such as a Taylor or Fourier series. Some physics constraints can be satisfied a priori by simple choices. While the fact itself that we seek a strain energy density function W already guarantees thermodynamic consistency, the objectivity requirement is a priori satisfied if we assume W to be a function of an objective strain measure, such as the right Cauchy-Green deformation tensor $\boldsymbol{C} = \boldsymbol{F}^T\boldsymbol{F}$. Finally, in the scope of isotropic materials, the strain energy density can be further reduced to a function of the principal invariants of $\boldsymbol{C}$ as

$$W(\boldsymbol{C}) \equiv W(I_1(\boldsymbol{C}), I_2(\boldsymbol{C}), I_3(\boldsymbol{C})) \tag{1}$$

with

$$I_1(\boldsymbol{C}) = \mathrm{tr}(\boldsymbol{C}),$$

$$I_2(\boldsymbol{C}) = \frac{1}{2}[\mathrm{tr}(\boldsymbol{C})^2 - \mathrm{tr}(\boldsymbol{C}^2)],$$

$$I_3(\boldsymbol{C}) = \det(\boldsymbol{C}).$$

To introduce a general parametric ansatz for W, the easiest option is to consider a large feature library $\mathbf{Q} : \mathbb{R}^3 \rightarrow \mathbb{R}^{n_f}$ of n_f non-linear functions of I_1, I_2, and I_3 and assume that the strain energy density can be expressed as their linear combination

$$W(I_1, I_2, I_3; \boldsymbol{\theta}) = \mathbf{Q}^T(I_1, I_2, I_3)\, \boldsymbol{\theta}, \tag{2}$$

where $\boldsymbol{\theta} \in \mathbb{R}^{n_f}$ is a vector of unknown real-valued material parameters, i.e. the material parameters to be estimated. This is the choice made in (Flaschel et al. 2021), where the vector $\boldsymbol{Q}$ contains features of the generalized Mooney-Rivlin model, a logarithmic feature as it appears in the Gent-Thomas model, and some volumetric terms to describe volumetric compressibility if needed. The stress follows as

$$\boldsymbol{P}(I_1, I_2, I_3; \boldsymbol{\theta}) = \frac{\partial \boldsymbol{Q}^T}{\partial \boldsymbol{F}}(I_1, I_2, I_3)\boldsymbol{\theta}, \tag{3}$$

hence, it depends linearly on $\boldsymbol{\theta}$.

A more flexible library is chosen in (Flaschel et al. 2023) to incorporate the Ogden family of hyperelastic models and possibly other models expressed in terms of the principal stretches rather than of the principal invariants of $\boldsymbol{C}$. For maximum flexibility, here the library includes a contribution $\tilde{W}_I$ that depends on the principal invariants of $\boldsymbol{C}$ and a contribution $\tilde{W}_\lambda$ that depends on the principal stretches, as follows

$$W = \tilde{W}_I(I_1, I_2, I_3; \boldsymbol{\theta}_I) + \tilde{W}_\lambda(\lambda_1, \lambda_2, \lambda_3; \mu_i, \alpha_i), \tag{4}$$

where

$$\tilde{W}_I(I_1, I_2, I_3; \boldsymbol{\theta}_I) = \boldsymbol{Q}^T(I_1, I_2, I_3)\, \boldsymbol{\theta}_I$$

$$\tilde{W}_\lambda(\lambda_1, \lambda_2, \lambda_3; \mu_i, \alpha_i) = \sum_{i=1}^{N_{\text{Ogden}}} \frac{2\mu_i}{\alpha_i^2}\left(\lambda_1^{\alpha_i} + \lambda_2^{\alpha_i} + \lambda_3^{\alpha_i} - 3\right).$$

Here $\boldsymbol{\theta}_I$ is again a vector of unknown real-valued material parameters, μ_i and α_i are further unknown real-valued material parameters, and N_{Ogden} is the desired number of terms in the generalized Ogden model. Importantly, unlike in the model library (1), in (4) the strain energy density depends *non-linearly* on the unknown material parameters, which in general complicates their identification through optimization (see Sections 3.5 and 4). To circumvent this issue, in (Flaschel et al. 2023) we assume an a priori fixed set of N_{Ogden} distinct candidate values of α_i. We show numerical evidence that the expressiveness of the model library increases upon increasing N_{Ogden}, and that (at least for the data at hand) choosing $N_{\text{Ogden}} = 2 \cdot 10^4$ values of α_i evenly distributed -100 and 100 excluding zero, i.e., $\alpha_i \in \{-100, -99.99, \ldots, -0.01, 0.01, \ldots, 99.99, 100\}$, provides a sufficiently fine realization of the candidate values of α_i. With this assumption, μ_i remain the only unknowns in the model ansatz for $\tilde{W}_\lambda$ which can now be written as a linear combination of non-linear feature functions

$$\tilde{W}_\lambda(\lambda_1, \lambda_2, \lambda_3; \boldsymbol{\theta}_\lambda) = \boldsymbol{Q}_\lambda^T(\lambda_1, \lambda_2, \lambda_3)\boldsymbol{\theta}_\lambda,$$

where the non-linear feature functions have been collected in the vector $\boldsymbol{Q}_\lambda$ with components

$$\{Q_\lambda(\lambda_1, \lambda_2, \lambda_3)\}_i = \lambda_1^{\alpha_i} + \lambda_2^{\alpha_i} + \lambda_3^{\alpha_i} - 3,$$

and $\boldsymbol{\theta}_\lambda$ is a vector of unknown real-valued material parameters that are related to the parameters μ_i through $\{\theta_\lambda\}_i = 2\mu_i/\alpha_i^2$. By defining

$$\boldsymbol{\theta} = \begin{bmatrix} \boldsymbol{\theta}_I \\ \boldsymbol{\theta}_\lambda \end{bmatrix}, \quad \boldsymbol{Q} = \begin{bmatrix} \boldsymbol{Q}_I(I_1, I_2, I_3) \\ \boldsymbol{Q}_\lambda(\lambda_1, \lambda_2, \lambda_3) \end{bmatrix},$$

the model library for the strain energy density can again be written as

$$W(I_i, \lambda_i; \boldsymbol{\theta}) = \boldsymbol{Q}^T(I_i, \lambda_i)\boldsymbol{\theta}. \tag{5}$$

For conciseness, we are using I_i as a short form for I_1, I_2, I_3 and analogously for λ_i. Clearly, the stress follows as

$$\boldsymbol{P}(I_i, \lambda_i; \boldsymbol{\theta}) = \frac{\partial \boldsymbol{Q}^T}{\partial \boldsymbol{F}}(I_i, \lambda_i)\boldsymbol{\theta}. \tag{6}$$

2.2 *Modeling space generated by a grammar*

The main drawback of the approach based on a library of candidate models is that a human expert needs to hand-design the primitives and assemble the library; this inevitably introduces a bias in the process of discovery and restricts the search space of possible candidate expressions to those included in the starting library. In (Kissas et al. 2024), we explore an alternative strategy: we develop formal grammars specifically designed to generate hyperelastic constitutive laws, so that the *language*(i.e. the ensemble of all hyperelastic strain energy density expressions) generated by the grammar can be adopted in place of a hand-crafted library. We ensure that the expressions derived by the grammar are valid elastic strain energy density functions by applying constraints either during the generation process or as acceptance conditions on the final expressions.

The grammar-based approach generates expressions which in general do not depend linearly on the unknown parameters θ_i; thus, the elastic strain energy density can no longer be expressed as in (5) but retains a more general expression

$$W(I_1, I_2, I_3; \boldsymbol{\theta}) \tag{7}$$

and the stress follows as

$$\boldsymbol{P}(I_1, I_2, I_3; \boldsymbol{\theta}) = \frac{\partial W}{\partial \boldsymbol{F}}(I_1, I_2, I_3; \boldsymbol{\theta}). \tag{8}$$

2.3 *Modeling space based on a neural network*

For both options of a hand-crafted feature library or of a library automatically generated from a formal grammar, the modeling space is limited to the space spanned by the linear (or non-linear) combination of the candidate functions in the library. On the other hand neural networks, being universal approximators, can, in principle, represent any generic constitutive behavior – with no limits on expressive power and no need to rely upon domain knowledge. In (Thakolkaran et al. 2022) we encode W using input convex neural networks. Their special architecture automatically satisfies some physics constraints, see Section 5. However, the larger modeling space goes at the expense of interpretability, since the outcome of the discovery process is no longer a closed-form mathematical expression but a trained neural network. In this case, the ansatz for the elastic strain energy density can still be written as in (7) and the stress still follows as in (8), with the difference that the unknown parameters θ_i are the weights and biases of the neural network for the chosen architecture.

2.4 *Summary*

Once the choice of the modeling space is performed, we have at hand an expression for $W(I_i; \boldsymbol{\theta})$ (or $W(\lambda_i; \boldsymbol{\theta})$ or $W(I_i, \lambda_i; \boldsymbol{\theta})$), which can be interpretable (stemming from a library of literature models, Section 2.1, or from the language generated by a grammar, Section 2.2) or non-interpretable (the ansatz of a neural network, Section 2.3). In the interpretable case, the dependency of W (hence, of $\mathbf{P}$) on $\boldsymbol{\theta}$ can be linear as in Equation (2) or non-linear as in Equation (7); this dependency is invariably non-linear in the non-interpretable case. The expression of W embeds a priori thermodynamic consistency, objectivity, isotropy, as well as additional physics constraints (see Section 5). Since the kinematic quantities are assumed to be known from experimental data, the only unknowns are the components of the real-valued parameter vector $\boldsymbol{\theta}$; they need to be found by minimization of some *objective function* (also denoted as *loss function* in machine learning jargon), which should measure the discrepancy between the model predictions and the experimental data. The choice of this function is discussed next.

3 OBJECTIVE FUNCTION AND PARAMETER IDENTIFICATION

In this section, we focus on the choice of the objective function (OF) whose minimization leads to model discovery. Rather than adopting the classification in supervised and unsupervised methods, which is classical of machine learning jargon, we distinguish between the OFs based on global quantities – typically one force component and one representative displacement component simultaneously recorded during a loading process – and those incorporating local quantities – i.e. quantities known on a fine grid, such as displacements delivered from digital image correlation (DIC) or, in the three-dimensional case, from digital volume correlation (DVC). In turn, as explained in the following, local quantities used to construct the OF can be directly the measured displacements, or they can be kinetic quantities computed by differentiation of the measured displacements (to obtain strains) passing through the assumed modeling space (to obtain stresses and thus nodal forces on a grid).

3.1 *OF based on global quantities*

The classical approach of material model identification relies on global force-displacement measurements and on test setups that are simple enough to enable the computation of the stress from the force and of the strain from the displacement. This approach is adopted in (Flaschel et al. 2023), where hyperelastic constitutive models are sought for human brain tissues and, due to the wet and shiny surface of these tissues, displacement field data are difficult to acquire. Thus, we rely on one-dimensional labeled data pairs stemming from uniaxial compression/tension and simple torsion tests on human brain tissue specimens, In the case of uniaxial compression/tension, the measurements are available in the form of stretch-stress pairs $(\hat{\lambda}^{(l)}, \hat{P}_{11}^{(l)})$ for load steps $l = 1, \ldots, n_{\text{UT}}$. Simple analytical derivations Brain23 provide the relation between λ and the stress P_{11} given by the model library, thus leading to n_{UT} equations

$$P_{11}(\hat{\lambda}^{(l)}; \boldsymbol{\theta}) = \hat{P}_{11}^{(l)}. \tag{9}$$

Analogously, in the case of simple torsion, the measurements are available in the form of twist-torque pairs $(\hat{\psi}^{(l)}, \hat{\tau}^{(l)})$ for load steps $l = 1, \ldots, n_{\text{ST}}$. Again analytical derivations lead to n_{ST} equations

$$\tau(\hat{\psi}^{(l)}; \boldsymbol{\theta}) = \hat{\tau}^{(l)}. \tag{10}$$

Since (Flaschel et al. 2023) adopts the modeling space in Section 2.1, stresses depend linearly on $\boldsymbol{\theta}$ and the above equations can be rewritten as linear systems

$$\boldsymbol{A}_{\text{UT}}\boldsymbol{\theta} = \boldsymbol{b}_{\text{UT}}, \quad \boldsymbol{A}_{\text{ST}}\boldsymbol{\theta} = \boldsymbol{b}_{\text{ST}}, \tag{11}$$

where $\boldsymbol{A}_{\text{UT}}$ and $\boldsymbol{A}_{\text{ST}}$ contain derivatives of the feature functions evaluated at the experimental stretches and twists, respectively, whereas $\boldsymbol{b}_{\text{UT}}$ and $\boldsymbol{b}_{\text{ST}}$ are the vectors of the experimental stresses and torques, respectively. The linear systems in (11) are concatenated to the system $\boldsymbol{A\theta} = \boldsymbol{b}$ with

$$\boldsymbol{A} = \begin{bmatrix} r_{\text{UT}}\boldsymbol{A}_{\text{UT}} \\ r_{\text{ST}}\boldsymbol{A}_{\text{ST}} \end{bmatrix}, \quad \boldsymbol{b} = \begin{bmatrix} r_{\text{UT}}\boldsymbol{b}_{\text{UT}} \\ r_{\text{ST}}\boldsymbol{b}_{\text{ST}} \end{bmatrix},$$

where $r_{\text{UT}} > 0$ and $r_{\text{ST}} > 0$ scale the contributions of the uniaxial compression/tension data and the simple torsion data, respectively. Finally, the total mismatch between the material model prediction and the experimental measurements is quantified by the squared L_2 -norm of the residual of the linear system

$$L_u(\boldsymbol{\theta}) = \| \boldsymbol{A\theta} - \boldsymbol{b} \|_2^2 . \tag{12}$$

Note that in (Flaschel et al. 2023) we further divide the OF by $n_{\text{UT}} + n_{\text{ST}}$. This is not done here for the sake of simplicity to unify the notation across different approaches.

With the modeling space generated by a grammar (Section 2.2) or based on a neural network (Section 2.3), the dependency on $\boldsymbol{\theta}$ in Equations (9) and (10) becomes non-linear. The OF can again be constructed from the squared L_2 -norm of the residuals of Equations (9) and (10), and results in a non-linear function $L_u(\boldsymbol{\theta})$.

3.2 *OF based on local unbalanced forces*

Global quantities such as representative force and displacement components deliver spatially integrated information on the material behavior, and it may easily happen that different models lead to a similar global response while the local behavior may be very different. For this reason, it is crucial to rely on local information, which is enabled by measurement technologies such as DIC or DVC and is an emerging paradigm in material testing (Pierron and Grédiac 2020). In (Flaschel et al. 2021, Joshi et al. 2022, Thakolkaran et al. 2022), we assume the availability of displacements measured on a fine grid (a mesh) on the surface of the specimen, from which the entire in-plane displacement field $\hat{\boldsymbol{u}}(\boldsymbol{X})$ can be reconstructed using a suitable set of shape functions. We then use this displacement field to construct the OF as follows.

Assuming negligible body forces, the weak formulation of linear momentum balance in the reference domain Ω under quasi-static conditions is given by

$$\int_{\Omega} \boldsymbol{P} : \nabla \boldsymbol{v} dV - \int_{\partial\Omega_t} \hat{\boldsymbol{t}} \cdot \boldsymbol{v} dS = 0, \quad \forall \quad \text{adm.}\boldsymbol{v}, \tag{13}$$

where $\hat{\boldsymbol{t}}$ is the surface traction acting on $\partial\Omega_t$ (zero in a displacement-controlled setting, but non-zero for force-controlled experiments), and the test function $\boldsymbol{v}$is admissible if it is sufficiently regular and it vanishes on the Dirichlet boundary.

Let $\mathcal{D} = \{(a,i) : a = 1, \ldots, n_n; i = 1,2\}$ denote the set of all nodal degrees of freedom. $\mathcal{D}$ is further split into two subsets of free and fixed (via Dirichlet constraints) degrees of freedom: $\mathcal{D}^{\text{free}} \subseteq \mathcal{D}$ and

$\mathcal{D}^{\text{fix}} = \mathcal{D} \backslash \mathcal{D}^{\text{free}}$, respectively. Using the same approximation basis as for $\hat{\mathbf{u}}$, the test functions are approximated as

$$\boldsymbol{v}(\mathbf{X}) = \sum_{a=1}^{n_n} N^a(\boldsymbol{X})\, \boldsymbol{v}^a, \tag{14}$$

with $v_i^a = 0$ if $(a,i) \in \mathcal{D}^{\text{fix}}$. Substituting the ansatz for the material model and noting that the discretized weak form must hold for all admissible $\{\boldsymbol{v}^a : a = 1, \ldots, n_n\}$, we obtain the following componentwise force balance equations at all free degrees of freedom:

$$\int_{\Omega} P_{ij}(I_1, I_2, I_3; \boldsymbol{\theta}) \nabla_j N^a dV - \int_{\partial\Omega_t} \hat{t}_i N^a dS = 0 \tag{15}$$

$\forall\, (a,i) \in \mathcal{D}^{\text{free}}$, where we use the Einstein summation convention. The integrals are computed via numerical quadrature on the mesh. The equations are ultimately assembled into the form

$$F_i^{\text{int},a}(\boldsymbol{\theta}) - F_i^{\text{ext},a} = 0 \tag{16}$$

$\forall (a,i) \in D^{\text{free}}$, where

$$F_i^{\text{int},a}(\boldsymbol{\theta}) = \int_{\Omega} P_{ij}(I_1, I_2, I_3; \boldsymbol{\theta}) \nabla_j N^a \, dV \tag{17}$$

$$F_i^{\text{ext},a} = \int_{\partial\Omega_t} \hat{t}_i N^a \, dS \tag{18}$$

can be interpreted as internal/constitutive and external forces respectively, acting on the degree of freedom denoted by (a,i). In the dynamic case, we have

$$F_i^{\text{int},a}(\boldsymbol{\theta}) - F_i^{\text{ext},a} + F_i^{\text{m},a} = 0 \tag{19}$$

where $F_i^{\text{m},a}$ are inertial forces. While the dynamic case is considered in Joshi22, in the following we assume quasi-static conditions for simplicity.

Note that the computation of the internal forces requires the local stress $\boldsymbol{P}$, which depends on I_i and needs to be evaluated at their experimental values, $\hat{I}_i$; this implies that we need to perform differentiation on the reconstructed displacement field, which may be challenging due to the inevitable presence of noise.

Additionally, the reaction forces are known at n_β Dirichlet boundaries. For $\beta = 1, \ldots, n_\beta$, let $\mathcal{D}^{\text{fix},\beta} \subseteq \mathcal{D}^{\text{fix}}$ (with $\mathcal{D}^{\text{fix},\beta} \cap \mathcal{D}^{\text{fix},\beta'} = \emptyset$ for $\beta \neq \beta'$) denote the subset of degrees of freedom for which the sum of reaction forces $\hat{R}^\beta$ is known. Each observed reaction force must balance the aggregated internal and (if present) inertial forces, i.e.,

$$\sum_{(a,i) \in \mathcal{D}^{\text{fix},\beta}} \left[F_i^{\text{int},a}(\boldsymbol{\theta}) + F_i^{\text{m},a} \right] = \hat{R}^\beta \tag{20}$$

When the ansatz for W (and thus $\boldsymbol{P}$) depends linearly on $\boldsymbol{\theta}$, such as in (2) or (5), the force balance equations in (16) and (20) deliver two linear systems of equations for $\boldsymbol{\theta}$, which can be assembled as

$$\boldsymbol{A}^{free} \boldsymbol{\theta} = \boldsymbol{b}^{free}, \quad \boldsymbol{A}^{\text{fix}} \theta = \boldsymbol{b}^{fix} \tag{21}$$

with $\boldsymbol{A}^{free} \in \mathbb{R}^{|\mathcal{D}^{free}| \times n_f}$, $\boldsymbol{b}^{free} \in \mathbb{R}^{|\mathcal{D}^{free}|}$, $\boldsymbol{A}^{fix} \in \mathbb{R}^{n_\beta \times n_f}$ and $\boldsymbol{b}^{fix} \in \mathbb{R}^{n_\beta}$. Note that in (Joshi et al. 2022), for computational efficiency, we randomly subsample $n_{\text{free}} < |\mathcal{D}^{\text{free}}|$ equations from $(21)_1$ such that $\boldsymbol{A}^{\text{free}} \in \mathbb{R}^{n_{\text{free}} \times n_f}$ and $\boldsymbol{b}^{\text{free}} \in \mathbb{R}^{n_{\text{free}}}$. Equation (21) can be written across all $t = 1, \ldots, n_t$ snapshots and concatenated into a joint system of equations as

$$\boldsymbol{A}\theta = \boldsymbol{b} \tag{22}$$

with

$$\boldsymbol{A} = \begin{bmatrix} \boldsymbol{A}^{\text{free},1} \\ \vdots \\ \boldsymbol{A}^{\text{free},n_t} \\ \lambda_r \boldsymbol{A}^{\text{fix},1} \\ \vdots \\ \lambda_r \boldsymbol{A}^{\text{fix},n_t} \end{bmatrix} \quad \text{and} \quad \boldsymbol{b} = \begin{bmatrix} \mathbf{b}^{\text{free},1} \\ \vdots \\ \boldsymbol{b}^{\text{free},n_t} \\ \lambda_r \boldsymbol{b}^{\text{fix},1} \\ \vdots \\ \lambda_r \boldsymbol{b}^{\text{fix},n_t} \end{bmatrix}, \tag{23}$$

where $\lambda_r > 0$ is a hyperparameter that controls the relative importance between the force balance of the free and fixed degrees of freedom. Finally, the OF can be expressed as

$$L_u(\boldsymbol{\theta}) = \|\boldsymbol{A\theta} - \mathbf{b}\|_2^2 \tag{24}$$

Once again, if using the modeling space from a grammar or a neural network, the dependency of P_{ij} on $\boldsymbol{\theta}$ becomes non-linear. The OF can again be constructed from the squared L_2 -norm of the residuals of Equations (16) and (20), and results in a non-linear function $L_u(\boldsymbol{\theta})$.

Summarizing, the construction of the OF based on local unbalanced forces requires i. the full displacement field reconstructed from the DIC (or DVC) measurements; ii. its differentiation to compute deformation measures such as I_i or λ_i, and iii. the ansatz for the material model. This methodology for construction of the OF is very similar to the one advocated by e.g. the equilibrium gap method (Roux and Hild 2020) and the virtual fields method (Grédiac et al. 2008). Since it requires no finite element simulations, it is computationally very efficient.

3.3 *OF based on displacement fields*

In the material model identification literature, the competitor of the equilibrium gap and virtual fields methods is also known as finite element model updating or FEMU (Kavanagh and Clough 1971). In FEMU, the OF is based directly on the discrepancy between the displacement field obtained from DIC, $\hat{\boldsymbol{u}}$, and the one computed by finite element simulations using the material model whose unknown parameters need to be identified, $\boldsymbol{u}^{FEM}(\theta)$, i.e.

$$L_u(\theta) = \left\| \boldsymbol{u}^{FEM}(\theta) - \hat{\boldsymbol{u}} \right\|_2^2 \tag{25}$$

The disadvantage of this approach is the significantly higher computational cost, due to the need for a finite element simulation at each iteration of the optimization process; the advantage is that the noisy displacement field does no longer need differentiation. Thus far, FEMU was only investigated in the traditional context of material parameter identification, as opposed to the model discovery advocated here. However, combining the OF in Equation (25) with the modeling space options in Section 2 using the discovery strategies in Section 4 would be conceptually straightforward.

3.4 *Non-dimensionalization*

As we have seen in this section, the OF may be constructed from global quantities, possibly measured from different tests, featuring in general different dimensions (Section 3.1), or from local unbalanced forces partly in the interior and partly at the restrained boundaries of tested specimens (Section 3.2) or from discrepancies in local displacements (Section 3.3). In principle, even mixed OFs can be conceived including information from different tests, some with global and some with local measurements. For all these reasons, it may turn out very useful to non-dimensionalize the OF, e.g. by normalization or by standardization as in (Flaschel et al. 2023).

3.5 *Parameter identification*

At this point, based on the modeling space in Section 2and on the experimental data, we can compute the OF $L_u(\theta)$ and, in principle, obtain the unknown vector θ by the minimization problem

$$\boldsymbol{\theta}_{opt} = \arg\min L_u(\boldsymbol{\theta}) \tag{26}$$

For the special cases of W depending linearly on $\boldsymbol{\theta}$, leading to an OF as in Equations (12) and (24), the optimization in (26) is convex and, assuming a non-singular $\boldsymbol{A}^T\boldsymbol{A}$ matrix, can be solved in closed form. In the general case, W depends non-linearly on $\boldsymbol{\theta}$, thus the optimization problem (26) can be highly non-convex and requires a suitable solution algorithm and a strategy to cope with possible multiple solutions. In all cases, including the convex case which may seem straightforward, the problem is likely to be significantly ill-posed, since the modeling space may easily contain several feature functions capable of interpreting the same data. This implies that (26) may be not solvable and, even if solvable, will deliver a solution that is highly sensitive to noise in the data (for the convex case, the condition number of $\boldsymbol{A}^T\boldsymbol{A}$ will be very high). Moreover, the solution vector $\boldsymbol{\theta}_{opt}$ will be in general a dense vector, i.e. a vector with all or most non-zero components. This implies that the model stemming e.g. from a library of 1000 feature functions will consist in an analytical expression with about 1000 components, which is clearly undesirable.

Note that the optimization problem (26) is conceptually nothing else but *parameter identification*- for the a priori chosen model $W(I_i;\boldsymbol{\theta})$ coinciding with our vast modeling space. In (26), the only difference with respect to conventional parameter identification is simply that the model we are starting from is significantly larger than usual.

4 FROM PARAMETER IDENTIFICATION TO MODEL DISCOVERY

In this section, we review the possible strategies to turn a parameter identification task into a model discovery task. As already introduced in Section 1, we define *model discovery*as the combination of *model selection*and *parameter identification*. Thus, we need a strategy for model selection to be combined with parameter identification. The guiding principle is that we seek a *parsimonious*material model, i.e. a model as simple as possible to explain the experimental data.

4.1 *Sparse regression*

In (Flaschel et al. 2021) and (Flaschel et al. 2023), we introduce L_p regularization in the OF, as follows

$$L_r(\boldsymbol{\theta}) = L_u(\boldsymbol{\theta}) + \lambda_p \|\boldsymbol{\theta}\|_p^p \tag{27}$$

and the minimization problem to be solved becomes

$$\boldsymbol{\theta}_{opt} = \arg\min L_r(\boldsymbol{\theta}) \tag{28}$$

This strategy is known as *sparse regression* and was first used for discovery of governing equations for dynamical systems in (Brunton et al. 2016). The L_p regularization term promotes sparsity, i.e. the presence of many vanishing components in the solution vector. Conceptually, this implies that the minimization problem now simultaneously performs model selection (i.e. it finds out which components of $\boldsymbol{\theta}$ should be zero, hence which features in the modeling space library should be discarded), and parameter identification (i.e. it determines the value of the non-zero components of $\boldsymbol{\theta}$, i.e. the material parameters). The hyperparameters $p \in [0, 1]$ and $\lambda_p > 0$ control the degree of sparsity desired in θ and thus the simplicity of the discovered model.

L_0 regularization ($p = 0$) is equivalent to penalizing the number of non-zero entries in the $\boldsymbol{\theta}$ vector. With this regularization, the optimization becomes a combinatorial subset selection problem which is computationally intractable for a large number of features. Traditionally, L_0 regularization is relaxed to the convex L_1 regularization ($p = 1$), which promotes sparsity by zeroing out some coefficients while shrinking the rest. Also known as LASSO (least absolute shrinkage and selection operator), L_1 regularization offers a good compromise between computational complexity and sparsity of the solution. Within $[0, 1]$, a smaller value of p more aggressively promotes sparsity at the cost of computational complexity, due to the higher degree of non-convexity in the penalty term. Clearly, the L_p term converges to the L_0 term as p approaches zero.

The choice of the weighting factor $\lambda_p \geq 0$ is crucial for striking a balance between data fitting accuracy and simplicity of the discovered model. If λ_p is too high, the resulting model will be too sparse and may not be rich enough to accurately fit the data. On the other hand, if λ_p is too low, the optimization problem (28) becomes highly ill-posed as the original unregularized problem.

Our choices for p and λ_p are summarized as follows. In (Flaschel et al. 2021), we adopt $p = 1/4$. As a result of this choice, the OF in (27) is non-convex and not continuously differentiable with respect to $\boldsymbol{\theta}$, thus, gradient-based optimization techniques such as gradient-descent and Newton (or Newton-like) methods do not perform well and we use a fixed-point iterative strategy instead. Due to the pronounced non-convexity, the fixed-point iteration scheme may converge to different local minima depending on the choice of the initial guess. For this reason, we perform multiple fixed-point iteration schemes, each with different randomly generated initial guesses. From all the converged solutions, we choose the one which leads to the smallest value of the OF. In order to estimate λ_p, we adopt an iterative strategy with the fixed-point algorithm nested inside it. We start with a small initial value of $\lambda_p = \lambda_p^0 > 0$ and perform sparse regression with the fixed-point algorithm. On the resulting model, we check some physics admissibility requirements (non-negativity and growth of W, see Section 5). If the checks are satisfied, the model is accepted. Otherwise, λ_p is multiplicatively increased by a factor $\kappa > 1$, i.e., $\lambda_p = \kappa\lambda_p^0$. The process is repeated until a physically admissible model is obtained. By gradually increasing λ_p, we ensure that the resulting model is not too sparse or simple.

With the computational setup in (Flaschel et al. 2023), we rely on the L_1 regularization term, thus, since in this study $L_u(\theta)$ is convex, we solve a convex minimization problem which admits a unique solution. We deploy the coordinate descent algorithm implemented in the *Lasso*subroutine of the open-source Python library *sklearn*, which is computationally very efficient as it leverages convexity. For the choice of λ_p we adopt a Pareto analysis, as qualitatively illustrated in Figure 1. A small value of λ_p yields a dense solution vector, thus a complex expression for the material model, that is in general associated with a high fitting accuracy, i.e., a small value of the unregularized OF, L_u. Increasing the value of λ_p results in a sparser solution vector, thus a simpler expression for the material model, at the cost of a reduced fitting accuracy, hence a larger L_u. In the limit of very large values of λ_p, the optimization problem returns a solution vector with only zero entries and L_u saturates. To strike a balance between the accuracy and the complexity of the discovered material model, we solve (28) multiple times using a set of different λ_p values, leading to a solution vector for every choice of λ_p. Then, we select from all solutions the one that exhibits an optimal combination of high fitting accuracy and high sparsity. To this end, L_u and the L_1 -norm of the solution vector are computed for each solution. The minimum and maximum L_u values among all solutions are denoted as $L_{u,\min}$ and $L_{u,\max}$, respectively. Further, we define a threshold value of L_u (see Figure 1) as

$$L_{u,\text{th}} = L_{u,\min} + r_\lambda(L_{u,\max} - L_{u,\min}), \qquad (29)$$

where r_λ is a small positive scalar, $0 < r_\lambda \ll 1$. All solutions for which L_u is larger than this threshold are expected to exhibit a low fitting accuracy and are thus discarded. From the remaining solutions, i.e., those for which L_u is below $L_{u,\text{th}}$, we select the sparsest one, i.e., the solution with the lowest value of the L_1 -norm term (see Figure 1). In this way, a sparse solution with a low L_u is obtained, whereby the amount of sparsity is dictated by the user-defined hyperparameter r_λ, which in (Flaschel et al. 2023) is set to $r_\lambda = 0.02$.

One more ingredient is needed in cases where a non-linear dependence of the model feature functions on the material parameters is circumvented by a fine discretization of the parameter space, such as for the Ogden model features in Section 2.1. Due to the fine discretization of the chosen range of α_i values, the feature library contains Ogden features with very similar values of α_i. For the solution vector obtained by sparse regression, the sparsity can be further improved by grouping together highly correlated features in the solution. To this end, we leverage a clustering algorithm that automatically groups together similar α_i values (Flaschel et al. 2023).

The last ingredients of the sparse regression strategy consist in a thresholding operation and in a final run of the optimization algorithm without regularization. The optimal solution obtained from sparse regression contains a large number of material parameters that are close or equal to zero. These "inactive features" can thus be permanently discarded from the initial library. This is performed through a simple thresholding operation, i.e., all parameters whose absolute value is below a threshold are set to zero. Subsequently, we deactivate the L_p regularization and solve the unregularized optimization problem (26) for the remaining parameters. If any of the resulting non-zero parameters is again below the defined threshold, the thresholding procedure is repeated until convergence is reached.

4.2 *Bayesian regression*

In (Joshi et al. 2022), we replace deterministic sparse regression with a probabilistic framework able to perform simultaneous model selection and parameter identification with uncertainty quantification. We use a modification of the hierarchical Bayesian model with discontinuous spike-slab priors proposed by (Nayek et al. 2021). These priors and their variants have a strong peak at zero and very shallow tails, representing the bias towards parsimonious solutions. To set up the discontinuous spike-slab priors on $\boldsymbol{\theta}$, we introduce a random indicator variable $\mathbf{z} \in [0,1]^{n_f}$ and auxiliary random variables $\nu_s \geq 0$ and $\sigma^2 \geq 0$, such that the prior probability distribution of each component θ_i is

$$p = \begin{cases} p_{\text{spike}}(\theta_i) = \delta(\theta_i) & \text{for } z_i = 0 \\ p_{\text{slab}}(\theta_i \mid \nu_s, \sigma^2) = \mathcal{N}_+(0, \nu_s\sigma^2) & \text{for } z_i = 1 \end{cases}, \tag{30}$$

with $i = 1, \ldots, n_f$.. When $z_i = 0$, the ith corresponding feature is deactivated (i.e., $\theta_i = 0$) and the probability density of that feature is given by a Dirac delta (*spike*) centered at zero. If $z_i = 1$, the corresponding feature is activated and θ_i is sampled from a truncated normal distribution $\mathcal{N}_+$ with zero mean and variance $\nu_s\sigma^2$ (*slab*), derived from a normal distribution by restricting the samples to non-negative values, i.e., $\theta_i \geq 0$ (see Section 5). With the prior assumption that the components of $\boldsymbol{\theta}$ are independent and identically distributed (i.i.d.), the joint prior distribution of $\boldsymbol{\theta}$ is written as

$$p_{joint} = p_{\text{slab}}\left(\boldsymbol{\theta}_r \mid \nu_s, \sigma^2\right) \prod_{i:z_i=0} p_{\text{spike}}\left(\theta_i\right) \tag{31}$$

with

$$p_{\text{slab}}\left(\boldsymbol{\theta}_r \mid \nu_s, \sigma^2\right) = N_+\left(\mathbf{0}, \nu_s\sigma^2\boldsymbol{I}_r\right), \tag{32}$$

where $\boldsymbol{\theta}_r \in \mathbb{R}_+^r$ is a reduced vector consisting of only slab/active components of $\boldsymbol{\theta}$ and $\mathbf{I}_r$ denotes the $r \times r$ identity matrix. We use the algorithm by (Botev 2016) for sampling from the multivariate truncated normal distribution. All features (active or not) are modeled as an i.i.d. Bernoulli trial with probability $p_0 \in [0,1]$,

$$z_i \mid p_0 \sim \text{Bern}(p_0). \tag{33}$$

The prior probability distributions for the variables ν_s, σ^2, and p_0 are modeled as

$$\begin{aligned} \nu_s &\sim \mathcal{IG}(a_\nu, b_\nu) && \text{(Inverse gamma distr.)} \\ \sigma^2 &\sim \mathcal{IG}(a_\sigma, b_\sigma) && \text{(Inverse gamma distr.)} \\ p_0 &\sim \text{Beta}(a_p, b_p) && \text{(Beta distr.)}, \end{aligned} \tag{34}$$

where (a_ν, b_ν), (a_σ, b_σ), and (a_p, b_p) are the parameters of the hyperpriors. Note that while both ν_s and σ^2 scale the variance of the slab, σ^2 also scales the variance of the likelihood.

In the Bayesian framework, Equation (22) defining the OF is better written as

$$\boldsymbol{b} = \boldsymbol{A\theta} + \boldsymbol{e}, \tag{35}$$

where $\mathbf{e}$ represents uncertainty in the experimental observations as well as model inadequacies. Given a constitutive model (represented by $\mathbf{q}$) sampled from the prior and the observed data (in the form of $\mathbf{A}$ and $\mathbf{b}$), we place an i.i.d. normal distribution likelihood on the residual as

$$\boldsymbol{b} \mid \boldsymbol{\theta}, \sigma^2, \boldsymbol{A} \sim \mathcal{N}(\boldsymbol{A\theta}, \sigma^2\boldsymbol{I}_N), \tag{36}$$

where $\boldsymbol{I}_N$ denotes the $N \times N$ identity matrix with $N = (n_{\text{free}} + n_\beta)n_t$. These dimensions stem from the definition of $\boldsymbol{A}$ and $\boldsymbol{b}$ in Section 3.2. However, generalization of the same approach to a different definition, or even to an OF depending nonlinearly on $\boldsymbol{\theta}$ would be conceptually straightforward.

The next step in the Bayesian learning process is to evaluate the joint posterior probability distribution for the random variables in the hierarchical model, i.e., $\{\boldsymbol{\theta}, \boldsymbol{z}, p_0, \nu_s, \sigma^2\}$, given the observed data and the consequent constraints embodied by $\boldsymbol{A}$ and $\boldsymbol{b}$. Using Bayes' theorem, the posterior is given by

$$p(\theta, \boldsymbol{z}, p_0, \nu_s, \sigma^2 \mid \boldsymbol{A}, \boldsymbol{b}) \propto \underbrace{p(\boldsymbol{b} \mid \theta, \boldsymbol{z}, p_0, \nu_s, \sigma^2, \boldsymbol{A})}_{\text{physics-constrainedlikelihood}} \underbrace{p(\theta, \boldsymbol{z}, p_0, \nu_s, \sigma^2)}_{\text{spike-slabmodelprior}} \tag{37}$$

For sampling from (37), Markov Chain Monte Carlo methods are needed. In (Joshi et al. 2022), we use a Gibbs sampler, which repeatedly and successively samples each variable in $\{\mathbf{q}, \mathbf{z}, p_0, \nu_s, \sigma^2\}$ using the respective conditional posterior distributions. This produces the following Markov chain of length N_G:

$$\begin{aligned} &\boldsymbol{\theta}^{(0)} \to \sigma^{2^{(0)}} \to \nu_s^{(0)} \to p_0^{(0)} \to \boldsymbol{z}^{(0)} \to \ldots\ldots \to \\ &\to \boldsymbol{\theta}^{(N_G)} \to \sigma^{2^{(N_G)}} \to \nu_s^{(N_G)} \to p_0^{(N_G)} \to \boldsymbol{z}^{(N_G)}, \end{aligned} \tag{38}$$

which empirically approximates the joint posterior distribution $p(\boldsymbol{\theta}, \boldsymbol{z}, p_0, \nu_s, \sigma^2 \,|\, \boldsymbol{A}, \boldsymbol{b})$. Note that the Markov chain states are only recorded after discarding the first N_{burn} burn-in samples to avoid the effects of a bad starting point. For the same reason, we also generate N_{chains} independent Markov chains (each with different starting points selected randomly) which are concatenated after discarding the burn-in states.

The average activity of the ith feature $(i = 1, \ldots, n_f)$ of the energy density function library is defined as

$$z_i^{\text{avg}} = \frac{1}{N_G \times N_{\text{chains}}} \sum_{k=1}^{N_G \times N_{\text{chains}}} z_i^{(k)}, \tag{39}$$

where $z_i^{(k)} \in \{0, 1\}$ is the activity of the ith feature in the kth sample of the Markov chain. Note that the average activity is a marginalization over the variables in the joint posterior distribution. Therefore, $z_i^{\text{avg}} = 1$ and $z_i^{\text{avg}} = 0$ imply that the corresponding feature is always active or inactive, respectively; whereas any value between zero and one implies that the joint distribution is likely multi-modal with the corresponding feature being both active and inactive in different modes.

4.3 *Symbolic regression*

In (Kissas et al. 2024), we propose a symbolic regression pipeline for constitutive law discovery. Due to the advantages of formal grammars (low complexity, ability to systematically embed constraints stemming from domain knowledge, and generalization capability) we explore grammar-based symbolic regression and, as already introduced in Section 2.2, develop formal grammars designed to generate hyperelastic constitutive laws. The proposed pipeline is composed by three steps: a library construction step, a pre-training step, and a data-driven discovery step. In the first step, we define a formal grammar that generates a library (a *language*) of mathematical expressions in the form of trees. This library is a useful result per se, as it may be deployed in place of a hand-crafted library (as discussed in Section 2.2) in the context of model discovery based on sparse or Bayesian regression, following Sections 4.1 or 4.2, respectively. In the second step, this library is used to train off-line a symbolic regression algorithm that takes advantage of the tree structure of the mathematical expressions and of the grammar of the constitutive laws. The pre-trained model is trained to encode a tree to a low-dimensional latent vector representations and decode this low-dimensional vector representation to a tree, creating a low-dimensional manifold of tree representations. In the third step, we perform gradient-free optimization to search the low-dimensional manifold for the vectorial representation of the model that best fits the given measurements.

Note that also in the grammar-based generation of the material law expressions we can embed the preference for a parsimonious model. In this case, the tree generation process can be manipulated so as to enforce or favor the generation of shallow trees, which results in parsimonious expressions, see (Kissas et al. 2024) for more details.

4.4 *Neural network training*

The approach in (Thakolkaran et al. 2022) is different from all others considered thus far in that it does not perform model discovery; instead, it performs what we can term *model encoding* through training of a neural network. In other words, the outcome of the method is not an intepretable mathematical expression, but a trained neural network that encodes the constitutive behavior which well interprets the data, whereby training minimizes an OF which in principle can be given by any of the choices in Section 3. In (Thakolkaran et al. 2022), this OF is given by (24). Within this type of approach, parsimonity is not explicitly enforced, with the exception that measures are adopted to prevent overfitting, as customary in regression tasks with neural networks. Further details are available in the original paper.

5 PHYSICS-BASED CONSTRAINTS

As follows, we briefly summarize the most important physics constraints that are (or are not) enforced by the various approaches overviewed in this paper.

Thermodynamic consistency, objectivity, and symmetry of the Cauchy stress tensor are guaranteed by construction by all overviewed approaches, as they all search for an elastic strain energy density of the form (7) (or of similar forms depending on the isochoric invariants or the principal stretches).

Normalization of energy and stress are automatically satisfied by the approaches based on handcrafted material model libraries, provided they are satisfied by all terms in the library. Instead, the grammar- and neural-network-based approaches need a posteriori corrections, which should be performed carefully in order not to violate other constraints, see also (Linden et al. 2023).

Non-negativity and volumetric growth are not guaranteed a priori by all approaches. Measures to at least empirically (thus approximately) enforce their satisfaction can be through the choice of a sufficiently strong sparse regularization (Flaschel et al. 2021) or through acceptance checks of the grammar-generated expressions (Kissas et al. 2024); other measures are the enforcement of non-negative components of the $\boldsymbol{\theta}$ vector (Joshi et al. 2022, Flaschel et al. 2023), which in presence of only coercive terms in the library guarantees coercivity of W, or details in the neural network architecture (Thakolkaran et al. 2022).

Polyconvexity is also not guaranteed a priori by all approaches. In principle, when using a hand-crafted material model library it is straightforward to guarantee polyconvexity by construction, by including only polyconvex terms in the library and enforcing non-negative components of the $\boldsymbol{\theta}$ vector. However, the libraries adopted so far include the popular Mooney-Rivlin model terms, which are known not to be polyconvex. Polyconvexity is not guaranteed in (Thakolkaran et al. 2022) nor in (Kissas et al. 2024).

In the above summary, we do not include material symmetry (isotropy in this paper) nor incompressibility, since we regard these more as modeling choices than as constraints. If desired, both of them can be straightforwardly embedded in the ansatz for the material model, as done in all reviewed approaches for isotropy, and in (Flaschel et al. 2023) for incompressibility.

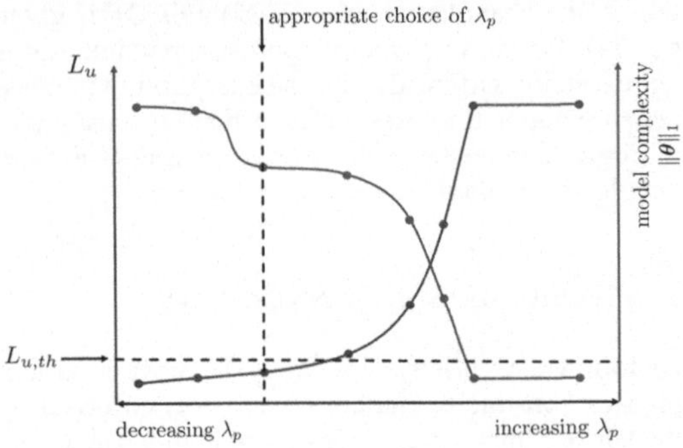

Figure 1. Pareto analysis for the choice of λ_p in sparse regression.

6 CONCLUSIONS AND OUTLOOK

We conclude this overview by highlighting open issues and promising directions for future research.

- The enforcement of physics-based constraints, while being a largely resolved issue for a modeling space based on a hand-crafted material model library, still relies on empirical checks for the more flexible grammar-based approach combined with symbolic regression. A priori satisfaction of all relevant physics constraints by grammar design is an interesting topic worth of further investigation;
- The assumption of W as a linear combination of the terms in the material model library is the most efficient from a computational standpoint, but also the most restrictive. For models not satisfying this assumption, such as the Ogden one, discretizing the parameter space to revert to a linear combination is an effective but not elegant solution which cannot be easily automatized; in general, sparse regression with loss functions which are non-linear in the unknown vector has never been investigated thus far and could significantly enlarge the spectrum of possibilities in the choice of the model ansatz;
- Experimental validation so far is limited to the brain tissue experiments in (Flaschel et al. 2023); the first validation with full-field measurements is in progress (Ricci et al. prep) but will need a significant effort in the future. A relevant related issue is that of optimal specimen design, in order to facilitate sampling the stress-strain space as extensively as possible with as few tests as possible. On questions such as the optimal amount of data as well as in general on uncertainty quantification, Bayesian approaches hold promise to deliver useful answers.
- Finally, extensions to more complex material behavior including large-strain viscoelasticity, heterogeneity and damage are still to be developed.

ACKNOWLEDGMENTS

This work is funded by the Swiss National Science Foundation (SNF) through grant N. 200021 204316 "Unsupervised data-driven discovery of material laws".

REFERENCES

Boddapati, J., M. Flaschel, S.Kumar, L.De Lorenzis, & C. Daraio (2023). Single-test evaluation of directional elastic properties of anisotropic structured materials. *Journal of the Mechanics and Physics of Solids 181*, 105471.

Botev, Z. (2016). The normal law under linear restrictions: simulation and estimation via minimax tilting. *J. R. Stat. Soc. Ser. B Stat. Methodol. 79*, 125–148.

Brunton, S., J.Proctor, & J.Kutz (2016). Discovering governing equations from data by sparse identification of nonlinear dynamical systems. *Proceedings of the National Academy of Sciences 113*, 3932–3937.

Flaschel, M., S.Kumar, & L.De Lorenzis (2021). Unsupervised discovery of interpretable hyperelastic constitutive laws. *Computer Methods in Applied Mechanics and Engineering 381*, 113852.

Flaschel, M., S.Kumar, & L.De Lorenzis (2022). Discovering plasticity models without stress data. *npj Computational Materials 8*, 91.

Flaschel, M., S.Kumar, & L.De Lorenzis (2023). Automated discovery of generalized standard material models with euclid. *Computer Methods in Applied Mechanics and Engineering 405*, 115867.

Flaschel, M., H.Yu, N.Reiter, J.Hinrichsen, S.Budday, P. Steinmann, S.Kumar, & L.De Lorenzis (2023). Automated discovery of interpretable hyperelastic material models for human brain tissue with euclid. *Journal of the Mechanics and Physics of Solids 180*, 105404.

Grédiac, M., F.Pierron, S.Avril, & E.Toussaint (2008). The virtual fields method for extracting constitutive parameters from full-field measurements: a review. *Strain 42*, 233.

Joshi, A., P.Thakolkaran, Y.Zheng, M.Escande, M.Flaschel, L.De Lorenzis, & S.Kumar (2022). Bayesian-euclid: discovering hyperelastic material laws with uncertainties. *Computer Methods in Applied Mechanics and Engineering 398*, 115225.

Kavanagh, K. T. & R. W.Clough (1971). Finite element applications in the characterization of elastic solids. *International Journal of Solids and Structures 7*, 11.

Kissas, G., S.Mishra, E.Chatzi, & L.De Lorenzis (2024). The language of hyperelastic materials. *arXiv:2402.04263*.

Lemaitre, J. & J.Chaboche(1994). Mechanics of solid materials. *Cambridge University Press*.

Linden, L., D.Klein, K.Kalina, J.Brummund, O.Weeger, & M.Kästner(2023). Neural networks meet hyperelasticity: A guide to enforcing physics. *J. Mech. Phys. Solids179*, 105363.

Marino, E., M.Flaschel, S.Kumar, & L.De Lorenzis (2023). Automated identification of linear viscoelastic constitutive laws with euclid. *Mechanics of Materials 181*, 104643.

Nayek, R., R.Fuentes, WordenK.C.E.J. (2021). On spike-and-slab priors for bayesian equation discovery of nonlinear dynamical systems via sparse linear regression. *Mech. Syst. Signal Process.107986*.

Pierron, F. & M.Grédiac (2020). Towards material testing 2.0. a review of test design for identification of constitutive parameters from full-field measurements. *Strain*.

Ricci, M., A.Abbasi, P.Carrara, M.Flaschel, S.Kumar, S. Marfia, & L.DeLorenzis (in prep.). Experimental validation of euclid for the automated discovery of hyperelastic material models.

Roux, S.& F.Hild (2020). Optimal procedure for the identification of constitutive parameters from experimentally measured displacement fields. *International Journal of Solids and Structures184*, 14.

Thakolkaran, P., A.Joshi, Y.Zheng, M.Flaschel, L.De Lorenzis, & S.Kumar (2022). Nn-euclid: deep-learning hyperelasticity without stress data. *Journal of the Mechanics and Physics of Solids 169*, 105076.

Constitutive Models for Rubber XIII – Dal (Ed.)
© 2025 The Author(s), ISBN 978-1-032-85138-9

Towards a Digital Twin of the tire

T. Berger, M. Böttcher, F. Hartung & M. Kaliske
Institute for Structural Analysis, Technische Universität Dresden, Dresden, Germany

ABSTRACT: In many application fields, the development of a Digital Twin is under way in order to open up new capabilities of generating information and steering a system via this digital approach. This methodology could achieve significant advantages as well for the use of tires. Based on sensor data, realistic information on the driving and loading history could be obtained, and with the help of simulation approaches, consequences for future characteristics and integrity could be predicted. These predictions could yield an active feedback to the driver in order to account for modified loading, i.e. driving conditions, to avoid for example excessive wear or failure.

Different sources of information need to be used like measured data as well as simulated physics. One core component is therefore the simulation capability which could help to develop a Digital Shadow which is a holistic but passive representation in comparison to the active Digital Twin.

The finite element method is a far developed methodology to represent the physical phenomena to be observed and evaluated. It can accompany all phases of the tire starting from production (unvulcanized tire), predicting the tire in use (friction, rolling resistance) and even different mechanisms of failure (damage, facture, wear) need to be considered. The production process influences clearly the final tire properties. Therefore, it is too much idealized to investigate the tire structure just based on the intended ideal design. Reality is different and these production features could be taken into account via a production simulation. The unvulcanized rubber is represented by an appropriate material formulation. The green tire is then placed in the mold and heated up, in order to modify the rubber into the cross-linked material. It ends up with a realistic tire structure incorporating the production features.

The working conditions of the tire are manifold. Essential are the frictional features in the contact patch because friction is fundamental for the handling of the vehicle. Braking and wear characteristics depend on it among others.

Failure and degradation properties have to be accounted for as well. Simulating wear is a challenging task but significant for the prediction of the properties of the worn tire and the lifetime. Even the interaction with its environment as the vehicle and the pavement are important. Predicting driving on snow or soft soil are specific challenges but in the reach of current simulation capabilities.

Nowadays, the simulation capabilities with respect to the tire are well developed. Thus, the presentation will focus on the simulation chain of the tire, which is a significant constituting element of the Digital Twin, which needs to be supplemented subsequently by measured data and fast reduced order models. Current simulation capabilities will be pointed out and illustrated by numerical examples.

1 INTRODUCTION

A holistic description of the tire lifetime and its steering is proposed with the concept of Digital Twining. This general approach uses the exponentially increasing amount of data created by digitalization and modern communication for a self-steering process that combines all available information (experimental, physical, sensors, …). This is achieved by three separate evolutionary steps that will be presented below. In the current state, the physical tire and its digital representation are only coupled with each other by manual data transfer. In other words, the information of the real tire obtained by experiments or sensors are gathered and processed by users, which apply these to individual digital descriptions of certain aspects of the tire lifetime. The obtained findings are used for the tire design to influence the physical model. The focus at this step is the reliable and efficient representation of the physical phenomena by appropriate numerical descriptions. This is used for the virtual tire design process. Due to the large amount of data, the manual transfer leads to a loss of information. Due to the individual description of certain aspects, the automation is difficult and errors could be introduced by manual data transfer. Therefore, the next step in the evolution

DOI: 10.1201/9781003516880-2

towards a Digital Twin is the so-called Digital Shadow. This means that the physical tire is constantly tracked via sensors and these information are transferred automatically to the digital model. This requires a reliable and holistic numerical representation of the tire during its lifetime from production to its end of service. The Digital Shadow is able to represent the physical model and predicts the tire behavior at the current conditions in an automatized way. The information of the real tire could be used to improve the tire and validate the numerical descriptions. In the final step, the digital and the real tire form a cyber-physical system with automatized interaction in both directions. This could be used for example to automatically detect possible failures of the tire due to the applied loading conditions and sending the driver an immediate warning message to change the tire or driving condition. In current applications, the concept of a Digital Twin is used for predictive maintenance of production processes at varying boundary conditions.

Numerical representations of the reality, like FEM or MPM simulations, have become an invaluable tool in engineering tasks which allow to predict structural responses under various loading scenarios. These methods are based on the formulation of constitutive material models and differential equations describing the physical behavior. Therefore, these models are categorized as physics-based methods, or white-box models. Analytical models describing physical phenomena or systems also fall into the same category. If properly applied, numerical simulations can yield very accurate, i.e. high fidelity, results. A main drawback of these models could be the high computational effort. Depending on the complexity of the structure and of the model, a simulation can take minutes, hours or days. Such a duration to obtain the result of a single simulation can make tasks like uncertainty analyses, sensitivity analyses, design optimizations or scenario analyses infeasible.

An increasingly popular solution approach is the usage of non-physics-based models. As the name suggests, these kind of models do not include any formulation of physical laws, but are constructed by analyzing data. Therefore, the term data-driven methods is used synonymously. These are inevitable to use in a Digital Twin for the real-time evaluation of driving conditions and immediate intervention from the digital to the physical model.

During their lifetime, tires undergo various loading conditions and changes in the geometry and material characteristics. Therefore, at each aspect of its service life, other features are of interest and needs to be described. In this contribution, each step will be presented and numerical descriptions are shown. From the design of the tire to its production and its service life until wear will end its service life. Tires are complex structures and all the presented aspects will influence each other. Therefore, a numerical representation of the tire lifetime in a holistic way is required. This contribution starts like the tire life with the vulcanization of rubber and how the production process influences the subsequent service life of the tire. Subsequently, an optimized tire design process in a robust way including uncertain loading conditions and varying material properties are presented. In such way produced and designed tires are investigated with respect to its service life and a thermo-mechanically coupled rolling simulation framework is shown. A special aspect of the tire service life is the description of friction and wear as both are strongly coupled with each other. Friction determines the tangential contact of the tire with the ground and is required for cornering and driving maneuvers. The non-catastrophic end of service of a tire is primarily determined by wear. Finally, data-driven approaches and metamodels are presented for a fast processing of real-time scenarios.

2 TIRE PRODUCTION SIMULATION

One key aspect of Digital Twinning or cyber-physical systems is to incorporate all available data and physical phenomena into the digital model. However, most tire models focus solely on the final product that leaves the factory. These tires already underwent a various and severe deformation history, which has to be taken into account for a holistic and realistic description of the tire lifetime. During the tire production, the green tire made of uncured rubber and reinforcements is placed into a heating press. By applying a large inner pressure, the green tire is formed into its final shape and the green rubber is exposed to large deformation. The temperature in the mold is high to start vulcanization. The long chain molecules of rubber build cross-links with each other to form a stiff matrix in its current position. Therefore, the shape of the tire in the mold is conserved and the tire receives its stiff viscoelastic mechanical properties. Finally, the tire is released from the mold and cools down to room temperature. Due to the shrinkage behavior of Nylon cords, this leads to prestress in the reinforcements and influences the tire dimensions. Therefore, this cooling step is carried out with an applied inner pressure. The contraction of the reinforcements is restricted and the tire dimensions are controlled. This is experimentally observed and numerically represented in (Kim, Berger, Paek, Zreid, & Kaliske 2022) and (Kim, Berger, Piao, Zreid, & Kaliske 2022).

The phase change of rubber is represented by a mixed material model that combines two arbitrary formulations for the uncured and cured phase, see Figure 1. By including an additional state variable, the state of cure c, the free energy function distinguishes between the uncured phase, the cured phase and any intermediate state,

$$\bar{\Psi}(c, \overline{\mathbf{F}}) = c\,\bar{\Psi}_u(\overline{\mathbf{F}}, P_u) + (1 - c)\,\bar{\Psi}_c(\overline{\mathbf{F}}, \overline{\mathbf{F}}_c, P_c). \quad (1)$$

By introducing a plastic deformation $\mathbf{F}_c$ on the cured rubber branch, the irreversible shape change during vulcanization is represented. The evolution law is not stated directly, it is derived by assuming a conservation of energy during vulcanization,

$$\frac{\partial \overline{\Psi}}{\partial c} = -\overline{\Psi}_u + \overline{\Psi}_c + c \frac{\partial \overline{\Psi}_c}{\partial \overline{\mathbf{b}}_e} \frac{\partial \overline{\mathbf{b}}_e}{\partial c} = 0, \quad (2)$$

using a multiplicative split of the deformation gradient $\overline{\mathbf{F}} = \overline{\mathbf{F}}_e \overline{\mathbf{F}}_c$ and the elastic left Cauchy-Green tensor $\overline{\mathbf{b}}_e = \overline{\mathbf{F}}_e \overline{\mathbf{F}}_e^T$. Therefore, the stiffness increase is taken into account in a thermodynamically consistent way by introducing the irreversible shape change. The main advantage of the proposed rubber curing description is the independent formulation of the curing kinematics and the uncured and cured rubber phase.

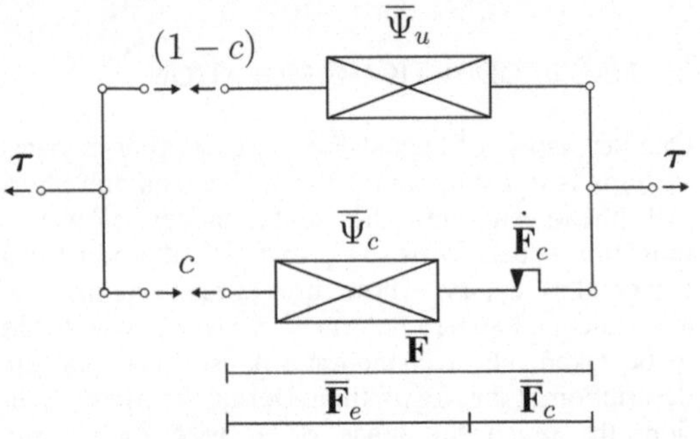

Figure 1. Generalized rheology of the mixed material model with the non-recoverable curing strain $\overline{\mathbf{F}}_c$.

In a numerical representation using the finite element method (FEM) and the LAGRANGE framework, the computational mesh distorts severely, which leads to artificial stiffness and convergence issues. Therefore, an arbitrary LAGRANGIAN EULERIAN (ALE) approach is proposed for an efficient and stable simulation framework (Berger & Kaliske 2022). An additional configuration is introduced to partially decouple the material and the spatial deformation. Therefore, it is possible to improve the quality of the computational mesh while tracking the deformation history of the material. Therefore, it is possible to represent large deformations in a numerically efficient way and to use the computational mesh for subsequent analyses. Therefore, it is possible to analyze the influence of the production process on the tire lifetime, for instance on the contact pressure distribution in the footprint (Berger, Kim, & Kaliske 2023) and on the tire rolling characteristics (Berger, Meyer, Uzeda, & Kaliske 2024).. Another possible approach to deal with large deformations is the use of a discretization by the material point method (MPM). In this approach, the deformable body is described by material points, which are not associated to a particular element, to track the deformation history. At each load increment, the material points are associated to a regular background mesh. The differential equations are evaluated using the background mesh and the movement of the material points is tracked and the computational mesh is reset after the load increment. This leads to a new distribution of the material points with respect to the computational mesh. Therefore, the elements cannot distort and the deformation history does not have to be projected. However, other challenges such as the consistent GAUSS integration or the description of tangential contact forces arise.

A first automatized approach of including sensor data in the production process is proposed in (Lee, Heo, Park, Piao, Boyadjian, Paek, Kim, Berger, & Kaliske 2021). Due to the real time validation of the simulation results and the update of the boundary conditions, possible defects of the tire caused by production can be detected. Additionally, the replacement cycles for wear parts, such as the bladder, can be optimized.

3 ROBUST TIRE DESIGN

As stated in the previous section, the tire production has a significant influence on the final tire characteristics and it is required to check in an automatized way that the produced tires are as intended. However, to minimize the possible defects of tires, a suitable design process in a robust manner is preferred, incorporating the variability, imprecision and incompleteness of material and geometry data as well as loading conditions. In tire design, these three uncertainty characteristics appear simultaneously in different combinations. Variability can be described by randomness, if a large number of statistical data is available. Imprecision and incompleteness are advantageously captured by the uncertainty model fuzziness (Götz, Graf, & Kaliske 2015). The mathematically construct of fuzzy sets based on a membership function μ is able to characterize small data sets and is applicable to asses experience and expert evaluations.

Robustness is a research field to utilize and quantify uncertainty in the multi objective optimization framework. Solutions in the design space are sought that are insensitive to perturbations caused by uncertain inputs. It may be less preferred to obtain a tire design that is identified as optimal for a certain loading condition and set of material parameters, while a small change, for example in the tire inner pressure or the YOUNG's moduli of the rubber compounds, leads to a distinctive change in the objective function. For evaluation of fuzzy sets, information reducing measures are introduced, which are classified into two classes. Reducing the uncertain quantities to a representative value M^R and measuring the uncertainty $M^{\mathcal{U}}$ (Schietzold, Leichsenring, G¨otz, Graf, & Kaliske 2021). For a robust design of the tire, the worst condition which could occur is investigated. Therefore, the lower bound of the fuzzy quantity support is of special interest,

$$M^R(\tilde{z}) = \min[z \in C(\tilde{z})]. \quad (3)$$

To quantify the amount of uncertainty in a fuzzy set, the area underneath the membership function,

$$M^{\mathcal{U}}(\tilde{z}) = \int_{-\infty}^{\infty} \mu_z(z_i)\,\mathrm{d}z, \tag{4}$$

is evaluated, which correlates with the zero-th moment of the fuzzy variable.

The optimal tire has low rolling resistance, high cornering stiffness, low wear, minimum noise and perfect grip at wet conditions. This has to be reached for a variety of loading conditions with as little material as possible, at low cost, and the tire should have a long lifetime. These design targets contradict each other and lead to a multiobjective optimization task with numerous conflicting objective functions. By taken into account the uncertainties of the loading and the material, the robustness of the design should be ensured as well. In the following example, the durability of the tire is improved while taken into account uncertainties. The main reason tires have to be replaced is due to wear or fracture. The first objective functions is the minimization of the difference of the contact pressure between the shoulder and the center,

$$\min f_1(\mathbf{x}_d, \tilde{\mathbf{p}}_a) = w_{eq}(\mathbf{x}_d, \tilde{\mathbf{p}}_a) = \frac{p_{N,S}(\mathbf{x}_d, \tilde{\mathbf{p}}_a)}{p_{N,C}(\mathbf{x}_d, \tilde{\mathbf{p}}_a)}. \tag{5}$$

This would lead to a homogeneous wear across the tire contact patch and minimizes the maximum contact pressure in the footprint and, consequently, the overall wear. The second target to increase the durability of the tire is to avoid fracture in areas of large deformation. This is usually found at the edges of the belt reinforcements near the tire shoulder. It is assumed that fracture is induced due to large amplitudes of the strain energy density during one revolution of the tire,

$$\Delta W = W_{\max} - W_{\min}. \tag{6}$$

Therefore, the minimization of the amplitude ΔW is the second objective function,

$$\min f_2(\mathbf{x}_d, \tilde{\mathbf{p}}_a) = \Delta W(\mathbf{x}_d, \tilde{\mathbf{p}}_a). \tag{7}$$

The most influential design parameters $\mathbf{x}_d$ regarding fracture and wear are identified as the belt angle, the number of capplies and the thickness of the tread layer. As stated above, the tire design should be less sensitive to changes in the loading or material variability. Therefore, some of them are introduced as fuzzy sets. The inner pressure of tires reduces over time and has to be controlled by the driver on a regular basis. However, in reality, this is not always the case. The tire is applied to different vehicles with variable weights and this yields to different guidelines for the inner pressure. The Nylon fibers used as reinforcements of tires are influenced by non-constant production conditions that causes variation in spacing. Finally, the varying stiffness of the tread compound is taken into account. A robust tire design is able to increase the durability at all possibly appearing inner pressures, at all admissible driving velocities and despite of geometrical production imperfections.

For an equal weighting of both objective functions and to take into account the robustness of the design, the uncertainty measure $M^{\mathcal{U}}$ and the information reduction measure M^R are used to form the aggregated target function,

$$\begin{aligned} f_{\text{aggr}} = \; & M^R(f_1(\mathbf{x}_d, \tilde{\mathbf{p}})) + M^{\mathcal{U}}(f_1(\mathbf{x}_d, \tilde{\mathbf{p}})) \\ & + M^R(f_2(\mathbf{x}_d, \tilde{\mathbf{p}})) + M^{\mathcal{U}}(f_2(\mathbf{x}_d, \tilde{\mathbf{p}})), \end{aligned} \tag{8}$$

using the design parameters $\mathbf{x}_d$ and take into account the uncertain parameters $\tilde{\mathbf{p}}$.

For an efficient optimization process, an artificial neural network (ANN) is utilized as a metamodel to reduce the required amount of FE solutions. Applications of data-driven models in general are discussed in Section 7. The three-dimensional tire model is simulated at 200 points with varying design and uncertainty parameters. For validation of the ANN, 40 points are stored and the remaining 160 points are used for training. The results of the optimization algorithm is shown in Figure 2. It can be seen, that the difference in the contact pressure as well as the strain energy density amplitude is decreasing during the optimization. The uniform contact pressure distribution is obtained for the worst and best case of the uncertainty parameters and shows only a small difference. This indicates a high robustness of the obtained design parameters with respect to this objective function. It can be seen from the optimization results that the main influence factor of the uniform contact pressure distribution is the thickness of the tread layer. The fatigue resistance however is mainly influenced by the number of capplies and the belt angle, with an almost constant robustness of the identified tire designs. The robust tire design is presented in more detail in (Serafinska, Kaliske, Zopf, & Graf 2013).

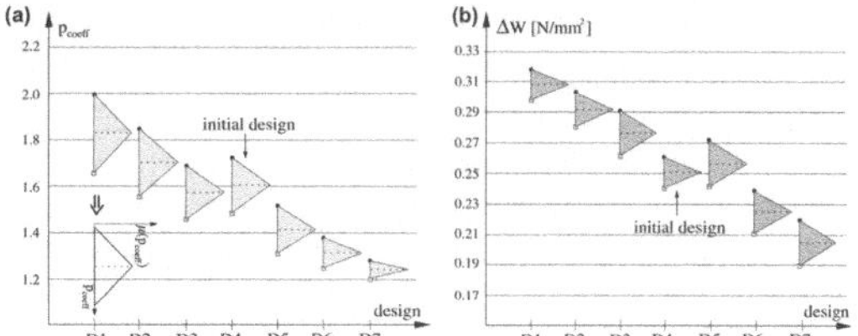

Figure 2. Improvement of the fuzzy results of (a) f_1 and (b) f_2 with the robust tire optimization.

4 TIRE IN SERVICE

For describing the motion of the tire, two completely different techniques are available, transient and stationary rolling. At transient approaches, the time scale is discretized and a time step is carried out tracking the motion of a material particle directly. Therefore, a non-axisymmetric tread pattern and various loading conditions can be represented. This usually leads to a large number of degrees of freedom, which increases the computational time. Therefore, it is used preferable for short time scales, for example a tire impact with a curb, and in combination with an explicit solution scheme. A numerically more efficient approach is realized by a stationary rolling framework. This is represented by an ALE approach, where the rigid body rotation of the tire is described in an EULERIAN frame and the deformation due to contact with the road in an LAGRANGIAN manner. Therefore, it is possible to use an adaptive discretization with more elements in the contact patch and less elements at the top of the tire. This leads to a time-independent formulation, where the behavior of the tire is represented as an observer sitting on the driving axes. The time-dependent material behavior of rubber is taken into account via so-called streamlines. The deformation history is not tracked at the integration point at a previous time step. Instead, it is evaluated at the previous integration point on the streamline. In other words, the material is flowing through the computational mesh to describe the time-dependent material behavior. In (Garcia, Kaliske, Wang, & Bhashyam 2016), a consistent derivation of the viscoelastic stress tangent is carried out that incorporates the whole streamline and leads to quadratic convergence.

In recent years, an other discretization method which is especially powerful in complex geometries as tires gets more popular. In the isogeometric analysis (IGA), the linear ansatzfunctions are replaced by non-uniform rational B-splines (NURBS). Therefore, the geometry is exactly described by control points and knot vectors. The geometry is represented in a smooth and continuous way even between element boundaries. It is possible to refine the discretization via element insertion and elevation or a combination of both (Israfilova, Garcia, & Kaliske 2021). Due to the approximation of the geometry via control points, special treatment of the reinforcements and contact has to be applied. However, as shown in (Garcia, Israfilova, Liang, Zhao, Wei, & Kaliske 2022), the velocity and acceleration fields in the tire can be described more accurately than with finite elements.

The viscoelastic material behavior of rubber influences not only the deformation during rolling. Due to the hysteresis observed by a loading and unloading behavior, energy is dissipated. During rolling, the rubber is exposed to a cyclic loading, which leads to a significant amount of dissipated energy and, consequently, to a significant temperature increase in the tire. Due to the highly temperature-dependent material behavior of rubber, this increase has to be taken into account for a realistic description of the tire rolling. In (Behnke & Kaliske 2015), a numerically efficient simulation framework is proposed to describe the temperature evolution in the tire over time.

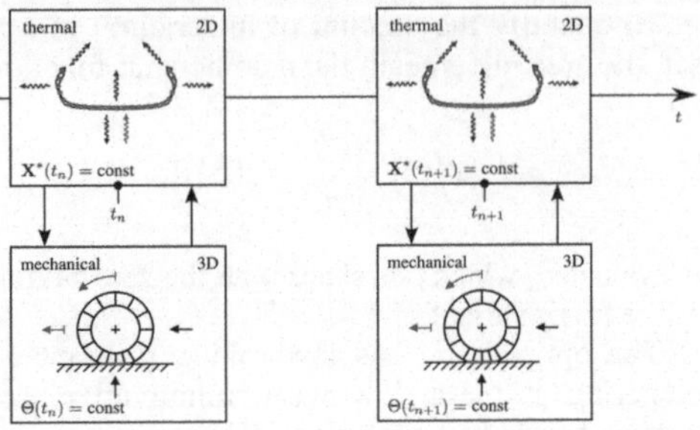

Figure 3. Simulation overview on the sequentially coupled tire rolling algorithm.

A staggered solution approach is proposed, where the mechanical and thermal representation are carried out subsequently. This is schematically shown in Figure 3. A steady state rolling simulation of a three-dimensional tire is carried out with a fixed temperature distribution. By applying the rotational and translational velocity as well as the vertical loading, a deformation state is computed. An averaged dissipated energy rate is obtained for each streamline according to the time-dependent viscoelastic material description. Therefore, the dissipated energy for a full rotation of the tire is obtained. It is assumed that the temperature evolves on a much larger timescale than one rotation and the temperature in the circumferential direction is constant. Therefore, the temperature increase is described using only the cross-section. In the transient thermal analysis, the displacement of the tire is neglected and a time step is represented. The energy dissipation rate, computed from the mechanical analysis, is taken into account as an external heat source. The temperature distribution is computed by taken into account the thermal conductivity, specific heat capacity and the surface heat exchange with the environment. After a predefined time step, the newly obtained temperature distribution in the tire cross-section is transferred back to the mechanical analysis and a new steady state rolling simulation is carried out. Due to the temperature-dependent material behavior, a new deformation state is computed and, consequently, the dissipated energy rate changes. This would influence the thermal analysis in the next time step and vice versa. At constant rolling conditions and by taken into account the heat exchange with the environment, the temperature state reaches a steady state after a certain time. For a numerically efficient simulation, the time step in the thermal analysis is adaptively increased based on the averaged temperature and dissipated energy rate,

$$\Delta t = t^{n+1} - t^n = s(t^n - t^{n-1}), \tag{9}$$

using the time step increase,

$$s = \frac{1}{2}\left(\left|\frac{\left[w_{\text{diss}}(t^{n-1}) - w_{\text{diss}}(t^{n-2})\right]\Delta t^n}{\Delta t^{n-1}\left[w_{\text{diss}}(t^n) - w_{\text{diss}}(t^{n-1})\right]}\right|\right). \tag{10}$$

Therefore, the time step is small at the beginning of the analysis, where the temperature and the dissipation rate change significantly. After several iterations, the temperature distribution approaches a steady state and the time steps increase accordingly, see Figure 4.

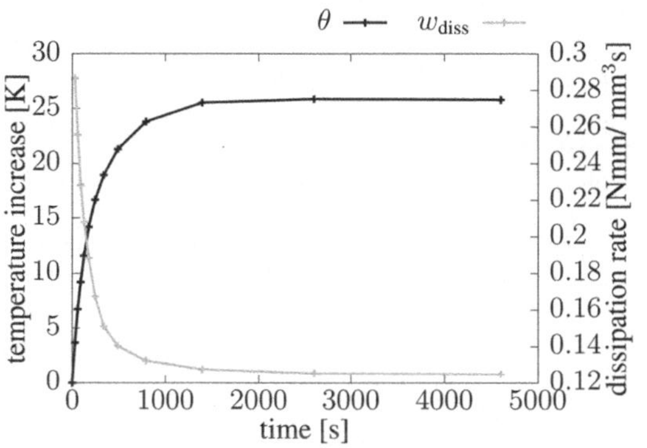

Figure 4. Average temperature and dissipation rate evolution over time using an adaptive time step size.

The material is described by the generalized MAXWELL model to take into account a wide range of frequencies. The time-independent elastic material behavior is captured using a hyperelastic formulation based on the extended tube model. The time-dependent behavior is described by a linear or a nonlinear viscoelastic formulation. The differences of the two descriptions as well as a comparison is presented in more detail in (Berger, Behnke, & Kaliske 2016).

5 FRICTION

Friction has a huge influence of almost every mechanical application and also on our daily life. Without friction, the tire would not move forward and could not transfer any forces from the vehicle to the road. However, at free rolling or sliding applications, the amount of friction is preferably small. An optimization of the friction contribution with respect to its application is required. Therefore, a deep understanding and characterization of rubber friction is needed for a holistic description of the tire.

Rubber friction can be described well with a multiscale method. The surfaces the tire normally rolls on appears only on the macroscale like a homogeneous surface. But in the range of cm, mm or even μm, the surface consists of several asperities and is a rough structure. The roughness of the surface can be approximated and described by the height difference correlation (HDC) function,

$$C_{\text{HDC}} = \langle (z(x+\zeta) - z(x))^2 \rangle. \tag{11}$$

This function compares the average height difference of two points with the same distance ζ. The height difference correlation function is linear in the logarithmic space, if the surface texture is self-affine. That means, that over all length scales, the ratio of amplitude and wavelength is constant (Wriggers & Reinelt 2009). Another approach of describing the surface texture is the power spectrum density (PSD),

$$C_{\text{PSD}}(\boldsymbol{q}) = \frac{1}{(2\pi)^2} \int \langle z(\boldsymbol{x})\, z(\boldsymbol{0}) \rangle\, e^{-i\boldsymbol{q}\cdot\boldsymbol{x}} d^2x, \tag{12}$$

which can be computed by FOURIER transformation (Persson, Albohr, Tartaglino, Volokitin, & Tosatti 2004). In Eq. (12), $\boldsymbol{q}$ represents the wave frequency vector. The PSD and the HDC function can be converted into each other. A third option is the bijective decomposition of the surface texture with a so-called bandpass filter (see Wagner, Wriggers, Veltmaat, Clasen, Prange, & Wies 2017). This is achieved by predefined lower and upper cutoff frequencies or wavelengths. For the proposed multiscale friction description, the height difference correlation function and the power spectrum density will be used.

Friction in general consists of different contributions, where the two predominantly ones are adhesion and hysteresis friction. Hysteresis friction stems from the viscoelastic material behavior of rubber. During sliding on a rough surface, rubber is exposed to a cyclic loading and due to the dissipation energy, a resistance against motion from the material arise. The hysteresis friction can be described analytically or numerically. In (Persson 2001), an analytical model to predict hysteresis friction is derived. Based on the linear elastic stiffness of rubber and the (isotropic) power spectrum density function $C_{\text{PSD}}(q)$, the real contact area is computed at each wavelength of the surface,

$$P(q) = \frac{A}{A_0} = \text{erf}\left(\frac{1}{2\sqrt{G}}\right), \tag{13}$$

using

$$G(q) = \frac{1}{8}\int_{q_l}^{q} q^3\, C_{\text{PSD}}(q) \int_0^{2\pi} \left|\frac{E(q\, v \cos\varphi)}{(1-\nu^2)\,\sigma_0}\right|^2 \mathrm{d}\varphi\, \mathrm{d}q. \tag{14}$$

Subsequently, the reaction force from the viscoelastic material contribution is obtained by taken into account the loss module and the sliding velocity,

$$\mu = \frac{1}{2}\int_{q_0}^{q_1} q^3\, C_{\text{PSD}}(q)\, P(q) \int_0^{2\pi} \cos\varphi\, \text{Im}\frac{E(q\, v \cos\varphi)}{(1-\nu^2)\sigma_0}\, \mathrm{d}\varphi\, \mathrm{d}q. \tag{15}$$

In (Persson 2006), the temperature-dependency of rubber material is included in the hysteresis friction formulation. During sliding over the rough surface, the dissipated energy of rubber leads to a significant temperature increase in the contact area. Therefore, the rubber material behavior changes, which yields to an increase of the real contact area. However, due to the softer behavior, the reaction forces decrease and the hysteretic friction contribution decreases. By taking into account thermal conductivity and heat exchange, the temperature at each length scale is derived. Therefore, the material behavior of each length scale can be described according to the temperature-dependent material characterization. Therefore, the for rubber typical decrease of the friction coefficient at large sliding velocities is represented. The analytical description of rubber friction using Eq. (15) has two main drawbacks. First, it is uncertain up to which length scale rubber can be represented as a bulk material and has a significant contribution to the hysteresis friction. Therefore, the smallest wavelength q_1, which has the largest influence on the friction is only guessed. The second drawback is related to the mechanical material description. The analytical derivation of PERSSON uses a linear elastic formulation. However, at smaller scales, rubber undergoes relatively large deformations and it is well known that rubber shows a highly nonlinear behavior at finite strains. This cannot be taken into account in the analytical formulation.

In a numerically description of the hysteretic rubber friction, the latter drawback could be overcome. Instead of an analytical derivation of the friction contribution, a multiscale homogenization approach is proposed. This is realized by splitting the surface into a finite number of sinusoidal waves with different length scales. At each scale, a friction map is constructed by numerically sliding a rubber block over the surface at a predefined contact pressure and velocity. The obtained reaction forces are used to construct a friction map, which will be used for the next larger scale. Therefore, the friction contribution from each length scale is taken into account to obtain the macroscopic friction map. The friction map for the smallest length scale is assumed by the adhesive friction contribution, which will be discussed in the following.

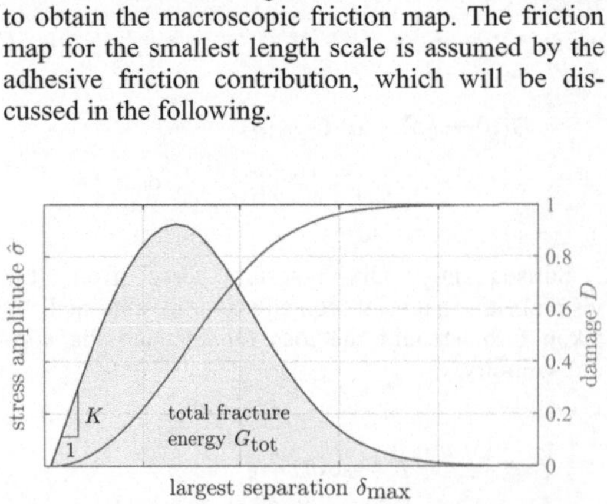

Figure 5. Traction-separation-law of adhesion model.

Adhesion arises from the molecular bonds between two different surfaces and gives and important contribution to rubber friction. Adhesive friction can e.g. be modeled by a nonlinear traction-separation-law with the adhesive stress vector,

$$\boldsymbol{\sigma} = (1 - D) \cdot K \cdot \boldsymbol{\delta}, \tag{16}$$

where $\boldsymbol{\sigma}$ relates on the damage function D, the initial adhesive stiffness K and the relative separation vector $\boldsymbol{\delta}$ between two contact points in normal and two tangential directions (Kaliske, Behnke, Hartung, & Wollny 2021). The devolution of the intensity of adhesion proposed in (Raous, Cang´emi, & Cocu 1999) is comparable to the evolution of the damage function D. A nonlinear approach for the damage function,

$$D = \frac{G}{G_{tot}} = \frac{1}{G_{tot}} \int_0^{\delta_{\max}} \hat{\sigma} \, \mathrm{d}\delta, \tag{17}$$

with only two unknown parameters (K, G_{tot}) is chosen to decrease the numerical effort of parameter identification. No stresses can be transferred between the contact points as soon as the damage value is zero (total debonding). Parameters G and G_{tot} in Eq. (17) represent the current and total fracture energy, accordingly. Symbol $\delta_{\max} = \max(\| \boldsymbol{\delta} \|)$ refers to the largest separation and symbol $\hat{\sigma} = \| \boldsymbol{\sigma} \|$ stands for the absolute adhesion stress. In Figure 5, the area underneath the transferred stress corresponds to the total fracture energy.

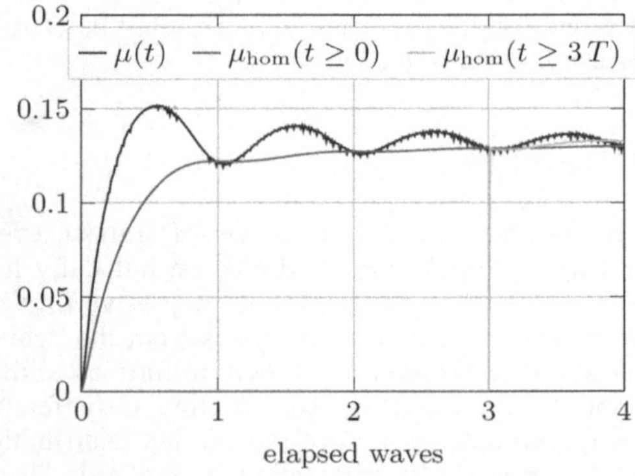

Figure 6. Time-homogenization of friction coefficient.

Friction features on a specific length scale can be homogenized to generate friction characteristics in order to apply them as a friction law for a next coarser length scale. As soon as a rubber block slides over a rough surface, a time-dependent friction coefficient results from the viscoelastic properties of the rubber material. The oscillation of friction features is smoothed by

$$\mu_t = \frac{1}{t_{\text{tot}} - t} \int_t^{t_{\text{tot}}} \mu(\tau) \, \mathrm{d}\tau, \tag{18}$$

where t and t_{tot} are referred to the starting time of homogenization and total simulation time, respectively. The outcome of the time homogenization is exemplarily presented in Figure 6, in which T stands for the periodic time of a sine wave function. Using a piecewise cubic spline interpolation, a load- and speed-dependent friction law for a coarser length scale is created by

$$\mu_{\text{spl}}(p, v) = \sum_{i,j=1}^{n_p, n_v} c_{i,j} \left(p - \xi_p\right)^{n_p - i} \left(v - \xi_v\right)^{n_v - j}, \quad (19)$$

with $n_p = n_v = 4$ (cubic spline generation), the breakpoints ξ_p and ξ_v (load and speed conditions in each block simulation of current length scale) and the spline coefficients $c_{i,j}$ (see Figure 7).

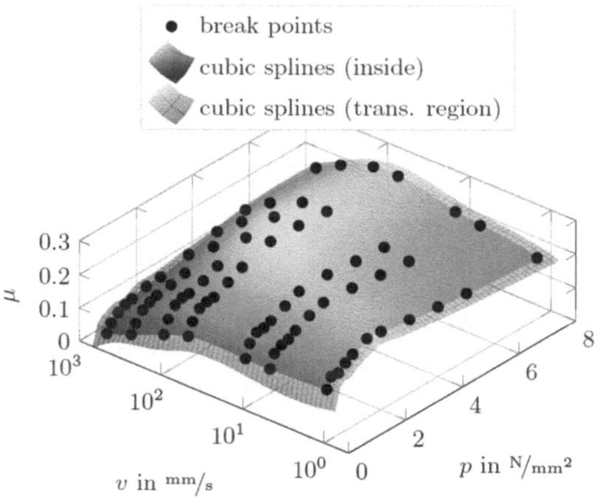

Figure 7. Derived friction map for next coarser length scale.

The combination of the length scale decomposition as well as the introduced time homogenization lead to a multi-scale friction procedure to compute macroscopic friction features (Hartung, Friederichs, Eckstein, & Kaliske 2022). However, numerical studies in (Kaliske, Behnke, Hartung, & Wollny 2021) have shown that a reliable multiscale rubber friction approach requires representative length scales, whose minimum wave length is a multiple of the maximum wave length of the next smaller length scale, i.e. a certain distance between neighboring scales has to be ensured to perform valid time homogenization with respect to friction.

6 WEAR

Wear as consequence of friction is divided into several types, such as adhesive, abrasive, corrosive and fatigue wear. On the other hand, abrasion (material loss due to wear) affects the friction characteristics. Following the findings of ARCHARD and SCHALLAMACH, see e.g. (Archard 1953) and (Schallamach & Turner 1960), wear directly correlates to frictional energy or frictional power. Furthermore, linear friction and wear tests on block level have stated that a nonlinear relationship between volume wear rate intensity,

$$\frac{\partial \dot{V}_w}{\partial A} = k_w \left(\frac{\partial \dot{E}_f}{\partial A}\right)^{a_w}, \quad (20)$$

and frictional power intensity is applicable, where k_w and a_w are wear model parameters to be identified. In Figure 8, the wear model proposed in Eq. (20) is fitted to the wear test results to obtain the wear model parameters.

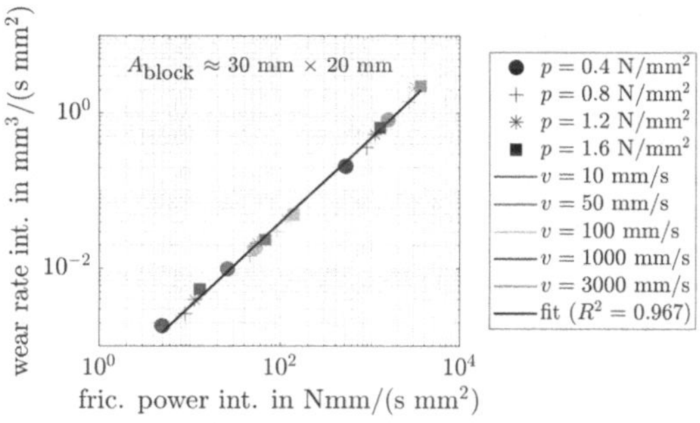

Figure 8. Identification of wear model parameters.

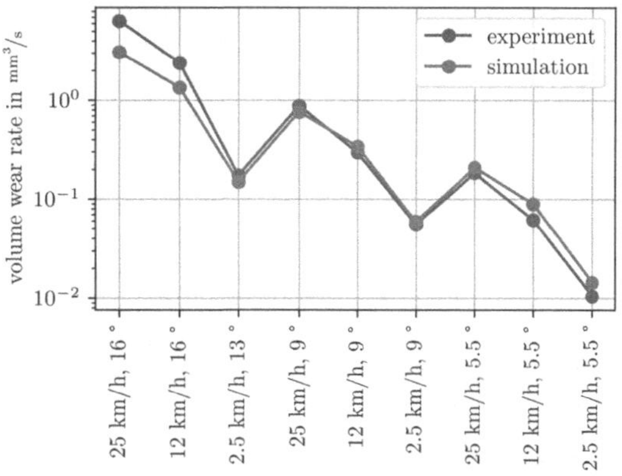

Figure 9. Validation of wear model using LAT 100 testing protocol.

The contact surface area of the investigated rubber blocks is approximately $A_{\text{block}} = [600]mm$. The resulting coefficient of determination of $R^2 = 0.967$ in Figure 8 proves again the direct correlation between frictional energy rate and wear rate.

There are different simulation frameworks to update the geometry of the underlying FE mesh as soon as the volume to be abraded is computed. For instance, a global-local approach is suitable for complex structures containing a large number of elements and nodes, e.g. rolling tires with detailed tread pattern. Typically, the wear volume is computed at FE nodes which is transferred to a wear height and a wear direction vector (Li, Li, & Wang 2020). Consequently, a sophisticated re-mesh algorithm is required to minimize the difference to the wear volume that results by moving the nodes in contact by the corresponding wear height. In addition, the geometry update frequency is a crucial factor towards the final worn shape. In (Hartung, Garcia, Berger, Hindemith, Wangenheim, & Kaliske 2021), a post-processing remeshing algorithm (mesh update after simulation) is compared to an adaptive meshing procedure (mesh is modified in every time step). Nevertheless, it is not advisable to use the findings on rubber blocks with respect to wear directly for rolling tires, since the wear mechanism as well as the energy intensity levels differ. The integration of a wear evolution law into the simulation framework may consider lower wear rates at small sliding distances within the contact patch of tires. The output of the developed post-processing wear analysis tool using an LAT testing protocol is presented in Figure 9. The speed and the angle indications in Figure 9 relate to the speed of the LAT 100 disc and the slip angle of the applied rubber wheel, accordingly (Salehi, Noordermeer, Reuvekamp, & Blume 2020).

7 DATA-DRIVEN MODELS

As a counterpart to physics-based methods, which are the focus of the previous sections, non-physics-based methods, or data-driven methods, provide a solution approach for tasks which require either a fast, close to real-time prediction of a systems response, or a large number of predictions with varying parameters. This is an essential functionality that a Digital Twin needs to possess, and a main advantage of data-driven models. Only then is a close to real-time scenario analysis possible, or a design optimization considering uncertainty as described in Section 3.

The subcategory of these methods on which this contributions is focused on, is that of model-based data-driven methods, or, in short, data-driven models. Mathematical formulations are trained, or fitted, on data to model the relationship between input and output parameters, without assigning any physical meaning. While simple data-driven models can be white-box models, e.g., a linear regression, it is more common to apply gray- or black-box models, such as neural networks, which are almost impossible to interpret, to more complex tasks, see, e.g., (Aldakheel, Satari, & Wriggers 2021), (Gentile & Galasso 2020) and (Serafinska, Graf, & Kaliske 2018).

Another subcategory which also falls under non-physics-based methods is the class of model-free data-driven methods. No underlying mathematical function is predefined and fitted to the data, but rather the data itself is utilized to make predictions for the unobserved parts of the domain. An application for engineering tasks is the prediction of stress-strain behavior based on existing data to avoid the necessity for a material model within the FE simulation, (Kirchdoerfer& Ortiz 2016) and (Zschocke, Graf, & Kaliske 2023).

A metamodel, which falls into the subcategory of data-driven models, is defined as a model of another, more complex model. From mathematical point of view, a metamodel provides a statistical approximation or, in some cases, interpolation of the response of the original model. The goal of the metamodel is to give a good approximation of the result that the computational simulation model would yield for the same values of the input parameters. To learn the relationship between input and output parameters present in the simulation model, a set of training data samples is required. While it is possible to train a metamodel using experimental results, this is mostly not practicable due to high costs and effort of performing the required number of experiments. Therefore, physics-based models, like FE models, are utilized in order to generate the data samples. A fundamental challenge is to create a data set which sufficiently describes this relationship throughout the entire domain. This is especially the case for a high number of input parameters, highly nonlinear functions and expensive, i.e. time-consuming, simulations. Different data sampling schemes to decrease the necessary number of data samples using sequential sampling strategies can be found, e.g., in (Böttcher, Fuchs, Leichsenring, Graf, & Kaliske 2021) and (Fuhg, Fau, & Nackenhorst 2021).

Another advantage of using metamodels is that no specific knowledge about the physical behavior of the system is needed by the user, making them widely applicable to a variety of tasks, provided that a deep understanding of the metamodeling procedure itself as well as high quality training data samples are available. Simultaneously, this lack of physical laws within the models, and in general the gray- or black-box nature of the models, is a point of criticism, since the result of the models are not verifiable if no data samples are available for a given parameter combination. For that reason, no reliable extrapolation is possible. In order to test the general applicability of the models apart from the training data samples, test data samples which are not included in the training data set are used to evaluate the models accuracy.

In general, a well executed metamodel training including a hyperparameter optimization of the model parameters, such as the number of hidden layers and neurons in an artificial neural network, is fundamental to create a reliable metamodel of sufficient accuracy, which can be used for predictions within the Digital Twin.

8 CONCLUSIONS

In this contribution, all the individual aspects of the tire lifetime from its production to its end of service are discussed and specific numerical representations of the single phases such as material descriptions and predictive methods are presented. It could be shown that these influence each other and cannot be considered separately. Suitable numerical descriptions for each step in the life of a tire are available and should be combined consistently for a holistic description of the tire lifetime. Furthermore, a robust optimization, which predicts an optimal tire design for a multicriterion objective function, while taken into account uncertainties, is discussed. For the efficient optimization and the real-time evaluation of sensor data, suitable data-driven or metamodels are required. Therefore, several classes of these methods as a counterpart to physic-based methods is presented. All the presented approaches are used to draft the vision of a Digital Twin, which combines all the available descriptions of the digital tire and real-time information of the physical tire to create a cyber-physical system. This could be used for an automatized evaluation and steering of the physical system based on changing conditions and digital representations. Therefore, all available information could be used to warn the driver and increase the safety and durability of tires.

REFERENCES

Aldakheel, F., R. Satari, & P. Wriggers (2021). Feed-Forward Neural Networks for Failure Mechanics Problems. *Applied Sciences 11*, 6483.

Archard, J. F. (1953). Contact and rubbing of flat surfaces. *Journal of Applied Physics 24*, 981–988.

Behnke, R. & M. Kaliske (2015). Thermo-mechanically coupled investigation of steady state rolling tires by numerical simulation and experiment. *International Journal of Non-Linear Mechanics 68*, 101–131.

Berger, T., R. Behnke, & M. Kaliske (2016). Viscoelastic linear and nonlinear analysis of steady state rolling rubber wheels: a comparison. *Rubber Chemistry and Technology 89*, 499–525.

Berger, T. & M. Kaliske (2022). An arbitrary lagrangian eulerian formulation for tire production simulation. *Finite Elements in Analysis and Design 204*, 103742.

Berger, T., S. H. Kim, & M. Kaliske (2023). A Microsphere-Based Rubber Curing Model for Tire Production Simulation. *Tire Science and Technology 51*, 82–113.

Berger, T., J. Meyer, S. Uzeda, & M. Kaliske (2024). Influence of the production process on tire rolling characteristics. *Constitutive Models for Rubber XIII – Proceedings of the 13th European Conference on Constitutive Models for Rubber.*

Böttcher, M., A. Fuchs, F. Leichsenring, W. Graf, & M. Kaliske (2021). Elsa: An efficient, adaptive ensemble learning-based sampling approach. *Advances in Engineering Software 154*, 102974.

Fuhg, J. N., A. Fau, & U. Nackenhorst (2021). Stateâ€'ofâ €'theâ€'Art and Comparative Review of Adaptive SamplingMethods for Kriging. *Archives of Computational Methods in Engineering 28*, 2689–2747.

Garcia, M., M. Kaliske, J. Wang, & G. Bhashyam (2016). A consistent implementation of inelastic material models into steady state rolling. *Tire Science and Technology 44*, 174–190.

Garcia, M. A., A. Israfilova, G. Liang, T. Zhao, Y. Wei, & M. Kaliske (2022). Isogeometric analysis for accurate modeling of rolling tires. *Computers & Structures 260*, 106717.

Gentile, R. & C. Galasso (2020). Gaussian process regression for seismic fragility assessment of building portfolios. *Structural Safety 87*, 101980.

Götz, M., W. Graf, & M. Kaliske (2015). Structural design with polymorphic uncertainty models. *International Journal of Reliability and Safety 9*, 112–131.

Hartung, F., J. Friederichs, L. Eckstein, & M. Kaliske (2022). Experimental and numerical investigation on multiscale hysteresis friction on artificial printed surfaces. *Tribology International*, 107999.

Hartung, F., M. A. Garcia, T. Berger, M. Hindemith, M. Wangenheim, & M. Kaliske (2021). Experimental and numerical investigation of tire tread wear on block level. *Lubricants 9*, 113–142.

Israfilova, A., M. Garcia, & M. Kaliske (2021). Isogeometric Analysis for Tire Simulations: From Mesh Generation to High Precision Results. *Tire Science and Technology 49*, 260–275.

Kaliske, M., R. Behnke, F. Hartung, & I. Wollny (2021). Multi-physical and Multi-scale Theoretical-Numerical Modeling of Tire-Pavement Interaction. In *Coupled System Pavement-Tire-Vehicle*, pp. 1–39. Springer, Cham.

Kim, S., T. Berger, J. Paek, I. Zreid, & M. Kaliske (2022). Numerical Modeling and Experimental Investigation of Thermal Shrinkage in Polymeric Cords. I: Constitutive Modeling. *Tire Science and Technology 51*, 39–62.

Kim, S., T. Berger, M. Piao, I. Zreid, & M. Kaliske (2022). Numerical Modeling and Experimental Investigation of Thermal Shrinkage in Polymeric Cords. II: Numerical Implementation. *Tire Science and Technology 51*, 63–80.

Kirchdoerfer, T. & M. Ortiz (2016). Data-driven computational mechanics. *Computer Methods in Applied Mechanics and Engineering 304*, 81–101.

Lee, W., S. Heo, D. Park, M. Piao, A. Boyadjian, J. Paek, S. Kim, T. Berger, & M. Kaliske (2021). Simulating tire manufacturing. *Tire Technology International 13*, 88–91.

Li, Z., Z. Li, & Y. Wang (2020). An integrated approach for friction and wear simulation of tire tread rubberr. part ii: Wear test, characterization and modeling. *Tire Science And Technology 48*, 146–165.

Persson, B. N. J. (2001). Theory of rubber friction and contact mechanics. *Journal of Chemical Physics 115*, 3840–3861.

Persson, B. N. J. (2006). Rubber friction: Role of the flash temperature. *Journal of Physics: Condensed Matter 18*, 7789–7823.

Persson, B. N. J., O. Albohr, U. Tartaglino, A. I. Volokitin, & E. Tosatti (2004). On the nature of surface roughness with application to contact mechanics, sealing, rubber friction and adhesion. *Journal of Physics: Condensed Matter 17*, R1–R62.

Raous, M., L. Cangémi, & M. Cocu (1999). A consistent model coupling adhesion, friction, and unilateral contact. *Computer Methods in Applied Mechanics and Engineering 177*, 383–399.

Salehi, M., J. W. M. Noordermeer, L. A. E. M. Reuvekamp, & A. Blume (2020). Parameter optimization for a laboratory friction tester to predict tire abs braking distance using design of experiments. *Materials & Design 194*, 108879.

Schallamach, A. & D. M. Turner (1960). The wear of slipping wheels. *Wear 3*, 1–25.

Schietzold, F. N., F. Leichsenring, M. Götz, W. Graf, & M. Kaliske (2021). Robustness versus Performance - Nested Inherence of Objectives in Optimization with Polymorphic Uncertain Parameters. *Advances in Engineering Software 156*, 102932.

Serafinska, A., W. Graf, & M. Kaliske (2018). Artificial neural networks based friction law for elastomeric materials applied in finite element sliding contact simulations. *Complexity 2018*, 1–15.

Serafinska, A., M. Kaliske, C. Zopf, & W. Graf (2013). A multi-objective optimization approach with consideration of fuzzy variables applied to structural tire design. *Computers & Structures 116*, 7–19.

Wagner, P., P. Wriggers, L. Veltmaat, H. Clasen, C. Prange, & B. Wies (2017). Numerical multiscale modelling and experimental validation of low speed rubber friction on rough road surfaces including hysteretic and adhesive effects. *Tribology International 111*, 243–253.

Wriggers, P. & J. Reinelt (2009). Multi-scale approach for frictional contact of elastomers on rough rigid surfaces. *Computer Methods in Applied Mechanics and Engineering 198*, 1996–2008.

Zschocke, S., W. Graf, & M. Kaliske (2023). Incorporating uncertainty in stress-strain data acquisition: extended model-free data-driven identification. *Proceedings in Applied Mathematics and Mechanics 23*(2), e202300008.

Constitutive Models for Rubber XIII – Dal (Ed.)
© 2025 The Author(s), ISBN 978-1-032-85138-9

Recent advances in magneto-active polymers: Experiments, modelling and simulations

M. Hossain & A. Bastola
Department of Mechanical Engineering, Zienkiewicz Institute for Modelling, Data and AI, Swansea University, UK

M.A. Moreno-Mateos & D. Garcia-Gonzalez
Department of Continuum Mechanics and Structural Analysis, Universidad Carlos III de Madrid, Madrid, Spain

C. Kadapa
Department of Mechanical Engineering, Edinburgh Napier University, UK

ABSTRACT: The last decade has witnessed the emergence of magneto-active polymers (MAPs) as one of the most advanced multi-functional soft composites. An MAP consists of a soft elastomeric matrix filled with micron- or nano-sized magnetisable rigid particles. These magneto-responsive particles are generally classified into two main groups: soft-magnetic particles and hard-magnetic particles. When soft-magnetic particles are used in an MAP, these can be magnetised by an external field. However, such a magnetisation will disappear as soon as the external magnetic field is switched off. In the case of hard-magnetic particles, their magnetisations will largely retain even after the removal of the applied magnetic field. When designing such responsive materials, the choice of both (soft matrix and rigid particles) phases is crucial. In this regard, the stiffness of the polymeric matrix determines the composite resistance to deform under an external magnetic field, i.e., the softer the matrix, the stronger the magnetostriction response. Polymeric materials are widely treated as fully incompressible solids that require special numerical treatment to solve the associated boundary value problem. Furthermore, both soft and hard magnetic particles-filled soft polymers are inherently viscoelastic. In this talk, we will at first present a wide range of experimental studies conducted on soft-and hard-magnetic polymers. Afterwards, we propose a unified simulation framework for magneto-mechanically coupled problems that can model hard and soft MAPs made of compressible and fully incompressible polymers, including the effects of the time-dependent viscoelastic behaviour of the underlying matrix. Finally, using a series of experimentally-driven examples consisting of beam and robotic gripper models under magneto-mechanically coupled loading, the versatility and benefits of the proposed framework are demonstrated. The effect of viscoelastic material parameters on the response characteristics of MAPs under coupled magneto-mechanical loading is also studied.

1 INTRODUCTION

In the mid of the last century, the notion of so-called smart or intelligent materials has emerged in the science and engineering fields. These materials can respond to more than one external fields. Such an external field may tune material properties and also may change their sizes or shapes. While in the early days, stimuli-responsive materials were largely metallic and hard composites in which an external field may generate small deformations or stresses, in the recent years, stimuli-responsive soft materials demonstrate large deformations resulting in gaining unprecedented attention in the research arena. These external fields could be electric field, magnetic field, temperature, moisture, pH, light, sound, chemical potentials or their combinations. Similarly, the bulk soft materials could either polymers, hydrogels and their composites. Soft responsive materials are ideal candidates that are in compliant with human bodies and are most suitable for developing human-machine interfaces. These newly emerged soft materials can be used in manufacturing actuators for soft robotics and stretchable electronics, energy harvesters from ambient motions, stretchable sensors for wearable devices and flexible sensors for structural health monitoring, see Figure 1.

2 MAGNETO-ACTIVE POLYMERS (MAPS): APPLICATIONS

Among various stimuli-responsive materials, magneto-active polymers (also known as magnetorheological polymers/elastomers, MRPs/MREs) attract a great attention in the research arena due to their untethered actuations, fast response and

DOI: 10.1201/9781003516880-3

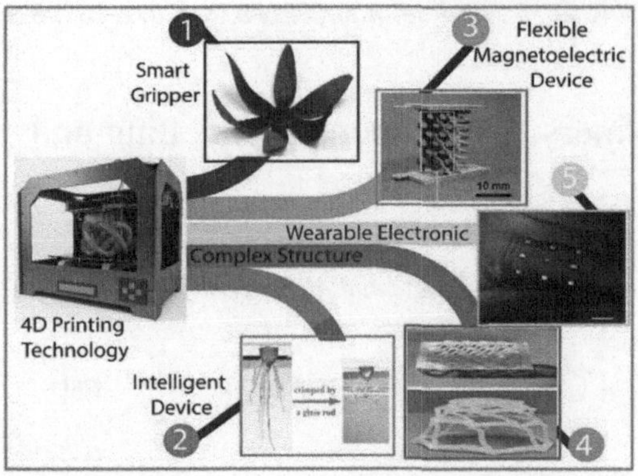

Figure 1. Recently 4D printing technology with magneto-active soft materials was used for various advanced sensors and robotics applications.

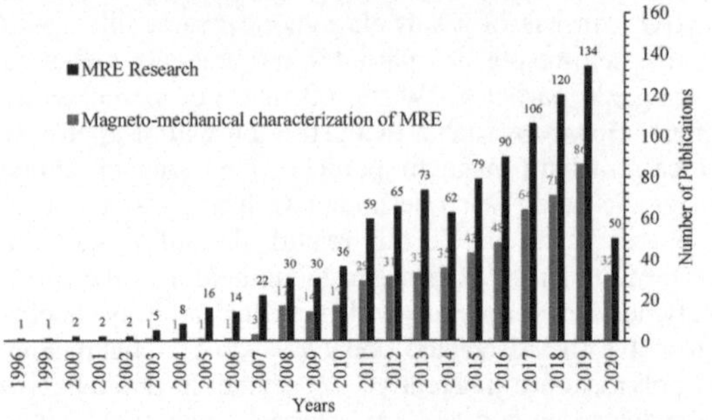

Figure 2. Publications on the magneto-mechanical characterizations of MAPs/MREs compared with various aspects of MRE-based research from 1996 to the mid of 2000. All Scopus archived documents (Journal articles, Conference proceedings, and books) are considered. The keyword "magneto-mechanical" is used to filter the research works that are focused on the magneto-mechanical characterizations of MAPs/MREs.

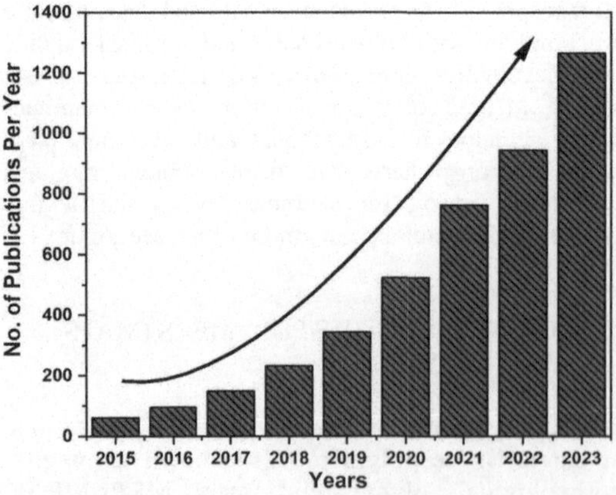

Figure 3. 3D Printing/additive manufacturing of magnetic actuated materials publication trends across the different years (Figure drawn based on the information from Scopus database using "3D printing", and "magnetic responsiveness" as keywords).

large actuation capabilities. Thanks to their untethered nature, magneto-driven actuations are best suitable for creating soft robots in applications for biomedical sciences such as controlled drug delivery, surgical robots, actuated cell culture or creating smart bone/cell scaffolds. Recent years have observed an unprecedented interest in MAPs with great potentials for a wide range of applications. Such growing interests are reflected in an exponential rise of the numbers of journal publications. For instance, the experimental studies on soft-magnetic MAPs are increased in ten to twenty folds in the last twenty years, see Figure 2 while the use of MAPs in the so-called 4D printing is also increased ten folds, see Figure 3.

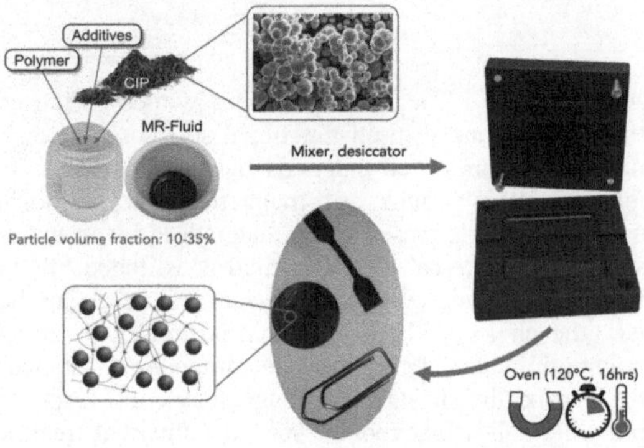

Figure 4. The key steps of an ideal/traditional synthesis process of soft-magnetic magneto-active polymers (MAPs).

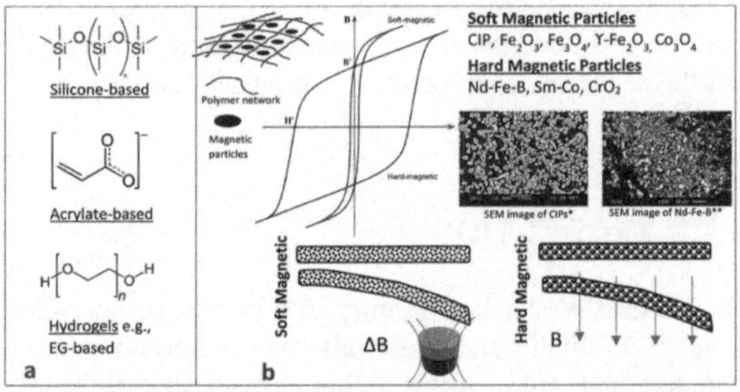

Figure 5. The key differences between hard-magnetic MAPs and soft-magnetic with respect to filler particles, residual magnetisation.

3 MAGNETO-ACTIVE POLYMERS (MAPS): SYNTHESIS

An MAP can be defined as a filled polymer in which magneto-responsive particles of nano to micro sizes are embedded in a soft polymeric matrix, cf. Figure 4. Based on the residual magnetisation characteristics, a MAP can be termed either a hard-magnetic or a soft-magnetic material. In a hard-magnetic MAP, when an external magnetic field is applied, particles are magnetised and a significant amount of residual magnetisation retains despite the withdrawal of the applied magnetic field. In contrast, in a soft-magnetic MAP, if an external magnetic field is applied, particles are magnetised and a complete demagnetisation occurs when the external field is switched off, cf. Figure 5.

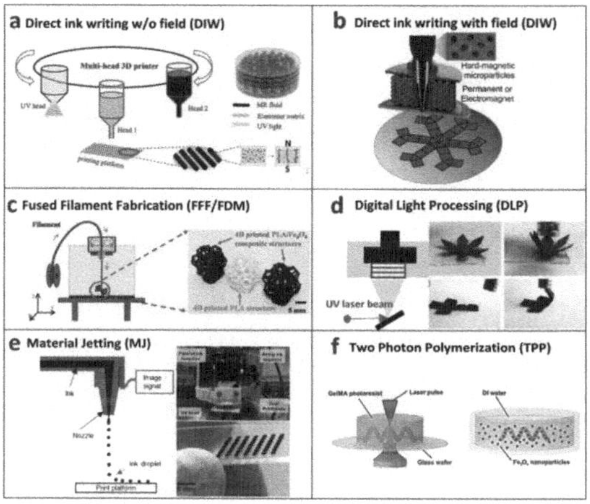

Figure 6. All major 3D printing processes can be applied to manufacture complex and intricate 3D magneto-active soft polymers.

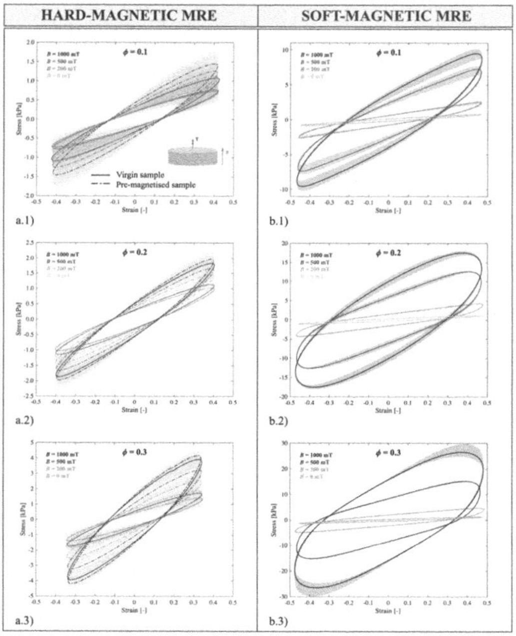

Figure 7. The change of rheological properties in the case of mixture of soft- and hard-magnetic particles in MAPs under a wide range of magnetic fields.

Both hard-magnetic and soft-magnetic MAPs have a wide range of applications with related merits and demerits. The synthesis process of an ideal soft-magnetic MAP is presented in Figure 4 in which carbonyl iron particles (CIP), widely considered as the soft-magnetic particles are used. In order to design more complex and intricate geometries/devices with MAPs, additive manufacturing (AM) aka 3D printing techniques are widely accepted. In this case, all major 3D printing processes can largely be applied to synthesis MAPs, e.g. Fused Filament/Deposition Modelling (FFM/FDM), Digital Light Processing (DLP), Stereolithography (SLA), Selective Laser Sintering (SLS) and Ink Jet Printing etc., see Figure 6. Similar to applications of MAPs, the experimental study in the literature is also increasing exponentially in the last decade, see Figure 2. One of the key characteristics of the addition of soft-magnetic particles is that they create tuneable mechanical properties such as the change of stiffness and loss moduli. In this case, we, for the first time, mix soft- and hard-magnetic particles to obtain the maximum tuneable properties, see Figure 7.

REFERENCES

1 Bastola A K, Hossain M, *The shape-morphing performance of magnetoactive soft materials performance*, Materials & Design, 211:110172, 2021.

2 Bastola A K, Hossain M, *A review on magneto-mechanical characterizations of magnetorheological elastomers*, Composites Part B: Engineering, 200: 108348, 2020.

3 Lucarni S, Hossain M, Gonzalez-Garcia D, *Recent advances in hard-magnetic soft composites: Syntheses, characterisations, computational modelling, and applications Ultra-stretchable and skin-mountable strain sensors using carbon nanotubes-Ecoflex nanocomposites*, Composite Structures, 279:114800, 2022.

4 Khalid M Y, Arif Z U, Tariq A, Hossain M, Umer R, *3D printing of stimuli-responsive polymeric hydrogel materials: Literature review and emerging applications*, GIANT, 17:100209, 2024.

5 Khalid M Y, Arif Z U, Tariq A, Hossain M, Khan K A, Umer R, *3D printing of magneto-active smart materials for advanced actuators and soft robotics applications*, European Polymer Journal, 205:112718, 2024.

6 Moreno-Mateos M A, Hossain M, Steinmann P, Gonzalez-Garcia D, *Hard magnetics in ultra soft magnetorheological elastomers enhance fracture toughness and delay crack propagation*, Journal of the Mechanics and Physics of Solids, 173:105232, 2023.

7 Moreno-Mateos M A, Hossain M, Steinmann P, Gonzalez-Garcia D, *Hybrid magnetorheological elastomers enable versatile soft actuators*, Nature Computational Materials, 8:162, 2022.

8 Moreno-Mateos M A, Hossain M, Steinmann P, Gonzalez-Garcia D, *Effects of soft and hard magnetic particles on the mechanical performance of ultra-soft magnetorheological elastomers*, Smart Materials and Structures, 31(6): 065018, 2022.

Constitutive models

Constitutive Models for Rubber XIII – Dal (Ed.)
© 2025 The Author(s), ISBN 978-1-032-85138-9

A data-driven constitutive model for compressible polymeric foams

A.K. Açan, Y. Gargı, S. Başdemir & H. Dal
Department of Mechanical Engineering, Middle East Technical University, Ankara, Türkiye
CAD/CAM Robotics Center, Middle East Technical University, Ankara, Türkiye

ABSTRACT: Compressible polymeric foams exhibit a highly non-linear mechanical behaviour. Classical constitutive models have a fixed mathematical expression. Degree of porosity and the cellular structure of the foam affects the mechanical response significantly. In this regard, phenomenological constitutive models may not be successful in modeling a variety of compressible polymeric foams. Data-driven constitutive modeling is a promising approach to solve this problem, here we propose an extended version of B-spline based data-driven framework (Dal, Denli, Açan, & Kaliske 2023) for compressible materials. The model adapts its control point values to reduce the error between data and prediction until a threshold is reached, and is thermodynamically consistent through the use of optimization constraints. The proposed model has been validated with closed-cell EPDM (ethylene propylene diene monomer) with three distinct foam densities. The uniaxial tension and the confined compression experiments were conducted to identify the mechanical behavior of the materials. A non-homogeneous uniaxial compression experiment was performed to validate the control points obtained through the simultaneous fitting of uniaxial tension and confined compression. Finite element analysis was conducted under conditions precisely replicated from the non-homogeneous uniaxial compression experiment. Furthermore, a comparison is performed between the classical constitutive models (Blatz & Ko 1962, Hill 1979) and the proposed data-driven approach. The results show that the data-driven approach is very promising to model the mechanical behavior of compressible polymeric materials.

1 INTRODUCTION

Considering that many polymer foams originate from rubber, and hyperelastic material models are crucial for understanding their mechanical properties. Dal et al. (Dal, Açıkgöz, & Badienia 2021) conducted a comprehensive examination of these models. Ricker and Wriggers (Ricker & Wriggers 2023) provide an examination of appropriate hyperelastic models designed to accurately represent the isochoric and volumetric characteristics of nine commonly used rubber compounds. The Ogden hyperfoam model (Hill 1979, Storåkers 1986), a modified version of the Ogden model (Ogden 1972) designed for incompressible materials, along with the Blatz-Ko material model (Blatz & Ko 1962), constitutes widely utilized and recommended hyperelastic material models for compressible polymeric foams in the literature. Yan et al. (Yan, Jia, Yu, Wang, Qiu, & Wan 2021) introduced a new approach for parameter fitting of the Ogden hyperfoam material model. By performing conducting uniaxial compression and simple shear tests, they obtained experimental data, which they then used to determine the material parameters. Effectively depicting significant volumetric deformations presents a considerable obstacle, often demanding the use of complex free energy functions. Alongside conventional material models found in current literature, scholars have introduced new free energy functions to improve accuracy in capturing notable volumetric alterations (Moerman, Fereidoonnezhad, & McGarry 2020), as well as new homogenization framework (Dal, Cansız, & Miehe 2018) to obtain compressible counterparts of any free energy function. However, utilizing the data-driven modeling approach, we can accurately obtain the mechanical behavior of compressible polymeric foams without the need for complex free energy functions. Nowadays, data-driven approaches gained attention in constitutive modeling. Using data-driven approaches to determine the form of the free energy function offers greater flexibility compared to employing fixed analytical forms. This flexibility enables capturing the mechanical behavior of various classes of materials more effectively. There are different approaches in data-driven modeling such as distance minimization based approaches (Kirchdoerfer & Ortiz 2016, Ibanez, Borzacchiello, Aguado, Abisset-Chavanne, Cueto, Ladeveze, & Chinesta 2017, Eggersmann, Kirchdoerfer, Reese, Stainier, & Ortiz 2019), neural-network based approaches (Linka & Kuhl 2023, Tac, Sree, Rausch,

DOI: 10.1201/9781003516880-4

& Tepole 2022, Ghaboussi, Garrett Jr, & Wu 1991), or B-spline based approaches (Sussman & Bathe 2009, Latorre & Montáns 2013, Dal, Denli, Açan, & Kaliske 2023, Tikenoğulları, Açan, Kuhl, & Dal 2023). A recent contribution introduced a parameter identification toolbox for incompressible materials, utilizing a B-spline-based data-driven approach (Durna, Açan, Tikenoğulları, & Dal 2024). However, according to our best knowledge, there is no data-driven model for highly compressible materials. For this purpose, in this contribution, we extend our previous B-spline-based data-driven model (Dal, Denli, Açan,& Kaliske 2023) to include compressible materials.

The paper is structured as follows: The concept of B-splines is elaborated in Section 2, while the data-driven hyperelasticity for compressible materials is presented in Section 3. The polyconvexity requirement is given in Section 4. Section 5 discusses experimental observations conducted on EPDM foam with three distinct foam densities. Numerical examples and predictions of the proposed model are provided in Section 6. Finally, the paper concludes with summarizing remarks in Section 7.

2 CONSEPT OF B-SPLINES

B-spline function and its derivative can be expressed as

$$\begin{aligned} C(\xi) &= \textstyle\sum_{k=1}^{n} \mathcal{N}_k^p(\xi)\mathcal{P}_k \\ C'(\xi) &= \textstyle\sum_{k=1}^{n} \mathcal{N}_k^{p'}(\xi)\mathcal{P}_k, \end{aligned} \tag{1}$$

where $\mathcal{N}_k^p(\xi)$ and $\mathcal{N}_k^{p'}(\xi)$ are the Bernstein basis functions and their derivatives, respectively. The parameter p denotes the degree of basis functions, while $\mathcal{P}_k$ is the control points associated to B-spline. The basis functions and their derivatives are obtained using Cox-de Boor recursive formula

$$\mathcal{N}_k^0(\xi) = \begin{cases} 1, & \text{if} \quad \xi_k \leq \xi < \xi_{k+1} \\ 0, & \text{otherwise} \end{cases} \tag{2}$$

along the given knot span. Basis functions $\mathcal{N}_k^p$ and their derivatives $\mathcal{N}_k^{p'}$ for $p > 0$ are computed through

$$\begin{aligned} \mathcal{N}_k^p(\xi) &= \tfrac{\xi-\xi_k}{\xi_{k+p}-\xi_k}\,\mathcal{N}_k^{p-1}(\xi) \\ &+ \tfrac{\xi_{k+p+1}-\xi}{\xi_{k+p+1}-\xi_{k+1}}\,\mathcal{N}_{k+1}^{p-1}(\xi), \\ \mathcal{N}_k^{p'}(\xi) &= \tfrac{p}{\xi_{k+p}-\xi_k}\,\mathcal{N}_k^{p-1}(\xi) \\ &+ \tfrac{p}{\xi_{k+p+1}-\xi_{k+1}}\,\mathcal{N}_{k+1}^{p-1}(\xi), \end{aligned} \tag{3}$$

where $\xi_{\hat{k}}$ stands for the knots, which are labeled by an index $\hat{k}$ ranging from 1 to m, where m represents the total count of knots. As per the Cox-de Boor formula in Equation 2, each basis function $\mathcal{N}_k^p(\xi)$ has its influence spread across $p+1$ adjacent knot spans. The quantity of knots, determined by the polynomial degree p and the number of control points n of the B-spline, is calculated as $m = p + n + 1$. For more detailed explanation of B-splines, see (Tikenoğulları, Açan, Kuhl, & Dal 2023).

3 DATA-DRIVEN HYPERELASTICITY FOR COMPRESSIBLE MATERIALS

Here, we utilize an alternative stretch measure instead of direct use of the first invariant of the right Cauchy-Green tensor for longitudinal stretch

$$\lambda_{\mathrm{n}} = \sqrt[2]{\frac{\lambda_1^2 + \lambda_2^2 + \lambda_3^2}{3}} = \sqrt[2]{\frac{I_1}{3}} \tag{4}$$

in the sense of (Arruda & Boyce 1993) and (Dal, Gültekin, & Açıkgöz 2020). The partial derivatives of the free-energy function with respect to the invariants are then modified as

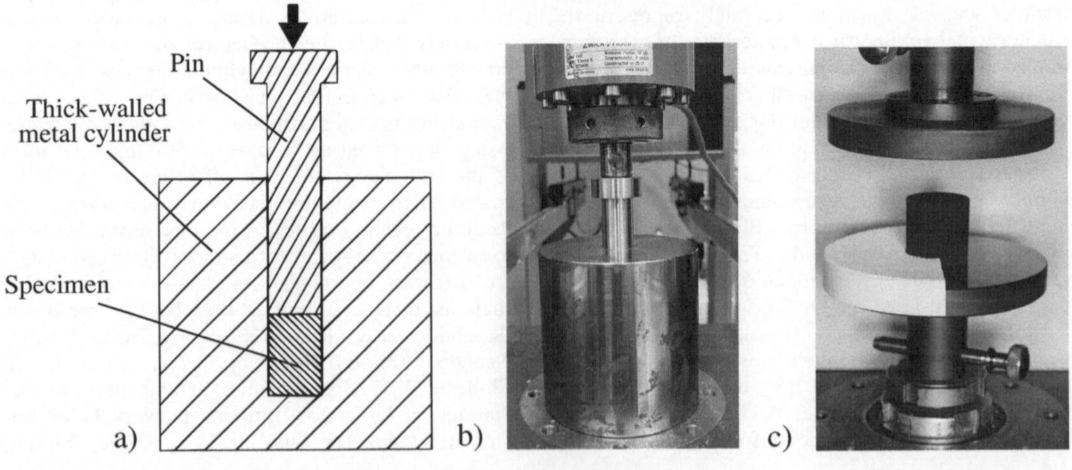

Figure 1. a) The schematic of the confined compression test, b) the confined compression test setup, and c) the apparatus used to centre the test specimen in the non-homogeneous uniaxial compression test.

$$\psi_1 \frac{\partial \psi}{\partial \lambda_n}\frac{\partial \lambda_n}{\partial I_1} \quad \text{where} \quad \frac{\partial \lambda_n}{\partial I_1} = \frac{1}{6}\frac{1}{\lambda_n}\,. \tag{5}$$

Finally, the partial derivative of free energy function can be expressed as

$$\psi_1 = \frac{1}{6}\frac{1}{\lambda_n}\hat{\psi}_1 \quad \text{where} \quad \hat{\psi}_1 = \frac{\partial \psi}{\partial \lambda_n}\,. \tag{6}$$

The Lagrangian and Eulerian stresses can be obtained by exploiting the chain rule

$$\begin{aligned} \boldsymbol{S} &= 2\partial_{\boldsymbol{C}}\psi &= 2\left[\frac{\partial \psi}{\partial I_1}\frac{\partial I_1}{\partial \boldsymbol{C}} + \frac{\partial \psi}{\partial I_3}\frac{\partial I_3}{\partial \boldsymbol{C}}\right], \\ \tau &= 2\partial_{g}\psi &= 2\left[\frac{\partial \psi}{\partial I_1}\frac{\partial I_1}{\partial g} + \frac{\partial \psi}{\partial I_3}\frac{\partial I_3}{\partial g}\right]. \end{aligned} \tag{7}$$

$$\begin{aligned} \boldsymbol{S} &= 2\psi_1 \boldsymbol{G}^{-1} - 2I_3\psi_3\boldsymbol{C}^{-1} + \boldsymbol{S}_0 \\ \tau &= 2\psi_1 \boldsymbol{b} - 2I_3\psi_3 g^{-1} + \tau_0 \end{aligned} \tag{8}$$

with

$$\psi_3 = \frac{\partial \psi}{\partial I_3}\,, \tag{9}$$

and the normalization terms

$$\boldsymbol{S}_0 = 2(\psi_1^0 + \psi_3^0)\boldsymbol{c}^{-1} \quad \tau_0 = 2(\psi_1^0 + \psi_3^0)g^{-1} \tag{10}$$

where ψ_i^0 with $i = \{1, 3\}$ are the initial values of the interpolation functions for $I_1 = 3$ and $I_3 = 1$, respectively. In the invariant-based data-driven formulation of hyperelasticity, the partial derivatives $\hat{\psi}_1$ and ψ_3 can be interpolated as

$$\hat{\psi}_1 = \sum_{i=1}^{n_s} \mathcal{N}_i^p(\lambda_n)\mathcal{P}_i^1 \quad \text{and} \quad \psi_3 = \sum_{i=1}^{n_s} \mathcal{N}_i^p(I_3)\mathcal{P}_i^3\,, \tag{11}$$

where $\mathcal{N}_i^p$ are the basis functions of order p and $\mathcal{P}_i^1$, $\mathcal{P}_i^3$ are the control points behaves as the material parameters in the constitutive equations. Parameter set $\mathcal{P}_i^1$ manages the functional form for the part of the free-energy function depending on the first invariant, whereas $\mathcal{P}_i^3$ controls the functional form for the part of the free-energy function depending on the third invariant.

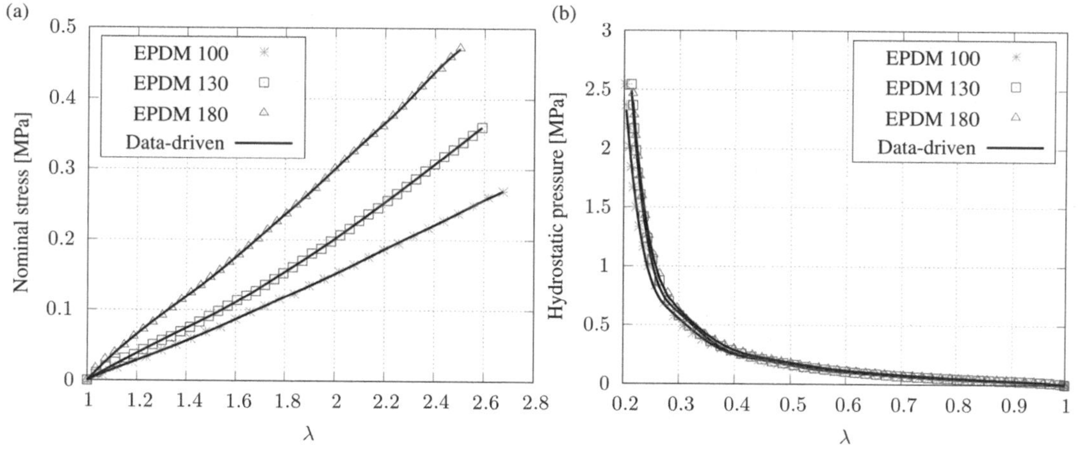

Figure 2. Data-driven model predictions for (a) uniaxial tension test and (b) confined compression test.

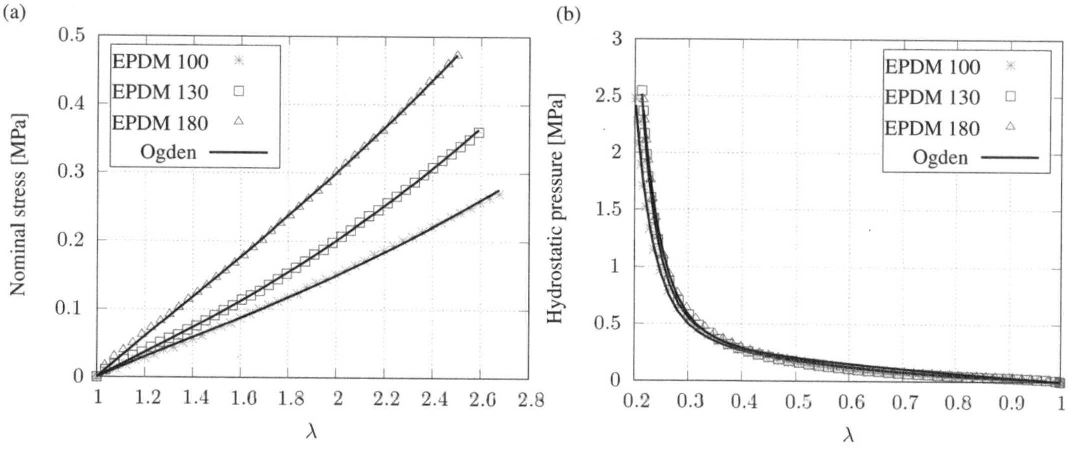

Figure 3. The third order Ogden hyperfoam model predictions for (a) uniaxial tension test and (b) confined compression test.

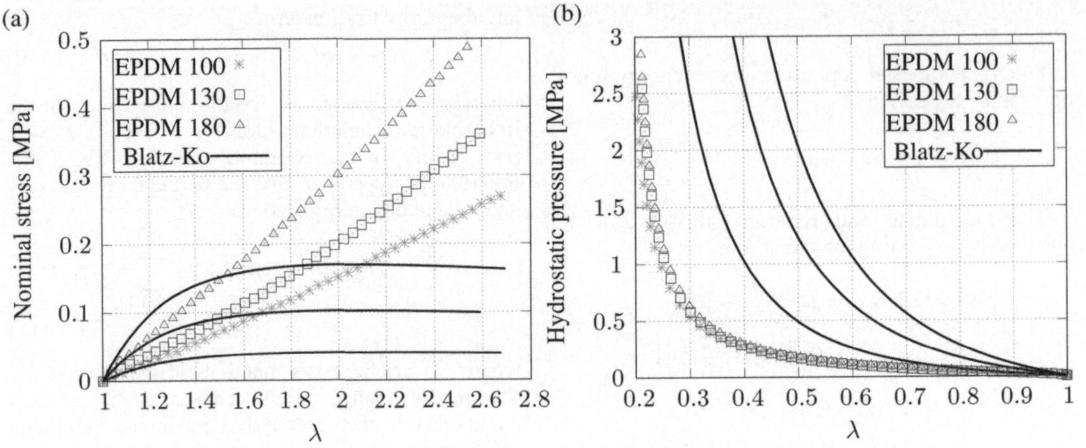

Figure 4. Blatz-Ko model predictions for (a) uniaxial tension test and (b) confined compression test.

4 POLYCONVEXITY OF THE COMPRESSIBLE FORMULATION

The polyconvexity requirement for the invariant based formulation can be expressed as

$$\psi(\boldsymbol{F}) = \psi(\boldsymbol{F}, \operatorname{cof}[\boldsymbol{F}], \det[\boldsymbol{F}]) = \psi(\boldsymbol{F}, \boldsymbol{H}, J) \quad (12)$$

be convex with respect to $\boldsymbol{F}$, $\boldsymbol{H} = \operatorname{cof}[\boldsymbol{F}]$, and $J = \det[\boldsymbol{F}]$, respectively (Ball 1976). Thus, for free energy functions expressed as polynomials or other forms involving I_1 and I_3, it is sufficient to demonstrate that

$$\psi_{11} \geq 0 \quad \text{and} \quad \psi_{33} \geq 0 \quad (13)$$

in order to ensure polyconvexity (Hartmann & Neff 2003). The polyconvexity relation for the modified invariant-based hyperelastic formulation can be expressed as

$$\psi_{11} \geq 0 \quad \rightarrow \quad \lambda_n \hat{\psi}_{11} - \hat{\psi}_1 \geq 0 \quad (14)$$

and

$$\psi_{33} \geq 0, \quad (15)$$

where ψ_{11}, ψ_{33} represent the second derivative of free energy function with respect to I_1 and I_3, respectively, while $\hat{\psi}_{11}$ denotes second derivative of free energy function with respect to modified invariant λ_n. During the data *data training/parameter identification* phase, the control points are subjected to the following inequality constraints

$$(\lambda_n^{i+1} - \Delta\lambda_n)\mathcal{P}_{i+1}^1 - \lambda_n^{i+1}\mathcal{P}_i^1 \geq 0, \quad (16)$$

and

$$\mathcal{P}_{i+1}^3 - \mathcal{P}_i^3 \geq 0. \quad (17)$$

in order to enforce polyconvexity.

5 EXPERIMENTAL STUDIES

In this study, uniaxial tension, non-uniform uniaxial compression and confined compression tests were conducted to model the mechanical behaviour of compressible polymeric foams using three different densities (EPDM100:100kg/m^3, EPDM130:130kg/m^3, EPDM180:180kg/m^3) with 100 % EPDM content. All tests were performed with 10 mm/min velocity and three specimens for each density.

The experimental setup utilized a Zwick/Roell Z010 universal testing machine fitted with two video extensometers and a Zwick/Roell XForceP series load cell capable of handling up to 10 [kN] nominal force. Experimental procedures, measurements, and control of the testing machine were performed using the TestXpert III software associated with the equipment. Image acquisition for the study was conducted using a Sony ZV-E10 video camera configured with 100 FPS capture settings.

5.1 *Uniaxial tension-UT*

Uniaxial tension tests were performed on three ISO37 Type 2 specimens per density, each triads corresponding to a distinct density (EPDM100, EPDM130 and EPDM180) manufactured from 5mm-thick EPDM foam sheets. Specimens were die-cut and loaded at a 10 mm/min velocity until rupture occurred.

5.2 *Confined compression-CC*

A confined compression test was performed on a cylindrical sample with 20 mm diameter and 30 mm height. A specimen is firmly placed in the cavity of the thick-walled cylinder. This setup effectively restricts lateral movement while applying compressive force, ensuring controlled volume changes. However, challenges arose due to friction

Table 1. The performance of the compressible hyperelastic models based on χ^2 quality of fit metric.

	Quality of fit (χ^2)								
	EPDM100			EPDM130			EPDM180		
	UT	CC	Total	UT	CC	Total	UT	CC	Total
Data-driven model	2.810e-04	0.0083	0.0086	0.0024	0.0242	0.0266	0.0022	0.0162	0.0184
3rd order Ogden hyperfoam model	0.0019	0.0705	0.0724	0.0047	0.0836	0.0883	0.0071	0.2186	0.2257
Blatz-Ko	3.0834	191.7302	194.8136	2.1693	1327.1966	1329.3659	3.2471	4308.4034	4311.6505

encountered during sample insertion, impeding accurate data collection during experimentation. LIQUI MOLY Silicone Spray, known for its compatibility with rubber materials, was utilized as a lubricating agent to address this issue. Due to the closed-cell nature of the foam used, penetration of the lubricant is difficult. Nevertheless, to avoid mechanical behaviour changes, only a minimal amount of lubricant was applied to create a thin film layer between the test specimen and the apparatus. The test apparatus and its schematic used in confined compression test can be seen in Figure 1-(a), (b).

5.3 *Non-homogeneous uniaxial compression-UC*

A non-homogeneous uniaxial compression test was conducted to verify the behaviour of closed-cell EPDM foams under compression. The test data were not utilized for determining material parameters due to their non-homogeneous nature. Instead, they were employed to validate FEA results obtained using material parameters that simultaneously fit uniaxial tension and confined compression tests.

The cylindrical specimen, measuring 30 mm in diameter and 30 mm in height, was positioned between two steel plates and affixed with instant adhesive to prevent surface slippage between the specimen and the plates. Ensuring coaxial alignment between the specimen and steel plates is crucial for evenly transferring the applied compression force. To achieve this, a 3D-printed sample centering device was developed, designed to be attached without causing imbalance. Describing its application process, the apparatus is first affixed to the metal part, following which a thin film of adhesive is applied to the lower surface of the specimen, ensuring that it does not change the mechanical behaviour of the specimen. Subsequently, the specimen is positioned within the device, and after a short waiting time, the device is removed. The adhesive is then applied to the upper surface of the specimen, and the metal piece is brought into contact and held briefly. The compression test setup is given in Figure 1-(c)

6 REPRESENTATIVE NUMERICAL ANALYSES

The performance of the proposed data-driven model is compared with Ogden hyperfoam model (Hill 1979) and Blatz-Ko model (Blatz & Ko 1962) based on the data from our experiments portrayed in Section 5. The uniaxial tension data and confined compression data have been utilized to calibrate the hyperelastic models, and a comparison has been conducted between the models based on χ^2 quality-of-fit

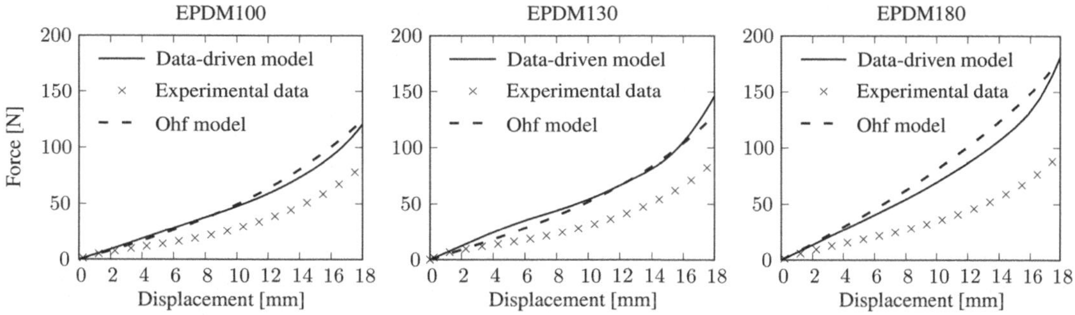

Figure 5. The finite element analysis results of the proposed data-driven model and Ogden hyperfoam model (Ohf model) for uniaxial compression experiment.

metric, see Table 1. 15 control points with the third order basis functions have been used to calibrate the data-driven model. The predictions of data-driven model for EPDM100, EPDM130 and EPDM180 foam can be seen in Figure 2. Moreover, the predictions of Ogden hyperfoam model (Hill 1979) and Blatz-Ko model (Blatz & Ko 1962) are given in Figures 3 and 4. It is noted that the data-driven approach demonstrates superiority in accurately capturing the mechanical behavior of EPDM material across various density ranges. Since the uniaxial compression experiment results a non-homogeneous deformation, the results of uniaxial compression is not utilized for calibration however it is used to validate the model. To this end, the proposed theory is implemented to general usage finite element analysis program FEAP 8.3 (Taylor 2011). To replicate the uniaxial compression experiment, the top and bottom surfaces are fixed, and a displacement is applied to the top surface. The simulation results for data-driven model and Ogden hyperfoam model 1979, (Hill Storåkers 1986). are given in Figure 5.

7 CONCLUSIONS

In this study, we extended our previous data-driven framework to encompass compressible materials. Instead of defining a volumetric free energy function directly, we utilized B-spline interpolation technique on the partial derivative of the volumetric free energy function concerning the third invariant of the right Cauchy-Green deformation tensor. To validate and train the model, we conducted uniaxial tension, confined compression, and uniaxial compression experiments on polymeric foam with three distinct densities. Our approach was compared with established models such as the Ogden hyperfoam model (Hill, 1979) and the Blatz-Ko model (Blatz & Ko 1962). Calibration of the models was performed using uniaxial tension and confined compression experiments, while testing was carried out using uniaxial compression experiments coupled with finite element analysis. Since there is friction in confined compression experiments, the model calibrated with confined compression test results exhibits overstiff behavior in uniaxial compression simulations. However, the overall results demonstrate the promising potential of proposed data-driven approach for accurately modeling the compressible behavior of polymeric materials.

ACKNOWLEDGEMENT

The financial support provided by the support of METU Scientific Research Project Coordination Unit under grant ADEP-312-2024-11490 is gratefully acknowledged.

REFERENCES

Arruda, E. M. & M. C. Boyce (1993). A three-dimensional constitutive model for the large stretch behavior of rubber elastic materials. *Journal of the Mechanics and Physics of Solids 41*(2), 389–412.

Ball, J. M. (1976). Convexity conditions and existence theorems in nonlinear elasticity. *Archive for Rational Mechanics and Analysis 63*, 337–403.

Blatz, P. & W. Ko (1962). Application of finite elastic theory to the deformation of rubbery materials. *Transactions of the Society of Rheology 6*, 223–251.

Dal, H., K. Açıkgöz, & Y. Badienia (2021). On the performance of isotropic hyperelastic constitutive models for rubber-like materials: A state of the art review. *Applied Mechanics Reviews 73*(2), 020802.

Dal, H., B. Cansız, & C. Miehe (2018). A three-scale compressible microsphere model for hyperelastic materials. *International Journal for Numerical Methods in Engineering 116*(6), 412–433.

Dal, H., F. A. Denli, A. K. Ac¸an, & M. Kaliske, (2023). Datadriven hyperelasticity, part i: A canonical isotropic formulation for rubberlike materials. *Journal of the Mechanics and Physics of Solids 179*, 105381.

Dal, H., O. Gültekin, & K. Açıkgöz (2020). An extended eightchain model for hyperelastic and finite viscoelastic response of rubberlike materials: Theory, experiments and numerical aspects. *Journal of the Mechanics and Physics of Solids 145*, 104159.

Durna, R., A. K. Açan, O. Z. Tikenoğulları, & H. Dal (2024). Hyper-data: A matlab based optimization software for datadriven hyperelasticity. *SoftwareX 26*, 101642.

Eggersmann, R., T. Kirchdoerfer, S. Reese, L. Stainier, & M. Ortiz (2019). Model-free data-driven inelasticity. *Computer Methods in Applied Mechanics and Engineering 350*, 81–99.

Ghaboussi, J., J., Garrett Jr, & X. Wu (1991). Knowledge-based modeling of material behavior with neural networks. *Journal of Engineering Mechanics 117*(1), 132–153.

Hartmann, S. & P. Neff (2003). Polyconvexity of generalized polynomial-type hyperelastic strain energy functions for near-incompressibility. *International Journal of Solids and Structures 40*(11), 2767–2791.

Hill, R. (1979). Aspects of invariance in solid mechanics. *Advances in applied mechanics 18*, 1–75.

Ibanez, R., D. Borzacchiello, J. V. Aguado, E. Abisset-Chavanne, E. Cueto, P. Ladeveze, & F. Chinesta (2017). Data-driven non-linear elasticity: constitutive manifold construction and problem discretization. *Computational Mechanics 60*(5), 813–826.

Kirchdoerfer, T. & M. Ortiz (2016). Data-driven computational mechanics. *Computer Methods in Applied Mechanics and Engineering 304*, 81–101.

Latorre, M. & F. J. Montáns (2013). Extension of the Sussman–Bathe spline-based hyperelastic model to incompressible transversely isotropic materials. *Computers & Structures 122*, 13–26.

Linka, K. & E. Kuhl (2023). A new family of constitutive artificial neural networks towards automated model discovery. *Computer Methods in Applied Mechanics and Engineering 403*, 115731.

Moerman, K. M., B. Fereidoonnezhad, & J. P. McGarry (2020). Novel hyperelastic models for large volumetric deformations. *International Journal of Solids and Structures 193*, 474–491.

Ogden, R. W. (1972). Large deformation isotropic elasticity–on the correlation of theory and experiment for incompressible rubberlike solids. Proceedings of the Royal Society of London. A. *Mathematical and Physical Sciences 326*(1567), 565–584.

Ricker, A. & P. Wriggers (2023). Systematic fitting and comparison of hyperelastic continuum models for elastomers. Archives of Computational Methods in Engineering, 1–32.

Storåkers, B. (1986). On material representation and constitutive branching in finite compressible elasticity. *Journal of theMechanics and Physics of Solids 34*(2), 125–145.

Sussman, T. & K.-J. Bathe (2009). A model of incompressible isotropic hyperelastic material behavior using spline interpolations of tension–compression test data. *Communications in Numerical Methods in Engineering 25*(1), 53–63.

Tac, V., V. D. Sree, M. K. Rausch, & A. B. Tepole (2022). Datadriven modeling of the mechanical behavior of anisotropic soft biological tissue. *Engineering with Computers 38*(5), 4167–4182.

Taylor, R. L. (2011). Feap: a finite element analysis program: version 8.3 usermanual. *University of California at Berkeley.*

Tikenoğulları, O. Z., A. K. Ac̦an, E. Kuhl, & H. Dal, (2023). Data-driven hyperelasticity, part ii: A canonical framework for anisotropic soft biological tissues. *Journal of theMechanics and Physics of Solids 181*, 105453.

Yan, S., D. Jia, Y. Yu, L.Wang, Y. Qiu, & Q. Wan (2021). Novel strategies for parameter fitting procedure of the ogden hyperfoam model under shear condition. *European Journal of Mechanics-A/Solids 86*, 104154.

Constitutive Models for Rubber XIII – Dal (Ed.)
© 2025 The Author(s), ISBN 978-1-032-85138-9

Characterization and modelling of the compressible hyperelastic behaviour of polyurethane foams: Influence of the density

J. Becker & J.B. Le Bail
ENSTA Bretagne, UMR CNRS 6027, IRDL, Brest, France
Vibracoustic – ESM Department, Carquefou, France

M. Le Saux, V. Le Saux & Y. Marco
ENSTA Bretagne, UMR CNRS 6027, IRDL, Brest, France

L. Maheo
Université Bretagne Sud, UMR CNRS 6027, IRDL, Lorient, France
Military Academy of Saint-Cyr Coetquidan, CReC Saint-Cyr, Guer, France

P. Charrier & W. Hervouet
Vibracoustic – ESM Department, Carquefou, France

ABSTRACT: In parts such as automotive jounce bumpers, microcellular polyurethane foams are subjected to multiaxial loading leading to very large deformations. The density of the material can vary from one part to another or within the same part. Thus, the numerical simulations used to design the parts require a model describing the mechanical behaviour of the material over a wide range of loading conditions, as a function of the material density. This paper presents an extensive experimental database of uniaxial tension and compression, simple shear and confined compression test results, obtained on elastomeric microcellular polyurethane foam samples with several densities between 380 and 600 kg.m^{-3}. The model proposed by Landauer et al. (2019), based on invariants of the Henck's strain tensor, is used to describe compressible hyperelasticity. The influence of the foam density is integrated in the model. A robust identification strategy is proposed. Once identified, the model is used to simulate the response of a jounce bumper under global axial compression. The results are compared to the experimental load-displacement curve and to the deformed shape of the jounce bumper obtained from X-ray tomography.

1 INTRODUCTION

Microcellular polyurethane foams are widely used in many industrial applications, including automotive industry, thanks to their interesting mechanical properties and their low density. In jounce bumpers, these foams are subjected to complex multiaxial loadings, leading to very large deformations and volume variations (several tens of percent). In addition, the foam density may vary over the volume of a jounce bumper. To correctly simulate the mechanical response of a jounce bumper in use, it is therefore necessary to get data from various loading conditions and a model covering the density range of interest.

The first step is to build up a relevant experimental database. Studies on the influence of the foam density on its mechanical behaviour are mostly based on the results of only one or two types of tests, for example, uniaxial compression (Bouix et al. 2009), uniaxial tension (Goods et al. 1998, Lin 1997),, simple shear (Landauer et al. 2019), hydrostatic compression (Viot 2009), confined compression (Liu and Subhash 2004) or combined compression-shear (Ling et al. 2018). The first objective of this paper is to present an extensive experimental database including uniaxial tension (UT), uniaxial compression (UC), simple shear (SS) and confined compression (CC) test results, obtained on microcellular polyurethane foams samples with several densities between 380 and 600 kg.m^{-3}. These data extend those presented in becker2022modelling,bail_modelisation_2022 obtained for a density of about 500 kg.m^{-3}.

The next step is to formulate a model capable of describing the mechanical behaviour of the material under multiaxial loading, in the loading range of interest here (strain and volume changes up to several tens of percent), and taking into account the effect of density. In the literature, relationships have been proposed to link certain remarkable quantities of the behaviour, like the Young modulus, the densification strain or the stress plateau, to the material density. The relationships proposed are linear (Di Landro et al. 2002),

DOI: 10.1201/9781003516880-5

based on power laws (Saha et al. 2005, Thirumal et al. 2008) or are more complex. They are sometimes based on microstructure considerations (Ashby and Gibson 1997, Saint-Michel et al. 2006, Chai et al. 2020). These relationships are limited to particular loading modes and do not a priori describe the response of the material under multiaxial loading.

Micro-mechanical models naturally depend on the density by considering porosity in the material (Lewis 2016, Danielsson et al. 2004, Rubin and Dorfmann 2020, Mihai et al. 2017). However, these models are often too computationally demanding to be used in finite element analysis (FEA) in an industrial context. Phenomenological models are better suited to this type of application. Few of the phenomenological models proposed in the literature explicitly take density into account. In these models, density dependence is considered by making certain parameters (Avalle et al. 2007, Wang et al. 2018, Goga 2011) or the energy potential (Rusch 1969, Sherwood and Frost 1992) density-dependent. The second objective of this paper is to propose a model, simple enough to be used to simulate complex parts, that describes the highly non-linear compressible elasticity of elastomeric polyurethane foams and its dependence on the initial density of the material. In this study, the phenomenological constitutive model from Landauer et al. (2019) has been chosen and identified to describe a large experimental database.

The experimental database is first presented in Section 2. Section 3 then describes the constitutive model and the identification procedure. Section 4 discusses the ability of the model to describe the response of the material to UT, UC, SS and CC tests. The model is finally applied to simulate, using the finite element method, the response of a jounce bumper in compression. The results are compared with experience.

2 EXPERIMENTAL DATABASE

2.1 *Material and procedure*

The material studied in this paper is a naphthalene diisocyanate-based cast elastomeric polyurethane micro-cellular foam, typical of the one used for automotive jounce bumpers. The behaviour of the material is assumed to be isotropic given its microstructure (Le Saux et al. 2023). Samples with different densities varying from 380 to 600 $kg.m^{-3}$ are used. As presented in Becker et al. (2022) for one material density, UC, UT, SS and CC tests have been performed, with nominal strain rate of about 0.01 s^{-1}. Four loading-unloading cycles were applied to each specimen to accommodate the cyclic softening, with a one-minute hold, between the third and the fourth cycle, at an imposed displacement for which the load is zero at the end of the third cycle, to recover the effects of short-term viscoelasticity. Only the response during the loading of the fourth cycle is considered in this paper, which focuses on non-linear elasticity. There is a residual deformation before the start of the fourth cycle. This deformation is taken into account to calculate the density of the material at the start of the fourth cycle (knowing the density before testing). The stress and strain are calculated relatively to this new configuration.

2.2 *Results*

The experimental curves are shown for each test in Figure 1. The results show that the elastic response and the compressibility of the material are highly non-linear whatever the material density. Density has a significant effect: the higher the density, the greater the apparent stiffness and the smaller the change in volume at which the densification phase occurs. Figure 1 shows the initial shear modulus determined from the SS test (between 0 and 0.2 of nominal shear strain), the volume ratio at densification determined from CC test (at a nominal stress of -10 MPa) and the initial bulk modulus determined using UC test (from the initial Young modulus and Poisson ratio determined over the first 5% of nominal strain). The dependence to the density of each of these quantities is approximately linear.

3 MODEL

3.1 *Model formulation*

The general curve profile (see Section 2.2) is about the same for the different densities. This means that the same model formalism can be used for the different densities and that the effect of density can be considered by introducing a dependence of certain parameters of the model on density. The behaviour law used in this paper is a hyperelastic model for compressible materials developed by Landauer et al. (2019). The potential uses the invariants (K_1, K_2, K_3) from Criscione et al. (2000), of the Hencky strain tensor, $\underline{\underline{E_H}}$, which can be decomposed as follow:

$$\underline{\underline{E_H}} = \frac{1}{3} K_1 \underline{\underline{I}} + K_2 \underline{\underline{N}} \tag{1}$$

where $\underline{\underline{I}}$ is the second-order identity tensor and $\underline{\underline{N}} = \frac{1}{K_2} \underline{\underline{dev(E_H)}}$ is a unit-magnitude and deviatoric tensor. $K_1 = tr(\underline{\underline{E_H}})$ describes the volume change, $K_2 = \sqrt{\underline{\underline{dev(E_H)}} : \underline{\underline{dev(E_H)}}}$ represents the magnitude of distortion and $K_3 = 3\sqrt{6}\det \underline{\underline{N}}$ represents the mode of distortion.

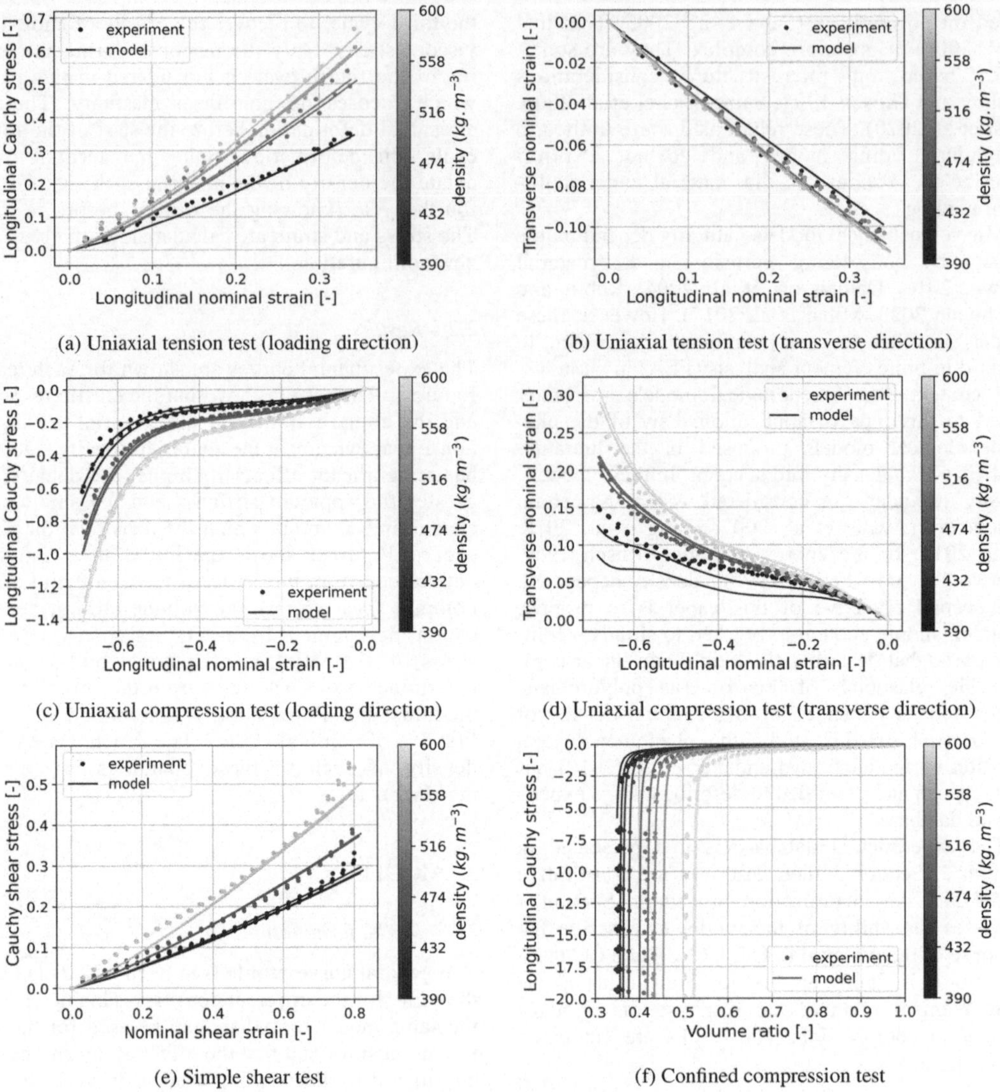

(a) Uniaxial tension test (loading direction)

(b) Uniaxial tension test (transverse direction)

(c) Uniaxial compression test (loading direction)

(d) Uniaxial compression test (transverse direction)

(e) Simple shear test

(f) Confined compression test

Figure 1. Experimental results and predictions of the multi density model (the stress values are normalised).

The free-energy potential writes:

$$W(K_1, K_2, K_3) = G_0[X(K_1)K_2^2 + L(K_2, K_3)] + Bf(K_1) \quad (2)$$

where:

$$X(K_1) = \frac{1}{2}\left(X_1' + X_2'\right)K_1 + \frac{\Delta_K}{2}\left(X_1' - X_2'\right) \times \ln\left(\frac{\cosh((K_1 - K_1^0)/\Delta_K)}{\cosh(K_1^0/\Delta_K)}\right) + 1 \quad (3)$$

$$L(K_2, K_3) = C_0 K_2^p + C_1(1 + K_3)K_2^q \quad (4)$$

$$f(J) = \frac{J^{C_2} - C_2 \ln J - 1}{C_2^2} + \frac{C_3}{r-1}\left[-J^{1-r} + \frac{(1 - J_{min})^r}{(J - J_{min})^{r-1}} + J_{min}\right] \quad (5)$$

where $J = \exp(K_1)$ is the volume ratio. This model has 14 parameters to identify $[G_0; B; J_{min}; C_1; K_1^0; \Delta_K; X_1'; X_2'; C_0; p; q; r; C_2; C_3]$.

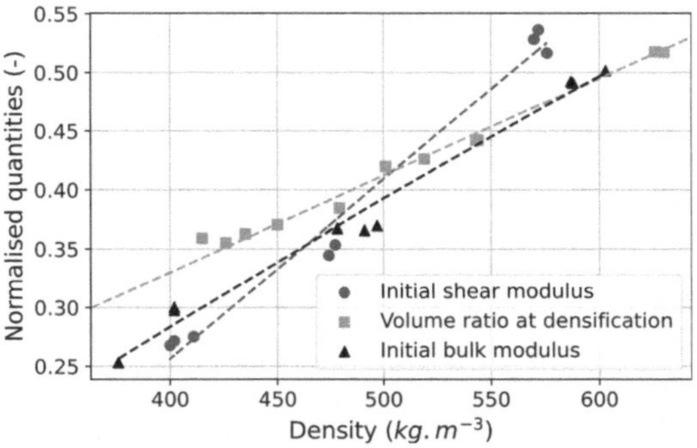

Figure 2. Experimental (normalised) values of the initial shear modulus, the volume ratio at densification and the initial bulk modulus as a function of the density.

The Cauchy stress $\underline{\underline{T}}$ can be calculated from the energy potential as follows:

$$\underline{\underline{T}} = J^{-1}\left[\frac{\partial W}{\partial K_1}\underline{\underline{I}} + \frac{\partial W}{\partial K_2}\underline{\underline{N}} + \frac{1}{K_2}\frac{\partial W}{\partial K_3}\underline{\underline{Y}}\right] \quad (6)$$

where $\underline{\underline{Y}} = 3\sqrt{6}\underline{\underline{N}}^2 - \sqrt{6}\underline{\underline{I}} - 3K_3\underline{\underline{N}}$

The f function proposed by Landauer et al. (2019) predicts an apparent compressibility modulus that tends to infinity in hydrostatic compression when J tends to J_{min}. This is not really realistic and in FEA, this can lead to convergence issues when the volume ratio is close from J_{min}. The f function is changed so that below a given value of volume ratio J_m, the hydrostatic Cauchy stress becomes a linear function of J:

$$f^*(J) = \begin{cases} f(J), & \text{if } J > J_m \\ \frac{B_{PU}}{2BJ_m}J^2 + aJ + b, & \text{otherwise} \end{cases} \quad (7)$$

where a, b and J_m are determined so that f^* is of class C^2 on the considered space.

J_m is then the solution of the equation:

$$\frac{d^2 f}{dJ^2}(J_m) - \frac{B_{PU}}{BJ_m} = 0 \quad (8)$$

This modification introduces an additional parameter, B_{PU}, which represents the bulk modulus of bulk polyurethane.

3.2 *Identification procedure*

3.2.1 *Determination of the polyurethane bulk modulus*

The bulk modulus of the bulk polyurethane, B_{PU}, is estimated using the results of CC tests. When the densification phase is reached, all the cells are closed and the material becomes nearly incompressible like bulk polyurethane. The CC test can then be approached as a hydrostatic compression test, where the hydrostatic stress is approximately equal to the longitudinal stress measured (Peng et al. 1994). The apparent foam bulk modulus B_{foam} is calculated as the slope of the CC hydrostatic stress-volume ratio curve from –40 MPa (corresponding to a volume ratio J_m) to the end of the test. Assuming that at $J = J_m$, the volume of the polyurethane matrix has changed very little and that the hydrostatic stress in the foam is approximately equal to that in the polyurethane matrix, B_{PU} can be related to B_{foam} with the relation:

$$B_{PU} \approx J_m B_{foam} \quad (9)$$

B_{PU} is calculated for each density: the average value is about 3711 MPa with a standard deviation of 147 MPa.

3.2.2 *Global identification strategy*

According to the experimental results in section 2.2, it is assumed that only the parameters $y = \{G_0, B, J_{min}\}$ in the model are density dependent, with a linear relationship between the parameter and the density ρ of the polyurethane foam, $y = a_y\rho + b_y$, with a_y and b_y new parameters of the model to be found.

Four groups of {UC, UT, SS, CC} tests with relatively close densities are constituted with mean density values of 404, 482, 492 and 580 kg.m^{-3}, respectively: {402; 399; 400; 415}, {478; 496; 477; 479}, {497; 496; 477; 501} and {588; 578; 576; 543}. Identification is carried out iteratively (at the first iteration, no parameters are fixed):

1. The parameters are identified using the protocol described in Section 3.2.3 for each group of material density.
2. For each parameter x, a coefficient of variation CV is defined: $CV(x) = \frac{\sigma_x}{\mu_x}$, where σ_x and μ_x are respectively the standard deviation and the mean value. Each parameter x for which $CV(x) \leq 20\%$ is fixed to its mean value: $x_{fixed} = \mu_x$. If no parameter fulfils the condition $CV(x) \leq 20\%$, the parameter with the lowest CV is fixed to its mean value.
3. Step 1 and 2 are repeated, identifying only the parameters not fixed in the previous step, until all parameters of the model are fixed (except G_0,B,J_{min}) using the method described in step 2.
4. Step 1 is repeated a last time. Coefficients a_x and b_x of the parameters G_0,B,J_{min} are found using linear regressions.

3.2.3 *Identification procedure for each density*

The identification procedure applied at the first step of the global identification strategy presented above is divided into a pre-identification and global identification. The pre-identification is a step-by-step process:

1. The minimal volume ratio J_{min} for which the densification occurs is directly determined from CC test data (see Figure 2).
2. The parameters of the X and L functions are identified using the UC and UT test data (as done in Landauer et al. (2019)). In addition, SS test is included.
3. The parameters of the f function are adjusted using the same approach as Landauer et al. (2019), from the UC and UT test data.

Then, a global identification is performed, using the UC, UT and SS test data, with the pre-identified set of parameters as starting point. The optimization is carried out using a Levenberg-Marquardt algorithm. The same objective function as Becker et al. (2022) is used. Finally, J_m is calculated using Eq. 8. It depends linearly on density.

4 RESULTS AND DISCUSSION

4.1 *Results on the identification database*

Once the parameters have been identified, the model is used to simulate the entire experimental database, considering the measured density of each sample (instead of the average density of the density group). The results are shown in Figure 1. For the CC test, the volume ratio corresponding to densification is well described (the rest of the response is not, as this test has not been considered in the identification). The response to UT, UC and SS tests is relatively well predicted, for all densities, over the entire wide range of deformations.

4.2 *Validation on a jounce bumper*

For a purpose of validation, the model is used to simulate a jounce bumper in compression, where the material is subjected to multiaxial loading. The finite element model used is similar to the one presented in Becker et al. (2022). The material of the jounce bumper is assumed to have a homogeneous density. Figure 3 shows the load-displacement response obtained according to the finite element model, compared to the experimental data for three densities. In addition, to have a more local validation, the simulated deformed shape is compared to the experimental one, obtained by X-ray tomography, in Figure 4. The left and right sides of the jounce bumper observed on an axial section of the tomography are represented by the solid and dashed black outlines (the deformation is not perfectly axisymmetrical). The K_1 field calculated by FEA shown in Figure 4 illustrates that the jounce bumper is subjected to large volume variations and heterogeneous mechanical fields. The results are in agreement with the experiments. This shows that the model is capable of correctly describing the non-linear compressible elastic response of the material subjected to multiaxial mechanical loading, resulting in significant strain and volume variations (several tens of percent), and the effect of density.

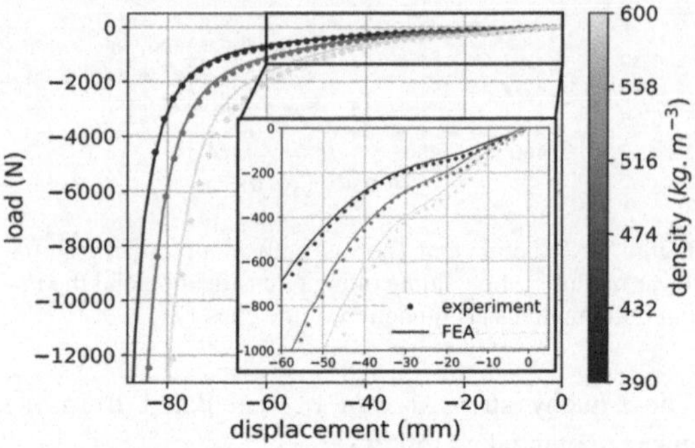

Figure 3. Experimental and finite element load-displacement curve of a jounce bumper, for three densities.

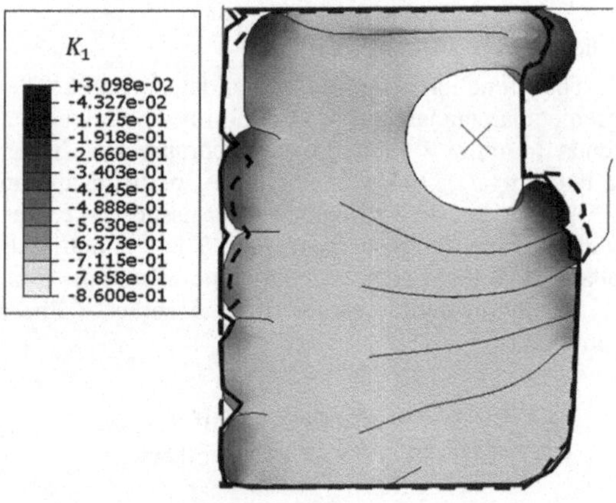

Figure 4. Comparison between the experimental (two deformed shapes from tomography shown by solid and dashed black lines) and the deformed shape (along with the field of K1) calculated by FEA of a jounce bumper with a mean density of 500 kg.m^{-3}.

5 CONCLUSIONS

In this paper, the phenomenological model from Landauer et al. (2019) has been adapted to consider the effect of the foam density on the non-linear elastic compressible response of an elastomeric polyurethane foam, for a density range of 380 to 600 kg.m^{-3}. The model includes 15 parameters. Only 3 of them depend on density, linearly. A robust identification process has been proposed based on an extensive experimental database composed of uniaxial compression, uniaxial tension, simple shear and confined compression tests. The constitutive model has been validated on a structure with the simulation of an automotive jounce bumper in compression. The results show a good correlation between the experiment and the simulation, for different densities. It is now possible to use the model to describe the behaviour of the material in parts with density gradients.

ACKNOWLEDGEMENT

The authors would like to thank Maëlys Pareilleux for her contribution to this study.

REFERENCES

Ashby, M. F. & L. J. Gibson (1997). Cellular solids: structure and properties. *Press Syndicate of the University of Cambridge, Cambridge, UK*, 175–231.

Avalle, M., G. Belingardi, & A. Ibba (2007). Mechanical models of cellular solids: Parameters identification from experimental tests. *International Journal of Impact Engineering 34*(1), 3–27.

Becker, J., J.-B. Le Bail, M. Le Saux, V. Le Saux, Y. Marco, L. Maheo, P. Charrier, & W. Hervouet (2022). Modelling of the compressible hyperelastic behavior of polyurethane foams: Influence of the identification strategy. In *Constitutive Models for Rubber XII*, pp. 20–25.

Bouix, R., P. Viot, & J.-L. Lataillade (2009). Polypropylene foam behaviour under dynamic loadings: Strain rate, density and microstructure effects. *International Journal of Impact Engineering 36*(2), 329–342.

Chai, H., Z. Xie, X. Xiao, H. Xie, J. Huang, & S. Luo (2020). Microstructural characterization and constitutive modeling of deformation of closed-cell foams based on in situ x-ray tomography. *International Journal of Plasticity 131*, 102730.

Criscione, J. C., J. D. Humphrey, A. S. Douglas, & W. C. Hunter (2000). An invariant basis for natural strain which yields orthogonal stress response terms in isotropic hyperelasticity. *Journal of the Mechanics and Physics of Solids 48*(12), 2445–2465.

Danielsson, M., D. M. Parks, & M. C. Boyce (2004). Constitutive modeling of porous hyperelastic materials. *Mechanics of Materials 36*(4), 347–358.

Di Landro, L., G. Sala, & D. Olivieri (2002). Deformation mechanisms and energy absorption of polystyrene foams for protective helmets. *Polymer Testing 21*(2), 217–228.

Goga, V. (2011). New Phenomenological Model for Solid Foams. In J. MurÃ-n, V. KompiÅ¡, and V. KutiÅ¡ (Eds.), *Computational Modelling and Advanced Simulations*, Computational Methods in Applied Sciences, pp. 67–82.

Goods, S. H., C. L. Neuschwanger, C. C. Henderson, & D. M. Skala (1998). Mechanical properties of CRETE, a polyurethane foam. *Journal of Applied Polymer Science 68*(7), 1045–1055.

Landauer, A. K., X. Li, C. Franck, & D. L. Henann (2019). Experimental characterization and hyperelastic constitutive modeling of open-cell elastomeric foams. *Journal of the Mechanics and Physics of Solids 133*, 103701.

Le Bail, J.-B. (2022). *Mod´elisation du comportement m ´ecanique sous chargement d'une but´ee d'amortisseur en mousse de polyur´ethane: vers une d´emarche*

Le Saux, M., J.-B. Le Bail, J. Becker, C. Caër,, P. Charrier, V. Le Saux, L. Maheo, & Y. Marco (2023). Statistical characterization of microcellular polyurethane foams microstructure based on 2D and 3D image analysis. *Journal of Cellular Plastics 59*(5-6), 395–417.

Lewis, M. (2016). A Robust, Compressible, Hyperelastic Constitutive Model for the Mechanical Response of Foamed Rubber. *Technische Mechanik - European Journal of Engineering Mechanics 36*(1-2), 88–101.

Lin, H.-R. (1997). The structure and property relationships of commercial foamed plastics. *Polymer Testing 16*(5), 429–443.

Ling, C., J. Ivens, P. Cardiff, & M. D. Gilchrist (2018). Deformation response of EPS foam under combined compression-shear loading. Part I: Experimental design and quasi-static tests. *International Journal of Mechanical Sciences 144*, 480–489.

Liu, Q. & G. Subhash (2004). A phenomenological constitutive model for foams under large deformations. *Polymer Engineering & Science 44*(3), 463–473.

Mihai, L. A., H. Wyatt, & A. Goriely (2017). Microstructure-based hyperelastic models for closed-cell solids. *Proceedings of the Royal Society A: Mathematical, Physical and Engineering Sciences 473*(2200), 20170036.

Peng, S. H., T. Shimbori, & A. Naderi (1994). Measurement of Elastomer's Bulk Modulus by Means of a Confined Compression Test. *Rubber Chemistry and Technology 67*(5), 871–879.

Rubin, M. B. & L. Dorfmann (2020). A Hyperelastic Model for Soft Polymer Foam Including Micromechanics of Porosity. *Journal of Elasticity 138*(2), 205–220.

Rusch, K. C. (1969). Load–compression behavior of flexible foams. *Journal of Applied Polymer Science 13*(11), 2297–2311.

Saha, M. C., H. Mahfuz, U. K. Chakravarty, M. Uddin, M. E. Kabir, & S. Jeelani (2005). Effect of density, microstructure, and strain rate on compression behavior of polymeric foams. *Materials Science and Engineering: A 406*(1), 328–336.

Saint-Michel, F., L. Chazeau, J.-Y. CavaillÃ©, & E. Chabert (2006). Mechanical properties of high density polyurethane foams: I. Effect of the density. *Composites Science and Technology 66*(15), 2700–2708.

Sherwood, J. A. & C. C. Frost (1992). Constitutive modeling and simulation of energy absorbing polyurethane foam under impact loading. *Polymer Engineering & Science 32*(16), 1138–1146.

Thirumal, M., D. Khastgir, N. K. Singha, B. S. Manjunath, & Y. P. Naik (2008). Effect of foam density on the properties of water blown rigid polyurethane foam. *Journal of Applied Polymer Science 108*(3), 1810–1817.

Viot, P. (2009). Hydrostatic compression on polypropylene foam. *International Journal of Impact Engineering 36*(7), 975–989.

Wang, S., Z. Zheng, C. Zhu, Y. Ding, & J. Yu (2018). Crushing and densification of rapid prototyping polylactide foam: Meso-structural effect and a statistical constitutive model. *Mechanics of Materials 127*, 65–76.

Constitutive Models for Rubber XIII – Dal (Ed.)
© 2025 The Author(s), ISBN 978-1-032-85138-9

Micromechanical modelling of the viscoelastic properties of filled elastomer blends

D. Juhre, J. Voges & M. Niemeyer
Chair of Computational Mechanics, Otto von Guericke University Magdeburg, Germany

M. Klüppel, A. Lang & M. Müller
German Institute of Rubber Technology, Hannover, Germany

ABSTRACT: Blending elastomers is an attractive method for achieving desired mechanical properties in different fields of application. While the experimental characterization of the viscoelastic properties is usually feasible for the pure phases, it can be difficult or even impossible for blends due to their components' interactions and the complex behavior resulting from their different glass transition temperatures. Typically, blending elastomers results in heterogeneous morphologies comprising regions with (almost) pure phases and finite interphases. The pure phases and interphases both significantly influence the viscoelastic properties. Material modeling and numerical simulations can be employed to understand the phase interactions better and predict the resulting viscoelastic properties. In this contribution, we investigate unfilled and filled binary blends consisting of natural rubber and styrene butadiene rubber, respectively. We use microscope images as the basis for the morphology that we consider in the numerical simulations. The morphology is stored within a phase parameter for each spatial point in the domain and is evolved in an Allen-Cahn framework to create differently sized diffuse interphases. These are subsequently used for mechanical simulations to investigate the impact on the storage and loss moduli.

1 INTRODUCTION

In order to meet the requirements of high-performance elastomeric materials, the method of blending elastomers is a crucial approach as it enables the creation of customized material properties. To achieve this, it is generally preferable to have morphologies in which the phases are separated rather than homogeneous. The interphases that arise due to the compatibility of the polymers can significantly impact the overall behavior of the blend, as has been investigated in several studies (Nelson et al. 1977, Hamed 1982, Lorenz et al. 2013).

To get insight into the viscoelastic material behavior, the blends are often investigated experimentally within the dynamic mechanical thermal analysis (DMTA), resulting in the storage moduli for information on stored elastic energy and the loss moduli quantifying the dissipative character. Both typically result as a function of the temperature and an excitation frequency. Applying the Williams-Landel-Ferry (WLF) approach, motivated by polymer physics, allows achieving such material behavior over a wide range of frequencies by only measurements in a small frequency interval at different temperatures (Williams et al. 1955). This works very well for many polymers. However, when blending these and dispersed phases form, the WLF procedure cannot be applied without restrictions, and the material behavior cannot be predicted for the full frequency range.

Conventionally, the DMTA is performed experimentally. However, due to the restrictions of the WLF approach and to make such results more easily accessible, the underlying steps are sometimes performed within computer simulations. To this end, molecular dynamics has been employed to investigate the interaction between different polymer chains (Saha & Bhowmick 2017, Ortiz-Serna 2020, Varghese 2020). Another approach of high practical interest is based on continuum mechanics via the finite element method (FEM). It is used in several works which simulate DMA measurements for filled elastomers (Haupt & Lion 2002, Jalocha 2015, Rendek & Lion 2010). However, besides our first study (Voges et al. 2022), approaches based on continuum mechanics for such simulations of blended elastomers cannot be found in the literature.

Therefore, this article contributes to the numerical investigation of the mechanical properties of dispersed elastomer blends with a specific interphase using the example of natural rubber (NR) and styrene

DOI: 10.1201/9781003516880-6

butadiene rubber (SBR) blends. In the theory section 2, first, a phase field framework is introduced to prepare the blend morphologies within representative volume elements (RVEs). Then, a rheological model including the phase parameter is proposed and comments on the numerical implementation are given. In section 3 the simulation results are presented and the impact of different interphase sizes and filler contributions on the material behavior is discussed.

2 THEORETICAL BACKGROUND

2.1 *Generating the blend morphology based on microscopic images*

The general idea of the numerical study within this contribution is to simulate the complex viscoelastic material properties of elastomer blends by a new approach in which the two polymer phases are modelled individually by an appropriate viscoelastic material model and to consider the interphase between the polymers by a mixture of both polymers following a mixture rule with a well-defined interphase width. This width is controlled by a phase parameter, which runs from $0 \leq \varphi \leq 1$, where $\varphi = 1$ signifies polymer A and $\varphi = 0$ polymer B, respectively (see Figure 1).

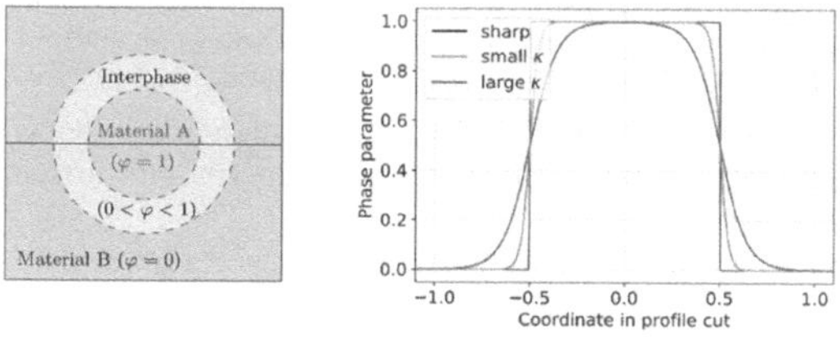

Figure 1. General idea of the description of interphase (in gray) between material A and material B by using a phase-field parameter φ (left). Influence of the interphase width κ (right).

Due to the highly complex morphology of the elastomer blends, a simulation of the entire body is computationally not feasible. Instead, only a representative volume element (RVE) is simulated, whose behavior accurately mimics that of the structure at the macro scale.

AFM measurements of DIK (see Figure 2) were used to prepare RVEs by an image-processing pipeline. Therefore, in a first step the pixel data of the measurement is modified, such that it contains black and white pixels only, representing the regions occupied by material A and B, respectively. Note that this step results in a sharp interface between the phases, and φ equals either 0 or 1. As a second step, the Allan-Cahn (AC) evolution equation is solved yielding a smooth interface region between the materials. Therefore, we consider the free energy functional

$$\Psi = \int_{\Omega} W\varphi^2(1-\varphi)^2 + \frac{1}{2}\,\kappa|\nabla\varphi|^2 dV \qquad (1)$$

where the second term accounts for the interphases. The first term, a double-well potential, represents the configurational energy. The parameters W and κ scale the energy contributions.

The evolution of the phase field yields

$$\frac{\partial\varphi}{\partial t} = -M\frac{\delta\Psi}{\delta\varphi} = M\left[\kappa\,\nabla^2\varphi - W\left(\varphi^3 - 6\varphi^2 + 2\varphi\right)\right] \qquad (2)$$

with the mobility M. Two diffuse interphase morphologies are created as illustrated in Figure 2. We label them as model 1 and model 2 and use $\kappa_1 = 2.5 \cdot 10^{-10}$ J/mm and $\kappa_2 = 2.5 \cdot 10^{-9}$ J/mm. Note that the interphase width also depends on the choice of W, which is set to $W = 1.0$ J/mm^3 in all simulations. Together with the sharp interface model 0 we provide three models for further investigation.

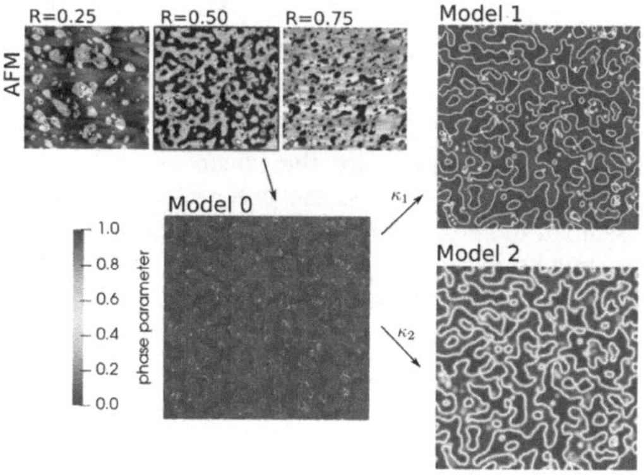

Figure 2. Creating diffuse interphases in blend morphologies from AFM images. Two different κ were used where $\kappa_2 = 10 \cdot \kappa_1$. The edge lengths of the microscope images are 5 μm.

2.2 *Mechanical model for the elastomer blend*

The viscoelastic behavior of the pure polymer phases is approximated by a generalized Maxwell model (GMM), for which the rheological model consists of a main elastic spring, and a parallel connection of n spring-dashpot-branches (see e.g., red model in Figure 3). While the standard GMM suffices to describe the viscoelastic response for pure materials (i.e., $\varphi = 0$ and $\varphi = 1$), it fails to provide an accurate description of the mixed material in the interphase region.

Therefore, we propose a material model for the elastomer blend that is based on a parallel connection

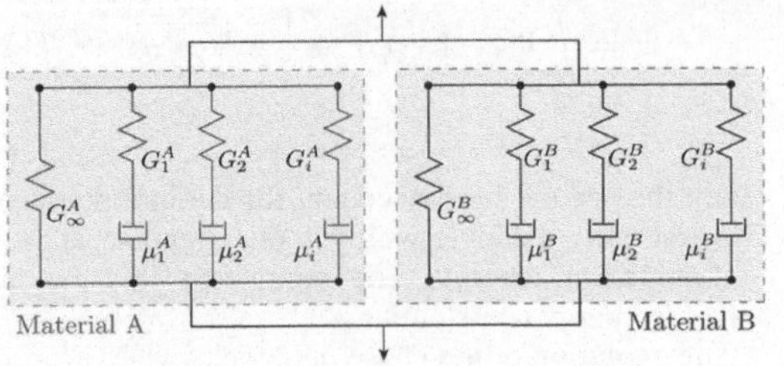

Figure 3. Structure of the generalized Maxwell model considering phase A and B in a mixture.

of two GMMs, as depicted in Figure 3, where the participation factor of branch A and B is expressed by the continuously changing phase field data φ.

The free energy of the system is defined by

$$\psi = (1-\varphi)\psi^A + \varphi\psi^B \tag{3}$$

where ψ^i is the free energy of each phase $i \in A, B$

$$\psi^i = G^i_\infty tr\ e^2 + \frac{K}{2}\ (tr\ \varepsilon)^2 + \sum_{n=1}^{N} G^i_n tr\left(e - e^i_{v,n}\right)^2. \tag{4}$$

Here, G^i_∞ and G^i_n are the shear moduli of the GMM (see Figure 2), K the compression modulus (assumed to be the same for phase A and B) and $\boldsymbol{e}$ is the deviatoric part of the strain ε. The evolution of the deviatoric viscous strain $\boldsymbol{e}_{v,n}$ is given by

$$\dot{e}^i_{v,n} = \frac{G^i_n}{\eta^i_n}\left(e - e^i_{v,n}\right). \tag{5}$$

Note that we assume that the total strain ε and $\boldsymbol{e}$ are the same for phase A and B.

The derivation of the required phase-field data, as well as the simulation of the elastomer blends' mechanical behavior, require elaborate numerical tools. Below, the key aspects are summarized that are essential for the successful numerical simulation.

2.3 *Numerical implementation and workflow for the simulations*

When simulating RVEs, the assumed repeating nature of the structure is typically achieved by periodic boundary conditions, which enforce a coupling of displacements on the boundary edges/faces opposite to each other. The actual loading is achieved by a prescribed uniform strain field $\tilde{\varepsilon}$ on the domain boundaries, which in our case, corresponds to a shear loading (see Figure 4).

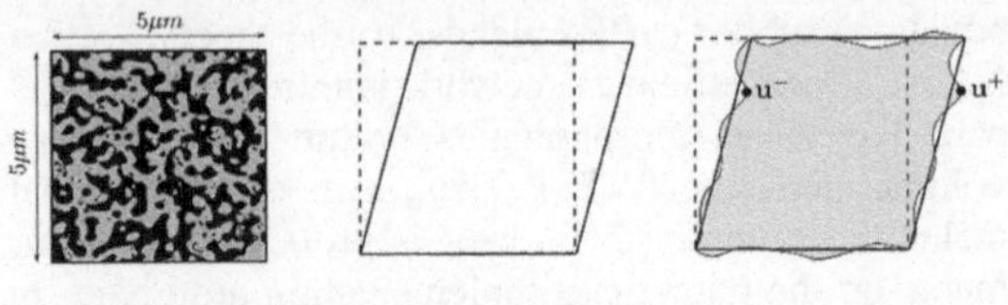

Figure 4. Loading of the RVE by cyclic simple shear considering periodic boundary conditions.

Note that viscous material behavior becomes apparent in transient cases, where the velocity dependent nature of the material response and relaxation effects are triggered. Therefore, instead of statically loading the structure, we use a sinusoidal loading $\tilde{\varepsilon} = \hat{\varepsilon}\sin(2\pi ft)$, where $\hat{\varepsilon}$ is the amplitude of the uniform strain field, and f the excitation frequency.

Due to the heterogeneous nature of the RVE, the strain and stress fields inherit local fluctuations. To formulate general statements regarding the macroscopic behaviour of the structure, we are interested in the effective properties of the spatially fluctuating fields and constitute laws. Due to the shear loading the average stress $\bar{\sigma}(t)$ and strain fields $\bar{\varepsilon}(t)$ are dominated by their shear components. Since the structure is loaded harmonically, the system response is also oscillatory, where due to the viscoelastic material, the stress and strain fields are not in phase. At the initial time steps, the phase shift is not constant, after a finite number of loading cycles however, it converges to a stationary value, which can be directly used to estimate the effective storage modulus G' and loss modulus G'' of the system for the current excitation frequency.

We perform our simulations using the Finite Element Analysis Program (FEAP). Due to the incompressibility of the elastomer blends, a nearly incompressible material behavior is assumed, which is characterized by the Poission's ratio of $\nu = 0.4995$. Since the standard displacement-based FE-formulation fails to accurately model such materials due to volumetric locking effects, we use a mixed u- p- θ formulation, where besides of the displacement field u, the pressure field p and volumetric strain field θ are also primary unknowns. All simulations are conducted in 2D using a Cartesian spatial discretization by 1000 × 1000 elements.

3 NUMERICAL RESULTS

3.1 *Simulation of unfilled blends*

The primary focus of the numerical experiments were unfilled elastomer blends consisting of NR and SBR materials. For the analysis of the viscoelastic behavior of the pure NR and SBR phases, and various blends of those, DMA was utilized.

In order to enable a smooth description for the storage and loss moduli of the pure NB and SBR phases in the simulation, the discrete experimental data measured by the DIK is interpolated via a Prony series, often used in linear viscoelasticity.

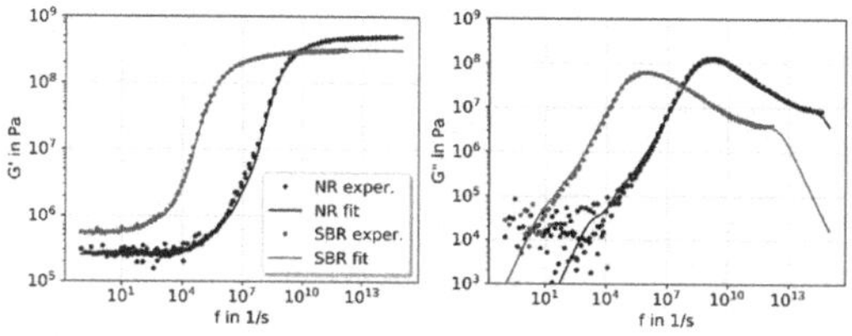

Figure 5. Fit of the pure phases NR and SBR by the general Maxwell model with 30 Prony series.

For both materials, the curve fitting was achieved via 30 Prony series terms, yielding an excellent fit as depicted in Figure 5. The downturn of the G'' curves at frequencies above 10^{12} Hz are due to lack of experimental data. Since we only investigate the model in the range up to 10^{12} Hz this shortfall is not harming the following results.

In the next step we investigate the elastomer blends with mixed phases. At first, we focus on the 50/50 blend and investigate the influence of the interphase width κ. In Figure 6, the effect of κ is compared to the sharp interface model, and additionally, the fitted curves corresponding to the pure NR and SBR materials are also depicted. In particular, we investigate the RVE of Figure 2 and the diffuse morphology is based on the two distinct interphase widths, κ_1 and κ_2.

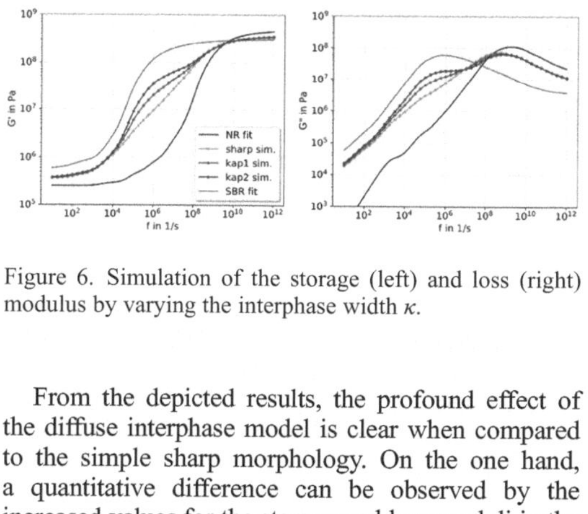

Figure 6. Simulation of the storage (left) and loss (right) modulus by varying the interphase width κ.

From the depicted results, the profound effect of the diffuse interphase model is clear when compared to the simple sharp morphology. On the one hand, a quantitative difference can be observed by the increased values for the storage and loss moduli in the mid-frequency range. On the other hand, a qualitative difference can be seen in the emerging stepped structure of the curves for increasing interphase widths, which manifests itself in an emerging second plateau for G' and second local maximum for G''. By comparing these results qualitatively with the blue lines in Figure 3 that represent the G' and G'' for the 50/50 blend via a temperature sweep, the pronounced step function of G' and the maxima of G'' are visible, too. Resulting that the consideration of the interphase in the simulation shows quite satisfying correlations to the experiment.

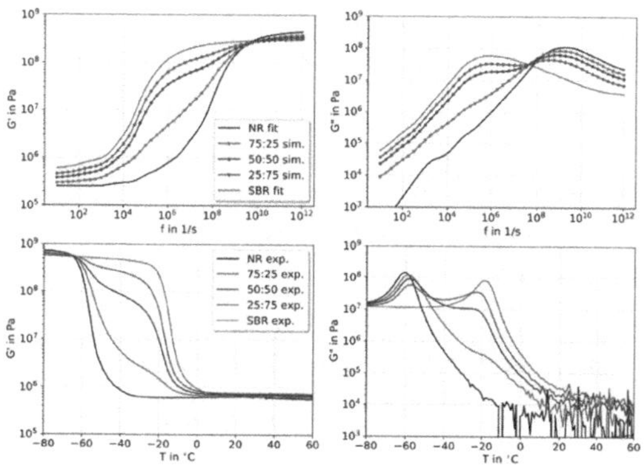

Figure 7. Plot of the storage moduli and the loss moduli determined via simulations for different blend ratios. All blend ratios use κ_2 for the interphase width. In the bottom line, experimental data is plotted over the temperature.

The same can be argued when simulating the different blend ratios by using the interphase width κ_2 (Figure 7) and comparing G' and G'' to the equivalent experiments in the temperature regime. As in the experiments the 75/25 blend shows a less pronounced step in G' and G'', whereas the 25/75 blend has a strong pronunciation of the two peaks in G''. Also, the switch of the G'' curves between the two glass transition zones can be represented very accurately by the simulations.

3.2 *Simulation of filled blends*

In a further step we also wanted to evaluate the model for filled elastomers. This step is much more complicated, since we have to introduce a third phase into our system, which is the filler. We have investigated two approaches, how to consider the filler content within the elastomer blends. Our approach is to obtain master curves for filled elastomer blends based on master curves of filled elastomers by means of simulation. Therefore, material data regarding the two-phase micro-structure and the average filler content of each polymer phase are necessary inputs. In Figure 8 two exemplary fits of the generalized Maxwell model to the experimental data of NR and SBR with 20 phr CB (N330), respectively, are shown.

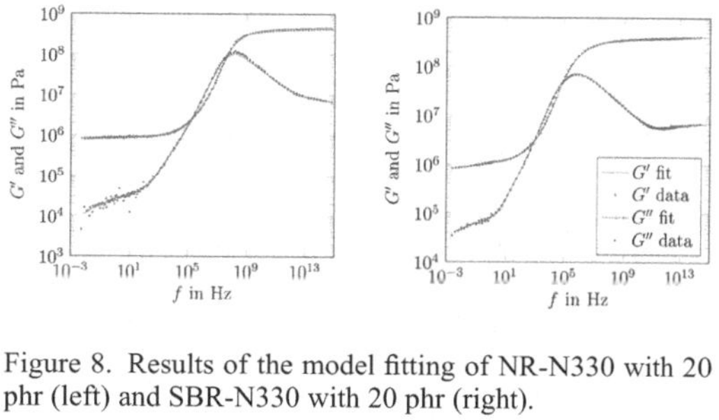

Figure 8. Results of the model fitting of NR-N330 with 20 phr (left) and SBR-N330 with 20 phr (right).

For considering the distribution of the filler in the two phases, it is worth mentioning that unfilled morphologies are not eligible for these filled elastomers, as the filler is distributed unevenly and the morphology differs significantly. If 100 % of the filler was in phase A in a A(50)-B(50) blend, then the filler content of phase A would be doubled compared to the total filler content of the blend. This implies that blends with filler content above 30 phr could be problematic, as only DMA data with filler content of up to 60 phr is available. And in fact, for all blends it is observed that carbon black preferentially accumulates in SBR, so that the filler content in the SBR phase is expected to be higher than in the NR phase. This is confirmed by the measured filler distribution of NR(50)-SBR(50) with varying filler content in Figure 9.

The presumed high filler content in SBR, exceeding 60 phr, for which the DMA data is available, is indeed found for the blends with a total filler content equal or greater than 40 phr, so that the evaluation of the model approach can only be reasonably evaluated for lower filler contents.

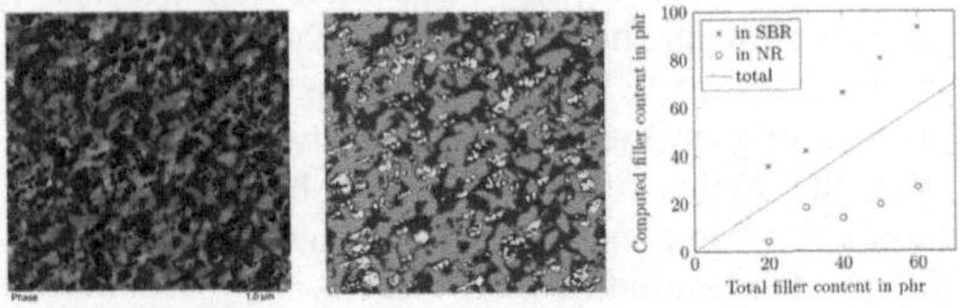

Figure 9. AFM measurement of NR(50)-SBR(50)-N330 (20) (left) and its description in the RVE (NR-light blue, SBR-dark blue, N330-yellow). Determined distribution of filler content in both phases (right).

The storage moduli of the reference data and those computed are compared in Figure 10. The results of simulations and measurements have in common that the blends with the same polymer fractions behave similarly. Furthermore, the NR(25)-SBR(75) curves are shifted towards pure SBR curves and the moduli rise with the filler content, except for the NR(50)-SBR(50)-N330(30)/N330(40) blends. The curve sets differ in the space between the groups of blends described above. In other words, there are more distinct differences between the groups in the reference data than in the simulation results. Also, the bends of the NR(50)-SBR(50) blends are clearer-cut in the reference data. This does not necessarily imply a problem attributed to the model, but simply due to the nonlinear relation between frequency and temperature do-main. The comparison of the loss modulus confirms the qualitative similarities within the groups and the characteristics of the NR-associated maximal G''.

The ascending order of the moduli with respect to the filler content is also confirmed, except for the SBR(50)-N330(30)/N330(40) blends.

The fact that this inconstancy occurs for both moduli suggests that there is a connection. Considering

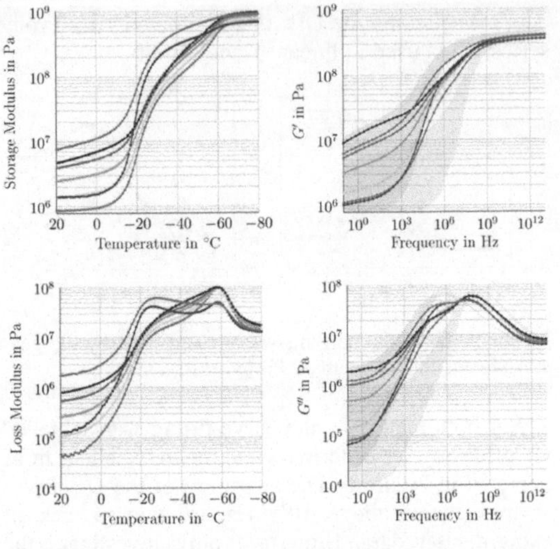

Figure 10. Experimental data (left) of storage and loss modulus in the temperature domain and simulation results (right) in the frequency domain.

Figure 9, it might be explained by the distribution of the filler to the respective polymer phases. Nevertheless, it is striking that there are two distinct maxima in the temperature domain, but only one broad or two merged maxima without minimum in-between in the frequency domain. This finding might be explained by or even indicates the nonlinear conversion between frequency and temperature domain.

4 CONCLUSIONS

In this contribution, we presented a workflow for deriving complex blend morphologies with smoothly varying phase-field data. The diffuse morphologies are used as an input for the material properties, for which we proposed a rheological model combining two generalized Maxwell models based on the phase-field data. Both the derivation of blend morphologies and the proposed material model were successfully implemented in an advanced FEA tool providing all the necessary computational features required for the simulations. Our numerical investigations of the viscoelastic material behavior of the complex phase morphologies reveal a good agreement with the experiments. It could be shown that the developed diffuse morphology is indeed superior to the sharp one, as the former leads to qualitatively more reasonable curves in the frequency domain, inhibiting two glass transitions. While the results reassure about the need for the elaborate interphase model having a great impact on the viscoelastic material behavior of the blend, the correct value for the interphase width is still under investigation.

The results obtained from image processing AFM data demonstrate that the developed method allows for preparing RVEs of filled elastomer blends, too. Since assigning the filler content to polymer phases is sparsely addressed in the literature, there is no standardized method on defining such blend structure. The filler dominantly distributes to the SBR phase, which is in line with the literature. Consequently, highly-filled blends consisting of low-filled NR and even higher-filled SBR exceed the limits of available DMA data, then causing extrapolation problems in computing dynamic moduli. The majority of results is physically reasonable, including the filler content-independent glass transition and the ascending moduli with increasing filler content. The behavior of dynamic moduli between the glass transitions is of particular interest when investigating elastomer blends and often attributed to the interphase, albeit not being comprehensively understood. Still, the dependence on the distribution of the filler to the polymer phases becomes apparent when comparing the NR(50)-SBR(50) blends, which is especially true for the glass transition.

REFERENCES

Hamed, G.R. 1982. Morphology-Property Relationships in EPDM-Polybutadiene Blends. In: Rubber Chem. Technol. 55, pp. 151–160.

Haupt, P. & Lion, A. 2002. On finite linear viscoelasticity of incompressible isotropic materials. In: Acta Mech. 159, pp. 87–124.

Jalocha, D. et al. 2015. Prestrained biaxial DMA investigation of viscoelastic nonlinearities in highly filled elastomers. In: Polym. Test. 42, pp. 37–44.

Lorenz, H. et al. 2013. Morphology and Micro-mechanics of Filled Elastomer Blends: Impact on Dynamic Crack Propagation. In: Fracture Mechanics and Statistical Mechanics of Reinforced Elastomeric Blends. Ed. by W. Grellmann et al. Berlin, Heidelberg: Springer Berlin Heidelberg, pp. 81–128.

Nelson, C.J. et al. 1977. The relationship between rheology, morphology and physical properties in heterogeneous blends. In: Macromol. Mater. Eng. 60, pp. 49–86.

Ortiz-Serna, P. et al. 2020. Exploring the role of lignin structure in molecular dynamics of lignin/bio-derived thermoplastic elastomer polyurethane blends. In: Int. J. Biol. Macromol. 158, pp. 1369–1379.

Rendek, M. & Lion, A.2010. Amplitude dependence of filler-reinforced rubber: Experiments, constitutive modelling and FEM – Implementation. In: Int. J. Solids Struct. 47, pp. 2918–2936.

Saha, S. & Bhowmick, A.K. 2017. Computer aided simulation of thermoplastic elastomer from poly (vinylidene fluoride)/hydrogenated nitrile rubber blend and its experimental verification. In: Polymer 112, pp. 402–413.

Varghese, G. et al. 2022. Experimental and Simulation Studies on Nonwoven Polypropylene–Nitrile Rubber Blend: Recycling of Medical Face Masks to an Engineering Product. In: ACS Omega 7, pp. 4791–4803.

Voges, J. et al. 2022. The influence of a diffuse interphase on the viscoelastic behavior of rubber blends". In: Constitutive Models for Rubber XII. CRC Press, pp. 127–132.

Williams, M.L. et al. 1955. The temperature dependence of relaxation mechanisms in amorphous polymers and other glass-forming liquids. In: J. Am. Chem. Soc. 77, pp. 3701–3707.

Constitutive Models for Rubber XIII – Dal (Ed.)
© 2025 The Author(s), ISBN 978-1-032-85138-9

Application of anisotropic hyperelastic model considering distributed fibers to artificial sports materials

A. Matsuda
Institute of Systems and Information Engineering, University of Tsukuba, Tsukuba, Ibaraki, Japan

ABSTRACT: In this paper, an anisotropic hyperelastic model considering distributed fibers to artificial sports materials is proposed. Also, an application of the proposed model to artificial leather materials for sports equipment are also shown. Artificial leathers which made from reinforcing fibers and resin matrix are some of the most important materials in sports equipment. However, the structure of both artificial leather materials is complex because artificial reinforcing fibers are contained in the body. Because of internal fibers, anisotropic mechanical characteristics of artificial leathers are observed. Also, the fiber orientations are dispersed, and fiber orientation dispersion and degree of anisotropy have correlations. Regarding the manufacturing process of sports equipment made with artificial leather materials, it is important to consider the anisotropic mechanical characteristics and fiber orientation dispersion to improve design performance. In this research, anisotropic hyperelastic model(Holzapfel and Gasser, 2000, 2001) was modified for artificial leather materials for sports equipment, and uniaxial tensile loading tests were conducted to investigate their anisotropic characteristics. Also, numerical simulations using the finite element method (FEM) were demonstrated. From tensile loading test results, the existence of one fiber family was revealed in the artificial leather. Regarding the FE simulations, the mechanical properties of the artificial of leathers were reproduced by the proposed model. The applicability of the proposed hyperelastic model to evaluate the mechanical properties of artificial leather materials was demonstrated.

1 INTRODUCTION

In this research, the mechanical characteristics of an artificial leather was investigated using uniaxial tensile loading tests and numerical simulations. The uniaxial tensile loading tests on the artificial leather was conducted to obtain the relationships between stress and strain. Also, numerical computations using the finite element method (FEM) were conducted to reproduce the mechanical properties of the artificial leather. In addition, a hyperelastic modelling of leather materials to reproduce the mechanical characteristics was investigated.

The artificial leather have fiber layers in the material. Chemical fibers such as polyethylene fibers, nylon fibers, and so on compose the fiber layer of artificial leather. The complex tangle of chemical fibers in artificial leather is made by reproducing the structure of the fiber layer of natural leather. Due to the fiber layers in the leather materials, artificial leather exhibit anisotropic mechanical characteristics.

The directions of internal fibers of the leather materials are assumed to be dispersed; this is defined as fiber orientation dispersion (Gasser et al., 2006). An anisotropic hyperelastic model, the Holzapfel-Gasser-Ogden form, was proposed by Holzapfel and Gasser et al 2006. The Holzapfel-Gasser-Ogden form was developed to evaluate the mechanical properties of arterial walls. Arterial walls have collagen fibers in the soft ground matrix, so they show anisotropic characteristics and the fiber directions are dispersed. The Holzapfel-Gasser-Ogden form could reproduce the anisotropic behaviors of arterial walls and consider the fiber orientation dispersion, but it is not suited to describe the mechanical characteristics of artificial leather because the modelling was conducted for a soft biomaterial. In order to reproduce the characteristics of artificial leathers, our proposed hyperelastic model was developed based on the Holzapfel-Gasser-Ogden form. Thus, the proposed model was developed by modifying the Holzapfel-Gasser-Ogden form. Using the test results and the proposed model, numerical simulations using the FEM were conducted and the applicability of the proposed model to evaluate mechanical properties of artificial leathers was investigated.

DOI: 10.1201/9781003516880-7

2 MODELING

In this section, the mechanical frameworks of continuum materials are explained. Fiber-reinforced materials, such as artificial leather, fiber-reinforced rubbers, and so on, are presumed to have one or two fiber families in the solid body. Therefore, the two unit vectors $\boldsymbol{a}_{01}$ and $\boldsymbol{a}_{02}$ (Gasser et al., 2000), which denote fiber orientations of the fiber families, are introduced as:

$$\boldsymbol{a}_{0i} = (\cos\theta_i, \sin\theta_i,\ 0) \tag{1}$$

where θ_i is the fiber orientation angle of the ith fiber family. By using orientation unit vector $\boldsymbol{a}_{0i}$, the ith structure tensor $\boldsymbol{A}_i$ is given by

$$\boldsymbol{A}_i = \boldsymbol{a}_{0i} \otimes \boldsymbol{a}_{0\mathbf{i}} \tag{2}$$

Then, the dispersion parameter κ $(0 \leqq \kappa \leqq 1/3)$ is prepared (Gasser et al., 2006). The dispersion parameter κ shows the degree of fiber orientation dispersion in fiber-reinforced materials and it can be determined experimentally. The lowest value of the dispersion parameter, i.e. $\kappa = 0$, denotes the ideal fiber alignment in the material. On the other hand, the highest value of the parameter, i.e. $\kappa = 1/3$, describes the dispersed fibers perfectly, and then the fiber reinforced material shows isotropic characteristics. The ith generalized structure tensor $\boldsymbol{H}_i$, i = 1,2, introduced by Holzapfel & Gasser (Gasser et al., 2006) is given by

$$\boldsymbol{H}_i = \kappa\boldsymbol{I} + (1 - 3\kappa)\boldsymbol{A}_i \tag{3}$$

where $\boldsymbol{I}$ denotes the identity tensor. The distribution of fiber orientations is based on a research of Lanir (Lanir, 1983). The ith anisotropic invariant I_{4i} is given by the structure tensor $\boldsymbol{A}_i$ with respect to the right Cauchy-Green deformation tensor $\boldsymbol{C}$.

$$I_{4i}(\boldsymbol{C}) = \boldsymbol{C} : \boldsymbol{A}_i = \mathrm{tr}(\boldsymbol{C}\boldsymbol{A}_i) \tag{4}$$

where the anisotropic invariant $\bar{I}_{4i} = 0$ is equal to the elongation squared in the direction of $\boldsymbol{a}_{0i}$. The second Piola-Kirchhoff stress tensor $\boldsymbol{S}$ of hyperelastic material is given by the strain energy function W with respect to the right Cauchy-Green deformation tensor $\boldsymbol{C}$ as follows:

$$\boldsymbol{S} = 2\frac{\partial W}{\partial \boldsymbol{C}} \tag{5}$$

The artificial leather which was applied to tensile loading tests showed different behaviors compared with the arterial layers. This materials showed strong anisotropic characteristics from the initial material states due to the manufacturing process. In order to represent the anisotropic behaviors of artificial leather for sports equipment, following strain energy function is proposed.

$$W = \bar{W}_{\mathrm{iso}}(\bar{\boldsymbol{C}}) + \bar{W}_{\mathrm{ani}}(\bar{\boldsymbol{C}}, \boldsymbol{H}'_1, \boldsymbol{H}'_2) + p(J-1) \tag{6}$$

Here $\bar{W}_{\mathrm{iso}}(\bar{\boldsymbol{C}})$ and $\bar{W}_{\mathrm{ani}}(\bar{\boldsymbol{C}}, \boldsymbol{H}'_1, \boldsymbol{H}'_2)$ are isotropic and anisotropic components, respectively. p is defined as the Lagrange multiplier, which corresponds to the hydrostatic pressure. J is the determinant of the deformation gradient tensor $\boldsymbol{F}$. For the isotropic components of the proposed model, incompressible isotropic Yeoh model was applied to represent the non-linearity as follows:

$$\bar{W}_{\mathrm{iso}}(\bar{\boldsymbol{C}}) = C_1(\bar{I}_1 - 3) + \frac{C_2}{2}(\bar{I}_1 - 3)^2 + \frac{C_3}{3}(\bar{I}_1 - 3)^3 \tag{7}$$

In order to represent the anisotropic behaviors in the large strain region of the artificial leather, following strain energy function for anisotropic material behavior is proposed in this paper.

$$\bar{W}_{\mathrm{ani}}\left(\bar{\boldsymbol{C}}, \boldsymbol{H}'_1, \boldsymbol{H}'_2\right) = \frac{\alpha_1}{2}\left\langle \bar{E}'^2_1 \right\rangle + \frac{\alpha_2}{2}\left\langle \bar{E}'^2_2 \right\rangle \tag{8}$$

This potential function include two linear components by means of $\bar{E}'^2_1$ and $\bar{E}'^2_2$ and are possible to represent anisotropic material behavior starting from the initial material state. Here, $\bar{E}'_1$ and $\bar{E}'_2$ are the modified Green-Lagrange strain-like quantities shown as follows:

$$\bar{E}'_1 = \kappa_1(\bar{I}_1 - 3) + (1 - 3\kappa_1)(\bar{I}_{41} - 1)$$

$$\bar{E}'_2 = \kappa_2(\bar{I}_1 - 3) + (1 - 3\kappa_2)(\bar{I}_{42} - 1) \tag{9}$$

3 RESULTS

In this section, the uniaxial tensile analyses of leather materials characterized by the proposed anisotropic hyperelastic models shown in Equation 7 and Equation 8 are discussed. The numerical simulations were performed using the finite element method (FEM). In this study, the mean orientations and the dispersion parameters approximated from the uniaxial tensile tests were applied to numerical simulations, and the applicability of the proposed hyperelastic model to the evaluation of the mechanical property of leather materials was investigated.

From the results of the tensile tests for the artificial leather, the material parameters of each material were approximated by the least-squares fitting between the test results and theoretical calculations. The material parameters of the artificial leather are defined as C_1 = 1.774, C_2 = - 1.431, C_3 = 3.943, $\alpha_1 = 2.236$, $\kappa_1 = 0.034$ and $\theta_1 = 0$.

The dispersion parameters of the first and second fiber families are described by κ_1 and κ_2. The mean fiber orientations of the first and second fiber families are denoted by θ_1 and θ_2. These are given by the results of the uniaxial tensile tests. From the tensile tests of the artificial leather, the artificial leather was estimated to have only one fiber family in the solid body. So, the parameters related to the second fiber family α_2 and θ_2 were set to zero in the analysis. An analysis model reproducing the test specimen and the fiber orientation vectors a_{01} and a_{02} in the model is shown in Figure 1.

The FE model for the observed section of the specimen was introduced, and the geometry is described as follows: length L = 20.0 mm and width W = 4.0 mm. The thickness of the FE model of the artificial leather T was set to 1.06 mm. Regarding the FE computations, the boundary condition was constrained to only the y-axis direction for the top and bottom surfaces to reproduce the tension tests. Tensile deformations reaching a strain of 0.3 were applied for the FE analysis.

The analysis results for the artificial leather are shown in Figures 2, 3, and 4. Relationships between the nominal stress and the nominal strain of the experimental results and simulated results in directions of 0, 45, and 90 degrees are shown. The experimental results and analysis results showed good agreement. As seen in the results of the FE computations, non-linear and anisotropic mechanical characteristics of the artificial leather could be described by the proposed hyperelastic model. In Figure 5, relationships between the nominal stress at 30% tensile strain and the cut-out angle of the experimental results and analysis results are shown. Equivalent stress distributions of FE computations for the artificial leather with the proposed hyperelastic model are shown in Figure 6. In the FE computation of the artificial leather, diagonal high stress distribution in the specimen was observed due to the one fiber family.

4 CONCLUSION

In this research, the mechanical characteristics of an artificial leather was investigated using uniaxial tensile tests, and an anisotropic hyperelastic model was proposed. Finite element computations with the proposed model were conducted to evaluate the mechanical properties of the leather materials. The artificial leather material, which are used for actual sports equipment, were used for the investigation of the mechanical properties. From the uniaxial tensile tests, the artificial leather was estimated to have a one-fiber family in the body.

The proposed model was able to describe the mechanical behaviors of the leather materials. Numerical simulations using the finite element method (FEM) for the artificial leather was conducted. For the analysis of the artificial leather, only the anisotropic component of the first fiber family was used for calculations and the symmetric anisotropic characteristics centered at the mean fiber orientation were obtained. The diagonal high stiffness distribution of the artificial leather, which had one fiber family, was observed in the material.

The numerical simulations with the proposed model reproducing the tensile loading test show the applicability of the proposed model for the evaluation of artificial leather for sports equipment.

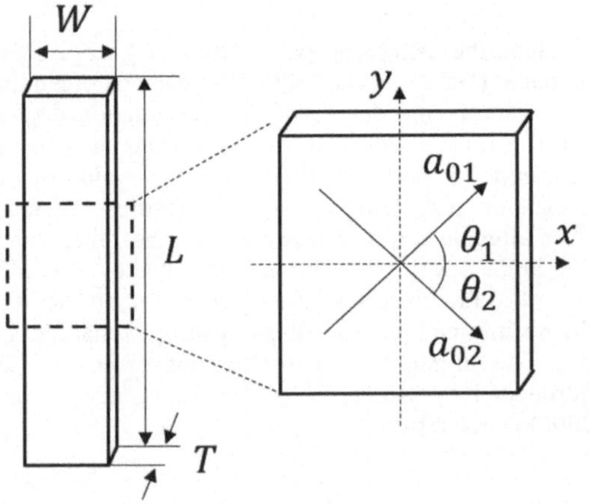

Figure 1. Analysis model of FE computations and distribution of fiber orientations vectors a_{01} and a_{02} in the FE model.

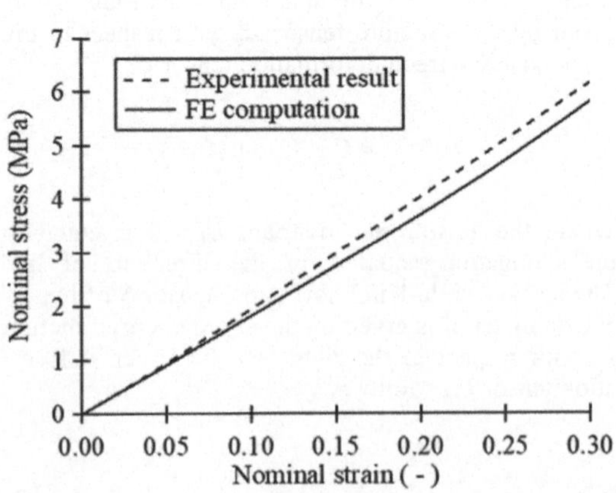

Figure 2. Relationships between nominal stress and nominal strain for the experimental results and FE analysis results in a direction of 0 deg.

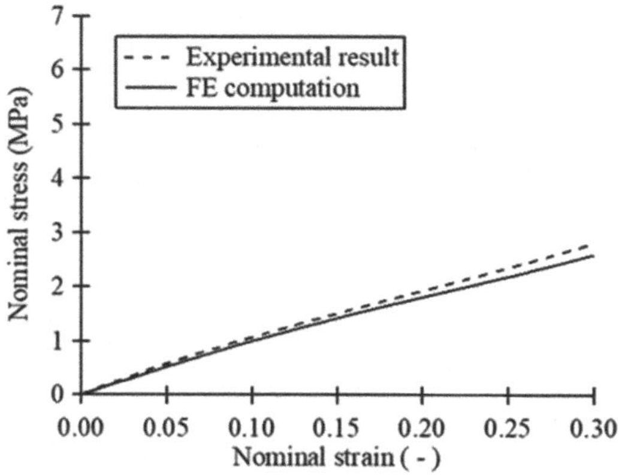

Figure 3. Relationships between nominal stress and nominal strain for the experimental results and FE analysis results in a direction of 45 deg.

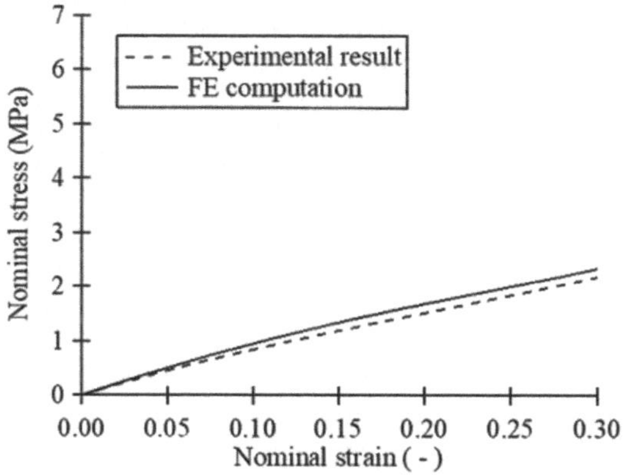

Figure 4. Relationships between nominal stress and nominal strain for the experimental results and FE analysis results in a direction of 90 deg.

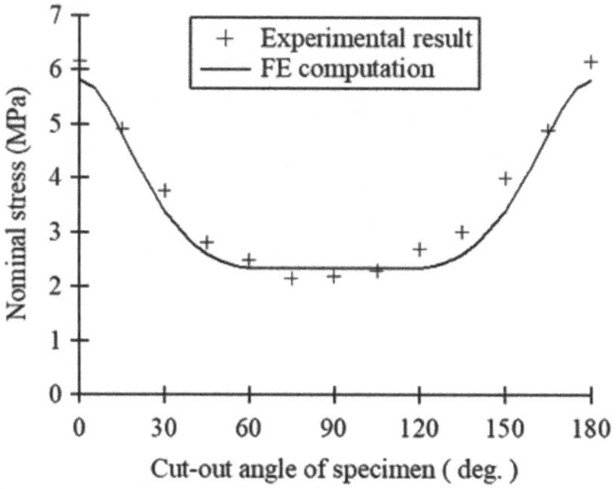

Figure 5. Relationships between nominal stress at 30% tensile strain and cut-out angle of specimen for the experimental results and FE analysis results for AL.

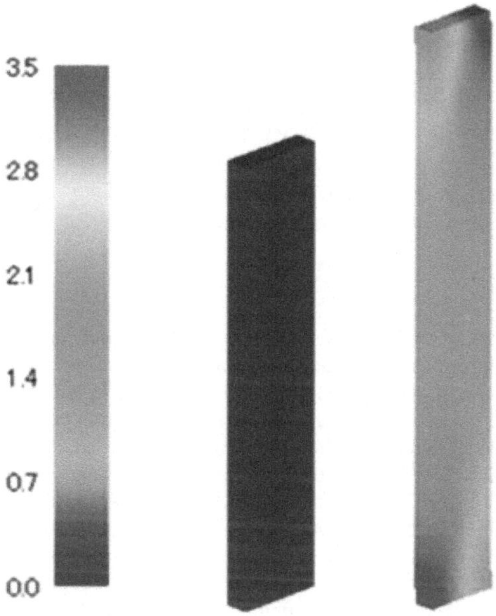

Figure 6. Finite elemental computations of a uniaxial test on specimens of the artificial leather.

REFERENCES

[1] Gasser, T. C., Holzapfel, G. A. and Ogden, R. W., A new constitutive framework for arterial wall mechanics and a comparative study of material models, J. Elaste, Vol. 61, No. 1 (2000), pp. 1–48.

[2] Gasser, T. C., Holzapfel, G. A. and Ogden, R. W., Hyperelastic modeling of arterial layers with distributed collagen fibre orientations. J. R. Soc. Interface, Vol. 3, No. 6 (2006), pp. 15–35.

[3] Holzapfel, G. A. and Gasser, T. C., A viscoelastic model for fiber-reinforced composites at finite strains: continuum basis, computational aspects and applications. Comput. Meth. Appl. Mech. Eng, Vol. 190, No. 34 (2001), pp. 4379–4403.

[4] Holzapfel, G. A. and Weizsäcker, H. W., Biomechanical behavior of the arterial wall and its numerical characterization. Computers in Biology and Medicine, Vol. 28, No. 4(1998), pp. 377–392.

[5] Lanir Y. Constitutive equations for fibrous connective tissues. J. Biomech, Vol. 16, Issue 1, No. 1 (1983), pp. 1–12.

[6] Li, Z., Paudecerf, D. and Yang, J., Mechanical behaviour of natural cow leather in tension. Acta Mechanica Solida Sinica, Vol. 22, No. 1 (2009), pp. 38–44.

[7] Martins, P. A. L. S., Natal, J. R. M. and Ferreira, A. J. M., A Comparative study of several material models for prediction of hyperelastic properties: Application to silicone-rubber and soft tissues. Strain (2006) Vol. 42 pp. 135–147.

Constitutive Models for Rubber XIII – Dal (Ed.)
© 2025 The Author(s), ISBN 978-1-032-85138-9

Numerical investigations on rubber-snow interaction utilizing an elastoplastic snow model with implicit gradient damage enhancement

A. Moeineddin, J. Platen & M. Kaliske
Institute for Structural Analysis, Technische Universität Dresden, Dresden, Germany

ABSTRACT: Interaction between rubber and snow surfaces yields intricate challenges that have been a subject of scientific investigations. To enhance tire traction in winter circumstances, a thorough understanding of the effects that take place in the contact patch between the tire tread and snow surface is required. Describing the effects of individual tire tread block parts during sliding on snow, and milling is the major goal of this study. Recognizing the inherent complexities posed by these interactions, especially under varying applied velocities, this study aims to deepen the understanding of these characteristics. To achieve this, a dual methodology is employed: detailed numerical simulations and rigorous experimental validations. Considering snow behavior to exhibit brittleness and transition to large deformations under high strain rates, an advanced elastoplastic constitutive model at finite strains, enriched further with an implicit gradient damage enhancement, to replicate the snow's brittle behavior is used. The simulations offer insight into distinctive deformation patterns of rubber. Furthermore, the strain-softening behavior exhibited by snow, especially at the range of applied velocities, introduces additional aspects of complexity. The rubber model is subjected to a range of computational scenarios, in parallel, experimental studies, conducting precise material testing on rubber interacting with snow, are employed to validate the numerical simulations, ensuring the reliability and applicability of the developed models.

1 INTRODUCTION

Ensuring the safety of all passengers and other vehicles is a crucial consideration when operating a vehicle in winter weather. The interaction between tires and road surfaces, particularly in snowy conditions, plays a critical role. As vehicles pass on snowy roads, the snow compacts under pressure, leading to a cycle of compression and release. Consequently, this compacted snow surface becomes notably stiff and slippery. Various factors, including snow characteristics and tire design, significantly influence traction on snow-covered roads. Environmental elements continually alter snow properties, such as density, with each fresh snowfall and passage of vehicles.

Consequently, the tire industry invests substantial resources into optimizing winter grip performance while ensuring compatibility with other tire characteristics. The allocation of outdoor test expenditures to snow and ice testing underscores the critical nature of this design parameter. However, conducting tests under variable outdoor conditions limits the precision of test outcomes, thereby impeding the detection of increasingly nuanced stages in development progress, often rendering them challenging, if not unattainable. Unlike summer tires, winter tires offer more than just different rubber compositions. They feature tread designs with sipes, small slits that form sub-blocks, enhancing their performance in winter conditions. There have been studies primarily investigating the frictional behavior of rubber on snow (Ella, Formagne, Koutsos, & Blackford 2013, Linke, Wiese, Wangenheim, Wies, & Wallaschek 2017, Hindemith, Heidelberger & Wangenheim 2022).

Snow, formed by the compacted clustering of ice crystals, has a complex nature that adds intrigue and challenge to its mechanical behavior. Snow becomes brittle when applied strain rates are high, but acts ductile when strain rates are low. The behavior of snow is numerically modeled using different methods. The Finite Element Method (FEM) offers a macroscopic understanding of continuum mechanics. (Meschke, Liu, & Mang 1996, Mundl, Meschke, & Liederer 1997, Haehnel, & Shoop 2004, Cresseri, & Jommi 2005, Shoop, Kestler, & Haehnel 2006, Cresseri, Genna, & Jommi 2010, Gaume, Gast, Teran, Van Herwijnen, & Jiang 2018, Vallero, Barbero, Barpi, Borri-Brunetto, & De Biagi 2023) are among the key contributors. Concurrently, studies have explored the microstructural level to bridge the gap between the emerging mechanical characteristics of snow and its intrinsic microstructure (Hansen & Brown 1988, Nicot 2004). There are several examples of discrete element methods that allow for detailed examination of snow behavior (Michael 2014, Gaume, van Herwijnen, Chambon, Wever,

DOI: 10.1201/9781003516880-8

& Schweizer 2017, Kabore, Peters, Michael, & Nicot 2021). When it comes to computer graphics, attempts to use the Material Point Method (MPM) to mimic snow have given more weight to visual appeal than scientific accuracy. For instance, MPM was used by (Stomakhin, Schroeder, Chai, Teran, & Selle 2013) to simulate snow dynamics, mostly for visual simulations.

Although previous studies have provided insight into snow behavior at low strain rates, there is a crucial knowledge gap concerning snow behavior at high strain rates, which is experienced in situations such as rolling tires on snow. The goal of this work is to model snow at high applied velocities, acknowledging the unique behavior that demands more investigation. Instead of trying to replicate tire-snow interactions exactly, the aim is to provide the groundwork for further studies in this area.

It is important to capture the transition of snow entering softening as it becomes brittle at high velocity. Damage mechanics is incorporated into a plasticity model in order to accurately describe this transition from the hardening phase, which is experienced at slower strain rates, to the softening and damage phases, which are observed at higher rates. Through this integration, the plasticity model may reflect the hardening features, and the implicit gradient damage model can represent the subsequent softening and damage. In this paper, fundamental concepts of the constitutive model, are summarized in Section 2. Simulation results of material calibration along with the numerical simulation of the rubber-snow friction are presented, followed by a comprehensive discussion of findings in Section 3. The exploration concludes with final remarks in Section 4.

2 METHODOLOGY

The principles of continuum mechanics are considered based on [Lee 1969, Miehe 1988, Miehe 1993, Waffenschmidt, Polindara, Menzel, & Blanco 2014]. In that sense, the deformation gradient is divided into parts describing distinct mechanisms for the purpose of material modeling. Different behaviors are displayed by many materials under pure isochoric and volumetric deformations. Furthermore, the deformation gradient is split into $\boldsymbol{F} = \boldsymbol{F}^e \cdot \boldsymbol{F}^p$, elastic $\boldsymbol{F}^e$ and plastic $\boldsymbol{F}^p$ portions. It is shown that the plastic deformations are irreversible. The intermediate configuration's additive split of the velocity gradient serves as the basis for the progression of the plastic deformations, where the time derivative of the displacements $\dot{u}$ is used as $\boldsymbol{L}^I = \boldsymbol{F}^{e^T} \cdot \text{grad}(\dot{u}) \cdot \boldsymbol{F}^e = \boldsymbol{L}^e + \boldsymbol{L}^p$.

The plastic evolution operator is therefore specified as $\boldsymbol{L}^p = \dot{\boldsymbol{F}}^p \cdot \boldsymbol{F}^{p^{-1}}$, compare [Miehe 1988, Miehe 1993]. Furthermore, elastic deformations are known to induce stress. Given the presence of rigid body motions in the deformation gradient, the Green-Lagrange strain tensor and the right Cauchy-Green tensor are stated as $\boldsymbol{C}^e = \boldsymbol{F}^{e^T} \cdot \boldsymbol{F}^e$, and $\boldsymbol{E}^e = \frac{1}{2}(\boldsymbol{C}^e - \mathbf{1})$.

The partial Legendre transformation of Gibbs equation yields the strain energy (Fleischhauer, Behnke, & Kaliske 2013). The energy that is retained in the material as a result of deformations is expressed by ψ. To derive stress measures, the dissipation inequality is connected to the strain energy. The first Piola-Kirchhoff stress is calculated as $\boldsymbol{P} = \frac{\partial \psi}{\partial \boldsymbol{F}}$. Furthermore, the Mandel stress tensor in the intermediate configuration is derived by expressing the dissipation inequality with respect to the intermediate configuration as $\boldsymbol{S}^I = \boldsymbol{C}^e \cdot \frac{\partial \psi}{\partial \boldsymbol{C}^e}$, (see Miehe 1993), where it is demonstrated to be $\boldsymbol{L}^p$'s work conjugate. Thus, in the intermediate configuration, the plastic evolution operator is defined in terms of the Mandel stresses ($\boldsymbol{L}^p = \dot{\gamma} \frac{\partial f_y}{\partial \boldsymbol{S}^I}$.), where f_y is the yield function and $\dot{\gamma}$ is the time derivative of the plastic multiplier. Finally, the Kuhn-Tucker loading conditions are used in the computational solution of the finite strain plasticity as $f_y \leq 0$, $\gamma \geq 0$ and $\gamma f_y = 0$.

It is demonstrated in Meschke, Liu, & Mang 1996, Cresseri & Jommi 2005, Gaume, Gast, Teran, Van Herwijnen, & Jiang 2018, Vallero, Barbero, Barpi, Borri-Brunetto, & De Biagi 2023 that the modified Cam-Clay yield function may capture the behavior of snow. This contribution presents a novel formulation of Cam-Clay, formulated in the plastic intermediate configuration. The suggested formulation is based only on the Mandel stresses and the plastic deformation gradient in the intermediate configuration. In that sense, Hencky strains of (Meschke, Liu, & Mang 1996, Gaume, Gast, Teran, Van Herwijnen, & Jiang 2018) are not included. The yield function can be formulated as $f_y = \frac{q^2}{M^2} + p(p - p_c)$.

The parameter M denotes the critical state line. q and p are scalar stress measures, which are derived from the Mandel stresses according to $q = \sqrt{\frac{3}{2}} \parallel \text{dev}(\boldsymbol{S}^{I,\text{eff}}) \parallel, p = \frac{1}{3}\text{tr}(\boldsymbol{S}^{I,\text{eff}})$.

The span of the yield function along the hydrostatic axis is determined by the parameter p_c. An exponential hardening law has been proven to be able to capture the properties of the soil [Borja & Lee 1990]. Therefore, the contribution at hand employs this broad methodology. It states that the hardening law is $\dot{p}_c = -\vartheta \dot{J}^p p_c$. The hardening law is associated with the time development of the determinant of the plastic component of the deformation gradient $J^p = \det(\boldsymbol{F}^p)$, as opposed to [Borja & Lee 1990]. It has been discovered that a hardening law based on the plastic deformation for snow [Meschke, Liu, & Mang 1996] is in good accordance with reality. Other constants that can be found through experimentation are connected to the material parameter ϑ by $\vartheta = \frac{1+e_0}{\lambda_s - \kappa_s}$, where, λ_s denotes the virgin compression index, κ_s is the swell index and the initial void ratio is denoted as e_0.

Localization in softening materials can cause numerical instabilities, mesh sensitivity, and simulation convergence problems. This problem is solved using the implicit gradient damage model proposed in

[Peerlings, de Borst, Brekelmans, & de Vree 1996]. This model introduces nonlocal fields based on the modified Helmholtz equation. Along with the local counterpart η, the gradient parameter c, and the nonlocal field variable η_{nl}, the equation also includes homogeneous Neumann boundary conditions. Research has shown that snow experiences softening at high loading rates, as mentioned in (Michael 2014, Wang, Fu, Han, Liu, Xiao, & Leng 2021). Capturing this behavior with local continuum mechanics leads to localization. This drawback is addressed by using a nonlocal damage formulation. Plastic deformations are the constitutively chosen driving force for the nonlocal field since Section 3 shows a good correspondence with reality. Over time, The local counterpart is evolved in time by $\dot{\eta} = |\dot{J}^p|$, where the absolute value guarantees a monotonic growth in the local counterpart. As a result, the damage variable is never reduced. Given that snow exhibits a considerable degree of hardening before softening, the variable γ is introduced as $\gamma = \begin{cases} \eta_{nl} - \gamma_0 & \forall \eta_{nl} > \gamma_0 \\ 0 & \forall \eta_{nl} \leq \gamma_0 \end{cases}$.

The volumetric plastic deformation threshold before damage appears is γ_0. For snow, this quantity varies with loading velocity. It is taken to be a material constant in the current contribution for the sake of brevity. The damage variable is then determined by $d = -\alpha + \alpha \exp[-\beta\gamma]$, where, β is the curvature parameter, which regulates the softening, and α is the maximum degradation of the material. The total stress tensors can be calculated as $\boldsymbol{P} = (1-d)\boldsymbol{P}^{\text{eff}}$ and $\boldsymbol{S}^I = (1-d)\boldsymbol{S}^{I,\text{eff}}$.

3 SIMULATION RESULTS AND DISCUSSION

This section outlines the procedure for determining the constitutive model's material parameters by comparing them to experimental data. The Modified Cam-Clay formulation, renowned for its capacity to represent the intricate mechanical behavior of soil under various loading conditions, serves as the foundation for the modeling of snow. To properly capture the quasi-brittle nature of snow softening and damage, damage mechanics with an implicit gradient is added to this model. Initially, basic tests emphasizing elasto-plastic behavior are the focus of calibration efforts. The hydrostatic compression and direct shear tests are specifically discussed. These tests not only provide a basic level of plausibility but also aid in the calibration of constitutive parameters that are necessary to guarantee the accuracy of the model. Furthermore, the uniaxial tests are carried out at varying loading rates which are included later to capture snow hardening and softening behavior by just calibrating the implicit gradient damage parameters. This part thoroughly assesses the model's capabilities by contrasting the model predictions with the experimental data.

As different types of snow require specific constitutive parameters, the hydrostatic compression test is replicated based on [Meschke, Liu, & Mang 1996]. In this test, a single finite element representing a cubic snow specimen with a length of 7.6 cm is subjected to entirely hydrostatic compression. It is typical practice to calibrate metrics like virgin compression and the loading-unloading indexes using the hydrostatic compression test. It also provides information about the pre-consolidation pressure. The simulation results are in agreement with the experimental data, which is shown in Figure 1 when the parameters suggested by [Meschke, Liu, & Mang 1996], are used: $\lambda = 0.38$ [-], $\kappa = 0.015$ [-], $M = 2.88$ [-], $G = 7500$ kPa, $p_0 = 70$ kPa.

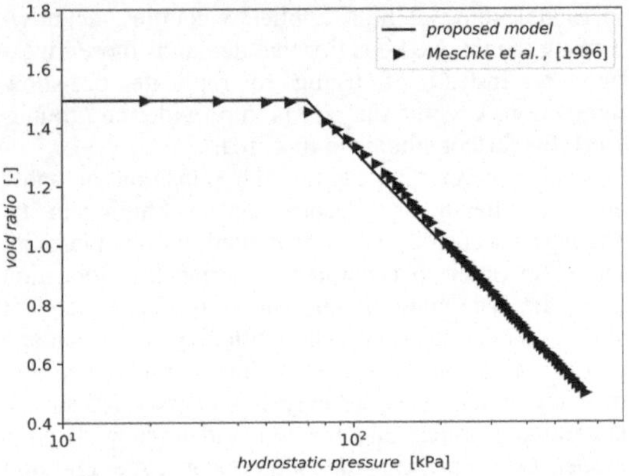

Figure 1. Results for isotropic compression test and proposed model.

Snow uniaxial compression tests provide important information about the behavior of materials under axial loads; stress-strain relationships highlight important characteristics like yielding, elasticity, and failure. Research such as [Wang, Fu, Han, Liu, Xiao, & Leng 2021] underlines the significance of using careful compaction techniques to guarantee consistent density throughout specimens. Tests carried out at varying applied rates, demonstrate varied behaviors. In Figure 2, at 2 mm/min, snow displays ductile hardening, while at 10 mm/min, it softens after reaching a peak force. At 100 mm/min, snow undergoes abrupt failure, without softening or hardening, indicating the significant impact of velocity on its mechanical response. Such findings underscore the need for careful consideration of compression conditions in snow material characterization. In these tests, the material parameters are $\rho = 600$ kg/m^3, G = 100E + 3 kPa, $\kappa = 0.015$ [-], $\lambda = 0.35$ [-], $M = 2.88$ [-]. Parameters including the damage parameters are shown as Table 1.

Table 1. Mateiral parameters.

$\dot{u}$ [mm/min]	p_0 [kPa]	α [-]	β [−]	γ [-]
2	450	-	-	-
10	1080	0.25	5E + 2	5E − 5
100	1000	0.93	6E + 3	5E − 5

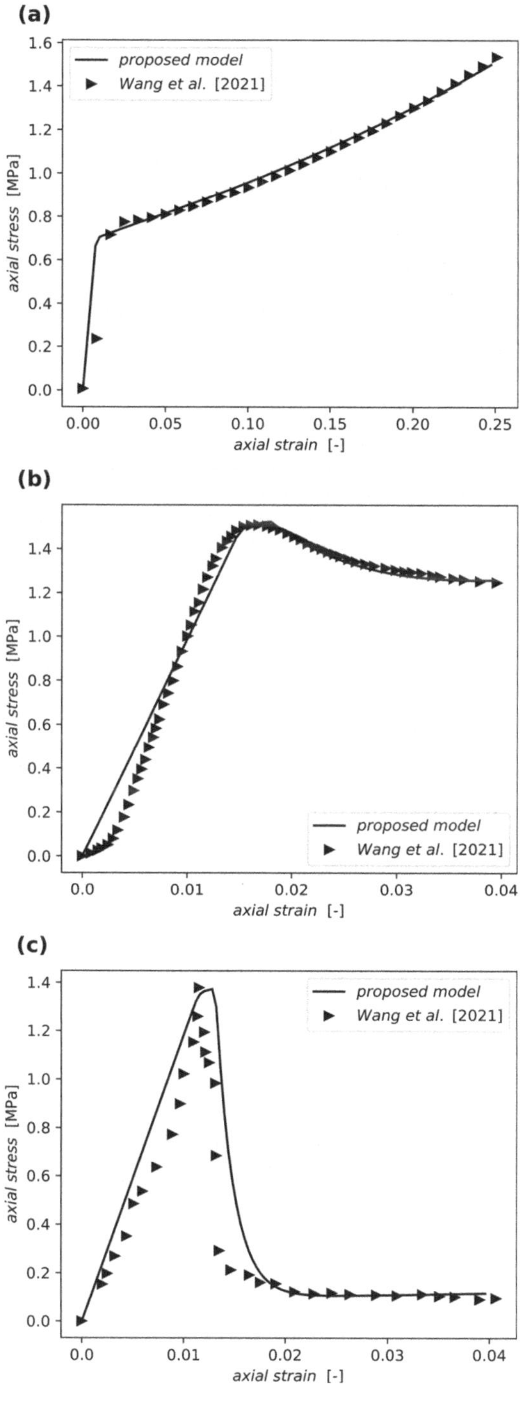

Figure 2. Relationship between axial stress and strain on uniaxial compression test at three different applied loading rates: a) 2 mm/min, b) 10 mm/min, and c) 100 mm/min.

A further study is undertaken to replicate and validate the behavior of individual tire tread blocks on snow-covered surfaces. The simulations are adjusted to match the experimental configuration, which is described in [Mundl, Meschke, & Liederer 1997]. Based on information from [Mundl, Meschke, & Liederer 1997], it is determined that the snowpack's density ranged from 450 to 750 kg/m^3. As a result, the snow in the simulations is assigned the material parameter from the hydrostatic compression test. The initial pre-consolidation pressure (p_0) is assumed to be 0.4 MPa due to the increased density of compacted snow. In the simulation, the snow's surface is covered by cubic rubber components. These rubber components are then subjected to a vertical pressure of 200 or 400 kPa, after which they are slid across the surface at a speed of 14.4 mm/min.

This simulation is primarily focused on measuring the exerted horizontal force (T) and displacement (u), to correlate these results with the experimental data to validate the model. A block with a width of 10 mm and a rubber hardness of rubber *ShA* 55 and *ShA* 65 is used for the comparative simulation. This model uses linear elements in a three-dimensional framework to discretize the rubber block and the snow. The thickness of the elements is 0.5 mm in the z-direction. The snow is modeled as a cubic shape, and the global element size is set to 1 mm for the rubber block and 0.5 mm for the snow. It is important to note that this model is fixed along the z-axis. Using the Mooney-Rivlin law, the rubber material's behavior is simulated, and the material characteristics are chosen based on [Lalo & Greco 2017]. The Coulomb friction model is used to represent the tangential force between the surface of the rubber block and the snow. The Coulomb friction coefficient is fixed at 0.55, following the suggestions made in [Mundl, Meschke, & Liederer 1997]. Figure 3 shows the comparison between numerical simulation and experimental data.

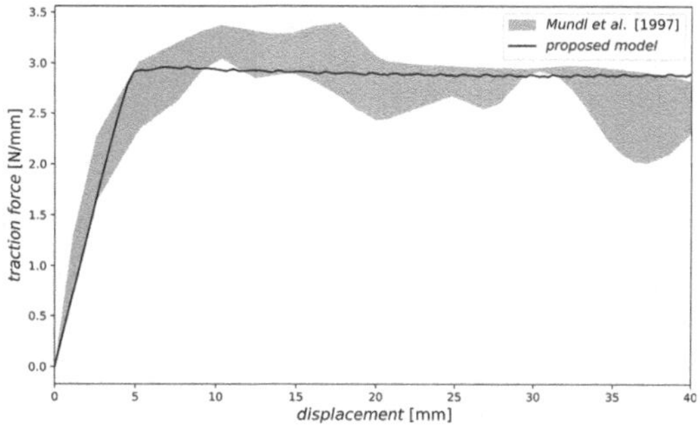

Figure 3. Relationship between the horizontal displacement and traction force from the FE simulation and experimental data.

Figure 4 illustrates the results of four distinct simulations involving the sliding of rubber on snow. Two rubber compositions, with rubber hardness of *ShA* 55 and *ShA* 65 are considered with applied loads of 200 kPa and 400 kPa. Figure 4 shows the relationship between relative traction force and displacement, where the relative traction force is defined as the traction force divided by the maximum traction achieved, specifically observed in the case of *ShA* 65 under 400 kPa pressure. Notably, the *ShA* 65 exhibits the highest friction coefficient among the tested materials, highlighting its superior performance in snow sliding scenarios under varying pressures. Figure 6 shows the plastic shear strain of snow for the rubber-snow interaction of a rubber with a hardness of *ShA* 55 under 400 kPa pressure, where the strain concentration is expanded on the leading edge of the rubber.

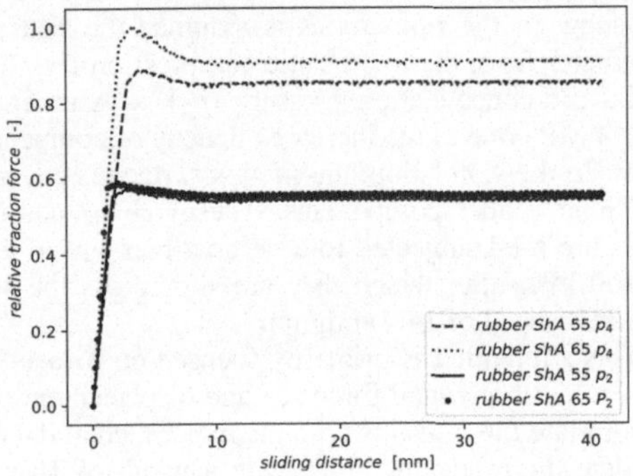

Figure 4. Relationship between relative traction force and displacement for two different rubber hardness and two different applied pressures.

Figure 5 compares the contact shear pressure at the leading edge of the rubber, and contact slip of the two rubber hardness under two different pressures. Here, the contact pressure is computed using the net contact normal force per unit area. At the same time, contact shear is frictional shear stress, and the contact slip is the amount of relative displacement of surface points that are in contact [Smith 2009].

4 CONCLUSIONS

This research is focused on modeling the quasi-brittle behavior of snow under high strain rates, addressing a crucial knowledge gap in snow mechanics relevant to engineering applications, especially on the sliding of

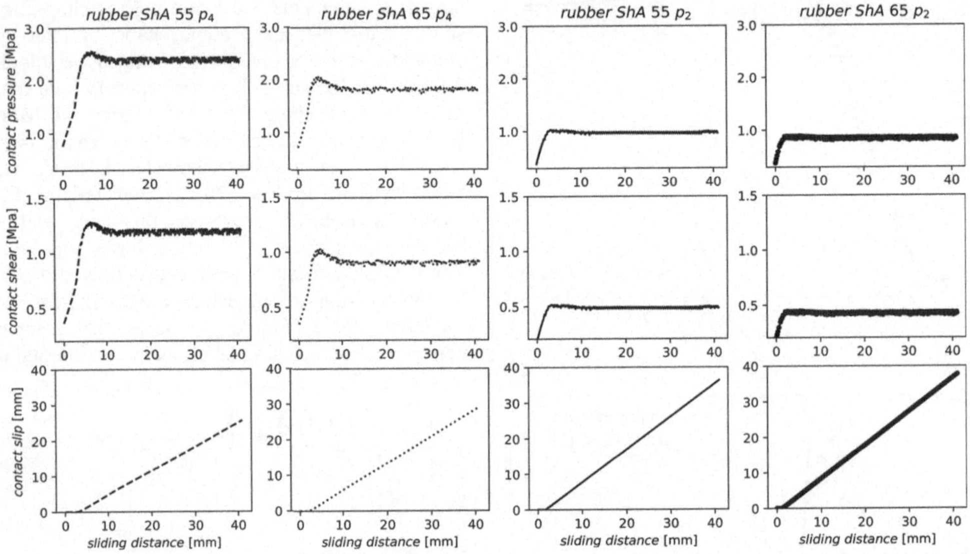

Figure 5. Comparison between contact pressure, shear and slip for two different rubber hardness and two different applied pressures.

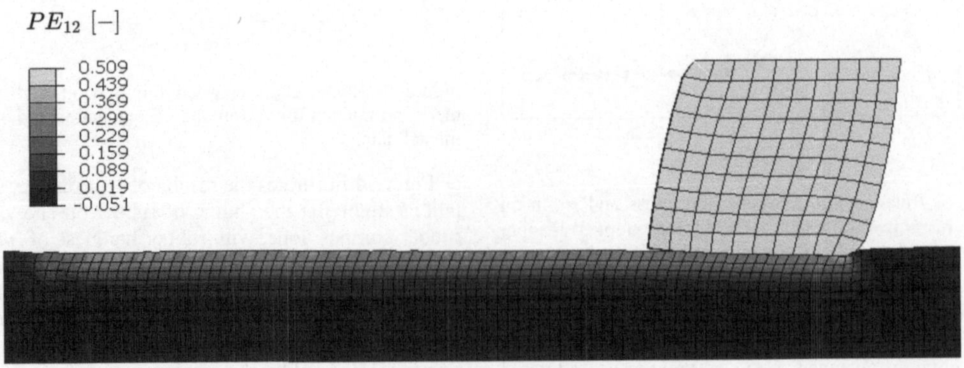

Figure 6. Plastic shear strain in snow under sliding of rubber.

rubber on snow. The study establishes a foundational model using an innovative constitutive description based on the modified Cam-Clay framework and multiplicative elasto-plasticity. The model incorporates implicit gradient enhanced damage, improving predictions and aligning with real-world observations. Additionally, the model can simulate dynamic transformations and strain rates, demonstrating its adaptability by calibrating damage parameters. This study delves into the intricate interactions between rubber and snow surfaces during tire sliding, specifically focusing on understanding the contact patch effects under various conditions. The simulation results provide valuable insights into the deformation patterns of rubber on snow. Notably, the comparison of different rubber hardness under varying pressures shows a trend. Overall, this research provides essential insights into snow mechanics, particularly under different applied velocities, laying the groundwork for further research.

ACKNOWLEDGMENTS

The scientific support by Continental Reifen Deutschland GmbH is sincerely appreciated. Additionally, gratitude is extended to both Continental Reifen Deutschland GmbH and the German Research Foundation (Deutsche Forschungsgemeinschaft) SFB/TRR 280, Sub-Project A02, Projekt-ID: 417002380, for their funding.

REFERENCES

Borja, R. I. & S. R. Lee (1990). Cam-clay plasticity, part 1: implicit integration of elasto-plastic constitutive relations. *Computer Methods in Applied Mechanics and Engineering 78*, 49–72.

Cresseri, S., F. Genna, & C. Jommi (2010). Numerical integration of an elastic–viscoplastic constitutive model for dry metamorphosed snow. *International Journal for Numerical and Analytical Methods in Geomechanics 34*, 1271–1296.

Cresseri, S. & C. Jommi (2005). Snow as an elastic viscoplastic bonded continuum: a modelling approach. *Rivista Italiana di Geotechnica 4*, 43–58.

Ella, S., P.-Y. Formagne, V. Koutsos, & J. R. Blackford (2013). Investigation of rubber friction on snow for tyres. *Tribology International 59*, 292–301.

Fleischhauer, R., R. Behnke, & M. Kaliske (2013). A thermomechanical interface element formulation for finite deformations. *Computational Mechanics 52*, 1039–1058.

Gaume, J., T. Gast, J. Teran, A. Van Herwijnen, & C. Jiang (2018). Dynamic anticrack propagation in snow. *Nature Communications 9*, 3047.

Gaume, J., A. van Herwijnen, G. Chambon, N. Wever, & J. Schweizer (2017). Snow fracture in relation to slab avalanche release: critical state for the onset of crack propagation. *The Cryosphere 11*, 217–228.

Haehnel, R. B. & S. A. Shoop (2004). A macroscale model for low density snow subjected to rapid loading. *Cold Regions Science and Technology 40*, 193–211.

Hansen, A. & R. Brown (1988). An internal state variable approach to constitutive theories for granular materials with snow as an example. *Mechanics of Materials 7*, 109–119.

Hindemith, M., J. Heidelberger, & M. Wangenheim (2022). Investigation of the influence of snow track density on tire tread block traction by experiments and discrete element method simulation. *Tire Science and Technology 50*, 316–336.

Kabore, B. W., B. Peters, M. Michael, & F. Nicot (2021). A discrete element framework for modeling the mechanical behaviour of snow - part l: Mechanical behaviour and numerical model. *Granular Matter 23*, 1–14.

Lalo, D. & M. Greco (2017). Rubber bushing hyperelastic behavior based on shore hardness and uniaxial extension. In *Proceedings of the 24th ABCM International Congress of Mechanical Engineering, COBEM-2017, Curitiba, PR, Brazil*, Volume 5280.

Lee, E. H. (1969). Elastic-plastic deformation at finite strains. *Journal for Applied Mechanics 36*, 1–6.

Linke, T., K. Wiese, M. Wangenheim, B. Wies, & J. Wallaschek (2017). Investigation of snow milling mechanics to optimize winter tire traction. *Tire Science and Technology 45*, 162–174.

Meschke, G., C. Liu, & H. A. Mang (1996). Large strain finite-element analysis of snow. *Journal of Engineering Mechanics 122*, 591–602.

Michael, M. (2014). *A discrete approach to describe the kinematics between snow and a tire tread*. Ph. D. thesis, University of Luxembourg, Luxembourg.

Miehe, C. (1988). *Zur numerischen Behandlung thermomechanischer Prozesse*. PhD thesis, Institut für Baumechanik und Numerische Mechanik, Universität Hannover, Hannover.

Miehe, C. (1993). *Kanonische Modelle multiplikativer Elasto-Plastizität: thermodynamische Formulierung und numerische Implementation*. Habilitation thesis, Institut für Baumechanik und Numerische Mechanik, Universität Hannover, Hannover.

Mundl, R., G. Meschke, & W. Liederer (1997). Friction mechanism of tread blocks on snow surfaces. *Tire Science and Technology 25*, 245–264.

Nicot, F. (2004). Constitutive modelling of snow as a cohesive-granular material. *Granular Matter 6*, 47–60.

Peerlings, R. H., R. de Borst, W. M. Brekelmans, & J. de Vree (1996). Gradient enhanced damage for quasi-brittle materials. *International Journal for Numerical Methods in Engineering 39*, 3391–3403.

Shoop, S., K. Kestler, & R. Haehnel (2006). Finite element modeling of tires on snow. *Tire Science and Technology 34*, 2–37.

Smith, M. (2009). *ABAQUS/standard user's manual, Version 6.9*. Dassault Systèmes Simulia Corp, United States.

Stomakhin, A., C. Schroeder, L. Chai, J. Teran, & A. Selle (2013). A material point method for snow simulation. *ACM Transactions on Graphics 32*, 1–10.

Vallero, G., M. Barbero, F. Barpi, M. Borri-Brunetto, & V. De Biagi (2023). *Some Improvements of a Visco-Plastic Constitutive Model for Snow.* Geotechnical Engineering in the Digital and Technological Innovation Era, 382–389, Springer Nature Switzerland.

Waffenschmidt, T., C. Polindara, A. Menzel, & S. Blanco (2014). A gradient-enhanced large-deformation continuum damage model for fibre-reinforced materials. *Computer Methods in Applied Mechanics and Engineering 268*, 801–842.

Wang, E., X. Fu, H. Han, X. Liu, Y. Xiao, & Y. Leng (2021). Study on the mechanical properties of compacted snow under uniaxial compression and analysis of influencing factors. *Cold Regions Science and Technology 182*, 103215.

Constitutive Models for Rubber XIII – Dal (Ed.)

A simple hyperelastic model for virtual product design

P. Schneider
Department of Simulation and Continuum Mechanics, Deutsches Institut für Kautschuktechnologie e. V., Hanover, Germany

S. Böttger
Continental AG, Hanover, Germany

ABSTRACT: Extensions for hyperelastic constitutive models towards an explicit continuous dependence of the temperature and the filler content are introduced in order to be used in virtual design optimization loops that also adjust the filler content. The basic models are the isochoric Neo-Hooke model and the volumetric standard model with constant bulk modulus, as well as the nonlinear compression model of Hartmann and Neff. Special emphasis is put on temperatures in the vicinity of the glas transition. The extended models are calibrated on a family of sulfur-crosslinked EPDM-mixtures with varying carbon black content that is tested in uniaxial tension and confined axial compression.

1 INTRODUCTION

In general virtual product design workflows are highly desirable especially due the potential to minimize cost intensive physical testing for interatively optimized prototypes. For example topology optimization can be used to minimize the weight (and used material) of a supporting structure under predefined design constrains, such as matching a target stiffness of the structure, without the need to test iterative prototypes. Therefore, such workflows are already applied especially for materials with established additive manufacturing technologies.

For the design of rubber parts the actual advantage of being a highly designable material is an obstacle in this regard: In a classical industrial workflow frequent alterations of the compound to match design goals are interwoven with changes of the design itself. For example, in order to match a target stiffness of a rubber part one could not only adjust the geometry, but also the material stiffness by altering the filler content. However, altering the filler content will change the material parameters and in turn a new material characterization based on physical testing is required, which hinders a virtual design workflow. Hence material models for whole "families" of compounds that have an explicit continuous dependence of the filler content and can be characterized beforehand are needed. In this contribution we also put a focus on an explicit continuous dependence of the temperature, especially close to the glas transition.

2 MATERIAL AND SETUP

Sulfur-crosslinked EPDM mixtures with varying carbon black content are investigated. The mixing process was carried out in an ideal manner. The EPDM was masticated at 50°C. The total mixing time for the unfilled and filled mixture was 5 and 7 minutes, respectively. The finished rubber compounds were rolled for an additional two minutes. The vulcanization took place in a heated press at 170°C.

Denotation	EPDM	\-25	\-50	\-75
Keltan 2450 [phr]	100	100	100	100
Zinc oxide [phr]	3	3	3	3
Stearic acid [phr]	1	1	1	1
Vulkanox 4010 [phr]	2	2	2	2
Corax N 550 [phr]	0	25	50	75
Rhenogran S-80 [phr]	2	2	2	2
CBS-80 [phr]	2	2	2	2
t_{90} at 170°C [min]	10.44	10.81	10.33	6.93
T_G [°C]	-49	-48	-49	-49
ϕ [%]	0	10.26	18.61	25.53

A universal tensile testing machine with an optical extensometer was used for the uniaxial tension tests on S2-type test pieces (cf. DIN 53504, din) with a preload of 0.1 N that were stored at the specific test temperature for at least 30 minutes before

DOI: 10.1201/9781003516880-9

testing. The deformation rate was 20 mm/min and data points up to 20% strain were used for the characterization of the material parameter by minimizing the quadratic error. For the volumetric characterization confined axial compression tests on the same machine with cylindrical specimen with a height of 3 mm and a diameter of 5.4 mm were performed with a prestress of 3 MPa. The deformation rate was 0.15 mm/min and data points up to 3% volumetric compression were used for the linear model and up to 6% for the Hartmann-Neff model.

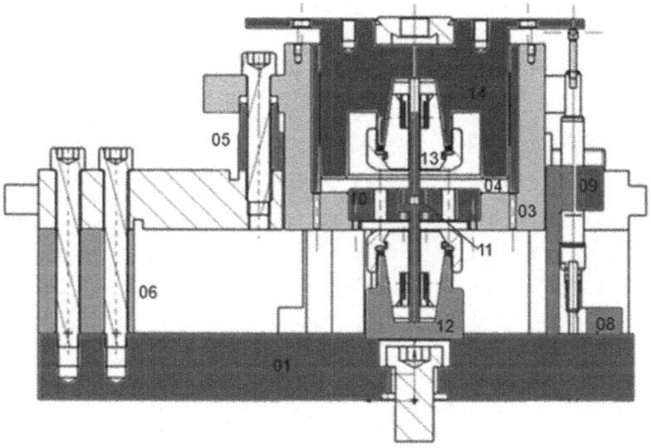

Figure 1. A test rig used to quantify the systematic error due to friction in confined axial compression testing.

The models were characterized by fitting the unloading curve.

The reason for this uncommon approach is that the systematic error due to friction of the test piece with the surrounding housing is minimized. In Figure 1 a special test rig design is displayed that was used to quantify the influence of the friction during confined axial compression measurements. It was inspired by a simpler design, Ilseng et al. (2017), that measured the force difference between an upper and lower load cell. However, due to measurement inaccuracies (noise) and the numerical concept of catastrophic cancellation a design to measure the friction force transferred by the housing directly seems preferable. The design illustrated in Figure 1 measures the force transferred from the compressed test piece 11 to the housing 10 via three load cells in 120°-steps around the circumference. A result of four sample measurements is illustrated in Figure 2. The relative error during the unloading is way smaller since the test piece has already settled.

The evaluation procedure used in this contribution corrects for the preload, machine stiffness and thermal expansion. All details of the setup and the evaluation procedure as well as the simpler design of the confined axial compression rig that was used to perform the tests in this contribution are documented in detail in Böttger et al. (2024).

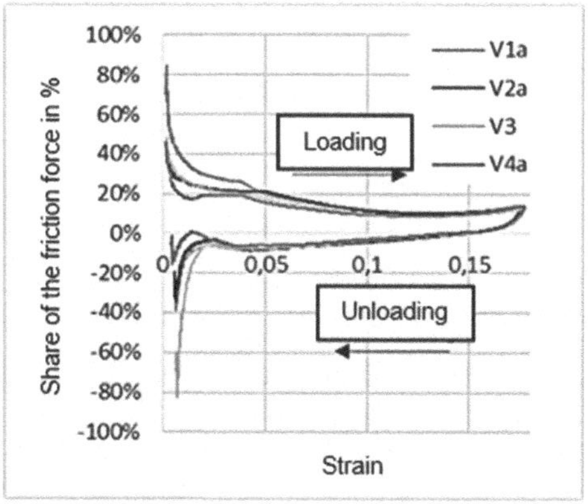

Figure 2. Frictional force in relation to applied pressure during confined axial compression measurements.

3 ISOCHORIC MODEL

For simplicity we start with the easiest useful model for the isochoric deformation of rubber which is the hyperelastic Neo-Hooke model which has only one material parameter. (Of course this model is limited to smaller strains and does not account for the complex phenomenology of filled rubber like viscous effects or softening.) The strain energy density is given by

$$\rho_0 \Psi^{iso} = c_{10}(\bar{I}_1 - 3), \tag{1}$$

where c_{10} is the stiffness parameter, which is in the original model proportional to the temperature $c_{10} = 1/2nk\vartheta$. However, for the unfilled EPDM this is only a good approximation, if the temperature is way higher than the glass transition temperature, cf. Figure 3.

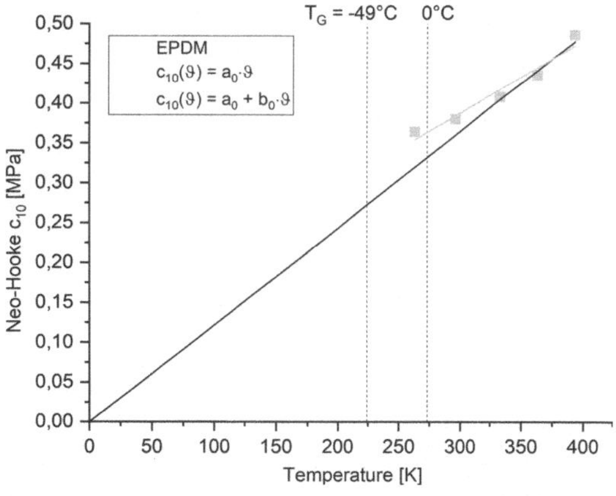

Figure 3. Temperature dependency of the Neo-Hooke parameter in the vicinity of T_G.

A pragmatic way to increase the accuracy for the in practice important regime close to the glas transition is to go over to an affine linear ansatz

$$c_{10}^{unfilled} = a_0 + \vartheta b_0, \tag{2}$$

cf. Figure 3. This linear behavior is attributed to the entropy related energy contribution of the polymers, cf. Treloar (1975) and Li et al. (2016).

Although the Neo-Hooke model was originally designed to model unfilled rubber, the model can also be fitted to stress-strain measurements of filled elastomers within the moderate strain regime. However, in general a superimposed enthalpy-related exponential dependence of the stiffness with respect to the temperature has to be anticipated, cf. Kraus (1971) and Guth (1945).

With respect to the filler volume concentration ϕ established approaches use a second-order polynomial ansatz, cf. Guth (1945). In general one would assume that the coefficients remain functions of the temperature

$$c_{10} = c_{10}^{unfilled}(\vartheta) + f(\vartheta)\phi + g(\vartheta)\phi^2. \tag{3}$$

If we subtract the unfilled stiffness and fit the parameters f and g freely for all temperatures, we see a reasonable prediction of the quadratic ansatz, cf. Figure 4.

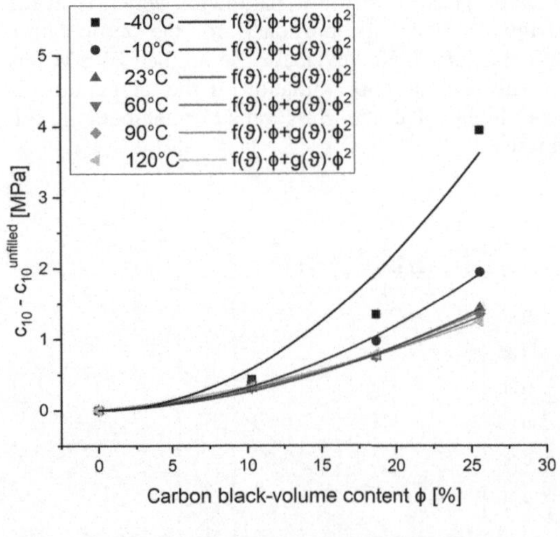

Figure 4. Filler network amplification of the stiffness.

and indeed the parameters f and g as functions of ϑ can be fitted well with an affine linear increasing (entropy-related) function for f and an (enthalpy-related) decreasing exponential ansatz for g, cf. Figure 5.

$$f(\vartheta) = a_1 + \vartheta b_1 \tag{4}$$

$$g(\vartheta) = a_2 + b_2 e^{-c_2\vartheta} \tag{5}$$

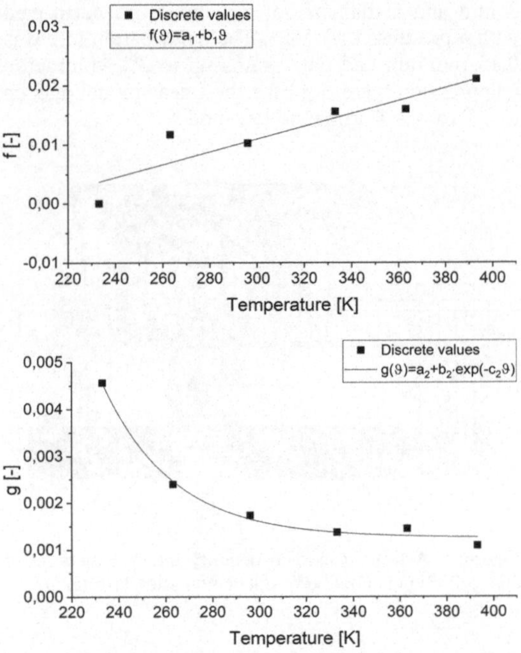

Figure 5. Entropy-related affine linear contribution f and enthalpy-related exponential contribution g of the filler network amplification of the stiffness.

For application the final 7-parameter model (1) to (5) should be fitted to the data of an unfilled and highly filled mixture in order to be predictive for intermediate values of ϕ and ϑ, cf. Figure 6.

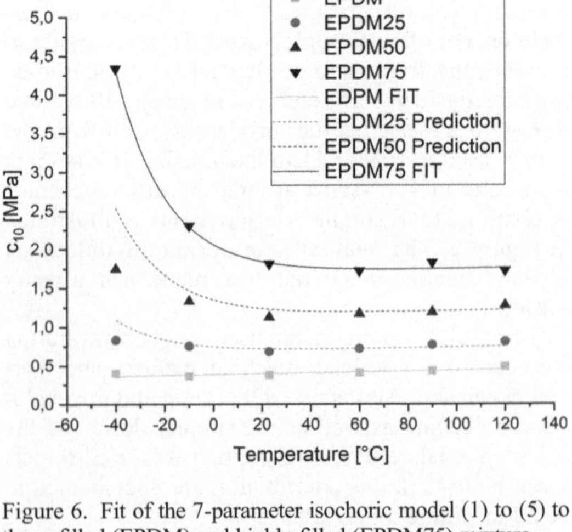

Figure 6. Fit of the 7-parameter isochoric model (1) to (5) to the unfilled (EPDM) and highly filled (EPDM75) mixture.

While the model is overall useful and interpolates very well in the higher temperature regime, we see that temperatures in the vicinity of the glas transition are challenging. Probably the pragmatic affine linear ansatz for the temperature is a bit to simple in the first place, however, it is also of interest to keep the number of material parameters low.

4 COMPRESSION MODEL

Technical rubber is quasi-incompressible, which means that its resistance to volume change is way higher than for volume preserving changes of its shape. In turn the isochoric contribution to the strain energy density dominates the deformation behavior of rubber for most load scenarios. Nevertheless, rubber is not ideally incompressible and the volumetric contribution can be pronounced in some load scenarios and undergoes strong changes in the vicinity of compressive load applications.

It is to be noted that in contrast to the isochoric part the compression behavior is indeed hyperelastic, i.e. strain rate dependencies, hysteresis or softening were not observed in our experiments, cf. B¨ottger et al.(2024),, whereas all of theses effects are observable for the isochoric part - here the Neo-Hooke model is just able to predict the moderate strain virgin response at quasi-static conditions.

The compression behavior is in general nonlinear. We investigate two models. The first one is the standard model

$$\rho_0 \Psi^{vol} = \frac{K}{2}(J-1)^2 \tag{6}$$

that implies a linear relation between the volumetric dilation J and the hydrostatic pressure. Hence, the compression modulus K interpreted as the current tangent is constant. The second model is a nonlinear model given by the equation

$$\rho_0 \Psi^{vol} = \frac{K_0}{2\beta^2}(J^\beta + J^{-\beta} - 2), \tag{7}$$

where K_0 is the initial tangent and β is the nonlinearity parameter. We denote it here as Hartmann-Neff model although the model was also proposed by different authors with different values for β, cf. Hartmann and Neff (2003).

With the standard model the temperature dependency for all filler contents follows in good approximation an (affine) linear trend, cf. Figure 7.

For the Hartmann-Neff model we first fitted the parameters K_0 and β freely, cf. Figures 8 and 9.

The behavior of β in the vicinity of the glas transition is complex, but seems to settle at a value around 20 for high temperatures. For pragmatic reasons and

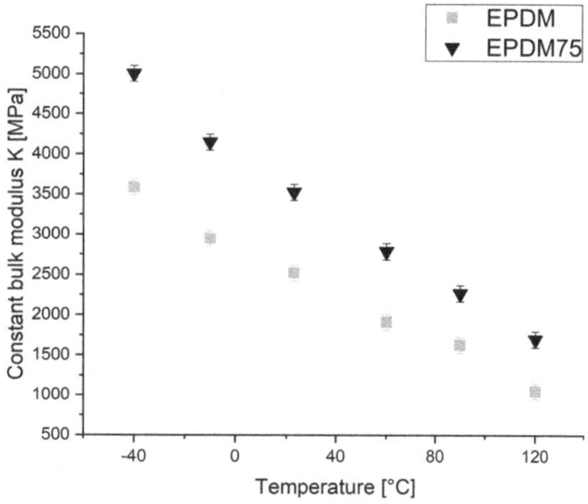

Figure 7. Temperature dependency of the bulk modulus K (standard model).

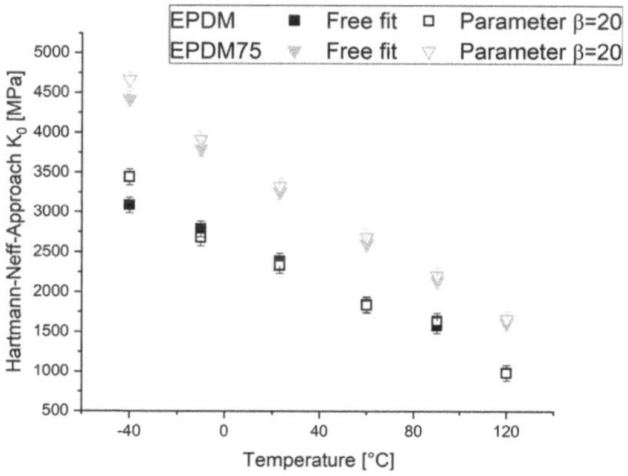

Figure 8. Temperature dependency of the initial bulk modulus K_0 (Hartmann-Neff model).

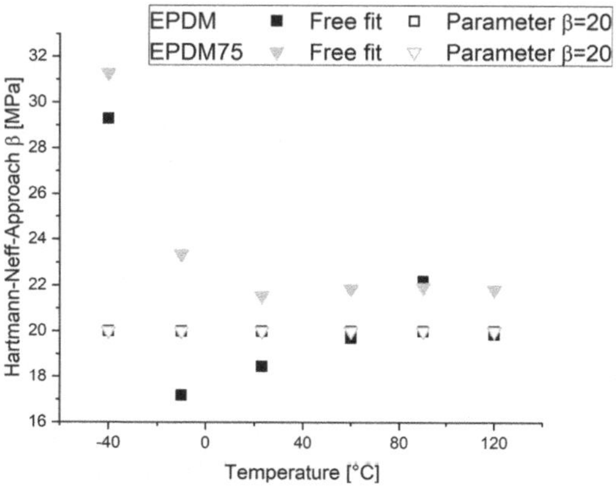

Figure 9. Temperature dependency of the nonlinearity parameter β of the Hartmann-Neff model.

to keep the number of parameters low, we choose to fixate $\beta = 20$, which does not affect the values for K_0 too much, cf. Figure 8. In both cases the temperature dependence follows in good approximation an (affine) linear trend as well. It is very noteworthy that the value $\beta = 20$ is way higher than fixed values proposed by other authors, whose model derivations are mostly based on theoretical arguments (such as polyconvexity). Also note that β is an exponent in model (7) which outlines that the compression behavior is indeed highly nonlinear.

Again, subtracting the unfilled values exemplarily for K (standard model) we see a linear trend for the filler volume dependency, cf. Figure 10,

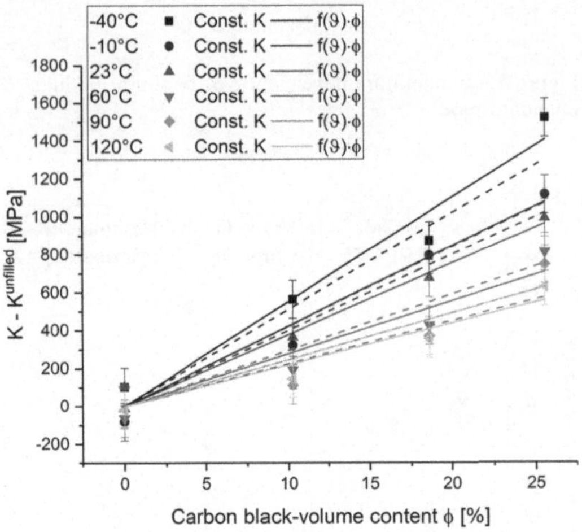

Figure 10. Filler volume dependency of the standard model.

whereas the proportionality parameter f follows an (affine) linear trend for the temperature dependency, cf. Figure 11.

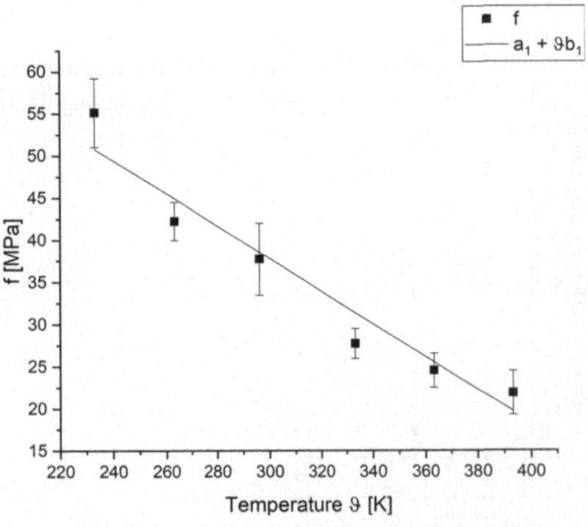

Figure 11. Temperature dependency of the proportionality parameter f.

For the Hartmann-Neff model with $\beta = 20$ the trends are similar. Hence we propose the models

$$K = (a_0 + b_0\vartheta) + (a_1 + b_1\vartheta)\phi \qquad (8)$$

$$K_0 = (a_0 + b_0\vartheta) + (a_1 + b_1\vartheta)\phi, \quad \beta = 20 \qquad (9)$$

(To avoid confusion: The denotations a_i, b_i are merely placeholders for parameters to be determined. They will not be equal for the standard and Hartmann-Neff model nor are they equal to parameters of the isochoric model.)

Fitting the final models to the data of the unfilled and highly filled (EPDM75) mixtures, we see that the model captures the behavior quiet good even in the vicinity of the glas transition, although here we also see the tendency for a slight upturn that is not covered by the model, cf. Figures 12 and 13.

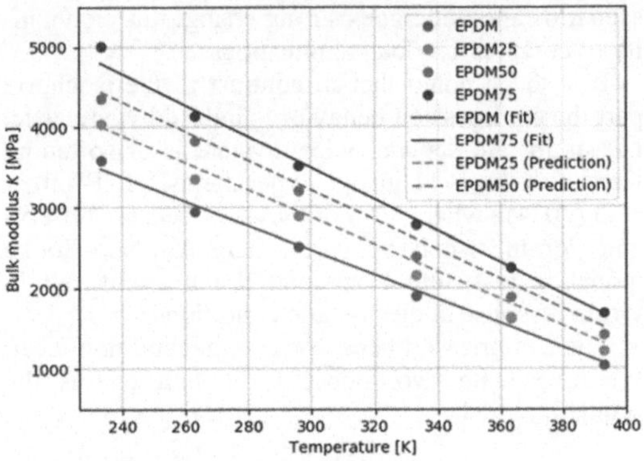

Figure 12. Fit of the 4-parameter standard model (8) to the unfilled (EPDM) and highly filled (EPDM75) mixture.

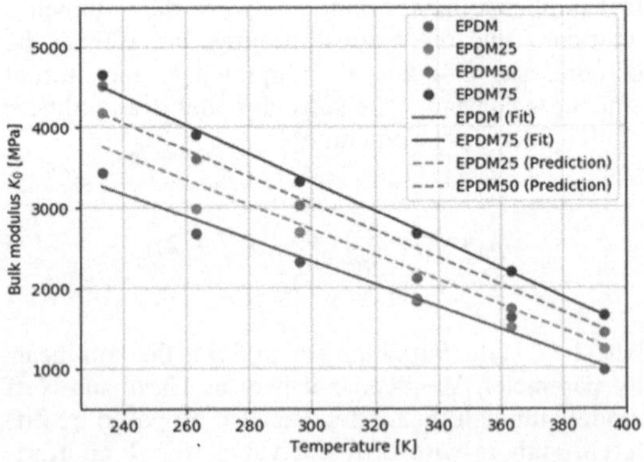

Figure 13. Fit of the 4-parameter Hartman-Neff model (9) to the unfilled (EPDM) and highly filled (EPDM75) mixture.

5 A REMARK ON THE ADDITIVE ISOCHORIC-VOLUMETRIC SPLIT

Usually an additive split of the strain energy density is assumed.

$$\rho_0 \Psi = \rho_0 \Psi^{iso} + \rho_0 \Psi^{vol} \tag{10}$$

Penn (1970) criticizes this split with the argument that under tensile load the stretch λ to dilation J curve shows the opposite curvature in experiments than predicted by a model using the split and the volumetric standard model (6). While we confirm his observations and DIC-measurements of the tested material show the same type of curvature as observed by Penn (1970), we could reproduce the correct type of curvature with the nonlinear Hartman-Neff model, cf. Figure 14.

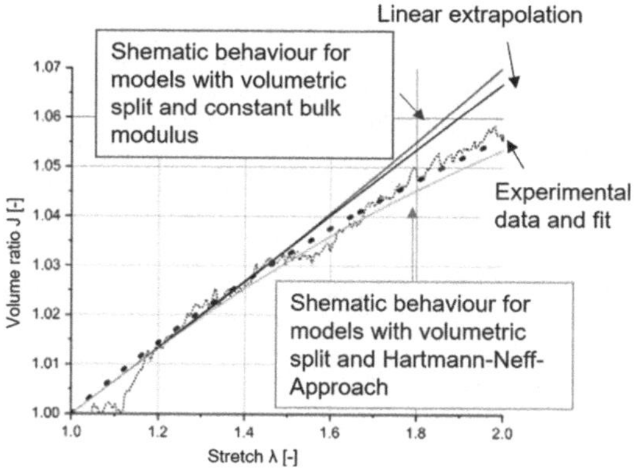

Figure 14. DIC-measurement based evaluation of the dilation J under tensile load with normal stretch λ in comparison to the type of curvature generated by the standard and the Hartmann-Neff model.

Hence, there is no principle limitation of the isochoric-volumetric split in this regard, if an adequate nonlinear compression model is used.

REFERENCES

Böttger, S., C. Peiler, J. Schilling, P. Schreiber, & P. Schneider (2024). Final report of IGF-Project 21767 N. Technical report, Deutsche Kautschuk-Gesellschaft e. V., Frankfurt. Deutsches Institut f'ur Normung e. V. (2017). Din 53504:2017-03, testing of rubber - determination of tensile strength at break, tensile stress at yield, elongation at break and stress values in a tensile test. Technical report, Beuth Verlag GmbH, Berlin.

Guth, E. (1945). Theory of filler reinforcement. Rubber Chemistry and Technology 18(3), 596–604.

Hartmann, S. & P. Neff (2003). Polyconvexity of generalized polynomial-type hyperelastic strain energy functions for near-incompressibility. International Journal of Solids and Structures 40, 2767–2791.

Ilseng, A., B. Skallerund, & A. Clausen (2017). An experimental and numerical study on the volume change of particle-filled elastomers in various loading modes. Mechanics of Materials 106, 44–57.

Kraus, G. (1971). Reinforcement of elastomers by carbon black.In Fortschritte der Hochpolymeren-Forschung, pp. 155–237. Springer.

Li, X., T. Bai, Z. Li, & L. Liu (2016). Influence of the temperature on the hyper-elastic mechanical behavior of carbon black filled natural rubbers. Mechanics of Materials 95, 136–145.

Penn, R. W. (1970). Volume changes accompanying the extension of rubber. Transactions of the Society of Rheology 14(4), 509–517.

Treloar, L. G. (1975). The physics of rubber elasticity. Oxford: Oxford university press.

Constitutive Models for Rubber XIII – Dal (Ed.)
© 2025 The Author(s), ISBN 978-1-032-85138-9

Constitutive modeling and FE-simulation of temperature dependent stator elastomers in positive displacement motors

A. Schöne, R. Landgraf & J. Ihlemann
Chair of Solid Mechanics, Chemnitz University of Technology, Germany

S. Zeller, C. Hohl & T. Regener
Baker Hughes, Celle Technology Center, Germany

ABSTRACT: In drilling operations the most commonly utilized drives for the drill bit are positive displacement motors (PDMs). They consist of rotor and stator and contain a rubber lining between the two parts, which is exposed to very high mechanical loads and volume changes. Moreover, due to the cyclic loading and the associated hysteresis heating of the elastomer material, the temperature inside the rubber lining increases inhomogeneously, which leads to local changes of the rubber's material behavior. In order to consider this effect in FE-simulations of the PDM, an appropriate material model is required. To realize this, an existing material model is extended to take the temperature dependency of the elastomer behavior into account. Although various approaches are conceivable, the use of power functions or tanh-functions has proven to be the best option for the investigated rubber material regarding the chosen optimization criteria. The extended material model is implemented and used to represent the material behavior of the rubber lining in FE-simulations with a short but representative section of the PDM. The simulation results give an insight into the distributions of strains and stresses inside the rubber lining and variations allow the investigation of the impact of different influencing factors.

1 INTRODUCTION

Positive displacement motors (PDMs) are the most used drives for the drill bit in oil drilling operations. The PDM is located downhole inside the bottom hole assembly closely above the drill bit and delivers the torque and rotation for the drill bit. The basis for the working principle of the motor are the helical geometries of rotor and stator, which form closed cavities between the two parts (Jaeger & Herlitzius 2016). When drilling fluid is pumped through the motor with high pressure, the cavities move alongside the motor axis, which also creates the rotation of the rotor and finally the drill bit. The vast majority of the motors use a rubber lining on the inside of the stator to seal the cavities from each other. In addition, it also represents a measure for compensating manufacturing inaccuracies. These factors make it an essential component of the motor.

However, the rubber lining is also the reason for most failures of the PDM (BA, Pushkarev, Kolyshkin, Song, & Yin 2016). The cause for this is the exposure of the elastomer material to high mechanical loads during the rotation of the rotor inside the stator. This is accompanied by volume changing effects due to high temperatures, high pressures and chemical swelling as a result to the contact with the drilling fluid. Furthermore, the hysteresis heating as a result of the cyclic load leads to an inhomogeneous temperature distribution inside the rubber lining, which is associated with local differences of the material behavior. Thus, computational investigations of the rubber lining inside the motor should consider this effect.

For the purpose of investigating the mechanical behavior of the elastomer lining inside the PDM, FE-simulations of the problem are suitable. In this case, the reduction of the problem size is necessary to limit the computation time of the simulation. Therefore, not the whole motor is simulated, but only a representative section of it. To ensure that the model behavior is not affected by the reduction of the model scope, periodic boundary conditions (PBCs) are applied to the bounding cross sections of the motor section. As shown in Schöne et al. (2022), this approach allows a detailed investigation of strains and stresses inside the rubber lining. The corresponding model is depicted in Figure 5.

Beside the FE-model itself, also an appropriate material model is required to represent the elastomer behavior accurately. This model should be able to reproduce the important effects of the rubber material, such as nonlinearity, hysteresis and permanent set. A constitutive model taking all these effects into account is the so-called MORPH, which stands for

DOI: 10.1201/9781003516880-10

Model of Rubber Phenomenology (Besdo & Ihlemann 2003, Landgraf & Ihlemann 2017). It acts as basic model and can be extended according to the requirements of the simulations. For example, an extension for volume changing effects can be realized as described in Schöne et al. (2022). Also curing phenomena (Landgraf & Ihlemann 2017) and time dependent effects (Landgraf & Ihlemann 2017) can be considered. Furthermore, the consideration of the temperature dependency is required. A first approach for this is given by Besdo et al. (2010). However, as the paper at hand shows, this can also be realized with different approaches.

The extended material model allows FE-simulations with inhomogeneous temperature distributions inside the rubber lining, that take varying material properties into account. In this way, a realistic investigation of the motor behavior and the resulting strains and stresses inside the rubber lining is enabled. This paper provides a brief insight into the available possibilities.

2 MATERIAL MODELING AND CHARACERIZATION

2.1 *Basic material model*

The modeling of the rubber material is based on the MORPH material model (Besdo & Ihlemann 2003, Landgraf & Ihlemann 2017). It includes eight free parameters $p_1 - p_8$ and considers typical effects of filled rubber materials, like nonlinearity, hysteresis, Mullins effects and permanent set. Only a brief description of the material model according to the implementation following Weiser et al. (2021) is given here. More extensive information can be found in the mentioned publications or Schöne et al. (2022).

The MORPH states that the total Kirchhoff stress $\underline{\underline{\tau}}$ results from the addition of a basic stress (a Neo-Hookean like part containing the deviator $()'$ of the isochoric part $\underline{\underline{\bar{b}}}$ of the left Cauchy-Green tensor $\underline{\underline{b}}$), the so-called additional stress $\underline{\underline{\tau}}^A$ and a volumetric part according to Hartmann and Neff (2003), which includes the bulk modulus K, the determinant J of the deformation gradient and the unit tensor $\underline{\underline{I}}$.

$$\underline{\underline{\tau}} = 2\alpha \underline{\underline{\bar{b}}}' + \underline{\underline{\tau}}^{A'} + \frac{K}{10}\left(J^5 - J^{-5}\right)\underline{\underline{I}} \tag{1}$$

The additional stress is given by an ordinary differential equation based on the Zaremba-Jaumann derivative (which is denoted by the *):

$$\overset{*}{\underline{\underline{\tau}}}{}^A = \beta \overset{*}{b}_F\left(\underline{\underline{\tau}}^L - \underline{\underline{\tau}}^A\right) . \tag{2}$$

This expression contains the limiting stress $2\tau^L$, which is defined as

$$\underline{\underline{\tau}}^L = \gamma \left[\exp\left(\left(p_7 \frac{\overset{*}{\underline{\underline{\bar{b}}}}}{\overset{*}{b}_F}\frac{b_F}{b_F^S}\right)\right) + p_8 \frac{\overset{*}{\underline{\underline{\bar{b}}}}}{\overset{*}{b}_F}\right]' . \tag{3}$$

It contains the very important variable b_F^S of the material model. This factor represents the maximum Frobenius norm of the deviator of the isochoric part of the left Cauchy-Green tensor during the loading history of the material point:

$$b_F^S = \max\left[b_F(s);\ 0 \le s \le t\right] , \tag{4}$$

$$\text{with } b_F = \sqrt{\underline{\underline{\bar{b}}}' \cdot\cdot \underline{\underline{\bar{b}}}'^T} . \tag{5}$$

This variable b_F^S is also part of the three material functions α, β and γ occurring in Equations (1)–(3):

$$\alpha = p_1 + \frac{p_2}{\sqrt{1 + \left(p_3 b_F^S\right)^2}} , \tag{6}$$

$$\beta = \frac{p_4}{\sqrt{1 + \left(p_3 b_F^S\right)^2}} , \tag{7}$$

$$\gamma = p_5 b_F^S \left(1 - \frac{p_6}{\sqrt{p_6{}^2 + \left(b_F^S\right)^2}}\right) . \tag{8}$$

In the context of the PDM, shrinkage during the vulcanization, temperature changes and chemical swelling due to the contact with the drilling fluid lead to volume changes of the elastomer. These aspects can be modeled with an approach based on the additive decomposition of the strain rate tensor, as described in more detail in Schöne et al. (2019).

2.2 *Extension to consider temperature dependent material behavior*

The basic MORPH as described in section 2.1 only models the material behavior at constant temperatures. Of course it is possible to identify different sets of material parameters at different temperatures. However, this would only generate discrete information at certain temperature levels and simulations with continuously changing temperatures would not be feasible. In order to consider the material behavior due to inhomogeneous and temporally changing temperature fields accurately, a temperature dependent formulation of the material model is required.

In the literature, such a modification of MORPH is already described by Besdo et al. (2010). Here, the eight original parameters of the material model are computed depending on the temperature change $\Delta\vartheta = \vartheta - \vartheta_0$ with the help of exponential functions (equation (9)). In the context of the paper at hand, the reference temperature is set to $\vartheta_0 = 0°C$.

$$p_k = A_k \cdot \exp(B_k \Delta\vartheta) \; for \; k = 1 \ldots 8 \tag{9}$$

This procedure establishes the new parameters A_k and B_k that have to be identified for a specific material. The number of additional parameters depends on the chosen variant. In the mentioned paper, three different variants are presented and the corresponding number of parameters ranges from 13 to 16. This results from setting some parameters p_k constant or adding a constant term to parameter p_8.

A detailed investigation of this approach reveals that a further reduction of the number of free parameters without significant loss of information is possible for the specific rubber material used in this context. The data basis are experimental data from tension and shear tests at 25°C, 50°C, 100°C and 150°C. The improvement of the modeling approach can be achieved by setting those parameters constant that show nearly no change over the investigated temperature range. The best compromise between a low number of free parameters and an accurate reproduction of the temperature dependency for the investigated rubber material is to keep the parameters p_3, p_4, p_5 and p_7 constant and to apply the exponential functions to the other four parameters. This leads to a total number of twelve free parameters and a normalized total error sum of 15.97. In comparison to the separate modeling of the four temperatures with 32 parameters and a normalized total error sum of 14.15 (see Table 1), the loss of modeling accuracy is absolutely acceptable for the reduction of free parameters.

Table 1. Comparison of different modeling approaches for the temperature dependency.

approach	number of free parameters	normalized total error sum
separate fits	32	14.15
exponential functions	12	15.97
power functions	12	15.21
tanh-functions	16	14.84

An even better fit of the experimental data can be achieved by the usage of alternative approaches. A very good compromise of a high modeling accuracy and a low number of parameters is given by the usage of power functions for the four non-constant parameters (with $\overline{\vartheta} = 1\,\mathrm{K}$ to ensure dimensional consistency):

$$\begin{aligned} p_k &= A_k && \text{for } k = 3,4,5,7 \\ p_k &= A_k \cdot \left(\frac{\Delta\vartheta}{\overline{\vartheta}}\right)^{B_k} && \text{for } k = 1,2,6,8. \end{aligned} \tag{10}$$

This version reduces the normalized total error sum compared to the exponential functions (see Table 1) and requires the same number of free parameters. However, the choice of this approach also restricts the usage of the model to temperatures above 0°C, because the power functions are not defined for a negative base with arbitrary exponents. Nevertheless, in the intended application in the context of PDM-simulations this is not important, because the temperatures investigated here are always above 0°C.

An alternative would be the use of sigmoid functions as ansatz, like for example the tanh-function:

$$\begin{aligned} p_k &= A_k && \text{for } k = 3,4,5,7, \\ p_k &= A_k \cdot \tanh(B_k \cdot \Delta\vartheta) + C_k && \text{for } k = 1,2,6,8. \end{aligned} \tag{11}$$

This model version allows an even more accurate reproduction of the experimental data, accompanied by a slightly higher number of free parameters. With this approach the consideration of negative temperatures is possible, but extrapolations should be checked for its physical plausibility. However, since the PDM-simulations are conducted at temperatures higher than 0°C, the version with power functions as given in equation (10) is used in the following.

2.3 *Parameter identification based on cyclic experiments*

To determine the free parameters of the MORPH material model for a specific carbon black filled rubber, cyclic tension and shear tests are conducted at different temperatures in the range from 25°C to 150°C. The tension tests are performed with standardized test specimens. By measuring the strain directly at the test specimens, specific strain levels can be defined, each of which is subjected to five loading cycles. The shear tests are conducted with special test specimens described in detail in Kanzenbach and Ihlemann (2021) and Kanzenbach et al. (2022). It is a rubber mat that can be fixed inside a special shear device with the help of small pins. The main benefit of this arrangement is the large homogeneous measuring area even at high shear deformations. As in the case of the tension tests, cyclic loading with increasing shear deformations is performed. The strain of the rubber mat is measured by digital image correlation (DIC), an optical measuring method.

The twelve parameters of the material model given in equation (10) are identified with the help of a nonlinear optimization using the least-squares method to consider the deviations of the stress values. Furthermore, also the hysteresis area is taken into account in the residuum. The balance between these two aspects can be regulated with a weighting factor. The stronger consideration of the stress values leads to a poorer reproduction of the hysteresis areas and vice versa. The product of the deviations of both criteria can be used to find an optimum weighting factor that allows a good replication of the stress-strain-behavior as well as an acceptable deviation of the hysteresis areas.

Since only the stationary behavior of the motor is of interest, the identification considers solely the stationary cycles of the tension and shear experiments. In order to capture the temperature dependency as good as possible, data from experiments at 25°C, 50°C, 100°C and 150°C are taken into account.

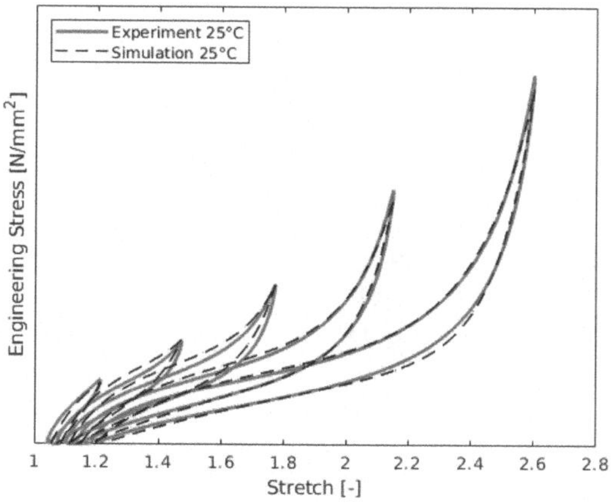

Figure 1. Comparison of simulation and experiment for the stationary cycles of a tension test at 25°C.

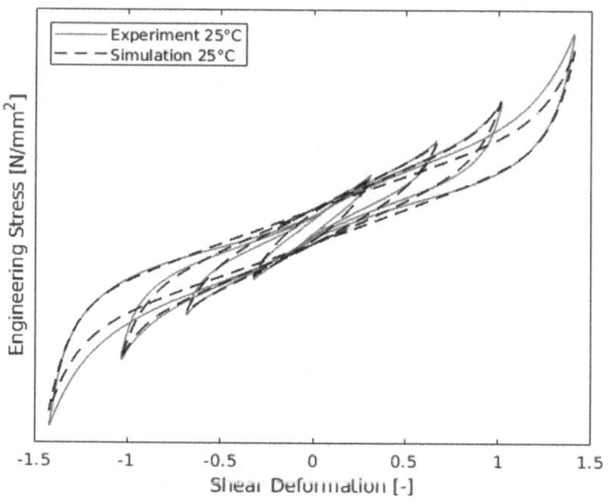

Figure 2. Comparison of simulation and experiment for the stationary cycles of a shear test at 25°C.

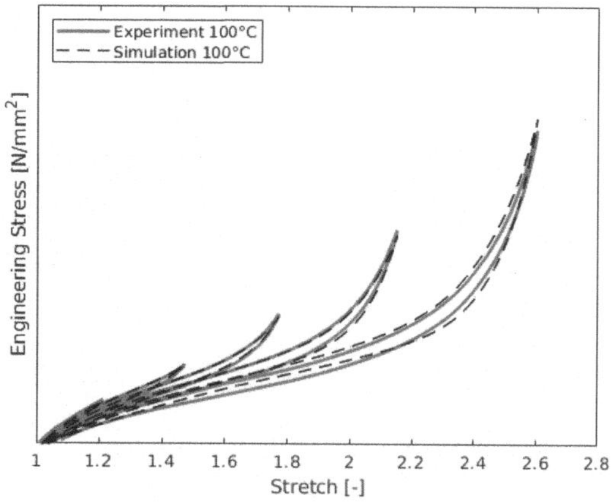

Figure 3. Comparison of simulation and experiment for the stationary cycles of a tension test at 100°C.

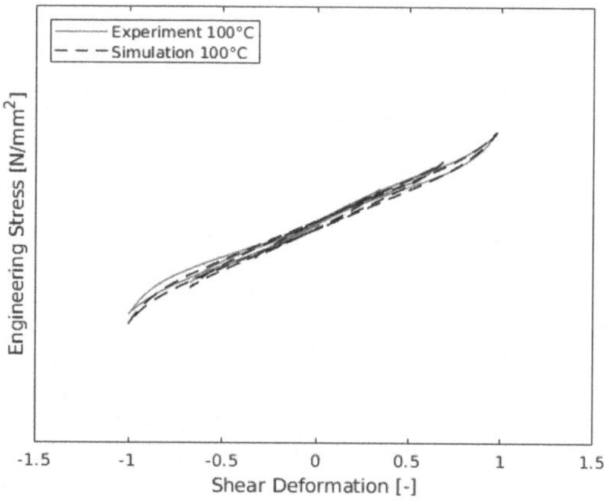

Figure 4. Comparison of simulation and experiment for the stationary cycles of a shear test at 100°C.

The resulting simulations are compared to the experimental data in Figures 1 to 4 for the tension and shear tests at 25°C and 100°C. All of the depicted simulated curves are generated with the same set of parameters. In general, the simulations reproduce the experimental curves very good. Moreover, the behavior change between the two temperatures is clearly visible both in the experimental and in the simulated curves.

3 FE-MODEL OF THE PDM

The FE-model to simulate the stationary operating state of the motor is the so-called representative model section (RMS) containing only a short section of the PDM, which is representative for the whole motor. The repetitive structure of the motor section is taken into consideration by multiple PBCs at the nodes of the opposing front faces. These PBCs link the degrees of freedom of two corresponding nodes, so that the two front faces experience the same deformation.

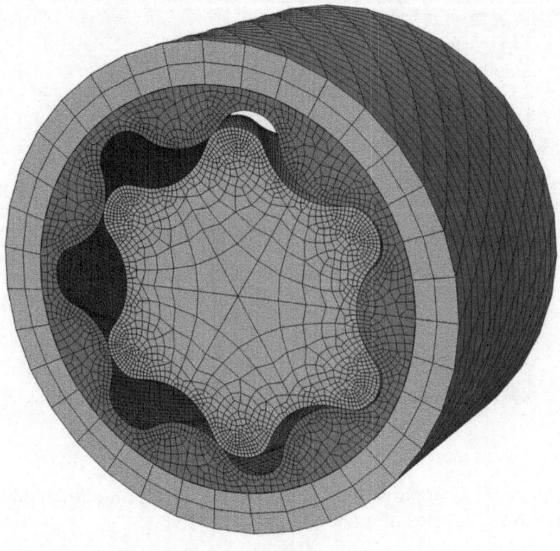

Figure 5. Meshed RMS consisting of outer stator, inner rotor and rubber lining in between.

Figure 5 shows an example of an RMS with six rotor lobes and seven stator lobes. In this case the characteristic shape of the stator is created solely by the rubber lining.

The eccentric rotation of the rotor in the stator is realized by the application of a rotational boundary condition on a reference point on the rotor axis. The second rotation of the rotor around the stator axis is not prescribed and results from the contact of rotor and stator.

A very important point of the model is the consideration of the fluid load. Since this model is not a coupled fluid-structure-interaction, the impact of the fluid on rotor and rubber lining is taken into account by static pressures, which follow the rotation of the rotor.

4 SIMULATION RESULTS

With the help of the RMS and the extended MORPH model, detailed FE-simulations of the strains and stresses inside the rubber lining can be performed. Moreover, the impact of variations of different boundary conditions, such as the operating state or various temperatures, can be investigated. In the following, an implementation of the MORPH version according to Equation (10) based on power functions with twelve parameters is used in the simulations.

An exemplary comparison of the impact of different temperatures on the resulting distribution of the variable b_F^S is given in Figure 6 (the rotor is hidden in the figure for better visibility). In this study, the single quarters of the rubber lining have a constant temperature of 25°C, 50°C, 75°C and 100°C, respectively. Consequently, also the distribution of the maximum equivalent strain b_F^S differs in the separate sections. Especially the quarter with a temperature of 25°C shows lower maximum values of this variable and also larger areas of very low values. The differences between the higher temperatures are less significant, but still visible.

Figure 7 shows a possible realistic temperature distribution inside the stator for a constant heat dissipation inside the rubber. On the surfaces of the rubber lining on the inside and the steel stator on the outside, heat is transported out of the system, which leads to lower temperatures in these areas. In contrast to that, in the interior of the rubber lining the heat is accumulated and high temperatures occur. This leads to high temperature differences inside the rubber lining (in the depicted example the difference is nearly 130 K).

The different temperatures have an impact on the elastomer properties. The varying material behavior is modeled with the temperature dependent MORPH presented in section 2.2. In this way, the current temperature at a material point directly defines the material behavior at this point.

As an example for the resulting distribution of the strains inside the rubber lining, a contour plot of the variable b_F^S is depicted in Figure 8. This variable represents the maximum equivalent strain of each material point throughout the whole simulation.

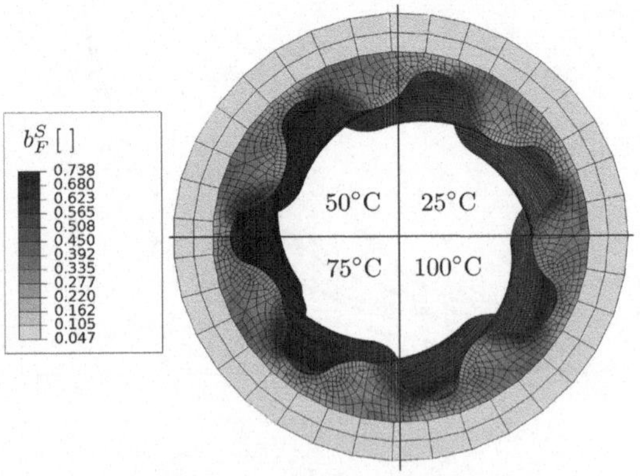

Figure 6. Example of the variable b_F^S inside the stator of the PDM at different temperatures.

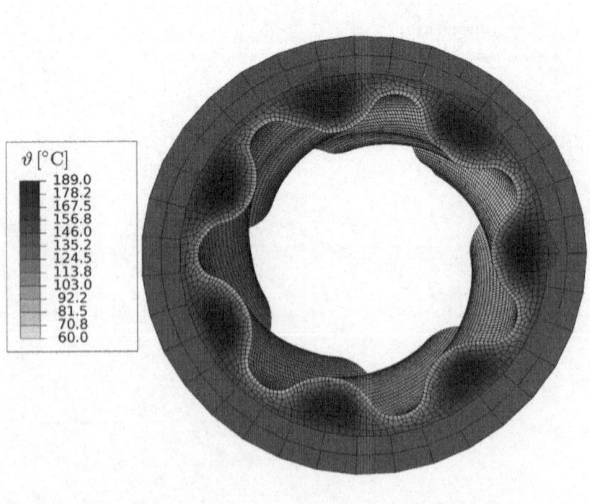

Figure 7. Example of the temperature distribution inside the stator of the PDM.

The distribution of b_F^S is inhomogeneous, but similar for the single lobes. Especially the strong difference between the loading and unloading flanks of the single lobes is very pronounced. The same effect can also be detected for the stresses inside the rubber material.

Furthermore, the model setup can be used to carry out variations and investigate the impact of different factors. This includes various thermal boundary conditions as well as multiple mechanical aspects, such as the rotational frequency, fluid pressures and the geometrical design of the motor. Moreover, also the material properties can be varied and the resulting impact on strains and stresses of the rubber lining can be determined.

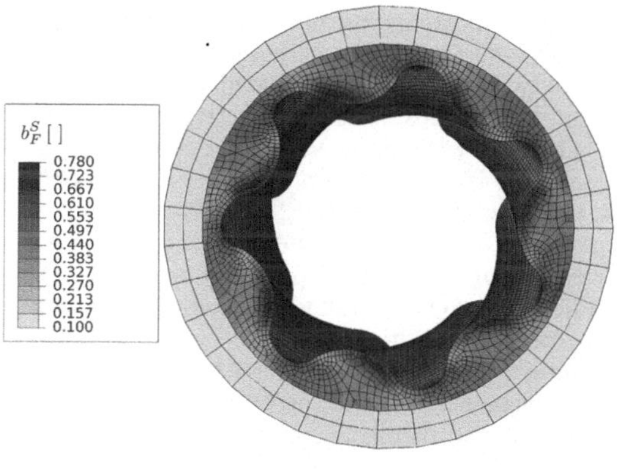

Figure 8. Example of the variable b_F^S inside the stator of the PDM.

5 CONCLUSIONS

In order to represent the material behavior of the elastomer in FE-simulations of PDMs accurately, an appropriate material model is required. In this context the MORPH material model can be extended to take the important temperature dependency into account. For the investigated rubber material and with respect to the optimization criteria applied, the best compromise between a low number of additional parameters and a high accuracy of the reproduction of the experimental data can be achieved with the usage of temperature dependent power functions or tanh-functions for four of the eight original parameters. The simulation results provide a detailed insight into the strains and stresses inside the rubber lining for a predefined temperature field. Additionally, different influencing parameters can be varied and their impact on the internal loading of the stator rubber can be investigated.

This offers the possibility to take measures in order to reduce strains and stresses of the elastomer material in future developments. This includes material properties, geometrical designs and the operating states of the PDM.

To be able to find the optimum of the single improvements, the simulations have to be as accurate as possible and should include as many aspects as possible. One important point for future developments of the model is to compute the heat originating from the hysteresis of the rubber material under cyclic loading. Since the thermal and mechanical processes occur at different time scales, thermomechanical coupled simulations are not an option for investigations that are restricted to the stationary state. Instead, an alternative algorithm has to be developed, which takes the temperature distribution due to hysteresis heating on one side and the mechanical loads in combination with the temperature dependent material behavior on the other side into account.

REFERENCES

Ba, S., M. Pushkarev, A. Kolyshkin, L. Song, & L. L. Yin (2016). Positive displacement motor modeling: Skyrocketing the way we design, select, and operate mud motors. In *Abu Dhabi International Petroleum Exhibition & Conference*. Society of Petroleum Engineers.

Besdo, D., N. Gvozdovskaya, & K. H. Oehmen (2010). Thermal effects and dissipation in a model of rubber phenomenology. *Lecture Notes in Applied and Computational Mechanics 51*, 95–121.

Besdo, D. & J. Ihlemann (2003). A phenomenological constitutive model for rubberlike materials and its numerical applications. *International Journal of plasticity 19*(7), 1019–1036.

Hartmann, S. & P. Neff (2003). Polyconvexity of generalized polynomial-type hyperelastic strain energy functions for near-incompressibility. *International Journal of Solids and Structures 40 (11)*, 2767–2791.

Jaeger, T. & J. L. Herlitzius (2016). *Navi-Drill Motor Handbook*. Baker Hughes Incorporated.

Kanzenbach, L. & J. Ihlemann (2021). Development of a shear device for precise simple shear measurements of rubber mats. *Strain 57*(2), e12376.

Kanzenbach, L., T. Lehmann, & J. Ihlemann (2022). Digital image correlation based characterization of rubber material at large shear deformations in an extended temperature range. *GAMM-Mitteilungen*, e202200009.

Landgraf, R. & J. Ihlemann (2017). Application and extension of the morph model to represent curing phenomena in a pu based adhesive. In *Constitutive Models for Rubber X*, pp. 137–143. CRC Press.

Landgraf, R. & J. Ihlemann (2023). A constitutive modeling approach to simulate time-dependent phenomena in rubber-like materials. *PAMM*, e202300277.

Schöne, A., R. Landgraf, & J. Ihlemann (2019). Simulation and modeling of stator elastomers in positive displacement motors. In *Proceedings of the 8th GACM Colloquium on Computational Mechanics for Young Scientists from Academia and Industry*, pp. 183–186. kassel university press.

Schöne, A., R. Landgraf, J. Ihlemann, S. Zeller, & C. Hohl (2022). Modeling and simulation of highly loaded rubber linings inside positive displacement motors. In *Constitutive Models for Rubber XII: Proceedings of the 12th European Conference on Constitutive Models for Rubber (ECCMR 2022)*, September 7–9, 2022, *Milano, Italy*, pp. 459–464. CRC Press.

Weiser, S., T. Lehmann, R. Landgraf, N. Goldberg, H. Donner, & J. Ihlemann (2021). Experimental and numerical analysis of cord–elastomer composites. *Journal of Rubber Research 24*(2), 221–225.

Constitutive Models for Rubber XIII – Dal (Ed.)

Constitutive modelling of amplitude dependent dynamic response of rubber-like materials

Ayşe Usta Yayla, Recep Durna & Hüsnü Dal
Department of Mechanical Engineering, Middle East Technical University, Ankara, Türkiye
Computational Micromechanics Laboratory, CAD/CAM Robotics Center, Middle East Technical University, Ankara, Türkiye

ABSTRACT: Rubber-like materials are widely used in the automotive, aerospace, and naval industries. The mechanical performance of rubbery polymers can be improved by incorporating filler particles. Filled rubber exhibits an amplitude-dependent viscoelastic response under dynamic loading, a phenomenon known as the "Payne effect," (Payne 1962). The Payne effect is characterized by a softening of the material as the amplitude of dynamic loading increases, resulting in a reduction in the storage modulus. Notably, this effect is reversible at room temperature, setting it apart from the Mullins effect. In this study, we present a novel constitutive model that accurately captures the Payne effect. The equilibrium response is modeled using an extended eight-chain framework (Dal, Gültekin, & Açıkgöz 2020), while the non-equilibrium behavior is described by quadratic free energy functions within the context of micro-sphere kinematics (Miehe & Göktepe 2005). We propose a nonlinear evolution equation for the viscous dashpot, incorporating amplitude dependency through an additional history variable based on the work of (Rendek & Lion 2010a). Our proposed evolution equation refines the Rendek-Lion formulation by accounting for loading rate-induced non-linearity in the storage modulus at high strain amplitudes. We demonstrate the model's ability to fit experimental data collected by our group. Additionally, we discuss the algorithmic considerations for the finite element (FE) implementation and showcase the model's numerical performance through the analysis of representative geometries.

1 INTRODUCTION

Due to their excellent sealing and damping properties, rubber-like materials are highly sought after in the automotive, naval, railway, and aerospace industries. These materials exhibit distinct mechanical response under static and dynamic loading conditions. The dynamic response of technical rubbers is highly sensitive to the filler ratio; i.e. such as silica or carbon black which are used to enhance their mechanical strength (Wolff 1996). Rubberlike materials exhibit large and nearly incompressible mechanical behaviour due to network of long polymer chains (Dal, Cansız, & Miehe 2018). Moreover, elastomers exhibit nonlinear viscoelastic behavior (Dal & Kaliske 2009) which is significantly influenced by rate, amplitude and temperature. To understand the dynamic behavior of rubber, the dependence of the material behaviour on temperature, loading frequency and amplitude, it is vital to perform Dynamic Mechanical Analysis (DMA) tests.

The Payne effect describes the amplitude-dependent softening behavior in filled elastomers under cyclic loading (Payne 1962). This effect, primarily attributed to the breakdown and reformation of active filler-filler networks (e.g., carbon black or silica) within the rubber matrix, results in a significant reduction in storage modulus at higher strain amplitudes. Over the decades, numerous experimental and theoretical studies have explored the mechanisms underlying this phenomenon. Kraus (1984) is credited with the creation of the first mathematical model in the frequency domain. This innovative model uses a conceptual framework that explains how the polymer network within the micro-structure can be destroyed and then reorganized. Ulmer (1996) adds an additional empirical factor to the model in which modifies it in response to Kraus's work. When compared to the original Kraus model, this augmentation greatly improves the model's ability to prediction of loss modulus values. Huber's study involves deriving the exponent from the empirical formulation by taking the filler particle connectivity into account (Huber, Vilgis, & Heinrich 1996). This approach aims to refine the understanding of material's behavior, particularly in the context of Payne effect. Lion & Kardelky (2004) payne proposed a 3D finite viscoelasticity model in the time domain. This model offers a more thorough knowledge of the dynamic behavior of the material on the stress response both under sinusoidal loading conditions and under random loading situations. Höfer &

DOI: 10.1201/9781003516880-11

Lion presents the results of an experimental examination into the dynamic material characteristics of elastomers loaded with carbon black in . This paper presents an innovative 3D constitutive modeling framework for finite non-linear viscoelasticity in addition to studying experimental findings. Interestingly, amplitude-dependent behavior is incorporated into the formulation by means of an intrinsic time scale expression. The study of constitutive modeling of amplitude dependency using experimental data is presented in Rendek & Lion (2010a). This work represents a step forward in the area since it integrates the 3D constitutive modeling framework that has been established for finite non-linear viscoelasticity, including amplitude-dependent behavior, into commercial software. Thermomechanical coupling effects such dissipative heat are examined and bimodal DMA and transient multistep tests are performed in Rendek & Lion (2010b).

2 CONSTITUTIVE MODELING OF RUBBER-LIKE MATERIALS

Rubber-like materials generally exhibit inelastic deformation under stress due to their polymer network structure. Their rate-dependent reaction is typically modeled separately. Put another way, it is separated into its viscoelastic and hyperelastic components. A free energy expression is given by

$$\Psi = \Psi^{e}(g;F) + \Psi^{ne}(g;F). \tag{1}$$

The equilibrium response is represented by Ψ^e in the equation, and the non-equilibrium response is represented by Ψ^{ne}. Within the framework of the generalized Maxwell model, the non-equilibrium response arises from the Maxwell branches, whilst the equilibrium response is assigned to the action of the single spring. It is possible to represent the hyperelastic response in this specific environment utilizing constitutive micro-mechanical or phenomenological approaches. A detailed account of phenomenological material modeling is omitted from this study, as its primary focus is on micro-structural changes in material networks. Comprehensive studies can be found in (Dal, Açıkgöz, & Badienia 2021). Equation (2) describes the dynamic modulus of the filled rubber-like materials, which explains the response under dynamic loading.

$$E^{*} = E' + \mathrm{i}E'' \tag{2}$$

The dynamic modulus E^* is made up of the storage and loss moduli, E' and E'', respectively. Loss modulus is out of phase with loading, whereas storage modulus is essentially in phase with the loading. The storage modulus curve has a sigmoidal shape in the frequency range. It rises in proportion to the frequency. At the maximum slope of the storage modulus, the loss modulus achieves its greatest value. Let the strain-controlled harmonic loading applied to the Maxwell elements be in the following form.

$$\varepsilon(t) = \Delta\varepsilon \sin(\omega t) \tag{3}$$

where $\Delta\varepsilon$ is the dynamic strain amplitude. The stress response for such loading is

$$\sigma(t) = \Delta\sigma \sin(\omega t + \delta) \tag{4}$$

Herein, $\Delta\sigma$ stands for the dynamic stress amplitude. Ratio between stress response to the strain input results in storage and loss modulus expressions for Maxwell branches. They can be defined in terms of phase difference as in

$$E' = \frac{\Delta\sigma}{\Delta\varepsilon}\cos(\delta) \quad \text{and} \quad E'' = \frac{\Delta\sigma}{\Delta\varepsilon}\sin(\delta) \tag{5}$$

Storage and loss modulus expressions in terms of frequency ω for generalized Maxwell model can be written as in Equation (6).

$$\begin{aligned} E'(\omega) &= E_0 + \sum_{i=1}^{k} E_i' = E_0 + \sum_{i=1}^{k} E_i \frac{(\omega\tau_i)^2}{1+(\omega\tau_i)^2} \\ E''(\omega) &= \sum_{i=1}^{k} E_i'' = \sum_{i=1}^{k} E_i \frac{(\omega\tau_i)}{1+(\omega\tau_i)^2} \end{aligned} \tag{6}$$

The spring coefficients in this case are $E_0, E_1, E_2, .., E_k$, and the relaxation times are τ_i. For incompressible material, the following relation between elastic modulus and shear modulus for the equilibrium response can be utilized: $E_0 \approx 3\mu$. Notably, Equation (6) is only able to describe the viscoelastic material's frequency dependency; it cannot handle the amplitude dependent behaviour. Therefore, it is essential that the Payne effect be taken into account for material modeling, particularly in loading scenarios where dynamic amplitudes are changing.

3 METHODOLOGY

This section explains the enhanced method for simulating amplitude dependent behaviour. Compared to models that employ invariant based formulations, the micro-sphere material model demands more processing power despite its ability to precisely capture the uniaxial and equibiaxial behaviour of the material response (Dal, Gültekin, & Açıkgöz 2020). Hence, utilizing the extended eight-chain model, (Dal, Gültekin, & Açıkgöz 2020) offers a thorough study of the equilibrium response. To further explore the non-equilibrium response under dynamic stress, the affine micro-sphere modeling approach—which has been well studied in (Miehe & Göktepe 2005)—is employed. By including an extra set of material parameters in the internal variable expressions, the modeling framework enhances the amplitude dependent characteristics of the non-equilibrium response. In this work, equilibrium response is simulated with the help of extended eight-chain material model. The theoretical features of the proposed material modeling approach for the non-equilibrium response is presented.

3.1 *Non-equilibrium response - Affine micro-sphere model*

This section explains the theoretical framework that describes how viscoelastic materials behave in response to amplitude changes. Affine micro-sphere material model aids in the modeling of non-equilibrium behaviour. Maxwell branches are used in each direction of the sphere to obtain the response. One can view the schematic representation in Figure 1.

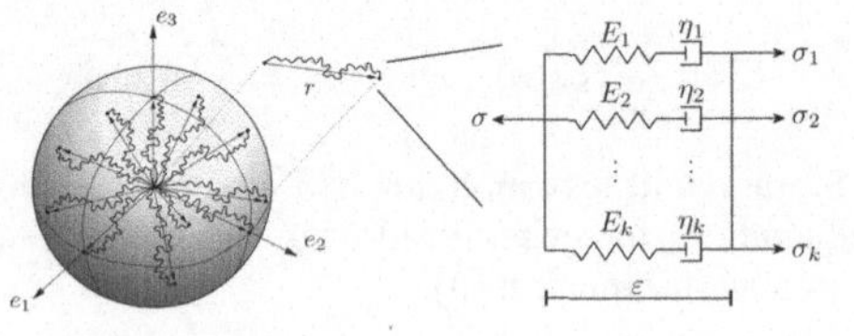

Figure 1. Demonstration of constitutive modeling rheology.

Affinity assumption $\lambda_f = \bar{\lambda}_f$ states that stretch due to free chain response is identical both in micro-structure and macro-structure of the polymer network.

$$\Psi^{ne} = \Psi^{ne}(\lambda_f, \varepsilon_f^1, \ldots, \varepsilon_f^{n_r}) \tag{7}$$

Free energy for the non-equilibrium response can be expressed as a function of free chain stretch and strain values, see Equation (7). In the equation, $\varepsilon_f^{n_r} := \ln \lambda_f$ states the strain in the n_r^{th} direction. Chain orientations and corresponding weights are tabulated in (2004). Free energy function for each Maxwell branch is defined as

$$\Psi_i^{ne}(\varepsilon_f^{n_r}, \alpha_i) = \frac{1}{2}\sum_{i=1}^{M} \mu_i (\varepsilon_f^{n_r} - \alpha_i)^2 \tag{8}$$

where M defines the number of Maxwell branches in one direction in the micro-sphere material model in Equation (8). Stress in the Maxwell branch is given in Equation (9).

$$\sigma_i = \partial_{\varepsilon_f^{n_r}} \Psi^{ne}(\varepsilon_f^{n_r}, \alpha_i) = \mu_i (\varepsilon_f^{n_r} - \alpha_i) \tag{9}$$

Force induced by internal variable β_i is also the same as stress relation; i.e. $\beta_i = -\partial_{\alpha_i} \Psi^{ne}(\varepsilon_f^{n_r}, \alpha_i) = \sigma_i$. Evolution of internal variable α_i is given in Equation (9).

$$\dot{\alpha}_i = \frac{1}{\hat{\eta}_i}\beta_i = \frac{1}{\hat{\eta}_i}\mu_i (\varepsilon_f^{n_r} - \alpha_i) \quad \text{with} \quad \hat{\eta}_i = \frac{\eta_i}{H_i} \tag{10}$$

where μ_i and η_i are the shear modulus and the dashpot coefficient in the corresponding Maxwell branch. Here, the viscosity in the dashpot element is defined with $\hat{\eta}_i$ which can be interpreted as the effective viscosity taking the material from initial viscosity level to the stationary viscosity level depending on the amplitude. $H_i(q_i)$ simulates amplitude dependent behaviour with the additional internal variable $q_i(t)$ for dashpot element. It is the structural variable describing the behaviour of micro-structure and dependent on the deformation process. Amplitude dependency is expressed as in Equation (11).

$$H_i(t) = 1 + q_i(t) \quad \text{with} \quad q_i(0) = 0 \tag{11}$$

with the improved structural variable evolution is given as in Equation (12).

$$\dot{q}_i(t) = \frac{1}{\tau_{st}} \left(p_c |\dot{\varepsilon}_f^{n_r}|^{p_\varepsilon} - q_i \right) \tag{12}$$

where p_c is the parameter introduced for unit consistency. Also, $p_\varepsilon \geq 1$ is the material parameter and τ_{st} can be interpreted as the saturation time of the micro-structure. Note that this type of evolution yields to the material model developed in (Rendek 2011) for $p_\varepsilon = 1$. For more information on the numerical implementation in orientation directions, we refer to (Miehe & Göktepe 2005, Dal, Zopf, & Kaliske 2018).

3.2 *Thermodynamic consistency*

For any constitutive model, the second law of thermodynamics should be verified. Inelastic materials' free energy functions are expressed in terms of internal variables. Incorporating the dissipation that occurred in the material response is the main reason. Here, Equation (13) provides the thermodynamic consistency statement. Expressions are provided in their general form, that is, without subscripts or superscripts, in order to maintain simplicity.

$$\mathcal{D} = \mathcal{P} - \dot{\Psi} \geq 0 \quad \text{and} \quad \mathcal{P} = S : \frac{1}{2}\dot{C} = \tau : d \tag{13}$$

where $\mathcal{P}$ is the stress power. Inserting the rate of free energy in Equation (13) gives the following relation.

$$\mathcal{D} = (\tau - 2\partial_g \Psi) : d - \partial_\alpha \Psi \dot{\alpha} - \partial_q \Psi \dot{q} \geq 0 \tag{14}$$

Then, the dissipation equation reduces to

$$\mathcal{D}_{red} = -\partial_\alpha \Psi \dot{\alpha} - \partial_q \Psi \dot{q} \geq 0 \tag{15}$$

In the constitutive model, non-equilibrium part of the free energy function is dependent on the internal variables. Therefore, the relations of the reduced dissipation are obtained by considering only non-equilibrium part of free energy.

$$-\frac{\partial \Psi}{\partial \alpha}\left(\dot{\alpha} + \frac{\partial \alpha}{\partial q}\dot{q}\right) \geq 0 \tag{16}$$

The term $\partial\alpha/\partial q$ is needed to be found in the equation. It can be obtained in discrete-time setting by utilizing Equations (10) and (11).

$$\alpha - \alpha_n = \frac{\Delta t}{\tau}(1 + q)(\varepsilon_f - \alpha) \tag{17}$$

In the above equation, α_n is the internal variable computed in the previous time step. After manipulations of Equation (17), derivative $\partial\alpha/\partial q$ becomes

$$\frac{\partial \alpha}{\partial q} = \frac{\Delta t/\tau}{\left(1 + \frac{\Delta t}{\tau}(1+q)\right)^2}\left(\varepsilon_f - \alpha_n\right) \geq 0 \quad (18)$$

It is seen that Equation (18) is always non-negative. Then, Equation (17) has its final form as

$$\sigma\left(\dot{\alpha} + \frac{\Delta t/\tau}{\left(1 + \frac{\Delta t}{\tau}(1+q)\right)^2}\left(\varepsilon_f - \alpha_n\right)\dot{q}\right) \geq 0 \quad (19)$$

From the equation, it can be inferred that the constitutive model is thermodynamically consistent for $\dot{\alpha}, \dot{q} \geq 0$.

4 MATERIAL IDENTIFICATION

In this part, the material identification procedure is explained and compared with experiments. Uniaxial and equibiaxial tension experiments are conducted under the support of the METU Computational Micro-Mechanics Laboratory (METU-CMML).

4.1 *Geometric factor identification*

This section explains the impact of specimen geometry on material behaviour. When boundary conditions during an experiment have an impact on the specimen reaction, it should be examined. For a long specimen that is unaffected by boundary conditions, material response should be obtained. According to Equation (20), a geometric factor is defined in this sense as the relation between two geometrically distinct specimen responses.

$$\alpha_g = \frac{m_{DMA}}{m_{UT}} \quad (20)$$

In the equation, geometric factor α_g is defined as the ratio of the initial slope of DMA specimen response m_{DMA} to the initial slope of sufficiently long specimen response m_{UT} under tensile loading. The shear modulus $\tilde{\mu}$ given in Dal, Gültekin, & Açıkgöz (2020) is utilized in closed form storage modulus equation as

$$\tilde{\mu}_{DMA} = \alpha_g \tilde{\mu} \quad (21)$$

Similarly, correction with geometric factor should be performed when overstress response is used in the uniaxial, equibiaxial tension response with the relation given in Equation (20).

$$\sigma_{i,UT,ET} = \frac{\sigma_{i,DMA}}{\alpha_g} \quad (22)$$

To apply the geometric factor in the material identification process, finite element modeling is carried out for various length dimensions under tensile loading. Initial slopes are derived from the simulations and normalized to the longest specimen. Dimensions are chosen as $2 \times 4 \times 4$, $2 \times 4 \times 8$, $2 \times 4 \times 16$, $2 \times 4 \times 64$ with thickness, width and length, respectively. From the simulations, it is concluded that the geometric factor α_g is 1.11 for the material investigated in this study (see Figure 2).

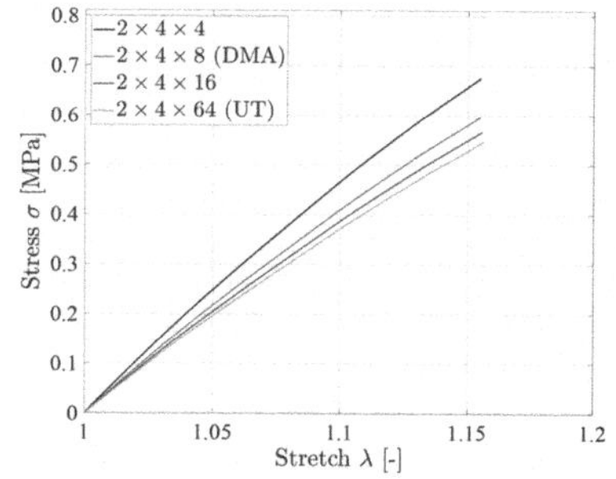

Figure 2. FEM simulation results for different specimen lengths.

4.2 *Material parameter identification*

Material parameters are obtained in two main steps. The material parameters for both the steady-state viscoelastic response and the hyperelastic response are simultaneously achieved in the first stage. The optimization process computes the transient response parameters after the material parameter has been established in the first step.

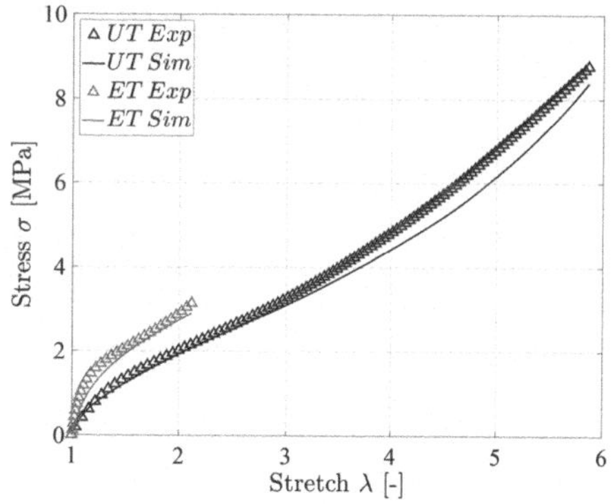

Figure 3. Uniaxial and equibiaxial tension test and material fit.

4.2.1 *Parameter identification for hyperelastic and steady-state viscoelastic response*

Hyperelastic material fit is performed for uniaxial and equibiaxial tensile experiments. For the steady-state viscoelastic material parameters, DMA test results are utilized.

- *Hyperelastic response material parameters:* The silicone-filled rubber material's hyperelastic behaviour is described by a parameter identification technique that is carried out based on the extended eight-chain material modeling approach. The optimization process makes use of the uniaxial and equibiaxial tension test results shown in Figure 3. Over the whole experiment range, a satisfactory simulation of the equibiaxial tension response is achieved. Furthermore, up to a stretch value of 3, the response produced with adjusted material parameters fully matches the data from the uniaxial stress test. In comparison to tests, material simulations produce a softer response for stretch values larger than 3.

- *Steady-state viscoelastic response material parameters:* The main idea for obtaining stationary response is that the micro-structural relaxation times $\tau_{st,i}$ are much larger than the loading periods; i.e. $\tau_{st,i} \gg 2\pi/\omega$. This assumption means that there is no change in the internal variable q_i with time; i.e. $\dot{q}_i(t) = 0$. Knowing that the loading is in the form of harmonic function such that

$$\varepsilon(t) = \varepsilon_0 + \Delta\varepsilon \sin(\omega t) \tag{23}$$

where, $\varepsilon_0, \Delta\varepsilon$ and ω are prestrain, dynamic strain amplitude and loading frequency, respectively. Then, its rate becomes

$$\dot{\varepsilon}(t) = \Delta\varepsilon\omega \cos(\omega t) \tag{24}$$

Inserting Equation (24) into Equation (12) yields

$$\dot{q}_i(t) = \frac{1}{\tau_{st}} \left(p_c|\Delta\varepsilon\omega|^{p_\varepsilon}|\cos(\omega t)|^{p_\varepsilon} - q_i\right) \tag{25}$$

The mean value should be obtained for steady-state value of internal variable $q_i(t)$. The mathematical procedure for expressing the steady-state value is explained below. Firstly, the expression over one period of loading is needed to be found.

$$\begin{aligned} \textstyle\int_0^T p_c|\dot{\varepsilon}_f^{n_r}(t)|^{p_\varepsilon} dt &= \textstyle\int_0^T p_c|\Delta\varepsilon\omega|^{p_\varepsilon}||\cos(\omega t)|^{p_\varepsilon}|dt \\ &= p_c|\Delta\varepsilon\omega|^{p_\varepsilon} \textstyle\int_0^T |\cos(\omega t)|^{p_\varepsilon} dt \end{aligned} \tag{26}$$

Utilizing the relation $\omega = 2\pi f$ in Equation (26) gives

$$p_c|\Delta\varepsilon\omega|^{p_\varepsilon} \int_0^T |\cos(\omega t)|^{p_\varepsilon} dt = 4p_c|\Delta\varepsilon\omega|^{p_\varepsilon} \frac{1}{f} I(p_\varepsilon) \tag{27}$$

where $I(p_\varepsilon)$ in Equation (27) is defined as the exact value of the corresponding integral. Exact result of the integral $I(p_\varepsilon)$ can be approximated with the exponential function in the form given in Equation (28) for small strains.

$$I(p_\varepsilon) \approx \hat{I}(p_\varepsilon) = a_1 \exp(a_2 p_\varepsilon) + a_3 \exp(a_4 p_\varepsilon) \tag{28}$$

The parameters in the equation are determined as $a_i = 0.4306, -1.176, 0.5560, -0.09901$ for $i = 1..4$ for $p_\varepsilon \in [0, 6]$. With the approximated expression of $\hat{I}(p_\varepsilon)$, Equation (27)

$$4p_c|\Delta\varepsilon\omega|^{p_\varepsilon} \frac{1}{f} \int_0^{1/4} (\cos(2\pi t))^{p_\varepsilon} dt \approx 4p_c|\Delta\varepsilon\omega|^{p_\varepsilon} \frac{1}{f} \frac{\hat{I}(p_\varepsilon)}{4} \tag{29}$$

Then, stationary value of the internal variable can be obtained with Equation (30).

$$q_{s,i} = p_c|\Delta\varepsilon\omega|^{p_\varepsilon} \hat{I}(p_\varepsilon) \tag{30}$$

Figure 4 provides a view of its simulation results. Estimating the elastic response of damper-like structures under harmonic loading—which is correlated with storage values—is essential. In terms of storage modulus, it is evident that amplitude dependent behaviour over a wide frequency range was successfully achieved. Conversely, the generated material model overestimates the material's loss modulus response. Since the absorbed energy cannot be detected during experiments, it cannot be accounted for in the developed material model. Additional cause for the variations in the loss modulus may come from thermal effects.

4.2.2 *Parameter identification for transient viscoelastic response*

In this part, transient behaviour parameters are obtained for storage modulus data at the loading frequency of $0.05 Hz$ and dynamic strain of 0.75%. Optimization region is depicted in Figure 5. After optimization process, the simulation is performed for the whole domain. In Figure 5, the difference between the stationary values of storage modulus comes from the results given in Figure 4. Optimized relaxation times $\tau_{st,i}$ for the internal variables $q_i(t)$ sufficiently cover the transient response.

5 NUMERICAL ANALYSIS

Numerical simulations are conducted in this chapter using a damper-like geometry presented in (Miehe & Göktepe 2005).

Side view of the benchmark geometry and dimensional information are presented in Figure 6. As a boundary condition, the structure is fixed at the bottom $x - z$ plane. Moreover, the inherent property of the benchmark issue of choice is that it has a non-homogeneous strain field under loading conditions. Within a specific frequency range, we examine the complicated nature of the Payne effect and show material behaviour that has a strong connection to the amplitude, see Figure 8. An additional examination of the behavior reveals a significant finding: as the amplitude of dynamic displacement rises, stress softening becomes apparent. The Payne effect is famously illustrated in this way. Payne effect must be considered for loading types with varying amplitudes of dynamic strain. The simulation result is shown where the Payne effect is not modeled in order to highlight its significance. The result regarding viscous stress in the loading direction is presented (see Figure 7), which is similar to Figure 8. It is investigated that larger overstress values are the outcome of increased loading amplitude. Nevertheless, increasing the dynamic strain amplitude does not result in any softening of the stresses.

Numerical simulations are conducted in this chapter using a damper-like geometry presented in (Miehe

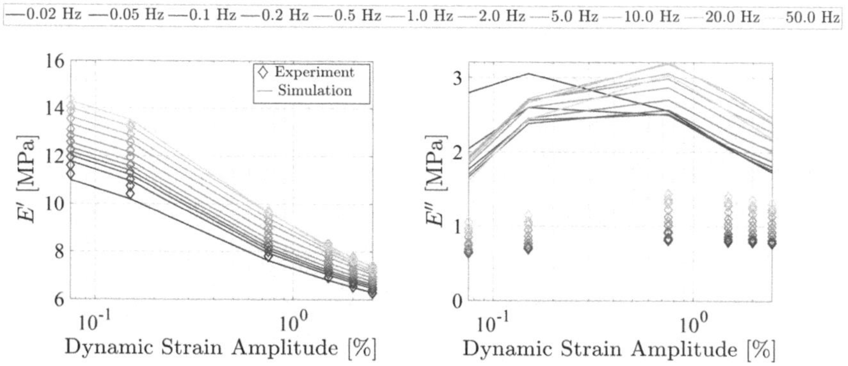

Figure 4. Material optimization performance for steady-state response - solid lines: simulation, diamond markers: experiments.

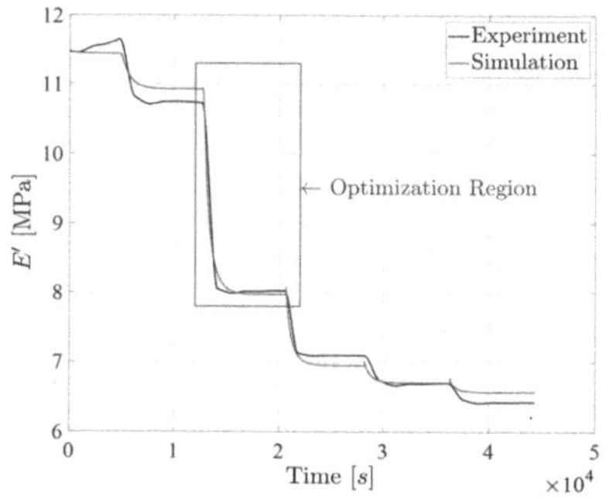

Figure 5. Performance of transient response parameters for whole range.

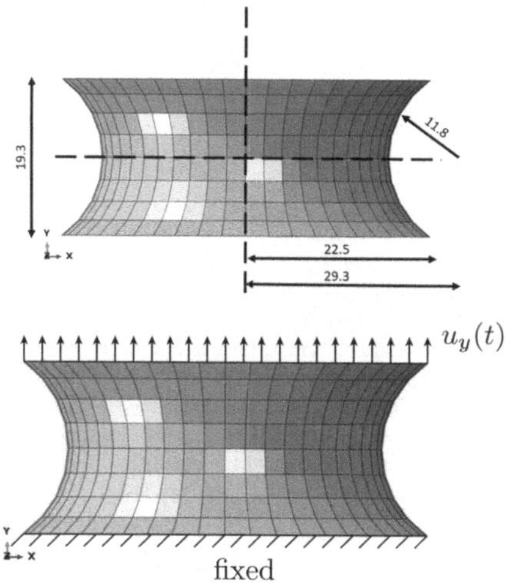

Figure 6. Dimensions of full geometry (all dimensions are in *mm*) (left) - illustration of boundary conditions (right).

& Göktepe 2005). Because it resembles vibration isolators, which are widely utilized in the automotive industry, this concept is very significant. Moreover, the inherent property of the benchmark issue of choice is that it has a non-homogeneous strain field under loading conditions. The numerical simulations cover a range of loading profiles with systematically varying amplitudes and frequencies in order to comprehensively evaluate the material behavior. The Finite Element Analysis Program *FEAP* is used to carefully document the complex investigations and simulation results (Taylor 2014).

6 CONCLUSION AND DISCUSSION

Rubber-like materials with filler particles that are dynamically loaded are the subject of this study. Viscoelastic and hyperelastic reactions to both dynamic and quasi-static stresses are investigated in the experiments. DMA tests reveal material behaviour variations with dynamic amplitudes. Constitutive modeling involves characterizing materials, where an improved evolution expression is proposed. The formula in question adresses the non-linear response commonly observed in filled rubbers by introducing an internal variable that is non-linearly dependent on loading rate. An extra material parameter, p_ε, needs to be optimized for the new strategy. The theory is notably applied for the first time in the micro-sphere material model framework.

The simulations highlight the model's increased fitting quality and show how well it captures the

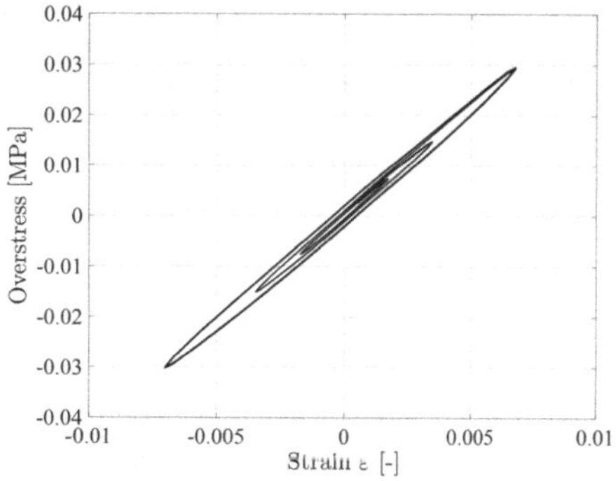

Figure 7. Classical viscoelasticity where amplitude dependence is excluded.

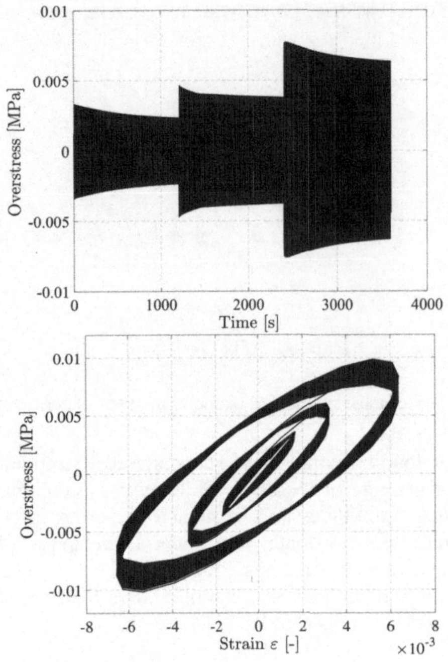

Figure 8. Quarter model simulation results: overstress change with time (left) and strain (right).

Payne effect through comparison with experimental data. In order to enhance comprehension for constitutive modeling, future research might concentrate on broadening material characterization to encompass a larger dynamic range. Clarifying material response could be achieved by looking at pre-strain effects and incorporating them into formulations. Improvements in the theoretical expressions at different amplitudes can improve the dissipative behaviour simulation accuracy. It is also essential to take into account and incorporate thermal effects, such as heat generated during DMA tests, into models.

ACKNOWLEDGEMENT

The financial support provided by the support of METU Scientific Research Project Coordination Unit under grant GAP-302-2020-10203 is gratefully acknowledged.

REFERENCES

Dal, H., K. Açıkgöz, & Y. Badienia (2021). On the performance of isotropic hyperelastic constitutive models for rubber-like materials: A state of the art review. *Applied Mechanics Reviews 73*(2).

Dal, H., B. Cansız, & C. Miehe (2018). A three-scale compressible microsphere model for hyperelastic materials. *International Journal for Numerical Methods in Engineering 116*(6), 412–433.

Dal, H., O. Gültekin, & K. Açıkgöz (2020). An extended eightchain model for hyperelastic and finite viscoelastic response of rubberlike materials: Theory, experiments and numerical aspects. *Journal of the Mechanics and Physics of Solids 145*, 104159.

Dal, H. & M. Kaliske (2009). Bergström–boyce model for nonlinear finite rubber viscoelasticity: theoretical aspects and algorithmic treatment for the fe method. *Computational mechanics 44*, 809–823.

Dal, H., C. Zopf, & M. Kaliske (2018). Micro-sphere based viscoplastic constitutive model for uncured green rubber. *International Journal of Solids and Structures 132-133*, 201–217.

Höfer, P. & A. Lion (2009). Modelling of frequency-and amplitude-dependent material properties of filler-reinforced rubber. *Journal of the Mechanics and Physics of Solids 57*(3), 500–520.

Huber, G., T. A. Vilgis, & G. Heinrich (1996). Universal properties in the dynamical deformation of filled rubbers. *Journal of Physics: Condensed Matter 8*(29), L409.

Kraus, G. (1984). Mechanical losses in carbon-black-filled rubbers. In Journal of Applied Polymer Science: *Applied Polymer Symposium*, Volume 39, pp. 75–92.

Lion, A. & C. Kardelky (2004). The payne effect in finite viscoelasticity: constitutive modelling based on fractional derivatives and intrinsic time scales. *International Journal of Plasticity 20*(7), 1313–1345.

Miehe, C. & S. Göktepe (2005). A micro–macro approach to rubber-like materials. part II: The micro-sphere model of finite rubber viscoelasticity. *Journal of the Mechanics and Physics of Solids 53*(10), 2231–2258.

Miehe, C., S. Göktepe, & F. Lulei (2004). A micro-macro approach to rubber-like materials— part I: the non-affine micro-sphere model of rubber elasticity. *Journal of the Mechanics and Physics of Solids 52*(11), 2617–2660.

Payne, A. R. (1962). The dynamic properties of carbon blackloaded natural rubber vulcanizates. part I. *Journal of applied polymer science 6*(19), 57–63.

Rendek, M. (2011). *Transient effects of filler-reinforced rubber with respect to the Payne effect: experiments, constitutive modelling and FEM implementation*. Ph. D. thesis, University of Federal Armed Forces.

Rendek, M. & A. Lion (2010a). Amplitude dependence of fillerreinforced rubber: Experiments, constitutive modelling and fem–implementation. *International Journal of Solids and Structures 47*(21), 2918–2936.

Rendek, M. & A. Lion (2010b). Strain induced transient effects of filler reinforced elastomers with respect to the payneeffect: experiments and constitutive modelling. *ZAMM Journal of Applied Mathematics and Mechanics/ Zeitschrift für Angewandte Mathematik und Mechanik: Applied Mathematics and Mechanics 90*(5), 436–458.

Taylor, R. L. (2014). Feap-a finite element analysis program.

Ulmer, J. (1996). Strain dependence of dynamic mechanical properties of carbon black-filled rubber compounds. *Rubber chemistry and technology 69*(1), 15–47.

Wolff, S. (1996). Chemical aspects of rubber reinforcement by fillers. *Rubber Chemistry and Technology 69*(3), 325–346.

Constitutive Models for Rubber XIII – Dal (Ed.)
© 2025 The Author(s), ISBN 978-1-032-85138-9

Zener model in Kelvin-Voigt representation with multiplicative split of the deformation gradient

Bülent Yagimli & Mohamed A. Abdelmoniem
Ostfalia University, Wolfenbüttel, Germany

ABSTRACT: In this paper, a new modeling approach for finite viscoelasticity is introduced based on a Zener model in Kelvin-Voigt representation. A novel methodology with new nomenclature for the definition of the deformation configurations is presented for the split of the deformation gradient tensor. Based on the thermo dynamic consistency, the evolution equation will be defined. The benefits of this model are discussed. The model is implemented in the finite element software Abaqus using a user material subroutine (UMAT). A practical example is provided involving an elastomeric puffer subjected to rotating shear load. The simulation predictions of the resulting mechanical behavior and heat dissipation agree very good with the experimental results. The model is affective in accurately capturing changes in the rubber transfer function induced by self-heating during operation.

1 INTRODUCTION

Elastomers can undergo very large elastic deformation. Depending on the application, the viscoelastic behaviour has to be considered. In the commercial finite element software Abaqus, finite viscoelasticity is handled by a convolutional integral function. The material model is suggested by Simo in (Simo 1987). The material model is based on a rheological Maxwell-element and can be implemented in user subroutine as presented in (Abdelmoniem & Yagimli 2021) to simulate the simple rotary dynamic testing machine from (Gent 1960). Analytical investigations in (Yagimli, Lion, & Abdelmoniem 2023) shows, that the suggested Simo-model leads to unphysical oscilations in certain frequency domain. The convolutional integral model in (Haupt & Lion 2002) leads to feasible stress response in all frequency domain. Another methodology, Zener model in Maxwell representation, is suggested by Lion in (Lion 1997) and (Reese & Govindjee 1998) with a multiplicative decomposition of the deformation gradient into an elastic and inelastic part. In that model the evolution equation for the inelastic deformation gradient is also based on a Maxwell element. A Zener model in Kelvin-Voigt representation is suggested in (Boyce, Weber, & Parks 1989). A new approach based on a rheological Kelvin-Voigt element will be presented in this work. Similar to Lion, the deformation gradient is split multiplicatively in two parts. Based on the thermo dynamic consistency, the evolution equation will be defined. The numerical integration algorithm will be presented and common testing scenarios will be simulated. The possibilities of both material models, the Zener model in Maxwell and the Zener model in Kelvin-Voigt representation will be discussed and validated with experimental results.

2 DEFORMATION NOMENCLATURE

Deformation gradient is split into an instantaneous and Kelvin-Voigt part. The meaning of the deformation gradients are presented with tangential vector

$$\mathrm{d}\underset{1}{x} = \bar{\mathbf{F}}_K \mathrm{d}\underset{0}{x} \quad \text{and} \quad \mathrm{d}\underset{2}{x} = \bar{\mathbf{F}}_I \mathrm{d}\underset{1}{x}. \tag{1}$$

The volumetric part is also split multiplicatively in a thermal expansion and a mechanical part as suggested in (Lu & Pister 1975)

$$\det(\mathbf{F}) = \mathbf{J} = \mathbf{J}_\theta \, \mathbf{J}_\mathrm{I} \tag{2}$$

so that the deformation gradient is the multiplication of all parts

$$\mathrm{d}\underset{C}{x} = \mathbf{F}\mathrm{d}\underset{0}{x} = J^{\frac{1}{3}}\bar{\mathbf{F}}\mathrm{d}\underset{0}{x} \tag{3}$$

$$\Rightarrow \quad \bar{\mathbf{F}} = \bar{\mathbf{F}}_I \bar{\mathbf{F}}_K \tag{4}$$

For the framework of the material model the following variables $\mathbf{F}$ deformation gradient and the internal variable $\bar{\mathbf{F}}_K$ are chosen. Based on the state variables the time derivative can be defined as:

DOI: 10.1201/9781003516880-12

$$\dot{\mathbf{F}} = \mathbf{L}\mathbf{F} \quad \text{withvelocitygradient} \quad \mathbf{L} = \dot{\mathbf{F}}\mathbf{F}^{-1}, \tag{5}$$

$$\dot{J} = (\mathbf{L} : \mathbf{1})J, \tag{6}$$

$$\dot{\bar{\mathbf{F}}} = \bar{\mathbf{L}}\bar{\mathbf{F}} \text{ with } \bar{\mathbf{L}} = \mathbf{L}^D \tag{7}$$

and the time derivative of the intantaneous part

$$\dot{\bar{\mathbf{F}}}_I = \bar{\mathbf{L}}\bar{\mathbf{F}}_I - \bar{\mathbf{F}}_I\bar{\mathbf{L}}_K \quad \text{with} \quad \bar{\mathbf{L}}_K = \dot{\bar{\mathbf{F}}}_K\bar{\mathbf{F}}_K^{-1} \tag{8}$$

This variables are applied to the definition of Helmholtz free energy.

3 ZENER MODEL IN KELVIN-VOIGT REPRESENTATION

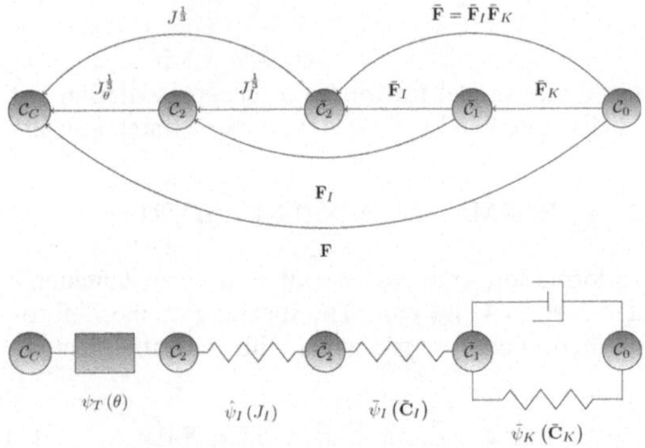

Figure 1. Energy storage for Zener model in Kelvin-Voigt representation.

Based on the rheological representation Figure 1 constitutive equations can be derived. The Helmholtz free energy is the addition of the energy storage function in each spring element and the thermal contribution ψ_T

$$\psi = \psi_T(\theta) + \hat{\psi}_I(J_I) + \bar{\psi}_I(\bar{\mathbf{C}}_I) + \bar{\psi}_K(\bar{\mathbf{C}}_K) \tag{9}$$

Herein $\bar{\mathbf{C}}_I$ is the right Cauchy-Green tensor between the intermediate configurations

$$\bar{\mathbf{C}}_I = \bar{\mathbf{F}}_I^T\bar{\mathbf{F}}_I \tag{10}$$

and the right Cauchy-Green tensor from initial to first intermediate configuration

$$\bar{\mathbf{C}}_K = \bar{\mathbf{F}}_K^T\bar{\mathbf{F}}_K \tag{11}$$

The time derivative of the Helmholtz free energy is

$$\dot{\psi} = \left(\frac{\partial\psi_T}{\partial\theta} - J_I J_\theta^{-1}\frac{\partial\hat{\psi}_I}{\partial J_I}\frac{\partial J_\theta}{\partial\theta}\right)\dot{\theta} +$$

$$+\left(J_I\frac{\partial\hat{\psi}_I}{\partial J_I}\mathbf{1} + 2\left(\bar{\mathbf{F}}_I^T\frac{\partial\bar{\psi}_I}{\partial\bar{\mathbf{C}}_I}\bar{\mathbf{F}}_I\right)^D\right) : \mathbf{D}$$

$$+\left(-2\bar{\mathbf{C}}_I\frac{\partial\bar{\psi}_I}{\partial\bar{\mathbf{C}}_I} + \bar{\mathbf{F}}_K^T\frac{\partial\bar{\psi}_K}{\partial\bar{\mathbf{C}}_K}\bar{\mathbf{F}}_K\right) : \bar{\mathbf{D}}_K \tag{12}$$

Within the restriction known as Clausius Duhem inequality in reduced form

$$\delta_{loc} = -\dot{\psi} + \mathbf{S} : \mathbf{D} - s\dot{\theta} \geq 0 \tag{13}$$

the relation for weighted Cauchy stress tensor

$$\mathbf{S} = J_I\frac{\partial\hat{\psi}_I}{\partial J_I}\mathbf{1} + 2\left(\bar{\mathbf{F}}_I^T\frac{\partial\bar{\psi}_I}{\partial\bar{\mathbf{C}}_I}\bar{\mathbf{F}}_I\right)^D \tag{14}$$

and the entropy function

$$s = \frac{\partial\psi_T}{\partial\theta} - J_I J_\theta^{-1}\frac{\partial\hat{\psi}_I}{\partial J_I}\frac{\partial J_\theta}{\partial\theta} \tag{15}$$

can be defined. The residual inequality is applied to define the evolution equation

$$\bar{\mathbf{D}}_K = \frac{1}{2\eta_K}\left(2\bar{\mathbf{C}}_I\frac{\partial\bar{\psi}_I}{\partial\bar{\mathbf{C}}_I} - \bar{\mathbf{F}}_K^T\frac{\partial\bar{\psi}_K}{\partial\bar{\mathbf{C}}_K}\bar{\mathbf{F}}_K\right)^D \tag{16}$$

Herein is η_K the viscosity parameter of the dashpot in Figure 1. Therefore the dissipation is given as

$$\delta_{loc} = \frac{1}{2\eta_K}\left\|\left(2\bar{\mathbf{C}}_I\frac{\partial\bar{\psi}_I}{\partial\bar{\mathbf{C}}_I} - \bar{\mathbf{F}}_K^T\frac{\partial\bar{\psi}_K}{\partial\bar{\mathbf{C}}_K}\bar{\mathbf{F}}_K\right)^D\right\|^2. \tag{17}$$

As long as the viscosity η_K has a positive value, the requirement from the Clausius Duhem inequality is fulfilled. A deformation dependent function can be defined to model the Mullins effect or Payne effect as suggested in [LionRendek and Lion2010].

For the numerical implementation Neo-Hook hyperelasticity is chosen

$$\hat{\psi}_I = \frac{\kappa}{2}(J_I - 1)^2 \tag{18}$$

$$\bar{\psi}_I = \frac{\mu_I}{2}(\bar{\mathbf{C}}_I : \mathbf{1} - 3) \tag{19}$$

$$\bar{\psi}_K = \frac{\mu_K}{2}(\bar{\mathbf{C}}_K : \mathbf{1} - 3) \tag{20}$$

Therefore the weighted Cauchy stress tensor is

$$\mathbf{S} = \kappa(J_I^2 - J_I)\mathbf{1} + \mu_I\left(\bar{\mathbf{F}}\bar{\mathbf{C}}_K^{-1}\bar{\mathbf{F}}^T\right)^D \tag{21}$$

and the evolution equation leads to time derivative of the internal variable

$$\dot{\bar{\mathbf{C}}}_K = \frac{1}{\eta_K}\left(\Delta\bar{\mathbf{T}}_K - \frac{1}{3}\left(\Delta\bar{\mathbf{T}}_K\bar{\mathbf{C}}_K^{-1} : \mathbf{1}\right)\bar{\mathbf{C}}_K\right) \tag{22}$$

with the delta stress tensor

$$\Delta\bar{\mathbf{T}}_K = \mu_I\bar{\mathbf{C}} - \mu_K\bar{\mathbf{C}}_K\bar{\mathbf{C}}_K \tag{23}$$

In case thermo-mechanical processes the dissipation is also needed to model self heating of visco-elastic deformation. The Neo-Hooke elasticity leads to the following dissipation function

$$\delta_{loc} = \frac{1}{2\eta_K}\left(\Delta\bar{\mathbf{T}}_K\bar{\mathbf{C}}_K^{-1}\right)^D : \left(\Delta\bar{\mathbf{T}}_K\bar{\mathbf{C}}_K^{-1}\right)^D \tag{24}$$

Euler forward integration is implemented for the numerical integration of Eq. (23), so that the variables are calculated with an explicit algorithm. To achieve a stable integration the time increment is limited to a small value.

4 ABAQUS IMPLEMENTATION

For the finite element implementation in Abaqus, the Zener model in Kelvin-Voigt representation is programmed in an user material subroutine (UMAT) using Fortran. The heat dissipation is considered in the model according to Eq. (25) and a transient thermal mechanical analysis is performed. To test the prediction capabilities of the model, the simulation results are compared with the experimental and simulation results from (Abdelmoniem & Yagimli 2024) for the same finite element model. The experimental setup is shown in Figure 2a.

(a) Experimental setup

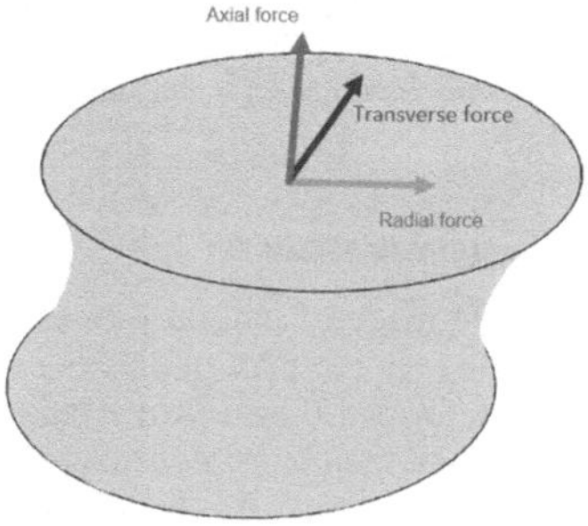

(b) Forces acting on the sample

Figure 2. Experimental setup and sample forces.

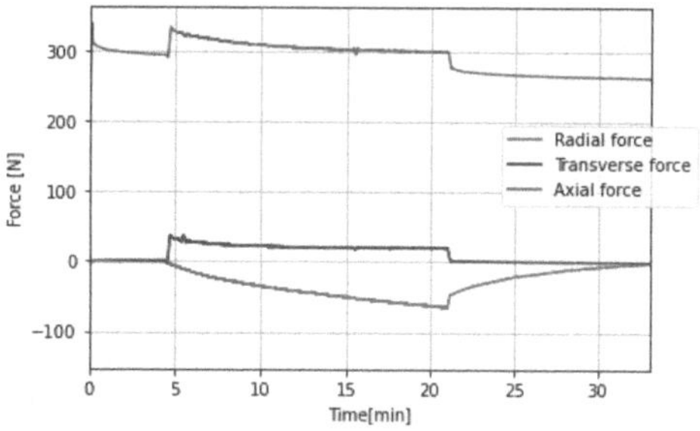

(a) Forces

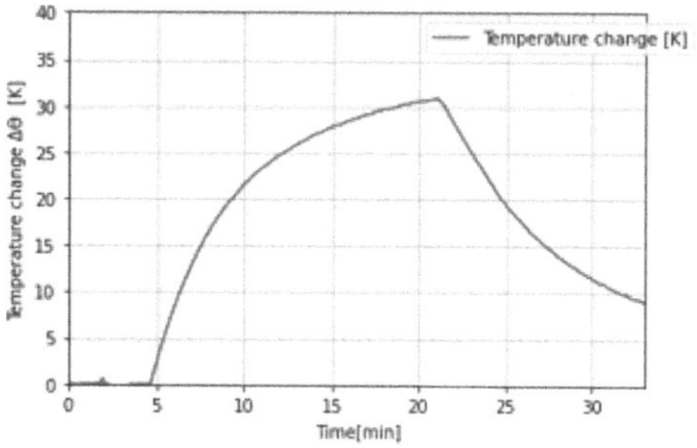

(b) Surface temperature

Figure 3. Experimental results.

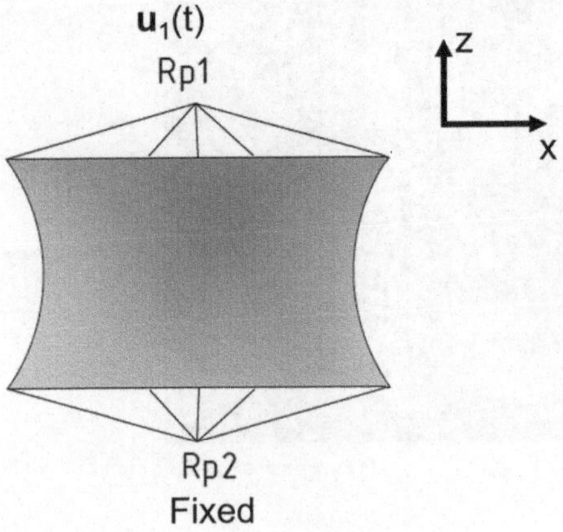

Figure 4. Boundary conditions for the simulation.

4.1 *Experimental procedure*

A rubber sample made of natural rubber is subjected to a shear loading. In the first step of the experiment, the sample is deformed 6 mm in the radial direction, the deformation is then kept constant so the radial forces relaxes with time. In the second step, the sample is rotated with a rotation frequency of 6 Hz and in the final step the rotation stops while keeping the deformation constant. A detailed demonstration of the set up Figure 2, sample geometry as well as the experimental procedure and the material properties is discussed in (Abdelmoniem & Yagimli 2024). The experimental results for the resulting forces and surface temperature development are shown in Figure 3.

4.2 *Simulation procedure*

The geometry of the rubber sample is discretised with 3377 C3D8HT elements (8 nodes linear brick, coupled temperature displacement hybrid elements with constant pressure). A displacement vector $u_1(t)$ is applied to the reference point (Rp_1) as in Figure 4 and according to equation 26. The angular frequency ω_1 is equal to $2\pi f$ where f is the rotation frequency and equal to 6Hz.

$$\mathbf{u_1(t)} = \begin{cases} \begin{bmatrix} 4u_1 t \\ 0 \\ 0 \end{bmatrix}, & 0 \leq t \leq 0.25\,\mathrm{s} \\ \begin{bmatrix} u_1 \\ 0 \\ 0 \end{bmatrix}, & 0.25 < t \leq 5\,\mathrm{s} \\ \begin{bmatrix} u_1 \cos(\omega_1 t) \\ u_1 \sin(\omega_1 t) \\ 0 \end{bmatrix}, & 5 < t \leq 15\,\mathrm{s} \\ \begin{bmatrix} u_1 \\ 0 \\ 0 \end{bmatrix}, & 15 < t \leq 20\,\mathrm{s} \end{cases} \tag{25}$$

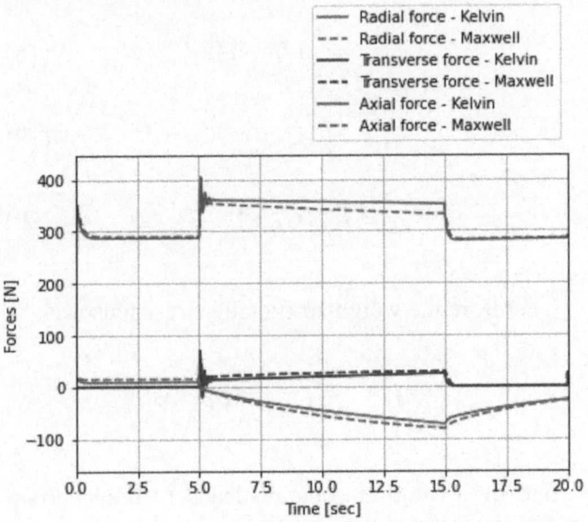

(a) Forces

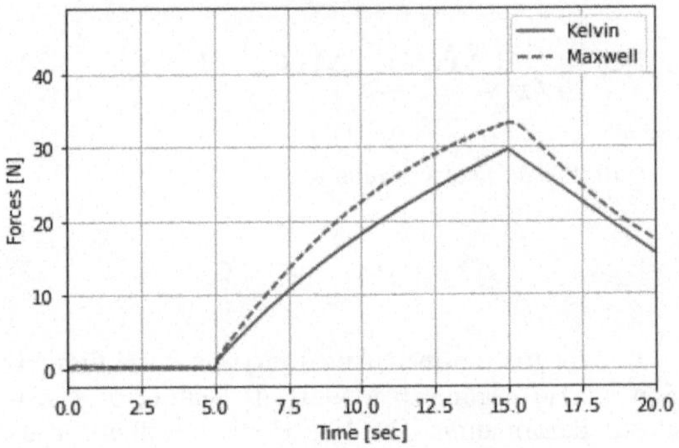

(b) Surface temperature

Figure 5. Simulation results.

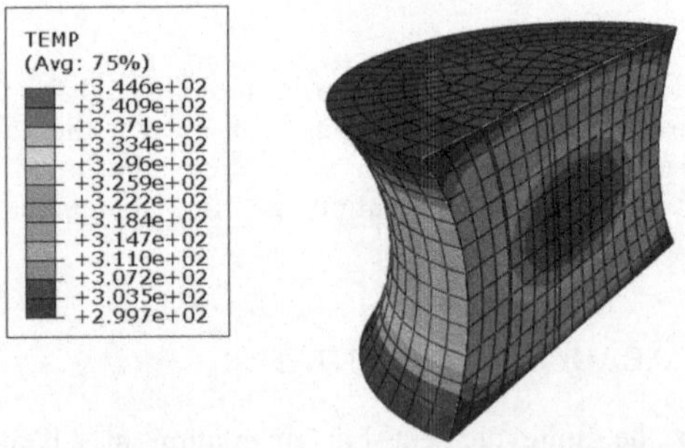

Figure 6. Temperature distribution along the sample cross section.

The Simulation results for the reaction forces and surface temperature development are compared with the results obtained in (Abdelmoniem & Yagimli 2024) using Maxwell model in Figure 5. The simulation results for the Kelvin-Voigt model agrees very good with the expermental and simulation results in (Abdelmoniem & Yagimli 2024). As the rotation step starts a force in transverse direction is induced due to viscoelasticity and dissipation. This transverse force does a mechanical work that is changed into heat energy which heats up the sample. The sample temperature increases till a steady state is reached. The self heating results in a thermal expansion of the sample, which is hindered due to the boundary conditions of the setup. That explains the increase of the axial force development with the temperature change. The overall increase in the surface temperature of the sample lies in the same range as experimentally observed (see Figure 3b). The temperature inside the sample was not measured, the predicition based on the applied thermal boundary conditions is shown in Figure 6.

5 CONCLUSION

The proposed modelling approach of finite viscoelasticity using Zenner model in Kelvin-Voigt representation is shown to provide good predictions of the viscoelastic and dissipation behaviour of the material under finite deformation. The model is based on the multiplicative split of the deformation gradient. The Simulation results shows very good agreement with the experimental ones, showcasing the model suitability to be adapted for the industrial applications, particularly in modeling the thermomechanical behavior of rubber components and their transfer function changes due to self-heating during operation. Future work will extend the model to incorporate multiple Kelvin-Voigt elements and additional phenomena, such as the Mullins effect and viscosity functions with deformation dependency. The implemented integration algorithm has to be improved to allow larger time increment.

REFERENCES

Abdelmoniem, M. & B. Yagimli (2021). Numerical studies on the heat dissipation process in elastomers under rotating loading direction. *Journal of Rubber Research 24*(5), 797–805.

Abdelmoniem, M. & B. Yagimli (2024). Self-heating in rubber components: Experimental studies and numerical analysis. *Journal of Rubber Research*.

Boyce, M. C., G. G. Weber, & D. M. Parks (1989). On the kinematics of finite strain plasticity. *Journal of the Mechanics and Physics of Solids 37*(5), 647–665.

Gent, A. (1960). Simple rotary dynamic testing machine. *British Journal of Applied Physics 11*(4), 165–167.

Haupt, P. & A. Lion (2002). On finite linear viscoelasticity of incompressible isotropic materials. *Acta Mechanica 159*(1-4), 87–124.

Lion, A. (1997). A physically based method to represent the thermo-mechanical behaviour of elastomers. *Acta Mechanica 123*(1-4), 1–25.

Lu, S. C. H. & K. S. Pister (1975). Decomposition of deformation and representation of the free energy function for isotropic thermoelastic solids. *Int. J. Solids Structures II*, 927–934.

Reese, S. & S. Govindjee (1998). A theory of finite viscoelasticity and numerical aspects. *International Journal of Solids and Structures 35*(26-27), 3455–3482.

Rendek, M. & A. Lion (2010). Strain induced transient effects of filler reinforced elastomers with respect to the Payne-Effect: Experiments and constitutive modelling. *ZAMM 90*(5), 436–458.

Simo, J. (1987). On a fully three-dimensional finite-strain viscoelastic damage model: Formulation and computational aspects. *Computer Methods in Applied Mechanics and Engineering 60*(2), 153–173.

Yagimli, B., A. Lion, & M. A. Abdelmoniem (2023). Analytical investigation of the finite viscoelastic model proposed by Simo: Critical review and a suggested modification. *Continuum Mechanics and Thermodynamics*.

Micro-structural investigations

Constitutive Models for Rubber XIII – Dal (Ed.)
© 2025 The Author(s), ISBN 978-1-032-85138-9

Development of a mesoscopic model for reinforcement in filled and strain-crystallizing elastomer networks

L. Tarrach & R. Hentschke
Fakultät für Mathematik und Naturwissenschaften, Bergische Universiät Wuppertal, Wuppertal, Germany

ABSTRACT: The model for strain-induced crystallization (SIC) in elastomers by Plagge & Hentschke [1] is combined with the model for generating filler network morphologies by Viktorova et al. [2] and extended by a rupture criterion, i.e. a critical free energy density, to analyze the rupture behavior of the model rubber networks. Here, 2D-networks have been studied. The mechanical behavior of non-crystallizing networks at small stretches shows the Payne effect, mainly caused by the reversible breaking of filler-filler bonds. The stress-stretch and crystallinity-stretch curves at larger deformations agree with experimental observations. Furthermore, the elongation at break dependence on filler content shows the expected decrease. For filler morphologies which contain larger agglomerates, the tensile strength decreases with increasing filler content, while it increases for randomly distributed or finely dispersed filler. Moreover, crystallizing networks possess smaller tensile strength than non-crystallizing networks. However, the networks do not fail beyond the percolation threshold. Altogether, the rupture behavior is only partly consistent with experimental observations.

1 INTRODUCTION

Filled rubber is widely used in industrial applications such as truck tires. In particular, Natural Rubber (NR) plays an outstanding role in this field due to its elasticity combined with its enormous tensile strength of up to 28MPa [3]. Wide Angle X-Ray Scattering (WAXS) experiments imply that its strength is related to its ability for strain-induced crystallization (SIC) [4]. SIC means that crystallites of linear dimension of around $10nm$ are reversibly formed at stretches of 300% or greater because polymer chains align along the stretching direction [5–7].

In addition, reinforcing fillers such as Carbon Black (CB) are added to the elastomer matrix to enhance the mechanical properties due to both hydrodynamic effects and surface properties [8]. Their impact depends on the properties of the filler itself [9–12]. The elastic modulus is elevated by filler at small deformations. Between $0.1 - 1\%$ to 10% of strain [12, 13], the Payne effect occurs which is indicated by a decrease of the modulus [8, 14]. Moreover, the stress in composites at fixed stretch is amplified also at higher deformations [15–20]. Since filler particles affect the local strain, the onset of SIC is shifted towards smaller stretches by the inclusion of filler into the rubber matrix [15–18]. In addition, the plateau of the stress-stretch curve, which is characteristic for SIC, becomes shorter, while the incline beyond it becomes more pronounced [15]. Furthermore, the tensile strength grows with increasing filler content until a maximum is reached [10, 11, 21–24]. The maximum depends on temperature [10], but coincides with the percolation threshold of the filler under certain conditions [23,24]. Beyond, the tensile strength decreases. The elongation at break decreases already for smaller filler content [21, 22, 25].

Although filler and SIC affect the mechanical behavior of rubber in a similar way, both effects must be quite different since filler particles restrict the mobility of the polymer chains [15], while SIC enhances the mobility of amorphous chain segments. Therefore, understanding of reinforcement requires for separating both aspects [19]. The aim of this work is to develop a model for an elastomer network which facilitates the simulation of both filler reinforcement and reinforcement by SIC as well as their interplay. For this purpose, the mechanical behavior of 2D-model networks is studied on the entire range of stretches until failure.

2 MODEL

For the simulation of SIC, the model proposed by Plagge & Hentschke [1] is applied. It is based on the theory for SIC suggested by Flory [26]. Detailed information on the model and the simulation parameters is provided in [27]. In order to extend the model to filled networks, the so-called morphology generator (MG) and the model for filler developed by Viktorova *etal.* [2] are

DOI: 10.1201/9781003516880-13

employed. Only essential aspects are summarized here. In the following, energies are given in units of $k_B T$ and lengths in units of the Kuhn length b (i.e. $b = 1$ and $k_B T = 1$).

An unfilled model elastomer network consists of links and nodes. While links represent polymer chains, nodes are chemical cross-links. For the initial configuration, N cross links are placed on a square lattice in 2D or a simple cubic lattice in 3D with lattice constant d_i. According to a given filler content ϕ, some of the cross links are replaced by filler at random. The MG [2], a nearest-neighbor site-exchange Monte Carlo (MC) algorithm, is applied to the configuration. For each node, a neighbor list is set up which contains all nodes located within a radius $r_{MG} = 2d_i$. At each MC step, a node is selected at random. Its position is interchanged with the position of a randomly chosen node from its neighbor list. If the Metropolis criterion is fulfilled, the new configuration is accepted, otherwise it is rejected:

$$\min(1, \exp[\Delta W]) \geq \text{random}(0, 1.0). \tag{1}$$

The change of the energy is determined by the polymer-filler interfaces:

$$\Delta W = \gamma_{pf} a \left(n_{pf}^{old} - n_{pf}^{new} \right), \tag{2}$$

where $a/(k_B T) = 0.421 m^2/mJ$ [2] is the effective contact area, n_{pf}^{old} and n_{pf}^{new} denote the number of interfaces in the old and the new configuration, respectively, and $\gamma_{pf} = 5.95 mJ/m^2$ [28,29] is the interface tension between polymer and filler. The latter is computed for NR filled with CB N339 from the disperse and the polar part of the contributing surface tensions according to OWRK-theory [30]. Since two-component systems are considered, the exact values of the surface tensions are of minor importance. It must only hold that $\gamma_{pf} > 0$. For counting the number of polymer-filler interfaces, only the neighbor lists of the two swapped nodes are considered. The algorithm favors the pairing of nodes of the same type.

Both the radius r_{MG} for setting up the neighbor lists and the number of MC steps determine the flocculation of filler. As depicted in Figure 1, the filler is finely dispersed in the initial random configuration. For up to 10. MC steps per node in 2D, connected agglomerates occur. Beyond that, the morphology becomes lumpy. Therefore, $10 \cdot N$MC steps are used in the following for 2D-networks. When ϕ exceeds the percolation threshold, a connected filler network forms which traverses the polymer.

In filled networks, links are replaced by filler-filler and polymer-filler bonds according to the filler morphology after an energy minimization which

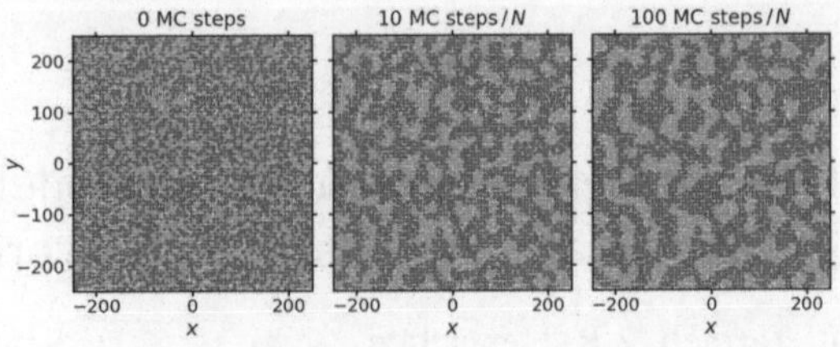

Figure 1. Evolution of the filler morphology in a network which consists of $N = 5041$ nodes and contains a fraction $\phi = 0.4$ of filler. The blue dots are cross links and the orange dots represent filler.

defines the equilibrium distances. The former bond type links two filler nodes while the latter connects a cross link and a filler. Afterwards, the network will be affinely deformed along the z-direction in small increments under volume conservation. The energy is minimized at each step by adjusting the positions of all nodes first and, then, considering energy minimization due to SIC for each link separately.

A link i is modeled as a Gaussian chain which consists of n_i Kuhn segments. Here, a single crystalline chain segment consists of l_c Kuhn segments and the crystalline segments are oriented along the stretching direction [1, 26]. The free energy of a semi-crystalline link i containing c_i crystalline segments is given as:

$$g(\vec{r}_i, c_i) = \frac{3}{2} \frac{(r_i - c_i l_c)^2}{n_i - c_i l_c} + c_i l_c \left(\theta_0 - \eta \sum_{i \neq j} c_j \exp\left(- \frac{|\vec{r}_{m,i} - \vec{r}_{m,j}|}{r_0} \right) \right), \tag{3}$$

where r_i is the end-to-end distance of the link, θ_0 is the energy which has to be overcome for SIC, η is a parameter for the interaction strength between links, r_0 is a characteristic distance and $\vec{r}_{m,i}$ denotes the position of the midpoint of link i [1]. The first term is because of the conformation of the chain according to the theory of rubber elasticity [31-33]. It is determined by the $n_i - c_i l_c$ amorphous chain segments. The last term accounts for the interaction of semi-crystalline links.

Filler-filler and polymer-filler bonds are modeled by harmonic potentials with variable spring constants:

$$u_i = \frac{1}{2} k_i \left(r_{i,0} - r_i \right)^2, \tag{4}$$

where $r_{i,0}$ denotes the equilibrium end-to-end distance of bond i. The spring constants are given as follows:

$$k_i = \begin{cases} 0, & r_i \geq R^{\mathrm{ff}} r_{i,0} \\ k_i^{\mathrm{ff}} = 5.0, & r_i < R^{\mathrm{ff}} r_{i,0} \end{cases} \text{filler} - \text{filler}, \quad (5)$$

$$\begin{cases} k_{i,\mathrm{weak}}^{\mathrm{pf}} = 1.0, & r_i \geq R^{\mathrm{pf}} r_{i,0} \\ k_i^{\mathrm{pf}} = 4.0, & r_i < R^{\mathrm{pf}} r_{i,0} \end{cases} \text{polymer} - \text{filler}, \quad (6)$$

where the cut-off radii are $R^{\mathrm{ff}} = 1.1$ and $R^{\mathrm{pf}} = 1.1$. The filler-filler bonds are defined as reversibly breakable depending on their extension. The values for the spring constants have been selected in order to agree with the phenomenology.

The forces $\vec{f_i}$ on the links and bonds are obtained by taking the derivative of Equations 3 and 4 respectively. At given stretch λ, they are related to the macroscopic mechanical properties of the entire network by the engineering stress σ [1]. Moreover, the crystallinity χ of the network can be computed at each stretch [1]. Only polymer chains contribute to χ because they are the only linkages which are able to crystallize.

3 RESULTS AND DISCUSSION

Subsequently, crystallizing (C) and non-crystallizing (NC) networks with variable filler content ϕ are simulated. The focus is on 2D-networks consisting of $N = 5041$ nodes. The remaining parameters are set as given in Tables 1 & 2 in [27]. Moreover, a detailed discussion of the stress-stretch and crystallinity-stretch behavior of unfilled model networks can be found in [1] and [27].

3.1 *Mechanical behavior of filled model networks at small deformations*

Experimentally, the mechanical behavior of rubber at small deformations is characterized by the dynamic moduli. Since the elastic modulus is proportional to the derivative of the stress-stretch curve, a numerical approximation of $\partial\sigma/\partial\lambda$ is employed. It is examined whether the model includes the Payne effect in filled networks. However instead of shearing, which is convenient for obtaining the moduli, a uniaxial stretching experiment is simulated. Therefore, the derivative of the stress for unfilled model networks is not constant but shows a decrease due to the geometry. It is also expected that the Payne effect in filled networks is indicated by a decrease of the derivative. In order to eliminate this superposition, the difference of the derivatives $(\partial\sigma/\partial\lambda)_\infty - \partial\sigma/\partial\lambda$ is analyzed. Here, $(\partial\sigma/\partial\lambda)_\infty$ is the derivative of the stress-stretch curve in the case of non-breakable filler-filler and polymer-filler bonds, i.e. $R^{\mathrm{ff}} = \infty$ and $R^{\mathrm{pf}} = \infty$, which only contains the systematic decrease. Since SIC is expected to start at stretches beyond the region in which the Payne effect occurs, NC-networks are simulated with variable filler content ϕ. Figure 2 shows plots of $\partial\sigma/\partial\lambda$ and of $(\partial\sigma/\partial\lambda)_\infty - \partial\sigma/\partial\lambda$ at small stretches λ for variable ϕ. For $\phi \leq 0.5$ which is below or at the percolation threshold, the data is approximately constant. For $\phi \geq 0.6$, the difference of the derivatives increases. The effect enhances with increasing filler content. Additionally, the onset of the increase shifts towards smaller stretches. In particular for $\phi = 0.9$, the difference tends to level off in a plateau beyond the increase. The increase indicates the occurrence of the Payne effect. Its stretch range, particularly for highly filled networks, agrees with the expectations [12, 13]. As depicted in Figure 3, varying the cut-off radii for polymer-filler and filler-filler bonds implies that this effect mainly arises from rupture of filler-filler bonds.

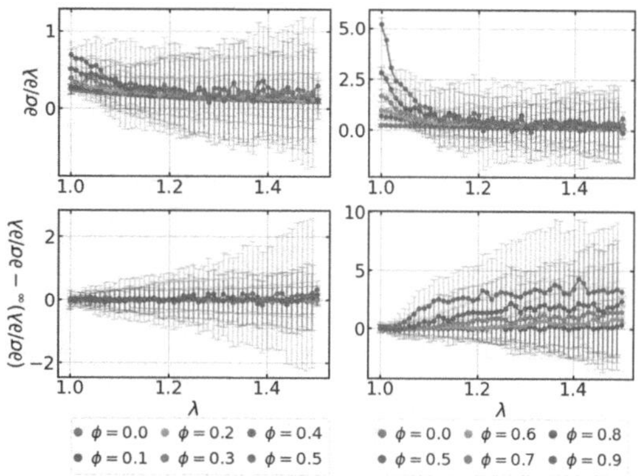

Figure 2. Plot of the derivative $\partial\sigma/\partial\lambda$ and of the difference of the derivatives $(\partial\sigma/\partial\lambda)_\infty - \partial\sigma/\partial\lambda$ versus λ for variable ϕ. The data points are connected for the eye, but this does not have any physical meaning.

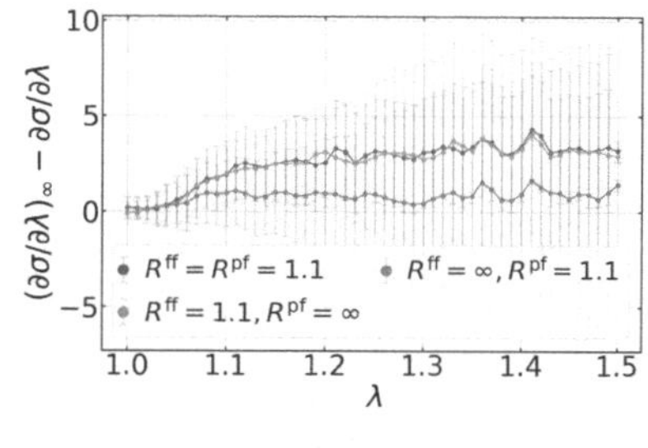

Figure 3. Plot of the difference of the derivatives $(\partial\sigma/\partial\lambda)_\infty - \partial\sigma/\partial\lambda$ versus λ with $\phi = 0.9$ and variable cut off radii R^{ff} and R^{pf}. The data points are connected for the eye, but this does not have any physical meaning.

3.2 *Mechanical behavior of the model networks at higher deformations*

For the sake of completeness, we summarize the essential results already presented in [34]. As depicted in Figure 4, the stress-stretch curves for NC-networks show stress-amplification in the range of intermediate and large stretches. The same holds for C-networks and is expected from experimental observations [15–20]. Figure 5 shows that the onset stretch for SIC shifts towards smaller stretches with increasing ϕ. In the unfilled case at the onset stretch of SIC, a distinct plateau forms in the stress-stretch curve which leads to softening. The plateau vanishes with increasing ϕ. Both the stress-stretch and the crystallinity-stretch curve show the for SIC characteristic hysteresis which tightens as the filler content increases. This agrees with experimental observations [15–18]. However, a steep incline beyond the plateau cannot be observed. For the unfilled network or small ϕ, the maximum stretch $\lambda_{max} = 5.0$ is too small to capture the plateau and, for higher filler content, the linear forces of filler-filler and polymer-filler bonds may dominate.

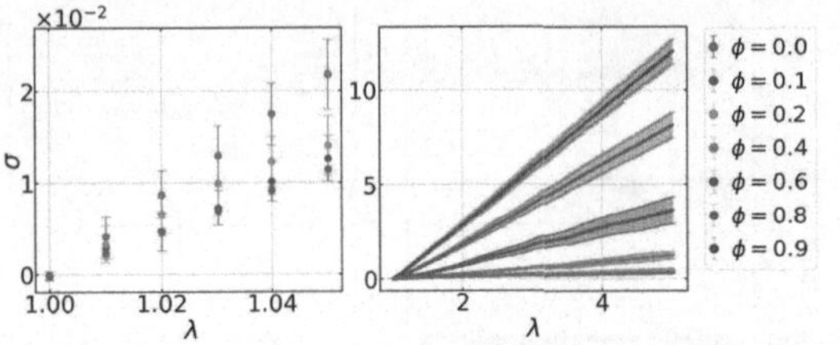

Figure 4. Stress-stretch curves of unfilled NC-networks with variable filler content ϕ. The left panel shows an extract of the right panel at small stretches.

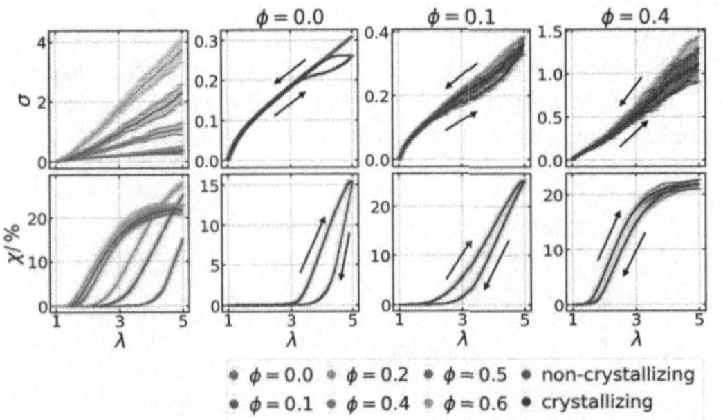

Figure 5. Stress-stretch and crystallinity-stretch curves for variable filler content ϕ. While the left panel only considers the stretching process in C-networks, the others represent entire strain cycles for C- and NC-networks. The arrows indicate stretching and contraction.

3.3 *Rupture behavior of the model networks*

In order to investigate the tensile behavior of the model networks, a rupture criterion for the links is introduced. It is defined by a critical free energy density $(g/n)_{crit}$, i.e. the free energy of a link per Kuhn segment. If the energy density of a link after energy minimization at fixed stretch exceeds this threshold value, the link breaks. A reasonable range for $(g/n)_{crit}$ has been obtained by comparison of free energy density histograms of an unfilled C- and of an unfilled NC-network. Due to SIC, the energies in the C-network are restricted to smaller values than those in the NC-network. Thus, $(g/n)_{crit}$ must be chosen from the range which is predominantly reached in the NC-network, but not in the C-network. For the definite determination, the fraction of broken links L_b/L, the tensile strength σ_t and the elongation at break λ_t dependent on $(g/n)_{crit}$, which is situated in the aforementioned range, are considered. For this purpose, unfilled C- and NC-networks are stretched up to $\lambda_{max} = 5.0$. The tensile strength is obtained as the maximum stress reached under the assumption that the stress monotonically increases until failure. The elongation at break is the stretch belonging to the tensile strength. A network is called stable if none of its links breaks during the deformation. A network is said to fail if it is destroyed by the formation of large holes such that the stress decreases towards zero.

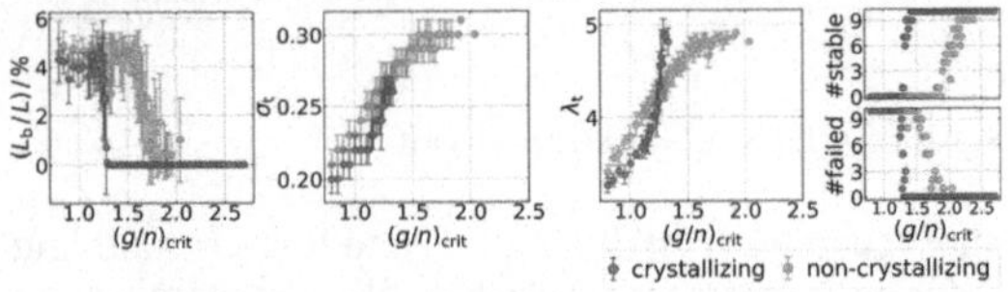

Figure 6. Plot of the fraction of broken links L_b/L, tensile strength σ_t, elongation at break λ_t, number of stable networks and number of networks which failed versus the critical energy density $(g/n)_{crit}$ for unfilled networks. For each $(g/n)_{crit}$, 10 networks have been simulated independently. The quantities L_b/L, σ_t and λ_t are computed as averages from the data for the networks which failed.

Figure 6 shows that a range of $(g/n)_{crit}$ occurs in which C-networks are stable, while NC-networks still break. For the purpose of doing tensile tests, rather both kind of networks must fail. Thus, the critical energy density is set to $(g/n)_{crit} = 1.25$. This value is situated at the falling edge of L_b/L for the C-case and in the region where tensile strength and elongation at break for the C-networks become larger than for the NC-networks.

In Figure 7a, L_b/L, σ_t, λ_t, the number of stable and the number of failed networks are plotted versus ϕ as results of a tensile test with $(g/n)_{crit} = 1.25$ and $\lambda_{max} = 5.0$. The tensile strength for C-networks is below that for the corresponding NC-networks until the percolation threshold $\phi \approx 0.5$ is reached. This is contradictory to the expectations, but may still originate from the choice of $(g/n)_{crit}$. At the percolation threshold, one of the NC-networks fails, while none of the C-ones does. Although σ_t should increase with increasing but small ϕ, it decreases until $\phi = 0.5$. An explanation may be that $10 \cdot N$MC steps generate badly dispersed filler. Simultaneously, λ_t becomes

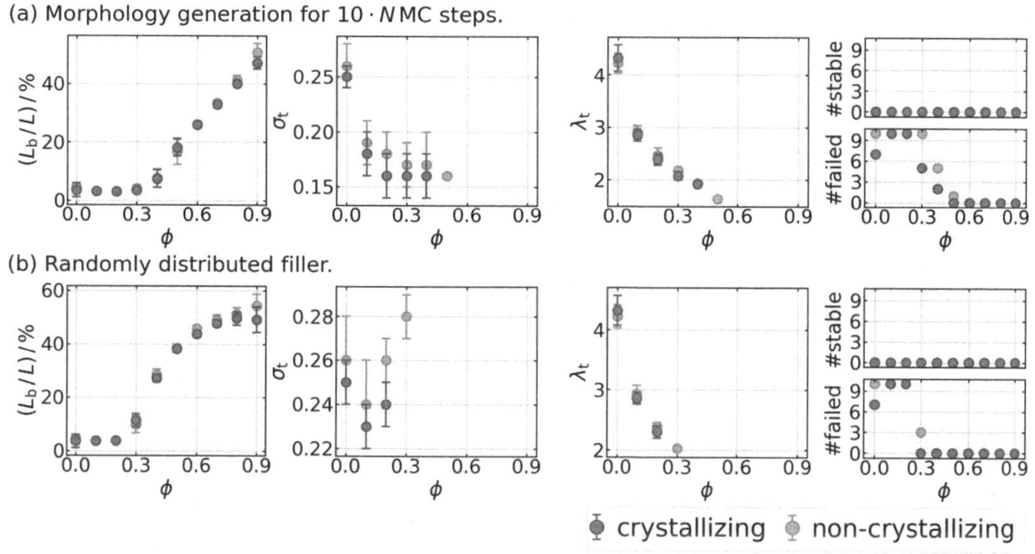

Figure 7. Plot of the fraction of broken links L_b/L, tensile strength σ_t, elongation at break λ_t, number of stable networks and number of networks which failed versus the filler content ϕ for $(g/n)_{crit} = 1.25$ and different filler morphologies. For each ϕ, 10 networks have been simulated independently. The quantities L_b/L, σ_t and λ_t are computed as averages from the data for the networks which failed.

smaller which, nevertheless, is consistent with experiments [21, 22, 25]. The behavior of the model networks may originate from an increasing number of polymer-filler and filler-filler bonds which appear to act as actuators for hole formation. These bonds weaken or break already at small deformations if they are aligned along the stretching direction. If links are located at the boundaries of a hole or adjacent to highly extended polymer-filler bonds and oriented along the stretching direction, the links are stretched even more such that the hole grows due to their rupture. Beyond $\phi = 0.5$, none of the networks fails, but they also do not become stable. Considering L_b/L, the reason might be that the filler agglomerates, and in particular bonds between them and the polymer, hold the network together. By construction, polymer-filler bonds only weaken during stretching but cannot break. Additionally, the applied maximum stretch may not be sufficiently high to generate failure. Nevertheless, the expected maximum and subsequent decrease of σ_t cannot be observed [10, 11, 21–23].

According to the dependence of the tensile behavior on the amount of the different linkages, there must also be a dependence on the filler morphology itself. Hence, simulations of networks with randomly distributed filler are done analogous to the previous. The results are presented in Figure 7b. The tensile strength of C-networks is still smaller than that of NC-networks. In addition, the behavior beyond the percolation threshold is not affected by the random filler morphology. In contrast to the previous case, the C-networks do not fail anymore already for $\phi = 0.3$. In agreement with experiments, λ_t declines, while σ_t shows an increase for filled networks. This observation is attributed to the filler structure. The filler in this random distribution is finely dispersed instead of building agglomerates. Therefore, polymer-filler bonds occur more often than in the previous morphology and compared to filler-filler bonds. Consequently, there is a larger number of bonds which could hold the network together and simultaneously a smaller number of bonds which already break at small stretches. Moreover since this model is coarse-grained, this implies that the filler nodes model larger filler particles than thought at first glance. Nevertheless, σ_t for the unfilled case is higher than for poorly filled networks. This may be related to the strain-amplification effect of the inclusion of filler.

4 CONCLUSIONS

A mesoscopic model which allows the simulation of filled and strain-crystallizing elastomer networks has been developed and 2D-networks have been studied. The mechanical behavior of unfilled NC-networks at small stretches features the Payne effect. Here, this phenomenon is generated by rupture of filler-filler bonds. At larger deformations, the stress-stretch curves of both C- and NC-networks and the crystallinity-stretch curves agree with experimental observations. By defining a critical energy density for the links, the rupture behavior has been investigated. In contrast to the expectation of an enhanced tensile strength due to SIC, it is smaller for C-networks than for NC-networks. Thus the critical energy density must be increased in the future. Furthermore, the tensile strength of networks containing filler agglomerates decreases with increasing filler content. This in contrast to experiments, but, as pointed out above, it may reflect the excessive 'lumpyness' of the filler morphology. Although the tensile strength is decreased compared to the unfilled case, it increases for randomly distributed filler. In general, the elongation at break decreases with increasing filler content as expected. However, none of the networks fails beyond the percolation threshold. Thus, the overall behavior of the tensile strength dependent on filler content does not yet agree with experiments. Prospectively, the impact of the filler morphology on the tensile strength must be studied further as well as the impact of SIC.

REFERENCES

1. Plagge, J. & Hentschke, R. Microphase Separation in Strain-Crystallizing Rubber. *Macromolecules* **54**, 5629–5635 (2021).
2. Viktorova, M. et al. Mesoscopic Model for the Simulation of Dynamic Mechanical Properties of Filled Elastomers: Model Construction and Parameterization. *ACS Applied Polymer Materials* **2**, 5521–5532 (2020).
3. R¨othemeyer, F. & Sommer, F. *Kautschuktechnologie: Werkstoffe - Verarbeitung - Produkte* (Carl Hanser Verlag Gmbh & Co. KG, 2013).
4. Huneau, B. STRAIN-INDUCED CRYSTALLIZATION OF NATURAL RUBBER: A REVIEW OF X-RAY DIFFRACTION INVESTIGATIONS. *Rubber Chemistry and Technology* **84**, 425–452 (2011).
5. Trabelsi, S., Albouy, P.-A. & Rault, J. Crystallization and Melting Processes in Vulcanized Stretched Natural Rubber. *Macromolecules* **36**, 7624–7639 (2003).
6. Chenal, J.-M. et al. Molecular weight between physical entanglements in natural rubber: A critical parameter during strain-induced crystallization. *Polymer* **48**, 1042–1046 (2007).
7. Candau, N. et al. Strain-Induced Crystallization of Natural Rubber and Cross-Link Densities Heterogeneities. *Macromolecules* **47**, 5815–5824 (2014).
8. Wang, M.-J. & Morris, M. *Rubber Reinforcement with Particulate Fillers* (Carl Hanser Verlag GmbH & Co. KG, 2021).
9. Payne, A. R. & Whittaker, R. E. Low Strain Dynamic Properties of Filled Rubbers. *Rubber Chemistry and Technology* **44**, 440–478 (May 1971).
10. Sambrook, R.W. Influence of Temperature on the Tensile Strength of Carbon Filled Vulcanizates. *Rubber Chemistry and Technology* **44**, 728–743 (July 1971).
11. Boonstra, B. Role of particulate fillers in elastomer reinforcement: a review. *Polymer* **20**, 691–704 (1979).
12. Tunnicliffe, L. B. et al. Flocculation and Viscoelastic Behaviour in Carbon Black-Filled Natural Rubber. *Macromolecular Materials and Engineering* **299**, 1474–1483 (2014).
13. St¨ockelhuber, K. W. et al. Impact of Filler Surface Modification on Large Scale Mechanics of Styrene Butadiene/ Silica Rubber Composites. *Macromolecules* **44**, 4366–4381 (2011).
14. Hentschke, R. The Payne effect revisited. *Express Polymer Letters* **11**, 278–292 (Apr. 2017).
15. Trabelsi, S., Albouy, P.-A. & Rault, J. Effective Local Deformation in Stretched Filled Rubber. *Macromolecules* **36**, 9093–9099 (2003).
16. Poompradub, S. et al. Mechanism of strain-induced crystallization in filled and unfilled natural rubber vulcanizates. *Journal of Applied Physics* 97, 103529 (2005).
17. Chenal, J.-M. et al. Parameters governing strain induced crystallization in filled natural rubber. *Polymer* **48**, 6893–6901 (2007).
18. Candau, N. et al. Compared abilities of filled and unfilled natural rubbers to crystallize in a large strain rate domain. *Composites Science and Technology* **108**, 9–15 (2015).
19. Rault, J. et al. Stress-Induced Crystallization and Reinforcement in Filled Natural Rubbers: 2H NMR Study. *Macromolecules* **39**, 8356–8368 (2006).
20. Spratte, T. et al. Investigation of strain-induced crystallization of carbon black and silica filled natural rubber composites based on mechanical and temperature measurements. *Polymer* **115**, 12–20 (2017).
21. Sadequl, A. M., Poh, B. T. & Ishiaku, U. S. Effect of Filler Loading on the Mechanical Properties of Epoxidized Natural Rubber (ENR 25) Compared with Natural Rubber (SMR L). *International Journal of Polymeric Materials and Polymeric Biomaterials* **43**, 261–278 (1999).
22. Savetlana, S. et al. The effect of carbon black loading and structure on tensile property of natural rubber composite. *IOP Conference Series: Materials Science and Engineering* **223**, 012009 (July 2017).
23. Azura, A. R., Ghazali, S. & Mariatti, M. Effects of the filler loading and aging time on the mechanical and electrical conductivity properties of carbon black filled natural rubber. *Journal of Applied Polymer Science* **110**, 747–752 (2008).
24. Plagge, J. Mastering of Filled Rubber Strength beyond WLF: Competition of Temperature, Time, Crack Deflection and Bond Breaking. *Polymers* **14** (2022).
25. Azura, A. & Leow, S. Effect of carbon black loading on mechanical, conductivity and ageing properties of Natural Rubber composites. *Materials Today: Proceedings* **17**, 1056–1063 (2019).
26. Flory, P. J. Thermodynamics of Crystallization in High Polymers. I. Crystallization Induced by Stretching. *The Journal of Chemical Physics* **15**, 397–408 (1947).
27. Tarrach, L., Hentschke, R. & Weilert, J. Modelling Study of Reinforcement and Rupture in Strain-Crystallizing Elastomer Networks. *Kautschuk Gummi Kunststoffe KGK*, 62–71 (3 2023).
28. St¨ockelhuber, K. W. et al. Contribution of physicochemical properties of interfaces on dispersibility, adhesion and flocculation of filler particles in rubber. *Polymer* **51**, 1954–1963 (2010).
29. Klat, D., Karimi-Varzaneh, H. A. & Lacayo-Pineda, J. Phase Morphology of NR/SBR Blends: Effect of Curing Temperature and Curing Time. *Polymers* **10** (2018).
30. Owens, D. K. & Wendt, R. C. Estimation of the surface free energy of polymers. *Journal of Applied Polymer Science* **13**, 1741–1747 (1969).
31. Rubinstein, M. & Colby, R. H. *Polymer Physics* (Oxford University Press, 2003).
32. Hentschke, R. *Thermodynamics* (Springer, Heidelberg, 2014).
33. Treloar, L. R. G. *The Physics of Rubber Elasticity* (Oxford University Press, 1975).
34. Tarrach, L. & Hentschke, R. *Model-Based Approach to Reinforcement by Filler and Rupture in Strain- Crystallizing Elastomer Networks* Presented at RubberCon2023, Edinburgh, UK. 2023.

Constitutive Models for Rubber XIII – Dal (Ed.)
© 2025 The Author(s), ISBN 978-1-032-85138-9

Predicting the effect of carbon black blends on compound properties by using a micro-structural simulation program

H. Wulf & J. Ihlemann
Department of Solid Mechanics, Chemnitz University of Technology

ABSTRACT: The amount, the size and the surface characteristics of the filler particles in a rubber compound influence hysteresis, permanent set and softening in a nonlinear way. To attain the desired combination of material properties, a blend of different filler types can be employed, but finding the optimal combination is difficult. Here, a method is proposed, that provides predictions concerning the mechanical properties of compounds with filler blends by simulation. To this end, the underlying abstract micro-structural model is extended to multiple filler types. A modification of the model parameters allows simulating a space of virtual materials, which is spanned by different ratios of the quantity of the individual filler types. A mapping algorithm translates this into predictions concerning the mechanical behavior for each point within the design space. It is demonstrated, that this approach can anticipate certain nonlinearities of the material properties. Comparisons with verification measurements demonstrate that typical properties like stiffness and hysteresis can be well predicted.

1 INTRODUCTION

Rubber materials are an extremely broad material class due to the immense number of combinations of ingredients and productions parameters. The main aspects of variation are polymer type, fillers, cross-linking, additives, mixing and the vulcanization process. All these aspects influence the ultimate properties of the material in a complex and interlocked manner. Therefore, fine-tuning the material for a specific application is difficult. Usually, the space of materials is explored by producing and testing material prototypes. In (Vijayabaskar et al. 2006), (Trebar et al. 2007) and (Sresungsuwan and Hansupalak 2013)such systematic measurements are extended to a model, that can predict the properties of an arbitrary material within the considered space of materials. Here, a material space spanned by mixture of different filler types is investigated. Such hybrid fillers further increase the range of possible rubber materials and can also increase the range of obtainable property combinations (for a review see Zainathul Akhmar Salim Abdul Salim and Ismail 2018)). Whereas the cited works on material prediction use linear interpolations or neural networks, this work employs a microstructurally motivated simulation program introduced in (Wulf & Ihlemann 2015) to generate the predictions. The core idea is, that due to the representation of the microstructure, an interpolation in the parameter space of the program will be more reliable and will yield good predictions even with a very small set of input data.

2 EXPERIMENTAL CHARACTERIZATION

In order to obtain the experimental data for the study, seven different material variations were produced and subjected to mechanical tests. All tested materials were CB-filled and sulfur-cured natural rubbers. The ingredients and the processing parameters for all materials are identical, except for the fillers. In order to study the influence of the filler type in isolation, the overall amount of filler is also identical at 45 pph (parts per hundred) for each material. The only variation is the filler type. Here, three different fillers types with different particle size and surface properties are employed, which are denoted by just CB1, CB2 and CB3. The most significant difference is the size with CB1 being the most fine-grained and CB3 the most coarse-grained filler. The first three compounds (named Exp1, Exp2 and Exp3) are the "pure filler" materials, filled with CB1, CB2 and CB3 up to the chosen pph value, respectively. The next three compounds are "hybrid filler" materials, obtained by mixing in two different filler types together, each with half the pph of the pure filler materials. This yields the materials Exp4 (CB1/CB2), Exp5 (CB1/CB3) and Exp6 (CB2/CB3). The final material Exp7 is obtained by mixing in all three filler types in combination, each with one third of the overall pph. Overall, the design space of materials can be best visualized in a ternary plot (Figure 1). Each point in the plot can be identified by its barycentric coordinates, which are the relative share of the three filler types.

DOI: 10.1201/9781003516880-14

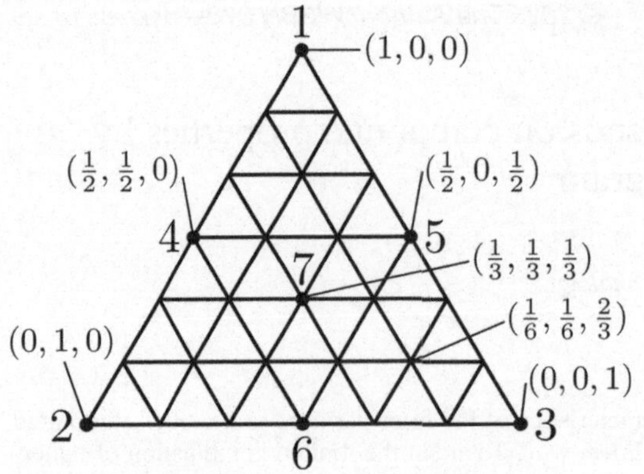

Figure 1. Setup of all the tested materials in a ternary plot. Additionally, the barycentric coordinates for selected points are given.

For the mechanical characterization, simple shear tests with load levels of 0.2, 0.4, 0.8, 1.0, 1.2 and 5 loading cycles per load level were conducted. Only the last cycle per load level, which is assumed to be stationary, is relevant for the further investigation. The results for the experiments Exp1, Exp2 and Exp3 (the pure filler materials) are shown in Figure 2. It can be observed that the material filled with CB3 is significantly softer. There is little difference between Exp1 and Exp2 concerning stiffness, but a slightly larger hysteresis for Exp1, especially for the larger strain cycles.

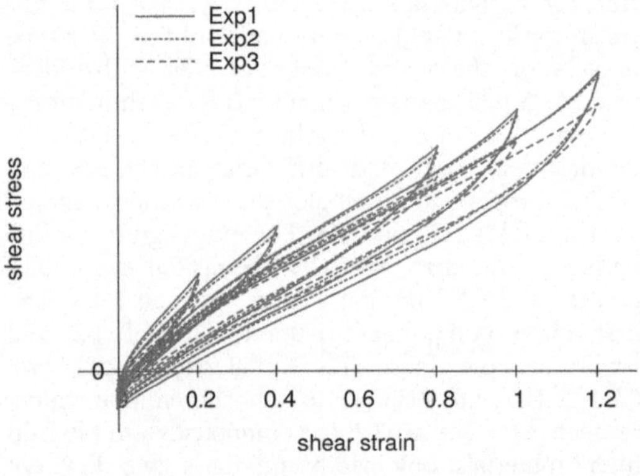

Figure 2. Mechanical behavior of the materials Exp0, Exp1 and Exp2.

Materials Exp4, Exp5 and Exp6 can be considered a middle point between two of the first three materials each (see 1). Therefore, they are compared with the corresponding pure materials in Figures 3, 4 and 5.

For Exp4 (CB1/CB2 hybrid) the properties do not deviate much from Exp1 and Exp2. For Exp5 (CB1/CB3 hybrid) and Exp6 (CB2/CB3 hybrid) however, it becomes obvious that the material properties are not linear along the edges of the material design space. The expected result would be stress levels for

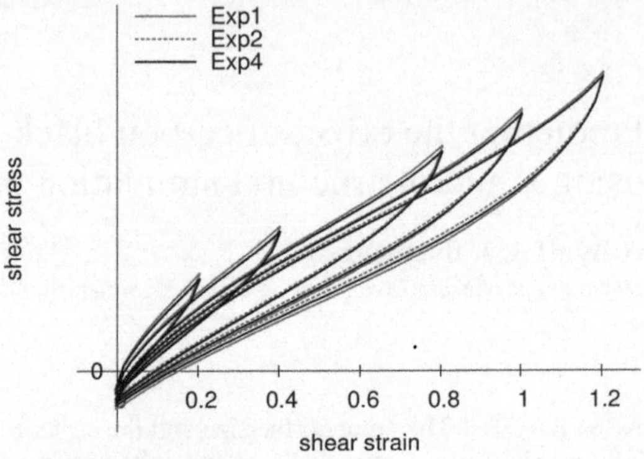

Figure 3. Mechanical behavior of the material Exp4 in comparison with Exp1, Exp2.

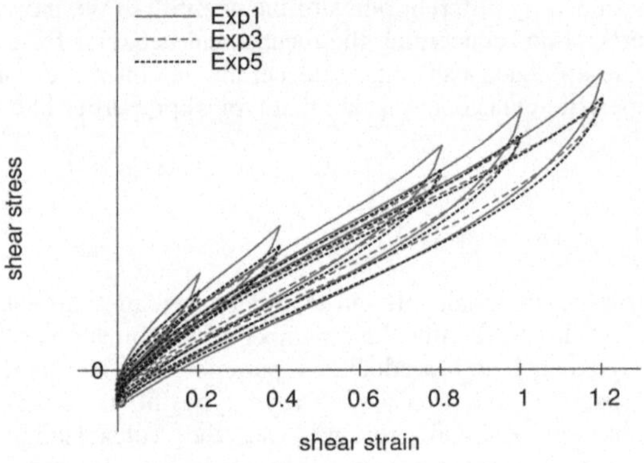

Figure 4. Mechanical behavior of the material Exp5 in comparison with Exp1, Exp3.

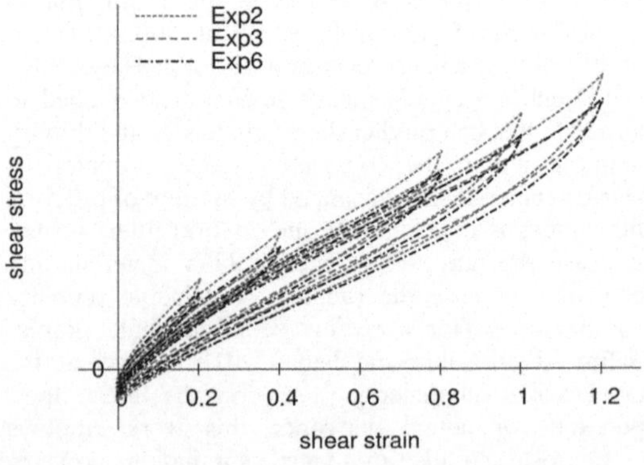

Figure 5. Mechanical behavior of the material Exp6 in comparison with Exp2, Exp3.

Exp5/Exp6, which are between Exp3 and Exp1/Exp2 respectively. However, the experimental data shows stress levels for Exp5 and Exp6 almost equal to Exp3 or even below (especially on the unloading paths). It appears that mix of two different fillers types is less reinforcing than each pure filler type.

A possible micromechanical explanation is that same type filler particles have better compatibility concerning their active surfaces and therefore can establish more and stronger bonds than different type filler particles.

Figure 6 shows the test results for Exp7 (the mix of all three filler types) in comparison with the three pure filler experiments. These results are definitely the most surprising. The stiffness of the triple filler blend material is significantly lower than each of the pure filler materials, despite having the same overall filler content. The general effect could be expected from the aforementioned micromechanical explanation, but the magnitude is still astonishing.

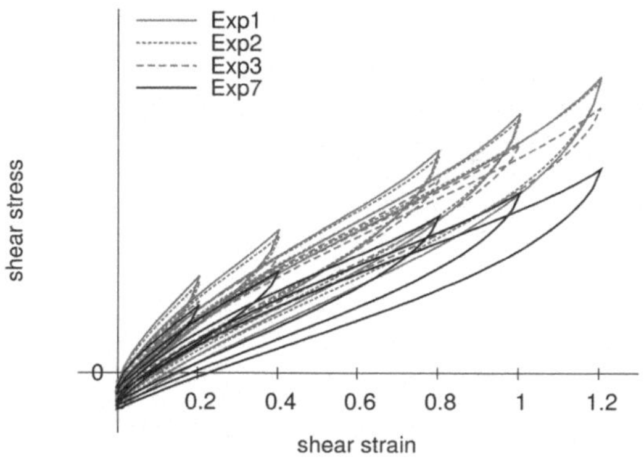

Figure 6. Mechanical behavior of the material Exp7 in comparison with Exp1, Exp2, Exp3.

3 SIMULATION MODEL

The simulation model and the theory behind it has been presented several time in literature, for instance Wulf & Ihlemann (2015). Hence, it is introduced very briefly here. The core idea is that various aspects of the material behavior can be explained by an inhomogeneous distribution of the density of weak physical links. The inhomogeneity emerges from a self-organization process. The central step for the implementation is to simplify any interactions as much as possible as long as the assumed self-organization process is still feasible. This is in stark contrast to methods like molecular dynamics, which aim at a precise representation of molecular interactions, but lead to excessive computational effort and very small simulated timeframes. In this work, molecular entities are abstracted and grouped to model elements, which rather represent basic properties of the micro-structure than micro-structural components. This means that the model elements cannot be directly identified as molecular entities and yields a more difficult interpretation of the model and its parameters. However, it enables significantly reduced simulation times.

The first model element is called strings and represents the flexibility and high number of configurations of the polymer chains. Each string consists of n_{seg} segments of equal length, which are beam elements connected at nodes at a random angle. A total number of n_n strings is generated and distributed equally in the simulation volume.

Next, the strings are connected by a total number of n_c crosslink connections. These are very stiff beam elements, that are inserted at random between the strings with a prescribed average length l_c. At this point, the model is purely elastic, but generates a variety of movements on deformation.

The final model element represents the properties of the physical links. First of all, a certain percentage p_a is marked as active nodes. Only these active nodes can establish physical links. Then, a critical distance l_{crit} is defined, which denotes the maximum length of the model element. During the deformation process, the distances between all active nodes is tracked. As soon as the distance between two such nodes drops below the critical distance, a physical link is inserted between the two nodes. These links are just linear springs with a stiffness of c_l. Any physical link that is extended above the critical length is removed. The stress response of the model changes abruptly, whenever large reorganization of links occur. In order to obtain a smooth and representative behavior, an average over a large number of model samples with identical parameters, but different random generation is computed.

Note that the model does not contain an extra element for the filler. Instead, the filler is acknowledged by its role as provider of active surface area as attachment sites for physical links. Hence, any active node in the model should interpreted as an area with a sufficient amount of filler present. The percentage of active nodes p_a is therefore directly linked to overall amount of filler in the material. The other two parameters of the physical links l_{crit} and c_l represent some averaged properties of the filler-filler and filler-polymer interactions.

The model can be easily extended to allow for a blend of different filler types in the same material. In the base model, each node only has the two stated of being inactive (presented by 0) or active (1). Now, each active node may be assigned a positive integer i from the range $1 \ldots n_f$ (with n_f different filler types) indicating, that it is active due to the presence of filler i. Obviously, the underlying assumption is that the dispersion of the filler is such that each node can be identified with one filler type exclusively. Instead of a single activity probability p_a, there is an activity probability $p_{a(i)}$ per filler type. As they are mutually exclusive, the sum of the $p_{a(i)}$ must be less than 1 and represents the probability, that a node is active.

When tracking the distance between all active nodes, the filler type assigned to each node must be considered. For a specific pair of nodes, these types can be denoted by i and j. It makes sense to define a separate parameter set $l_{crit(ij)}$ and $c_{l(ij)}$ for each combination i,j. The parameter sets with $i = j$ specify the behavior of the pure filler and its interaction with itself, whereas the parameter sets with $i \neq j$ specify the interaction of different filler types.

4 PROPERTY PREDICTION

4.1 *Parameter adaption*

The concept of the work is that the pure filler experiments are provided as knowledge base, whereas the behavior of the hybrid filler materials is to be predicted. In general the materials for Exp1, Exp2 and Exp3 are different and require different parameter sets. Especially for phenomenological material models, completely separate parameter sets would have to be identified. For the presented model with microstructural interpretation however, common parameter can be used for all string and crosslink properties, as they are not connected to the filler. The filler-related parameters are the activity probability $p_{a(i)}$, the critical length $l_{crit(ii)}$ and link stiffness $c_{l(ii)}$, which are identified per material $i = 1 \ldots 3$. Overall, this yields model 14 parameters to identify. Another challenge for the identification process is the fact, that the simulated stress response is prone to some systematic errors. For instance, the permanent set is usually too large. This is remarkably hard to avoid, because all nonlinear properties are not introduced by carefully designed mathematical terms in a material model, but instead emerge on their own from the model dynamics. Therefore, the model is good at predicting the trends of material properties and their relationships, but not that good at directly fitting stress-strain curves. Instead, the model parameters are fit to a dimensionless hysteresis measure H_i and a dimensionless softening measure S_i as defined in [IhlemannWulf and Ihlemann2019]. Furthermore, each stationary cycle is expressed by the minimum stress T_{yx}^{min}, maximum stress T_{yx}^{max} and a dimensionless shape function $g(\xi) \in [0, 1]$ as:

$$T_{yx}(\xi) = T_{yx}^{min} + (T_{yx}^{max} - T_{yx}^{min})g(\xi). \qquad (1)$$

At the end, the objective function for the parameter adaption is squared error sum that incorporates the softening and hysteresis measures, the maximum stress values and the shape functions $g(\xi)$ for all stationary cycles. For the optimization process, a Downhill-Simplex method is used. Gradient-based methods cannot be used due to the random number-controlled model generation and the discontinuous model behavior whenever links are inserted or removed. The main interesting result are the physical link parameters, which are separate for each material. The results are shown in table:

	Exp1 (CB1)	Exp2 (CB2)	Exp3 (CB3)
$p_{a(i)}$	0.5761	0.4633	0.3339
$l_{crit(ii)}$	0.2743	0.3231	0.2751
$c_{l(ii)}$	0.3154	0.2454	0.2683

The most obvious trend is the reduction of the activity from CB1 to CB3, which corresponds to the fact that smaller the particle size leads to larger active surface area at similar amounts of filler. For CB2 the increased interaction length partly balances out the reduced activity, whereas CB3 is just least reinforcing as easy to see from the model parameters alone. In order to define the complete model behavior for hybrid filler materials, the interaction parameters $l_{crit(ij)}$ and $c_{l(ij)}$ for each combination $i \neq j$ are still missing. It is impossible to identify these from the first three experiments. Of course, they could be identified from the experiments Exp4, Exp5 and Exp6 - but this leaves almost no experimental data for verification of the procedure. This would defeat the main purpose of the paper, which is generating property predictions for materials, which have not been tested so far. As a solution, a simple estimate for the interaction parameters is computed by taking the average for each combination of the pure filler parameters. The obtained link stiffness is reduced by a factor of $\alpha = 0.5$ to account for the fact that each filler interact more strongly with itself, than with other filler types.

4.2 *Prediction algorithm*

In order to generate a prediction for a material reinforced with a mix of the three different fillers, the model parameter set for this material is defined by multiplying the activity probability parameters $p_{a(i)}$ with the barycentric coordinates (which are the fraction of each filler type compared to the pure filler materials). The reason is the simple assumption, that the available active surface area (which is modeled by the $p_{a(i)}$) is directly proportional to the amount of filler for each type. Then, a simulation with this modified parameter set is carried out. The properties derived from that simulation are denoted by $P_{sim,varied}$. As explained, the model is not that good at predicting the stress-strain curves directly, but it good at predicting trends. Therefore, the change in the model response between the simulations for the pure filler materials and the varied simulation are computed and applied to the measurements by simple multiplication. There are contributions from the three pure filler Experiments, with associated properties $P^{(k)}$, and corresponding simulation results $P_{sim}^{(k)}$, which are added up with weighting factors $w^{(k)}$:

$$P_{predict} = \sum_{k=1}^{3} w^{(k)} P^{(k)} \frac{P_{sim,varied}}{P_{sim}^{(k)}}. \qquad (2)$$

A natural choice for the weighting factors $w^{(k)}$ are the barycentric coordinates of the material to predict. In comparison to the previously proposed prediction method presented in (Wulf & Ihlemann 2019) this is an extension to multiple reference measurements and a change from extrapolation to interpolation.

4.3 *Prediction results*

With the shown prediction method, predicted stress-strain curves for the material corresponding to experiments 4 to 7 can be generated. These can be compared with the actual measurements with these materials. The results are shown in Figures 7, 8, 9 and 10.

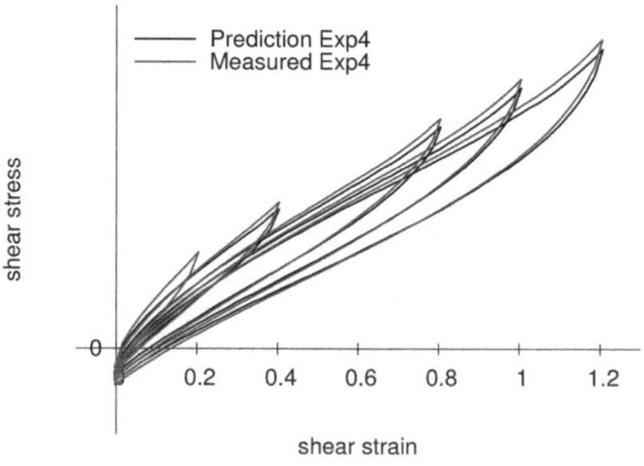

Figure 7. Predicted stress-strain curve for experiment 4 and actual validation measurement.

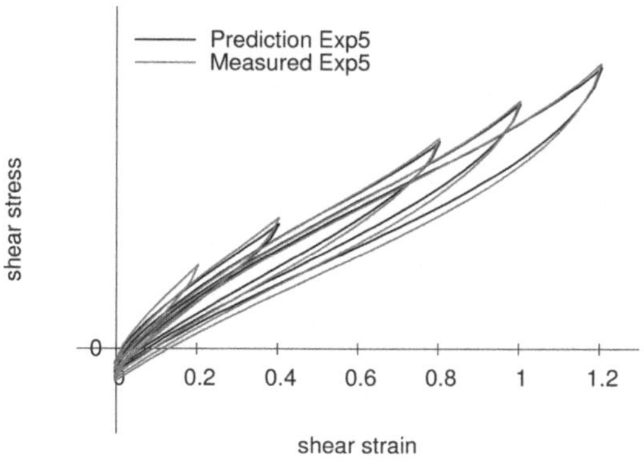

Figure 8. Predicted stress-strain curve for experiment 5 and actual validation measurement.

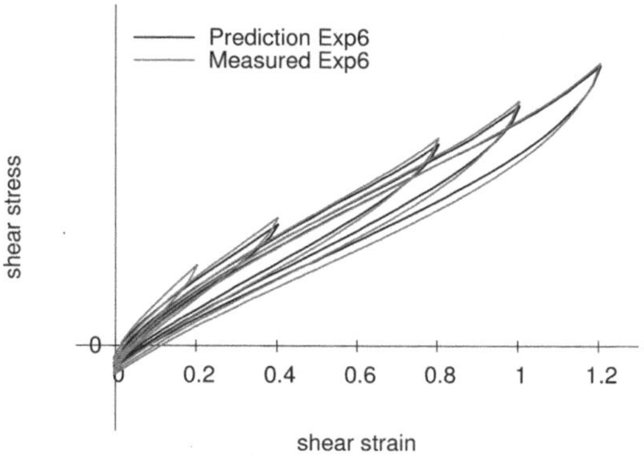

Figure 9. Predicted stress-strain curve for experiment 6 and actual validation measurement.

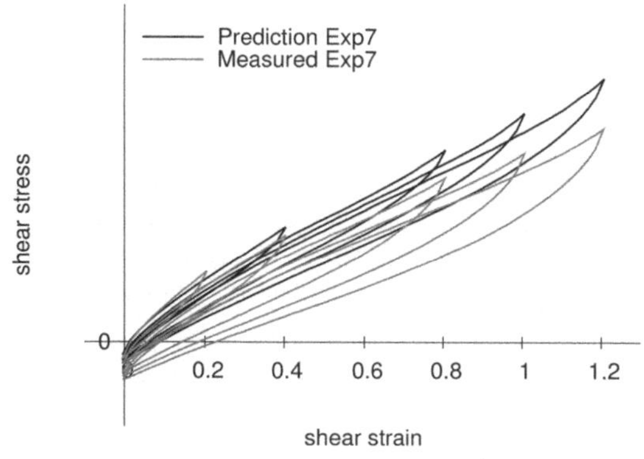

Figure 10. Predicted stress-strain curve for experiment 7 and actual validation measurement.

It can be seen that the overall stress level for Exp4 (CB1/CB2 hybrid) is slightly under-predicted. This is due to assumption, that different filler types have less interaction than the pure filler with itself. For the combination of CB1 and CB2 the assumed interaction factor of $\alpha = 0.5$ was to small, apparantly. The prediction of the hysteresis measure is very good. For Exp5 and Exp6 the overall stress level is predicted very well. In particular, the prediction is much better than a simple linear interpolation of the stress levels of the involved pure filler materials. The hysteresis measure is slightly underestimated for the larger strain cycles. Unfortunately, the prediction for the last material Exp7 (equal mix of all filler types) is not good. The stresses are severely overestimated. The simulation correctly predicts that this material has the smallest stiffness, but the magnitude of this effect is clearly underestimated. The reason for the considerable reduction in stiffness when mixing all three filler types remains unclear. Apparently some additional micromechanical processes occur, which are not present yet for the dual-filler materials. These processes are not fully covered in the model, leading to the deviations. However, the trends in stiffness and hysteresis are correct - this is already a success considering the unexpected and extreme nature of the last materials behavior.

The presented method can generate predictions for materials, which have been actually produced and measured - but it can also generate predictions for any other material in the triangular material space. For the final results, predictions for 28 virtual materials equally spaced in the triangle have been computed. In Figure 11 the distribution of the predicted maximum stress over the material space is shown. It can be clearly seen, that the method predicts a minimum in peak stress in the inside of the triangle. The minimum is located at the coordinates $(\frac{1}{6}, \frac{1}{6}, \frac{2}{3})$ and it seems logical to mix in a larger fraction of the third filler to this end, considering it is the least reinforcing filler type.

In Figure 11 the distribution of the relative hysteresis measure for the largest strain cycle is shown. Again, a nonlinear distribution over the material space is obtained. Unfortunately, there is no further experimental data inside the triangle to validate these predictions. Overall, these results demonstrate the essential advantage of the model-based prediction over purely data-driven methods: With measured data available only at the corners there is hardly any other choice but linear interpolation when relying only on the raw data. The model-based approach however, can correctly predict nonlinearities.

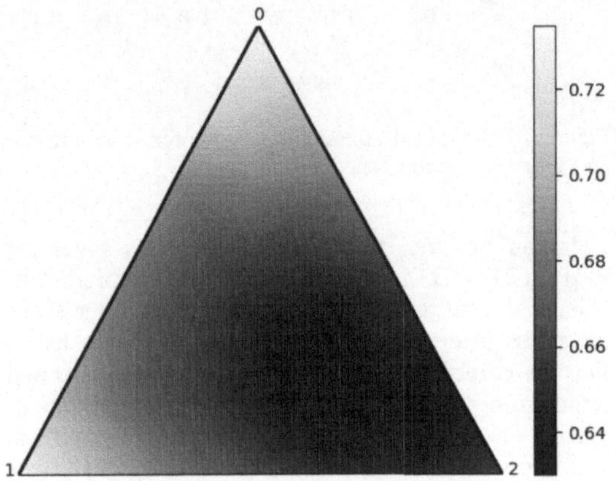

Figure 11. Distribution of the predicted maximum stress in the material space.

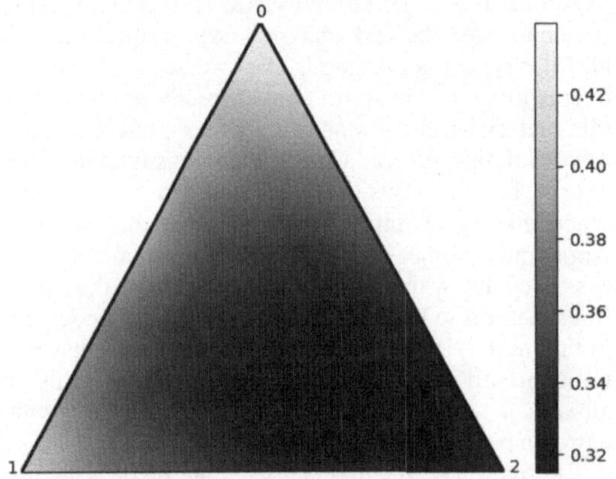

Figure 12. Distribution of the predicted relative hysteresis in the material space.

5 CONCLUSION

A method for predicting the properties of rubber materials is presented. It relies on experimental data for a set of reference materials and generates predictions for any materials in the space spanned by these reference materials. It relies on fitting the parameters of a microstructural simulation program to the supplied measurements using a specialized fitting procedure. Thereafter, interpolated materials can be simulated by interpolating the model parameters. Predicted stress-strain curves for corresponding material are generated by a transfer algorithm. The method is applied to predict the behavior of materials with hybrid filler typed based on measurements on pure filler type materials. Validation measurements reveal that this method provides good results for materials with two filler types. For the triple filler material, significant deviations occur, indicating the need for further model refinement. Overall, the capability of the method of generating correct trends for non-linear property distributions in the considered material space is demonstrated.

REFERENCES

Sresungsuwan, N. & N. Hansupalak (2013). Prediction of mechanical properties of compatibilized styrene/natural-rubber blend by using reaction conditions: Central composite design vs. artificial neural networks. *Journal of Applied Polymer Science 127*(1), 356–365.

Trebar, M., Z. Susteric, & U. Lotric (2007). Predicting mechanical properties of elastomers with neural networks. *Polymer 48*(18), 5340–5347.

Vijayabaskar, V., R. Gupta, P. Chakrabarti, & A. K. Bhowmick (2006). Prediction of properties of rubber by using artificial neural networks. *Journal of applied polymer science 100*(3), 2227–2237.

Wulf, H. & J. Ihlemann (2015). Simulation of mullins effect and relaxation due to self-organization processes in filled rubber. In *Constitutive Models for Rubber IX*, Volume 9, pp. 305–310. Balkema.

Wulf, H. & J. Ihlemann (2019). Predicting rubber compound properties by using a micro-structural simulation program. In *Constitutive Models for Rubber XI*, Volume 11, pp. 104–109. Balkema.

Zainathul Akhmar Salim Abdul Salim, A. H. & H. Ismail (2018). A review on hybrid fillers in rubber composites. *Polymer-Plastics Technology and Engineering 57*(6), 523–539.

Experimental methods and characterization

Constitutive Models for Rubber XIII – Dal (Ed.)

Innovations in rubber compounds crack growth testing

Mathieu Badard
Metravib Material Testing, Lissieu, France

Nathan Selles
Elanova, Paris, France

ABSTRACT: Polymers exposed to high stress or at harsh environmental conditions, are subject to crack propagation, which limits quickly their lifespan. In order to quantify this phenomenon, Metravib Material Testing developed a method to quantify the cracks growth, by combining the mechanical solicitation capability of DMA and a mobil camera. This paper describes this new method and how it can be used to determine the effect of the temperature and of frequency of the solicitation on the cracks growth. Tests were performed on Styrene-Butadiene Rubber (SBR) and elastomeric silicone. The results showed that for a fixed number of excitation cycles, the crack propagation was increasing at low frequency. This phenomenon at low frequency has been attributed to a temporal mechanical contribution in addition to the cyclic mechanical contribution. This presentation also depicts that crack propagation rate increases with increasing temperature. This effect is more pronounced for SBR material than for silicone material, due to their high thermal stability.

1 INTRODUCTION

In recent decades, the use of elastomeric material increased significantly due to their unmatched performances. While it is already tremendous materials, many scientists and engineers are continuously working on their optimization, especially their lifespan. One of the main challenges limiting the lifespan of rubber is the apparition and the propagation of cracks when material are submitted to intense energies level or/and exposed to rough surfaces.

In order to study crack propagation, only few techniques of crack monitoring have been developed. Most of these methods consist in applying a monotonic or a dynamic strain over time on a pre-cut rub-ber layer and following the crack propagation at a millimeter scale (Heinrich 2021). The original point of the present work is the use of a new method developed by Metravib Material Testing, which combines a Dynamic Mechanical Analyzer (DMA) and an accurate optical system detecting automatically the crack tip position changes with a great accuracy. This part is further developed in the Methodology part.

In some cases, the main challenges of the crack propagation tests is to imitate the crack propagation in real conditions. Then, the study of critical parameters on the crack growth rate and the understanding of the physical laws that govern this process are mandatory. The aim of the paper is to determine the effect of the temperature and the frequency of the mechanical excitation on the crack growth behavior of SBR and silicone rubbers.

Lake and Lindley (Lake 1964) have studied the frequency effect on the fatigue cracking behavior of elastomers. They demonstrated that for amorphous elastomers, the frequency of solicitation has a non-negligible in-fluence and they proposed a model in which crack growth occurs from two contributions: a mechanical contribution, related to the sinusoidal excitation and a temporal contribution linked to the duration of the cycles. Baumard (Baumard 2017) confirmed these results in his thesis work by showing that the cracking curve shifts upwards as the frequency decreases.

Temperature plays a key role on the viscoelastic behavior of polymer materials. Young examined the influence of temperature on fatigue crack propagation within a NR material, over a temperature range be-tween the glass transition temperature and the degradation temperature (Young 1986).

This work shows that for a given tearing energy, the speed of propagation in-creased with temperature. This influence of temperature on the cracking behavior of elastomeric materials has also been highlighted previously in the literature for samples of NBR and HNBR (Beranger 2018) or CIIR (Young 1986).

DOI: 10.1201/9781003516880-15

2 METHODOLOGY

2.1 *Materials*

SBR layer were provided by Elanova. The second material is a silicon rubber based on silicium and provided by Silcomp Company. All specimen were cut as rectangular band specimen (80mm width, 3.5 mm thickness and 15 mm height) and fixed in a dedicated specimen holder (see Figure 1).

Figure 1. Specimen fixed in the specimen holder.

2.2 *Fatigue crack growth system*

The crack propagation was measured by the crack growth-testing module designed by Metravib Material Testing. This system consists of a DMA instrument coupled to a digital camera fixed on a micrometric motorized table controlled by a dedicated software ensuring a real time image-processing (see Figure 2 below).

Figure 2. A picture of the crack growth system, the optical system is on the right, the DMA part on the center.

2.3 *Stabilization step*

Prior to the crack growth monitoring, two preliminary tests are performed. The first one is the stabilization test, which consist of applying high strains on the specimen. This test aims to verify if the specimen is properly fixed in the specimen holders, especially to avoid any slippering effects. The other objective is to cancel the Payne and /or Mullins effect that can occurs for the material, especially for elastomeric polymers.

2.4 *Characterization step*

The second preliminary test is the characterization, which is performed to establish a correlation between the strain applied on the specimen and the corresponding energy G (J/m^2, the surface being the section of the specimen). Because the correlation between the strain applied and the energy provided vary a lot depending on material, it is more relevant to compare the crack growth propagation on material using the same level of the energy. Here we note that Metravib software allows to works on the provided energy Gp or the restituted energy Gr (Rivlin 1953) (Kadir 1984). The test itself consists of applying different level of strain on the specimen. For each strain, a sinusoidal stress is applied over 500 periods, and the energy is calculated from the load and unload curve.

2.5 *Crack growth testing*

Then, four cracks are initiated with a blade on the specimen. Figure 3 below.

Figure 3. Crack initiation on the specimen using a scalpel blade.

The crack growth process can described with the following steps:

1) The 4 crack tips position are automatically monitored by the optical system
2) The DMA instrument apply a significant number of cycles (set by the operator) of mechanical stress on the specimen. Consequently, the cracks growth in the material
3) The DMA stops and the optical system monitors the new crack tips position
4) Return to the step 2, the DMA apply a periodic mechanical stress over a large number of cycles.

Compare to other systems, the fact that the camera stops between each measurement allows the optical to take picture of the crack tips with a micrometric accuracy on each four cracks.

2.6 *Tests parameters*

As explained in the introduction part, frequency effect and temperature effects were studied. About frequency tests, three frequencies were tested, 1, 10 and 95 Hz. The temperature was stabilized at 23°C during 30 minutes before the tests start. For both materials SBR and Silicone, the crack growth rate was studied for 10 levels of energy.

About the temperature tests, three temperatures were tested, -10°C, 23°C and 100°C. The frequency was set to 10 Hz for all the tests, and 10 level of energy G were tested.

3 RESULTS AND DISCUSSION

3.1 *Frequency effects*

Crack growth testing of SBR compound and silicone compound were achieved at 1 Hz, 10 Hz and 95 Hz at 23°C. Figures below (Figures 4 and 5) presents the crack growth rate as a function of G energy (J/m²) for SBR material and silicone material respectively. For each frequency, a power law shows a clear dependence between the crack growth rate (nm/cycle) and the energy G. Those results are common among different paper in the literature (Heinrich 2021).

The data also shows a low data dispersion, thus highlighting the reliability of the method. For both materials, results are close to the ones mentioned in the literature: for a given value of energy supplied, the propagation speed is higher as the frequency is lower (Lake 1964) (Baumard 2017). This phenomenon can be explained by the fact that at low frequency, the rubber is stretched during a longer time at maximum strain. Consequently, the crack get a longer time to progress during this temporal contribution.

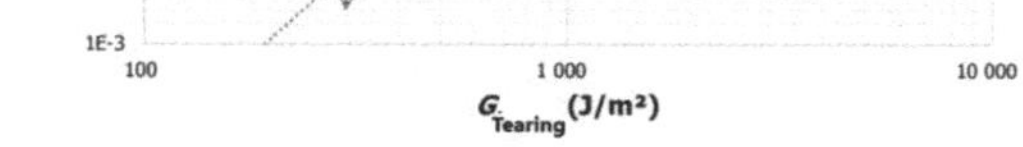

Figure 4. Crack growth rate as a function of the provided energy, tests were performed at 1, 10 and 95 Hz. SBR materials.

However, the exponents of the power laws between materials are not equal and the lines are not parallel. In addition, frequency has a less significant influence on the behavior of Silicone than on that of SBR. These results depicts different crack propagation mechanisms in the material, correlated to their different chemical structure and their different processing.

Finally, it is important to consider the influence of self-heating when analyzing these results. Due to greater stress per second, the cracking curves at 1 Hz, 10 Hz and 95 Hz present different self-heating potentials and this phenomenon should be considered in a further study.

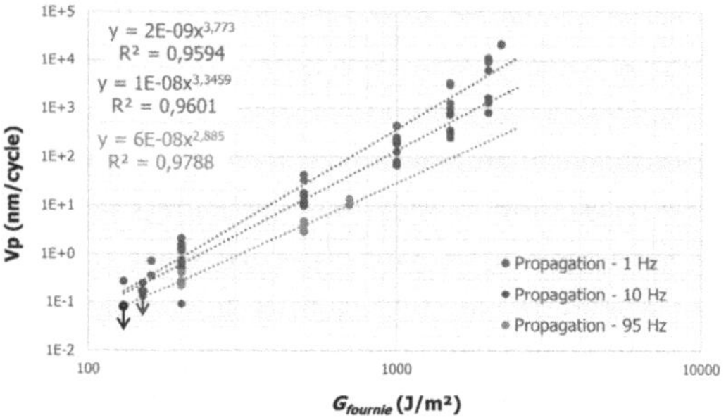

Figure 5. Crack Growth rate as a function of the provided energy, tests were performed at 1, 10 and 95 Hz. Silicone materials.

3.2 *Temperature effects*

For both materials, the crack growth tests have been performed at three temperature stages: -10°C, 23°C and 100°C at 10 Hz. The effect of temperature on crack growth rate is depicted in Figure 6 and Figure 7 below.

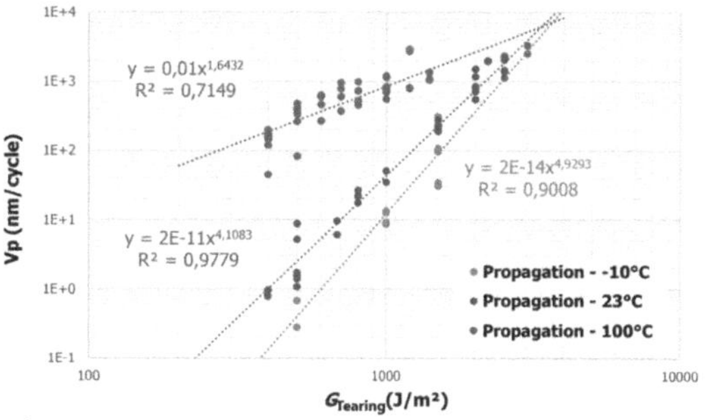

Figure 6. Crack growth rate as a function of the provided energy, tests were performed at -10, 23 and 100 °C. SBR materials.

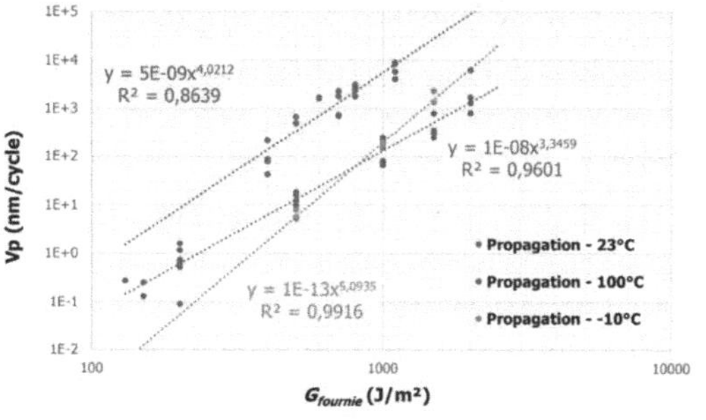

Figure 7. Crack Growth rate as a function of the provided energy, tests were performed at -10, 23 and 100 °C. Silicone materials.

As observed in the literature (Beranger 2018) (Young 1986) for a given value of energy supplied, the speed of propagation is higher as the temperature is high, this clearly shows that temperature increases the propagation rate of the crack tip. This behavior is explained by the weakening effect of the temperature on the polymer chains constituting the rubber material.

For the SBR material, the comparison of the results at 23 and -10°C is close: the curves are almost parallel and the exponents of the power law are similar. On the other hand, the behavior at 100°C is quite different: the exponent of the power law representing the behavior is very different from those obtained at -10 and 23°C and depicts a faster crack growth propagation for the same level of energy. This result can be explained by a phenomenon of degradation of the material due to holding time at 100°C for a long time.

The following graph shows (Figure 8), for a given energy of 1000 J/m², the change of the crack growth rate for a SBR material exposed to a high temperature of 100°C. It seems that the crack growth rate increase with the number of cycle, but actually, these results highlight the fact that a long exposure to a high temperature weaken the material.

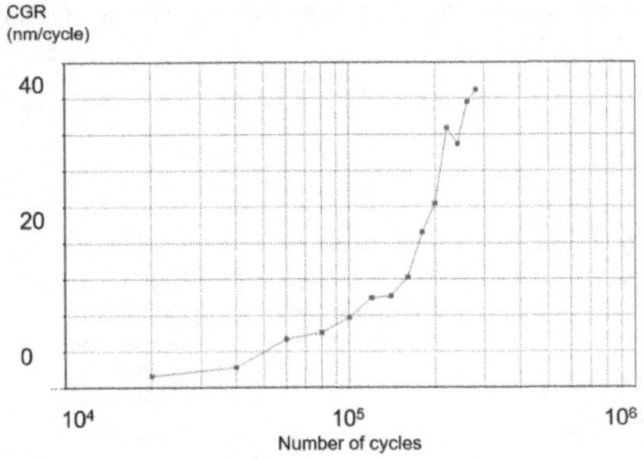

Figure 8. Crack Growth rate as a function of the time at 100 °C for SBR materials.

On the other hand, silicone materials are characterized by better stability of mechanical behavior when the temperature increases but also by better resistance to thermal degradation. The energy cracking results are therefore little influenced by temperature: the behaviors at 23 and -10°C are almost similar, as for the characterization and energy restitution results. The cracking curve at 100°C is slightly shifted towards higher values of propagation speed, while keeping a power law exponent value of the same order of magnitude.

4 CONCLUSION

In this paper, we exposed a new method combining an automated optical system and a DMA to monitor the crack propagation. This new system allowed to measure simultaneously up to four cracks position with a micrometric accuracy.

Due to this technology, we were able to determine how the frequency of the mechanical solicitation does affect the crack propagation for both material. It was shown that at low frequency, the compound is stretched during a longer time at maximum length, which results into a more important crack propagation.

The temperature effect was also investigated and it was shown that crack growth rate increase with temperature. Moreover, those tests highlight different behaviors between the two materials. While the crack growth rate increases significantly with the temperature for SBR material, it is more stable for silicone polymer. It was also shown for SBR compound that a long exposure to high temperatures lead to a thermal degradation of the polymer structure and leads to an increase of the crack growth rate.

Overall, this study shows how a new technique of characterization of the crack growth testing can bring new insights to study the frequency effect and the temperature effect on the crack growth propagation in SBR.

REFERENCES

Baumard, T. L. M. (2017). Investigation on Fatigue Failure in Tyres (Doctoral dissertation, Queen Mary University of London).

Béranger, A. S., Qin, J., Heuillet, P., & Baurier, H. (2018). Fatigue crack growth behavior of NBR, HNBR, HNBR ZSC compounds. Procedia engineering, 213, 145–152.

Heinrich, Kipscholl, Stocek (2021). Fatigue Crack Growth in Rubber Materials. Advances in Polymer Science, 286.

Kadir, A., & Thomas, A. G. (1984). Tearing of unvulcanized natural rubber. Journal of Polymer Science: Polymer Physics Edition, 22(9), 1623–1634.

Lake, G. J., & Lindley, P. B. (1964). Cut growth and fatigue of rubbers. II. Experiments on a noncrystallizing rubber. Journal of Applied Polymer Science, 8(2), 707–721.

Rivlin, R. S., & Thomas, A. G. (1953). Rupture of rubber. I. Characteristic energy for tearing. Journal of Polymer Science Part A: Polymer Chemistry, 10(3), 291–318.

Young, D. G. (1986). Fatigue crack propagation in elastomer compounds: effects of strain rate, temperature, strain level, and oxidation. Rubber chemistry and technology, 59(5), 809–825.

Constitutive Models for Rubber XIII – Dal (Ed.)
© 2025 The Author(s), ISBN 978-1-032-85138-9

Time, amplitude, and frequency dependence of the Payne effect a study on silica-filled styrene-butadiene rubbers

J. Cervi, C. Marano & F. Briatico Vangosa
Dipartimento di Chimica, Materiali e Ingegneria Chimica "G. Natta", Politecnico di Milano, Milano, Italy

ABSTRACT: The study explores the Payne effect in silica-filled styrene-butadiene rubbers (SBR) and acrylonitrile-butadiene rubber (NBR). Dynamic mechanical tests in torsion were performed on SBR, unfilled and containing 25, 50 and 75 phr silica particles, and on a 60 phr silica-filled NBR to assess the influence of filler content, strain amplitude, and frequency on the Payne effect. The results indicates that a linear analysis of the nonlinear response is an acceptable approximation in the range of strain amplitudes considered here. Further, the Payne effect is reversible in nature, as the materials can recover their initial storage modulus after deformation. However, the recovery is slow, and the effect of multiple strain sweeps does not accumulate. The also study confirms the separability of strain amplitude and frequency effects, allowing for the construction of a Payne effect master curve.

1 INTRODUCTION

Numerous applications including tires, damping devices, and belts for transmitting power utilize reinforced rubbers. These are typically vulcanized elastomers reinforced by the addition of inorganic particles like carbon black and silica. When exposed to oscillatory inputs, these materials exhibit a nonlinear response to increasing amplitudes in oscillation. This phenomenon, known as the "Payne effect," prominently occurs within the amplitude range typical of these applications.

The Payne effect takes its name from the first scientist who brought to the attention the softening shown by filled rubber under increasing strain amplitudes in 1962 [Payne, 1962]. This softening occurs as a reduction of the storage component of the dynamic modulus from a linear region value, G'_0 in shear conditions. Instead the loss component increases from the linear region at low amplitudes value, G''_0, up to a maximum and then it decreases toward values lower than G''_0. Despite a plethora of studies has been conducted on this phenomenon because of its central role in many fields, its structural interpretation is not yet completely understood. A first debated point is the representation of the Payne effects in terms of the dynamic modulus components: indeed it's important to point out that they are defined based on the application of a linear analysis, which leads to the constraint that the response to sinusoidal inputs must remain sinusoidal. As a matter of facts, the Payne effect manifests itself at rather high shear strain levels, where non-linear behavior may be dominant. To take it into account the data analysis, the response should be expressed as a linear combination of sinusoidal functions (Fourier series) and analyzed using methods appropriate to the response to Large Amplitude Oscillation in Shear (LAOS) [Wilhelm 2002, Ewoldt 2008]. Literature discussion on this linear/non-linear dichotomy [Randall, 2014] is still open.

A second point of debate is about the reversible/irreversible nature of the Payne effect, already investigated by Chazeau at al. in 2000 [Chazeau, 1962]. According to the study that filled rubber would recover in time the G'_0 value after being subjected to a strain amplitude sweep which induced the Payne effect. Therefore, it seems that the softening of G' related to the Payne effect is reversible in nature. However, other works [for example Rendek, 2010] report that applying a series of subsequent amplitude sweeps changes the material's response to oscillatory inputs. This change seems to be irreversible, in a fashion similar to what observed in some instances for the Mullins effect.

Finally, another interesting point is the combined effect of the sinusoidal strain amplitude and frequency. Indeed in [Chazeau, 2000] and in [Li, 2017] observed that frequency affects the dependence of the components of the dynamic modulus on the amplitude. Further, they suggested the separability of the two variables according to equation 1 in a broad range of frequencies (0,004 – 10 Hz)

$$\begin{cases} G'(\gamma_0, \omega) = G'_0(\omega) f(\gamma_0) \\ G''(\gamma_0, \omega) = G''_0(\omega) g(\gamma_0) \end{cases} \quad (1)$$

DOI: 10.1201/9781003516880-16

The current work aims at contributing to these three points presenting the results of an experimental campaign carried by dynamic mechanical tests in torsion performed on crosslinked SBR with varying silica contents (0, 25, 50, and 75 phr). In order to widen the scope of the work, also a 60 phr silica filled NBR was taken into consideration.

2 MATERIALS AND METHODS

The investigation of Payne effect was carried out on a styrene-butadiene rubber purposely prepared by Pirelli S.p.A. differing in silica particles content (0, 25, 50, and 75 phr), indicated as SBR 0, SBR 25, SBR 50 and SBR 75. E-SBR 1500 by Versalis, containing 23% styrene and Zeosil 1165 MP by Solvay were used. A commercial acrylonitrile-butadiene rubber (NBR) with 60 phr of silica, provided by Arlanxeo Germany Gmbh, indicated as NBR60, was further considered. The SBR compounds were sulfur cured in a press under 10 MPa at 170 °C for 10 minutes to produce rubber sheets from which rectangular samples with dimensions 34x9x3 mm^3 were die cut. The specimen dimensions were set after a preliminary study aimed at maximizing the applicable strain amplitude. NBR was sulfur cured at 170°C for 5 minutes to mold sheets from which rectangular 34x9x2 mm^3 were die cut.

The oscillatory experiments are performed using the Anton Paar MCR502 Rheometer in strain control with a torsional setup.

The tests are mainly divided in three typologies:

1. Amplitude sweep tests, consisting in a stepwise logarithmically increasing strain amplitude (from 0,01 to 30 %) performed at constant frequencies (0,1; 1; 5; 10; 20 rad/s). These tests have been conducted on unstrained (virgin) specimens to investigate the silica reinforcing effect, and to check the frequency-insensitive feature of the Payne effect.
2. A sequence of amplitude sweep tests consisting in three consecutive strain amplitude sweeps, separated by a set rest time. The sequence was applied starting from a virgin sample, and the results will be indicated as #0, #1 an #2 meaning that the material was virgin or preprocessed once or twice. This test aims at investigating irreversible feature of the Payne effect.
3. At the end of the previous test a constant low strain amplitude (0,01%) sinusoidal stimulus is applied to the sample for a long time, spanning several time decades (up to 10^5s). This analysis has the goal to understand if the nature of the structural changes induced by the amplitude sweeps are reversible or irreversible.

The last two tests were carried out at 5, 10 or 20 rad/s.

3 EXPERIMENTAL RESULTS

3.1 *Linearity of the oscillatory response*

To assess the degree of non-linearity due to the large shear strain amplitude applied to the sample, the FT-rheology was applied. In this technique the stress response is described as a linear combination of sinusoidal functions. Considering the ratio between the intensities of the third to the first harmonic, $I_{(3\omega)}/I_{(1\omega)}$ it's possible to verify the higher harmonics significance in the shear strain amplitude range considered here. Results of $I_{(3\omega)}/I_{(1\omega)}$ were obtained (Figure 1) performing a single shear strain amplitude sweep test (for SBR 0, 25, 50, 75 phr).

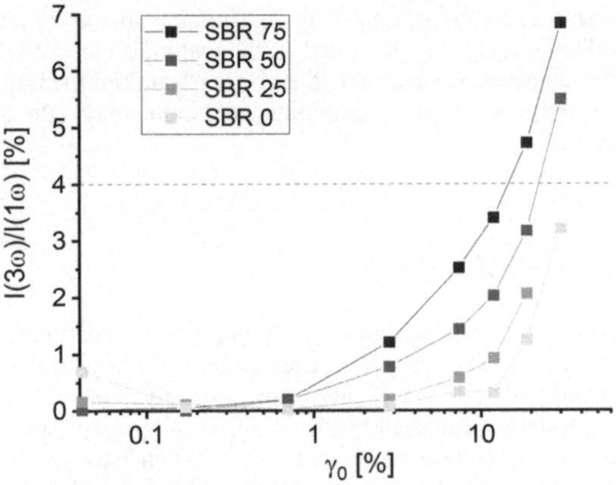

Figure 1. Third to first harmonic intensity ratio, $I_{(3\omega)}/I_{(1\omega)}$, vs shear strain amplitude for filled SBR with different amount of silica.

Literature [Li, 2017] reports that non-linear effects can be neglected for $I_{(3\omega)}/I_{(1\omega)}<4\%$. The limit is exceeded at high amplitude for the SBR50 and SBR75, indicating that the linear analysis is questionable. However, the values are still lower than 10%, a value indicating a limited contribution of the higher harmonics modes. Based on these results, the hypothesis of linear behavior can be taken as an acceptable approximation, and the results will presented based on the linear viscoelasticity theory.

3.2 *Filler effect*

The results from the single amplitude sweep test on each material are shown in Figure 2 in terms of $G'(\gamma_0)$ (a) and $G''(\gamma_0)$ (b) functions.

The typical trend of the Payne effect can be observed along with the expected reinforcing effect of silica [Payne, 1962].

3.3 *Shear strain amplitude effects*

The results of the consecutive amplitude sweeps (sweep #0-#1-#2) are shown in Figure 3.

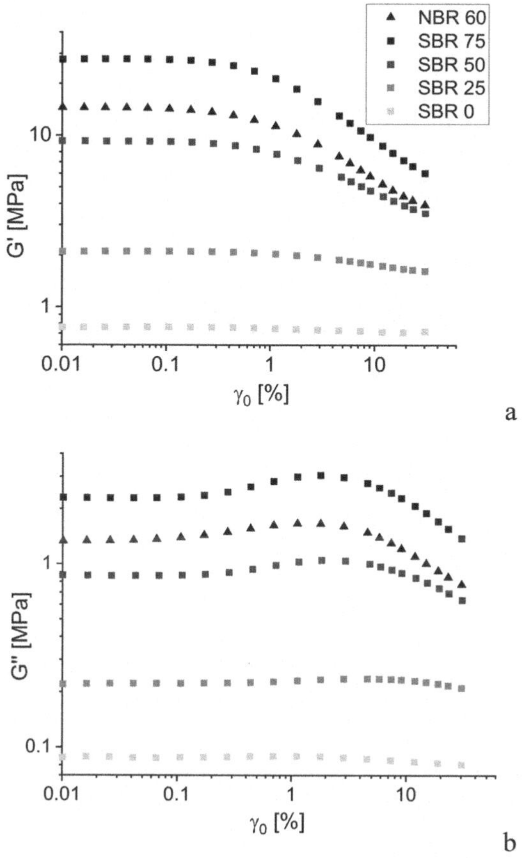

Figure 2. (a) G' and (b) G'' vs. γ_0 for filled SBR and NBR 60.

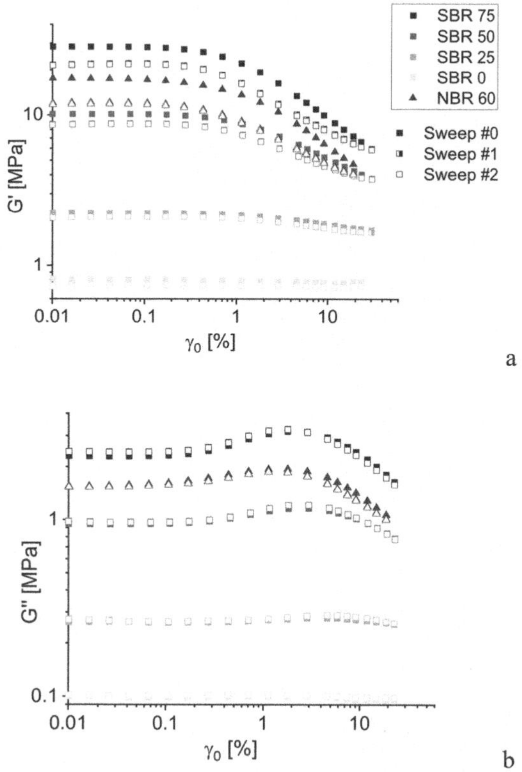

Figure 3. (a) G' and (b) G'' vs. γ_0 at 20 rad/s for SBR materials and NBR 60; the three sweeps are represented as full symbols (sweep #0), half-full symbol (sweep #1) and hollow symbols (sweep #2).

From figure 3, for each filled material the G' curves corresponding to strain sweeps #1 and #2 overlap, and are both lower than that corresponding to the strain sweep #0 applied to the virgin material. Notably #1 and #2 curves seem to display a second plateau at the highest amplitude, which is not present in the #0 one. As for G'', the effect of previous strain application is neglectable.

These results suggest a permanent structural change occur upon the first cycle, and then the materials structure is not further modified.

3.4 *Time effect – initial structure recovery*

In the present paragraph, the constant amplitude and frequency test performed subsequently the three previous sweeps is aimed at monitoring the low amplitude modulus evolution in time. The analysis considers only the storage component, $G'_{0\#3}(t)$, because the loss component variation is very small. To compactly report the results for each material, the ratio between the recovering storage modulus and the virgin material's storage modulus is calculated ($G'_{0\#3}(t)/ G'_{0\#0}(t)$) and shown in Figure.

Figure 4 shows that the materials are able to recover their pristine low amplitude storage modulus. This suggests that the structural changes related to the Payne effect are reversible for these silica filled rubbers. However the recovery kinetics remains slow.

In the case of NBR50, which was observed to fully recover the initial value of G'_0, a further shear strain amplitude sweep was applied, and the overlapping of the response for the virgin material and the same sample after the recovery, in Figure 5, confirms the reversibility of the Payne effect.

3.5 *Frequency effects*

The frequency and amplitude separability has been assessed by performing the single amplitude sweep test for different frequencies (0,1; 1; 5; 10 and 20 rad/s). The superposition of the curves would demonstrate that the separability of the variables holds, so the frequency does not affect the amplitude dependence of the moduli. The data from the amplitude sweeps at different frequencies are shown in Figure 6.

Observing the results, it becomes clear that the separation of frequency and amplitude variables

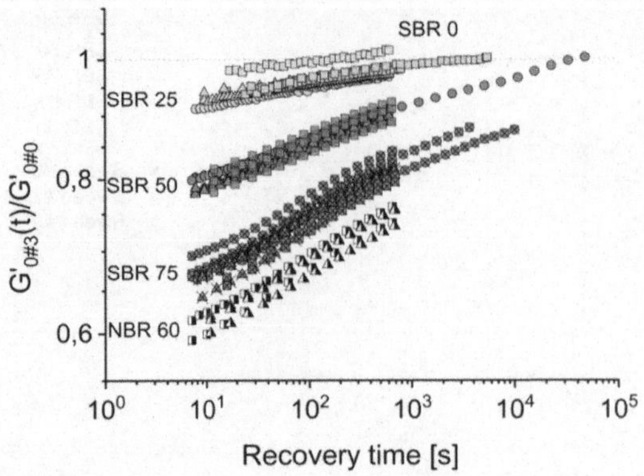

Figure 4. $G'_{0\#3}(t)/G'_{0\#0}(t)$ for the considered materials. Square represent results of tests performed at 5 rad/s, triangles results of tests performed at 10 rad/s, circles results of tests performed at 20 rad/s.

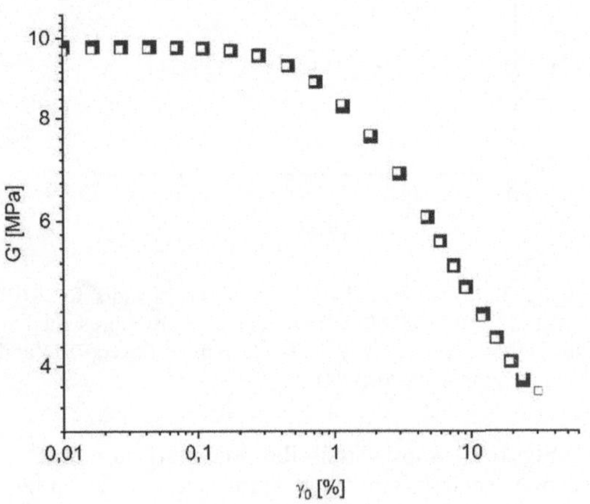

Figure 5. G' vs. γ_0 for filled SBR 50 measured at 20 rad/s. Filled squares refer to a test performed on a virgin material specimen, hollow squares refer to the same specimen tested after 1 month recovery.

seems valid especially at lower shear strain amplitudes. Additionally, at the lowest frequencies (0,1 and 1 rad/s) it seems not to hold, especially for SBR 50 and NBR 60 materials. This discrepancy from the overall trend might be due either to the application of the linear analysis in a region in which we showed the nonlinear effects may have an effect or the recovery of structure during the long testing times required for the lower frequency tests.

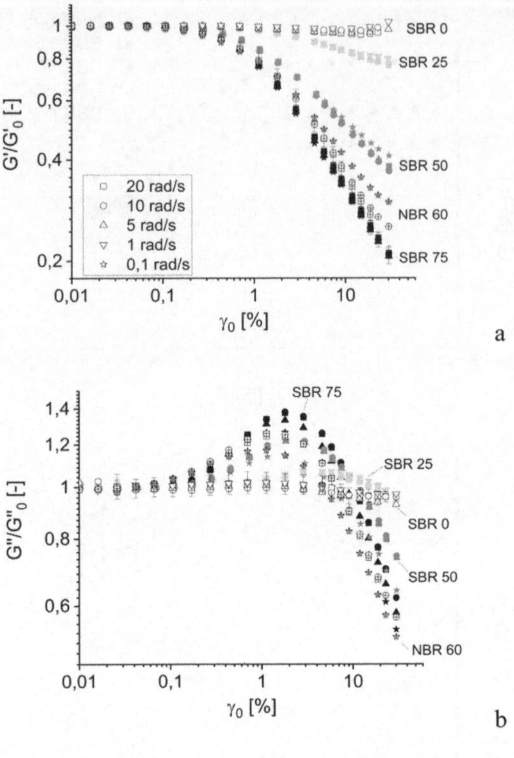

Figure 6. (a) G'/G'_0 and (b) G''/G''_0 for SBR 0, 25, 50 and 75 and NBR 60 as a function of the shear strain amplitude.

4 CONCLUSIONS

The present work was aimed at the investigation of the Payne effect for silica filled SBR and NBR.

The non-linear analysis of the mechanical response to sinusoidal shear strain confirmed the possibility to adopt the traditional description of the Payne effect as a decrease of G' with shear strain amplitude, given the fact that the level of nonlinearity, described in terms of the ratio of third harmonic to the first one, is modest.

Resorting to this method, the effect of filler content was described, both in terms of reinforcing ability and of compound softening upon straining in shear. Independent of the filler content and strain application frequency, an apparently permanent change in properties – suggesting a modification in materials structure – is observed after a first strain sweep. However, the effect of applying multiple strain sweeps to the same specimen is not cumulative, as the Payne effect signature remains unchanged after the first one. Furthermore, the change in properties – and in structure – is only apparently irreversible, as the components of the complex modulus at small strain amplitude are observed to revert to those of the virgin materials in a sufficiently long time-interval.

Finally, the possibility to separate the effects of strain amplitude and frequency in the analysis of Payne effect was confirmed allowing to build a "Payne effect" master curve . The limited deviation from the master curve may be attributed either to non-linearity at the largest strain amplitudes or – more likely – to recovery during the long lasting tests performed at the lower frequencies.

ACKNOWLEDGMENTS

We wish to thank Dr. Luca Giannini from Pirelli S.p.A. for kindly preparing and providing the SBR silica filled materials considered in this research.

REFERENCES

Payne A.R. 1962 The dynamic properties of carbon black-loaded natural rubber vulcanizates. Part I *Journal of Applied Polymer Science, vol. 6, no. 19, pp.* 57-63.

Wilhelm M. 2002 Fourier-Transform Rheology. *Macromolecular Materials and Engineering, vol 287, no. 2, pp.* 83-105.

Ewoldt R. H., Hosoi A. E. & McKinley G. H. 2008 New measures for characterizing nonlinear viscoelasticity in large amplitude oscillatory shear, *Journal of Rheology, vol. 52, p.* 1427-1458.

Randall A. M., Robertson C. G. 2014, Linear-Nonlinear Dichotomy of the Rheological Response of Particle-Filled Polymers. *Journal of Applied Polymer Science, 40818.*

Chazeau, L., Brown, J. D., Yanko L. C. & Sternstein, S. S. 2000 Modulus Recovery Kinetics and Other Insights Into the Payne Effect for Filled Elastomers, *Polymer Composites, vol 1, no. 2, pp.* 202-222.

Rendek M. & Lion A. 2010. Strain induced transient effects filler reinforced elastomers with respect to the Payne Effect: experiments and constitutive modelling, *Journal of Applied Mathematics and Mechanics, vol. 90, no. 5, pp.* 436-458.

Li S., Mi Y. & Wang X. 2017. Superposed nonlinear rheological behavior in filled elastomers *Journal of Rheology, vol. 61, no. 3, pp.* 409-425.

Constitutive Models for Rubber XIII – Dal (Ed.)
© 2025 The Author(s), ISBN 978-1-032-85138-9

Strain-hardening of hydrogenated nitrile-butadiene rubber and its similarities to natural rubber

C. Gögelein
Arlanxeo Deutschland GmbH, Leverkusen, Germany

M. Hemstede-van Urk
Arlanxeo Netherlands B.V., Geleen, The Netherlands

ABSTRACT: We examine the effect of polymer composition, network density, and temperature on the tensile properties of hydrogenated nitrile-butadiene rubber (HNBR). Seven HNBR co- and terpolymers with varying degrees of hydrogenation and with 21-49 wt% acrylonitrile (ACN) contents are considered. For crosslinked samples with ≤43 wt% ACN, the stress-strain curves reveal that strain-hardening (SH) disappears below 60 °C, and for 49 wt% ACN, below 80 °C. At 23 °C, we find that the tensile strength (TS) decreases with increasing network density for 34 wt% ACN, but for 49 wt% ACN, TS undergoes a maximum. For carbon black (CB) filled samples tested at 23 °C, it is found that for ACN ≤39 wt%, TS is significantly reinforced by the addition of CB. For ≥43 wt% ACN, the filler reinforcing effect is less pronounced. These results are rationalized by the crystallization properties of HNBRs and comparison to natural rubber.

1 INTRODUCTION

Nitrile-butadiene rubber (NBR) consists of unsaturated butadiene (BD) and acrylonitrile (ACN) units. Upon hydrogenation, BD units are converted into two ethylene groups, which are typically referred to as hydrogenated butadiene (HBD) units. The degree of hydrogenation (DH) can be adjusted precisely and is expressed in terms of the residual double bond (RDB) content, given in units percent. For example, for NBR, RDB is 100%, and for a complete hydrogenation of all BD units, RDB is 0%.

Partially hydrogenated (PH) polymers with 4-18% RDB can be vulcanized with sulfur as well as with peroxide. Fully hydrogenated (FH) HNBR with <0.9% RDB are used in conjunction with peroxide crosslinking.

The chemical and physical properties of HNBR depend upon the ACN content and DH. In general, HNBRs have better oil- and chemical-resistance than NBRs and can withstand much higher temperatures. The working temperature range is -25 to 160 °C. Moreover, HNBRs have excellent mechanical properties like tensile and tear strength, elongation, and abrasion resistance. The polymers are typically used in the automotive and oil and gas industries, e.g. timing belts (Bradford & Wood, 1994) and seals (He, et al., 2016), but also for the production of lithium-ion batteries (Sedlmeier, et al., 2022).

In previous studies, it was found that the ACN content determines the glass transition temperature, T_g, as well as the polymer crystallization behavior (Obrecht, et al., 1986). Differential scanning calorimetry (DSC) measurements show that T_g increases with increasing ACN content, in accordance with the free volume theory (Wrana, 2014). In addition, these experiments reveal that HNBR polymers with 19 wt% ACN content have 14% degree of crystallinity (DC), determined from the first DSC heating curve (Obrecht, et al., 1986). With increasing ACN content, DC decreases. For example, polymer samples with 34 wt% ACN content have DC = 1.6%. At ~39 wt% ACN, the polymer samples are amorphous, and for >39 wt% ACN content, DC increases. For HNBR polymers with <39 wt% ACN content, the melting enthalpies obtained from the first and second heating curves, ΔH_{1st} and ΔH_{2nd}, respectively, are similar or even identical, indicating that crystallization is fast. By contrast, for >39 wt% ACN, no melting peak is observed during the second heating curves, suggesting slow crystallization.

Wide angle X-ray scattering (WAXS) experiments reveal that for <39 wt% ACN content, crystallinity is caused by ethylene sequences (Obrecht, et al., 1986). Like other ethylene-containing copolymers such as ethylene-vinyl acetate, ethylene acrylates, and ethylene propylene rubber (EPM), crystals are formed when the polymer melt is cooled, or when the polymer material

DOI: 10.1201/9781003516880-17

is heated within a temperature range below its melting temperature. As a results, HNBRs with <39 wt% ACN content are semi-crystalline polymers.

For EPM copolymers, it was found that the crystallinity increases upon stretching (Auriemma, et al., 2020). Moreover, for EPM copolymers an incremental crystallinity increase is observed after stretching and retraction. Interestingly, for EPM samples with high ethylene content, the initial crystals break up or melt, and act as nuclei. In addition, new crystals are formed from the highly oriented amorphous phase.

For HNBRs with >39 wt% ACN content, alternating HBD-ACN sequences crystallize (Kobatake, et al., 1997). As with natural rubber (NR), the crystallization kinetics of HNBRs with high ACN content is slow, but upon stretching, crystallization is strongly enhanced, and during retraction, crystal melting is observed. Thus, HNBRs with >39 wt% ACN content exhibit strain-induced crystallization (SIC) that is similar to NR.

The effects of network density on sulfur-vulcanized partially hydrogenated HNBR samples with 34, 36, and 44 wt% ACN contents were recently studied (Femina, et al., 2023) (Femina, et al., 2024). Samples with 44 wt% ACN are amorphous at room temperature (23 °C). The tensile strength at break (TS) reaches a maximum at intermediate network densities, and WAXS measurements show that this maximum is associated with SIC. In contrast, the samples with 34 and 36 wt% ACN contents do not exhibit SIC, and consequently, no maximum in TS is observed for varying network density. Stretching only induces the orientation of the amorphous chain segments in the direction of strain.

Until now, however, it is unclear up to what temperature self-reinforcing SIC occurs in HNBR networks. In addition, the extent to which SIC occurs in filler-reinforced networks has not yet been systematically investigated. Since filled HNBR networks are often used for high temperature applications, these two questions are of interest not only for academic reasons, but also for technical applications. This is particularly the case because per- and polyfluoroalkyl substances, such as fluoroelastomer, may be banned (European Chemicals Agency, 2023), and HNBR is regarded as an alternative polymer for some applications.

For this study, we selected FH and PH HNBR polymers with 21 to 49 wt% ACN content. The study considers the effect of temperature on the tensile properties of (non-vulcanized) polymer samples at 23 °C and vulcanized HNBR networks at 23-100 °C. In addition, for vulcanized HNBR networks, the effects of varying network density and carbon black (CB) reinforcement are investigated at 23 °C.

2 EXPERIMENTS

2.1 *Polymers*

Polymers were provided by Arlanxeo Deutschland GmbH, Dormagen, Germany. The ACN and RDB contents were taken from the manufacturers specifications and are listed in the Table 1. Note that polymer sample 1 contains ACN and HBD units plus a third monomer that is not specified by the manufacturer. Polymer samples 1-5 have RDB <0.9 % and, thus, are FH. In contrast, for polymer samples 6 and 7, RDB = 5.5 %, showing that these samples are PH.

Table 1. Content of acrylonitrile (ACN) and residual double bonds (RDB) of polymers.

Polymer sample #	ACN [wt%]	RDB [%]
1	21	<0.9
2	34	<0.9
3	39	<0.9
4	43	<0.9
5	49	<0.9
6	34	5.5 ± 1
7	43	5.5 ± 1

Mooney viscosities (MV) were measured according to DIN 53523 at ML_{1+4} at 100 °C, with viscosity results given in Mooney units (MU).

To study the polymer thermal properties, a Q200 differential scanning calorimeter (DSC) from TA Instruments, Newcastle, DE, USA was used. First, the polymer samples were cooled ballistically from room temperature to -100 °C, and then heated to 100 °C, and cooled to -100 °C at 10 K/min heating and cooling rates, respectively. Thereafter, the heating and cooling cycle was repeated. T_g was determined from the second heating curve, and the peak maximum melting temperature, T_m, from the first heating curve. In addition, ΔH_{1st} and ΔH_{2nd} were determined from the first and second heating curves, respectively.

2.2 *Compounds*

Peroxide, triallyl isocyanurate (TAIC), 4,4′-Bis (α, α-dimethylbenzyl)diphenylamine (CDPA), magnesium oxide (MgO), and CB were purchased from VWR International GmbH, Darmstadt, Germany.

Mixing was carried out in a Werner & Pfleiderer GK 1.5E internal mixer (1.5 l volume and PS5 intermeshing rotor blades). The flow temperature of the mixer was set to 40 °C, the rotor speed to 40 rpm, and 70% fill factor was chosen. Following a conventional mixing sequence, first, the polymer was added to the internal mixer. Second, after 1 min mixing time, CDPA, MgO, and, if used, CB were added. After 3 min the ram was raised, swept, and lowered again, and after 5 min total mixing time the batch was transferred to a Troester L 300-MA 01 two-roll mill

(200 mm diameter, 1:1.21 friction ratio, 14 rpm front and 17 rpm rear roll). The mill rolls were preheated to 40 °C. Peroxide and TAIC were added, and the compound was refined by cutting the banded sheet five times from the left- and right-hand sides, respectively. Finally, the compound was rolled and passed through the mill nip five times. Total milling time was 10 min.

The compound formulations are listed in Table 2 showing unit parts per hundred rubber (phr), giving the active amounts of the pure ingredients. The volume fraction of CB is in the range of 0-13 vol%, being below the mechanical percolation threshold of 22 vol% for CB in HNBR (Liu, et al., 2022).

Table 2. Compound formulations in units parts per hundred rubber (phr).

Ingredient	Amount [phr]
HNBR	100
CB	0-30
CDPA	1.5
MgO	3
TAIC	1.5
Peroxide	2.4
Total phr	108.4-138.4

To test the effect of the network density on the tensile properties, additional compounds were prepared with 0.2-5.6 phr peroxide content, keeping the amounts of all other ingredients constant.

Vulcameter curves were measured with a moving die rheometer (MDR) from Alpha Technologies, Hudson, OH, USA, according to DIN 53529 at 180 °C.

DIN S2 test specimens were cut from 2 mm thick slabs, which were compression molded in polytetrafluoroethylene (PTFE) foils. Polymer samples were molded for 10 min at 105 °C, and compounds were vulcanized for times equivalent to 2 x t90 + 10% time plus a mold specific preheating at 180 °C. Clamping pressure was 120 bar.

Tensile stress-strain properties of polymer samples were tested at 0.2 N preload and 1000 mm/min crosshead speed. Stress-strain curves of vulcanized compounds were measured according to DIN 53504 with 200 mm/min crosshead speed. Tensile tests at 23 ± 2 °C were carried out on a Zwick ZN 5 tensile test machine, and tests at 40-100 °C were done on a Zwick ZN10 tensile tester equipped with an environmental chamber (±1 K accuracy). The temperature of the sample was equilibrated in the chamber for 3 min before the measurements were started. Tensile measurements were repeated three times per sample and median values of the tensile strength at break, TS, were calculated.

3 RESULTS AND DISCUSSION

3.1 *Polymer properties and green strength*

MV and DSC results are summarized in Table 3. The Mooney viscosities of polymer samples 1-6 are in the range from 70 to 81 MU. By contrast, polymer sample 7 has a significantly lower MV (43 MU). Differences in MV, or likewise molar mass, may affect the green strength and tensile properties of rubbers (Wrana, 2014).

Table 3. Polymer Mooney viscosity (MV), glass transition temperature, T_g, obtained from second heating curve, peak maximum melting temperature, T_m, from first heating curve, and melting enthalpies from first and second heating curves, ΔH_{1st} and ΔH_{2nd}, respectively.

Polymer sample #	MV [MU]	T_g [°C]	T_m [°C]	ΔH_{1st} [J/g]	ΔH_{2nd} [J/g]
1	76	-40	-26	2	3
2	74	-27	-6	5	7
3	72	-25	-	0	0
4	68	-20	42	4	0
5	81	-12	43	13	0
6	70	-29	-15	2	3
7	43	-21	42	2	0

Considering polymer sample series 2-5, and polymer samples 6 and 7, respectively, in Table 3 reveals that T_g decreases with increasing ACN content, as expected (Obrecht, et al., 1986).

For the polymer 1, which contains ACN and HBD units plus a third monomer, T_g = -40 °C. To compare this value with the glass transition temperature of a HNBR polymer with 21 wt% ACN and 79 wt% HBD, we used the Gordon-Taylor expression obtained by Obrecht, et al., 1986, for <39 wt% ACN HNBR polymers and adjusted the intercept to match the results for polymers 2 and 3 (Gordon & Taylor, 1952). Applying this expression and estimating the glass transition temperature of a HNBR polymer with 21 wt% ACN gives $T_g \approx$ -29 °C. This value is significantly higher than T_g of polymer 1, showing that the glass transition of polymer 1 is shifted by 11 K to lower temperatures by the third monomer. Furthermore, for a FH HNBR polymer with a lower ACN content than polymer 2, a higher DC and therefore, higher T_m-, ΔH_{1st}-, and ΔH_{2nd}-values than for polymer 2 are expected. In lieu, polymer 1 (21 wt% ACN) has a lower melting temperature and melting enthalpy than polymer 2 (34 wt% ACN), T_m = -26 vs. -6 °C, ΔH_{1st} = 2 vs. 5 J/g, and ΔH_{2nd} 3 vs. 7 J/g, respectively. Hence, for polymer 1, the crystallization of HBD sequences is hindered by the third monomer.

Likewise, polymer 6 has the same ACN content as polymer 2 (34 wt%), but due to partial hydrogenation, the crystallization of the HBD sequences of polymer 6 is disturbed by BD sequences. As a result, polymer 6 has lower T_m-, ΔH_{1st}- and ΔH_{2nd}-values than polymer 2, -15 vs. -6 °C, 2 vs. 5 J/g, and 3 vs. 7 J/g, respectively.

Polymer 3 with 39 wt% ACN content has no melting peak, indicating that the polymer is amorphous.

For polymers 4 and 5 with 43 and 49 wt% ACN, T_m = 42 and 43 °C, and ΔH_{1st} = 4 and 13 J/g, respectively. Hence, T_m of polymers 4 and 5 are almost identical, but ΔH_{1st} of polymer 5 is higher than that of polymer 4. Comparing these polymers with the polymer 2 (34 wt% ACN) reveals that polymers 4 and 5 have higher T_m-values than polymer 2, i.e. 42 and 43 °C vs. -6 °C, respectively. These differences in physical properties of polymers with high (≥39 wt%) and low (<39 wt%) ACN contents are due to the different crystallization properties of HBD and alternating HBD-ACN sequences. It was also noted that the enthalpy of fusion of polymer 4 with 43 wt% ACN (ΔH_{1st} = 4 J/g) is almost identical to the ΔH_{1st} value of polymer 2 with 34 wt% ACN content (5 J/g). Since the crystallization of alternating HBD-ACN sequences is slow, no crystallization and melting are observed during the second DSC cooling and heating cycle for polymers 4, 5, and 7.

The stress-strain curves of the polymer samples 1-7 were carried out at 23 °C and the results are shown in Figure 1. Note that the y-axis denotes the engineering stress. Furthermore, note that the curves are labeled with the ACN contents of the polymers in units weight percent, but for simplicity, we shortened the labels to the percent sign. When comparing the shape of the stress-strain curves in Figure 1, for polymer 5 with 49 wt% ACN content, the pronounced stress increase at strains >500% is striking. As shown in previous studies, this pronounced strain-hardening (SH) effect is caused by SIC of the alternating HBD-ACN units (Obrecht, et al., 1986). With decreasing ACN contents, the SH effect decreases, and consequently, TS decreases and the elongation at break, E_b, increases; cf. samples 3, 4, and 5 with 39, 43, and 49 wt% ACN, respectively.

Taking a closer look at the stress-strain curves of samples 2, 3, 4, and 5 with 34, 39, 43, and 49 wt% ACN contents respectively, shows that all samples exhibit an almost linear stress-strain progression before the test specimens break (Figure 1). Moreover, the data reveal that the slope of the linear regimes decreases with decreasing ACN content. In addition, comparing the shapes of the stress-strain curves shows that for sample 2, the onset of the SH regime is more uniform (less steep) than for samples 3, 4, and 5.

Comparing the stress-strain curves of the FH polymers 2 and 4 with the PH samples 6 and 7, respectively, reveals that partial hydrogenation decreases TS and increases E_b, as expected.

For sample 1 (21% ACN), crystallization is disturbed and as a result, no SH is observed.

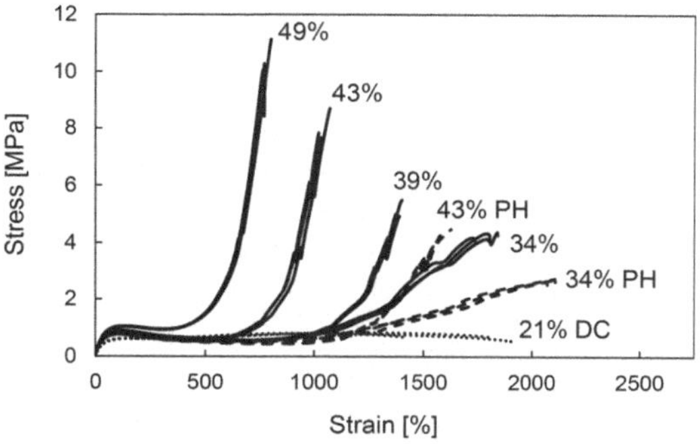

Figure 1. Stress-strain curves of polymer samples (green strength) at 23 °C. The curves are labeled with the polymer acrylonitrile (ACN) contents in units weight percent. In addition, partially hydrogenated (PH) and sam-ples with disturbed crystallization (DC) are indicated.

3.2 *Vulcanized polymer networks*

Next, the effect of the sample temperature on TS for vulcanized polymer networks at T = 23, 40, 60, 80, and 100 °C is investigated (Figure 2). TS decreases with increasing temperature, as expected. The stress-strain curves of vulcanized networks of polymer 1 show no SH for T ≥ 40 °C (data not shown). Therefore, the TS-values measured for polymer 1 networks are used as (approximate) threshold values, indicating the absence of SH. Applying these criteria, it is found that for samples with ≤43 wt% ACN content, SH disappears above 60 °C, and for 49 wt% ACN above 80 °C.

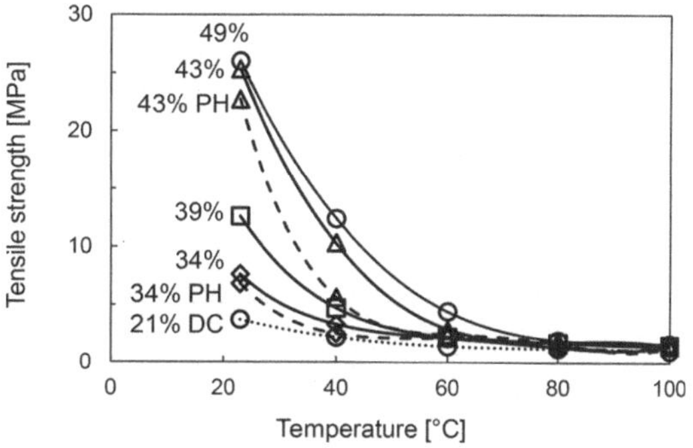

Figure 2. Effect of temperature on tensile strength for vulcanized polymer networks with 21-49 wt% ACN content. Note that the labeling of the curves is analogous to Figure 1.

Having obtained for high ACN HNBRs information on both the melting temperatures and the temperatures above which SH disappears, T_l, it is revealing to compare these temperatures with those of NR. For polymers 4 and 5, values of T_m = 42 and 43 °C, and T_l ≈ 60 and 80 °C, were found respectively, and for NR, T_m = 49 °C (Gögelein & van

Duin, 2023) and $T_l \approx 80$ °C (Huneau, 2011), showing that both temperatures are similar for NR and 43 wt% ACN polymer networks, and almost identical for NR and 49 wt% ACN networks.

The tensile test results for samples with different peroxide contents are presented in Figure 3, showing TS in relation to the rheometer maximum torque, S'_{max}. The latter quantity is proportional to the network density (Gögelein & van Duin, 2023). Tensile tests were carried out at 23 °C.

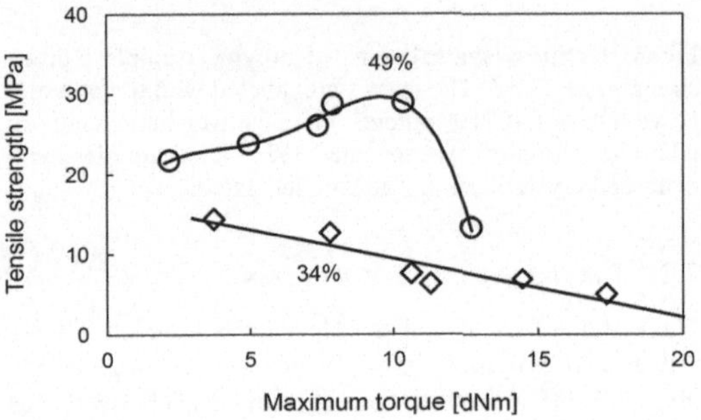

Figure 3. Tensile test results for vulcanized polymer networks with different rheometer maximum torque values (network densities). Shown are data for networks that were vulcanized from polymers with 34 and 49 wt% ACN contents.

For 49% ACN samples with increasing network densities, TS undergoes a maximum at an intermediate network density. By comparison, for 34% ACN, TS decreases with increasing network density. For the 49 % ACN samples and low network densities, the network connectivity increases with increasing network density (Gögelein & van Duin, 2023). As a result, the polymer network strands become increasingly better aligned upon stretching and crystallization (SIC) of the alternating HBD-ACN sequences sets in. On the other hand, at high network densities, the network strands are short, and the network junctions hinder their ability to crystallize, resulting in a TS maximum occurring at intermediate network densities. For 34 % ACN content, HBD sequences crystallize and presumably, a TS maximum occurs at network densities below the network densities tested here.

3.3 *Reinforced polymer networks*

With the intention to study the effect of reinforcing CB filler on the tensile strength of vulcanized networks, TS versus the ACN content was plotted in Figure 4. For non-reinforced polymer networks (0 vol% CB content), it was found that samples with high ACN content have significantly higher tensile strength values than samples with low ACN content, TS = 4-7 MPa for <39 wt% ACN, and TS = 23-27 MPa for >39 wt% ACN.

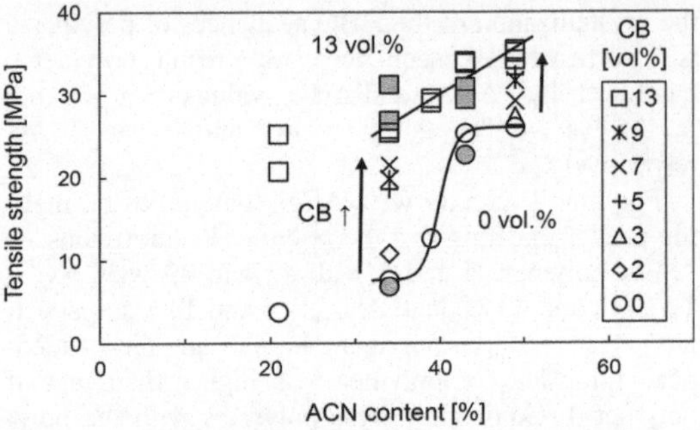

Figure 4. Tensile strength of reinforced networks of vulcanized and reinforced polymer networks with 21-49 wt% ACN content and 0-13 vol% carbon black (CB) filler loading at 23 °C. The gray filled symbols mark the results for the partially hydrogenated samples.

Next, for networks with 34 wt% and 49 wt% ACN content, the CB filler loading was increased stepwise from 0 to 13 vol%. For the 34 wt% ACN samples, TS increases from 7 MPa at 0 vol% to 27-31 MPa at 13 vol% CB content. For the 49 wt% ACN samples, TS increases from 26 MPa at 0 vol% to 29-35 MPa at 13 vol% CB content. Therefore, by adding 13 vol% CB, TS of the 34 wt% ACN sample is increased by a factor of 3 compared to the unfilled system, whereas TS of the 49 wt% ACN sample increases only by a factor of 1.2. Overall, it was found that TS increases with increasing reinforcing CB content, but the reinforcing effect is more pronounced for samples with 34% than 49% ACN content.

These results show that networks consisting of HNBR polymers with low ACN contents are effectively reinforced by CB fillers. By contrast, unfilled networks of high ACN HNBR polymers already have high TS properties because of SIC. For >39 wt% ACN networks, TS increases by the addition of filler as well, but the reinforcing effect is comparatively low. Interestingly, similar observations are made for NR (Röthemeyer & Sommer, 2013). Because of SIC of isoprene sequences, NR polymer networks have per se excellent tensile and tear strength properties. For this reason, to adjust the hardness to a desired value, only small amounts of non-reinforcing CB (≤30 phr) are usually added to NR compounds. By contrast, reinforcing CB is added to compounds with synthetic rubbers to adjust the hardness and to increase the strength of the material.

4 CONCLUSIONS

The tensile properties of HNBR polymer melts with low and high ACN contents were tested, and pronounced SH is found for FH polymer samples with ≥39 wt% ACN content.

Furthermore, the stress-strain curves of vulcanized HNBR polymer networks were measured at various temperatures. For samples with ≤43 % ACN content, SH disappears above 60 °C, and for 49% ACN, above 80 °C. These data and the results from DSC measurements on the polymer samples reveal that for HNBRs with 43 and 49 wt% ACN, the melting temperatures as well as the temperature ranges up to which SH occurs are similar and almost identical, respectively, to those of NR.

For polymer networks with 49% ACN content, a maximum in TS occurs at intermediate network densities due to SIC. In lieu, for 34% ACN samples, the crystallization is hindered by network junctions, and therefore TS decreases with network density.

Tensile tests on samples with varying filler loadings at 23 °C reveal for 34 wt% ACN content that TS increases significantly due to the reinforcing effect of CB particles. By adding 13 vol% CB, the TS-value of the 34 wt% sample has reached the TS level of the non-reinforced samples with ≥43 wt% ACN content. For ACN ≥ 43 wt%, TS also increases with filler loading, but the reinforcing effect of CB is less pronounced than for samples with ≤39 wt% ACN content.

The investigation of the crystallization and mechanical properties of HNBR polymers and their vulcanizates is of fundamental interest, but the findings of this study might also be of relevance for practical applications in the rubber industry. For example, the reinforcing effect of SIC of high ACN polymers could be utilized for applications (e.g. seals) in chemically aggressive environments where NR cannot be used.

REFERENCES

Auriemma, F. et al., 2020. Effect of stretching on the crystallization of un-crosslinked ethylene/propylene(/diene) random copolymers. *Polymer*, Band 199, p. 122540.

Bradford, W. & Wood, M., 1994. High strength HNBR - The new benchmark elastomer for automotive synchronous and serpentine belts. *SAE Technical Paper 940551*.

European Chemicals Agency, 2023. [Online] Available at: https://echa.europa.eu/de/-/echa-publishes-pfas-restriction-proposal [Zugriff am 6 3 2024].

Femina, G. et al., 2023. X-ray diffraction study of strain-induced crystallization of hydrogenated nitrile-butadiene rubbers: Effect of crosslink density. *Polymer*, Band 271, p. 125782.

Femina, G. et al., 2024. Molecular chain orientation and mechanical properties of hydrogenated nitrile-butadiene rubbers with different acrylonitrile content and crosslink density. *Polymer*, Band 295, p. 126749.

Gögelein, C. & van Duin, M., 2023. Sulfur vulcanization of low- versus high-unsaturated rubbers (IIR and EPDM versus NR and BR): part II - network structure and tensile properties. *Rubber Chem. Technol.*, 96(3), p. 400–421.

Gordon, M. & Taylor, J. S., 1952. Ideal copolymers and the second-order transitions of synthetic rubbers. i. non-crystalline copolymers. *J. Appl. Chem.*, 2(9), pp. 493–500.

He, X., Shi, X., Hoch, M. & Gögelein, C., 2016. Mechanical properties of carbon black filled hydrogenated acrylonitrile butadiene rubber for packer compounds. *Polym. Test.*, Band 53, pp. 257–266.

Huneau, B., 2011. Strain-induced crystallization of natural rubber: a review of X-ray diffraction investigations. 84(3), p. 425–452.

Kobatake, T., Kodama, K., Hayashi, S. & Yoshioka, A., 1997. Improvement of low-temperature flexibility of hydrogenated nitrile—butadiene rubber. *Rubber Chem. Technol.*, 70(5), pp. 839–854.

Liu, J. et al., 2022. Investigation of filler network percolation in carbon black (CB) filled hydrogenated butadiene-acrylonitrile rubber (HNBR). *Polym. Bull.*, Band 79, p. 87–96.

Obrecht, W. et al., 1986. Hydrierter Nitrilkautschuk ein Wekrkstoff mit neuen Eigenschaften. *Angew. Macromol. Chem.*, Band 145/146, pp. 161–179.

Röthemeyer, F. & Sommer, F., 2013. *Kautschuk Technologie*. 3 Hrsg. Munich: Carl Hanser Verlag.

Sedlmeier, C. et al., 2022. From powder to sheets: a comparative electrolyte study for slurry-based processed solid electrolyte/binder-sheets as separators in all-solid-state batteries. *J. Electrochem. Soc*, Band 169, p. 070508.

Wrana, C., 2014. *Polymerphysik*. Heidelberg: Springer Spektrum.

Constitutive Models for Rubber XIII – Dal (Ed.)

A comprehensive exploration of strain-induced crystallization through surface calorimetry and thermodynamics analysis

V.N. Khiêm & J.-B. Le Cam
Department of Continuum Mechanics, RWTH Aachen University, Germany
Institut de Physique UMR 6251 CNRS/Université de Rennes, Rennes, France

ABSTRACT: To date, investigations of strain-induced crystallization are out of reach for most mechanical laboratories due to costly X-ray diffraction facilities. Making use of the exothermic nature of such phase transition, infrared thermography based quantitative surface calorimetry has provided an alternative approach to investigate strain-induced crystallinity in conventional mechanical laboratories. Moreover, this calorimetric approach provides continuum quantities of interest for enriching and validating constitutive models. Nevertheless, its application has been confined to the evaluation of crystallinity under only the loading case. In this contribution, strain-induced crystallization is investigated in various types of natural rubbers by coupling data analysis with domain knowledge of thermodynamics. It pushes the boundaries of the current quantitative surface calorimetry methodology to explore not only the crystallinity during loading-unloading cycles but also different compositions of the internal energy.

1 INTRODUCTION

The transformation of amorphous polymer chains under deformation into ordered molecular structures is known as strain-induced crystallization (SIC). This process holds significant role in understanding the resistance of natural rubber (NR) to crack propagation. However, real-time observations of strain-induced crystallite formations have predominantly relied on X-ray diffraction (XRD) techniques.

From a thermodynamic standpoint, SIC in NR represents a highly exothermic phenomenon. Consequently, conventional X-ray diffraction techniques offer limited insights into this process, failing to capture critical aspects such as energy transfer during crystallization and melting. Misinterpretation of stress-strain hysteresis in NR diagrams can easily occur due to this limitation, as noted by previous studies (Samaca Martinez et al. 2013b, Le Cam et al. 2015, Khiem and Itskov 2018b).

Recent advancements have introduced infrared thermography in combination with mechanical measurements to monitor temperature evolution during SIC in NR (Samaca Martinez et al. 2013a, Candau et al. 2015, Spratte et al. 2017). Moreover, calorimetric analyses during cyclic loading have provided profound insights into crystallization kinetics and the thermomechanical behavior of NR (Samaca Martinez et al. 2013b). Recent research efforts (Le Cam 2018) have integrated infrared thermography-based quantitative surface calorimetry to accurately quantify strain-induced crystallinity in unfilled NR under the loading case. These findings show strong agreement with crystallinity measurements obtained via XRD on identical NR samples (Le Cam et al. 2020).

Nevertheless, crystallization kinetics under loading and unloading are significantly different (Khiêm and Itskov 2018, Khiêm et al. 2022b). Thus, initial endeavors to apply the aforementioned method to quantify strain-induced crystallinity under unloading condition yielded unsatisfactory results. In this paper, we investigate SIC in various types of NR by coupling data analysis with domain knowledge of thermodynamics. This results in an excellent alignment with the crystallinity observed via X-ray diffraction throughout loading-unloading cycles, thereby facilitating the derivation of novel insights into the internal energy of NR.

2 EXPERIMENT SETUP

The material considered here is an unfilled NR. The test was carried out with a rectangular specimen with dimensions 24x10x2mm. Figure 1 presents an overview of the experimental setup composed of a biaxial testing machine, an infrared (IR) camera and the rectangular-shaped specimen. It should be noted that an optical camera and a home-made LED light appear in this figure.

DOI: 10.1201/9781003516880-18

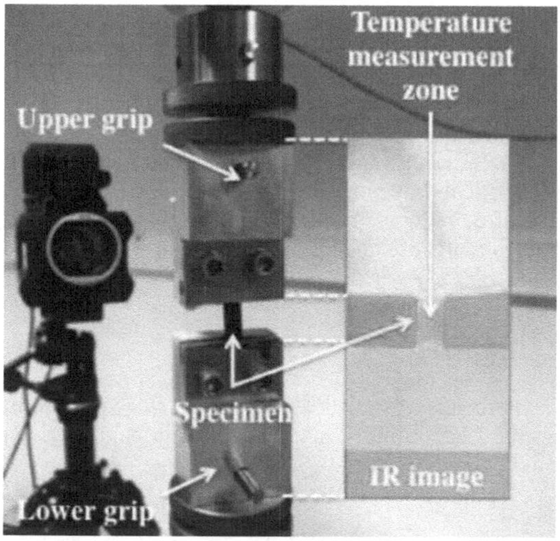

Figure 1. Overview of the experimental setup.

3 THERMODYNAMICS OF STRAIN-INDUCED CRYSTALLIZATION

The local form of the first law of thermodynamics is given in the Lagrangian description by (Khiêm and Itskov 2018)

$$\dot{U}_F + \dot{U}_\theta = \mathbf{P} : \dot{\mathbf{F}} - \operatorname{Div} \boldsymbol{Q} + R + \hat{R}, \qquad (1)$$

where $\boldsymbol{Q}$ is the heat influx per unit referential area, R is the external radiation, $\hat{R}$ denotes the internal heat supply per unit referential volume of the body, $\mathbf{P}$ is the first Piola-Kirchhoff stress tensor and the superposed dot indicates the material time derivative. The heat flux can be written as $\boldsymbol{Q} = -\mathbf{F}^{-1}\mathbf{k}_0\mathbf{F}^{-\mathrm{T}}$ det $\mathbf{F}$ Grad T, where $\mathbf{k}_0$ is the spatial thermal conductivity tensor and $\mathbf{F}$ is the deformation gradient. The first term in the right hand side of (1) is the deformation-dependent part of the internal energy. The second term U_θ due to the rise of temperature is given by

$$U_\theta = \rho_0 c(T - T_0), \qquad (2)$$

where c is the heat capacity and ρ_0 is the density of NR in the reference configuration.

In the case of NR, the internal heat supply resulting from crystallization is given by (Khiêm and Itskov 2018)

$$\hat{R} = \Delta h \dot{\zeta}, \qquad (3)$$

where Δh is the latent heat of fusion per unit volume (Roberts and Mandelkern 1955) and ζ is the total crystallinity.

In view of (2), one can reformulate (1) as

$$\hat{S} = \rho_0 c\dot{T} + \operatorname{Div} \boldsymbol{Q} - R = \mathbf{P} : \dot{\mathbf{F}} - \dot{U}_{\mathbf{F}} + \hat{R}. \qquad (4)$$

Since the heat source $\hat{S}$ and stress power $\mathbf{P} : \dot{\mathbf{F}}$ can be measured, the crystallization degree ζ can be quantified once the prior knowledge of the internal energy $\dot{U}_{\mathbf{F}}$ is given.

4 QUANTITATIVE SURFACE CALORIMETRY

Quantitative surface calorimetry (QSC) involves the identification of the heat source through temperature field measurement, employing the heat diffusion equation. The temperature field, captured utilizing an infrared (IR) camera at the flat surface of a specimen, is inherently two-dimensional. Consequently, to accommodate this dimensional constraint, the above three-dimensional heat diffusion Equation (4) should be simplified. In pursuit of this simplification, a series of assumptions are introduced.

- The heat conduction is considered isotropic so that $\boldsymbol{Q} = -k_0\mathbf{C}^{-1}$ det $\mathbf{F}$ Grad T, where k_0 denotes the thermal conductivity coefficient and $\mathbf{C} = \mathbf{F}^{\mathrm{T}}\mathbf{F}$ is the right Cauchy-Green tensor.
- The material is incompressible so that $\det \mathbf{F} = 1$.
- The temperature is assumed constant along the specimen thickness. This indicates that the specimen must be relatively thin and the stress state is plane-stress. Thus, (4) can be integrated over the thickness as

$$\begin{aligned} &\rho_0 c\left(\dot{T} + \frac{T - T_{amb}}{\tau_{2D}}\right) \\ &- \operatorname{Div}_{2D}(k_0\mathbf{C}^{-1}\operatorname{Grad}_{2D}T) - R = \hat{S}, \end{aligned} \qquad (5)$$

where T_{amb} is the ambient temperature. τ_{2D} is a characteristic time of the heat exchanges along the thickness direction by convection with the air at the specimen surface.

- The external radiations R are assumed to remain constant during the deformation process. In this case, R can be removed from the heat diffusion equation by using the temperature variation $\theta(X, Y, t) = T(X, Y, t) - T_{ref}(X, Y)$ instead of the temperature itself. It should be noted that T_{ref} is the ambient temperature if it is constant over the test. In this case, (5) reads

$$\rho_0 c\left(\dot{\theta} + \frac{\theta}{\tau_{2D}}\right) - \operatorname{Div}_{2D}(k_0\mathbf{C}^{-1}\operatorname{Grad}_{2D}\theta) = \hat{S}. \qquad (6)$$

For materials with low thermal diffusivity, such as rubber, the thermal gradient tends to be predominantly localized within the vicinity of the grips of the testing machine. Consequently, under conditions where either the heat conduction within the specimen plane is negligible or the heat exchange with the grips does not generate temperature gradient in the measurement zone, the two-dimensional heat diffusion equation can be further reduced (Chrysochoos 1995, Berthel et al. 2008) to a "0D" formulation of the heat diffusion equation (Khiêm et al. 2022a)

$$\rho_0 c\left(\dot{\theta}+\frac{\theta}{\tau}\right)=\hat{S}. \tag{7}$$

This equation holds for homogeneous heat source field. Furthermore, it is assumed that $\tau_{2D}=\tau$, where τ represents the characteristic time of a natural return to room temperature after a heating (or a cooling) for each testing configuration. In this case $\theta=\theta_0 e^{\frac{-(t-t_0)}{\tau}}$, and τ can be determined at different incremental stretches, as discussed in Samaca Martinez et al. 2013b.

5 CRYSTALLIZATION KINETICS

The success of our theories on SIC (Khiêm and Itskov 2018b, Khiêm et al. 2022b) relies on a postulation of two crystallite types. Here, we shortly summarize the kinetics of these crystallites as it serves as the domain knowledge for the analysis of internal energy (Section 6).

According to our equilibrium theory of SIC (Khiêm and Itskov 2018), the crystallization in NR contains two steps: formation of a solid cluster of polymer segments (nucleus) in the amorphous network and growth of such cluster. The former is referred to as crystallization nucleation which can occur in both loading and unloading, while the latter is referred to as crystallization growth solely taking place in unloading.

The primary nucleus is assumed to be bundle-like with a cylindrical geometry. Each primary nucleus is built by C neighbouring polymer strands. The change in the free energy of a strand due to the formation of a nucleus in a direction i of the microsphere is given by

$$\Delta G=\gamma_t 2Cs+\gamma_s 2\sqrt{\pi Csl}-Csl\Delta\Psi_{\mathrm{i}}, \tag{8}$$

where γ_t is the surface tension at the top surfaces of the crystallite. γ_s is the surface tension at the side of the crystallite. $\Delta\Psi_{\mathrm{i}}$ is the difference in bulk free energies between the semi-crystalline and the amorphous strands. s is the cross-section area of a single polymer chain, and l is the length of the crystallite.

Thus, according to the Lauritzen-Hoffman nucleation theory (Hoffman and Lauritzen, Jr. 1961), the critical number of molecules C^* in the nucleus can be obtained by maximizing (8) with respect to C

$$C^*=\frac{4\pi\gamma_s^2}{s\Delta\Psi_{\mathrm{i}}^2}. \tag{9}$$

Nuclei with the number of molecules larger than this critical value are likely to form solid crystallites. The free energy barrier required for the crystal nucleation is obtained by substituting C^* into (8) as

$$\Delta G^*=\frac{8\pi\gamma_t\gamma_s^2}{\Delta\Psi_{\mathrm{i}}^2}. \tag{10}$$

By means of statistical mechanics, the equilibrium primary crystallinity in each direction can be given by (Khiêm and Itskov 2018)

$$\zeta^{\mathrm{I}}=\sum B_1\exp\left(-\frac{\Delta G^*}{k_{\mathrm{B}}T_0}\right), \tag{11}$$

where B_1 is the primary crystallinity constant.

Under unloading, mobility of molecules reduces and the secondary crystallites grow by aggregation of polymer segments into the critical nuclei (Khiêm and Itskov 2018). The number of secondary crystallites in the direction i is a multiplication of the number of nuclei and the number of grown crystallites n_G from each nucleus.

$$\zeta^{\mathrm{II}}=\sum B_2\exp\left(-\frac{\Delta\bar{G}^*}{k_{\mathrm{B}}T_0}\right)\cdot n_G \tag{12}$$

where B_2 is the secondary crystallinity constant.

Thus, the total crystallinity is the sum of the primary and secondary crystallinity

$$\zeta=\zeta^{\mathrm{I}}+\zeta^{\mathrm{II}}. \tag{13}$$

This theory has been successfully applied to both unfilled and filled NR and facilitates elucidation of the non-monotonic crystallinity (Candau et al. 2014) and misorientation angle of the crystallites (Trabelsi et al. 2003) during unloading.

6 INTERNAL ENERGY

Motivated by the above kinetics of SIC, the internal energy $U_{\mathbf{F}}$ also contains two parts

$$U_{\mathbf{F}}=U_{\mathbf{F}}^{\mathrm{I}}+U_{\mathbf{F}}^{\mathrm{II}}. \tag{14}$$

$U_{\mathbf{F}}^{\mathrm{I}}$ results from the internal energy due to deformation of the primary crystallites, while $U_{\mathbf{F}}^{\mathrm{II}}$ is the internal energy due to deformation of the secondary crystallites. Since the former can be easily quantified from the calorimetry analysis (see e.g. Le Cam 2018), hereafter we focus on the definition of $U_{\mathbf{F}}^{\mathrm{II}}$.

To this end, it is assumed that there are ζ^{II} secondary crystallites at a material point of the sample. For the sake of simplicity, a single secondary crystallite is considered as a deformable solid with Young modulus E. Thus, its internal energy can be expressed as

$$u = \frac{1}{2}E\varepsilon^2, \tag{15}$$

where ε is the strain due to deformation of the secondary crystallite. Thus, the internal energy of secondary crystallites with the crystallinity ζ^{II} are given by

$$U_{\mathbf{F}}^{\mathrm{II}} = \zeta^{\mathrm{II}}\frac{1}{2}E\varepsilon^2. \tag{16}$$

Due to the two-type crystallization kinetics, the internal energy $U_{\mathbf{F}}^{\mathrm{II}}$ is only activated during unloading.

7 RESULTS

In this section, calorimetric data including heat source and stress power (Figure 2) of an unfilled NR (Section 2) is used to quantify crystallinity and internal energy of the sample.

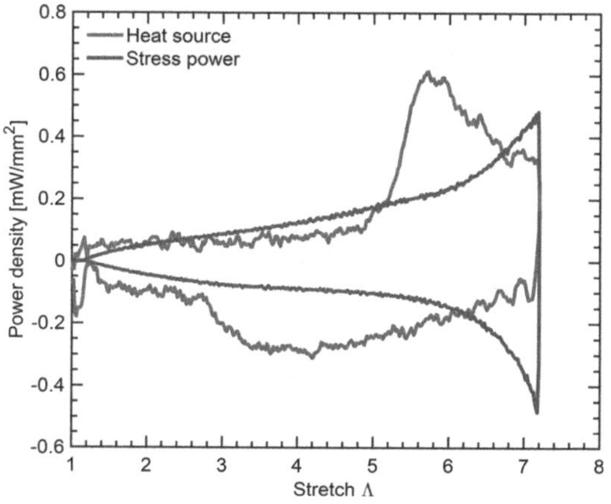

Figure 2. Heat source and stress power of unfilled NR versus stretch.

In view of (14), the first law of thermodynamics can be given by

$$\hat{S} = \mathbf{P} : \dot{\mathbf{F}} - \dot{U}_{\mathbf{F}}^{\mathrm{I}} - \dot{\zeta}\left(\Delta h - \frac{1}{2}E\varepsilon^2\right) - (\zeta - \zeta^{\mathrm{I}})E\varepsilon\dot{\varepsilon} - \dot{\zeta}^{\mathrm{I}}\frac{1}{2}E\varepsilon^2. \tag{17}$$

Since the primary crystallinity and its internal energy $U_{\mathbf{F}}^{\mathrm{I}}$ can be directly extracted from the calorimetric analysis, the unknown total crystallinity can be integrated from (17). For validation of our thermodynamics-informed quantitative surface calorimetry analysis, the quantified total crystallinity from (17) is plot against that detected by X-ray diffraction technique on the same NR (Figure 3).

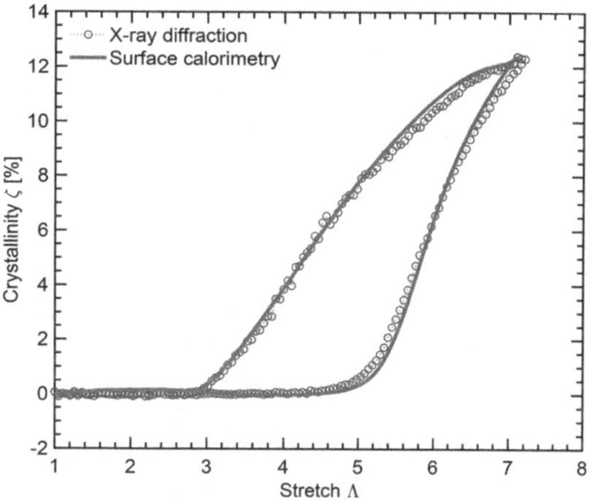

Figure 3. Crystallinity discovered by thermodynamics-based quantitative surface calorimetry versus X-ray diffraction measurements.

It can be seen that the discovered crystallinity shows excellent agreement with X-ray diffraction measurements. We can also extract the internal energy due to deformation of NR under any loading condition, a quantity that has not been made accessible by any measurement technique (Figure 4).

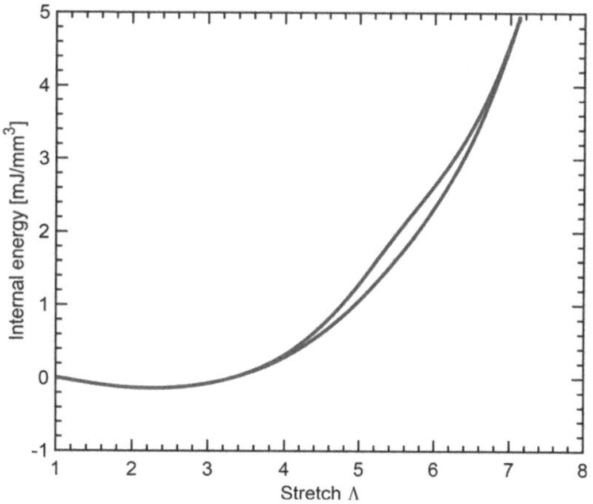

Figure 4. Internal energy due to deformation in NR.

8 CONCLUSION

In the present study, we introduce a novel thermodynamics-based quantitative surface calorimetry technique to discover strain-induced crystallinity in NR.

Through the integration of domain knowledge into the current quantitative surface calorimetry analysis, our methodology achieves the SIC disclosure. Most remarkably, our approach demonstrates a high degree of concordance with XRD measurements conducted on identical NR specimens in both loading and unloading conditions, thereby affirming its reliability and validity. Furthermore, our methodology allows for quantification of internal energy within NR specimens, facilitating understanding of their compositions. This discovery enables more comprehensive understanding of SIC, a perspective not attainable through any other data acquisition technique.

REFERENCES

Berthel, B., A. Chrysochoos, B.Wattrisse, & A. Galtier (2008). Infrared image processing for the calorimetric analysis of fatigue phenomena. *Experimental Mechanics 48*, 79–90.

Candau, N., R. Laghmach, L. Chazeau, J.-M. Chenal, C. Gauthier, T. Biben, & E. Munch (2014). Strain-induced crystallization of natural rubber and cross-link densities heterogeneities. *Macromolecules 47*(16), 5815–5824.

Candau, N., R. Laghmach, L. Chazeau, J.-M. Chenal, C. Gauthier, T. Biben, & E. Munch (2015). Influence of strain rate and temperature on the onset of strain induced crystallization in natural rubber. *European Polymer Journal 64*, 244–252.

Chrysochoos, A. (1995). Analyse du comportement des matériaux par thermographie infra rouge. In *Colloque Photomécanique*, Volume 95, pp. 201–211.

Hoffman, J. D. & J. I. Lauritzen, Jr. (1961). Crystallization of bulk polymers with chain folding: Theory of growth of lamellar spherulites. *Journal of Research of the National Bureau of Standards-A. Physics and Chemistry 65A*(4), 297.

Khiêm, V. N., J.-B. L. Cam, S. Charlès, & M. Itskov (2022a). Thermodynamics of strain-induced crystallization in filled natural rubber under uni- and biaxial loadings, part i: Complete energetic characterization and crystallinity evaluation. *Journal of the Mechanics and Physics of Solids 159*, 104701.

Khiêm, V. N., J.-B. L. Cam, S. Charlès, & M. Itskov (2022b). Thermodynamics of strain-induced crystallization in filled natural rubber under uni- and biaxial loadings, part II: Physically-based constitutive theory. *Journal of the Mechanics and Physics of Solids 159*, 104712.

Khiêm, V. N. & M. Itskov (2018). Analytical network-averaging of the tube model: Strain-induced crystallization in natural rubber. *Journal of the Mechanics and Physics of Solids 116*, 350–369.

Le Cam, J.-B., P. Albouy, & S. Charlès (2020). Comparison between X-ray diffraction and quantitative surface calorimetry based on infrared thermography to evaluate strain-induced crystallinity in natural rubber. *Review of Scientific Instruments*, in publication.

Le Cam, J.-B., J. R. Samaca Martinez, X. Balandraud, E. Toussaint, & J. Caillard (2015). Thermomechanical analysis of the singular behavior of rubber: Entropic elasticity, reinforcement by fillers, strain-induced crystallizaation and the Mullins effect. *Experimental Mechanics 55*(4), 771–782.

Le Cam, J.-B. (2018). Strain-induced crystallization in rubber: A new measurement technique. *Strain 54*(1), e12256. e12256 10.1111/str.12256.

Roberts, D. E. & L. Mandelkern (1955). Thermodynamics of crystallization in high polymers: Natural rubber. *Journal of the American Chemical Society 77*(3), 781–786.

Samaca Martinez, J. R., J. B. Le Cam, X. Balandraud, E. Toussaint, & J. Caillard (2013a). Mechanisms of deformation in crystallizable natural rubber. Part 1: Thermal characterization. *Polymer 54*(11), 2717–2726.

Samaca Martinez, J. R., J. B. Le Cam, X. Balandraud, E. Toussaint, & J. Caillard (2013b). Mechanisms of deformation in crystallizable natural rubber. Part 2: Quantitative calorimetric analysis. *Polymer 54*(11), 2727–2736.

Spratte, T., J. Plagge, M. Wunde, & M. Klüppel (2017). Investigation of strain-induced crystallization of carbon black and silica filled natural rubber composites based on mechanical and temperature measurements. *Polymer 115*, 12–20.

Trabelsi, S., P. A. Albouy, & J. Rault (2003). Effective local deformation in stretched filled rubber. *Macromolecules 36*(24), 9093–9099.

Constitutive Models for Rubber XIII – Dal (Ed.)
© 2025 The Author(s), ISBN 978-1-032-85138-9

Adhesive properties of tyre tread smear wear and its link to transfer layer deposition onto road surfaces

E. Koliolios & J.J.C. Busfield
Queen Mary University of London, London, UK

S. Nakano & T. Kawamura
Sumitomo Rubber Industries Ltd., Kobe, Japan

ABSTRACT: Rubber wear is typically categorised into three different processes known as abrasive wear, fatigue wear, and smear wear. The mechanisms of the abrasive and fatigue wear have been extensively studied in the past. There are much fewer reports into the smear wear mechanism. Smear wear has also been identified as 'sticky debris' by several authors, but there have been no attempts to quantify and characterise this so called 'stickiness'. Additionally, the 'stickiness' could potentially be linked to the transferring of rubber material to a surface when it is abraded against it. This transferred rubber debris is commonly referred to as a transfer layer. This research therefore aims to initially quantify the adhesion of smear wear debris produced from a variety of tyre tread compounds composed of different polymers and fillers. A new experimental method was used to evaluate adhesion energy. This new technique evaluated the adhesive properties of smear wear generated from different tyre tread compounds during abrasion. These results were linked to independent surface abrasion experiments to provide in-depth understanding of smear wear adhesion and how it influences the amount of transfer layer deposited on the countersurface during abrasion.

1 INTRODUCTION

Measuring the adhesive forces between interfaces has been of interest in several scientific fields including rubber tribology. The methods of evaluating adhesion range from mechanical to chemical. One of the most common mechanical methods uses a peel test. This involves different geometrical setups where an adhesive is applied between two interfaces and then the energy required to peel the interface is measured (Bartlett (2023)). Currently, standard methods to characterise adhesion use atomic force microscopy (AFM) or a nano-indenter. The idea of using the AFM as a tool to measure adhesion for polymers was discussed by Mizes (1991) and Creuzet (1992). It involves the loading and unloading of a nano-indenter on a surface of interest to evaluate the energy required for the two surfaces to be fully detached after unloading (Mizes 1991). Although AFM measured adhesion provides repeatable results, a faster and simpler alternative macroscopic approach was required here. Liquid droplet contact angles can also be used as a chemical method to evaluate surface energy of a material which can then be linked to adhesive force, as performed by Maeda (2007) for rubber compounds. Nevertheless, the rough surface produced during rubber abrasion, and the use of hazardous chemicals makes this method of quantifying adhesion surfaces much more difficult. Such limitations in the available methodologies have led to the development of a new method for adhesion characterisation.

In this investigation, the adhesive forces created through the existence of a so called 'smear wear layer' which is formed during abrasion on the outer surface of tyre tread were the main point of interest. Rubber failure can divide to three different categories, abrasive wear, fatigue wear and smear wear (Southern and Thomas 1973), (Gent and Pulford 1989). Abrasive wear and fatigue wear have no apparent effect on altering the surface adhesion of rubber although both play a significant role on the resulting surface roughness (Schallamach 1958). However, when a smear wear layer forms during abrasion, the rubber surface has been characterised as sticky (Gent and Pulford 1989).

As a result, the smear wear layer has been a subject of investigation for numerous researchers (Gent and Pulford 1989), (Wu 2017 and 2023) (Huang 2018), (Koliolios 2022). Tiwari (2021) discussed extensively both the formation and the transferring of the smear wear film from a rubber surface to a hard solid surface as well as how this transferred layer influences friction and adhesion. However, the extent of this transfer layer was not evaluated and the detailed effect of compound design and how this influenced the transfer effect.

Therefore, the goal of this study was to understand this smear wear adhesion to a substrate with a new robust experimental method; using a screw driven

DOI: 10.1201/9781003516880-19

Instron universal test machine. Additionally, the elucidation of the adhesion mechanisms was performed to establish links with the mechanism of transferring of the smear film onto a substrate through an independent surface abrasion experiment.

2 MATERIALS AND METHODS

2.1 *Adhesion instrumentation*

The schematic of the instrumentation used for the adhesion measurement is presented in Figure 1.

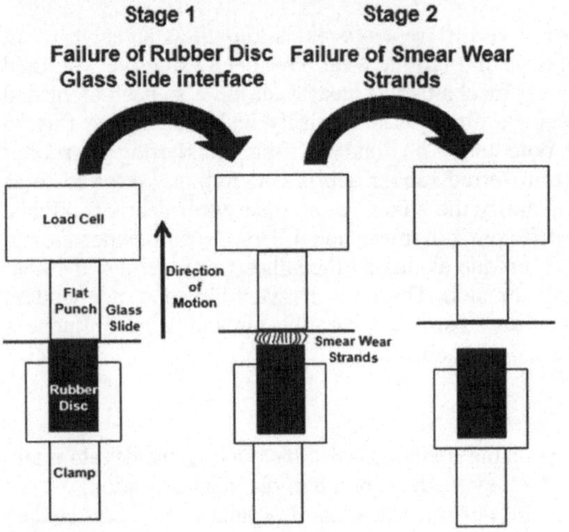

Figure 1. Adhesion method measurement schematic.

The setup was composed of an upper part, which included a 100 N load cell, a flat punch which was attached to the load cell and a VWR Superfrost Plus microscope adhesion slide which was attached to the flat punch using double-sided tape. The lower part of the setup consisted of an abraded rubber disc which was fixed in place using a clamp. The upper part of the setup was brought into contact with the lower part as shown in Figure 1 until the load reached a compressive force of 2.0 N and an interface was formed between the glass slide and the rubber disc. The upper part then moved upwards at a displacement rate of 1 mm/min as shown in Figure 1 until the interface between the glass slide and the rubber disc failed. The energy released during this process was labelled as total energy of adhesion and it is composed of two failure processes. The first failure of the interface is of a purely interfacial nature; the glass slide is detached from the interface. The second failure is more cohesive in nature; since smear wear strands attach to the glass slide and failure occurs at a much larger displacement of the upper part. Regardless, both stages of failure contribute to the adhesion of the smear wear layer to a substrate and thus an additional contribution of friction between the rubber and the road surface.

2.2 *Sample preparation*

To evaluate how different compound recipes influence the adhesion of the smear wear layer to a substrate, four tyre tread compounds were selected. Two synthetic polyisoprene rubber (IR) compounds one filled with carbon black (CB), and one filled with silica and two styrene-butadiene rubber (SBR) compounds, similarly one filled with CB and the other filled with silica. The total filler loading in all the compounds was 70 parts per hundred rubber (phr). The naming convention of the compounds as well as the main polymer and filler types used is presented in Table 1.

Table 1. Rubber compounds.

Compound Name	Polymer	Filler
IRN220	IR	CB N220
IRVN3	IR	UltrasilVN3
SBRN220	SBR	CB N220
SBRVN3	SBR	UltrasilVN3

The compounds were moulded into rubber discs with a diameter of 54 mm and a thickness of 13 mm. The compounds were then abraded using the blade abrasion machine presented in Figure 2.

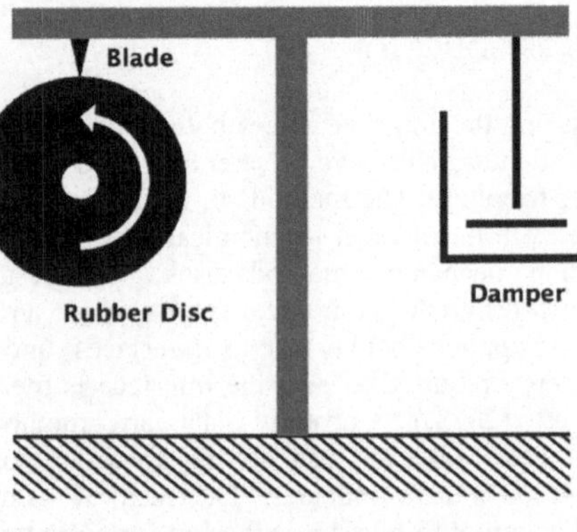

Figure 2. Blade Abrasion Machine Schematic (Adapted from Wu (2023)).

The blade abrasion machine consists of a blade, a damper filled with oil which reduces the vibration, and different dead weights which are applied on top of the blade. The rotational speed of the disc can be varied. For this study, two rotational speeds of 100 rpm and 75 rpm were used, as well as two different loads of 10 N and 6.87 N. After the discs were abraded for 30 minutes, they were immediately placed on the adhesion setup and an infrared thermometer was used to measure their temperature during testing.

2.3 *Surface abrasion and transfer layer measurement*

A surface abrasion instrumentation was used to evaluate and measure the abrasion behaviour and the amount of debris transferred as a film on an abrasive surface. The surface abrasion equipment consists of a motor, an applied load in the form of a dead weight, a 12-grit silicon carbide grinding block, a clamped rubber disc and a thermal camera which measured the temperature of the disc during abrasion. A schematic of the surface abrasion machine is presented in Figure 3.

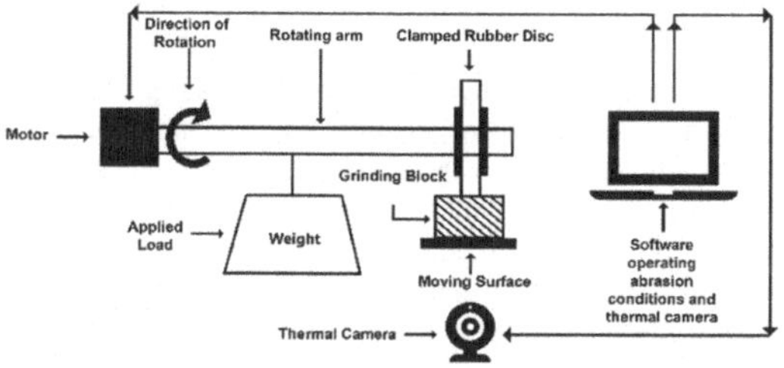

Figure 3. Surface Abrasion Machine Schematic (Adapted from Kyei-Manu (2022)).

Three different motor rotational speeds were used for this investigation, 60 rpm, 120 rpm and 180 rpm. The applied load was kept constant at 20 N and the grinding block surface moved continuously forwards and backwards at a speed of 5 mm/min. The dimensions of the clamped rubber disc are the same as the discs used during the blade abrasion experiment. Each surface abrasion experiment lasted for 30 minutes.

The mass of the rubber disc was measured before and after the experiment. The mass difference obtained is symbolised with ΔM_{disc}. The mass of the grinding block was also measured before and after the experiment. The grinding block's mass increased during the experiment and the mass difference is symbolised by $\Delta M_{grinding\ block}$. The transfer layer percentage of mass loss ($TL\%$) was calculated using the following equation:

$$TL\% = 100\% \times \frac{\Delta M_{grinding\ block}}{\Delta M_{disc}} \qquad (1)$$

2.4 *Characterisation of smear wear layer samples*

Characterisation was also performed on the smear wear layer samples. The samples were collected during blade abrasion by pressing the adhesive slides on the abrading rubber disc, resulting in a smear wear layer sample attaching on the surface of the glass slide. Similar characterisation was also performed previously by Koliolios (2022) and Kyei-Manu (2022).

A TA instruments DSC 25 was used to measure the glass transition temperature (T_g) smear wear layer samples generated. Approximately 5 mg of each specimen was hermetically sealed in an aluminium pan and subjected to a heat-cool-heat-cool cycle in a nitrogen atmosphere. The samples were heated from room temperature to 100°C, cooled to −70°C, heated to 100°C and finally cooled to 25°C at a rate of 10°C/min. The glass transition temperature of each specimen was calculated as per ISO11357 standard procedure.

An Agilent 1260 Infinity Gel Permeation Chromatography (GPC) system was used to evaluate the molecular weight distribution of smear wear layer samples. A minimum of 10 mg of smear wear samples were dissolved in tetrahydrofuran (THF) for a minimum of 48 hours. The resulting solutions were filtered through a 0.2 μm polytetrafluoroethylene syringe filter and then GPC runs were performed with a constant flow rate of 1 mL/min for 30 minutes. The GPC was equipped with Refractive Index Detector (RID), 2 PLgel 5 μm mixed-C columns (300mmx 7.5 mm), a PLgel 5 mm guard column (50 mm x 7.5 mm) and an autosampler. The columns and the RID were operated at 40°C. The instrument was calibrated with linear narrow poly (methyl-methacrylate) standards in a range of 550 to 2136000 g/mol.

2.5 *Processing of experimental data*

The data generated from the adhesion experiment is of the form of load versus extension. An example of the adhesion profiles which were encountered during initial tests are presented in Figure 4.

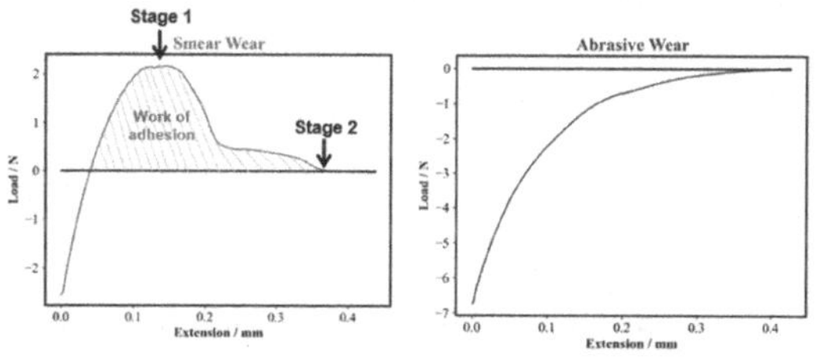

Figure 4. Example of adhesion profile created by a compound which has a smear wear layer on its abraded surface (left) and one that does not have one (right).

Figure 5. Surface or rubber disc tested with smear wear (left) and abrasive wear (right).

The compound in the left of the figure has produced a smear wear layer during abrasion and as a result a work of adhesion can easily be measured during the test. The compound on the right produced abrasive wear which results in no work of adhesion. The surfaces of the two rubber discs tested can also be seen in Figure 5. There was an evident difference in the surfaces of the two specimens, since the surface which has generated smear wear was smooth with no visible saw-tooth pattern whereas the surface which generated abrasive wear was rough with a clearly defined saw-tooth pattern.

Since specimens which have produced abrasive wear during blade abrasion do not generate any adhesive forces, this research only focused on the compounds and conditions which generated a smear wear layer and thus adhesive forces during the experiment. The resulting adhesion profiles were processed as shown in Figure 6, where an example test of sample IRVN3 is presented.

The adhesion profile is divided into two different stages as also explained in Figure 1. Stage 1 failure, which is the failure of the interface between the rubber disc surface and the glass slide surface, occurs when the measured force reaches a maximum. Stage 2 failure, which is the failure of the smear wear strands which have been attached to the glass slide, occurs when the measured force reaches the value of zero. Therefore, three energy values of interest were generated during this experiment, according to the equations presented below:

$$E_{IA} = \int_{x_1 \text{ at } F=0}^{x \text{ at } F=max} F(x)\mathrm{d}x \quad (2)$$

Where E_{IA} is defined as the energy of interfacial adhesion, x is defined as the extension, F is defined as the load and x_1 is the lowest value of the extension where $F = 0$, as shown in Figure 6.

$$E_{SF} = \int_{x \text{ at } F=max}^{x_2 \text{ at } F=0} F(x)\mathrm{d}x \quad (3)$$

Where E_{SF} is defined as the strand failure energy and x_2 is the highest value of the extension where $F = 0$, as shown in Figure 6.

$$E_{TEA} = E_{IA} + E_{SF} \quad (4)$$

Where E_{TEA} is defined as the total energy of adhesion. The task of integral calculation has been automated using a Python script for consistency. The results of E_{IA}, E_{SF} and E_{TEA} were thus obtained for each test of the four compounds of interest.

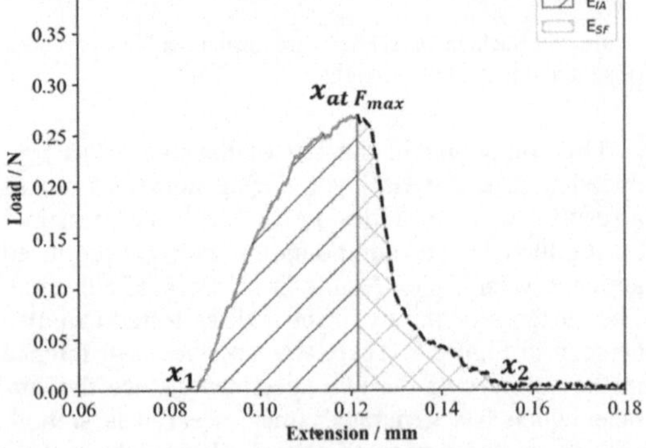

Figure 6. Methodology of adhesion energy calculation for an example test of compound IRVN3.

3 RESULTS AND DISCUSSION

3.1 *Initial adhesion results*

After the processing discussed in Figure 6, an average value for the adhesion energies were obtained for each abrasion conditions for all four different compounds. Three repeats were performed in total for each condition and sample. The measured energies were divided by an approximated area of contact which is equal to 13 mm x 13 mm. The area of contact being the average value of the 'fingerprint' dimensions that each specimen deposits on the glass slide. The values were then plotted against the measured temperature of the rubber discs. The results obtained are presented in Figure 7, Figure 8, and Figure 9 respectively. From the figures, it can be observed that the disc temperature influences the adhesive interactions between the rubber disc and glass slide.

Additionally, a significant difference between SBR and IR compounds can be seen, since IR compounds have lower values for both E_{IA}, E_{SF} and E_{TEA} as a result. This potentially implies that in the case where a smear wear layer is formed during tyre operation, SBR compounds will benefit further from additional adhesion and thus grip. In the case of compound IRN220 it appears that the total energy of adhesion is minimal even compared to IRVN3. Therefore, both the main polymer type and the filler play an important role in this type of interaction between surfaces. Moreover, there are also differences between the two SBR compounds as at certain temperatures it appears that SBRN220 provides stronger E_{TEA} especially at higher operating temperatures.

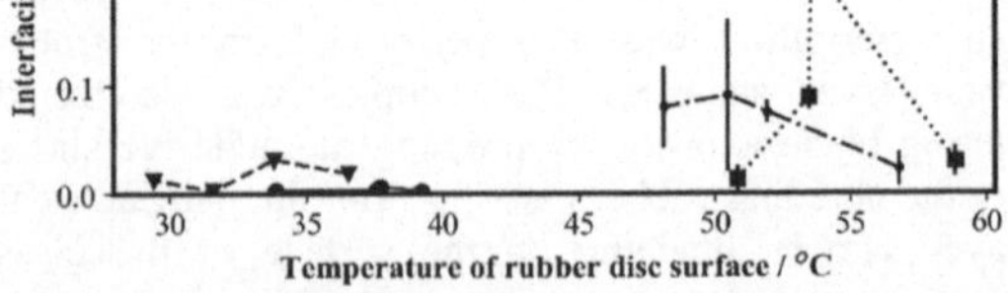

Figure 7. Interfacial Adhesion Energy (E_{IA}) against disc temperature.

SBRN220 also exhibited a peak of adhesion for E_{IA}, E_{SF} and E_{TEA}, which could mean that there are different types of mechanisms in smear wear layer adhesion; one of which increases adhesion with temperature and one of which decreases adhesion with temperature. The occurrence of the peak was also noticed for SBRVN3. However, the peak only occurred in the E_{IA} component, which hints to the

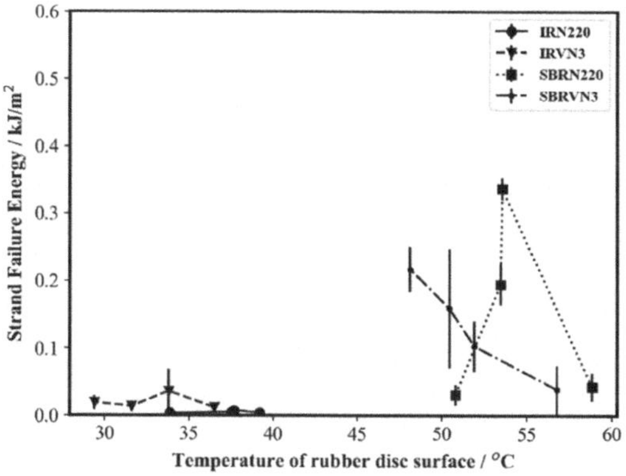

Figure 8. Strand Failure Energy (E_{SF}) against disc temperature.

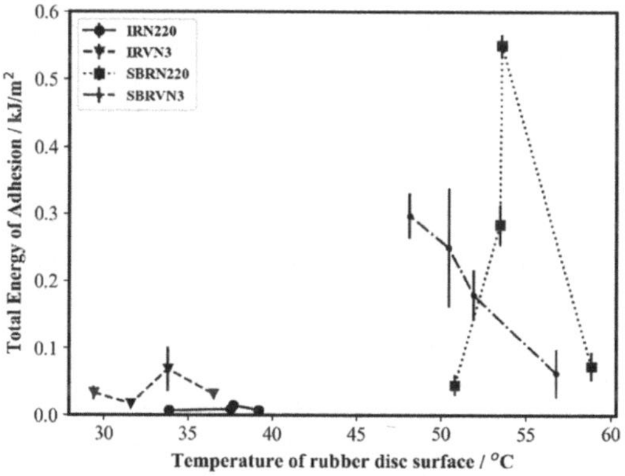

Figure 9. Total Adhesion Energy (E_{TEA}) against disc temperature.

competing mechanism being related to the separation of the interface. Potentially the mechanism which increases E_{IA} with temperature is related to the real area of contact of the two interfaces, which was assumed to be an average value for all compounds, at all measured temperatures. Otherwise, it might also be an effect of the thickness of the smear layer which could also be changing with different abrasion conditions. Therefore, a hypothesis could be made for the nature of the two competing mechanisms. The first mechanism decreases with temperature and relates to the increase in chain mobility and smear strand strength. The second mechanism increases E_{TEA} with temperature and relates either to the unaccounted increase in the real area of contact or a change in the thickness of the smear wear layer.

3.2 *Adhesion and glass transition temperature*

Considering the effect that the glass transition temperature has on the molecular interactions which govern adhesion is crucial when evaluating it. Therefore, considering approaches of other researchers such as Morita (2005), E_{TEA} was plotted against reduced temperature where reduced temperature is defined as (Morita 2005):

$$T_{reduced} = \frac{T - T_g}{T_g} \tag{5}$$

Where T is the operating temperature of the sample with all the temperature values being in Kelvin.

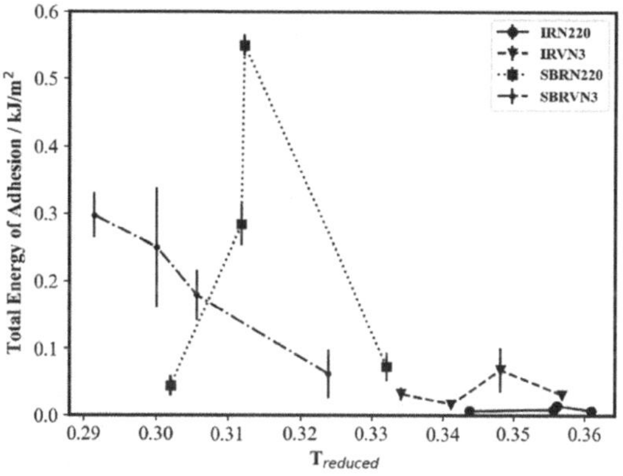

Figure 10. Total Adhesion Energy (E_{TEA}) against reduced temperature.

From Figure 10, it is evident that the lower the reduced temperature the higher the E_{TEA}, like the findings of Morita (2005), due to implications on both polymer chain mobility, filler interactions and material strength. Morita (2005) also mentioned that different separation mechanisms occur at different reduced temperatures. These mechanisms were also identified during these experiments. The three different separation modes mentioned are (a) strand failure dominated separation at lower $T_{reduced}$ (found in SBRVN3 samples) (b) cavitation dominated separation at intermediate $T_{reduced}$ (found in SBRN220 samples) (c) interface dominated separation at high $T_{reduced}$ (found in IRN220 samples). Therefore, the existence of different separation modes could also provide an additional explanation to various different peaks occurring in certain samples. Since, multiple types of separation modes could exist simultaneously along the area of contact.

3.3 *Transfer layer and its links with adhesion*

To link the mechanism of the transferring of the smear film to a countersurface, additional considerations should be made. Successful transfer of material does not necessarily require strong adhesion, but molecules which are mobile enough to transfer from the rubber disc to the grinding block. As a result, both a measure of adhesion strength and molecular mobility must be considered to explain the transfer layer mechanism. Okuno (2001) discussed how the 'stickiness index' and thus the adhesive force of adhesive wear of chloroprene rubber is proportional to the number average molecular weight (M_N) of the uncured rubber samples as well as its viscoelastic properties. The viscoelastic properties of the smear wear samples are potentially accounted for by the reduced temperature. Thus, the number average molecular weight of the smear wear layer samples (M_N) was used to further evaluate the nature of the adhesive interaction. To link the measured adhesion and the $TL\%$, $T_{reduced}$ and M_N will therefore be used.

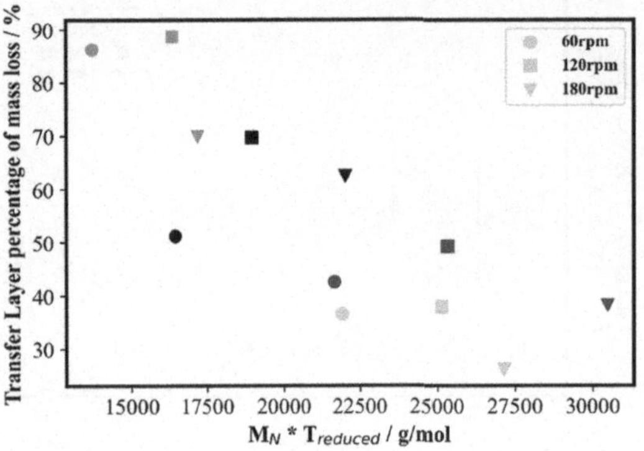

Figure 11. Transfer layer percentage of mass loss against reduced temperature times number average molecular weight.

From Figure 11, an inverse proportional relationship was observed between $TL\%$ and $T_{reduced} * M_N$, similar to the one seen between E_{TEA} and $T_{reduced}$ and as a result the percentage of generated wear material which is transferred is both a function of adhesion energy, through its similar link with and chain mobility through its dependency to M_N.

4 CONCLUSION

To conclude, the energy of adhesion between a rubber sample and a glass slide was evaluated using a new approach. The adhesive energy was evaluated and divided into components for tyre tread compounds abraded under different conditions and thus at different temperatures. The measured adhesive energy was found to be influenced by the main polymer type and the filler type of the elastomer sample as well as the sample temperature during abrasion. Finally, an understanding of the smear layer adhesion obtained through this experiment was used to evaluate the effect that adhesion and compound properties have on the amount of debris that are transferred on a rigid abrasion surface during abrasion at different speeds.

REFERENCES

Bartlett, M. D., Case, S. W., Kinloch, A. J., & Dillard, D. A. 2023. Peel tests for quantifying adhesion and toughness: A review. *Progress in Materials Science, 137*, 101086.

Creuzet, F., Ryschenkow G., Arribart H., 1992. A New Tool for Adhesion Science: The Atomic Force Microscope, *The Journal of Adhesion, 40:1*, 15–25.

Gent, A. N., Pulford, C. T. R., 1983. Mechanisms of rubber abrasion. *Journal of Applied Polymer Science 28*, 943–960.

Huang, M., Guibert, M., Thévenet, J., Fayolle, C., Chaussée, T., Guy, L., Vanel, L., Loubet, J.L., Sotta, P., 2018. A new test method to simulate low-severity wear conditions experienced by rubber tire materials. *Wear 410-411*, 72–82.

Koliolios, E., Nakano, S., Kawamura, T., Tsumori, I., Busfield, J., 2022. Determining the smear wear mechanism in filled SBR tyre tread compounds. *Constitutive Models for Rubber XII. pp.* 483–488.

Kyei-Manu, W.A., Hurrell, P., Akutagawa, K., Busfield, J., Herd, C.R., Tunnicliffe, L.B., 2022. Effect of carbon black properties on the abrasion resistance of rubber compounds at low sliding speed. *Constitutive Models for Rubber XII. pp.* 489–494.

Maeda, K., Bismarck A., Briscoe B., 2007, Effect of bulk deformation on rubber adhesion, *Wear 263*, 1016–1022.

Mizes, H. A., Loh, K.-G., Miller, R. J. D., Ahuja, S. K., Grabowski, E. F. 1991. Submicron probe of polymer adhesion with atomic force microscopy: Dependence on topography and material inhomogeneities. *Appl. Phys. Lett. 25 November 59 (22)*: 2901–2903.

Morita, H., Yamada, M., Yamaguchi, T., & Doi, M. (2005). Molecular Dynamics Study of the Adhesion between End-grafted Polymer Films. *Polymer Journal, 37*, 782–788.

Okuno, S. 2001. Evaluation of adhesive wear and dynamic viscoelasticity of chloroprene rubber. *Nippon Gomu Kyokaishi, 74*(6), 248–252.

Schallamach, A., 1958. Friction and abrasion of rubber. *Wear*, 1(5):384–417.

Southern, E., Thomas, A.G., 1979. Studies of rubber abrasion. *Rubber Chemistry and Technology 52*, 1008–1018.

Tiwari, A., Miyashita, N. & Persson, B.N.J. 2021. Rubber Wear and the Role of Transfer Films on Rubber Friction on Hard Rough Substrates. *Tribol Lett 69*, 42.

Wu, G., 2016. The mechanisms of rubber abrasion. *Queen Mary University of London, London.*

Wu, G., Sotta, P., Huang, M., Tunnicliffe, L.B., Busfield, J.J., 2023. Characterisation of sticky debris generated during smear wear. *Rubber Chemistry and Technology 96*, 588–607.

Constitutive Models for Rubber XIII – Dal (Ed.)
© 2025 The Author(s), ISBN 978-1-032-85138-9

Swelling coefficient of CO_2 gas sorption versus temperature: Arrhenius equation

E. Lainé & J.-C. Grandidier
Institut Pprime, CNRS, ISAE-ENSMA, Université de Poitiers, F-86961 Futuroscope, France

J. Felten, B. Omnès, F. Destaing & H. Lejeune
Centre Technique des Industries Mécaniques, F-44308 Nantes Cedex 3, France

ABSTRACT: Despite its importance in the sealing field, fundamental knowledge of the behavior of elastomer O-rings in an environment of carbon dioxide (CO_2) still needs to be completed. It is essential to identify the behavior of the O-ring under the coupled effect of sorption and temperature to carry out predictive simulations. In addition, the coupling of volume change due to CO_2 uptake with thermal dilatation in the context of materials with entropic or enthalpic behavior should be identified. A non-contact experimental measurement technique is used to identify swelling coefficients during CO_2 pressurization of an HNBR O-ring. Tests were performed at three pressures from 0 to 6MPa under isothermal conditions at 10 to 129°C. Measurements show that the CO_2 swelling coefficients are independent of pressure but dependent on temperature, decreasing with increasing temperature. An Arrhenius-type law (like the solubility of CO_2) perfectly describes this evolution.

1 INTRODUCTION

Carbon dioxide (CO_2) is a natural gas often found in hydrocarbon environments; even at relatively low CO_2 concentrations, it can cause seals to swell significantly and induce a plasticizing effect (Ahmed et al. 2019). However, seals must withstand it and remain tight throughout their service life.

Hydrogenated nitrile-butadiene rubber (HNBR) is well known for its resistance to rapid gas decompression and, more generally, to chemical aggression and thermal degradation (Alcock et al. 2015, Grelle et al. 2015, Qamar et al. 2013). Saturation conditions, particularly in CO_2, are known to modify gas permeability due to plasticization effects in elastomers (Lainé et al. 2022). Nevertheless, the lifetime sealing performance can be considerably affected when subjected to the gaseous environment over a long period, combined with changing temperatures.

There are numerous studies in the literature on the swelling behavior of elastomers under CO_2. (Alcock et al. 2015, Briscoe et al. 1996, Davies et al. 1999 and other authors: see Lainé et al. 2022). The solubility of CO_2 in polymers has been identified by measuring gas sorption (Hilic et al. 2001) using gravimetric (Felder 1980, Haroonabadi et al. 2018, Kamiya et al. 1998) or spectroscopic methods (Brantley et al. 2000). It has been quantified by direct observation of polymer swelling (Kamiya et al. 1998, Royer et al. 1999, Wissinger & Paulaitis 1987).

Nevertheless, identifying the diffuso-mechanical properties of O-rings is difficult. One of the first obstacles is related to the geometry of the sample, which needs to be better suited to conventional mechanical testing. While there are many works in the literature, they generally concern bulk or plate materials. The properties of O-rings, developed using an industrial process that may be slightly different from those used in the laboratory, are still a matter of concern.

Consequently, the present work is motivated by the absence of systematic studies of CO_2 sorption through sealing materials and dedicated applications. The challenge is to characterize behavior directly on the seal and avoid problems associated with sample microstructure compared with those generated by the industrial process.

This article concentrates on the study of HNBR filled with carbon black. This work completes the studies of the articles (Lainé et al. 2019, 2020). Swelling coefficients at different temperatures are extracted from in situ measurements during solubility (sorption) tests under CO_2 gas at different pressures. Section 2 describes the material and the test bench used for sorption tests under CO_2 gas. Next, the in situ measurement technique is recalled (Lainé et al. 2020) and the calculation of CO_2 sorption strain, in volumetric test conditions is explained. Finally, the two test protocols are presented. Section 3 presents all the results of CO_2 sorption tests under isothermal conditions in the range 10 to 129°C range (seven different

DOI: 10.1201/9781003516880-20

temperatures). Finally, the values of the swelling strains due to CO_2 sorption under the different conditions and the temperature dependence of the swelling coefficient translated by Arrhenius' law are presented and commented. Section 4 concludes this work.

2 EXPERIMENTAL

2.1 *Material and sample*

One elastomer type, a hydrogenated nitrile rubber (HNBR, Zetpol 2010, Zeon) was selected for these investigations. This compound was produced with a nominal hardness of 80 Shore A. HNBR was compounded as summarized in Table 1, using a base polymer that is 96% saturated with 36% acrylonitrile content and was vulcanized with peroxide curing system, dicumyl peroxide. The quantity of reinforced fillers (Carbon Black) is fixed at 70 phr.

The test samples used in the CO_2 gas sorption tests are HNBR O-rings. The nominal dimensional characteristics of the O-rings are 50.17 x 5.33mm, respectively the inner diameter and the cross-section diameter.

Table 1. Composition of the HNBR.

	HNBR
Material	Phr*
HNBR	100
N-330 HAF carbon black	70
Antioxidant agent Left	1.5
Vulcanizing agent Right	8
Vulcanizing accelerator	2

* Parts per hundred rubber parts in weight.

2.2 *Experimental device*

The HNBR O-rings were positioned on a flat support placed in a cylindrical high-pressure chamber. This chamber is pressure-regulated up to 40MPa and temperature-regulated up to 150°C.

Tests can be conducted under nitrogen (N_2), hydrogen (H_2), carbon dioxide (CO_2) or certain mixtures. The chamber volume is limited (1.77liters for a diameter of 150mm and a length of 100mm) to ensure hydrogen safety. The device is described in detail in previous articles (Lainé et al. 2020, 2022).

The special feature of this chamber is that it has optical access with a central cylindrical sapphire window 40 mm in diameter at the front, enabling the use of non-contact optical measurement methods. Temperature is measured using a thermocouple placed inside the chamber. Three nitrogen purges were systematically carried out at pressures of up to 1MPa after the chamber was closed to remove oxygen from the ambient air and before introducing carbon dioxide.

2.3 *Non-contact extensometry (markers tracking method)*

The metrology used to perform the measurements is a non-contact extensometer system (Benoit et al. 2015, Lainé et al 2019). This marking method, which was developed in the laboratory, has the advantage of being applicable under extreme conditions (very high temperatures, high pressures). Figure 1 shows a local view of an O-ring with four markers placed on the generatrix of the O-ring (sealing plane of the O-ring). The conventional technique consists of tracking the markers' barycenter displacements in the plane of a pair of them. As a result, these markings enable horizontal measurement of dimensional variations by taking them two at a time.

The camera is centered between markers 3 and 4 (Figure 1). Markers 1 and 2 are furthest away. Horizontal deviations along the generatrix at the parting line are measured between markers 1 and 4 and markers 2 and 3, respectively. In the following, these horizontal distances are noted as *DevH12* and *DevH34*, respectively. Measuring the distances between the furthest (1 and 2) and nearest (3 and 4) markers enables values to be framed and assess the quality of measurements with these markers. In addition, knowledge of the distances between the object and the camera must be measured accurately, which is difficult in the pressure machine environment.

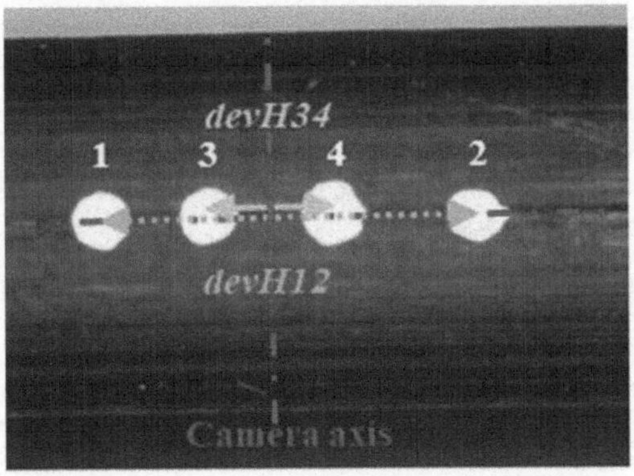

Figure 1. Measurement horizontal deviations (*devH12* & *devH34*) - Markers tracking method (Lainé et al. 2021).

With this experimental technique, a direct link between marker alignment errors and O-ring expansion measurement errors can be expected. Numerical simulations of the thermal expansion of an O-ring were carried out to confirm the capability of the proposed analytical methodology (gap measurement) to identify a coefficient of thermal expansion, and to evaluate the effect of an error in the positioning of one or more markers.

The numerical simulations' results demonstrated that the distance between the horizontal markers or positioning errors had no influence on the result. In addition, horizontal deviation (*devH*) measurement was validated for quantifying expansion in a pressurized environment.

2.4 *Measuring CO_2 sorption strain*

As the strain are small, this measurement deviation (noted: *devH*) corresponds to the sum of the mechanical (noted: ε_{mech}), thermal (noted: ε_{th}), due to CO_2 sorption (noted: $\varepsilon_{sorption\_CO_2}$) strains and an optical corrective value due to the gas phase changes that disrupt the measurement:

$$devH = \varepsilon_{sorption\_CO_2} + \varepsilon_{mech} + \varepsilon_{th} + \delta_{op} \quad (1)$$

where δ_{op} = optical corrective value. All tests are conducted at constant temperature, and CO_2 pressure is applied only after the O-ring has stabilized thermally (at least 1 hour). Thus, in the initial state of the measurements, thermal strain is null.

The CO_2 test is a volumetric test, the strain is assumed isotropic. The strain component ε_v is identical in each direction and

$$\varepsilon_v = \frac{1}{3}\frac{\Delta V}{V_0} \quad (2)$$

where ΔV is the volume change, and V_0 the initial volume. This one has been numerically verified. From linear elastic law, the hydrostatic pressure *(P)* in the O-ring is estimated as follows:

$$P = -K\frac{\Delta V}{V_0}. \quad (3)$$

Hence in direction of measurement

$$\varepsilon_{mech} = \varepsilon_v = -\frac{1}{3K}P \quad (4)$$

with the bulk modulus *K* was identified at 20°C over the pressure range 2.4 to 39.3MPa, with a value of 3234MPa. This value is constant over the temperature range 10 to 130°C as the module increases.

This last equation is used to correct the experimental data based on the imposed pressure *P* and knowing the material parameter *K*. Thus, the mechanical strain is of the order of 0.01% per 1MPa.

2.5 *Optical corrective value*

Initial feasibility tests revealed a change in marker sharpness as early as 2MPa CO_2 gas pressure, leading to a jump in deviation measurement. A test was carried out on a cubic metal sample, as this material does not swell under CO_2 pressure at 10°C.

Four markers were placed on the front face of the cube, two in horizontal direction and two in vertical direction, in order to measure horizontal (*devH12*) and vertical (*devV34*) deviation respectively. This sample is placed in the chamber under a CO_2 pressure of approximately 4.55MPa.

Figures 2a-2b respectively present images of the face of the metal sample observed optically in the initial state (under air) and after 1500s under CO_2 pressure. The test was run for 30 minutes, as the measurements stabilize fairly quickly.

Figure 3 shows measurements of CO_2 pressure, temperature and deviations in the gas. Switching from the initial air phase to a pressurized CO_2 gas phase results in an optical change that overestimates the horizontal and vertical deviations by 0.28 and 0.18% respectively. These two values (δ_{op}) obtained in this test on a metal sample are virtually identical to those obtained in the various O-ring tests, whatever the pressure and temperature conditions.

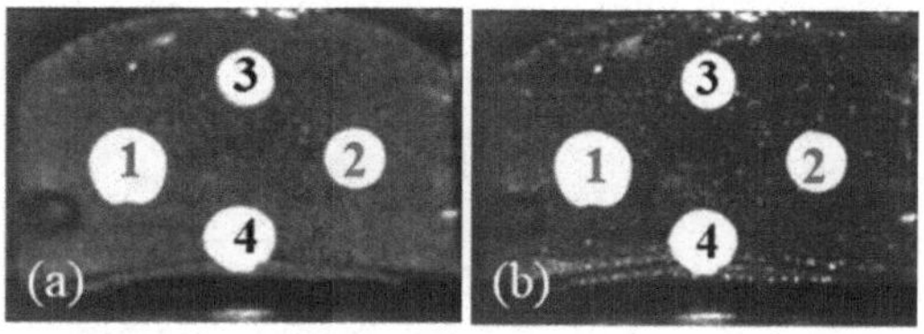

Figure 2. Tracking of markers on metal sample at 10°C – (a) Air phase – (b) Gas phase at a CO_2 pressure of 4.55MPa at 1500s.

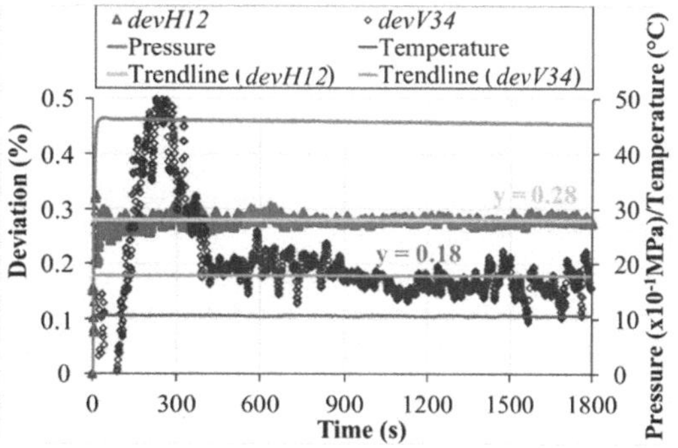

Figure 3. Deviations measured versus time on a metal sample at a CO_2 pressure of 4.55MPa and at 10°C.

2.6 *Protocol*

An identical protocol for sorption tests for one condition (pressure, temperature) (Lainé et al. 2020) has been defined to characterize HNBR O-rings.

The O-ring is placed in the chamber at the test temperature. A minimum of one hour is required to ensure thermal stability. Thus, in equation 1, the thermal strain is zero. The pressure is then built up slowly enough to maintain equilibrium. The deviation is measured and corrected by δ_{op}, the mechanical strain is calculated knowing the imposed mechanical pressure, and it is then possible to quantify the strain due to CO_2 sorption. On the other hand, only one seal per temperature was used for the three temperatures (44.6, 84.7, and 110°C), and three pressure steps were imposed. Measurement reversibility for HNBR has been highlighted, making this protocol possible.

Figure 4 shows the phase diagram for pure CO_2 and the different test conditions (CO_2 pressure-temperature) performed. Colored dots represent conditions, each color corresponding to a given temperature.

The pressures and temperatures considering the presence of impurities in the CO_2 stream can slightly modify the phase diagram as a function of concentration (Aursand et al. 2016).

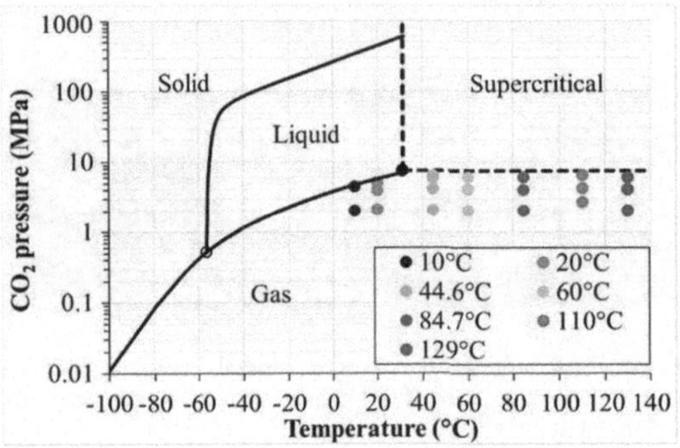

Figure 4. CO_2 phase diagram. Dots indicate tests conducted under different conditions (CO_2 pressure-temperature).

3 RESULTS AND DISCUSSION

3.1 *CO_2 pressure sorption tests under isothermal conditions*

Different HNBR O-rings were tested under different temperature (10, 20, 60 and 129°C) and CO_2 pressure (approx. 2, 4 and 6MPa) conditions.

Figures 5-8 show the CO_2 sorption strains calculated according to equation 1 as a function of time during CO_2 pressurization, at 10, 20, 60, and 129°C, respectively. At 10°C, only two pressures (2.02 and 4.44MPa) were tested. At 20, 60, and 129°C, the three conditions of imposed pressure in the gas phase are indicated in the corresponding figures.

The transient curves, i.e. before stabilization, have identical characteristic times to diffusion and give the same response. A comparison of characteristic times reveals that the higher the pressure, the shorter the transient phase. In addition, as the temperature rises, the stabilization time at a given pressure decreases.

After several tests, it was found that measurement errors on the strains were in the order of 2 to 5%, depending on the pressure level applied. Despite the complexity of the experiment, image analysis provides fairly accurate measurements, since it enables us to quantify O-ring expansion fluctuations generated by the slight temperature oscillations imposed by the enclosure during the test. These variations are due to the regulation of the thermal system. They lead to a modulation of solubility coupled with thermal gradients.

By comparing the responses of the four temperatures tested, the increase in strains during the pressurization phase under CO_2 perfectly reflects the effect of temperature on the absorption phase and solubility.

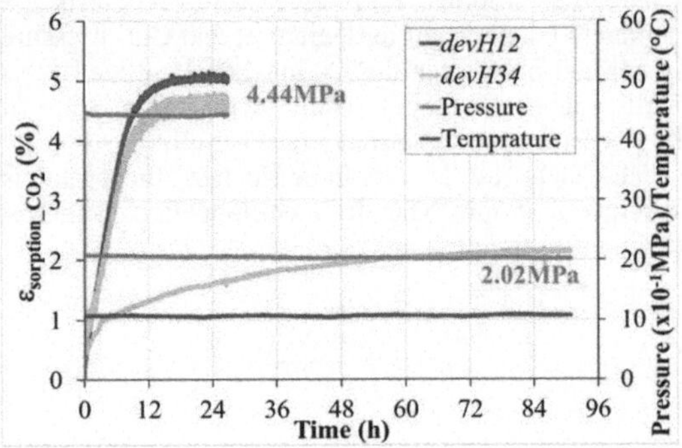

Figure 5. Sorption tests under different CO_2 pressures at 10°C.

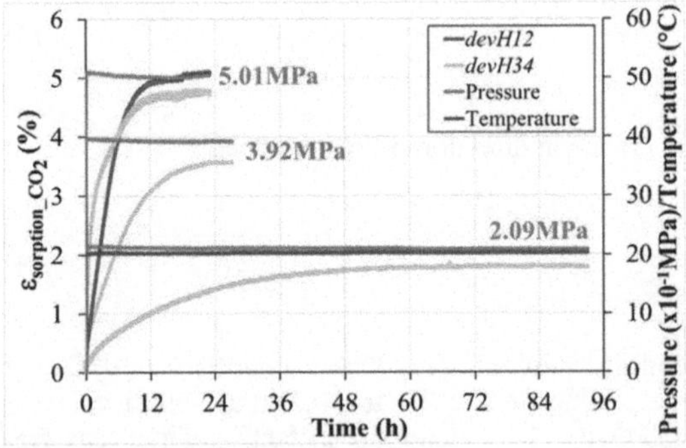

Figure 6. Sorption tests under different CO_2 pressures at 20°C.

3.2 *CO_2 sorption tests at different pressure levels under isothermal conditions*

Three different HNBR O-rings were used to conduct these sorption tests under CO_2 at the following three isothermal temperatures: 44.6, 84.7, and 110°C. As explained in the protocol section, the three CO_2

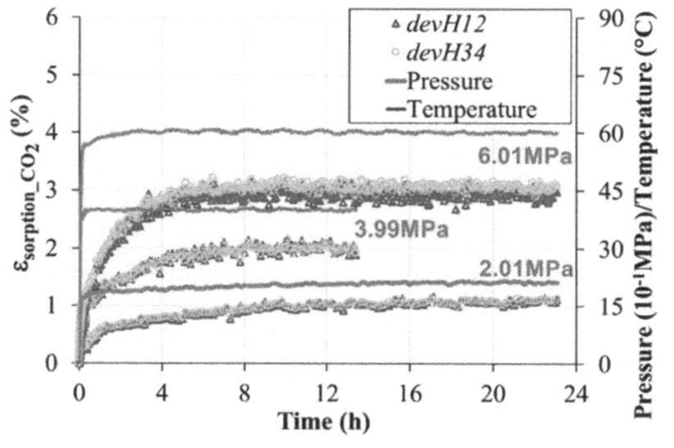

Figure 7. Sorption tests under different CO_2 pressures at 60°C.

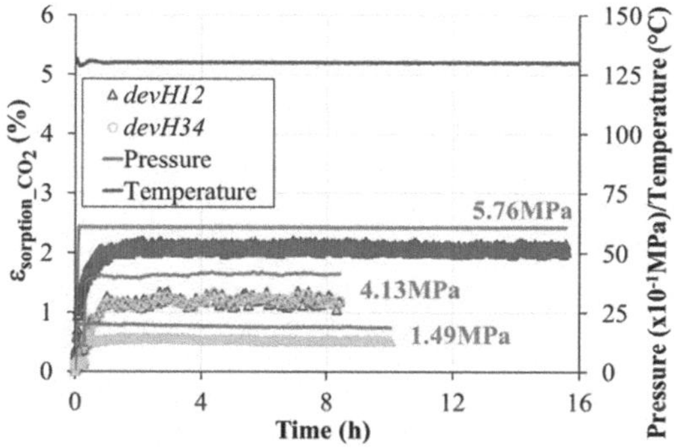

Figure 8. Sorption tests under different CO_2 pressures at 129°C.

pressures (approx. 2, 4, and 6MPa) were applied successively after a stabilization time.

Figures 9-11 show the evolution of pressure, temperature, and CO_2 sorption strain calculated according to equation 1 over time.

3.3 *Swelling as a function of CO_2 pressure*

The volume change (swelling-Sw) of the O-ring under CO_2 pressure is related to the calculated CO_2 sorption strain by the relationship (Lainé et al. 2022):

$$Sw_{CO_2} = \left(1 + \varepsilon_{sorption\_CO_2}\right)^3 - 1 \qquad (5)$$

Figure 12 plots the swelling strains calculated according to equation 5 as a function of conditions (pressure of CO_2-temperature) and represented by a colored square with a black outline. Each color corresponds to a given temperature.

For a given temperature, a linear line passing through 0 and passing as close as possible between the measurement points can be used to identify the CO_2 swelling coefficient, corresponding to the slope of this line, expressed by the following equation:

$$Sw_{CO_2} = \alpha_{Sw\_CO_2}\left(P_{CO_2} - P^0_{CO_2}\right). \qquad (6)$$

3.4 *CO_2 swelling coefficient as a function of temperature*

Transport coefficients are commonly expressed as temperature-dependent according to Arrhenius' law, so solubility is written as (Flaconnèche et al. 2001, Signorini et al. 2024):

$$S(T) = S_0 e^{-\frac{\Delta H_s}{RT}} \qquad (7)$$

where S_0 is the pre-exponential factor, and ΔHs is the apparent heat of the solution, and R the universal gas constant. The logarithmic representation of S versus 1/T allows us to obtain a straight line, the slope of which is proportional to ΔHs. Klopffer et al. (2001) have indicated that for more condensable gases and vapors (e.g. CO_2, SO_2), ΔH_S is negative, and, for a given system, a decrease in solubility will be observed with increasing temperature. This expresses that the penetrant finds condensing increasingly difficult in the polymer as temperature increases.

By analogy, the CO_2 swelling coefficient can be written as:

$$\alpha_{Sw\_CO_2}(T) = \alpha^0_{Sw} e^{-\frac{\Delta H_s}{RT}} \qquad (8)$$

where α^0_{Sw} is the pre-exponential factor.

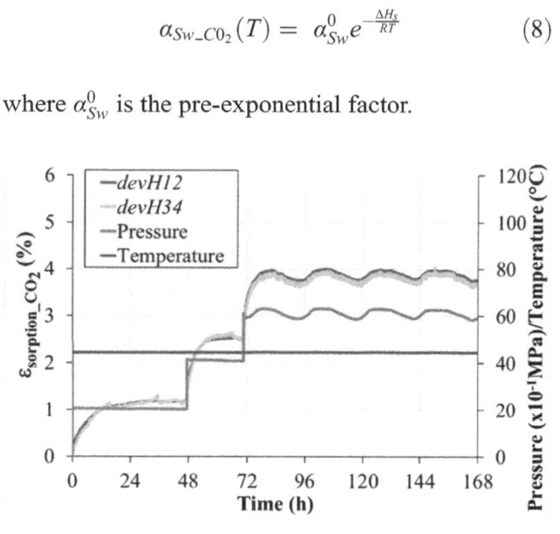

Figure 9. Sorption tests with three pressure levels at 44.6°C.

Figure 13 shows CO_2 swelling coefficients as a function of reciprocal temperature, plotted as black dots. Since the x-axis is logarithmic, the dotted

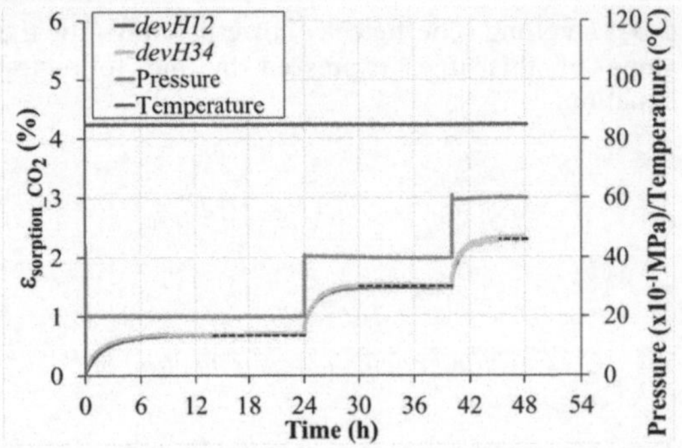

Figure 10. Sorption tests with three pressure levels at 84.7°C.

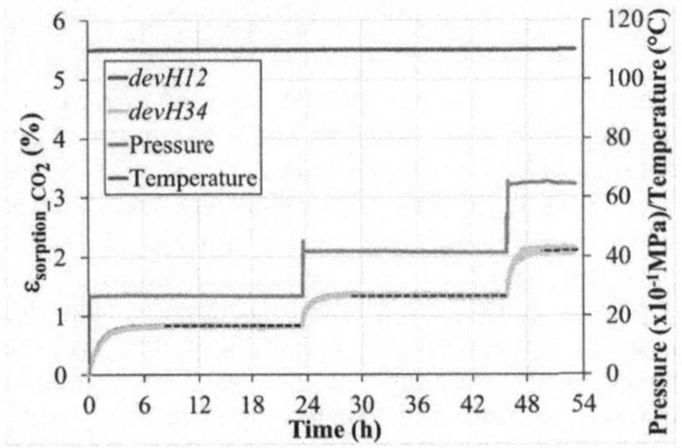

Figure 11. Sorption tests with three pressure levels at 110°C.

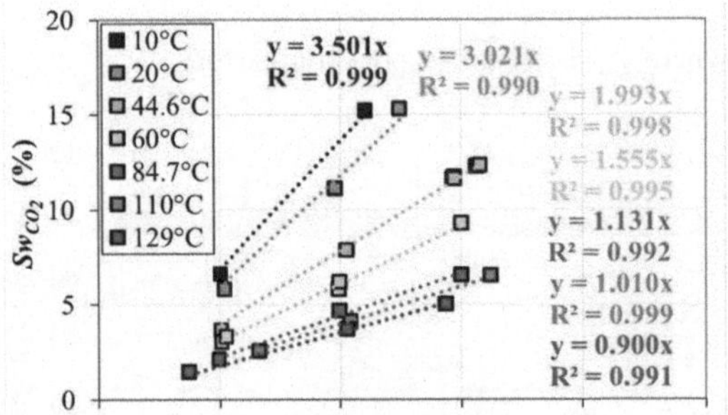

Figure 12. CO_2 sorption strains versus CO_2 pressure at different temperatures.

linear line passing through the set of points has Arrhenius' law as its equation. By multiplying the exponential term (-1.359) by the value of the universal gas constant, the apparent heat is equal to -11.3kJ/mol for HNBR.

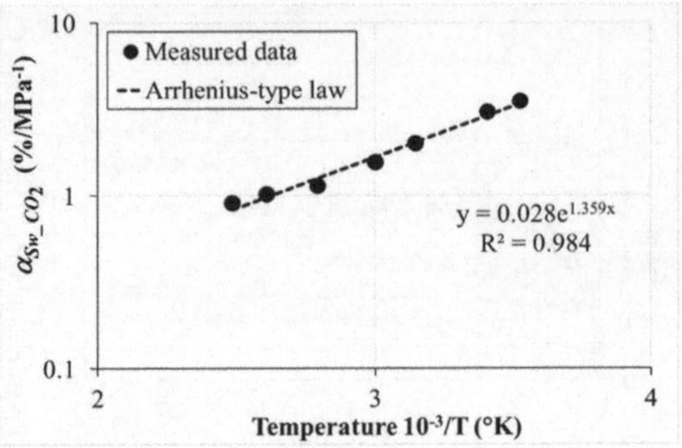

Figure 13. CO_2 swelling coefficient vs reciprocal temperature.

4 CONCLUSIONS

Non-contact optical measurements are not always easy to implement on structures in a gaseous environment. The method developed enabled us to measure the strain of an O-ring subjected to CO_2 gas pressure during a sorption test. All the tests carried out at different temperatures and pressures demonstrated that:

- CO_2 sorption Strain increases as CO_2 pressure increases at a given temperature and decreases as temperature increases at a given pressure.
- Stabilization time after CO_2 absorption decreases with increasing CO_2 pressure at a given temperature, and with increasing temperature at a given pressure.
- CO_2 swelling coefficient is constant as a function of pressure at a given temperature.
- CO_2 swelling coefficient is temperature-dependent, decreasing with increasing temperature.
- The evolution of the CO_2 swelling coefficient as a function of temperature is expressed by an Arrhenius' law similar to solubility.
- The determined ΔHs value of -11.3kJ/mol for HNBR seems to be in agreement with the values obtained for polymers by Signorini et al. 2024.

ACKNOWLEDGMENTS

We thank TotalEnergies EP Company and Saipem Company for their Sponsoring in the Citeph project PIPESEALCO2 (CITEPH2022 - 53) and led by the Cetim.

This work was supported by the CPER FEDER project of Région Nouvelle Aquitaine.

REFERENCES

Ahmed, S. & Salehi, S. et al. 2019. Experimental investigation of elastomers in downhole seal elements: Implications for safety. *Polym. Test.* 76: 350–364.

Alcock, B. & Peters, T.A. et al. 2015. The effect of hydrocarbon ageing on the mechanical properties, apparent crosslink density and CO_2 diffusion of a hydrogenated nitrile butadiene rubber (HNBR), *Polym. Test.* 47: 22–29.

Aursand, E. & Aursand, P. et al. 2016. The influence of CO_2 mixture composition and equations of state on simulations of transient pipeline decompression. *Int. J. Greenh. Gas Control* 54: 599–609.

Benoit, G. et al. 2015. Mechanical testing in pressurized hydrogen and carbon dioxide, in: The 10th BSSM International Conference on Advances in Experimental Mechanics, Edinbourg, Ecosse, 1-3 September.

Brantley, N.H. et al. 2000. In situ FTIR measurement of carbon dioxide sorption into poly(ethyleneterephthalate) at elevated pressures. *J. Appl. Polym. Sci.* 77: 764–775.

Briscoe, B.J. & Kelly, C.T. 1996. The effect of structure on gas solubility and gas induced dilation in a series of poly(urethane) elastomers, *Polymer* 37: 3405–3410.

Davies, O.M. et al. 1999. The mechanical properties of elastomers in high-pressure CO_2, *J. Mat. Sci.* 34: 417–422.

Felder, R.M.H. 1980. *Permeation, diffusion, and sorption of gases and vapors* in: R.A. Fava, ed., Methods Exp. Physics, Part C: Polymer Physics, Academic Press, New York.

Flaconnèche, B. & Martin, J. & Klopffer, M.H. 2001. Permeability, Diffusion and Solubility of Gases in Polyethylene, Polyamide 11 and Poly(vinylidene fluoride). Oil & Gas Science and Technology – Rev. IFP 56: 261–278.

Grelle, T. & Wolff, D. & Jaunich, M. 2015. Temperature-dependent leak tightness of elastomer seals after partial and rapid release of compression, *Polym. Test.* 48: 44–49.

Haroonabadi, L. et al. 2008. Investigation of the effect of thermal aging on rapid gas decompression (RGD) resistance of nitrile rubber, *Polym. Test.* 67: 37–45.

Hilic, S. & Boyer, S.A.E. et al. 2001. Simultaneous measurement of the solubility of nitrogen and carbon dioxide in polystyrene and of the associated polymer swelling, *J. Polym. Sci. Part B: Polym. Phys.* 39: 2063–2070.

Kamiya, Y. et al. 1998. CO_2 sorption and dilation of poly(methyl methacrylate), *Macromolecules* 31, 472–478.

Klopffer, M.H. & Flaconnèche, B. 2001. Transport Properties of Gases in Polymers: Bibliographic Review. *Oil& Gas Science and Technology* 56: 223–224.

Lainé, E. & Grandidier, J.-C. et al. 2019. Effects of sorption and desorption of CO_2 on the thermomechanical experimental behavior of HNBR and FKM O-rings – Influence of nanofiller reinforced rubber. *Polym. Test.* 75: 298–311.

Lainé, E. & Grandidier, J.-C. et al. 2020. Non-contact method used to determine the swelling/shrinking coefficients under CO_2 sorption/desorption on an HNBR O-ring – Study of coupling with temperature and pressure. *Polym. Test.* 85: 1–16.

Lainé, E. & Grandidier, J.-C. et al. 2022. Identification of swelling/shrinking coefficients under CO_2 on an FKM O-ring – Comparison with HNBR and influence of reinforcements on the matrices – Influence of nanofiller reinforced rubber. *Mat. & Techniques* 110: 303–319.

Qamar, S.Z. & Akhtar, M. et al. 2013. Mechanical and structural behavior of a swelling elastomer under compressive loading. *Materials & Design* 45: 487–496.

Royer, J.R. et al. 1999. Carbon dioxide induced swelling of poly(dimethylsiloxane), *Macromolecules* 32: 8965–8973.

Signorini, V. et al. 2024. Characterization and modeling of CO_2 transport through Fluorinated thermoplastics. *ACS Appli. Polym. Mater.* 6: 379–389.

Wissinger, R.G. & Paulaitis M.E. 1987. Swelling and sorption in polymer-CO 2 mixtures at elevated pressures, *J. Polym. Sci Part B: Polym. Phys.* 25: 2497–2510.

Constitutive Models for Rubber XIII – Dal (Ed.)
© 2025 The Author(s), ISBN 978-1-032-85138-9

Fundamental approach to build viscoelastic master curves for heterogeneous elastomer blends

A. Lang, M. Müller, M. Klüppel & U. Giese
Deutsches Institut für Kautschuktechnologie e. V., Hannover, Germany

J. Voges & D. Juhre
Otto-von-Guericke-Universität, Magdeburg, Germany

ABSTRACT: The time-temperature superposition principle is used as standard procedure for various elastomers to extend the viscoelastic spectrum limited by the measurement apparatus to higher and lower frequencies. Heterogeneous blend systems represent a challenge that has not yet been clarified in this regard. These blends consist of at least two different polymers and do not exhibit a homogeneous phase, i.e. the individual phases are located in separated domains allowing the coexistence of individual properties. Thus, the generation of so-called viscoelastic master curves is not directly possible. In this work an approach is presented to generate such master curves by decomposing the spectrum within the temperature domain and projecting the individual phase contributions onto the frequency domain using defined shift factors.

1 INSTRUCTION

In rubber industry, customizing properties are achieved by blending carefully chosen polymers to create rubber blends with distinct phases, meeting the diverse requirements of the elastomer market. Over the years, it's been understood that besides the phase morphology and reinforcing filler particles, the physical properties of rubber blends are significantly influenced by the quality of phase bonding in the interphase and the successful co-vulcanization of both phases. This necessitates some level of compatibility among the often-immiscible blend components. Phase bonding notably affects key properties like wear and crack growth rates, closely related to the high-frequency viscoelastic response of the blend. Another motive for blending different rubbers e.g. in tire treads is to enhance wet traction properties by optimizing viscoelastic losses in the kHz to MHz range. This means that the high-frequency range should be well characterized already at the laboratory level to have a rough prediction of the performance indicators. However, the determination of the viscoelastic properties in this frequency range is quite difficult or even impossible. There is no suitable experimental method that allows direct access to reliable data for high-frequency excitations.

The only concept, which is often used as standard for various elastomers to extend the measured viscoelastic spectrum limited by the measurement apparatus to higher and lower frequencies or relaxation times, is the time-temperature superposition principle (TTS) [1] based on the assumption of thermo-rheologic simplicity. But in this regard, heterogeneous blend systems represent a challenge that has not yet been clarified. These blends consist of different polymer types located in phase separated domains, which in turn have different time-temperature dependent relaxation relations. Thus, the generation of viscoelastic master curves is not directly possible. A first approach to evaluate viscoelastic master curves for heterogeneous blend elastomers is presented here.

2 MATERIALS AND EXPERIMENTAL

2.1 *Materials*

For preliminary studies styrene-butadiene rubber (SBR) with higher glass transition temperature T_g of about -20°C and natural rubber (NR, T_g~-60°C) blends with blend ratios varying in steps of 25 phr were chosen. The detailed compound formulations are listed in Table 1.

Before mixing of compounds the Pale Crepe (NR) was masticated using a two roller mill for 15 minutes to reduce viscosity and improve processability.

The ingredients were processed using a tangential internal mixer Rheomix 3000 (Haake) with initial temperature of 50°C and rotational speed of 50 min^{-1}. First the NR was mixed for 2 min following by blending for additional 2 min with SBR.

DOI: 10.1201/9781003516880-21

Table 1. Compound formulations of studied elastomer blend materials.

Components	Content [phr]
SBR (Nipol NS 616)	0/25/50/75/100
NR (Pale Crepe)	100/75/50/25/0
ZnO	1.0
Stearic Acid	0.5
IPPD	1.0
CBS	1.0
Sulfur	1.0

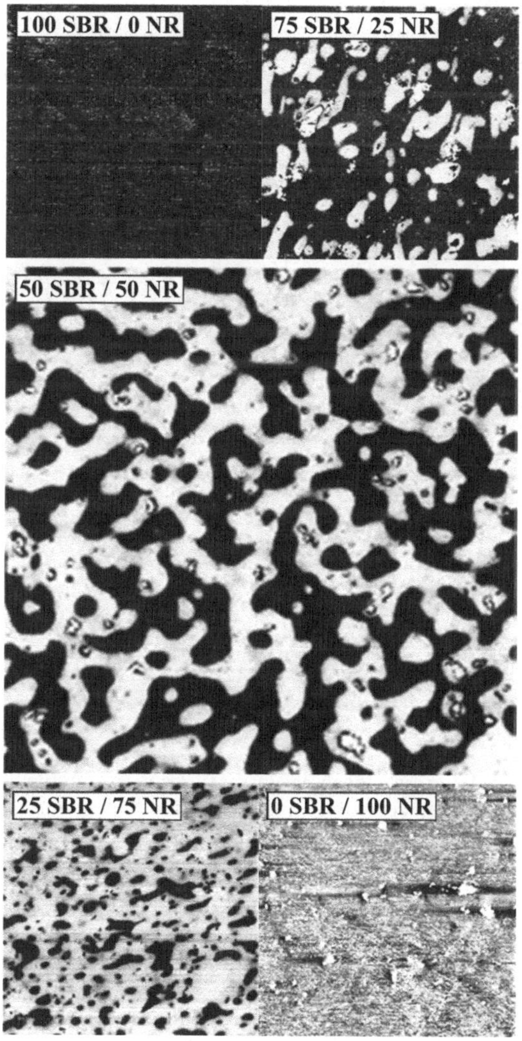

Figure 1. Blend morphology measured by AFM for different blend ratios varying from 100 SBR/0 NR (top left) to 0 SBR/100 NR (bottom right) in 25 phr steps with bright areas correspond to NR and dark regions related to SBR phase. Shown area corresponds to 5μm by 5μm [18].

Subsequently, all additives including the curing chemicals were added for further 2 min and the melt was ejected. As a final step, the compounds were passed through the roller mill five times upside down. The vulcanization was carried out using a heat press at temperature of 160°C. The curing parameters were based on the rheometer curve, with a curing time selected as $t_{90\%}$ + 1 min per 1 mm sample thickness.

The blend morphology was verified using an AFM Dimension Icon from Fa Bruker with tapping mode, as can be seen in Figure 1. For compounds with different blend ratios (25/75 and 75/25) the subordinate phase structure is well distributed in the main matrix. The typical size of the islands is about 100 nm to 500 nm in diameter. For the 50/50-blend a more interpenetrating structure can be observed. It should be mentioned that the interphase can only be seen insufficiently or not at all. On the one hand, this is due to the measuring method itself, as the tip diameter is in the range of 20 nm and therefore finer structures cannot be resolved. On the other hand, it cannot be ruled out that layers close to the measured surface also have an undefined influence on the results.

2.2 *Experimental setup*

To determine viscoelastic properties, an ARES rheometer from Rheometric Scientific^{TM} was used. The samples with a geometry of 40 x 10 x 2 mm³ were dynamically excited in torsion mode. The temperature sweeps were performed at temperatures ranging from -80°C to 60°C applying a frequency of 1 Hz and a deformation amplitude of 0.5 %. For construction of viscoelastic master curves, the temperature-frequency sweeps were performed by varying temperature from -100°C to 60°C and frequency from 0.1 Hz to 10 Hz. The deformation amplitude was 0.5 % as above.

3 RESULTS AND DISCUSSION

3.1 *Conventional evaluation of viscoelastic master curves*

Elastic as well as viscous properties of elastomeric materials are mainly based on the molecular structure of polymers. The rotational and translational degrees of freedom of the chain segments determine the mobility of the whole chain. One principle measure of this is the glass transition temperature. The position of the glass transition temperature depends on the arrangement and nature of the chain segments as well as the presence of side groups in the polymer chain. By changing the microstructure, the polymer dynamics can be strongly modified. The movement of the chains or chain segments is only possible if there is sufficient free volume in which the chains can relax. This "special volume" was first measured by T. G. Fox (Jr.) and P. J. Flory as a function of molar mass and temperature for polystyrene [2]. As the temperature decreases, less free volume is available, which reduces the mobility of the polymer chains until the free volume reaches a critical value, so that the chain segments are "frozen". This occurs at the glass transition temperature. A semi-empirical description was provided by H. Vogel, G. S. Fulcher, G. Tammann and W. Hesse [3, 4, 5].

$$f = f_\infty \cdot \exp\left(-\frac{E_A}{k_B \cdot (T - T_{VF})}\right) \quad (1)$$

The equation links the frequency f with the temperature T starting from a so-called Vogel-Fulcher temperature T_{VF}, at which the free volume disappears completely, and a relaxation rate for infinitely high temperatures f_∞. E_A is an apparent activation energy that corresponds approximately to the van der Waals interaction between neighboring chains [97]. k_B is the Boltzmann constant.

Based on the theoretical description, it can be deduced that the same rotation and translation relaxation modes of the chain segments exist for the same temperature differences, namely $\Delta T = T - T_g$, where T is the service temperature and T_g is the glass transition temperature [6]. An equivalent relationship between the temperature and excitation frequencies was determined by M. L. Williams, R. F. Landel and J. D. Ferry using various polymers [1, 7, 8].

$$\log(a_T) = \log\left(\frac{f}{f_{ref}}\right) = \frac{-C_1 \cdot (T - T_{ref})}{C_2 + (T - T_{ref})} \quad (2)$$

This so-called WLF equation establishes the relationship for the chain mobility, where a temperature change corresponds to a shift on the frequency or time axis. The shift takes place on the logarithmic frequency or time scale multiplying the reference frequency f_{ref} by the temperature-dependent shift factors $a_T = f/f_{ref}$, which transform the measured data to a master curve. The parameters C_1 and C_2 are generally free parameters that depend on the polymer type and the reference temperature T_{ref} (e.g. application temperature). For various rheological simple polymers, universal parameters $C_{1,u} = 17.44$ and $C_{2,u} = 51.6\,K$ were determined in case the reference temperature corresponds to the glass transition temperature [1, 8]. For most polymers with a linear structure, C_1 agrees quite well, whereas larger deviations are found for C_2 [9]. So, these parameters should be determined individually for every elastomeric material. Additionally, the validity of this relationship is limited to the temperature range $T_g < T < T_g + 100°C$ [1, 8]. Nevertheless, this technique allows to extend the instrumentally limited measuring range for isothermal processes by several decades and thus to combine individually measured data for obtaining a viscoelastic master curve, as shown in Figure 2.

For technical applications, this method is used to characterize the material behavior especially for high-frequency excitations. Good agreement can be found using alternative measurement methods such as dielectric relaxation spectroscopy or ultrasonic analysis [10-15].

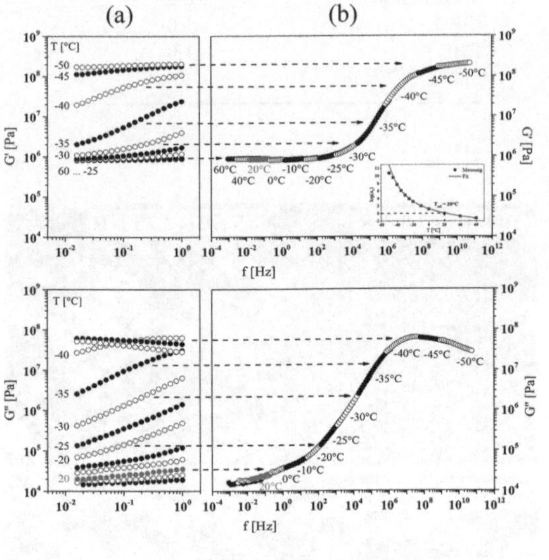

Figure 2. Application of the time-temperature superposition principle to the measurement data (a) of an unfilled SBR sample to create a master curve (b) with the associated factors a_T (inset) for a reference temperature of 20°C at a deformation amplitude of ε=0.2%.

It should also be noted that this procedure cannot be used for common technically relevant elastomers without further modification, as the influence of the reinforcing filler on the dynamics is not taken into account, which in turn can lead to larger errors [16, 17].

3.2 *Viscoelastic master curves for blends*

To determine viscoelastic master curves for blend materials, contributions from individual polymers (here SBR and NR) should be taken into account. Therefore, the individual materials should be characterized in a proper way. Additionally, the temperature dependence of the blend materials should be determined. The results are depicted in Figure 3. As can be seen, with increasing SBR amount an additional contribution to the storage modulus as well as to the loss modulus of the NR phase in the temperature range between -60°C and 0°C can be found. Here, the SBR-phase is nearly-rigid and behaves like a reinforcing filler, which leads to higher stiffness and higher dissipation of the full phase [18].

In order to explain the master procedure in a simplified way, we will focus on the loss modulus of the compounds. By using the procedure of Klüppel et al. [19-21], the individual contributions are adapted to the blend spectrum by considering the glass transition of both polymers. This is intended to determine the interphase contribution and split the full spectrum into individual contributions, which can be transferred into the frequency dependent spectrum by the TTS and reconstruct the full master curve.

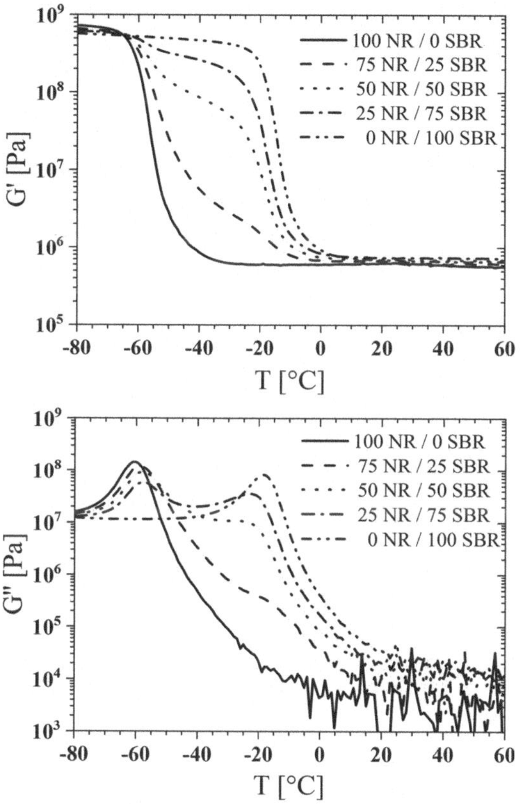

Figure 3. Storage (top) and loss (bottom) moduli for pure materials as well as for blends in dependence of temperature measured at 1 Hz and deformation amplitude of 0.5%.

In Figure 4 an example for the 75SBR/25NR-blend reveals a significantly gap (grey region) between the blend data (measurement points) and the additive contributions from individual elastomers (dashed line). This indicates the existence of an interphase, which additionally contributes to the spectrum of the individual pure phases (doted lines). However, it cannot be completely excluded that the morphological structure does not have an additional effect on this gap. For the sake of simplicity, it is assumed here that the two phases act independently of each other (since these are not miscible) and that the morphology only changes the two individual contributions not the interphase, which is already been taken into account by the adjustment according to procedure of Klüppel et al. [19-21].

So, the interphase contribution (IPh) can be determined by subtracting the individual polymer contributions from the measured curve of the blend. As can be seen in Figure 5, the interphase represents a broad spectrum, which can be interpreted as a gradient of different concentrations of the individual polymers in the interphase. In this case we used three interphase contributions each represented by a Gaussian distribution having different concentrations of both polymers. IPh-1 has higher concentration of NR molecules, IPh-3 is higher filled with SBR, whereas IPh-2 represents a homogeneous mixture with 50% SBR and 50% NR. The sum of all three interphase contributions (IPh-Fit) matches quite well the resulting total interphase spectrum.

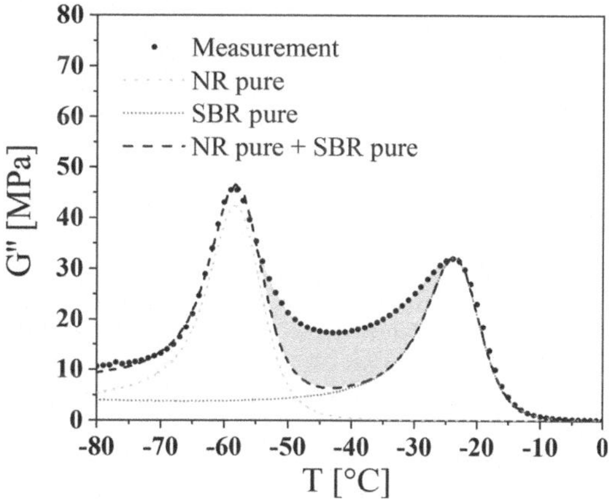

Figure 4. Loss modulus comparison of measured data resulting from blend material (75 SBR/25 NR) and adapted contributions of pure polymers.

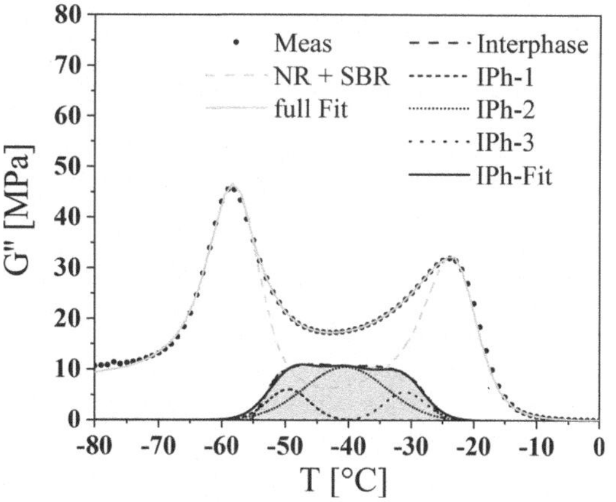

Figure 5. Loss modulus comparison of measured data resulting from blend material (75 SBR/25 NR) and adapted contributions of pure polymers.

Having split the individual spectra resulting from the two domains and three interphase proportions, this enables us to transfer the individual contributions to the frequency axis using the time-temperature superposition principle. The only question is, which shift parameter set results in a reliable transfer?

It's quite evident that the two separated domains should be transferred using the corresponding shift parameters $C_{1,SBR}$, $C_{2,SBR}$ and $C_{1,NR}$, $C_{2,NR}$, like it is shown in Figure 6 (top). The contribution of the interphase is taken into account by adding the interphase components, which is depicted in Figure 6 (bottom). The shift factors $C_{2.Iph\text{-}i}$ for individual interphase components can be calculated using the universal parameter $C_{2,u} = 51.6\ K$ according

$$C_{2,IPh-i} = C_{2,u} + T_{ref} - T_{g,IPh-i} \qquad (3)$$

with $T_{g,IPh-i}$ being the glass transition temperature for each interphase component, whereas $C_{1,IPh-i}$ is more difficult to determine. With the assumption that C_1 is proportional to a kind of apparent activation energy for viscoelastic relaxations [1] and the possibility to superpose linearly the individual activation energy contributions in a miscible blend, $C_{1,IPh-i}$ are calculated by averaging of the individual C_1-Parameters from pure SBR and pure NR scaling with the assumed amount of the individual polymer types in the respective interphase, i.e. IPh-2 consists of 50 % SBR and 50 % NR, which results in $C_1 = 0.5 \cdot C_{1,SBR} + 0.5 \cdot C_{1,NR}$. Ratios of 0.25/ 0.75 and 0.75/ 0.25 were selected for IPh-1 and IPh-3 respectively.

It is obvious that for a reference temperature of -20°C an additional contribution between both loss peaks from SBR and NR can be observed, which leads to more realistic data compared to pure components. Shifting the data to a refence temperature of 20°C, the spectrum merges and becomes wider due to different shift parameters. This already allows a rough estimation of the viscoelastic properties, e.g. for tire applications and the dissipation range on the frequency axis.

Nevertheless, a comparison with finite element (FE)-simulations – depicted in Figure 7 (bottom, dotted line) – shows that on a qualitative level the approach can be used to determine a viscoelastic master curve of non-miscible blend materials, but on a quantitative level a gap can be observed. It is still not clear whether the shift procedure or the FE-simulations misleading here.

For the storage modulus an equivalent procedure can be applied considering the individual contributions, which is not shown here since more physical motivated clarification in this regard is needed.

4 CONCLUSION

This work has shown that the determination of master curves for blend systems is still not fully clarified.

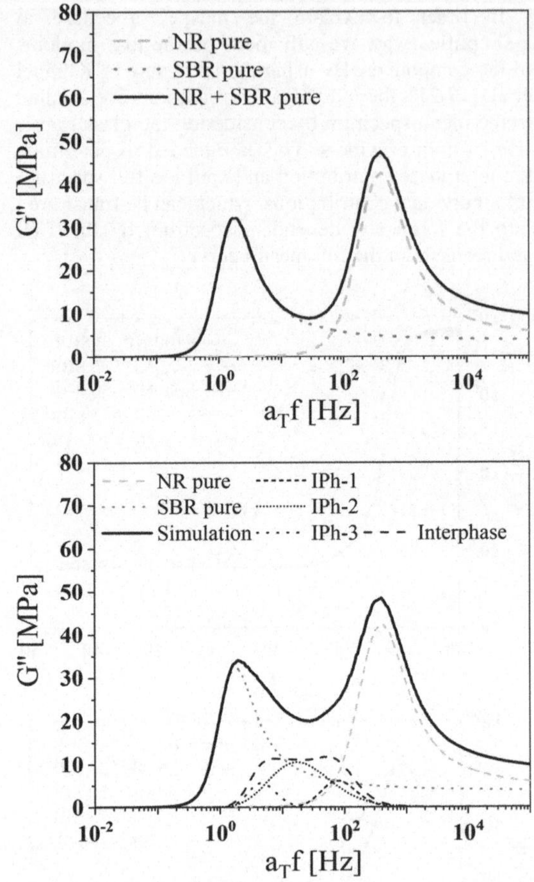

Figure 6. Loss modulus of the viscoelastic master curve consists only on both main phase contributions (NR and SBR) for a reference temperature of -20°C (top) and the full combination including the interphase components (bottom).

Although the presented concept can deliver qualitative results, quantitatively the concept still needs to be checked more closely and possibly improved in some parts. Unfortunately, there are no known methods that allow a blend system to be tested at high excitation frequencies without significant dissipation and corresponding self-heating. This means that there is no suitable validation of the concept. The only way to make a comparison is to use numerical simulation, but there are only preliminary models that need to be validated too. Thus, the topic of mastering of viscoelastic properties of heterogeneous blend elastomers still remains a challenge for future work.

ACKNOWLEDGEMENTS

The funding by the Deutsche Forschungsgemeinschaft (DFG, German Research Foundation) is appreciated with the grant 426080407.

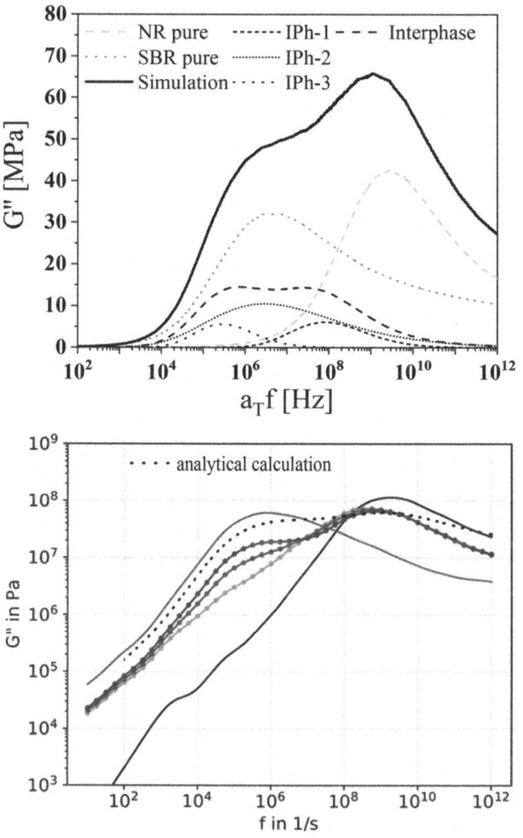

Figure 7. Loss modulus of the viscoelastic master curve for a reference temperature of 20°C (top) and the resulting curve on a logarithmic scale in comparison with simulations (bottom, dotted line) [22, 23].

REFERENCES

[1] J. D. Ferry, Viscoelastic Properties of Polymers, New York: John Wiley & Sons, Inc., 1980.

[2] T. G. Fox and P. J. Flory, Journal of Applied Physics, 21, 581–591, 1950.

[3] H. Vogel, Physikalische Zeitschrift, 22, 645, 1921.

[4] G. S. Fulcher, Journal of the American Ceramic Society, 8, 339, 1925.

[5] G. Tammann and W. Hesse, Zeitschrift für anorganische und allgemeine Chemie, 156, 245–257, 1926.

[6] F. Röthemeyer and F. Sommer, Kautschuktechnologie, München, Wien: Hansen, 2001.

[7] M. L. Williams and J. D. Ferry, Journal of Polymer Science, 11, 169–175, 1953.

[8] M. L. Williams, R. F. Landel and J. D. Ferry, Journal of the American Chemical Society, 77, 3701–3707, 1955.

[9] M. T. Shaw and W. J. MacKnight, Introduction to Polymer Viscoelasticity, 3. Hrsg., New Jersey: John Wiley & Sons, Inc., 2005.

[10] J. Fritsche, Leibniz Universität Hannover, Dissertation, 2009.

[11] S. Eckert-Kastner, T. Alshuth and R. H. Schuster, Kautschuk Gummi Kunststoffe, 57, 423-429, 2004.

[12] J. Kroll, T. Alshuth and R. H. Schuster, Kautschuk Gummi Kuststoffe, 59, 33–40, 2006.

[13] J. Kroll, T. Alshuth and R. H. Schuster, Katschuk Gummi Kunststoffe, 59, 184–189, 2006.

[14] C. A. Oprisoni, T. Alshuth and R. H. Schuster, Kautschuk Gummi Kunststoffe, 61, 301–305, 2008.

[15] J. G. Meier, J. W. Mani and M. Klüppel, Physical Review B, 75, 054202, 2007.

[16] M. Klüppel, Journal of Physics: Condensed Matter, 21, 035104, 2009.

[17] A. Lang and M. Klüppel, Wear, 380, 15-25, 2017.

[18] M. Müller, A. Lang, M. Klüppel and U. Giese, Constitutive Models for Rubber XII, CRC Press, 121–126, 2023

[19] R. Schuster, J. Meier and M. Klüppel, Kautschuk Gummi Kunststoffe, 53, 663–674, 2000.

[20] M. Wunde and M. Klüppel, Rubber Chem. Technol., 89, 588–607, 2016.

[21] M. Wunde and M. Klüppel, Continuum Mech. Thermodyn., 29, 1135–1148, 2017.

[22] J. Voges, D. Juhre, M. Müller, A. Lang and M. Klüppel, Constitutive Models for Rubber XII, CRC Press, 127–132, 2023.

[23] J. Voges, M. Müller, A. Lang, M. Klüppel and D. Juhre, Archive of Applied Mechanics (submitted) 2024.

Constitutive Models for Rubber XIII – Dal (Ed.)

About mechanically induced crystallization kinetics in filled natural rubber

A. Le Bihan, Y. Marco & V. Le Saux
ENSTA Bretagne, UMR CNRS 6027, IRDL, Brest, France

L. Chazeau & J.-M. Chenal
INSA Lyon, UMR CNRS 5510, MATEIS, Villeurbanne, France

I. Warneboldt, P. Charrier & C. Champy
Vibracoustic Nantes, CAE and Durability Prediction Department, Carquefou, France

ABSTRACT: In this work, we provide information about mechanically induced crystallization (MIC) kinetics. First, we show that the usual definition of threshold elongations for crystallization and melting does not always make sense. We then show the effect of the loading path on the crystallization levels reached. Finally, we define a quantity representative of the crystallization kinetics and try to describe it using quantities as elongation or stress. At the same time, we illustrate that crystallization is a self-regulating phenomenon: a saturation of crystallization kinetics can be reached.

1 INTRODUCTION

Natural Rubber (NR) is widely used in automotive anti-vibration parts due to its exceptional damping and vibration resistance properties: it's an ideal material for reducing unwanted vibrations from the road, engines, suspension or other components. NR is also known for its excellent fatigue properties.

Automotive parts are subjected to very complex loading conditions. For NR, so-called non-relaxing conditions lead to local reinforcement of the material, resulting in increased fatigue (Cadwell et al., 1940). This increase in service life can be graphically illustrated by a HAIGH diagram: iso-curves of lifetime in a mean-amplitude load plane. Such diagram is characterized by a bell curve (Champy et al., 2021).

This local reinforcement effect is characteristic of NR and its ability to crystallize under tension due to its stereoregularity (Huneau, 2011). In reality, the increase in lifetime duration is merely a consequence of other phenomena that occur during cracking. One of the first investigations into the effect of non-relaxing loading on crack propagation was that of (Busse, 1935). He observed that the main crack, initially orthogonal to the loading direction, gives rise to two new cracks. These tend to orient themselves parallel to the loading direction or turn back for highly non-relaxing loading. A second effect related to crack propagation was observed by (Saintier, Cailletaud, and Piques, 2011; Warneboldt et al., 2022): crack branching *i.e.* ramification near the main crack. The crack propagates in a direction perpendicular to the loading direction.

Crystallization kinetics is also fundamental to explain the hysteretic behaviour observed under dynamic conditions which leads to the reinforcement and the lifetime increase. Usually, to better understand crystallization kinetics, the main idea is to stretch the material to a given strain and follow the evolution of the crystallinity index, usually by Wide Angle X-Ray Diffraction (WAXD) measurements. (Brüning et al., 2015) studied the evolution of crystallization during relaxation, at different temperatures and for different imposed strains. This crystallization follow-up was carried out over short times, of the order of a minute. More recently, (Albouy and Sotta, 2020) performed the same exercise over longer times (one hour) at ambient temperature. In particular, they have shown that, over a given imposed elongation, there is an independence of crystallization kinetics to the imposed elongation. They also proved the existence of a time-strain equivalence that may be used to model crystallization (Sotta and Albouy, 2020).

These first data provide a better understanding of the kinetic effects and also raise some questions: how does crystallization kinetics evolve? It is possible to easily relate it to simple quantities such as elongation or stress? Kinetics established by (Albouy and Sotta, 2020) are given for a monotonic loading, but how would kinetics vary if the loading path is different? In order to answer some of these questions, we chose to realize an anhysteretic (ANH) test and simple tension tests under synchrotron beam. The results of the two kinds of tests are analyzed and compared. The paper is organized as follows. Section 2 describes the studied material and the experimental set-up. Section 3 presents the experiments and the data analysis procedure. Section 4 is devoted to the results discussion. Section 5 concludes and gives perspectives for this work.

DOI: 10.1201/9781003516880-22

2 MATERIALS AND EXPERIMENTAL SET-UP

2.1 *Materials and specimen*

The material of interest is an industrial filled at 42 phr of carbon black and vulcanized natural rubber. The basic formulation is given in Table 1.

Table 1. Recipe of the material.

Component	Content [phr]
Matrix	100
Fillers	42
Zinc Oxyde	4
Plasticizer	5
Stearic acid	2
Antioxidant	2
Antiozonant	2
Accelerator	0,6
Sulfur	2,8

The specimens were all manufactured under identical molding process settings and from a single material batch to ensure equal mixing and curing conditions. Samples used are called mini PureShear, see Figure 1. The 1mm thickness is a good compromise for WAXD experiments, between X-ray absorption and X-ray scattering. The specimen lengths used vary between 5 and 10 mm to ensure uniaxial tension conditions.

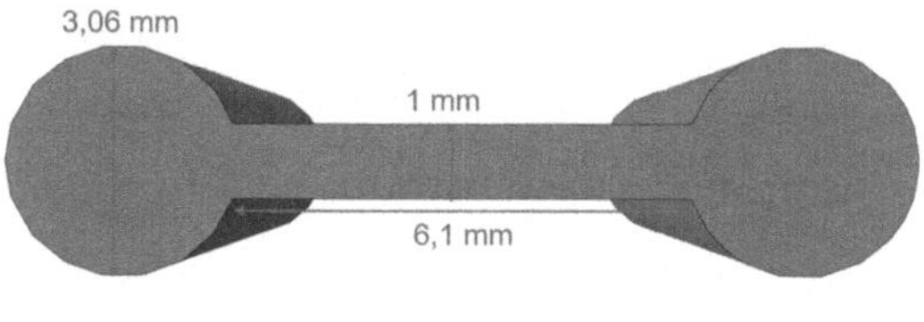

Figure 1. Mini-PureShear sample dimensions.

2.2 *Synchrotron beamline*

The in-situ WAXD experiments were performed on the D2AM beamline at the European Synchrotron Radiation Facility. The X-Ray energy is 8 keV corresponding to a wavelength of 1, 54 Å... and the beam size is 300 μm$\times$ 200 μm.

Two important Positive Intrinsic Negative (PIN) diodes, called PM, are visible on the Figure 2. The PM3 is located at the exit of the optical cabin and is used to measure the intensity of the incident beam. The PM1 is located at the center of the WAXD detector, and is used to measure the intensity of the transmitted beam.

The WAXD sensor used is a strip sensor and its surface area is 1340 $\times$ 700 pixels. Each pixel size is 13μm and measures a quantity proportional to the number of photons received by that pixel.

The used tensile machine LINKAM MODULE FORCE STAGE allowing displacement driving. The specificity of this machine is a space along the axis perpendicular to the sample: the machine lets an X-ray beam pass through itself as well as through the sample.

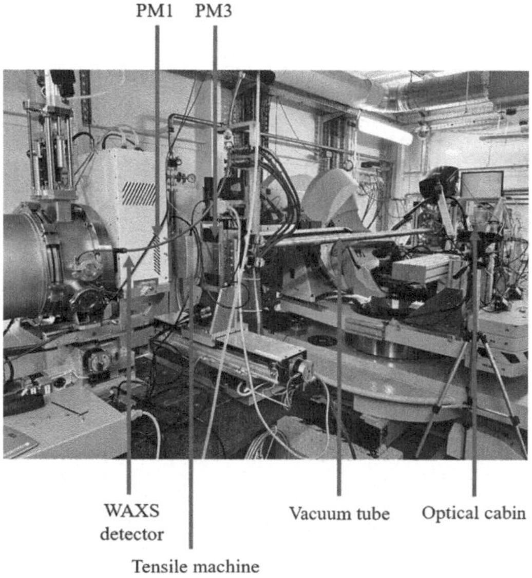

Figure 2. Line disposition: beam, machine and WAXD detector.

3 TEST PROCEDURE

3.1 *Mechanical signals*

All the tests are driven by crosshead displacement. Several types of tests are considered. First, classic load-unload tests were carried out between $\lambda = 1$ and $\lambda = 5.5$ where λ is the stretch of the material in the tensile direction. The three elongation rates investigated are: $\dot{\lambda}_1 = 5 \times 10^{-1}\text{s}^{-1}$, $\dot{\lambda}_2 = 5 \times 10^{-2}\text{s}^{-1}$, $\dot{\lambda}_3 = 5 \times 10^{-3}\text{s}^{-1}$.

Another test performed is the anhysteretic (ANH) or multi-relaxation test. The principle is to apply to the sample a low stretch, then relax it at this constant stretch, then repeat the process for increasing elongations until the maximum elongation. Stretching rate and relaxation time are kept the same during the full test. Once maximum elongation has been reached, the same process is repeated during the unload. The elongation increment is $\Delta\lambda = 0.3$, the phase duration of each relaxation is $\Delta t = 60$ seconds, the incremental strain rate is $\dot{\lambda}_2 = 5 \times 10^{-2}\text{s}^{-1}$, the minimal elongation $\lambda_{\min} = 1.0$ and the maximal one is $\lambda_{\max} = 5.5$. A representative scheme of these two tests is proposed in Figure 3.

3.2 *Background correction and post-treatment*

A raw diffraction pattern issued from the WAXD sensor cannot be analyzed as it stands. Although a diffraction pattern $I_{d,raw}$ and a PM1 value, $I_{1,raw}$,

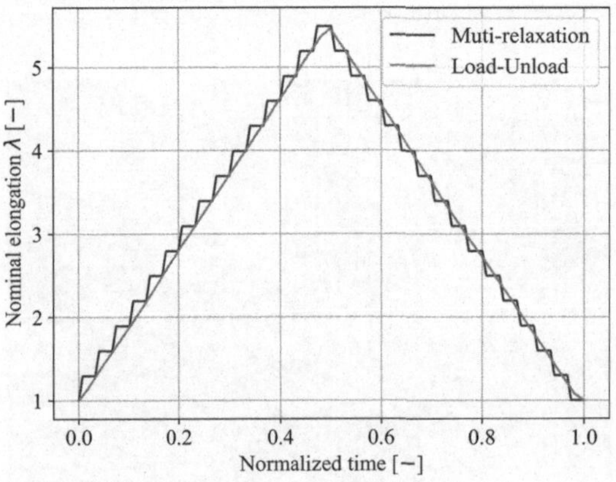

Figure 3. Test procedures for performed tests (normalized time).

can be obtained from a measurement, several factors hamper its analysis. First, the air between the exit of the optical cabin and the WAXD detector is diffracted by the electrons. Then, during a mechanical test, the specimen is stretched and its thickness varies.

To subtract the air diffraction, it is common to measure the *background* diffraction: a measurement without any specimen, where only the air is diffracted. A diffraction pattern, $I_{d,bck}$, and a PM1 value, $I_{1,bck}$, are therefore recorded. Then, in order to overcome all the biases mentioned above, it is necessary to apply successive BEER-LAMBERT laws and, assuming the incompressibility of the material when uniaxially stretched, it is then possible to obtain the following correction formula (Ran et al., 2001; Beurrot-Borgarino et al., 2013):

$$\frac{1}{\sqrt{\lambda}}\left(\frac{I_{d,raw}}{I_{1,raw}} - \frac{I_{d,bck}}{I_{1,bck}}\right) \quad (1)$$

Applying this correction formula to a raw diffraction pattern obtained from the WAXD detector typically produces the pattern shown in Figure 4.

It shows the three ZnO rings at large angles, the amorphous halo and the classic crystalline peaks found in industrially formulated natural rubbers. (120) and (200) peaks in the direction orthogonal to the tensile direction and the (201) peak between these two directions (Beurrot-Borgarino et al., 2013; Candau et al., 2012; Huneau, 2011; Albouy and Sotta, 2020). A post-processing based on deconvolution is then carried out to distinguish the diffraction of the crystalline phase from the amorphous one: the areas diffracted by these phases are called $\mathcal{I}_{\text{crist}}$ and $\mathcal{I}_{\text{amo}}$ respectively.

The next step is to estimate a crystallinity index: the volumetric fraction of the crystallized phase within the material. Several definitions exist in the literature. The definition used here is (Beurrot-Borgarino et al., 2013):

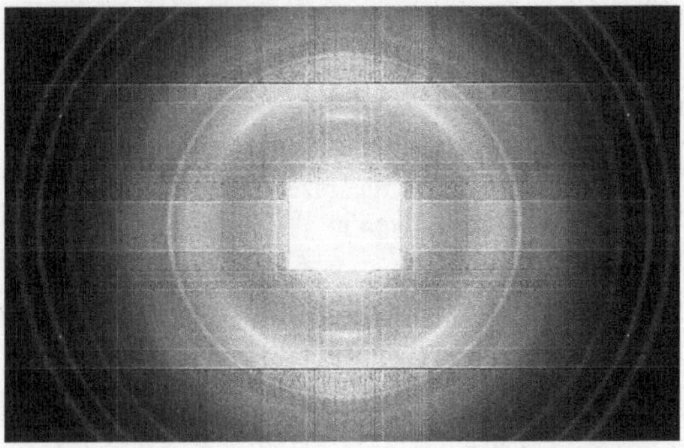

Figure 4. Typical diffraction pattern (tension direction is horizontal).

$$\chi_c = \frac{\mathcal{I}_{crist}}{\mathcal{I}_{crist} + \mathcal{I}_{amo}} \quad (2)$$

4 RESULTS

The results are all extracted from the same ESRF campaign (Le Bihan et al., 2023).

4.1 *Load-unload tests*

The three load-unload curves in the elongation *vs.* cristallinity index plane (λ, χ_c), are plotted in Figure 5.

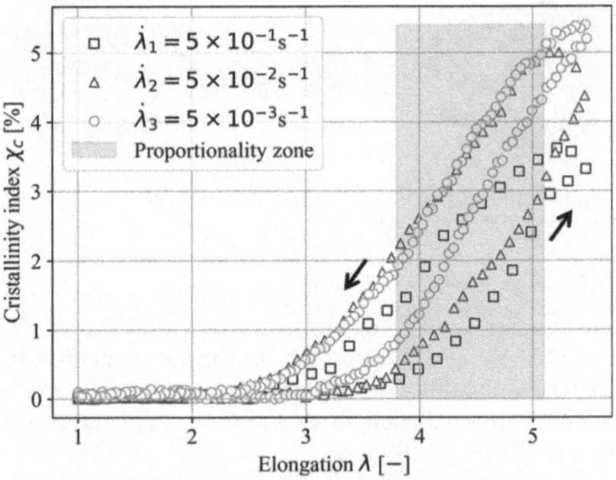

Figure 5. Load-unload curves for the three different strain rates.

It is possible to extract the elongations at the onset of crystallization, λ_C and melting λ_M. These are presented in the Table 2.

The melting threshold elongation λ_M of crystallites varies little or not at all with the strain rate, matching with (Albouy and Sotta, 2020) and seems to converge towards 2.2 . On the other hand, the elongation at the onset of crystallization increases with $\dot{\lambda}$, in line with (Albouy and Sotta, 2020; Candau et al., 2015).

Table 2. Evolution of threshold elongations with strain rate.

Strain rate [s^{-1}]	$\dot{\lambda}_3$	$\dot{\lambda}_2$	$\dot{\lambda}_1$
λ_C	2,7	3,0	3,1
λ_M	2,3	2,2	2,2

The Figure 5 shows that crystallization is "delayed": a maximum elongation does not correspond to a maximum of crystallinity. In fact, the crystallinity index continues to increase during the beginning of the unload phase, then decreases. The higher the strain rate, the more pronounced the overshoot. For a given stretch, a slower strain rate leads to a higher crystallinity index.

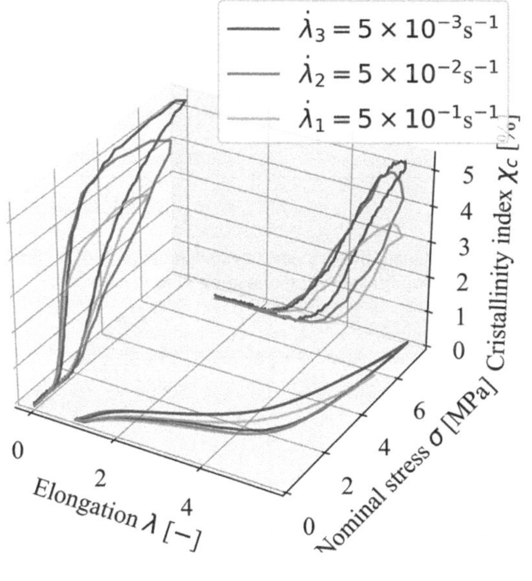

Figure 6. Point cloud $(\lambda, \sigma, \chi_c)$ projections for the 3 strain rates.

Figure 6 highlights three coupling hysteresis loops: stress to strain, crystallization to strain and crystallization to stress.

It is possible to describe cristallinity *versus* time or cristallinity *versus* stress curves by linear approximation. On some well chosen elongation ranges (grey zone of the Figure 5), this linear approximation is justified: when MIC appears, it seems to be proportional to elongation. The performed approximations are:

$$\chi_c = \frac{\partial \chi_c}{\partial t} \times t + \chi_c(t_0) \tag{3}$$

$$\chi_c = \frac{\partial \chi_c}{\partial \sigma} \times \sigma + \chi_c(\sigma_0) \tag{4}$$

where slopes $\frac{\partial \chi_c}{\partial t}$ and $\frac{\partial \chi_c}{\partial \sigma}$ are representative of crystallization kinetics (regarding time or nominal stress) and $\chi_c(t_0)$ and $\chi_c(\sigma_0)$ describe the cristallinity index at the beginning of the step. Conventional cycles at the three strain rates permits to define crystallization kinetics for load and melting kinetics for unload. These kinetics are presented in Table 3.

Table 3. Kinetics of crystallization (or melting) for simple cycles.

	$\frac{\partial \chi_c}{\partial t}$ [%/s]		$\frac{\partial \chi_c}{\partial \sigma}$ [%/MPa]	
Strain rate	Load	Unload	Load	Unload
$\dot{\lambda}_1$	1,440	−1,159	0,898	3,099
$\dot{\lambda}_2$	0,155	−0,149	0,830	3,174
$\dot{\lambda}_3$	0,018	−0,016	0,987	3,180

4.2 *Comparison of anhysteretic test and conventional cycles*

Figure 7 shows evolution of nominal stress and crystallinity index *versus* elongation for ANH tests. Be careful which way these curves are read. For stress, the load curve is "above" the unload curve. For crystallization, it's the opposite.

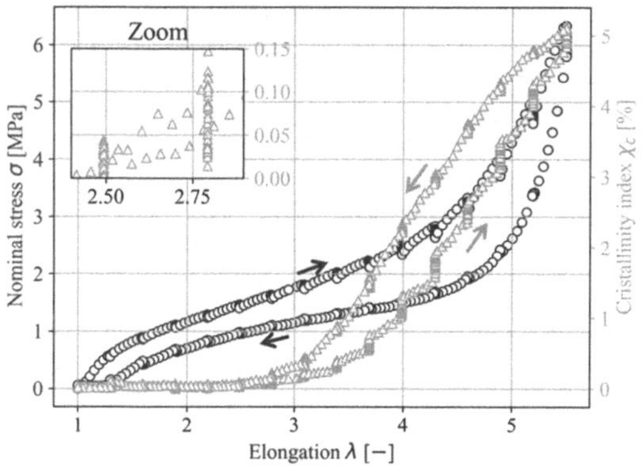

Figure 7. Stress and cristallinity index curves for ANH test.

At each stress relaxation step, whether load or unload, crystallinity increases. At first sight, one might wonder what effect the stress relaxation steps have on crystallization *i.e.* the effect of the test duration. A suitable comparison test would be a load-unload at the same draw ratio rate, $\dot{\lambda}_2$. We use the test performed in section 4.1, at $\dot{\lambda}_2$ and $\dot{\lambda}_3$ and compare both types of tests on Figure 8.

At each relaxation stage of the load, the ANH curve starts from the simple curve at $\dot{\lambda}_2$ and ends at the simple curve at $\dot{\lambda}_3$. On the other hand, the overall melting kinetic does not seem to be affected by time effect. Still on the global considerations, it was previously established that the onset elongation for the start of crystallization, λ_C, was 3.0 for a strain rate $\dot{\lambda}_2 = 5,0 \times 10^{-2}s^{-1}$ (see Table 2). For the ANH

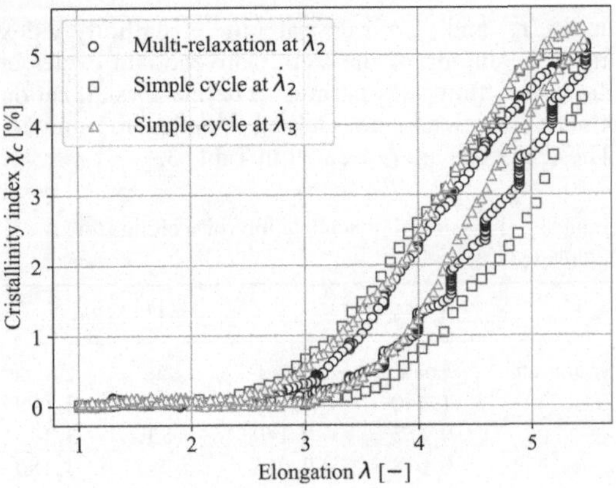

Figure 8. Comparison of ANH test with simple cycles.

test, the onset of crystallization is detected at $\lambda = 2.6 < \lambda_C$. The relaxation time allows the material to crystallize before reaching the threshold elongation λ_C. The same phenomenon occurs at the end of the test, when crystallisation is detected for $\lambda < \lambda_M$. This directly questions the relevance of the definition of threshold elongations. For dynamic test, it is well known that MIC only appears when $\lambda > \lambda_C$ and disappears when $\lambda < \lambda_M$. However, thanks to time effects, this ANH test shows that it is possible to induce mechanical crystallization outside the interval $[\lambda_M; \lambda_C]$.

This test also provides an answer about the effect of thermomechanical history on crystallization kinetics: for a given elongation, which loading path leads to the highest crystallinity indices: continuous loading (conventional cycle) or with intermediate relaxation steps (ANH test)? It appears that giving the material more time does not lead to a significant increase of the crystallinity index. The main difference is the absence of overshoot in the ANH test, since the relaxation steps have modified the crystallization kinetics during the loading phase, especially at the highest stretching levels before.

4.3 *MIC kinetics during ANH test*

Figures 9 to 12 represents cristallinity index χ_c and crystallization kinetics evaluated by equation (3) or (4) during loading phase (from $\lambda_{\min}$ to $\lambda_{\max}$) or unloading phase (from $\lambda_{\max}$ to $\lambda_{\min}$).

On the temporal curves of Figures 9 and 10, it can be seen that the crystallization kinetics increase with elongation until a saturation is reached, as classically observed on AVRAMI curves (Avrami, 2004): MIC kinetics is a self-regulating phenomenon. If we look at the values of these kinetics, there is a clear difference between the loading (or unloading) phases and the relaxation phases: we can clearly see a difference of a factor of 10 regarding kinetics between the two types of steps. We can also compare the kinetics obtained during the ANH test with those obtained for simple cycles (see Table 3). In terms of crystallisation kinetics, relaxation steps appear to be negligible, but in terms of crystallinity index, they still lead to a significant increase. On the other hand, the crystallization kinetics values of the loadings steps of the ANH test correspond to those obtained for a simple cycle performed under a ten times lower tension speed ($\partial\chi_c/\partial t = 0.018\%/\text{s}$ for $\dot{\lambda}_3$ in Table 3): it show that, although at first sight relaxation steps have almost no effect on the final crystallinity indexes reached but they have a significant impact on the crystallisation kinetics.

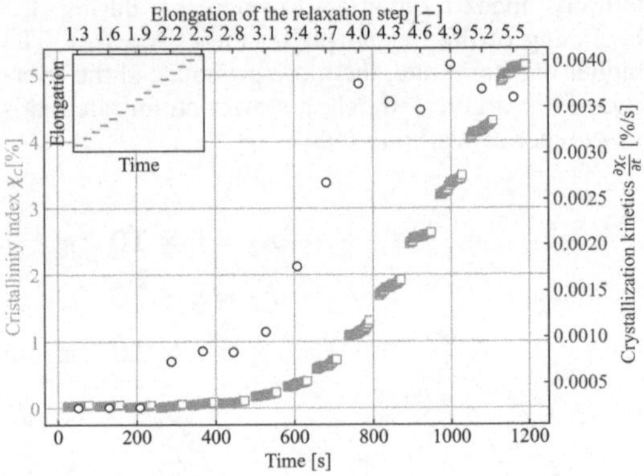

Figure 9. Cristallinity index (□) and kinetics (○) *vs.* time during relaxation steps of the loading phase (ANH test).

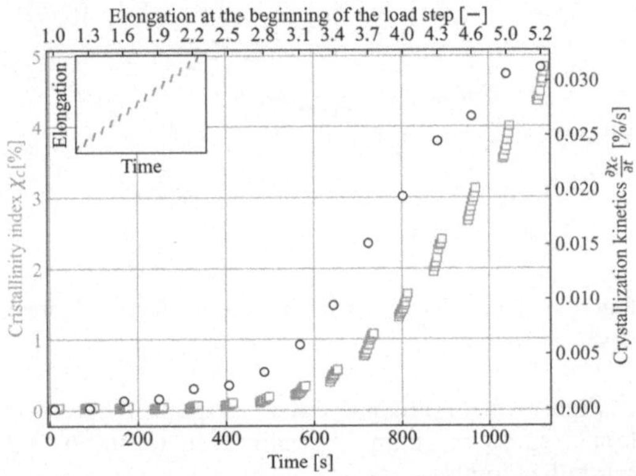

Figure 10. Cristallinity index (□) and kinetics (○) *vs.* time during loading steps of the loading phase (ANH test).

Figures 9 and 10 indirectly show that elongation is not an unifying variable for crystallization kinetics. Would nominal stress be? Figures 11 and 12 show that the nominal stress cannot unify the kinetics, otherwise the slope $\partial\chi_c/\partial\sigma$ would be constant. What is worth noting, however, is that it exists a nominal stress that leads to an optimum of crystallization or melting kinetics, in both the loading (Figure 11) and unloading (Figure 12) phases.

Regarding melting kinetics values, while those obtained on the simple cycles seemed to stagnate at around 3.1 %/MPa (see Table 3), those obtained on the ANH test are clearly dependent on the nominal stress, although of the same order of magnitude (see Figure 12).

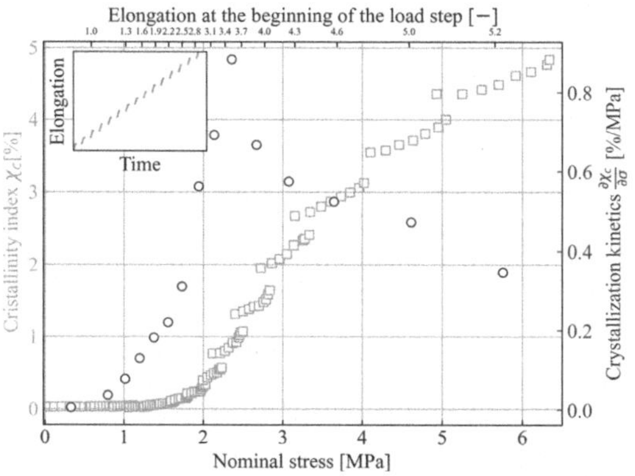

Figure 11. Cristallinity index (□) and kinetics (○) *vs.* nominal stress during loading steps of the loading phase (ANH test).

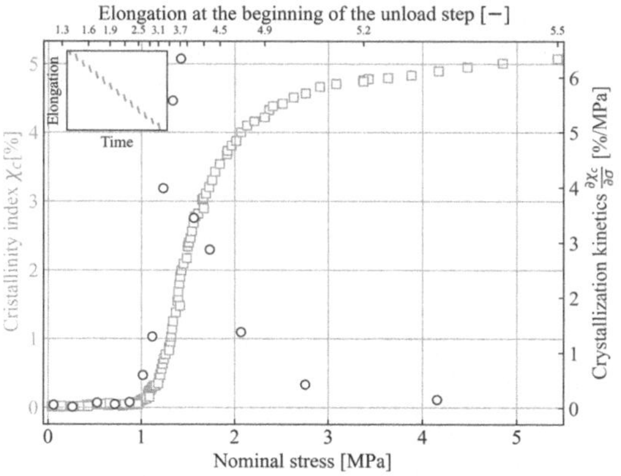

Figure 12. Cristallinity index (□) and kinetics (○) *vs.* nominal stress during unloading steps of the unloading phase (ANH test).

5 CONCLUSIONS AND PERSPECTIVES

The ANH test confirms that the classic definitions of threshold elongations for crystallization and melting should be treated with caution: it is possible to obtain crystallization at very small stretches by giving time to the material to crystallize.

The ANH test also shows that the crystallization kinetics are higher during the loading or unloading phases than during the relaxation phases. For the strain rates tested, the test also demonstrates a saturation of the crystallization kinetics. The various graphical representations show that the unification of kinetics (by time, strain or stress) is not straight forward. The comparison of a simple cycle (monotonic loading) with an ANH test (interrupted loading) shows that the loading path has almost no effect on the final crystallinity indexes.

This ANH test has provided a lot of answers about the notions of crystallization kinetics, but it also raises many questions: for the ANH test, what would happen if the draw ratio were different? if there were fewer relaxation steps? if there were longer relaxation steps? or creep steps instead of relaxation ones?

REFERENCES

Albouy, P.-A. and Sotta, P. "Draw Ratio at the Onset of Strain-Induced Crystallization in Cross-Linked Natural Rubber". *Macromolecules* 53, no. 3 (2020): 992–1000.

Avrami, M. "Kinetics of Phase Change. II Transformation-Time Relations for Random Distribution of Nuclei". *The Journal of Chemical Physics* 8, no. 2 (2004): 212–224. Visited on 01/23/2024.

Beurrot-Borgarino, S. et al. "Strain-Induced Crystallization of Carbon Black-Filled Natural Rubber during Fatigue Measured by in Situ Synchrotron X-ray Diffraction". *International Journal of Fatigue* 47 (2013): 1–7.

Brüning, K. et al. "Kinetics of Strain-Induced Crystallization in Natural Rubber: A Diffusion-Controlled Rate Law". *Polymer* 72 (2015): 52–58.

Busse, W. F. "Tear Resistance and Structure of Rubber". *Rubber Chemistry and Technology* 8, no. 1 (1935): 122–137.

Cadwell, S. M. et al. "Dynamic Fatigue Life of Rubber". *Industrial & Engineering Chemistry Analytical Edition* 12, no. 1 (1940): 19–23.

Candau, N. et al. "Characteristic-Time of Strain Induced Crystallization of Crosslinked Natural Rubber". *Polymer* 53, no. 13 (2012): 1423–1464.

Candau, N. et al. "Influence of Strain Rate and Temperature on the Onset of Strain Induced Crystallization in Natural Rubber". *European Polymer Journal* 64 (2015): 244–252.

Champy, C. et al. "Fatigue of Crystallizable Rubber: Generation of a Haigh Diagram over a Wide Range of Positive Load Ratios". *International Journal of Fatigue* 150 (2021): 106313.

Huneau, B. "Strain-Induced Crystallization of Natural Rubber: A Review of X-ray Diffraction Investigations". *Rubber Chemistry and Technology* 84, no. 3 (2011): 425–452.

Le Bihan, A. et al. (Influence of the Loading History on Strain Induced Cristallisation of Elastomers with Original Architecture of Crosslinks Network - https://doi.esrf.fr/10.15151/ESRF-ES-1191261152. In collab. with Gilbert Chahine and Isabelle Morfin). 2023.

Ran, S. et al. "Structural Changes during Deformation of Kevlar Fibers via On-Line Synchrotron SAXS/WAXD Techniques". *Polymer* 42, no. 4 (2001): 1601–1612.

Saintier, N., Cailletaud, G., and Piques, R. "Cyclic Loadings and Crystallization of Natural Rubber: An Explanation of Fatigue Crack Propagation Reinforcement under a Positive Loading Ratio". *Materials Science and Engineering*: A 528, no. 3 (2011): 1078–1086.

Sotta, P. and Albouy, P.-A. "Strain-Induced Crystallization in Natural Rubber: Flory's Theory Revisited". *Macromolecules* 53, no. 8 (2020): 3097–3109.

Warneboldt, I. et al. "Experimental Investigations about Complex Non-Relaxing Fatigue Loads for Carbon-Black Filled Natural Rubber". *International Journal of Fatigue* 156 (2022): 106696.

Constitutive Models for Rubber XIII – Dal (Ed.)
© 2025 The Author(s), ISBN 978-1-032-85138-9

Frequency-domain analysis of elastomeric vibration isolators: Experimental and numerical investigations under dynamic loading

T. Gejgus, O. Farkas, K. Loos, A. Lion & M. Johlitz
Institute of Mechanics, University of the Bundeswehr Munich, Neubiberg, Germany

ABSTRACT: When dynamic loads act upon rubber isolators used in electric vehicles, the response of the material is highly nonlinear, and can vary significantly in the region of structural resonances. For this reason, it is of interest to be capable of dynamic measurements to determine the dynamic stiffness on both the sample and the component level in the relevant frequency range. In addition, the density of the industrial rubber blend has a tremendous impact on the response in the resonance region. Since this dependence on the material response is strongly affected by the temperature of the elastomer as well, further experiments, such as temperature-dependent pycnometry or thermal mechanical analysis are performed to quantify these aspects. Moreover, further enhancements to the Test Bench for Dynamic Stiffness (TBDS) introduced by Gejgus and Farkas in 2022 are presented, with a specific focus on investigating the influence of thermal effects. The subsequent emphasis shifts to developing a master curve for the elastomer. Initially, the geometrical approach of curve generation is compared with established shift principles, exploring the relevance of the free choice of the reference temperature. Subsequently, the master curve is approximated via a Prony series, and both the raw master curve and the identified Prony series are used to describe the viscoelastic material behaviour within two different FE-simulations. The deviations in the simulation results caused by the defined viscoelasticity and the influence of the rubber compound compressibility are presented and discussed. Lastly, torque and vibration profiles during the run-up of the permanent magnet synchronous motor (PMSM) are measured under various boundary conditions. This method enables the determination of relevant excitation amplitudes for finite element simulations by analysing the frequency spectrum of a PMSM during its operating phases.

1 INTRODUCTION

An improvement in the driving comfort, especially at lower frequencies of the vibration range of interest ($\leq$ 500Hz), is an undeniable consequence of the increase of electric vehicles available on the market. Despite the overall improvement in the acoustic qualities of the vehicles, a number of negative experiences have been reported by the users over millions of operating hours. Driven by the desire to continuously improve the technical solutions, this study focuses on the material behaviour of elastomeric engine mounts used as vibration isolators for electric vehicle drives.

Firstly, the Noise-Vibration-Harshness (NVH) behaviour of the tested Global Vehicle Motor (GVM) shown in Figure 1 is investigated and the causes of several effects are analysed and interpreted.

Next, the material parameters of the selected industrial rubber compound are determined. The temperature-dependent density of the blend is investigated experimentally using gas pycnometry tempered by a Peltier element. The hyperelastic material behaviour is then defined with help of a two-parametric Mooney-Rivlin free energy function. To

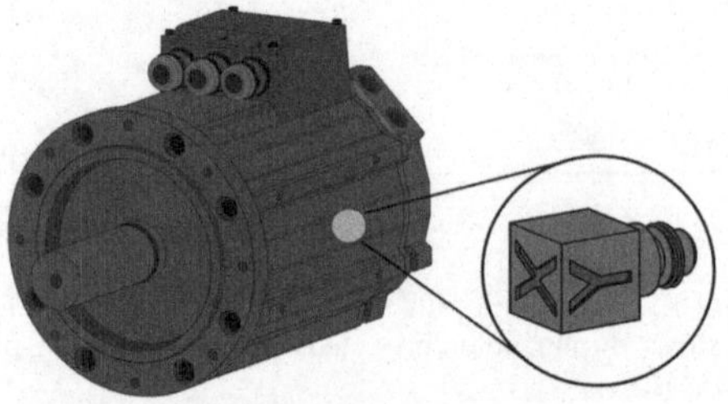

Figure 1. Global vehicle motor produced by Parker Hannifin and the accelerometer used to measure the response during the engine run-up.

determine the viscoelastic behaviour of the material under investigation, the standardised dynamic mechanical thermal analysis is performed under consideration of the Mullins and Payne effects. The result of the temperature-frequency sweep is shifted vertically and extrapolated horizontally with respect to the time-temperature superposition principle. The details of this principle are further elaborated in this study. The focus of this contribution then shifts to the

DOI: 10.1201/9781003516880-23

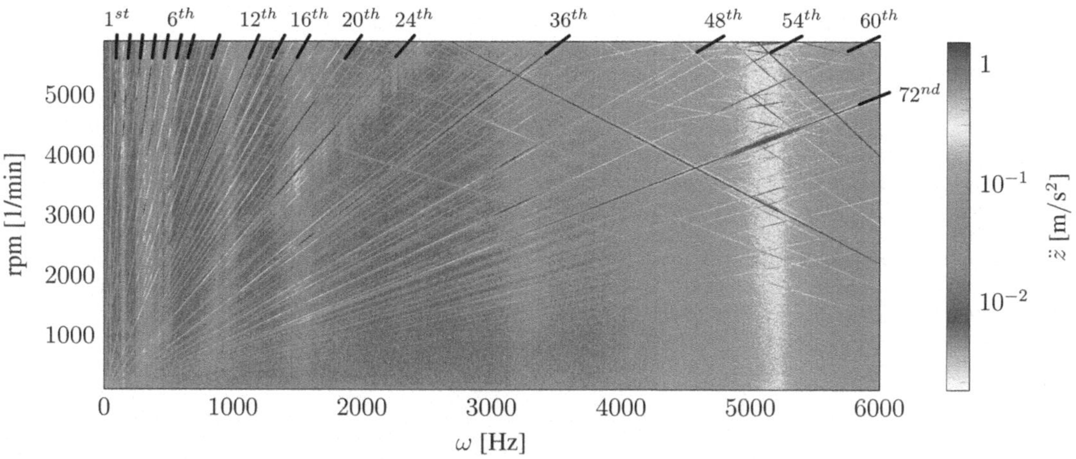

Figure 2. Campbell diagram displaying acceleration magnitudes measured on the PMSM housing with respect to the increasing rotational speed in rpm (created with PAK 6.1).

high-frequency dynamic measurement presented in Gejgus et al. (2022), where its improvements, advantages and weaknesses are discussed. The frequency sweep conducted up to 3000Hz is compared with the standard dynamic measurement and the FE-simulation results presented in the next chapter. Finally, possible sources of error are discussed and future objectives are outlined.

2 EXPERIMENTAL INVESTIGATIONS OF THE E-DRIVE

Electric vehicles can be powered by a variety of conceptionally different electric motor designs. The majority consists of a permanent magnet synchronous machine (PMSM) and a fixed-ratio gearbox. This solution gained its popularity mainly because of PMSM's ability to recuperate mechanical energy, its efficiency, and its compelling power-to-weight ratio.

The analysed synchronous machine provided by Parker Hannifin GmbH is equipped with 12 poles on the rotor and 18 slots on the stator. The drive employs 3-phase voltage controlled by pulse width modulation (PWM). Based on the conclusions of Kazmierkowski et al. (2002), a space vector modulation (SVM) algorithm is used to minimize the total harmonic distortion caused by the rapid switching of the motor control.

The NVH behaviour of the PMSM was investigated during controlled run-ups from 100rpm to 5900rpm over a period of 600s. A triaxial IEPE (integrated electronics, piezoelectric) accelerometer was accurately placed on PMSM's side as marked in Figure 1. The typical spectral effects and specific pronounced orders associated with the design and control of the tested AC motor are shown in Figure 2 via the Campbell diagram dependent on the rotational speed measured with the optical tachometer sensor. The Campbell diagram allows the identification of structural resonances of the tested drive, as these remain independent of the motor speed. The 6th order is linked to the so-called slot locking and thus to the rotor pole count. The spectrum includes higher harmonics of the 6th order up to the 72nd order, which is of significant interest. Slightly above 5000Hz, the 72nd order intersects with the off-zero harmonics of the PWM carrier frequency. Furthermore, an order crossing occurs in the region where one of the structural resonances of the drive is located. This phenomenon is responsible for an enormous amplification of the excitation amplitudes. A similar amplification can be observed around 1500Hz, where the 24th order intersects one of the structural resonances.

While designing engine mounts for electric vehicles, it is crucial to avoid the presence of structural resonances of the elastomeric isolator close to the areas described above.

3 DETERMINATION OF MATERIAL PROPERTIES

The material under examination is a compound of natural and butadiene rubber (NR-BR) filled with 22phr of carbon black supplied by BOGE Elastmetall GmbH. A single measurement cannot fully characterise the material in its entirety, especially when subjected to a complex deformation state. Instead, every aspect of the material behaviour needs to be isolated and experimentally characterised individually. This section presents the experiments including their results used to determine both the temperature-dependent and the mechanical properties, briefly describing the methods employed.

3.1 *Density via gas pycnometry*

Since the density of the material acts as a source of inertia within the numerical simulation, it is important to determine it with as much accuracy as possible. Fortunately, gas pycnometry provides a quick and precise method for determining the density. According to Tamari (2004), gas pycnometry compares the pressure difference between the reference chamber and the chamber containing the sample whose density is of interest. The material's density is then calculated using Boyle's law. For the NR-BR compound, the density at three different temperatures at a pressure of approximately 124kPA is determined using a Paan Ultrapyc 5000 gas pycnometer by Anton Paar GmbH. The measurement results displayed in Table 1 are presented with a low mean variance of 0.0052% and show that the dependence between density and temperature can be considered as linear.

Table 1. NR-BR density at different temperatures via gas pycnometry.

Temperature θ [°C]	*Density* [g/cm^3]
20	1.0255
35	1.0174
50	1.0097

3.2 *Hyperelastic material behaviour*

Moving on to mechanical properties of the material, the next step is to determine the hyperelastic behaviour. To examine the stress response of the NR-BR compound, one has to establish a uniform state for the samples via preconditioning, which eliminates the presence of the Mullins effect. Consequently, a stepwise stress-relaxation guided by ISO527-1, respectively ISO7743 leads to the determination of equilibrium stresses under uniaxial tension (UT), uniaxial compression (UC) and pure shear (PS) at an ambient temperature of 20°C.

The stress response is then approximated to define the rate-independent part of material behaviour within the FE-simulation. The selected material model yields a two-parametric hyperelastic model, as outlined by Mooney-Rivlin, with the subsequent strain energy function

$$\rho_0\psi = C_{10}(\mathbf{I_C} - 3) + C_{01}(\mathbf{II_C} - 3). \quad (1)$$

The proposed model contains both the first $\mathbf{I_C}$ and the second $\mathbf{II_C}$ invariant of the right Cauchy-Green strain tensor $\mathbf{C}$. The resulting equations governing the first Piola-Kirchhoff stress response for UT, UC and PS are defined as functions of the stretch λ

$$P_{\text{UT,UC}} = 2C_{10}\left(\lambda - \frac{1}{\lambda^2}\right) + 2C_{01}\left(1 - \frac{1}{\lambda^3}\right), \quad (2)$$

$$P_{\text{PS}} = 2(C_{10} + C_{01})\left(\lambda - \frac{1}{\lambda^3}\right). \quad (3)$$

Here, the constant material parameters C_{10} and C_{01} are to be identified with the help of a particle swarm optimisation algorithm (PSO), which is inspired by a collective behaviour of natural systems, such as flocks of birds searching for a source of food. The optimisation is based on the minimisation of the residual between the measured and the calculated stress for each of the deformation modes. The aim to identify such a set of material parameters for with the goodness of fit (GoF), determined by the root-mean-square deviation, is the lowest. The non-linear quasi-static measurement results (data points) and their approximation (lines) displayed in Figure 3 show a high accuracy of fit.

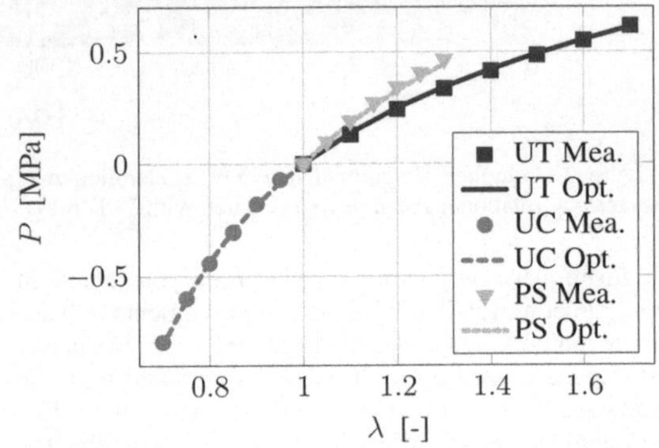

Figure 3. Results of quasi-static measurements along with the approximation identified via the PSO algorithm.

3.3 *Viscoelastic material behaviour*

Considering the application of the elastomer as an isolator of vibrations generated by the PMSM, a characterisation of the rate-dependent material properties must be performed. For this purpose, a small NR-BR cylindric sample with dimensions according to DIN 815 is examined. Generally, there are two well-known and established ways to determine the rate-dependent material response: a relaxation test, which defines the behaviour over a longer time period, or a dynamic mechanical thermal analysis (DMTA), which, in contrast, determines the material response as a function of frequency. Since this study concerns the investigation of elastomers in the frequency domain, the latter method based on the superposition of a static, compression preload and a dynamic oscillation is chosen. The DMTA yields results in form of the storage and the loss modulus denoted by $G'(\omega)$ and $G''(\omega)$, respectively. These are, according to Ehrenstein et al. (2012), proportional to the amount of fully recoverable stored energy and the energy dissipated as heat, and can be used to calculate the dynamic response represented by the complex dynamic modulus $G^*(\omega)$ or the delay between the excitation and the material response given by the phase angle $\delta(\omega)$.

Furthermore, the carbon black filler of the NR-BR compound causes the so-called Payne effect, which must be taken into account in the characterisation. The Payne effect is interpreted as a breakdown of the

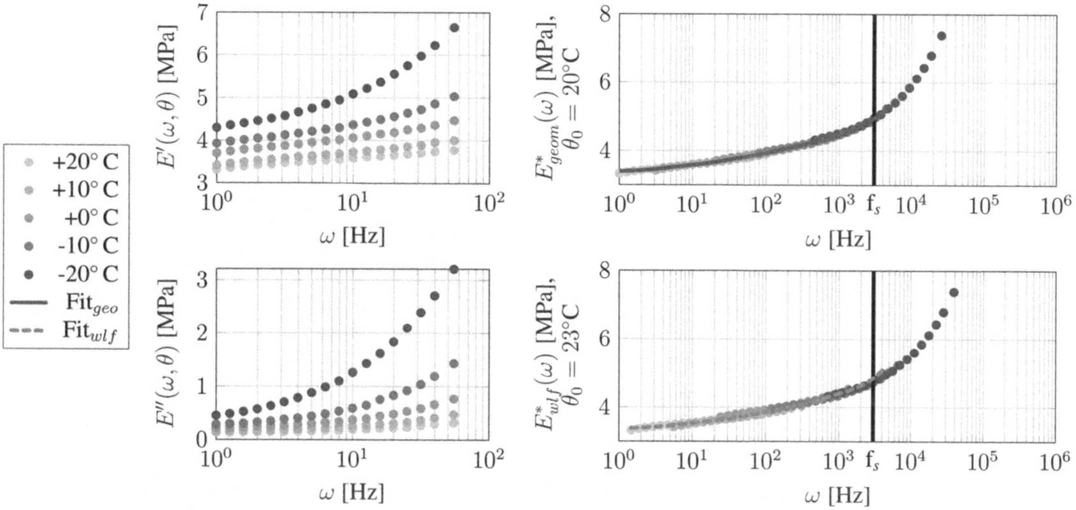

Figure 4. Storage and loss dynamic moduli obtained via temperature-frequency sweep and master curves of complex dynamic modulus at different reference temperatures θ_0 created via geometrical and WLF approach with fitted Prony series.

filler network leading to a pronounced non-linearity of the material response even at relatively small excitation amplitudes, see Raghunath et al. (2016). To define the linear viscoelastic region (LVR), i.e. the region in which the stress response is not affected by network breakdown, a dynamic sweep of the excitation amplitude must be performed. As with the static predeformation of the material, the sample exhibits a negligible dependence in the range of the relevant pre-deformation amplitudes.

Finally, the exact parameters of the temperature-frequencysweep are to be defined. Firstly, the frequency within the measurement is distributed logarithmically between 1Hz and 55Hz. The frequency sweep is executed in the temperature range from -20°C up to 20°C with steps of 10°C. The results of the static and dynamic sweeps are then used to define the excitation amplitudes ensuring that the response remains within the framework of linear viscoelasticity. The temperature-frequency sweep is thus successfully defined, and its results are presented in the left part of Figure 4. It is crucial to note that the results do not account for the mass of the sample used for the measurement.

However, the results of the temperature-frequency sweep alone are not sufficient to describe the viscoelastic part of the material response, since the upper boundary of the vibration range of interest for the PMSM is far beyond the measured 55Hz. Fortunately, the time-temperature superposition principle outlined by van Gurp and Palmen (1998) can be of help. The presented procedure assumes that the effects of both frequency and temperature on the material are in fact equal. One can therefore select a reference temperature θ_0 and shift the rest of the frequency sweeps horizontally to produce one continuous curve, thus extrapolating the results far beyond the measurable range to cover the range of interest for the PMSM. This curve is referred to as the master curve, and, as indicated in the abstract of this contribution, there are multiple ways to obtain it. The first, namely the geometrical approach, determines the shift factors a_θ based on the overlap windows. These depend on the measurement results and, of course, on the selected reference temperature. On the other hand, the approach presented by Williams et al. (1955) is governed by the following equation

$$\log a_\theta = -\frac{C_1(\theta-\theta_0)}{C_2+\theta-\theta_0}. \tag{4}$$

Here, the shift factors a_θ are determined using the constants C_1 and C_2 which vary with the examined material. The WLF approach requires significantly more computing power, since the material parameters $C_{1,2}$ need to be determined using a suitable optimisation algorithms, however, it offers the possibility of shifting the master curve to an arbitrary temperature outside of the measurement range. In both shifting methods, a vertical shift compensating for the changes in volume and therefore density of the material is applied. More details and studies to generate master curves from temperature-frequency sweeps are published in Farkas et al. (2024).

Upon establishing a master curve through one of the two outlined methods, vectors corresponding to the storage and loss moduli are derived. These serve as direct input variables for the analysis of the viscoelastic material behaviour using FE simulation. Optionally, input data for the FE simulation can be obtained by specifying parameters within a Prony series, whereby a simplified material model based on the generalised Maxwell model consists of a singular nonlinear hyperelastic spring and a predefined number of linear Maxwell elements. Using this representation, the storage and loss moduli are explicitly defined as follows:

$$G'(\omega) = E_\infty + \sum_{k=1}^{n} \mu_k \frac{(\omega\tau_k)^2}{1+(\omega\tau_k)^2}, \tag{5}$$

$$G''(\omega) = \sum_{k=1}^{n} \mu_k \frac{\omega\tau_k}{1+(\omega\tau_k)^2}. \tag{6}$$

The relaxation times τ_k and the stiffnesses of the springs μ_k are determined using a nonlinear least squares algorithm. In this example, the master curve is approximated using 6 Maxwell elements.

Finally, the right part of Figure 4 shows the master curve obtained via both the geometrical and the WLF approach extrapolating to 40,000Hz, i.e. far beyond the measured maximum frequency of 55Hz. In addition, the Prony series calibrated to approximate the master curves (Fit) shows a high level of agreement. Both the master curves and the parameters identified for the Prony series are used in the FE-simulations to define the viscoelastic properties of the material.

4 HIGH-FREQUENCY DYNAMIC MEASUREMENTS

As presented in Section 3.3, standardised dynamic measurements are incapable of reaching high frequencies without the time-temperature superposition principle. For instance, the measurement setup Eplexor 500N by NETZSCH-Geratebau GmbH used to create the temperature-frequency sweep shown in Figure 4, achieves frequencies up to 200Hz at different ambient conditions and provides precise control of static and dynamic quantities. Nevertheless, based on the stiffness of the measured sample, the results might be corrupted by resonances of the measurement setup itself. The measurement of the material behaviour therefore contains a slight uncertainty, which can lead to incorrect results in numerical simulations, especially at high frequencies. On the other hand, high-frequency test rigs developed by ACOEM France or Muller-BBM GmbH require a considerable investment, or they are not available for purchase at all. Due to these reasons, the frequency sweep from 100Hz to 3000Hz in this study was performed using TBDS, a high-frequency analysis setup shown in Figure 5, developed at the University of the Bundeswehr Munich and first presented by Gejgus et al. Since last mentioned in 2022, multiple design upgrades eliminate the disturbances caused by the frame oscillation. Additionally, the support of the hanging mass, which acts as a source of inertia, was exchanged to one with lower stiffness. Lastly, an accelerometer observing the oscillation of the electrodynamic vibration exciter was exchanged to another one with higher sensitivity.

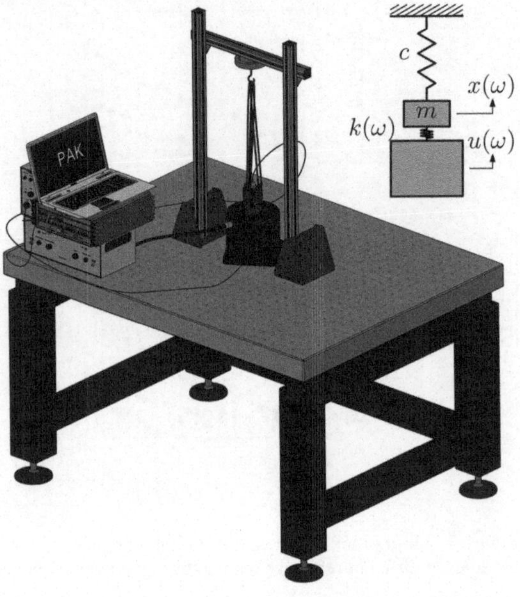

Figure 5. The high-frequency analysis setup along with the equivalent mathematical model for sample response presented by Gejgus et al. in 2022.

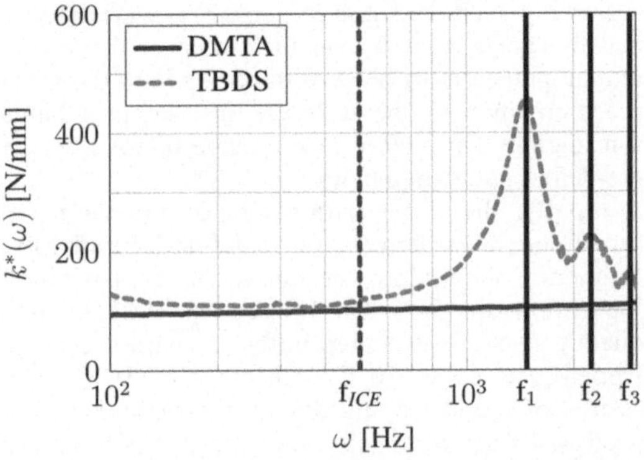

Figure 6. Dynamic stiffness magnitude calculated from the DMTA measurement and measured stiffness using TBDS.

Moving on, the preload acting upon the cylindrical sample with the diameter of about 25mm was set to match the preload conditions for the DMTA analysis. The oscillation of the sample is regulated with the excitation velocity amplitude v_a of 5mm/s with the tolerance of 2%. The regulation is done in closed loop via PAK MKII, a data acquisition unit available from Muller-BBM VAS GmbH.

After the measurement is completed, the next step is to determine the dynamic stiffness of the sample. To this end, the analytical model displayed at the upper left corner of Figure 5 is solved to determine the frequency-dependent dynamic stiffness of the cylindrical sample. The magnitude of dynamic stiffness $k^*(\omega)$ then reads as

$$\mathrm{k}^*(\omega) = \frac{(c - \omega^2 m)\,\widehat{x}(\omega)}{\widehat{u}(\omega) - \widehat{x}(\omega)}, \tag{7}$$

where $\widehat{u}$ and $\widehat{x}$ are frequency-dependent magnitudes of the excitation, respectively response displacement amplitudes. The constant c represents the linearised stiffness of the elastomeric supports at the operating point, which prevent the horizontal movements of the seismic mass. Since the TBDS was performed at ambient temperature of 23°C, the magnitude of dynamic stiffness can be directly compared to the stiffness obtained from the DMTA measurement with the cylindrical sample. The results of the temperature-frequency sweep are mastered to $\theta_0 = 23$°C and represented via complex dynamic stiffness. This quantity is calculated using the magnitude of the complex dynamic modulus $G^*(\omega)$ and the dimensions of the sample. The comparison of the two measurements is then presented in Figure 6. Looking closely at this figure, one can see that the curves at and below the frequency f_{ICE} differ very little from each other. This frequency corresponds to 500Hz and represents the vibrational region of interest for vehicles with internal combustion engine. However, considering the application of engine mounts for vehicles with PMSM, the relevant region is much wider. Since the upper limit of the frequency spectrum of the electric motor studied in this paper can easily reach ten times the value of f_{ICE}, it is necessary to analyse the results above this frequency as well to ensure the desired level of vibroacoustic passenger comfort. By doing so, a significant difference between the two measurements arises. The TBDS is able to detect three pronounced peaks of the stiffness, which are believed to be the structural resonances of the elastomeric material. These peaks, however, stay undetected by the extrapolation of results obtained with the standardised DMTA measurement. The frequencies at which the structural resonances occur are presented in Table 2. With the increase of more than 350%, the sudden change in the dynamic stiffness can amplify the structure-borne noises, thereby greatly affecting the NVH characteristics of the observed system. Since resonances can only be detected if the mass of the sample is taken into account, it is crucial to consider the material density when dealing with the problems of NVH and acoustics of vehicles with electric engines. For example, comparing the first resonance frequency of the sample tested with TBDS with the Campbell diagram shown in Figure 2, it can be seen that one of the structural resonances of the PMSM is located in roughly the same frequency region. The vibrations of the 24th order amplified by the structural resonance of the electric motor housing would be transmitted even stronger due to the sudden increase of the dynamic stiffness of the engine mount.

Lastly, due to heat transfer from the body of the electromagnetic exciter into the sample, the high-frequency measurement was analysed with the thermal imaging camera. The results from the thermography measurement without the cooling of the exciter show that the bottom of the sample increased its temperature by approximately 12.5°C in the span of 650s. Afterwards, the measurement was repeated with assisted air cooling of the exciter. This simple step lowered the temperature change to 4°C. The images captured by the camera before and after the measurement with the exciter cooling are displayed in Figure 7. Looking at the temperature profile of the sample after the measurement, which is displayed in the right part of the figure, it can be noticed that the heat conduction only affects the bottom part of the sample. Since the average temperature change along the sample height is 1.3°C and the ambient temperature during TBDS measurement was 21.5°C, the master curve generated using the WLF approach needed to be shifted to $\theta_0 = 23$°C.

Table 2. Frequencies of structural resonances with their respective dynamic stiffnesses.

Frequency [Hz]	*Dynamic stiffness* [N/mm]
$\mathrm{f}_1 = 1475$	460.3
$\mathrm{f}_2 = 2236$	226.0
$\mathrm{f}_3 = 2882$	169.6

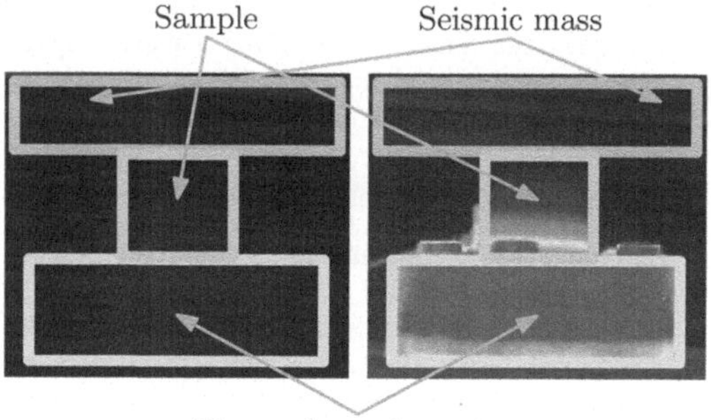

Figure 7. Temperature profile before and after the TBDS measurement.

5 FINITE ELEMENT SIMULATION

Finally, the issues described, investigated and presented in previous sections can be represented within one complex mathematical model - a digital representation of the cylindrical elastomeric sample. This representation is known as a digital twin, and it can be used to reduce the research expenses by observing the behaviour of the specimen with only a few standardised measurements and a finite element model. The digital twin of the cylindrical sample is developed using the Abaqus simulation environment - this software provides a wide range of material descriptions, it is therefore more than suitable to compare different definitions of the viscoelastic behaviour of the material. The performance of the digital twin is evaluated upon the results of the TBDS simulation. Thereby, the boundary and loading conditions are set to match the initial state of the TBDS measurement. The NR-BR blend is defined using three material properties:

1. **Density** The definition of the density and thus the inertia of the material is rather simple: the temperature-dependent data measured using a gas pycnometer, which is displayed in Table 1, are used directly as input.
2. **Hyperelastic behaviour** The hyperelastic behaviour was approximated with a two-parametric Mooney-Rivlin free energy function. The parameters C_{10} and C_{01} were identified using PSO and implemented directly into the material definition module in Abaqus. The material parameter D_1, which is related to the compressibility of the NR-BR mixture, is defined to satisfy the Poisson's ratio of $\nu = 0.4995$.
3. **Viscoelastic behaviour** With respect to the previous statements, the definition of viscoelastic material properties differs in each of the performed FE-simulations. The basis for the material definition is the master curve at the reference temperature of θ_0= 23°C obtained with uniaxial DMTA. The first simulation uses the input outlined by Bahuguna et al. (2007): the frequency dependent material properties are calculated using the dynamic moduli and the preload of the sample. In the second approach, the material is defined using the relaxation times τ_k and the dimensionless constants g_k determined by calibration of the Prony series.

To reduce the computational power requirements, the body of the cylindrical sample is represented through an axisymmetric deformable solid. This solid is discretised using linear axisymmetric elements whose size is estimated to 0.4mm. These parameters of the mesh ensure that the propagation of sound waves generated by the electromagnetic exciter is guaranteed without distortion. Further refinement of the mesh only increases the demands on the computing power, but has no influence on the simulation results.

In the first step of the simulation, the initial conditions of the TBDS measurement are replicated. Both the top and the bottom of the sample are connected to two remote points, which are used to define the input and to observe the output, respectively. A predefined temperature field used for the interpolation of the density data is defined for each node of the mathematical model. Next, the simulation procedure performs the static step - the top of the sample is preloaded with the weight of the seismic mass. That being said, the gravity generated by this mass must be compensated by a concentrated force to match the preload in the TBDS measurement. Finally, the state of predeformation is passed to the steady-state dynamic step responsible for the dynamic excitation of the bottom of the sample. Considering that the displacement input $\hat{u}(\omega)$ is identical for both the measurement and the FE-simulations, it is desired to compare the displacement amplitudes $\hat{x}(\omega)$ at the top of the sample, since these are proportional to the transfer stiffness of the investigated elastomeric isolator. The frequency in the dynamic simulation step ranges from 100Hz to 3kHz and the displacement at the top of the sample is logged with the rate of 2Hz. With this definition and the available computational power of a standard desktop computer, the numerical simulation is completed in roughly 10 minutes. The comparison of the simulations and the TBDS measurement is shown in Figure 8. The results of the numerical simulation clearly show that the material model is able to capture the structural resonances. In the range up to 1250Hz, both simulations show high concordance with the measurement. The simulation with the tabular definition of the viscoelastic properties agrees significantly better with the measurement curve than that with the Prony series definition. Despite the obvious shift of structural resonances to higher frequencies, the simulation is viewed as successful - the

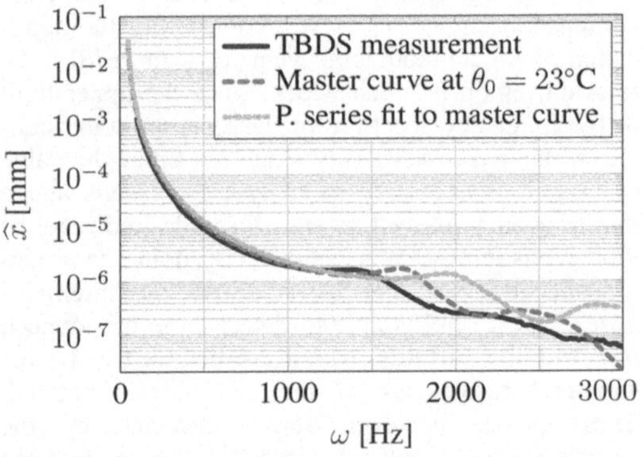

Figure 8. Comparison between TBDS measurement and FE-simulations.

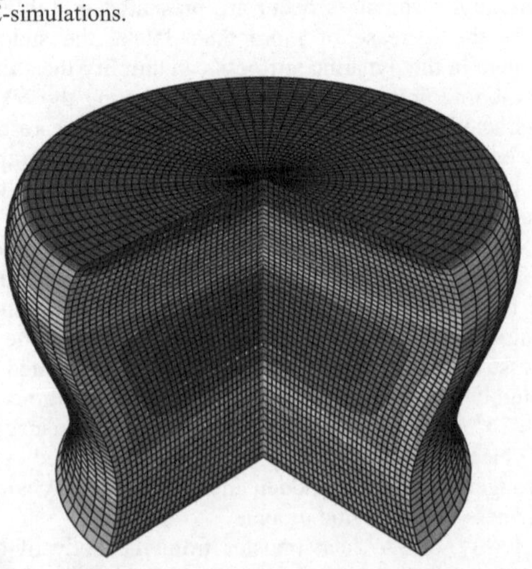

Figure 9. Longitudinal wave passing through the sample at first calculated resonance frequency of 1650Hz.

structural resonances are identified with the material definition, which consists of standardised measurements and their results only. It is believed that the reason for this difference in the resonance frequencies lies in the definition of the viscoelastic material properties. The temperature change at the bottom of the sample during the measurement could contribute to this difference as well. To summarise this section, Figure 9 shows a structural wave passing through the sample at the first resonance frequency of the simulation. The material model works with tabular viscoelastic input of the master curve.

6 CONCLUSION

As presented at the beginning of the study, despite the fact that the operation of electric vehicles emits less noise than that of the vehicles powered by an internal combustion engine, some undesired noises are still generated. The areas of the Campbell diagram highlighted in Section 2 are especially critical and should be treated with utmost attention.

After successful determination of the material properties, the finite element simulations were performed to validate the high-frequency measurement conducted with the help of TBDS. The heat transfer between the electromagnetic exciter and the tested cylindrical sample was investigated and has been taken into consideration while defining the material model. Moreover, several definitions of the viscoelastic material behaviour were specified and compared to gain knowledge regarding model sensitivity. Finally, two representative simulation results were compared with the high-frequency measurement conducted with TBDS.

A satisfactory correlation of the simulation and the measurement data was achieved using the tabular definition of the viscoelastic properties within the Abaqus simulation environment. On the contrary, while using parameters of Prony series, a significant shift of the resonance behaviour was observed. Clearly, this phenomenon needs to be investigated further.

In the future, an extended NVH characterisation of the tested electric drive and its dependence on control strategy, loading conditions and mounting is certainly of interest. Lastly, a complex simulation of E-drive for different constellations of soft mounting would be contributory.

REFERENCES

ACOEM France (13/06/2022). High Frequency DMTA testing, https://www.metravib-design.com/.

Anton Paar GmbH (23/03/2023). Ultrapyc 5000, https://www.anton-paar.com/.

Bahuguna, S., R. Marlow, & T. Dalrymple (2007). Understanding Frequency Domain Viscoelasticity in Abaqus. Elastomere oder thermoplastische Elastomere - Bestimmung des Druckverformungsrestes - Teil 1: Bei Umgebungstemperaturen oder erhöhten Temperaturen.

Ehrenstein, G., G. Riedel, & P. Trawiel (2012). *Thermal Analysis of Plastics: Theory and Practice*. Carl Hanser Verlag GmbH & Company KG.

Farkas, O., T. Gejgus, A. Lion, & M. Johlitz (2024). *From Raw Data to Master Curve - Improving the Time-Temperature Superposition Process Using Geometrical Properties of Temperature-Frequency Sweeps*, pp. 35–56. Cham: Springer Nature Switzerland.

Gejgus, T., O. Farkas, A. Lion, & M. Johlitz (2022). Experimental and numerical investigation of NVH behaviour of NR - BR blend used for engine mounts of electric vehicles. In L. Andena, F. B. Vangosa, R. Frassine, and C. Marano (Eds.), *Constitutive Models for Rubber XII*, pp. 193–198. London: CRC Press.

Gejgus, T., J. Schröder, K. Loos, A. Lion, & M. Johlitz, (2022, 01). Advanced characterisation of soft polymers under cyclic loading in context of engine mounts. *Polymers 14*, 429.

ISO 527-1 (2019). Plastics - determination of tensile properties - part 1: General principles.

ISO 7743 (2016). Rubber, vulcanized or thermoplastic—determination of compression stress-strain properties.

Kazmierkowski, M. P., R. Krishnan, F. Blaabjerg, & J. D. Irwin (2002). *Control in Power Electronics: Selected Problems*. Academic Press Series in Engineering. Elsevier Science.

Mooney, M. (1940). A Theory of Large Elastic Deformation. *Journal of Applied Physics 11*(9), 582–592.

Müller-BBM GmbH (13/06/2022). Measurement rig for the determination of the dynamic stiffness of resilient elements, https://www.muellerbbm.com/.

Müller-BBM VAS GmbH (26/04/2022). PAK MKII.

NETZSCH-Gerätebau GmbH (27/03/2023). High-Force DMTA Gabo Eplexor Series up to ± 500 N.

Raghunath, R., D. Juhre, & M. Klüppel (2016). A physically motivated model for filled elastomers including strain rate and amplitude dependency in finite viscoelasticity. *International Journal of Plasticity 78*, 223–241.

Rivlin, R. S. (1948). Large elastic deformations of isotropic materials. I. Fundamental concepts. *Philosophical Transactions of the Royal Society of London. Series A, Mathematical and Physical Sciences 240*(822), 459–490.

Tamari, S. (2004). Optimum design of the constant-volume gas pycnometer for determining the volume of solid particles. *Measurement Science and Technology 15*(3), 549–558.

van Gurp, M. & J. Palmen (1998). Time-temperature superposition for polymeric blends. *Rheology Bulletin 67*, 5–8.

Williams, M. L., R. F. Landel, & J. D. Ferry (1955). The temperature dependence of relaxation mechanisms in amorphous polymers and other glass-forming liquids. *Journal of the American Chemical Society 77*(14), 3701–3707.

Williams, M. L., R. F. Landel, & J. D. Ferry (1955). The temperature dependence of relaxation mechanisms in amorphous polymers and other glass-forming liquids. *Journal of the American Chemical Society 77*(14), 3701–3707.

Constitutive Models for Rubber XIII – Dal (Ed.)

Exploring the stress-softening effects in filled elastomers: Beyond uniaxial stretching

Thanh-Tam Mai & Kenji Urayama
Department of Material Chemistry, Kyoto University, Kyoto, Japan

Katsuhiko Tsunoda
Sustainable and Advanced Material Division, Bridgestone Corporation, Tokyo, Japan

ABSTRACT: Our study investigates the stress-softening effect, also known as the Mullins effect, in particulate filler-reinforced elastomers by conducting loading-unloading stretching cycles under various deformation modes, including unequal biaxial stretching. We utilize conductive carbon black as a filler to monitor strain-induced changes in electrical surface resistivity. This approach offers insights into how the filler network disconnects under complex deformations. Our finding reveals that both the loss factor, which measures the ratio of energy dissipated to energy input, and the changes in electrical surface resistivity in softened matrix are simply determined by the first invariant of the deformation tensor for the maximum applied deformation, regardless of the deformation mode. Our research further delves into the impact of stress-softening on crack propagation dynamics. We compare three types of specimens: (I) one without stress-softening, (II) one with significant stress-softening, and (III) one displaying directional softening. Through this analysis, we elucidate the influence of directional stress-softening on key factors such as tearing energy, crack-growth velocity, crack-tip shape, and local strain fields near the crack tip.

1 INTRODUCTION

In the realm of materials science, particulate filler-reinforced elastomers/rubbers are prized for their advanced mechanical strengths and are integral to numerous everyday life and industrial applications (Donnet & Custodero 2013). These filled rubbers are distinguished by their stress-softening behavior, known as the Mullins effect (Mullins 1948), which manifests as a notable stress reduction upon cyclic stretching cycles. This phenomenon is linked to microstructural changes (Diani *et al.* 2009), including the breakdown of filler aggregates, rupture at filler-rubber interfaces, and chain slippage, with the destruction of filler networks being crucial. These changes suggest a redistribution of filler particles within the rubber matrix, along with other dissipative phenomena. Despite advancements in analytical techniques such as X-ray scattering (Takenaka 2013), which indirectly assess these microstructural shifts during stretching, the intricate nature of the Mullins effect and its underlying causes remain elusive, hindered by the difficulty of directly observing these microstructural evolutions.

Carbon black (CB) stands out as the primary reinforcing filler in rubber products (Rahaman *et al.* 2019), particularly within the tire industry. Beyond enhancing physical properties, CB confers electrical conductivity to elastomers through the generation of percolated CB aggregates or networks with a sufficient CB concentration. This electrical behavior offers a unique avenue to merge mechanical and electrical investigations to probe the complex relationship between mechanical energy loss and structural changes in CB filled rubber system.

However, it is noteworthy that previous assessments have predominantly been confined to uniaxial stretching conditions, representing merely one specific deformation type among the array of deformations elastomers can undergo (Figures 1a).

Addressing the gap in stress-softening research, particularly for deformations beyond uniaxial stretching, is critical for understanding the behavior of elastomers used in practical applications like tires, which are subject to complex deformations (Figures 1b). To this end, we have been developed a custom-built biaxial tester (Mai *et al.* 2017a) to simulate a broad range of deformations through two orthogonal strains (Figure 1c), expanding the scope of Mullins effect analysis in filled elastomers to include a variety of biaxial stretching modes.

Our recent study (Taniguchi *et al.* 2022) sheds light on stress-softening phenomenon focusing on both mechanical and electrical resistivity measurements under various stretching conditions, including

DOI: 10.1201/9781003516880-24

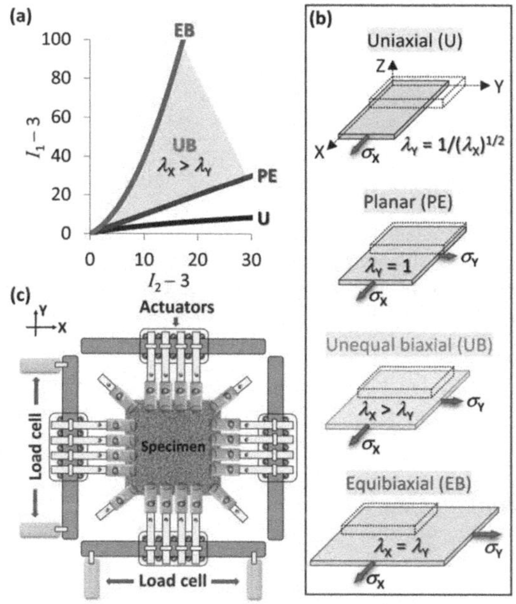

Figure 1. (a) Map of principal invariants (I_1 Vs. I_2) of deformation tensor for an incompressible material ($\lambda_x\lambda_y\lambda_z = 1$, where λ_i is the elongation in i-direction): The first and second invariants are evaluated as $I_1 = \lambda_x^2+\lambda_y^2+\lambda_z^2$ and $I_2 = \lambda_x^{-2}+\lambda_y^{-2}+\lambda_z^{-2}$. (b) Schematic of the custom-built biaxial tensile tester (Mai *et al.* 2017a). (c) Several deformation types for a sheet specimen.

uniaxial (U) and biaxial extensions with planar (PE), unequal biaxial (UB), and equibiaxial (EB) geometries, as shown in Figure 1b with a sheet specimen. In EB, the specimen is equally stretched in x- and y-directions at the same stretching speed ($V_x = V_y$), whereas UB involves unequal stretching speed ($V_x > V_y$). PE distinguishes itself by only stretching in the x-direction, keeping the y-dimension fixed ($V_y = 0$), thus setting it apart from common uniaxial stretching.

Additionally, in filled elastomers, stress-softening resulting from anisotropic pre-strain exhibits notable directional dependence (Machado *et al.* 2012, Mai *et al.* 2017b, 2023). Specifically, when triggered by uniaxial or unequal biaxial stretching, the magnitude of stress reduction in the direction of greater pre-strain significantly surpasses that in other directions. What happens to cracks in filled rubbers when they grow at different speeds, and how does the Mullins effect—where materials soften in specific directions—affect this process?

Our research (Mai *et al.* 2021a, b) seeks answers to these above questions by examining the impact of directional stress-softening on crack propagation in filled elastomers, focusing on various critical aspects: tearing energy, crack-growth velocity, crack-tip shape and local strain fields. We employed the digital image correlation (DIC) technique (Schreier *et al.* 2009), aided by a high-speed camera with a

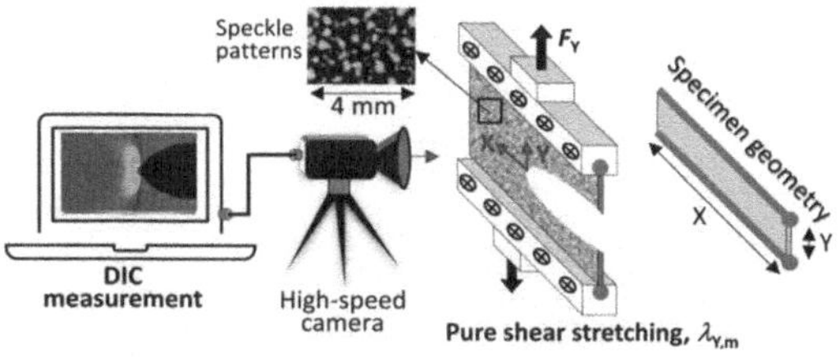

Figure 2. A custom-designed system that incorporates the digital image correlation (DIC) technique into crack-growth studies (Mai *et al.* 2020), focusing on specimens with a pure shear geometry.

phenomenal speed of up to 540000 frames per second, to analyze the local strain fields during dynamic crack propagation. Incorporating DIC into crack-growth experiments (Figure 2) provides a deeper understanding of the complex strain fields at the crack tip (Mai *et al.* 2020, 2021a, b, Osumi *et al.* 2022), unlocking new insights into their significant role in various fundamental aspects of crack propagation, especially under the effects of strain-induced softening.

2 CHARACTERIZING STRESS-SOFTENING EFFECT BY BIAXIAL STRETCHING

2.1 *Specimens*

Specimens of styrene-butadiene rubber (SBR) reinforced with conductive carbon black (CB) across various CB volume fractions (ϕ) were used in our experimental and analytical work. The sample preparation details can be found in reference (Taniguchi *et al.* 2022).

2.2 *Stress-softening characterized by cyclic tensile stretching*

Several deformation modes (U, PE, UB with a V_y/V_x ratio of 0.5, and EB) were utilized to explore the Mullins effect, as illustrated in Figure 1b. In experiment, we subjected sheet specimens ($65 \times 65 \times 1$ mm³) to loading-unloading cycles with various $\lambda_{x,m}$ (maximum x-elongation) for each stretching type.

Figure 3 compares the nominal stresses (σ_x and σ_y) against elongations (λ_x and λ_y) during virgin loading and reloading, under UB ($V_y/V_x = 0.5$) at $\lambda_{x,m} = 2$ and corresponding $\lambda_{y,m} = 1.5$. A notable reduction in stress upon reloading is observed in both directions. This phenomenon is consistent across diverse deformations for various types of filled rubbers (Mai *et al.* 2017a, Taniguchi *et al.* 2022).

Stress-softening at each $\lambda_{x,m}$ is quantified by the energy dissipation per unit volume (D), calculated by the difference in mechanical work per unit volume between the virgin loading (W_0) and the reloading (W_r). The calculation of W_0 or W_r employs the formula $W = \int \sigma_x d\lambda_x + \int \sigma_y d\lambda_y$ for general biaxial

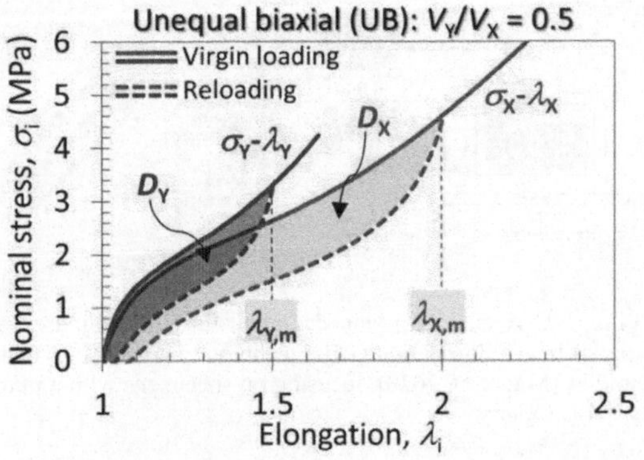

Figure 3. The biaxially nominal stresses (σ)-elongations (λ) relation under virgin loading and reloading for UB with $V_y/V_x = (\lambda_y-1)/(\lambda_x-1) = 0.5$. The graphical evaluation of energy dissipation (D) in each direction is also indicated in the figure. The data of SBR/CB with $\phi = 0.21$ are shown in the figure.

deformation. Figure 3 illustrates the measurement of D for a typical unequal biaxial deformation, represented as: $D = D_x + D_y$. Additionally, we introduce the concept of the "dissipation factor" (Δ), defined as the ratio of D to the W_0: $\Delta = D/W_0$. This factor facilitates a comparison of energy dissipation across different deformations, calibrated by W_0.

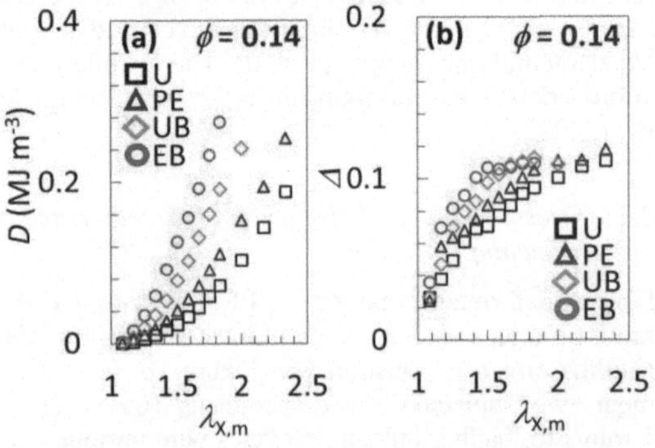

Figure 4. (a) Energy dissipation (D) and (b) dissipation factor (Δ) as a function of $\lambda_{x,m}$ for SBR/CB with $\phi = 0.14$.

As shown in Figures 4a and 4b, the dependency of D and Δ on $\lambda_{x,m}$ varies across deformation types for SBR/CB (ϕ = 0.14). Their values are highest for EB, followed by UB, PE, and U, at the same $\lambda_{x,m}$. They also escalate with increasing CB content (Taniguchi *et al.* 2022). These characteristics are shared by particulate filled rubbers, including silica-filled SBR (Mai *et al.* 2017a) and CB-filled natural rubber (NR) (Taniguchi *et al.* 2022). In the subsequent sections, we will present a detailed discussion on the variables that determine Δ, paired with an analysis of electrical resistivity.

2.3 *Stress-softening characterized by electrical resistivity*

In this investigation, SBR/CB specimens with ϕ = 0.14 were employed. This ϕ level, indicative of the initial CB percolation stage, is highly deformation-sensitive, enabling detailed study of electrical resistivity changes during deformation. Other specimens with higher or lower ϕ level were also investigated in reference (Taniguchi *et al.* 2022).

The evaluations of electrical resistivity, under different types of stretching, were specifically conducted under two conditions: (1) the deformed state (ρ_D), and (2) the unloaded and relaxed state (ρ_S), as illustrated in Figure 5. Comprehensive details on the methodology of these electrical resistivity measurements are provided in reference (Taniguchi *et al.* 2022).

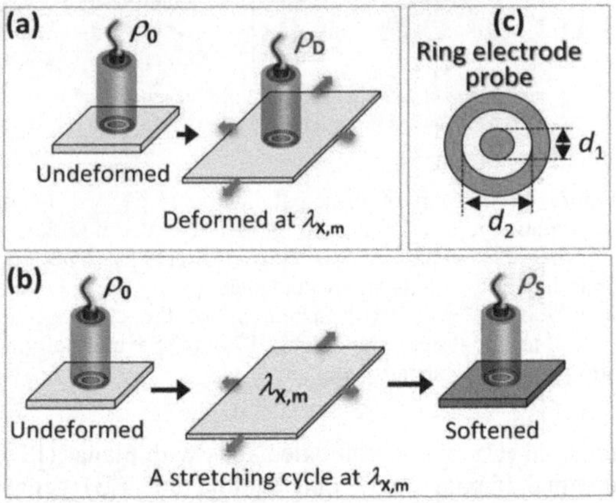

Figure 5. Evaluations of electrical resistivity under (a) the deformed state (ρ_D), and (b) the unloaded and relaxed state (ρ_S). (c) A ring electrode probe contains inner and outer-ring electrodes with diameters of d_1 = 5.9 and d_2 = 11 mm, respectively.

Figure 6a demonstrates that the normalized electrical resistivity of deformed state ($\rho_D^* = \rho_D/\rho_0$) (Figure 5a) increases with $\lambda_{x,m}$ and varies significantly with the deformation types, following the order: ρ_D^*(U), ρ_D^*(PE), ρ_D^*(UB), to ρ_D^*(EB). This trend, mirroring mechanical energy dissipation (Figure 4a), underscores a consistent link between deformation and electrical/mechanical behaviors.

The ρ_D^* variation during deformation reflects two opposing effects: stretching in the x- and y-directions increases ρ_D^* by breaking CB particles connections, while contraction in the z-axis (stress-free) reduces ρ_D^* by bringing particles closer. Initially, minimal stretching prominently increases ρ_D^*, but as stretching continues, this rate decreases, indicating densification of CB particles. Enhanced stretching in the y-direction further elevates ρ_D^*, emphasizing the filler-disconnection effect in the x-y plane under biaxial strains.

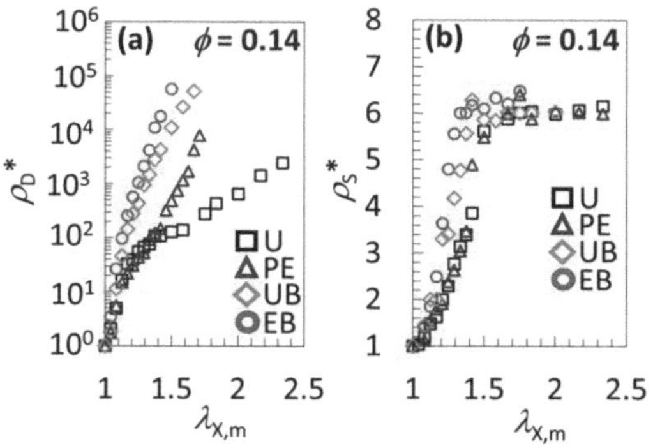

Figure 6. The normalized electrical resistivity under (a) deformed state ($\rho_D^* = \rho_D/\rho_0$) and (b) unloaded-relaxed state ($\rho_S^* = \rho_S/\rho_0$) as a function of $\lambda_{x,m}$ for SBR/CB with $\phi = 0.14$.

After a loading–unloading cycle (Figure 5b), the electrical resistivity in the softened and relaxed state (ρ_S) shows a residual change in the conductive path. Figure 6b illustrating that ρ_S^* (= ρ_S/ρ_0) increases with $\lambda_{x,m}$ for different deformation types. Notably, ρ_S^* rises sharply until $\lambda_{x,m}$ reaches 1.5, then levels off, showing a similar trend to energy dissipation factor Δ, which will discuss thoroughly in the following section.

2.4 *Determining variable of mechanical and electrical behaviors*

In the comparison of the values of Δ and ρ_S^* among various types of extension, it is more plausible to employ quantities other than $\lambda_{x,m}$ as a measure of the degree of general deformation (Figures 1b). For instance, the first principle invariants ($I_{1,m}$) of deformation tensor for the finite deformation theory. The value $I_{1,m}$ at each $\lambda_{x,m}$ is evaluated by the relation of $I_{1,m} = \lambda_{x,m}^2 + \lambda_{y,m}^2 + \lambda_{z,m}^2$, where $\lambda_{i,m}$ (i = x, y, z) is a elongation in i-direction and $\lambda_{z,m} = 1/(\lambda_{x,m}\lambda_{y,m})$ derived from the incompressible assumption of rubber-like materials (Treloar 1975).

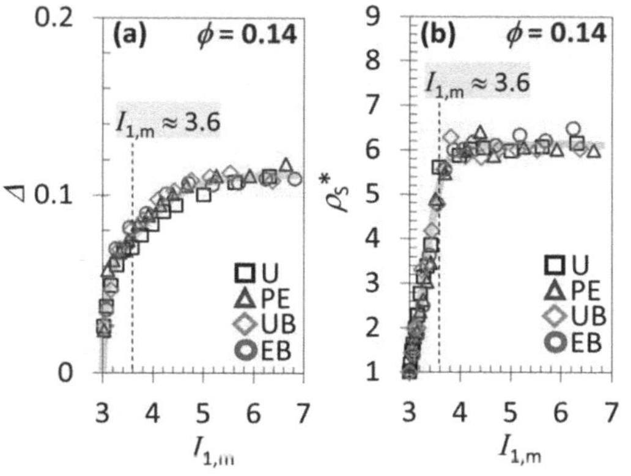

Figure 7. (a) Δ and (b) ρ_S^* as a function of $I_{1,m}$ for SBR/CB with $\phi = 0.14$.

In Figures 7a and 7b, we see that Δ and ρ_S^* align across various types of deformation when analyzed using $I_{1,m}$ as a variable. The finding highlights that the impact of deformation on Δ and ρ_S^* is consistent, regardless of the type of deformation. Similar patterns are observed in different filled elastomers, like SBR/Si (Mai *et al.* 2017a) and NR/CB (Taniguchi *et al.* 2022), where energy loss mainly occurs through the breaking of physical bonds, such as Van der Waals and hydrogen bonds, contrasting with energy dissipation through covalent bond breaking seen in double network hydrogels (Mai *et al.* 2018).

Furthermore, Figure 7b show a sharp rise in ρ_S^* with increasing $I_{1,m}$ up to a critical point ($I_{1,m} \approx 3.6$), after which it plateaus, indicating no further loss in CB conductivity for larger deformation. This suggests the filler network reaches a stable structure beyond critical point that doesn't change with more deformation, which is consistent with Δ behavior. Sharp increases in Δ and ρ_S^* for small deformations ($I_{1,m} < 3.6$) highlight filler network breakdown as the main cause of early mechanical damage. On the other hand, at larger deformations ($I_{1,m} > 3.6$), damage primarily comes from breaking down rubber-filler interfaces, like the detachment and/or slippage of rubber chains from filler particles.

3 EFFECT OF STRESS-SOFTENING ON FAST DYNAMIC CRACK GROWTH

3.1 *Specimens*

The examination of stress-softening and anisotropic damage effects on rapid crack propagation, orthogonal to the pre-stretching axis (Figure 2), is facilitated through a comparative analysis of three distinct specimen types: (I) SBR/CB with $\phi = 0.05$ (SBR5), (II) a pristine SBR/CB with $\phi = 0.21$ (SBR21), and (III) SBR21-SY, an elastomer, derived from SBR21 subjected to pre-uniaxial loading-unloading cycles with extensive maximum stretching $\lambda_{y,m} = 4.5$.

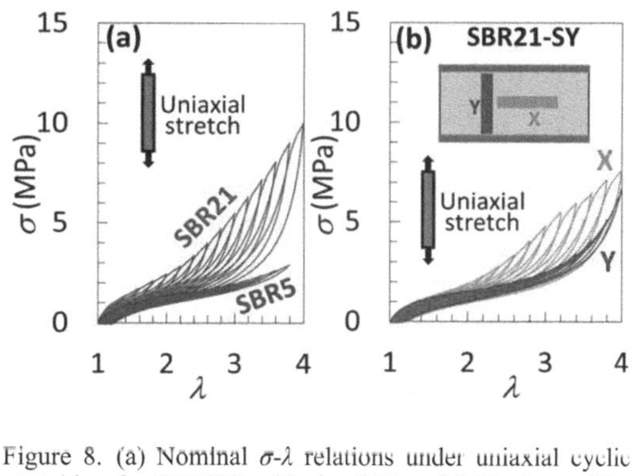

Figure 8. (a) Nominal σ-λ relations under uniaxial cyclic stretching for SBR/CB with $\phi = 0.05$ and 0.21, denoted by SBR5 and SBR21, respectively. (b) σ-λ data of subsamples in x- and y- directions cut from SBR21-SY.

Figure 8a compares the σ-λ curves of SBR5 and SBR21 during stepwise cyclic stretching with $\lambda_m = 1.2 \sim 4.0$. The high-filled SBR21 shows clear stress-softening, unlike the low-filled SBR5, which lacks this feature. Significantly, directional softening in SBR21-SY, induced by pre-stretching with $\lambda_{y,m} = 4.5$, is analyzed using subsamples stretched uniaxially, SBR21-SY-x and SBR21-SY-y. Figure 8b reveals that damage is mostly directional. SBR21-SY-x shows significant extra stress-softening, whereas SBR21-SY-y shows less due to previously softening.

3.2 *Experimental investigation of crack-growth and strain fields*

A bespoke experimental apparatus, as illustrated by Figure 2, was employed to investigate both crack propagation and local strain fields. A sheet specimen was pre-stretched to certain elongation in the y-direction (denoted λ_y^{pre}). In the pre-stretched state of λ_y^{pre}, speckle patterns were imposed on the sample surface using a custom-made speckle stamp. Crack was initiated by cutting the specimen's edge midway along the y-axis, and a high-speed camera captured the crack movement across the specimen's center.

The steady-state crack-growth velocity (V_C) at each λ_y^{pre} was evaluated from the images captured during the crack propagation. The local displacement field around the crack tip was measured from the captured speckle images using the DIC method with the software of Correlated Solutions (Vic-2D). Detailed experimental procedures are available in reference (Mai *et al.* 2021a, b).

3.3 *Effect of stress-softening on input tearing energy and crack-growth velocity*

Input tearing energy (Γ) at different λ_y^{pre} levels, as depicted in Figure 9a, is calculated using the relationship of $\Gamma = WL_{y,0}$, where W is the elastic energy stored during the unloading curves for strain-induced stress-softening materials, and $L_{y,0}$ is initial y-length of sample. The Γ values for SBR21-SY are significantly lower than SBR21 but similar to SBR5 at the same λ_y^{pre}, indicating prior y-direction damage in SBR21-SY.

Figure 9b represents V_C as a function of Γ across various λ_y^{pre} levels. Notably, the figure delineates a marked, discontinuous transition in V_C demarcating two distinct behavioral regimes: a "slow mode" characterized by $V_C < 10^{-5}$ m/s and a "fast mode" where $V_C > 10^{-1}$ m/s, intersecting at a critical tearing energy threshold (Γ_C).

The crack velocity transition shows SBR21 with the highest Γ_C, followed by SBR21-SY, and the lowest for SBR5, as seen in Figure 9b. Previous research indicates Γ_C is closely linked to fracture toughness (Morishita *et al.* 2019). This pattern is confirmed by our results, where fracture toughness under uniaxial stretching ranks as SBR5 (4.6 MJ/m^3), SBR21-SY (16 MJ/m^3), and SBR21 (23.3 MJ/m^3), matching the order of Γ_C values. The difference in

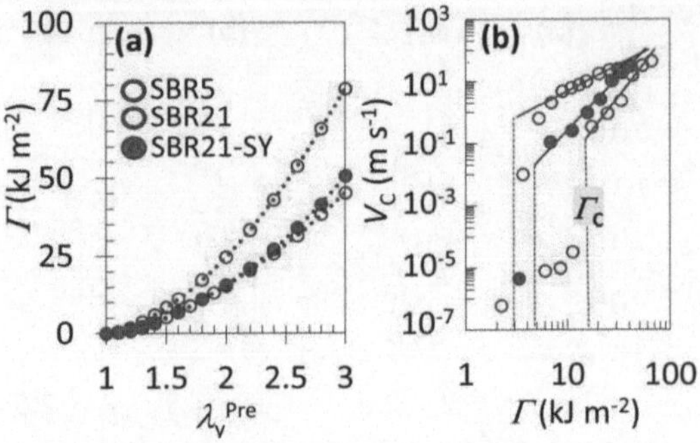

Figure 9. (a) The λ_y^{pre} dependence of the input tearing energy (Γ) for SBR5, SBR21 and its softened SBR21-SY. (b) The steady-state crack-growth velocity (V_C) as a function of Γ for three specimens.

toughness between SBR21 and SBR21-SY is due to prior y-softening as discussed above.

Furthermore, within the fast mode, the V_C–Γ relationship adheres to a power-law, expressed as $\Gamma \sim (V_C)^{\beta}$, a phenomenon corroborated by analogous findings in prior research for various rubber-like materials (Persson *et al.* 2005). This β value is linked to the power-law exponent (κ) for the time-dependent linear relaxation modulus ($G(t) \sim t^{-\kappa}$) in the glass-rubber transition. The difference in β between SBR5 and SBR21 is due to the different κ values, which are affected by filler content (Morishita *et al.* 2019). Additionally, prior softening treatments on CB-filled elastomers have no appreciable effect on $G(t)$ in this regime (Diaz *et al.* 2014), explaining why SBR21-SY and SBR21 share similar β values.

3.4 *Effect of stress-softening on crack-tip shape and local strain fields*

Figure 10 illustrates the comparison of strain fields in specimens under identical input tearing energy

Figure 10. 2D crack-tip strain fields of $\Delta\varepsilon_{yy}$ and $\Delta\varepsilon_{xx}$ at $\Gamma = 35$ kJ/m^2 for the specimens of SBR5 ($V_C = 25$ m/s), SBR21 ($V_C = 5$ m/s), and SBR21-SY ($V_C = 19$ m/s).

(approximately 35 kJ/m^2). SBR21 and SBR21-SY exhibit similar in $\Delta\varepsilon_{xx}$ (strain in the x-direction) but show significant differences in $\Delta\varepsilon_{yy}$ (strain in the y-direction). The strains $\Delta\varepsilon_{xx}$ and $\Delta\varepsilon_{yy}$ are those measured from the pre-stretched state in the x- and y-directions, i.e., the strain increments due to crack growth. SBR21-SY displaying a higher $\Delta\varepsilon_{yy}$, attributed to prior softened in the y-direction with $\lambda_m = 4.5$ (Figure 8). Conversely, $\Delta\varepsilon_{xx}$ remains similar between SBR21-SY and SBR21 due to comparable stress-softening characteristics along the x-axis (Figure 8). SBR5 and SBR21-SY exhibit similar in $\Delta\varepsilon_{yy}$ but marked differences in $\Delta\varepsilon_{xx}$, reflecting a directional pre-softening in SBR21-SY.

Notably, the similarity in $\Delta\varepsilon_{xx}$ (also shear strain $\Delta\varepsilon_{xy}$) just behind the crack tip correlates with the observed similarity in crack-tip shape between SBR21-SY and SBR21. In contrast, variations in $\Delta\varepsilon_{yy}$ within the same area do not affect the crack-tip shape, as the contraction behind the crack tip results from unloading of materials. Furthermore, we observed that the strain singularity phenomenon at the crack tip, characterized by the relationship $\Delta\varepsilon_{yy} \sim (1/r)^{\alpha}$ where r is the distance from the crack tip, persists without alteration with prior softening.

4 CONCLUSIONS

Our study demonstrates that both the mechanical loss factor (Δ; $\Delta = D/W_0$) and changes in electrical surface conductivity (ρ_s*) in carbon black-filled rubbers are primarily dictated by the first invariant of deformation tensor (I_1) for the maximum applied deformation, independent of the deformation types. This finding provides an important basis for understanding the correlations between filler network disconnection and nonlinear mechanical properties, and for developing constitutive models suited to complex deformations.

Furthermore, our research uncovers the effects of prior softening along the y-axis on crucial crack-growth factors such as tearing energy, crack-growth speed, and crack-tip strain field. We find that softening notably affects properties along the y-axis, yet it has a minimal effect on properties along the x-axis, including the crack-tip shape. This selective influence elucidates the relationship between material pre-conditioning and its subsequent response to crack-growth.

ACKNOWLEDGEMENT

This work was partly supported by JST, CREST (grant number JPMJCR2091), Japan.

REFERENCES

Diani, J., Fayolle, B., and Gilormini, P., 2009. A review on the Mullins effect. *European Polymer Journal*, 45 (3), 601–612.

Diaz, R., Diani, J., and Gilormini, P., 2014. Physical interpretation of the Mullins softening in a carbon-black filled SBR. *Polymer*, 55 (19), 4942–4947.

Donnet, J.-B. and Custodero, E., 2013. Reinforcement of Elastomers by Particulate Fillers. In: J.E. Mark, B. Erman, and M. Roland (eds.), *The Science and Technology of Rubber*. Waltham: Elsevier, 367–400.

Machado, G., Chagnon, G., and Favier, D., 2012. Induced anisotropy by the Mullins effect in filled silicone rubber. *Mechanics of Materials*, 50, 70–80.

Mai, T.-T., Matsuda, T., Nakajima, T., Gong, J.P., and Urayama, K., 2018. Distinctive Characteristics of Internal Fracture in Tough Double Network Hydrogels Revealed by Various Modes of Stretching. *Macromolecules*, 51 (14), 5245–5257.

Mai, T.-T., Morishita, Y., Tsunoda, K., and Urayama, K., 2021a. Experimental Analysis of Fast Crack Growth in Elastomers. *Advances in Polymer Science*, 289, 239–272.

Mai, T.-T., Morishita, Y., and Urayama, K., 2017a. Novel features of the Mullins effect in filled elastomers revealed by stretching measurements in various geometries. *Soft Matter*, 13 (10), 1966–1977.

Mai, T.-T., Morishita, Y., and Urayama, K., 2017b. Induced anisotropy by Mullins effect in filled elastomers subjected to stretching with various geometries. *Polymer*, 126, 29–39.

Mai, T.-T., Okuno, K., Tsunoda, K., and Urayama, K., 2021b. Anisotropic stress-softening effect on fast dynamic crack in filler-reinforced elastomers. *Mechanics of Materials*, 155, 103786.

Mai, T.-T., Taniguchi, Y., Tsunoda, K., and Urayama, K., 2023. Comparison of Mullins Effect Anisotropy of the Elastomers Reinforced by Carbon-Black and Silica Filler. *Nihon Reoroji Gakkaishi*, 51 (1), 33–39.

Mai, T., Okuno, K., Tsunoda, K., and Urayama, K., 2020. Crack-Tip Strain Field in Supershear Crack of Elastomers. *ACS Macro Letters*, 9 (5), 762–768.

Morishita, Y., Tsunoda, K., and Urayama, K., 2019. Universal relation between crack-growth dynamics and viscoelasticity in glass-rubber transition for filled elastomers. *Polymer*, 179, 121651.

Mullins, L., 1948. Effect of Stretching on the Properties of Rubber. *Rubber Chemistry and Technology*, 21 (2), 281–300.

Osumi, R., Yasui, T., Tanaka, R., Mai, T.-T., Takagi, H., Shimizu, N., Tsunoda, K., Sakurai, S., and Urayama, K., 2022. Impact of Strain-Induced Crystallization on Fast Crack Growth in Stretched cis -1,4-Polyisoprene Rubber. *ACS Macro Letters*, 11 (6), 747–752.

Persson, B.N.J., Albohr, O., Heinrich, G., and Ueba, H., 2005. Crack propagation in rubber-like materials. *Journal of Physics: Condensed Matter*, 17 (44), 1071–1142.

Rahaman, M., Khastgir, D., and Aldalbahi, A.K. (eds.) 2019. *Carbon-Containing Polymer Composites*. Singapore: Springer Singapore.

Schreier, H., Orteu, J.-J., and Sutton, M.A., 2009. *Image Correlation for Shape, Motion and Deformation Measurements*. Boston, MA: Springer US.

Takenaka, M., 2013. Analysis of structures of rubber-filler systems with combined scattering methods. *Polymer Journal*, 45 (1), 10–19.

Taniguchi, Y., Mai, T.-T., Yamaguchi, M., Tsunoda, K., and Urayama, K., 2022. Investigating Multiaxial Mullins Effect of Carbon-Black-Reinforced Elastomers Using Electrical Resistivity Measurements. *ACS Applied Polymer Materials*, 4 (2), 1139–1149.

Treloar, L.R.G., 1975. *The physics of rubber elasticity*. New York: Oxford University Press Inc.

Constitutive Models for Rubber XIII – Dal (Ed.)
© 2025 The Author(s), ISBN 978-1-032-85138-9

Stretch induced anisotropy and fracture toughness: An experimental study on a silica- filled NBR compound

I. Denora, A. Bianucci, G. Caccia, M. Contino & C. Marano
Dipartimento di Chimica, Materiali e Ingegneria Chimica "Giulio Natta", Politecnico di Milano, Milano, Italy

A. Kaiser
ARLANXEO Deutschland GmbH

ABSTRACT: Anisotropic pre-stretching of a filled rubber compound is responsible for the anisotropic behaviour of the stretched material, which is commonly evidenced by evaluating only the tensile response of the pre-stretched material. In this work, an evaluation of the effects of stretch-induced anisotropy on the fracture behaviour of the material was also carried out. Preliminary pre-stretching of silica-filled SBR-based compounds was performed by deforming the material under uniaxial or equibiaxial tensile conditions or under pure shear deformation conditions, up to different strain levels. The tensile and fracture behaviour of the softened material was evaluated along two perpendicular directions and correlated with both the principal strains applied and the biaxiality index defined to characterize the anisotropy of preliminary stretching.

1 INTRODUCTION

Filled elastomers are commonly used in products, such as tires, flexible hoses and seals, which are subjected to complex loading conditions during their life and could undergo failure. It is reported in literature that, when a rubber compound is stretched under uniaxial or unequal biaxial tensile loading conditions, the mechanical response of the resulting softened material is anisotropic (a review of these studies is reported in (Denora & Marano, 2024)). Any difference between the tensile response of the softened material tested in the direction of maximum pre-stretching and along any other different direction has been commonly considered as index of anisotropy of the softened material. However, as far as the authors know, the effect of such anisotropy has been evaluated just on the tensile response of the material and not yet on the fracture behavior, which is an interesting issue from an application point of view.

The research work has been performed within the Polymers4Hydrogen COMET Project on silica filled NBRs supplied by Arlanxeo Germany GmbH. The aim is to develop a methodology that allows to evaluate the anisotropy in the fracture response of a preliminarily stretched material and correlate it with the anisotropy of the applied pre-stretching (uniaxial/equibiaxial (EB) tension or pure shear (PS) deformation).

2 MATERIALS AND METHODS

2.1 *Materials and specimens' geometry*

NBR-based compounds filled with 60 phr of silica have been supplied by Arlanxeo Germany GmbH, as 2 mm thick sheets. Non silanised (Si) and in situ-silanised silica (Sil-Si) were used. The NBR/Si compound was tested to define a methodology for the characterization of the preliminarily stretched material's anisotropy. This methodology was than applied to testing the compound NBR/Sil-Si.

Sulphur-based curing was performed by compression moulding at 170°C for 5 minutes.

Rectangular large specimens (100 mm wide, 125 mm long and 1.5 mm thick) and square shaped large specimens (128 mm wide and long and 1.5 mm thick) have been obtained through home-made moulds. Double-T specimens with a 25 mm gauge length and 65 mm long and 5 mm wide rectangular specimens were die-cut from the large specimens.

A 2 mm notch was introduced, with a razor blade, on one side and at mid height of the rectangular specimens for "single-edge notched tension" (SENT) fracture testing: a gauge length of 35 mm was used in fracture tests.

2.2 *Strain induced softening*

A loading-unloading cycle under tensile conditions, up to a strain of 200%, has been performed on large

DOI: 10.1201/9781003516880-25

rectangular specimens to preliminarily pre-stretch the NBR/Si compound.

A loading-unloading cycle under equibiaxial tension or pure shear deformation conditions, respectively up to 30% and 75% of deformation, have been performed on large square shaped specimens to preliminarily pre-stretch the NBR/Sil-Si compound. The values of the strain applied in the pre-stretching have been selected to avoid the failure of the specimens.

The tests have been video-recorded to evaluate strain distribution via Digital Image Correlation (DIC) analysis.

2.3 *Biaxiality index*

In order to characterize the deformation state applied at the end of the pre-stretching loading step, the index of biaxiality was defined as follows

$$BI = \frac{(e_1 - e_2)}{e_1} \tag{1}$$

where e_1 and e_2 are respectively the major and minor principal strains, evaluated by means of DIC analysis.

Since deformation resulted to be non-homogeneous in the large specimens tested, the BI was evaluated locally in 25 subregions (18 x 25.4 mm for rectangular specimens and 21.6 x 25.4 mm for square shaped ones) identified on these specimens: a map of *BI* was thus obtained for each of the different loading conditions considered.

2.4 *Mechanical testing*

For a preliminary evaluation of the anisotropic behavior of the pre-stretched softened material, uniaxial tensile tests were performed, at room temperature and at a nominal strain rate of 1 min^{-1}, on double-T specimens, cut from the large rectangular specimens uniaxially pre-stretched. The double-T specimens were cut at some of the subregions defined above, along a direction parallel or perpendicular to the preliminary pre-stretching direction. The anisotropic material response was related to the value of both the maximum principal strain locally applied and the corresponding biaxiality index. Similarly, rectangular specimens have been cut from pre-stretched large rectangular specimens and notched: SENT tests have been then performed to evaluate the effect of pre-stretching on material's fracture behaviour. Fracture tests were performed at room temperature at a nominal strain rate of 1 min^{-1}. Also in this case, the specimens have been cut with their length parallel or perpendicular to the pre-stretching direction. Any single specimen was identified by the pre-stretching conditions locally applied in the zone ahead of the crack tip, that, again, will be described through the value of the maximum principal strain locally applied and the value of the local biaxiality index. This preliminarily work was carried out on the NBR/Si compound.

The same procedure was adopted to evaluate the anisotropy of the softened NBR/Sil-Si: uniaxial tensile tests and SENT tests were carried out on specimens that have been cut from large square-shaped specimens, preliminarily stretched in biaxial loading conditions (PS or EB tension).

2.5 *Fracture toughness*

The tearing energy was evaluated analytically as:

$$T_c = \frac{2\pi}{\sqrt{\lambda_c}} \frac{U_c a}{twh} \tag{2}$$

where U_c is the input energy at crack onset, λ_c is the draw ratio (or stretch) at crack onset, t is the specimen's thickness, w its width, h the gauge length and a the crack length (Greensmith, 1963; Rivlin & Thomas, 1953).

3 RESULTS

In paragraph 3.1 and 3.2 the results of the test campaign carried out on the NBR/Si compound, aimed also to define a testing methodology, are reported. The methodology was then applied to the NBR/Sil-Si compound, on which pre-stretching was performed under EB tension or PS deformation conditions. The latter campaign results are reported in paragraph 3.3.

3.1 *Anisotropic response of softened NBR/Si: Uniaxial tensile tests*

In Figure 1, the stress strain curves obtained from the uniaxial tensile tests performed on the NBR/Si compound preliminarily stretched in uniaxial tension conditions up to 200% are reported both for specimens tested in a direction perpendicular (a) and parallel (b) to the pre-stretching direction. The biaxial index evaluated in correspondence of each specimen gauge-length is reported to distinguish between the tested specimens. The stress-strain curves of the virgin material (no pre-stretched) are reported for comparison: the virgin material is slightly anisotropic, and this can be related to the calendaring process adopted for material preparation. As expected, the anisotropic pre-stretching is responsible of the anisotropy in the tensile response of the material for deformation lower than 200%. In particular, a significant softening is observable when the material is tested in the same direction of the pre-stretching, while in the perpendicular direction the behavior is not very different from that of the virgin material. It seems that there is not any significant effect of the *BI* on the softened material behaviour.

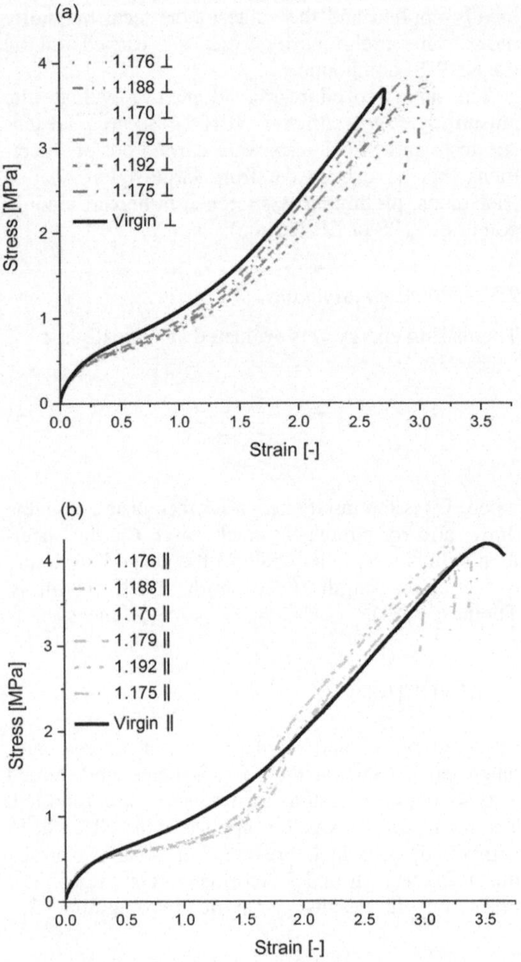

Figure 1. Uniaxial tensile stress-strain curves of NBR/Si: black curves refer to the virgin material, grey curves refer to the material preliminarily stretched in uniaxial tension up to 200% and tested in direction perpendicular (a) and parallel (b) to the pre-stretching direction. The value of *BI* evaluated in correspondence of each specimen is reported.

3.2 *Anisotropic response of softened NBR/Si: Fracture tests*

In Figure 2, the force-displacement curves obtained from the fracture tests performed on the NBR/Si compound preliminarily stretched in uniaxial tension conditions up to 200% are reported both for SENT specimens tested in a direction perpendicular (dark grey) and parallel (light grey) to the pre-stretching direction, together with that of the virgin material. The dots correspond to the crack onsets.

In Figure 3 the value of NBR/Si tearing energy is plotted versus *BI*. No correlation between these two quantities can be observed, at least in the range of *BI* explored. Nevertheless, the tearing energy results to be lower when fracture test was performed loading the material in direction perpendicular to the pre-stretching one, with respect to the tests performed in the parallel direction. This result indicates that the softened material is anisotropic in its fracture behaviour too ($T_{softened\_//} = 9.10 \pm 1.83$, $T_{softened\_\perp} = 4.77 \pm 1.60$). As expected, $T_{softened\ \perp}$ is lower than that of the virgin material ($T_{virgin} = 6.44 \pm 1.72$). Differently, $T_{softened\ //}$ is higher than T_{virgin}: this result could be related to the larger strains at which crack onset occurs in the softened material.

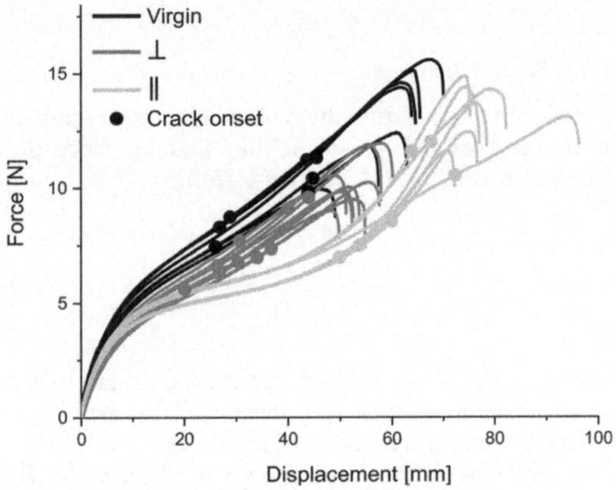

Figure 2. Force displacement curves for fracture test on NBR/Si. (Black: virgin material; light grey: softened material tested along the pre-stretching direction; dark grey: softened material tested in the direction perpendicular to the pre-stretching direction).

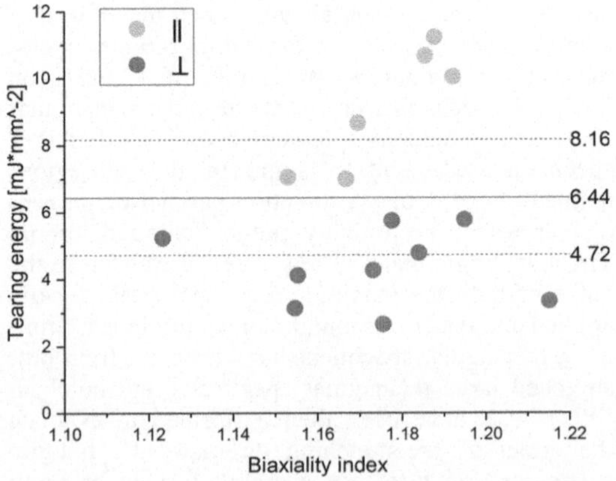

Figure 3. Tearing energy versus the index of biaxiality of NBR/Si preliminarily stretched in uniaxial tension. Light grey: SENT test loading direction parallel to the pre-stretching. Dark grey: SENT test loading direction perpendicular to the pre-stretching. The continuous black line corresponds to the mean value of T for the virgin material. The dotted black lines identify the standard deviation range.

It must be highlighted that the pre-stretching has been performed at a strain value that is higher than the strains experienced by the virgin material at crack onset, as it can be observed in Figure 4, that reports the strain as a function of the distance from the crack tip on the notch plane, as measured, by DIC analysis, in a virgin NBR/Si SENT specimen at the onset of the fracture.

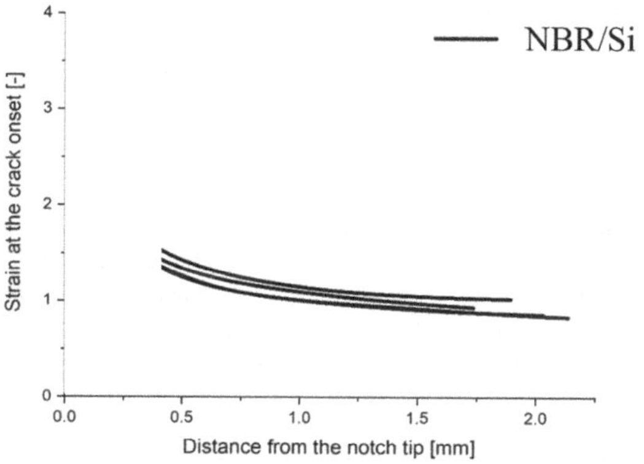

Figure 4. Strain as a function of the distance from the crack tip on the notch plane, as measured, by DIC analysis, in a virgin NBR/Si SENT specimen at the onset of the fracture.

3.3 *Fracture behaviour of NBR/Sil-Si pre-stretched in PS deformation conditions and equibiaxial tension*

Figure 5 reports the force displacement curves relevant to the SENT tests carried out on NBR/Sil-Si: (a) refers to the compound preliminary stretched in PS deformation conditions up to a nominal strain of 75 % and then tested along a direction parallel (light gray) or perpendicular (dark gray) to the pre-stretching one, (b) refers to the compound preliminary stretched in equibiaxial tension up to a nominal strain of 30 % and then tested along the two pre-stretching directions.

In Figure 6 the tearing energies are reported as a function of the index of biaxiality that was locally measured close to the notch tip of each specimen.

The force-displacement curves for the softened materials are very similar to that of the virgin material, irrespective of the testing direction, the pre-stretching loading conditions and the applied deformation. These results are consistent with the strain distribution on the notch plane of a SENT specimen obtained by DIC analysis for virgin material (Figure 7): at crack onset the strains are larger than the pre-stretching ones, and the material behaves as the virgin one in the strain range between the pre-stretching strain and that at the fracture onset. Because of that, the tearing energy of the softened material is almost the same of the virgin material and independent of both the pre-stretching loading conditions (*BI*) and the pre-stretching strain level. It was not possible to reach higher strain values in the pre-stretching for premature failure of the specimens.

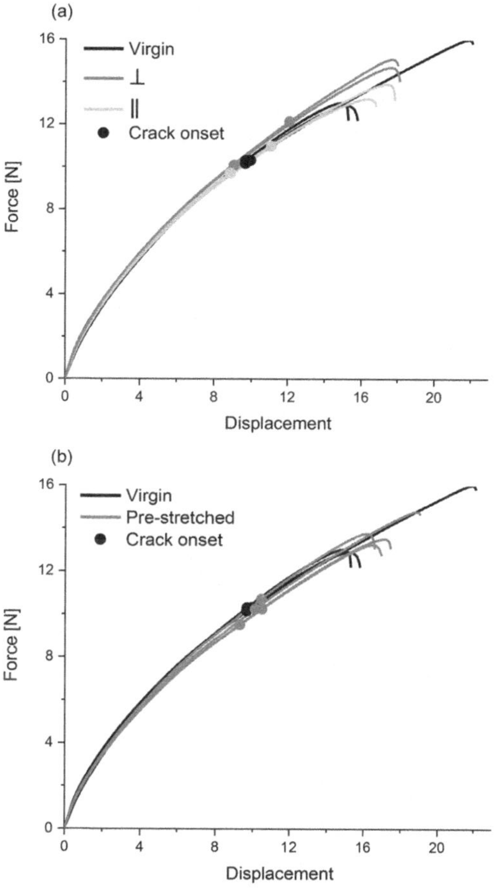

Figure 5. Force displacement curves for fracture test on NBR/Sil-Si preliminary pre-stretched up to 75% in PS deformation conditions (a) or to 30% in equibiaxial tension (b). In (a): black - virgin material; light grey - softened material tested along the pre-stretching direction; dark grey - softened material tested in the direction perpendicular to the pre-stretching direction. In (b): black - virgin material, dark grey: pre-stretched material.

4 CONCLUSIONS

A methodology to evaluate the effects of the Mullins effect anisotropy on the fracture behaviour of a filled rubber has been defined. The pre-stretching conditions are defined through the value of the maximum principal pre-strain applied under different loading conditions and a proper index of biaxiality. The difference in the tearing energy evaluated performing SENT tests along different direction with respect to the pre-stretching direction was considered as an index of the Mullins effect anisotropy.

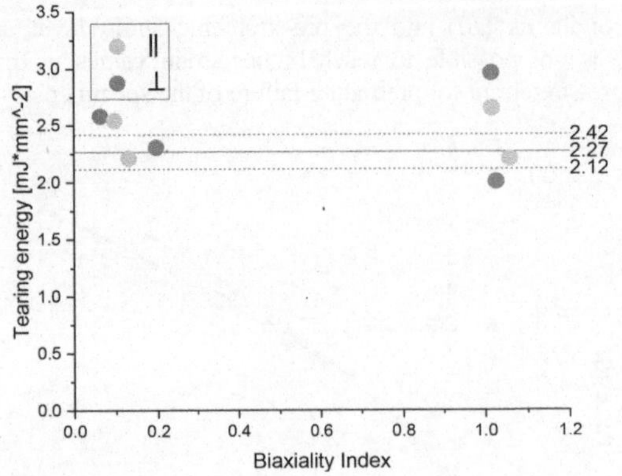

Figure 6. Tearing energy versus the index of biaxiality of NBR/Sil-Si preliminarily stretched in PS deformation conditions ($BI \approx 1$) or equibiaxial tension ($0 \leq BI \leq 0.2$). Light grey: SENT test loading direction parallel to the pre-stretching. Dark grey: SENT test loading direction parallel to the pre-stretching. The continuous black line corresponds to the mean value of T for the virgin material. The dotted black lines identify the standard deviation range.

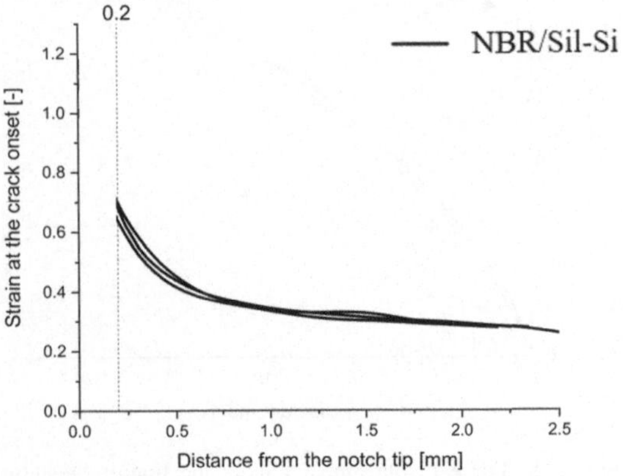

Figure 7. Strain distribution on the notch plane of a SENT specimen obtained by DIC analysis for virgin NBR/Sil-Si material evaluated at crack onset.

In the case of the pre-stretching performed in uniaxial tension conditions, at a strain higher than the strain at fracture onset of the NBR/Si compound, the softened material resulted to be anisotropic both in its uniaxial tensile and fracture response.

In the case of the pre-stretching performed in PS deformation conditions or EB tension, at strains lower or close to the strain at fracture onset for the NBR/Sil-Si compound, the material has not been affected by the pre-stretching, for what concerns both its uniaxial tensile and fracture response. To observe an effect on toughness and possibly even anisotropic behavior, an anisotropic pre-stretching should be applied and up to larger strains. For this purpose, specimens' geometry should be improved to reduce the stress concentration close to the clamping system, which is responsible for specimens' premature failure.

ACKNOWLEDGEMENTS

The research work was performed within the COMET-project "Polymers4Hydrogen" (project-no.: 21647053) at Politecnico di Milano within the framework of the COMET-program of the Federal Ministry for Climate, Action, Environment, Energy, Mobility, Innovation and Technology and the Federal Ministry for Digital and Economic Affairs, Austria with contributions by Polymer Competence Center Leoben (PCCL), Montanuniversität Leoben (Department Polymer Engineering and Science, Chair of Chemistry of Polymeric Materials, Chair of Materials Science and Testing of Polymers), Technical University of Munich, Tampere University of Technology, Bundesanstalt für Materialforschung und -prüfung (BAM) and Arlanxeo Deutschland GmbH, ContiTech Rubber Industrial Kft., Peak Technology GmbH, SKF Sealing Solutions Austria GmbH, Faurecia.

REFERENCES

Denora, I., & Marano, C. (2024). Stretch-induced softening in filled elastomers: A review on Mullins effect related anisotropy and thermally induced recovery. *Polymer Testing*, *133*, 108399. https://doi.org/10.1016/J.POLYMERTESTING.2024.108399

Greensmith, H. W. (1963). Rupture of rubber. X. The change in stored energy on making a small cut in a test piece held in simple extension. *Journal of Applied Polymer Science*, *7*(3), 993–1002. https://doi.org/10.1002/app.1963.070070316Rivlin

Rivlin, R.S., & Thomas, A. G. (1953). Rupture of rubber. I. Characteristic energy for tearing. *Journal of Polymer Science*, *10*(3), 291–318. https://doi.org/10.1002/pol.1953.120100303

Constitutive Models for Rubber XIII – Dal (Ed.)
© 2025 The Author(s), ISBN 978-1-032-85138-9

Crystallization of natural rubber under non-uniform deformation

D. Nozaki, T.-T. Mai & K. Urayama
Department Materials Chemistry, Graduate School Engineering, Kyoto University, Kyoto, Japan

R. Osumi
Department Macromolecular Engineering Science, Kyoto Institute of Technology, Kyoto, Japan

K. Tsunoda
Sustainable Advanced Materials Division, Bridgestone Corporation, Tokyo, Japan

ABSTRACT: Natural rubber (NR) exhibits strain-induced crystallization (SIC) which enhances significantly mechanical strength and fracture toughness. Most of previous studies have studied SIC and the resulting strain hardening under uniaxial elongation, but rubber products often undergo complex deformation in service. SIC under complex and non-uniform deformation has not been explored thoroughly, despite its importance. This work has characterized SIC in NR driven by a non-uniform deformation during the uniaxial stretching of a wide sheet along the short axis. DIC analysis reveals that the resultant non-uniform deformation includes a range of deformation spanning from planar to uniaxial elongation. In parallel, thermographic analysis uncovers a heterogeneous distribution of crystallinity, reflecting the non-uniform deformation. These independent methodologies underscore a crucial role of strain biaxiality in SIC behavior.

1 GENERAL INSTRUCTIONS

Natural rubber (NR) is widely used in applications that require high reliability, such as tires for heavy-duty vehicles and aircraft. One of the reasons is that NR possesses a capability of strain-induced crystallization (SIC) (Katz, 1925), a self-reinforcing mechanism that significantly improves mechanical strength and fracture toughness, distinguishing it from other non-SIC rubber materials (Candau et al. 2016). There has been much research on SIC, especially focusing on the correlation between SIC and the resulting hardening (Toki et al. 2002, Trabelsi et al. 2003, Tosaka et al. 2004, Tosaka et al. 2006, Brüning et al. 2012, Candau et al. 2015, Le Cam et al. 2020). Most of these previous studies have employed simple and uniform uniaxial elongation as a trigger for SIC. Rubber products in service are generally exposed to complicated and non-uniform deformation. However, the SIC behavior triggered by complex and non-uniform deformation is not fully understood despite its importance (Chen et al, 2019, Khiêm et al. 2022, Hiraiwa et al. 2023).

In this work, we characterize the non-uniform SIC driven by the inhomogeneous strain field in a wide NR sheet subjected to "pseudo-planar" elongation achieved by uniaxial elongation in its short axis. In this pseudo-planar elongation, the strain field varies continuously from planar elongation in the center to nearly uniaxial elongation in the free edges. We quantify the non-uniform strain field by digital image correlation (DIC) method, and the resulting non-uniform SIC from the heat generation associated with crystallization. We explore the relationship between local strain and crystallinity, and the influence of strain biaxiality on SIC behavior.

2 EXPERIMENTAL

A wide sheet sample ($x \times y \times z = 175 \times 20 \times 1$ mm) of vulcanized natural rubber (NR: RSS#3, sulfur: 1.12 phr) was used for pseudo-planar elongation experiments. The wide sheet sample was subjected to uniaxial stretching along the short (y-) axis (Figure 1a) at a crosshead speed of 15 mm/s, resulting in what we term "pseudo-planar" elongation with finite strain inhomogeneity across the material. In the context of pseudo-planar elongation, the central region of the rubber sheet underwent planar stretching without width change, because the width shrinkage was suppressed due to the geometric constraint effect. In contrast, the regions in the vicinity of the free edges experience approximate uniaxial stretching. The key characteristic of this stretching geometry is the gradual variation of strain biaxiality from planar in the center to an approximate uniaxial stretching in the free edges along the x-axis. This variation allows us to analyze the impact of strain biaxiality on SIC in a single experiment.

DOI: 10.1201/9781003516880-26

The non-uniform strain distribution in pseudo-planar elongation was characterized by DIC method from the displacements of a random surface pattern captured by a high-speed camera (FASTCAM Mini AX10, Photon, Frame rate: 500 Hz). The surface temperature variation of the sample during the pseudo-planar elongation process was tracked using a high-speed infrared camera (Telops: SPARK M150, Frame rate: 200 Hz), with an ambient baseline of 25 °C.

3 RESULTS

3.1 *Strain distribution*

We evaluate the non-uniform distributions of the true strains in the y- and x-directions (designated ε_y and ε_x, respectively) in pseudo-planar elongation by the DIC method. Figure 1b displays the distributions of ε_y at a macroscopic true strain of E_y = 1.94. The figure shows a quoter section of the sample because the data were symmetric. Figure 1c illustrates the distribution of the strain ratio (μ_{yx}), defined by, $\mu_{yx} = -\varepsilon_x/\varepsilon_y$, in the corresponding state. Figures 2a-c display the 1D plots of ε_x, ε_y and μ_{yx} against distance (d) from the center ($d = 0$) to the edge at various E_y values.

In Figure 2a, ε_y at each E_y is independent of d across the entire sample, and approximately equal to E_y. In Figure 2b, ε_x at each E_y significantly depends on d. In a wide region near the center, ε_x is zero, indicating planar elongation. In the region of $d \geq 40$ mm, ε_x becomes increasingly negative with approaching the free edge, highlighting a continuous increase in width shrinkage. In Figure 2c, μ_{yx} varies from 0 in the center to approximately 0.5 near the edge, independent of E_y. These features of the μ_{yx} - d curves indicate that the strain field in the pseudo-planar elongation varies continuously from planar elongation to approximate uniaxial elongation.

3.2 *Surface temperature distribution*

When deformed by pseudo-plane elongation at sufficiently high E_y, the sample experiences SIC, resulting in a rise in surface temperature. Figure 3a displays a 2D distribution of surface temperature at E_y = 1.94. While every position in the stretched state undergoes a rise in surface temperature from the initial state ($T_i \approx 25$ °C), the magnitude of the temperature rise (ΔT: $\Delta T = T - T_i$) significantly depends on the position, and ΔT increases towards the edge. Figure 3b shows the corresponding 1D distribution of ΔT along the x-axis at various E_y. At each E_y, ΔT remains almost constant at d < 40 mm whereas it increases with approaching the edge.

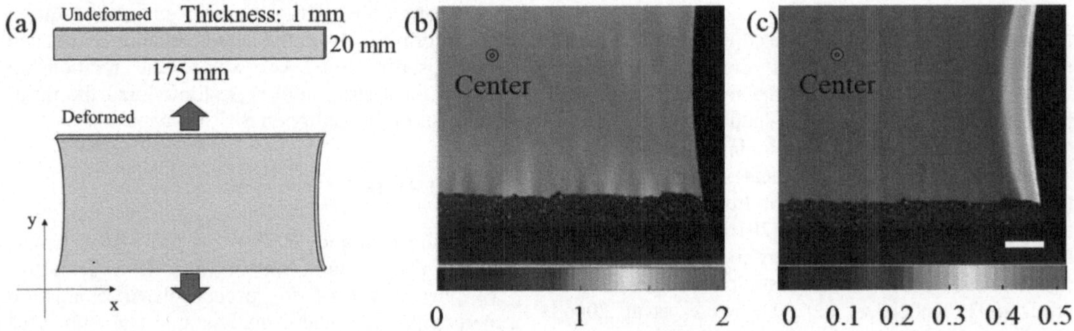

Figure 1. (a) Illustration of "pseudo-planar elongation" of a wide NR sheet specimen, (b)-(c) 2D distributions of (b) ε_y, (c) μ_{yx} at a macroscopic true strain of E_y = 1.94 evaluated with DIC technique. Scale bar indicates 15 mm.

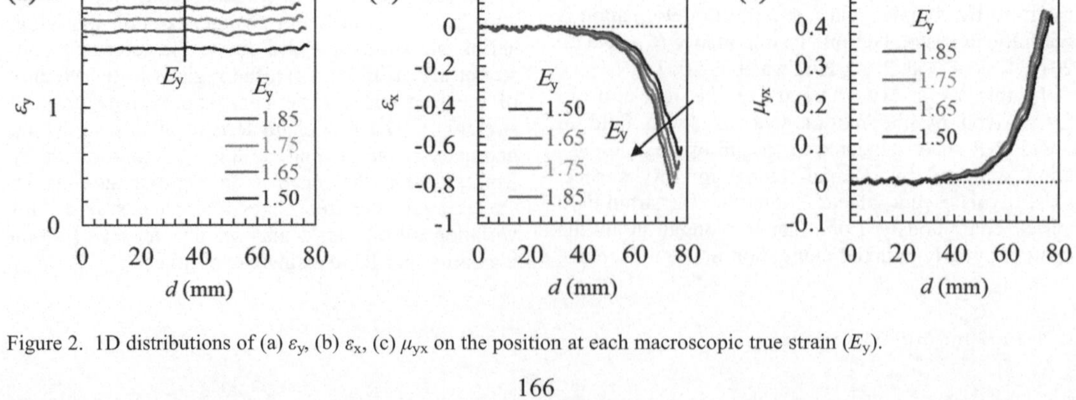

Figure 2. 1D distributions of (a) ε_y, (b) ε_x, (c) μ_{yx} on the position at each macroscopic true strain (E_y).

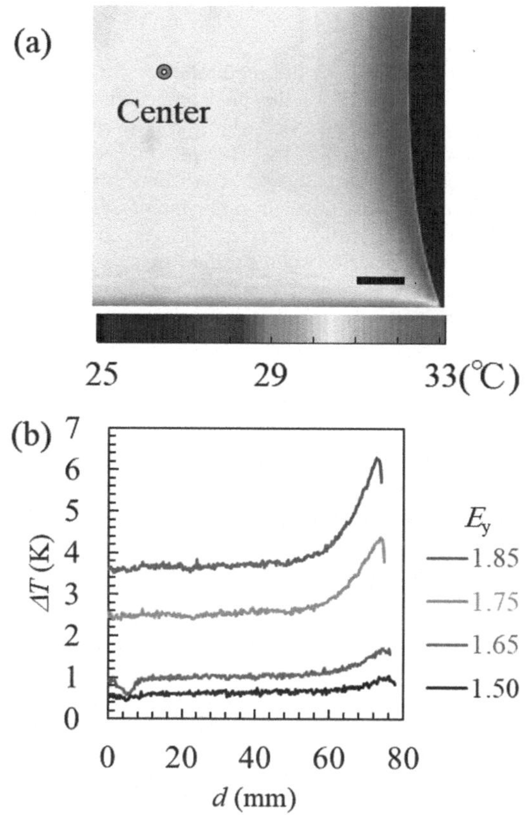

Figure 3. (a) 2D distributions of T at a macroscopic true strain of $E_y = 1.94$ observed by infrared camera. Scale bar indicates 15 mm. (b) 1D distributions of ΔT on the position at each macroscopic true strain (E_y).

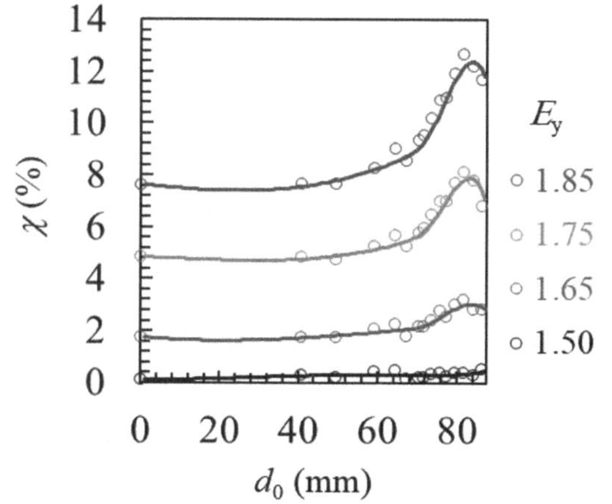

Figure 4. 1D distributions of χ on the position at each macroscopic true strain (E_y). The positions d_0 are defined in the undeformed state.

3.3 *Crystallinity distribution*

The finite spatial ΔT distribution in Figure 3a indicates non-uniform crystallinity (χ) due to varying strain fields. We assessed the χ value at each point by analyzing the corresponding ΔT variation as a function of ε_y by applying the methodology proposed by Le Cam (Le Cam, 2018). They evaluated χ from ΔT in NR samples subjected to uniform uniaxial stretching, by considering the contributions of SIC, entropy elasticity, and heat dissipation to the total heat emission. In this study, we apply this

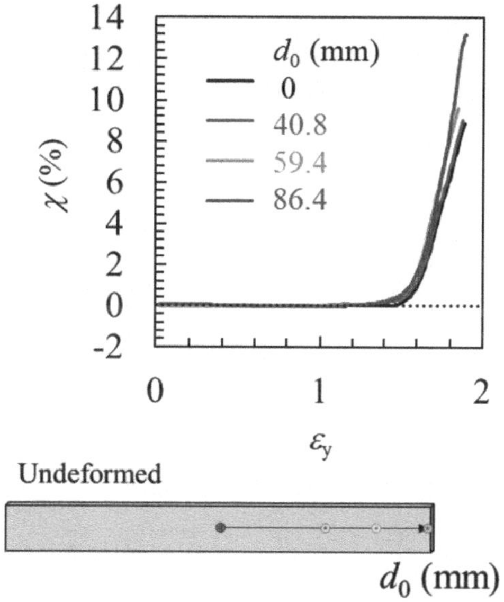

Figure 5. Tracking crystallinity variations at 4 points in the undeformed state.

approach to 15 horizontal positions across the sample to assess the spatial χ distribution. We outline this method in Appendix.

Figure 4 shows the χ values for each position at various degrees of E_y. The position at $d_0 = 0$ corresponds to the center of the sample. It should be noted that d_0 values are defined in the undeformed state. At each E_y, χ has finite distribution: In the central region of $0 < d < 40$ mm, χ is insensitive to position, while χ increases with approaching the free edge in the region of $d > 40$ mm. Evidently, the spatial distribution of χ reflects the strain field gradient from planar to uniaxial elongation, as shown in Figure 2(c).

4 DISCUSSION

Figure 5 plots χ against ε_y for four specific positions, integrating the ε_y - E_y data in Figure 2a and χ - E_y data in Figure 4. Remarkably, the onset of SIC denoted as ε_y^*, is approximately 1.5 and remains largely unaffected by the sample's position. This observation suggests that the critical strain for initiating SIC is almost insensitive to strain biaxiality within the examined range.

Beyond ε_y^*, an increase in ε_y leads to a rise in χ. However, this rise in χ is not uniform, but varies depending on the position, more specifically, on μ_{yx}, when compared at the same ε_y.

Figure 6 delves into this relationship by plotting χ against μ_{yx} across various ε_y values for the examined 15 positions. At each ε_y, χ increases with an increase in μ_{yx}, suggesting that the variation from planar to uniaxial elongation significantly influences the degree of crystallinity. These results in Figure 6 obviously demonstrate the pronounced effect of strain biaxiality on χ. It should be noted that the classical theoretical models, which attribute SIC primarily to entropy reduction following deformation, expects an opposite trend (Flory, 1947). According to these models, χ is expected to increase as the deformation shifts from uniaxial to planar elongation at an equivalent ε_y. This expectation arises from the assumption that planar stretching, which involves finite lateral extension, resulting in a greater overall entropy reduction than uniaxial stretching. Our finding indicates a need for a revised understanding of the effect of strain biaxiality on the crystallization behavior of rubber under deformation.

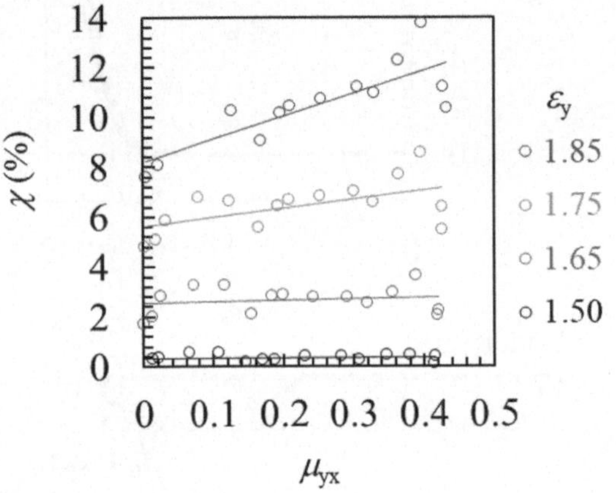

Figure 6. The μ_{yx} dependence of χ compared at the same ε_y.

5 CONCLUSION

Pseudo-planar elongation with non-uniform strain field for a NR sheet undergoes non-uniform SIC once stretching exceeds a critical threshold. The central portion of the sample experiences planar extension, while the strain field continuously varies from planar to uniaxial stretching with approaching the free edges. Notably, local crystallinity increases towards the free edges, even under consistent applied strain levels. Crucially, crystallinity is reduced when the deformation shifts from uniaxial to planar elongation, underscoring the role of strain configuration in SIC behavior.

6 APPENDIX

This section outlines the methodology for evaluating crystallinity from the heat emission observed during stretching of NR, which was proposed by Le Cam (Le Cam, 2018). Temperature variations observed under nonadiabatic conditions are affected by thermal dissipation and other factors during the test. Here, the following equation is adopted as a model to describe such thermal dissipation phenomena in which heat generation and dissipation are combined.

$$s = \rho C\left(\frac{dT}{dt} + \frac{\Delta T}{\tau}\right) \tag{1}$$

where ρ (= 0.936 kg/dm^3) and C (= 1768 J/kg/K) are the sample density and the heat capacity, respectively, and are assumed to be independent of strain and temperature, using literature values (Chrysochoos, 1995). The time constant for heat dissipation of the measurement system (τ) can be separately obtained from the recovery behavior of the surface temperature to room temperature for a preheated specimen. The overall heat power density triggered by the deformation process (s) includes not only heat generation due to SIC but also due to entropy elasticity (Gough-Joule effect). The temperature variation data in the low-strain region without SIC effect were used to obtain a phenomenological equation describing the strain dependence of the contribution of the Gough-Joule effect (s_{GJ}). The temperature variation due to SIC (ΔT_{SIC}) at each strain (or t) was evaluated using the corresponding heat power density s_{SIC} ($= s - s_{GJ}$) and the following equation.

$$\Delta T_{SIC}(t) = \frac{s_{SIC}(t)\Delta t}{\rho C} + \Delta T_{SIC}(t - \Delta t) \tag{2}$$

where $\Delta T_{SIC}(0) = 0$. The crystallinity χ as a function of t was determined by the following equation.

$$\chi(t) = \frac{\rho C \Delta T_{SIC}(t)}{\Delta H} \tag{3}$$

where ΔH (= 6.2×10^4 J/dm^3) is the equilibrium enthalpy of melting (Kim et al. 1972), which is assumed to be independent of strain and temperature.

The time constant for heat dissipation of the measurement system (τ) was estimated at each point from the cooling process following the temperature rise mainly due to SIC, until the temperature stabilized at ambient temperature after the stretch was held constant. The maximum time for this process (τ_{max}) at each position was determined after rapid stretch to a nominal strain of 6. We also accounted for the local strain dependence of τ, which results from thickness reductions during stretching, according to the method of Le Cam et al (Le Cam et al.

2020). They consider this effect through the deformation biaxiality ratio in the following.

$$\tau = \frac{\tau_0}{\lambda_x \lambda_y} = \frac{\tau_0}{\exp(\varepsilon_x)\exp(\varepsilon_y)} \tag{4}$$

Where τ_0 is τ in the undeformed state. τ_0 was obtained by substituting the experimental value τ_{max} at the maximum measured elongation into Equation (4).

ACKNOWLEDGEMENTS

This work was carried out with the support of JST, CREST grant number JPMJCR2091, Japan.

REFERENCES

Brüning, K. Schneider, K. Roth, S. V. & Heinrich, G. 2012. Kinetics of Strain-Induced Crystallization in Natural Rubber Studied by WAXD: Dynamic and Impact Tensile Experiments. *Macromolecules*. 45(19): 7914–7919.

Candau, N. Laghmach, R. Chazeau, L. Chenal, J.-M. Gauthier, C. Biben, T. & Munch, E. 2015. Influence of strain rate and temperature on the onset of strain induced crystallization in natural rubber. *Eur. Polym. J.* 64: 244–252.

Candou, N. Chazeau, L. Chenal, J.-M. Gauthier C. & Munch, E. 2016. A comparison of the abilities of natural rubber (NR) and synthetic polyisoprene *cis*-1,4 rubber (IR) to crystallize under strain at high strain rates. *Phys. Chem. Chem. Phys*. 18: 3472–3481.

Chen, X. Meng, L. Zhang, W. Ye, K. Wang, D. Chen, W. Nan, M. Wang, S. & Li, L. 2019. Frustrating Strain-Induced Crystallization of Natural Rubber with Biaxial Stretch. *ACS Appl. Mater. Interfaces*. 11: 47535–47544.

Chrysochoos, A. 1995. Analyse Du Comportement Des Matériaux Par Thermographie Infra Rouge. *Colloq. Photomécanique*. 95: 201–211.

Flory, P. J. 1947. Thermodynamics of Crystallization in High Polymers. I. Crystallization Induced by Stretching. *J. Chem. Phys*. 15: 397–408.

Hiraiwa, S. Mai, T.-T. Tsunoda, K. & Urayama, K. 2023. Strain hardening in biaxially stretched elastomers undergoing strain-induced crystallization. *RCS Adv.* 13: 34630–34636.

Katz, J. R. 1925. Röntgenspektrographische Untersuchungen am gedehnten Kautschuk und ihre mögliche Bedeutung für das Problem der Dehnungseigenschaften dieser Substanz. *Die Naturwissenschaften*. 13(19): 410–416.

Khiêm, V. N. Le Cam J.-B. Charlès, S. & Itskov, M. 2022. Thermodynamics of strain-induced crystallization in filled natural rubber under uni- and biaxial loadings, Part I: Complete energetic characterization and crystallinity evaluation. *J. Mech. Phys. Solids*. 159: 104701.

Khiêm, V. N. Le Cam J.-B. Charlès, S. & Itskov, M. 2022. Thermodynamics of strain-induced crystallization in filled natural rubber under uni- and biaxial loadings, Part II: Physically-based constitutive theory. *J. Mech. Phys. Solids*. 159: 104712.

Kim, H.-G. & Mandelkern, L. 1972. Multiple Melting Transitions in Natural Rubber. *J. Polym. Sci. Part A2 Polym. Phys*. 10(6): 1125–1133.

Le Cam, J.-B. 2018. Strain-induced crystallization in rubber: A new measurement technique. *Strain*. 54: e12256.

Le Cam, J.-B. Albouy, P.-A. & Charlès, S. 2020. Comparison between x-ray diffraction and quantitative surface calorimetry based on infrared thermography to evaluate strain-induced crystallinity in natural rubber. *Rev. Sci. Instrum*. 91: 044902.

Toki, S. Sics, I. Ran, S. Liu, L. Hsiao, B. S. Murakami, S. Senoo, K. & Kohjiya, S. 2002. New Insights into Structural Development in Natural Rubber during Uniaxial Deformation by In Situ Synchrotron X-ray Diffraction. *Macromolecules*. 35(17): 6578–6584.

Tosaka, M. Kawakami, D. Senoo, K. Kohjiya, S. Ikeda, Y. Toki, S. & Hsiao, B. S. 2006. Crystallization and Stress Relaxation in Highly Stretched Samples of Natural Rubber and Its Synthetic Analogue. *Macromolecules*. 39(15): 5100–5105.

Tosaka, M. Murakami, S. Poompradub, S. Kohjiya, S. Ikeda, Y. Toki, S. Sics, I. & Hsiao, B. S. 2004. Orientation and Crystallization of Natural Rubber Network As Revealed by WAXD Using Synchrotron Radiation. *Macromolecules*. 37(9): 3299–3309.

Trabelsi, S. Albouy, P.-A. & Rault, J. 2003. Crystallization and Melting Processes in Vulcanized Stretched Natural Rubber. *Macromolecules*. 36(20): 7624–7639.

Constitutive Models for Rubber XIII – Dal (Ed.)
© 2025 The Author(s), ISBN 978-1-032-85138-9

Effects of loading conditions, temperature, and magnetic field on the dynamic compressive behavior of an isotropic magnetorheological elastomer

T.H. Nam, I. Petríková, B. Marvalová & M. Sivčák
Technical University of Liberec, Liberec, Czech Republic

ABSTRACT: An isotropic magnetorheological elastomer (MRE) has been prepared by filling magnetically sensitive micro-sized carbonyl iron particles (CIPs) into silicone rubber. The dynamic mechanical behavior of the isotropic MRE was investigated through cyclic compression tests using the Instron Electropuls E3000 testing system. Effects of pre-strains (5-20%), strain amplitudes (0.1-3%), frequencies (1-50 Hz), magnetic field intensities (0-0.296 T), and temperatures (RT-60 °C) on the dynamic compressive properties of the isotropic MRE were examined. Although the storage and loss moduli of the isotropic MRE increased with increasing the pre-strain and magnetic field intensity, they decreased with rising strain amplitude and temperature. The storage modulus of the isotropic MRE was enhanced considerably with increasing frequency to about 20 Hz, which then increased gradually. The loss modulus of the isotropic MRE increased significantly with increasing frequency to about 20 Hz and then changed slightly. The dependence of the isotropic MRE dynamic moduli on the frequency at different pre-strains, strain amplitudes, magnetic field intensities, and temperatures was numerically studied using a fractional derivative viscoelastic model. The model was fitted well to the dynamic moduli of the isotropic MRE. The model parameters were identified by fitting the dynamic moduli to the measured data of the isotropic MRE. The dynamic moduli of the isotropic MRE estimated from the investigated model with fitted parameters were in good agreement with the measured data.

1 INTRODUCTION

Magnetorheological elastomers (MREs) are considered smart composite materials fabricated by filling ferromagnetic particles into a non-magnetic elastomeric matrix. The typical behavior of MREs is the change in their shape and mechanical properties under the application of an external magnetic field (Bastola et al. 2020). Besides, MREs have shown other interesting capabilities, such as dynamic locomotion, programmable shape morphing, manipulation and assembly of objects, and reconfigurable structures (Bastola & Hossain 2021). Therefore, MREs have potential applications in damping systems, vibration absorbers and isolators, sensors and actuators, soft robotics, morphable structures, metamaterials, minimally invasive surgery, controlled drug delivery, and other biomedical devices (Nam et al. 2024, Díez et al. 2021).

MREs have been created on elastomeric matrices. Therefore, they inherit the main properties of these matrices, such as large deformation, viscoelastic behavior, stress softening effect, amplitude and frequency dependency, stiffness reduction under cyclic loading, and time-dependent response (Nam et al. 2020). Besides, MREs contain magneto-sensitive particles. As a result, the mechanical properties of MREs vary when the magnetic field is applied. Therefore, the effects of loading conditions, magnetic field, and temperature on the dynamic mechanical properties of MREs need to be investigated.

The dynamic mechanical behavior of MREs can be simulated using fractional derivative viscoelastic models. Agirre-Olabide et al. (2017) used a four-parameter fractional derivative model to fit the dynamic complex shear modulus of isotropic MREs. The fitting error did not exceed 1% for the storage and loss moduli. Nam et al. (2020) investigated the dependence of dynamic shear properties of isotropic and anisotropic MREs on the frequency and magnetic field intensity (MFD) using the fractional derivative Zener model. Results showed that the model was fitted well to measured data for both isotropic and anisotropic MREs. Zhu et al. (2021) developed a nonlinear dynamic model with a fractional element for MREs in a linear viscoelastic regime. The proposed model can be used for dynamic mechanical analysis and simulation of MRE-based vibration systems.

For this paper, the isotropic MRE was fabricated by mixing micro-sized carbonyl iron particles (CIPs) with a volume fraction of 27% into a silicone rubber matrix. The dynamic compressive behavior of the isotropic MRE was investigated through the cyclic compressive test. Effects of different frequencies, pre-strains, strain amplitudes, MFDs, and temperatures on

DOI: 10.1201/9781003516880-27

the dynamic compressive properties of the isotropic MRE were examined. Besides, a fractional derivative viscoelastic model with seven parameters was used to simulate the dynamic compressive behavior of the isotropic MRE at various pre-strains, strain amplitudes, MFDs, and heating temperatures.

2 EXPERIMENTAL STUDY

2.1 *Materials*

The raw materials used to produce the isotropic MRE are micro-sized CIPs (type 44890, Sigma-Aldrich, USA), silicone rubber ZA13, and its catalyst (Zhermack SPA, Italy). The CIPs with a grain size of 5-9 μm have a purity of 99.5% and a density of 7.86 g/cm^3. The isotropic MRE specimens with a 27% volume fraction of CIPs were fabricated in a solid cylindrical shape with about 13 mm in height and 28 mm in diameter according to the standard ISO D575-91:2018. Detailed procedures for processing the isotropic MRE samples are described in our earlier article (Nam et al. 2021). Microstructural morphologies of the isotropic MRE can be found in the paper (Nam et al. 2021).

2.2 *Dynamic compressive test*

The dynamic mechanical behavior of the isotropic MRE was studied via the cyclic compressive test using the Instron Electropuls E3000 testing system (Instron Corp., MA, USA). The isotropic MRE samples were subjected to cyclic harmonic compressive loading in the displacement control. The experimental setups of the dynamic compressive tests for isotropic MRE samples with and without the magnetic field and at elevated temperatures are described in Figure 1.

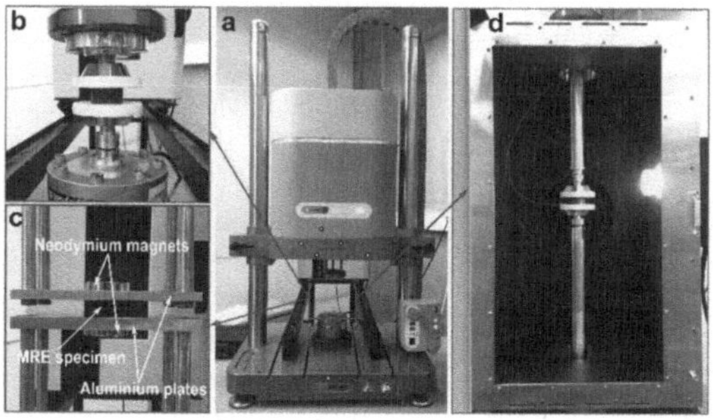

Figure 1. (a) The experimental setup of dynamic compression tests for isotropic MRE samples: (b) without the magnetic field, (c) with magnetic fields, and (d) at elevated temperatures.

Two fixtures in the frame structure with aluminum plates are used to insulate neodymium magnets from contacting the MRE sample during the dynamic test (see Figure 1c). The average MFDs were measured at several points between neodymium magnets by a digital magnetometer TECTRA (Model 5180 Gauss/Tesla meter). At a pre-strain of 10%, the external magnetic fields with mean MFDs of 0.132, 0.206, and 0.296 T corresponding to 1, 2, and 6 pairs of neodymium magnets were applied to the MRE specimens.

2.3 *Characterizations*

The experimental study was conducted to characterize the dynamic compressive properties of the isotropic MRE with the change of frequencies (1-50 Hz), pre-strains (1-20%), strain amplitudes (0.1-3%), and various MFDs (0-0.296 T). The isotropic MRE sample subjected to a sinusoidally varying compressive strain $\varepsilon(t)$ will reach the steady state after several cycles, and the resulting stress $\sigma(t)$ is also sinusoidal with the same angular frequency ω. However, the compressive stress and strain will lag by a phase angle δ (loss angle). The loss angle was determined using the discrete Fourier transform (DFT) at the main excitation frequency in Matlab. The dynamic complex modulus (E^*), defined as the ratio of the stress amplitude ($\Delta\sigma$) and the strain amplitude ($\Delta\varepsilon$), is represented in the form:

$$E^* = E' + iE'' \tag{1}$$

where E' is the storage modulus, and E'' is the loss modulus. The storage and loss moduli and the loss angle are expressed as follows:

$$E' = \frac{\Delta\sigma}{\Delta\varepsilon}\cos\delta,\ E'' = \frac{\Delta\sigma}{\Delta\varepsilon}\sin\delta,\ \tan\delta(\omega) = \frac{E''}{E'} \tag{2}$$

3 NUMERICAL SIMULATION

A fractional derivative viscoelastic model with seven parameters described in Figure 2 was used to simulate the dynamic compressive behavior of the isotropic MRE. This model was composed of an elastic spring E_0 and two fractional Maxwell elements in parallel with six parameters (E_α, α, c_α, E_β, β, c_β) (Nadzharyan et al. 2018).

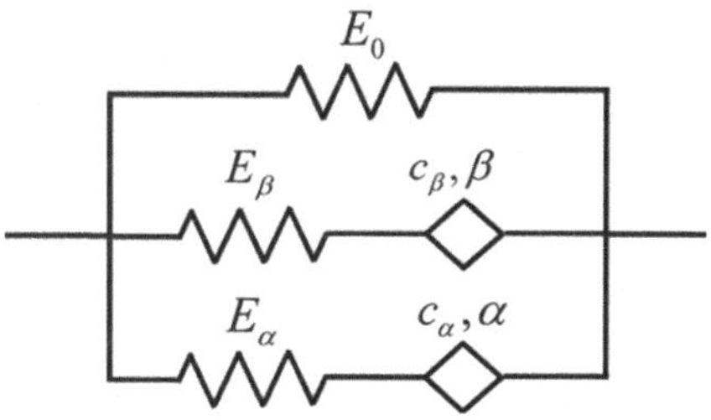

Figure 2. A fractional derivative viscoelastic model.

The constitutive equation for the fractional derivative viscoelastic model in the time domain is written as follows:

$$\frac{1}{E_\alpha}\frac{d^{\alpha+\beta}\sigma}{dt^{\alpha+\beta}}+\frac{1}{c_\alpha}\frac{d^\beta\sigma}{dt^\beta}+\frac{E_\beta}{c_\beta E_\alpha}\frac{d^\alpha\sigma}{dt^\alpha}+\frac{E_\beta}{c_\alpha c_\beta}\sigma(t)$$
$$=\left(1+\frac{E_0+E_\beta}{E_\alpha}\right)\frac{d^{\alpha+\beta}\varepsilon}{dt^{\alpha+\beta}}+\frac{E_0+E_\beta}{c_\alpha}\frac{d^\beta\varepsilon}{dt^\beta} \tag{3}$$
$$+\frac{E_\beta}{c_\beta}\left(1+\frac{E_0}{E_\alpha}\right)\frac{d^\alpha\varepsilon}{dt^\alpha}+\frac{E_0E_\beta}{c_\alpha c_\beta}\varepsilon(t)$$

where E_0, E_α and E_β are the elastic moduli of corresponding springs; c_α and c_β are viscoelasticity coefficients of the fractional elements; α and β are the order parameters of the fractional elements.

The dynamic complex modulus of the fractional derivative viscoelastic model in the frequency domain takes the form of Equation 3. The expressions for the storage and loss moduli obtained for this model are as follows:

$$E' = \frac{N'_0+N'_\alpha+N'_\beta+N'_{\alpha\beta}}{D_0+D_\alpha+D_\beta+D_{\alpha\beta}} \tag{4}$$

$$E'' = \frac{N''_\alpha+N''_\beta+N''_{\alpha\beta}}{D_0+D_\alpha+D_\beta+D_{\alpha\beta}} \tag{5}$$

The terms in the numerator and denominator of the storage and loss moduli are expressed as follows:

$$N'_0 = E_0E_\alpha^2E_\beta^2,$$

$$N'_\alpha = (E_0+E_\alpha)E_\beta^2c_\alpha^2\omega^{2\alpha} + (2E_0+E_\alpha)E_\alpha E_\beta^2c_\alpha\omega^\alpha\cos(\alpha\pi/2)$$

$$N'_\beta = (E_0+E_\beta)E_\alpha^2c_\beta^2\omega^{2\beta} + (2E_0+E_\beta)E_\alpha^2E_\beta c_\beta\omega^\beta\cos(\beta\pi/2)$$

$$\begin{aligned}N'_{\alpha\beta} &= (E_0+E_\alpha+E_\beta)c_\alpha^2c_\beta^2\omega^{2(\alpha+\beta)}\\ &+(2E_0+E_\alpha+2E_\beta)E_\alpha c_\alpha c_\beta^2\omega^{\alpha+2\beta}\cos(\alpha\pi/2)\\ &+(2E_0+2E_\alpha+E_\beta)E_\beta c_\alpha^2c_\beta\omega^{2\alpha+\beta}\cos(\beta\pi/2)\\ &+(2E_0+E_\alpha+E_\beta)E_\alpha E_\beta c_\alpha c_\beta\omega^{\alpha+\beta}\cos(\alpha\pi/2)\cos(\beta\pi/2)\end{aligned}$$

$$N''_\alpha = (E_0+E_\alpha)E_\alpha E_\beta^2c_\alpha\omega^\alpha\sin(\alpha\pi/2)$$

$$N''_\beta = (E_0+E_\beta)E_\alpha^2E_\beta c_\beta\omega^\beta\sin(\beta\pi/2)$$

$$\begin{aligned}N''_{\alpha\beta} &= E_\alpha^2c_\alpha c_\beta^2\omega^{\alpha+2\beta}\sin(\alpha\pi/2)\\ &+E_\beta^2c_\alpha^2c_\beta\omega^{2\alpha+\beta}\sin(\beta\pi/2)\\ &+E_\alpha E_\beta c_\alpha c_\beta\omega^{\alpha+\beta}[E_0\sin((\alpha+\beta)\pi/2)\\ &+2E_\alpha\sin(\alpha\pi/2)\cos(\beta\pi/2)\\ &+2E_\beta\cos(\alpha\pi/2)\sin(\beta\pi/2)]\end{aligned}$$

$$D_0 = E_\alpha^2E_\beta^2$$

$$D_\alpha = E_\beta^2c_\alpha^2\omega^{2\alpha}+2E_\alpha E_\beta^2c_\alpha\omega^\alpha\cos(\alpha\pi/2)$$

$$D_\beta = E_\alpha^2c_\beta^2\omega^{2\beta}+2E_\alpha^2E_\beta c_\beta\omega^\beta\cos(\beta\pi/2)$$

$$\begin{aligned}D_{\alpha\beta} &= c_\alpha^2c_\beta^2\omega^{2(\alpha+\beta)}+2E_\alpha c_\alpha c_\beta^2\omega^{\alpha+2\beta}\cos(\alpha\pi/2)\\ &+2E_\beta c_\alpha^2c_\beta\omega^{2\alpha+\beta}\cos(\beta\pi/2)\\ &+4E_\alpha E_\beta c_\alpha c_\beta\omega^{\alpha+\beta}\cos(\alpha\pi/2)\cos(\beta\pi/2)\end{aligned}$$

4 RESULTS AND DISCUSSION

The storage and loss moduli of the isotropic MRE are determined from experimental data at different pre-strains, strain amplitudes, MFDs, and temperatures. The parameters of the fractional derivative viscoelastic model are obtained by fitting equations 4 and 5 to themeasured results at various pre-strains, strain amplitudes, MFDs, and temperatures. The fitting was done simultaneously for the storage and loss moduli using nonlinear least-squares multiple data-fitting in Matlab.

4.1 *Effects of pre-strains*

The model parameters fitted with experimental data of the isotropic MRE at different pre-strains are given in Table 1. As Table 1 shows, the parameters E_0 and E_β increase with increasing pre-strain. Besides, the parameters E_α and c_α are corresponding much lower than the parameters E_β and c_β. This can indicate that the effect of a fractional Maxwell element in the investigated model on the compressive dynamic behavior of the isotropic MRE is not large. However, when authors use the four-parameter fractional viscoelastic model presented in the article (Nam et al. 2020), the fitting of the loss modulus is not very accurate at high frequencies, leading to a significant difference between the fitted and measured loss angle. Therefore, the authors applied the seven-parameter model with two fractional Maxwell elements in this study.

Table 1. Fitting parameters of the model with experimental data of the MRE at different pre-strains.

	Pre-strains (%)			
Parameters	5	10	15	20
E_0 (MPa)	2.39	2.92	3.68	4.97
E_α (MPa)	3.2E-05	7.1E-05	7.8E-05	6.4E-05
E_β (MPa)	2.08	2.65	3.18	3.43
c_α (MPa s$^\alpha$)	3.4E-05	6.4E-05	7.1E-05	6.0E-05
α (rad)	0.128	0.170	0.202	0.206
c_β (MPa s$^\beta$)	0.509	0.541	0.562	0.579
β (rad)	0.368	0.383	0.397	0.420

The measured and model-fitted results of the storage and loss moduli for the isotropic MRE at various pre-strains are presented in Figure 3. Results showed that the storage and loss moduli increase with increasing pre-strain. The storage modulus enhances with increasing frequency to 50 Hz, while the loss modulus increases to about 20 Hz and then changes slightly. Besides, a very good agreement between fitted and measured moduli of the isotropic MRE is clearly observed in Figure 3.

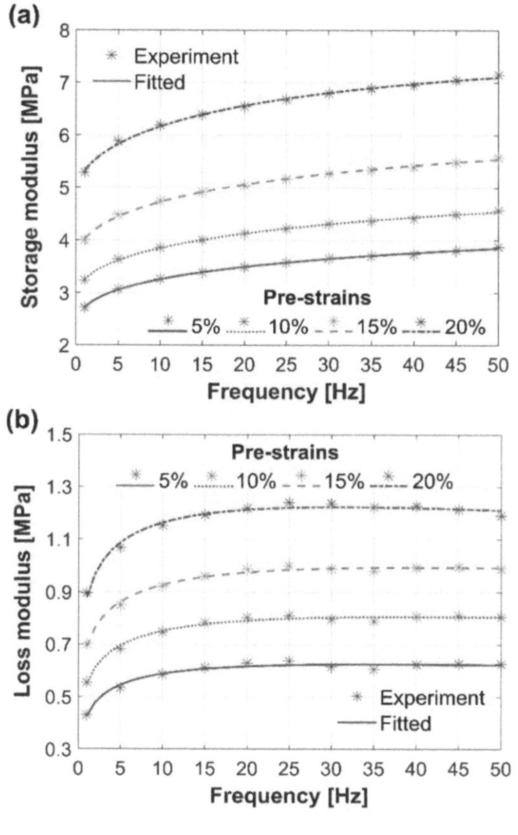

Figure 3. (a) Storage and (b) loss moduli of the isotropic MRE loaded at a strain amplitude of 1% and different pre-strains. The star markers show the storage and moduli ascertained from measured data. The solid and dash lines show model-fitted storage and loss moduli.

4.2 *Effects of strain amplitudes*

The fitted parameters of the investigated model with experimental data of the isotropic MRE at different strain amplitudes are given in Table 2. As seen in Table 2, the parameter E_0 decreases with increasing strain amplitude. Like above, the parameters of a fractional Maxwell element (E_β, c_β, and β) are much higher than those of the other (E_α, c_α, and α). The variation of storage and loss moduli with frequency at different strain amplitudes is described in Figure 4. The model-fitted and measured results of the storage and loss moduli for the isotropic MRE at various strain amplitudes as functions of frequency are shown in Figure 4. As observed in Figure 4, the storage and loss moduli decrease with increasing strain amplitude. Besides, the storage modulus

Table 2. Fitting parameters of the model with experimental data of the MRE at different strain amplitudes.

	Strain amplitudes (%)					
Parameters	0.1	0.25	0.5	1	2	3
E_0 (MPa)	3.54	3.38	3.16	2.95	2.69	2.49
E_α (MPa)	2E-05	4E-05	5E-05	6E-05	6E-05	1E-04
E_β (MPa)	2.65	2.68	2.61	2.53	2.51	2.66
c_α (MPa s$^\alpha$)	3E-05	5E-05	5E-05	6E-05	5E-05	9E-05
α (rad)	0.133	0.159	0.154	0.151	0.173	0.232
c_β (MPa s$^\beta$)	0.544	0.547	0.537	0.532	0.531	0.535
β (rad)	0.407	0.384	0.384	0.383	0.381	0.380

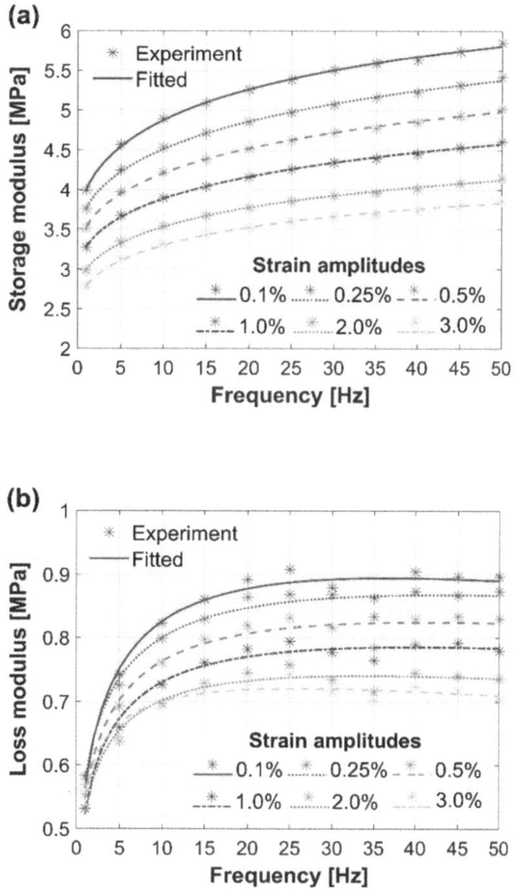

Figure 4. (a) Storage and (b) loss moduli of the isotropic MRE loaded at the pre strain of 10% and different strain amplitudes. The star markers show the storage and moduli determined from measured data. The solid and dash lines show model-fitted storage and loss moduli.

increases with rising frequency, whereas the loss modulus enhances to about 20 Hz frequency and then changes irregularly. There is a slight decrease in the modulus loss at frequencies of 30 Hz and 35 Hz in the measured data. Moreover, at the strain amplitude of 3%, there is an irregular variation in the modulus loss of the isotropic MRE at frequencies below 10 Hz.

4.3 *Effects of magnetic fields*

The parameters of the fractional derivative model for the isotropic MRE fitted to experimental data under different MFDs were obtained, with results presented in Table 3. The parameters E_0 and E_β increase with the rise of MFD. The increase in these elastic moduli proves that the isotropic MRE stiffness is enhanced with increasing MFD, which is in accordance with the magnetic-control stiffness of the MRE.

Table 3. Fitting parameters of the model with experimental data of the MRE under different MFDs.

	MFDs (T)			
Parameters	0	0.132	0.206	0.296
E_0 (MPa)	2.98	3.25	3.45	3.76
E_α (MPa)	6.1E-05	7.4E-05	8.3E-05	1.1E-04
E_β (MPa)	2.01	2.21	2.37	2.39
c_α (MPa s$^\alpha$)	5.8E-05	6.4E-05	6.9E-05	7.8E-05
α (rad)	0.124	0.147	0.166	0.182
c_β (MPa s$^\beta$)	0.533	0.550	0.562	0.586
β (rad)	0.375	0.374	0.371	0.366

The fitted and experimental results of the loss and storage moduli as frequency functions under different MFDs for the isotropic MRE are presented in Figure 5. As mentioned above, the storage modulus tends to boost with increasing frequency, while the loss modulus only increases with the frequency rise to about 25 Hz. Besides, the measured loss modulus reduces at 30 Hz and 35 Hz frequencies. The model-fitted loss modulus shows an insignificant change at frequencies above 20 Hz.

4.4 *Effects of temperatures*

The model parameters obtained from fitting the loss and storage moduli of the isotropic MRE to measured data at various temperatures are listed in Table 4. The model-fitted and measured values of loss and storage moduli for the isotropic MRE at different temperatures are described in Figure 6.

As Table 4 shows, the parameter E_0 and E_β of the model decreases slightly with increasing temperature. Besides, the fraction parameters of a fractional Maxwell element (E_α, c_α, and α) are much lower than those of the other element (E_β, c_β, and β). As observed

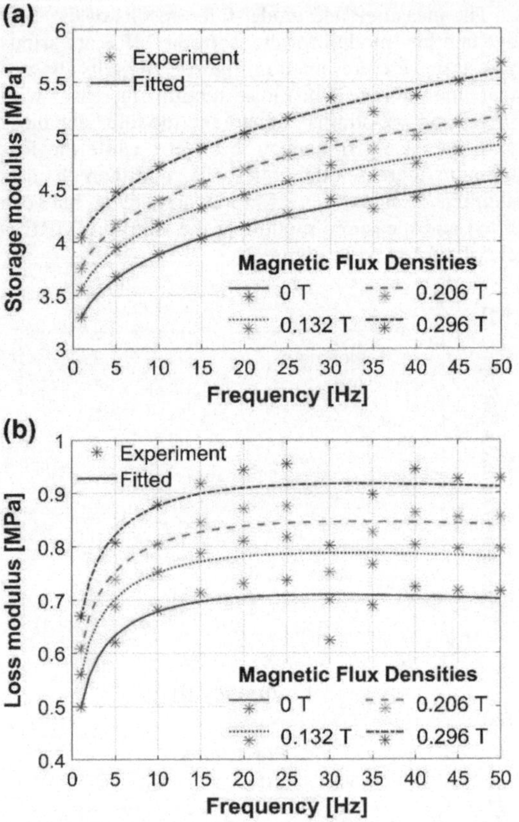

Figure 5. (a) Storage and (b) loss moduli of the isotropic MRE loaded at the pre-strain of 10%, the strain amplitude of 1% under different MFDs. The star markers show the storage and moduli ascertained from measured data. The solid and dash lines show model-fitted storage and loss moduli.

Table 4. Fitting parameters of the model with experimental data of the MRE at different temperatures.

	Temperature (°C)				
Parameters	RT	30	40	50	60
E_0 (MPa)	2.91	2.84	2.77	2.72	2.71
E_α (MPa)	6.4E-05	5.8E-05	5.6E-05	5.1E-05	5.3E-05
E_β (MPa)	2.64	2.55	2.29	2.14	2.00
c_α (MPa s$^\alpha$)	6.1E-05	5.5E-05	5.4E-05	5.1E-05	5.3E-05
α (rad)	0.165	0.161	0.145	0.138	0.132
c_β (MPa s$^\beta$)	0.533	0.528	0.512	0.500	0.495
β (rad)	0.380	0.373	0.362	0.353	0.344

in Figure 6, the storage and loss moduli decline with rising temperature. The results indicated that the higher the temperature, the lower the dynamic moduli. Besides, the loss modulus determined from experimental data shows a reduction at a frequency of 30 Hz. Moreover, the model-fitted loss modulus indicates almost unchanged at frequencies above 20 Hz.

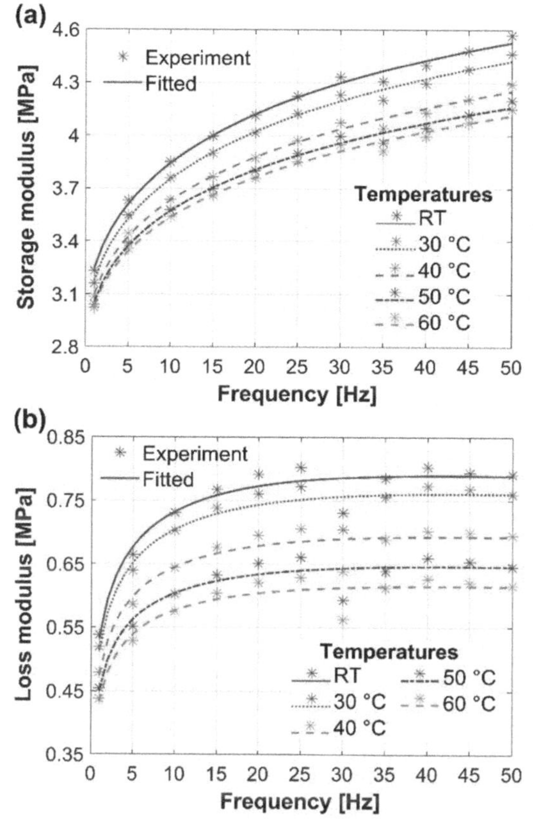

Figure 6. (a) Storage and (b) loss moduli of the isotropic MRE loaded at the pre-strain of 10%, the strain amplitude of 1%, and at different temperatures. The star markers show the storage and moduli determined from measured data. The solid and dash lines show model-fitted storage and loss moduli.

5 CONCLUSIONS

The dynamic compressive behavior of the isotropic MRE made of silicone matrix and magnetically sensitive micro-sized CIPs was investigated in this study. The impacts of pre-strains, strain amplitudes, frequencies, MFDs, and temperatures on the dynamic compressive properties of the isotropic MRE were examined. The storage modulus of the isotropic MRE increased with increasing frequency, while the loss modulus enhanced to about 20 Hz and then changed slightly above 20 Hz. Although the storage and loss moduli of the isotropic MRE were boosted by rising pre-strain and MFD, they were reduced by increasing strain amplitude and temperature. The rise in the temperature resulted in the decrease of the dynamic compressive properties of the isotropic MRE.

The dynamic moduli of the isotropic MRE as frequency functions at different pre-strains, strain amplitudes, MFDs, and temperatures were numerically simulated using a seven-parameter fractional derivative viscoelastic model. The fractional derivative viscoelastic model was fitted well to the dynamic moduli of the isotropic MRE. The model parameters were identified by fitting the dynamic moduli to the measured data of the isotropic MRE. The dynamic moduli of the isotropic MRE estimated from the investigated model with fitted parameters were in good agreement with the measured data.

ACKNOWLEDGEMENT

This work was supported by the Ministry of Education, Youth and Sports of the Czech Republic and the European Union – European Structural and Investment Funds in the frame of Operational Program Research, Development and Education – project Hybrid Materials for Hierarchical Structures (HyHi, Reg. No. CZ.02.1.01/0.0/0.0/16_019/0000843).

REFERENCES

Agirre-Olabide I., Lion A. & Elejabarrieta, M.J. 2017. A new three-dimensional magnetoviscoelastic model for isotropic magnetorheological elastomers. *Smart Materials and Structures* 26: 035021.

Bastola, A. K., & Hossain, M. 2021. The shape–morphing performance of magnetoactive soft materials. *Materials & Design* 211: 110172.

Bastola, A.K., Paudel, M., Li, L. & Li, W. 2020. Recent progress of magnetorheological elastomers: a review. *Smart Materials and Structures* 29(12): 123002.

Díez, A.G., Tubio, C.R., Etxebarria, J.G. & Lanceros-Mendez, S. 2021. Magnetorheological elastomer-based materials and devices: State of the art and future perspectives. *Advanced Engineering Materials* 23(6): 2100240.

Nadzharyan, T.A., Kostrov, S.A., Stepanov, G.V., & Kramarenko, E.Y. 2018. Fractional rheological models of dynamic mechanical behavior of magnetoactive elastomers in magnetic fields. *Polymer* 142: 316–329.

Nam, T.H., Petríková, I. & Marvalová, B. 2020. Experimental characterization and viscoelastic modeling of isotropic and anisotropic magnetorheological elastomers. *Polymer Testing* 81: 106272.

Nam, T.H., Petríková, I., & Marvalová, B. 2024. Effects of applied strain, magnetic field, and temperature on the compressive stress relaxation behavior of magneto-sensitive elastomers. *Mechanics of Time-Dependent Materials*, 1–20. https://doi.org/10.1007/s11043-023-09654-4

Nam, T.H., Petríková, I., Marvalová, B. & Hdaib, M.Y. 2021. Self-heating and dynamic mechanical behavior of silicone rubber composite filled with carbonyl iron particles under cyclic compressive loading. *Journal of Composite Materials* 55(28): 4273–4292.

Zhu, G., Xiong, Y., Li, Z., Xiao, L., Li, M., & Bai, X.F. 2021. A nonlinear dynamic model of magnetorheological elastomers in magnetic fields based on fractional viscoelasticity. *Journal of Intelligent Material Systems and Structures* 32(2): 228–239.

Constitutive Models for Rubber XIII – Dal (Ed.)

Exploring non-uniform strain-induced crystallization around a stationary crack tip in natural rubber

T.-T. Mai & K. Urayama
Department of Material Chemistry, Kyoto University, Kyoto, Japan

T. Yasui, R. Tanaka & S. Sakurai
Department of Biobased Materials Science, Kyoto Institute of Technology, Kyoto, Japan

H. Masunaga & T. Kabe
Japan Synchrotron Radiation Research Institute, Hyogo, Japan

K. Tsunoda
Sustainable and Advanced Materials Division, Bridgestone Corporation, Tokyo, Japan

ABSTRACT: This research investigates the relationship between non-uniform strain and the induced non-uniform crystallization near a stationary crack-tip in natural rubber, employing micro-scale DIC and X-ray scattering techniques. We demonstrate that most of the area undergoing crystallization near the crack-tips is subjected to a limited range of the deformations between planar and uniaxial extension. Our finding shows that maximum tensile strain in local principal strain tensor governs local crystallinity, while shear strain dictates crystal orientation.

1 INTRODUCTION

While many studies (Huneau 2011) have investigated strain-induced crystallization (SIC) in natural rubber (NR) (Hirata *et al.* 2014), focusing on its relationships with variables like strain and temperature, most of them have used unnotched samples under uniform uniaxial extension. SIC not only stiffens the rubber at high elongation but also enhances tear strength and suppresses the crack growth (Persson *et al.* 2005, Osumi *et al.* 2022). Generally, the strain steeply increases with approaching the crack-tip. The local strains near crack-tips are enhanced sufficiently to induce SIC. Characterizing SIC around a crack front is challenging due to its confined and heterogeneous nature. Previous studies (Trabelsi *et al.* 2002, Brüning *et al.* 2013, Rublon *et al.* 2014, Demassieux *et al.* 2021) showed the occurrence of SIC in the region expanding several hundred micrometers from the crack-tip in the neat and carbon-black filed NR samples. Some researchers (Brüning *et al.* 2013, Rublon *et al.* 2014, Demassieux *et al.* 2021) investigated the heterogeneous strain field near the crack-tip using digital image correlation (DIC) (Schreier *et al.* 2009) to discuss the relationship with the spatial distributions of SIC properties. However, a comprehensive understanding of these relationships remains incomplete due to the inherent localized and heterogeneous features of the SIC near a crack tip.

We use a combination of two-dimensional wide angle X-ray diffraction (WAXD) scan with a 10 μm square beam at 40 μm intervals, and micro-scale DIC techniques with a high resolution of approximately 4 μm, to delve into the relationships between non-uniform strain and crystallization around a crack-tip of NR (Mai *et al.* 2024). This potent combination of these two techniques unravels a governing quantity in the local principal strain tensor that influences local SIC near the crack tip. This information enables us to infer the spatial distribution of crystallinity at any degree of crack tip opening, using solely the associated strain field data.

2 EXPERIMENTS

2.1 *Material*

The NR specimens were prepared using natural rubber gum (Ribbed Smoked Sheets RSS#3). Vulcanization was conducted by using sulfur (1.4 g per 100 g of rubber) as a cross-linking agent (Mai *et al.* 2024).

2.2 *Measurements*

Micro-scale DIC measurements quantify the 2D displacement of the micro-speckle patterns on the

DOI: 10.1201/9781003516880-28

specimen surface caused by crack tip opening. We employed a rectangular sheet sample with a dimension of 40 mm × 10 mm × 0.25 mm which has a 10 mm pre-crack along the X-axis on one side of the edge (Figure 1). A random speckle pattern with a typical speckle size of 20 μm was applied on the specimen surface. The pre-notched specimen was elongated in the Y-axis with a true strain of 0.62 (nominal strain of 0.85) which is below the threshold strain for crack growth (approximately 0.79 and 1.2 in true and nominal strain, respectively. Prior to the measurement, the sample was subjected to preliminary loading-unloading cycles with the same degree of the maximum strain as in the DIC/WAXD tests, in order to exclude influence of manufacturing history and achieving strain-hysteresis stability.

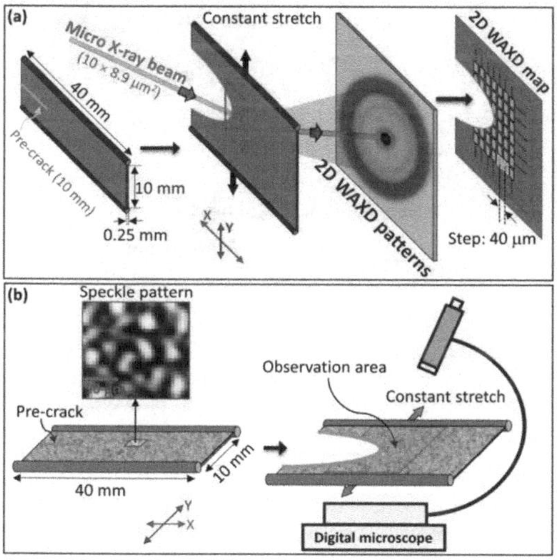

Figure 1. Complementary analysis using (a) 2D scanning micro-beam WAXD and (b micro-DIC measurements. Both the experiments were performed for the identical specimen, enabling precise data correlation.

DIC measurements (Figure 1b) were conducted for a pre-notched specimen under a digital microscope by capturing a set of the micro-speckle images with a resolution of 1200 pixels × 1600 pixels (pixel size of approximately 4 μm). The 2D local strain tensor at each position measured in laboratory (X, Y) coordinates was calculated from the corresponding displacement gradient tensor. The DIC analysis also identifies the major and minor principal strains (ε_1 and ε_2, respectively, in true strain) both in magnitudes and directions (Schreier *et al.* 2009).

A micro-beam scanning WAXD experiment (Figure 1a) was conducted at the BL-15A2 beamline of the Photon Factory at the High Energy Accelerator Research Organization (Tsukuba, Ibaraki, Japan). The identical pre-cracked sample and same custom-made tensile instrument were employed in both the micro-DIC and the micro-beam WAXD measurements, thereby enabling the precise correlation between the results of the two independent tests. The crack contour of the stretched specimen was identical with that in the micro-DIC test, confirming the same strain field in the two independent tests. After the quasi-equilibrium of the tensile stress, the narrow x-ray beam with a size of x × y = 10 × 8.9 μm^2 was irradiated for 0.2 seconds at each position and scanned over a region around the crack tip, covering a grid of 20 points along the x-axis and 72 points on the y-axis, with each point separated by a distance of 40 μm.

3 RESULTS AND DISCUSSION

The micro-beam scanned in the X-Y plane around the crack tip yielded a map of the 2D WAXD patterns over an area of approximately 3000 μm × 800 μm around the crack tip. This map provides a definite basis for characterizing the heterogeneous SIC (Figure 2) (Mai *et al.* 2024).

Occurrence of SIC at each point is revealed based on the diffractions from the crystallographic planes (201), (200), and (120) of NR. The figure displays the diffraction patterns along selected six lines along the X-axis. The crystal diffractions are observed near the crack profile, while the positions significantly depart from the crack contour exhibit no crystal diffractions, thereby assigned as the totally amorphous state. We evaluate a boundary between SIC and amorphous regions by assuming the positions with a crystallinity of less than 1% to be amorphous. The SIC region is almost symmetric, and extends 160 ± 20 μm along the Y = 0 line, and about 1600 μm in the Y-direction apart from the crack tip when the applied tensile true strain in the Y-direction ($\varepsilon_{Y,macro}$) is 0.62.

Both the crystal diffraction intensities and the crystal *c*-axis direction significantly vary depending on their positions. These distributions in crystal characteristics stem from the considerably heterogeneous strain field which induces the crystallization. The micro-scale DIC analysis characterizes the heterogeneous strain field: By the crack-tip opening, positions near the Y = 0 line undergo stretching along the Y-axis with small contraction along the X-axis, without any body rotation.

In contrast, positions significantly distant from the Y = 0 line (where |Y| ≫ 0) experience more complicated displacements including body rotation and shear in the XY plane. Crystallization is expected to be predominantly driven by the maximum tensile component in the total displacement by the crack-tip opening. DIC analysis can evaluate the principal strain tensor with no diagonal component at each position. As positions depart from the Y = 0 line, the principal strain axis has a noticeable tilt relative to the Y-axis (Figure 3a), while it is parallel to the Y-axis near the Y = 0 line. A close correlation

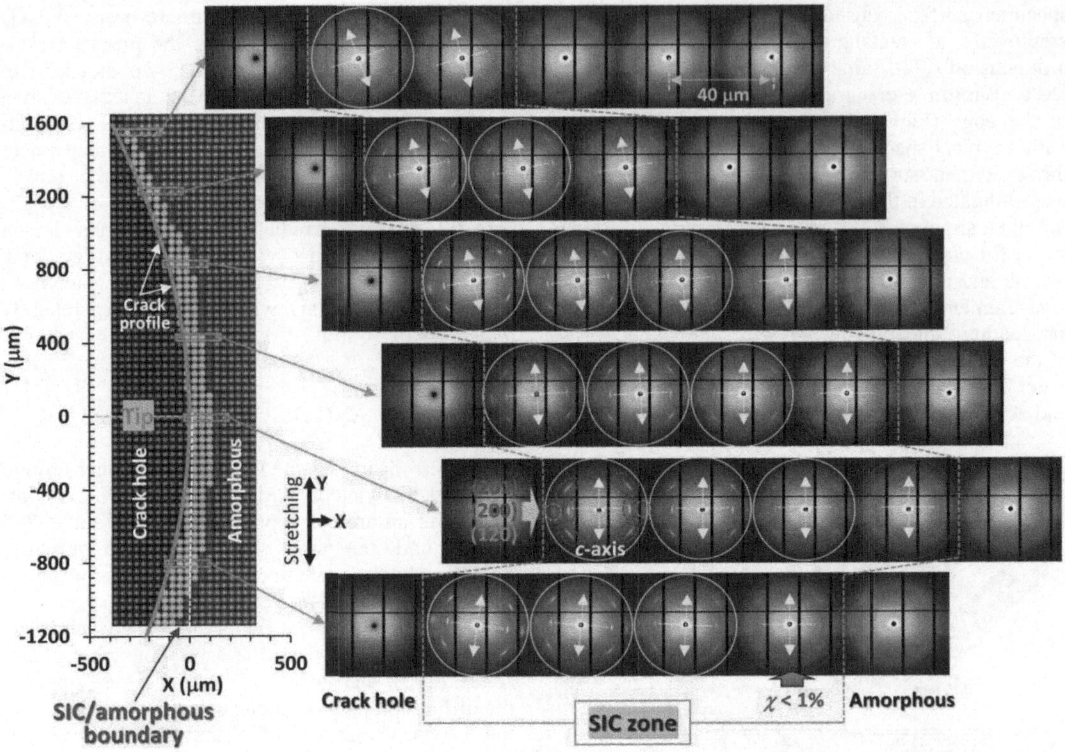

Figure 2. 2D mapping of WAXD patterns around the crack tip. Each pattern is separated by 40 μm. The SIC zone, highlighted in yellow, is defined by a boundary with <1% crystallinity, marked by red circle patterns. The orientation of the crystallites (c-axis) is shown by yellow arrow at positions along seven specific lines along the X-axis.

between the tilt angles of the principal strain axis (ϕ_p) and the crystal *c*-axis (ϕ_c) concerning the Y-axis is shown in Figure 3b. The relationship is approximated by a linear correlation with a proportional factor of about 0.77.

This indicates that finite tilting of the crystal *c*-axis results from the tilting of the principal strain axis, reflecting finite 2D shear deformation occurring at the corresponding position. This observation indicates the crucial role of shear deformation in the orientation of SIC microcrystallites. The slope slightly less than unity is attributed to the complex deformation composed of uniaxial and shear strains.

To correlate the local strain with local crystallinity, we utilize the major and minor components of the principal strain tensor [ε_1 and ε_2, expressed in true (Hencky) strain, respectively] due to finite ϕ_p, instead of the YY- and XX-components of the strain tensor measured in laboratory coordinates (ε_{YY} and ε_{XX}, respectively) (Mai *et al.* 2024). Due to significant strain concentration, ε_1 exceeds 2 (or nominal strain of 6) near the crack tip, while the applied global true strain ($\varepsilon_{Y,macro}$) is only 0.62 (or nominal strain of 0.85) (Figure 4B). The strain ε_1 increases as the position approaches the crack tip, while ε_1 becomes similar to the $\varepsilon_{Y,macro}$ value at the positions far from the crack tip. In the peripheral area far from the crack tip where considerable body rotation occurs by crack-tip opening (Figure 3A), ε_1 significantly differs from ε_{YY} (Figures 4A and 4B).

The type of deformation is characterized by the strain biaxiality $\varepsilon_2/\varepsilon_1$. Specifically, the ratio $\varepsilon_2/\varepsilon_1$ in the SIC area lies within a narrow range, approximately -0.25 ± 0.1, as shown in Figure 4C. The corresponding deformations are intermediate between planar ($\varepsilon_2/\varepsilon_1 = 0$) and uniaxial stretching ($\varepsilon_2/\varepsilon_1 \approx -0.5$). Interestingly, the SIC/amorphous boundary is approximated by the iso-strain line of $\varepsilon_1 = 1.3$ (Figure 4D). This unexpected simplicity is attributed to the limited variation in deformation type within the SIC region. This SIC/amorphous boundary ($\varepsilon_1 = 1.3$) is close to the critical strains ($\varepsilon_1 \approx 1.4$) for SIC onset under both planar and uniaxial stretching obtained for the unnotched NR sheets. Chen et al. (Chen *et al.* 2019) also reported that the critical major strains for SIC onset remained almost constant between uniaxial and planar extension. These consistencies imply that the SIC onset strain around crack tips is comparable to those in unnotched samples under uniform strain, because of the relatively narrow range of local deformation types in the SIC area.

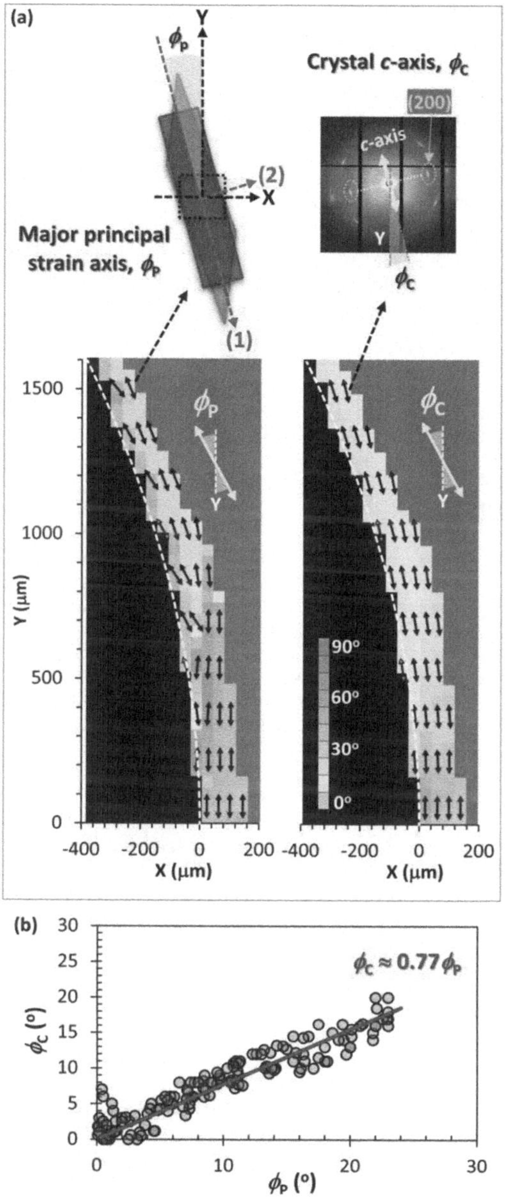

Figure 3. (a) Spatial distributions of the angles of major principal strain axis (ϕ_P) and crystal c-axis (ϕ_C) relative to the Y-axis near a crack tip. To avoid data overlap in the visualization, arrows indicate data points spaced at 160 μm intervals, equivalent to every fourth measuring point. (b) Relationship between ϕ_C and ϕ_P for all measured positions. The solid line represents the result by a linear regression with a slope of 0.77.

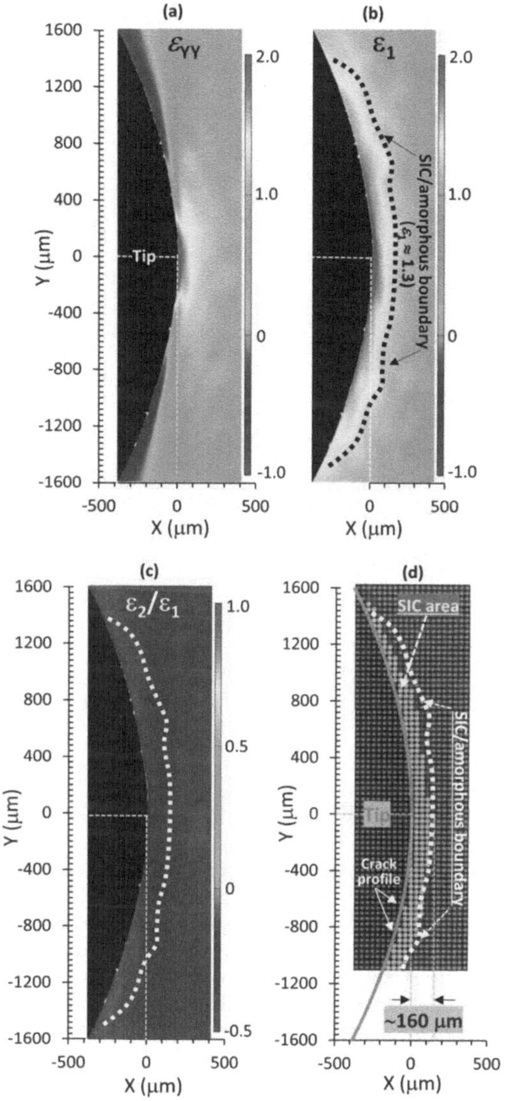

Figure 4. 2D distribution of local strains near a crack tip. (a) true strain ε_{YY} in laboratory coordinate, (b) major principal true strain ε_1, and (c) strain biaxiality of principal strains $\varepsilon_2/\varepsilon_1$ for a pre-notched specimen subjected to an applied macroscopic tensile true strain of 0.62. The dotted line represents the iso-ε_1 line of $\varepsilon_1 = 1.3$. (d) Comparison between the SIC zone (as shown in Figure 2), and the area of $\varepsilon_1 > 1.3$. The iso-ε_1 line of $\varepsilon_1 = 1.3$ almost aligns with the boundary between SIC and amorphous states.

The spatial distribution of degree of crystallinity (χ_c) has been obtained by analyzing 2D WAXD patterns at various positions. We investigated the dependence of χ_c on ε_1 at different positions on six specific lines including both the peripheral area of |Y| ≫ 0 and the area near the Y = 0 line (Figure 5A). Notably, the χ_c data for various positions fall on a single line (depicted by a violet line) when plotted against ε_1 at the corresponding positions (Figure 5B). This result demonstrates that the heterogeneous distribution of χ_c near the crack tip can be described by a function only dependent on ε_1, irrespective of ε_2. The χ_c-ε_1 data are fitted by a quartic function of ε_1 (X, Y) for χ_c at the respective position:

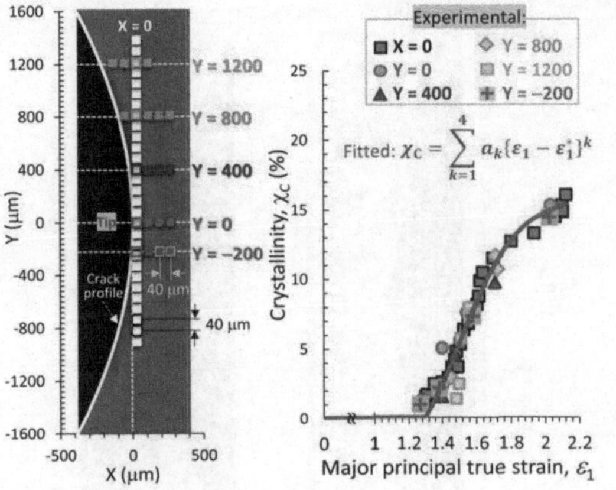

Figure 5. Correlations between local strains and local crystallinity in the SIC zone. The various data at different positions collapse into a fitted curve, depicted by the violet line. The fitted curve is given by a quartic function of ε_1, χ_c (X, Y) = $\sum a_k$ $[\varepsilon_1 - \varepsilon_1^*]^k$ for $\varepsilon_1 \geq \varepsilon_1^*$ with k = 1, 2, 3, 4, and χ_c (X,Y) = 0 for $\varepsilon_1 < \varepsilon_1^*$. The SIC onset strain ε_1^* is 1.3, and the coefficients are a_1 = 17.85, a_2 = 70.02, a_3 = −148.55, and a_4 = 78.62.

$$\chi_c(\mathrm{X,Y}) = \sum a_k[\varepsilon_1(\mathrm{X,Y}) - \varepsilon_1{}^*]^k \qquad (1)$$

with k = 1, 2, 3, 4 when $\varepsilon_1 > \varepsilon_1^*$, and χ_c (X, Y) = 0 at $\varepsilon_1 < \varepsilon_1^*$. The value of ε_1^* is the ε_1 value at the SIC/amorphous boundary and set at 1.3 here. It should be noted that no master function for χ_c like Equation (1) can be obtained when using ε_{YY} measured in the laboratory coordinate as a variable. This result indicates that χ_c is governed by the maximum tensile local strain derived in the principal axes coordinates, i.e., the local strain along the local crystal *c*-axis. Significantly, Equation (1) enables the computation of the 2D distribution of χ_c around the crack tip at any degree of crack tip opening, only using the corresponding ε_1 field data (Mai *et al.* 2024).

4 CONCLUSION

We reveal the correlation between the heterogeneous strain field and SIC near a crack tip and its peripheral zone in notched NR under constant applied stretching. In peripheral region apart from the crack tip, the local principal strain axis has finite tilt relative to the applied stretching direction due to finite shear strain caused by crack-tip opening. The *c*-axis of the resultant SIC crystal also shows tilting. The tilt angles for these two quantities exhibit a linear correlation, indicating that finite tilting of the crystal *c*-axis is attributed to shear strain caused by crack-tip opening. Local crystallinity (χ_c) and the initiation of SIC near the crack tip and its peripheral zones are governed by the maximum tensile true strain (ε_1) within the tensor of local principal strains, devoid of diagonal components. The boundary separating SIC and amorphous regions is approximated by an iso-ε_1 line, and the spatial distributions of χ_c can be accurately described using a function dependent only on ε_1. This simplicity is due to the limited variation in types of deformation (characterized by a narrow $\varepsilon_2/\varepsilon_1$ range of approximately −0.25 ± 0.1) within the SIC region, with corresponding distortions falling between planar and uniaxial stretching.

ACKNOWLEDGEMENT

This work was partly supported by JST, CREST (grant number JPMJCR2091), Japan.

REFERENCES

Brüning, K., Schneider, K., Roth, S. V., and Heinrich, G., 2013. Strain-induced crystallization around a crack tip in natural rubber under dynamic load. *Polymer (United Kingdom)*, 54 (22), 6200–6205.

Chen, X., Meng, L., Zhang, W., Ye, K., Xie, C., Wang, D., Chen, W., Nan, M., Wang, S., and Li, L., 2019. Frustrating Strain-Induced Crystallization of Natural Rubber with Biaxial Stretch. *ACS Applied Materials and Interfaces*, 11 (50), 47535–47544.

Demassieux, Q., Berghezan, D., and Creton, C., 2021. Microfocused Beam SAXS and WAXS Mapping at the Crack Tip and Fatigue Crack Propagation in Natural Rubber. *Advances in Polymer Science*, 286, 467–491.

Hirata, Y., Kondo, H., and Ozawa, Y., 2014. Natural rubber (NR) for the tyre industry. *In*: Shinzo Kohjiya; Yuko Ikeda, ed. *Chemistry, Manufacture and Applications of Natural Rubber*. Waltham: Elsevier, 325–352.

Huneau, B., 2011. Strain-induced crystallization of natural rubber: A review of X-Ray diffraction investigations. *Rubber Chemistry and Technology*, 84 (3), 425–452.

Mai, T., Yasui, T., Tanaka, R., Masunaga, H., Kabe, T., Tsunoda, K., Sakurai, S., and Urayama, K., 2024. Unraveling Non-Uniform Strain-Induced Crystallization Near a Crack Tip in Natural Rubber. *Advanced Science*.

Osumi, R., Yasui, T., Tanaka, R., Mai, T.-T., Takagi, H., Shimizu, N., Tsunoda, K., Sakurai, S., and Urayama, K., 2022. Impact of Strain-Induced Crystallization on Fast Crack Growth in Stretched cis -1,4-Polyisoprene Rubber. *ACS Macro Letters*, 11 (6), 747–752.

Persson, B.N.J., Albohr, O., Heinrich, G., and Ueba, H., 2005. Crack propagation in rubber-like materials. *Journal of Physics: Condensed Matter*, 17 (44), 1071–1142.

Rublon, P., Huneau, B., Verron, E., Saintier, N., Beurrot, S., Leygue, A., Mocuta, C., Thiaudière, D., and Berghezan, D., 2014. Multiaxial deformation and strain-induced crystallization around a fatigue crack in natural rubber. *Engineering Fracture Mechanics*, 123, 59–69.

Schreier, H., Orteu, J.-J., and Sutton, M.A., 2009. *Image Correlation for Shape, Motion and Deformation Measurements*. Springer. Boston, MA: Springer US.

Trabelsi, S., Albouy, P.A., and Rault, J., 2002. Stress-induced crystallization around a crack tip in natural rubber. *Macromolecules*, 35 (27), 10054–10061.

Advanced computational techniques for elastomers

Constitutive Models for Rubber XIII – Dal (Ed.)
© 2025 The Author(s), ISBN 978-1-032-85138-9

Influence of the production process on tire rolling characteristics

T. Berger, J. Meyer, S. Uzeda & M. Kaliske
Institute for Structural Analysis, Technische Universität Dresden, Dresden, Germany

ABSTRACT: During the tire production process, a green tire consisting of uncured rubber is placed inside a heating press. By applying a large inner pressure, it is formed into its final shape. Subsequently, the temperature is increased and rubber vulcanization starts, which significantly changes the material behavior. Finally, the tire is extracted from the mold and cools down, which leads to shrinkage of the reinforcement layers of the tire. During the production of the tire, it undergoes a complex deformation history, which leads to prestress in the rubber compounds and reinforcements. This process and its influence on the final tire are usually neglected in the tire design process. However, it could be shown that the production process has an influence on the final tire shape.

In a previous contribution, a finite element (FE) material formulation, which reliably describes the phase change from uncured to cured rubber, has been presented. It has been formulated in combination with the micro-sphere approach. An arbitrary Lagrangian-Eulerian (ALE) formulation is applied to take into account the severe distortions, which are present during the tire production process. In the presented framework, the results of the tire production process are directly transferred to a footprint analysis and a subsequent steady state rolling investigation. In this contribution, the visco-elastic material description of cured rubber is adopted to the mixed Lagrangian-Eulerian steady state rolling approach. Therefore, the same material description is used for the analysis of the production process and the tire in service. This is a further step towards a closed numerical representation of the whole tire lifetime, which could be used as a digital twin of the tire.

In this contribution, the influence of the tire production process is investigated with respect to the tire driving performance. Therefore, a numerical comparison of a rolling tire is carried out with respect to rolling resistance and its temperature increase. By neglecting the tire production process, prestress in the rubber and the geometrical change due to the in-molding step are not taken into account, which leads to a different driving behavior of the final tire.

1 INTRODUCTION

A tire is made out of various rubber compounds and reinforcement materials that are shaped in a sophisticated geometry to withstand various loading conditions. Tire manufacturers design their products to have low rolling resistance, high driving safety at minimum wear. However, these targets contradict each other and are influenced further by additional requirements as for example low noise and wet driving behavior. Due to increasing knowledge on material modeling and advances in computational resources, the tire design process is more and more aided by numerical methods such as the finite element method (FEM). Therefore, different designs and combinations of various rubber compounds can be evaluated without building a tire prototype. In most cases, small changes in tire geometry or initial distribution of rubber compounds lead to significant changes in tire performances. Tire designers are usually focused on the final cured tire geometry. However, the production process of the tire leads to changes in the rubber compound distribution in the tire and, consequently, in the mechanical tire behavior. In Berger, Kim, & Kaliske (2023), the influence of the production process on the footprint of the tire is investigated. Due to the description of the production process by the FEM, the optimal tread geometry of the green tire is identified to match the interaction with the mold. Therefore, the distribution of the contact pressure in the footprint could be improved solely by changing the green tire.

In this study, the influence of the production process on the rolling tire performance is investigated. Therefore, the arbitrary Lagrangian Eulerian (ALE) approach is used in the production process to increase the quality of the discretization and to remove mesh distortions that stem from large deformations of the uncured rubber during in-molding (Berger & Kaliske 2022). Rubber in its uncured and cured phase and the transition is described by a mixed material model in combination

DOI: 10.1201/9781003516880-29

with the micro-sphere approach by MIEHE (Miehe, Göktepe, & Lulei 2004). For a thermodynamically consistent formulation and a permanent shape change during curing, plastic strain is introduced. The evolution law of this strain depends on the state of cure and is derived from the conservation of energy. Due to the proposed material description by the micro-sphere approach, the thermo-mechanical simulation framework of a steady state rolling tire is modified in this contribution. A numerical investigation of the tire driving performance with and without considering the influence of the production process is presented as well. Due to the application of the ALE framework and the modification of the steady state rolling formulation, the results of the production process are applied directly to subsequent simulations. This leads the way to a complete numerical description of the tire from its production to its end of service. If the numerical tire description is supported by sensor data, a digital shadow of a real tire is obtained. Finally, a digital twin of the tire is reached, if the real tire is adjusted by the results of the numerical simulations.

2 THERMO-MECHANICALLY COUPLED STEADY STATE ROLLING

For the thermo-mechanical investigation of the tire, the sequentially coupled approach by Behnke & Kaliske (2015) is adopted to the material description based on the micro-sphere formulation.

2.1 *Sequential simulation approach*

A brief overview on the thermo-mechanically coupled simulation framework is given in this section. Due to the cyclic deformation in the tire and the viscoelastic behavior of cured rubber, its temperature increases over time. The different rubber compounds and the heat exchange with the environment lead to non-uniform temperature distribution in the tire cross-section. The rate of the evolution of the temperature is assumed to be several orders slower than one rotation of the tire. Therefore, the mechanical and thermal field are computed sequentially. It is assumed that the temperature is constant in circumferential direction. In the mechanical analysis, the temperature of the tire remains constant and the deformation is computed based on the steady state rolling ALE formulation by Nackenhorst (2004). According to the viscoelastic material description presented in the next section, the dissipation energy is derived that serves as the heat influx in the thermal analysis. The temperature distribution in the tire is obtained by a transient analysis, which takes into account the conductivity and specific heat capacity of the rubber and the heat exchange with the environment. The anisotropic conductivity of the rebars is taken into account via homogenization of the reinforcement layers, see (Berger & Kaliske 2017). Subsequently, the updated temperature distribution is transferred to the mechanical analysis. Due to the increased temperature, the mechanical properties of rubber change, which leads to a different deformation state and a decreased energy dissipation rate. These two analyses are carried out subsequently until a certain time step is reached or a steady state temperature distribution is obtained. Only in the thermal simulation, an actual time increment is carried out. Therefore, one mechanical simulation represents this certain time step. For an efficient implementation of the approach, this time step is adaptively controlled by the change of temperature over time. Therefore, at the beginning of the analysis, when the temperature evolves quickly from room temperature, a small time increment is chosen. While the tire is rolling and the temperature distribution approaches its steady state, the time increment increases. With the presented thermo-mechanically coupled simulation framework, the evolution of the temperature over time can be efficiently predicted at steady state rolling and under certain driving maneuvers. An option is included to automatically estimate the free rolling velocity of the rolling tire based on the reaction moment at the rim. By a linear approximation between rotational velocity Ω and reaction moment M_y, the free rolling velocity is estimated as

$$\Omega^{t_n} = -\frac{M_y - m\Omega^{t_{n-1}}}{m}, \tag{1}$$

with

$$m = \frac{M_y^{t_{n-2}} - M_y^{t_{n-1}}}{\Omega^{t_{n-2}} - \Omega^{t_{n-1}}}. \tag{2}$$

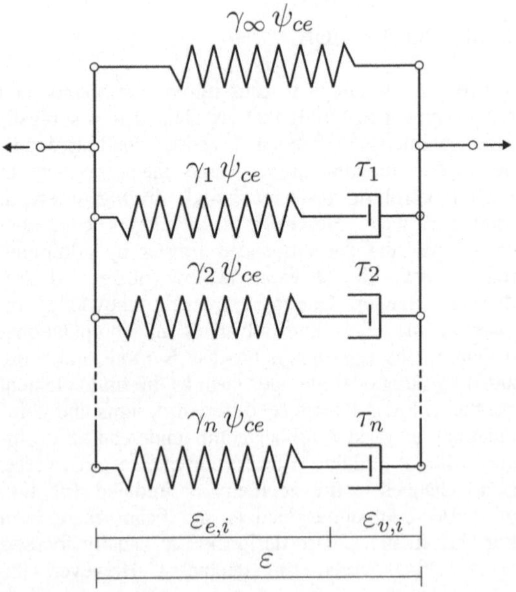

Figure 1. Rheology of the generalized MAXWELL model used in the micro-sphere approach.

2.2 *Viscoelastic material description*

After the production of the tire, it is assumed that all rubber compounds are fully cured. Therefore, only the cured rubber branch of the mixed material model is activated in the following description. Assuming an incompressible behavior, the deformation and the mechanical description are split into a volumetric and an isochoric part. The volumetric part is described by a penalty function to ensure a constant volume, while the complex isochoric behavior is described by a generalized MAXWELL element, which is the standard approach to cover a wide range of frequencies. However, due to the application of the micro-sphere model, it is defined with respect to the micromechanical stretches λ in each direction of the discretized sphere, see Figure 1. The free energy function reads

$$\psi_c = \gamma_\infty \psi_{ce}(\lambda) + \sum_{i=1}^{N_{\text{br}}} \gamma_i \psi_{ce}(\lambda_{e,i}), \tag{3}$$

for N_{br} MAXWELL branches. The same free energy function for each spring is used,

$$\psi_{ce}(\lambda) = \frac{\mu_c}{\delta_c}\left[\lambda^{\delta_c} - 2 + \lambda^{-\delta_c}\right]. \tag{4}$$

For the dashpots, the stretch is split into an elastic and a viscous part,

$$\lambda = \lambda_{e,i}\lambda_{v,i}. \tag{5}$$

The evolution law of the viscous stretch is derived from the rate-dependent back stress,

$$\beta_{v,i} = -\frac{\partial \gamma_i \psi_{ce}}{\partial \varepsilon_{v,i}}, \tag{6}$$

using the logarithmic strain $\varepsilon = \ln\lambda$. From (Dal, Zopf, & Kaliske 2018), the nonlinear flow rule for the viscous strain is defined as

$$\dot{\gamma}_{v,i}(\beta_{v,i}) = \frac{1}{\eta_{v,i}}\left|\beta_{v,i}\right|^{m_{v,i}}. \tag{7}$$

By using the nonlinear flow rule, the evolution law of the viscous strain is defined as

$$\dot{\varepsilon}_{v,i} = \dot{\gamma}_{v,i}(\beta_{v,i})\beta_{v,i}. \tag{8}$$

Due to the steady state rolling assumption with the ALE framework, the deformation history $\varepsilon_{v,i}^{t_n}$ is taken from the previous integration point on the so-called streamline. The internal time step is derived from the distance to the predecessor integration point Δd_i and the rotational velocity Ω,

$$\Delta t_i = \frac{\Delta d_{l,i}}{\sum_j^n \Delta d_{l,j}} \frac{2\pi}{\Omega}. \tag{9}$$

Due to a NEWTON-RAPHSON iteration, the logarithmic viscous strain is obtained using the residual equation,

$$r_{v,i} = \varepsilon_{v,i} - \varepsilon_{v,i}^{t_n} - \dot{\gamma}_{v,i}(\beta_{v,i})\beta_{v,i} = 0. \tag{10}$$

For more information on the derivation of the stresses and the algorithmic tangent of the viscous part, the reader is referred to (Berger & Kaliske 2020).

During rolling, the rubber exhibits a cyclic deformation, which leads to a temperature increase due to the viscoelastic mechanical behavior. Assuming a constant temperature field in the mechanical analysis, the dissipated energy is derived from the CLAUSIUS-PLANCK inequality (Holzapfel 2000),

$$\mathcal{D}_{\text{int}} = \tau : \mathbf{d} - \dot{\Psi}. \tag{11}$$

Due to the micro-sphere model, Eq. (11) has to be fulfilled at each micromechanical direction,

$$\mathcal{D}_{\text{int}} = \tau : \dot{\lambda} - \dot{\psi} \geq 0. \tag{12}$$

By applying the multiplication rule to Eq. (12),

$$\tau : \dot{\lambda} - \frac{\partial \psi}{\partial \lambda}\dot{\lambda} - \frac{\partial \psi}{\partial \varepsilon_{v,i}}\dot{\varepsilon}_{v,i} \geq 0, \tag{13}$$

the stress response is obtained as the derivative of the free energy function with respect to the stretch. Therefore, the energy dissipation rate of the viscoelastic material reads

$$\mathcal{D}_{\text{int},v} = \beta_{v,i}\dot{\gamma}_{v,i}\beta_{v,i}. \tag{14}$$

Due to the time-independent steady state rolling approach, the dissipation rate is obtained by Eq.

(14). The temperature is assumed to be constant in circumferential direction and, therefore, the dissipation rate is averaged over the circumferential streamline with the respective distance,

$$\overline{w}_{\text{diss}} = \frac{1}{\sum_{i}^{n} \Delta d_{l,i}} \sum_{j}^{N_{\text{Stream}}} w_{\text{diss},j} \, \Delta d_{l,j}. \tag{15}$$

The averaged dissipation rate of one streamline computed by Eq. (15) is used as the internal heat source in the two-dimensional cross-section of the thermal analysis. Additionally, this can be used to compute the total loss power,

$$P_{\text{RR}} = \int_{\varphi=0}^{2\pi} \int_{A} \overline{w}_{\text{diss}} \, r \, \mathrm{d}A \, \mathrm{d}\varphi. \tag{16}$$

By a numerical integration over the cross-section and an analytical integration in circumferential direction, the total loss power is obtained and used to compute the rolling resistance force

$$F_{\text{RR}} = \frac{P_{\text{RR}}}{v}, \tag{17}$$

with the translational velocity v. The rolling resistance obtained by Eq. (17) yields the force stemming from the viscoelastic material during rolling. This can be used to identify rubber compounds that cause the largest rolling resistance. Additionally, it can be used to distinguish between other sources of rolling resistance, such as friction that is present in non-free rolling conditions.

The identification of the time-independent and time-dependent parameters are carried out separately. From so-called multistage tests, where the strain is applied and held for a period of time, the long-term response of the rubber is captured. Based on these experiments, the stiffness μ_c and the power term δ_c are identified via a gradient based optimization to describe the nonlinear mechanical behavior at large strains, (Berger, Kim, & Kaliske 2023). Via dynamical mechanical analysis (DMA), the time-dependent behavior of the rubber is experimentally captured. The DMA tests are carried out at small strains and the linear PRONY series is applied to identify the scaling factors and relaxations times of the generalized MAXWELL element. Therefore, the storage and loss moduli are identified via,

$$E' = \gamma_\infty E_0 + \sum_{i=1}^{N_{\text{br}}} \gamma_{v,i} E_0 \frac{\omega^2 \tau^2}{1 + \omega^2 \tau^2}, \tag{18}$$

and

$$E'' = \gamma_\infty E_0 + \sum_{i=1}^{N_{\text{br}}} \gamma_{v,i} E_0 \frac{\omega \tau}{1 + \omega^2 \tau^2}. \tag{19}$$

However, in the micro-sphere model, the viscosity is used instead of the relaxation times. According to (Behnke 2015), the viscosity is computed by the initial stiffness and the relaxation times,

$$\eta_{v,i} = \gamma_{v,i} E_{\text{ini}} \tau_i. \tag{20}$$

For the proposed free energy function, the initial stiffness is obtained as the consistent tangent at zero strain,

$$E_{\text{ini}} = \mu_c \left[(\delta_c - 1) \, 1^{\delta_c - 2} - (-\delta_c - 1) \, 1^{-\delta_c - 2} \right]. \tag{21}$$

Therefore, the well established parameter identification can be applied to identify the relaxation times and scaling factors of the PRONY series and transfer them to the micro-sphere approach. In Figure 2, the macroscopic stress-strain relation of the micro-sphere model is compared to the multistage relaxation test. The long-term response and the hysteresis of the mechanical behavior are well captured.

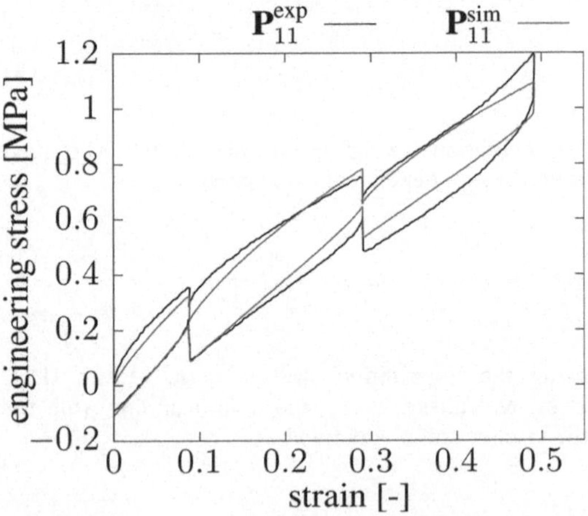

Figure 2. Stress response of the micro-sphere model compared to the multistage experiment.

3 NUMERICAL INVESTIGATIONS

In the following investigation, the impact of different outcomes of the tire production process on the final tire characteristics are studied.

3.1 *Influence of the green tire geometry*

In a first analysis, two different green tire geometries are studied. One with a non-optimal location of the circumferential grooves and one with an optimized geometry. In Figure 3, the location of the rubber compounds after the production is shown. With the initial geometry, the pre-cut edges for the circumferential grooves are not aligning with the mold. Therefore, the subtread compound is not equally distributed in the tire cross-section, which leads to a non-uniform contact pressure distribution in the footprint of the tire. For more information, the reader is referred to (Berger, Kim, & Kaliske 2023).

The position of the rubber compounds is not as intended and the tire has worse driving characteristics, which can be seen in the evolution of the temperature and the rolling resistance force. Both tires are rolling at a low speed of $v = 10$ kph and the temperature evolution is recorded for 2000 s in Figure 4. At the beginning of the analysis, the temperature increase is similar and reaches almost a steady state condition at 2000 s. At this point, the heat source from the dissipated energy is balanced with the heat exchange with the environment. However, due to the slow rolling velocity, the average temperature increase of $2K$ is almost negligible. However, even at these small changes, an influence of the green tire geometry is visible. The modified geometry has a better contact pressure distribution and, therefore, less dissipation from the rubber material. This leads to a lower temperature increase and rolling resistance force. It is expected that this effect is amplified at higher velocities, which will be investigated in subsequent studies. The temperature difference is clearly visible in the tread compound of the two tires, see Figure 5. Due to the non-uniform distribution of the subtread compound, the dissipation energy is higher in the initial geometry and, therefore, the temperature increase is more severe. Even though the absolute difference is low at 0.5 K, this is an approximately 30% increase in the tread rubber, which leads to larger differences at higher velocities.

A small reduction of the rolling resistance is visible during rolling due to the softer material at higher temperatures. However, the sharp decrease of the rolling resistance in the first three steps occurs due to the poor initial guess of the rotational velocity Ω. After three mechanical simulations, the free rolling velocity Ω_{free} is obtained and the influence of the larger deformations resulting from an inappropriate free rolling velocity vanishes.

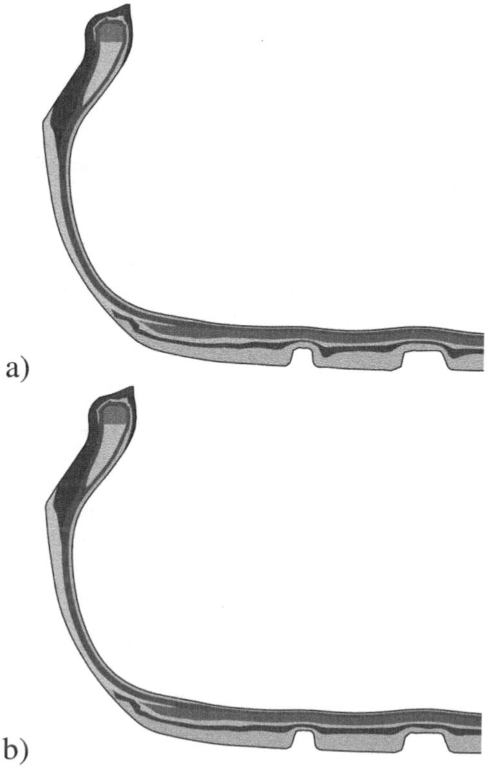

Figure 3. Location of the rubber compounds in the cured tire using a) the initial green tire geometry and b) the modified geometry.

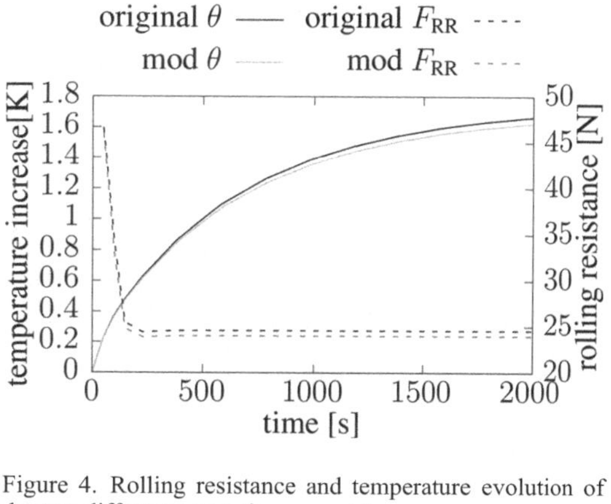

Figure 4. Rolling resistance and temperature evolution of the two different green tire geometries.

3.2 *Influence of the prestress*

During the production, the rubber compounds are cured in a deformed position, which leads to a new equilibrium configuration. This in combination with the reinforcements leads to a stress distribution in the tire from the production process. In most cases, these prestresses are neglected by tire designers. To show the influence of these prestresses, the tire geometry after inflation is studied and depicted in Table 1. It could be seen that the radial deformation of the tire does not change if the prestress of the rubber is included. However, taking into account the prestress of the rebars, the radial and lateral deformation of the tire changes. It is clearly visible that the prestress of the rebars influence the deformation

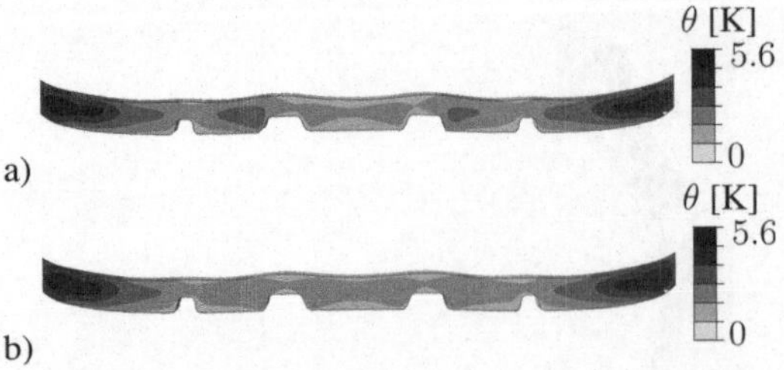

Figure 5. Temperature distribution in the tread at free rolling for a) the initial green tire geometry and b) the modified geometry.

more compared to the rubber prestress. In (Chen 2002), the influence of the shrinkage behavior of the nylon cords in the tire is investigated. To reduce the lateral deformation, an additional post-cure inflation (PCI) step is added to the production process. Therefore, the tire is immediately after curing put on a rim and inflated. Due to the applied inner pressure, the nylon cords cannot shrink, which leads to less prestress in the rebar. This leads to less lateral deformation at the inflated tire. In Kim, Berger, Paek, Zreid, & Kaliske (2022) and Kim, Berger, Piao, Zreid, & Kaliske (2022), a numerical description of the thermal shrinkage behavior of nylon cords is proposed and applied to numerical simulations.

Table 1. Influence of the prestress on the deformation of the tire.

prestress	no	rubber	rebar	all
radial [mm]	359.93	360.31	362.85	362.86
lateral [mm]	278.51	278.27	280.82	280.82

3.3 *Influence of the rebar angle*

During the tire production, the green tire undergoes various deformation modes. During the in-molding step, the tire is laterally compressed and then exposed to a large inner pressure to obtain its final shape. The rubber that shows an isotropic behavior is not significantly influenced by the complex deformation. However, the reinforcements, such as steel belts and nylon carcass are usually made of layers of bars or fibers. During the deformation, the angle and spacing of the fibers in the layer change, which is normally not included in the tire design process. The influence of the changed rebar geometry is investigated at the static loading of the tire. By assuming the initial rebar angle and spacing, the force is 200 N larger than using the rebar information obtained from the tire production simulation. The angle of the belt rebar layers that are placed diagonal in the tire increases around 2°. This leads to lower stress in the reinforcements and the stiffness reduces. This can influence the driving behavior as well and it is assumed that these changes of the rebar angle significantly influences the cornering characteristics of the tire.

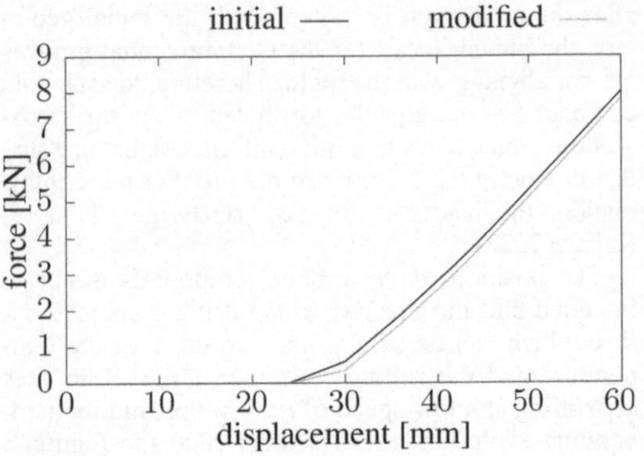

Figure 6. Force over displacement of the tire with different definitions of the rebar angle.

4 CONCLUSIONS

In this study, certain effects of the tire production process on the tire are studied. Therefore, a thermomechanically coupled steady state rolling framework with a sequential simulation flowchart is presented. Two different arbitrary LAGRANGIAN EULERIAN (ALE) frameworks are introduced. The first ALE is applied to the axisymmetric tire production simulation to overcome large mesh distortions and to obtain a tire discretization, which can be used in a subsequent tire rolling simulation. The ALE framework for steady state rolling is adapted to directly use the tire discretization obtained from the production simulation in any subsequent analysis. Therefore, the viscoelastic material description coupled to the micro-sphere approach is adapted to this ALE framework. Especially, the computation of the dissipated energy is presented as input for the thermal analysis to predict the temperature evolution during rolling. In a free rolling simulation of the tire, the forces and moments of the tire are investigated by taking into account certain results of the tire production. It could be seen that an insufficient tire production process has an influence on the rolling behavior. If the outcome of the production does not match with the intended geometry by the tire designers, the rolling resistance and the temperature evolution are changed which could lead to a lower lifetime of the tire. Additionally, the influence of the prestress in the rubber from the production process is investigated. The inflated tire dimensions change if the prestresses are considered. Above all, the prestress in the rebar together with its spacing and angle of the belt layer have the largest influence on the tire dimensions.

In subsequent studies, the influence of the change of the rebar angle due to the deformations in the production process should be investigated for cornering maneuvers. It is expected that the changed rebar angle has an influence on the cornering behavior, as the belt reinforcements have a significant influence on the cornering behavior. As concluded, the rebar prestress has an influence on the tire dimension. Therefore, the thermal shrinkage behavior of nylon cords with and without a post-cure inflation step should be investigated as well. Also the possible scenario and the consequence of undercured rubber compounds in the tire could be studied. The combination of a tire production simulation with a steady state rolling framework leads a way to a complete description of the tire lifetime. Therefore, possible influences on the tire driving performance can be related to the production process. By including sensors in the tire and coupling the information gathered from this framework with the real tire, the vision of a digital twin can be achieved.

REFERENCES

Behnke, R. (2015). *Thermo-mechanical Modeling and Durability Analysis of Elastomeric Components under Dynamic Loading*. Ph. D. thesis, Technische Universität Dresden, Institut für Statik und Dynamik der Tragwerke.

Behnke, R. & M. Kaliske (2015). Thermo-mechanically coupled investigation of steady state rolling tires by numerical simulation and experiment. *International Journal of Non-Linear Mechanics 68*, 101–131.

Berger, T. & M. Kaliske (2017). Advanced rebar formulation within a thermo-mechanically coupled tire simulation. *Proceedings in Applied Mathematics and Mechanics 17*, 289–290.

Berger, T. & M. Kaliske (2020). A Thermo-mechanical Material Model for Rubber Curing and Tire Manufacturing Simulation. *Computational Mechanics 66*, 513–535.

Berger, T. & M. Kaliske (2022). An arbitrary lagrangian eulerian formulation for tire production simulation. *Finite Elements in Analysis and Design 204*, 103742.

Berger, T., S.-h. Kim, & M. Kaliske (2023). A Microsphere-Based Rubber Curing Model for Tire Production Simulation. *Tire Science and Technology 51*, 82–113.

Chen, B. (2002). Influences of thermo-mechanical properties of polymeric cords on tires with and without post-cure inflation. *Tire Science and Technology 30*, 156–179.

Dal, H., C. Zopf, & M. Kaliske (2018). Micro-sphere based viscoplastic constitutive model for uncured green rubber. *International Journal of Solids and Structures 133*, 201–217.

Holzapfel, G. A. (2000). *Nonlinear Solid Mechanics: A Continuum Approach for Engineering*. Chichester: Wiley.

Kim, S., T. Berger, J. Paek, I. Zreid, & M. Kaliske (2022). Numerical Modeling and Experimental Investigation of Thermal Shrinkage in Polymeric Cords. I: Constitutive Modeling. *Tire Science and Technology 51*, 39–62.

Kim, S., T. Berger, M. Piao, I. Zreid, & M. Kaliske (2022). Numerical Modeling and Experimental Investigation of Thermal Shrinkage in Polymeric Cords. II: Numerical Implementation. *Tire Science and Technology 51*, 63–80.

Miehe, C., S. Göktepe, & F. Lulei (2004). A micro-macro approach to rubber-like materials, Part I: The non-affine micro-sphere model of rubber elasticity. *Journal of the Mechanics and Physics of Solids 52*, 2617–2660.

Nackenhorst, U. (2004). The ALE-formulation of bodies in rolling contact: Theoretical foundations and finite element approach. *Computer Methods in Applied Mechanics and Engineering 193*, 4299–4322.

Constitutive Models for Rubber XIII – Dal (Ed.)
© 2025 The Author(s), ISBN 978-1-032-85138-9

A hybrid meta-FEM approach for numerical computation of wear parameters

S. El Masri, F. Hartung, T. Berger & M. Kaliske
Institute for Structural Analysis, Technische Universität Dresden, Dresden, Germany

P. Dubey, V. Bansal, S. Goyal & S. Nair
CEAT Tyre Ltd., Halol, India

ABSTRACT: Due to the interlocking of asperities, rubber friction is a dissipative process that converts mechanical energy into heat and produces material wear. Furthermore, wear is the loss of material under the frictional sliding of two bodies against each other. The wear parameters calibrated by using Linear Friction Tester (LFT) experiments lead to an overestimation of wear when used in Laboratory Abrasion & Skid Tester 100 (LAT100) simulations. Furthermore, calibration of wear parameters by LAT100 experiment uses assumptions regarding the contact area and slip velocity which cannot be measured directly. Therefore, numerical calibration can be one way to obtain reliable wear parameters. To numerically calibrate the wear parameters, a hybrid meta-finite element method (FEM) approach is developed. For the FEM representation of the LAT100 experiment, a steady state rolling framework is adopted and an arbitrary LAGRANGIAN-EULERIAN(ALE) adaptive meshing procedure is used for wear calculation. Due to the mesh dependency of the ALE adaptive meshing, a failure surface algorithm is developed. The workflow is composed of a loop of three main steps. First, an initial set of wear parameters is identified by the minimization algorithm. Next, the failure surface algorithm decides if these wear parameters are passed to the FEM or the meta-model. Finally, the FEM or the meta-model predicts the wear mass, and a new iteration is started from step one. The loop ends once optimal wear parameters are found. Both approaches are able to predict the amount of wear of the LAT100 with a similar accuracy.

1 INTRODUCTION

Wear and tear of tires results in 9768 tons of yearly pollution which ends up in the environment (Kole, Löhr, Van Belleghem, & Ragas 2017). Heavier wear particles pollute land and water, while lighter particles remain airborne. The rise of electrical vehicles in the future will also contribute to an increase in tire wear and pollution. Electrical vehicles are on average heavier than their combustion engine counterpart and have much larger torsional moment which leads to increasing wear. The weight increase is attributed to the heavy weight of the batteries. Therefore, it is clear that proper modeling of wear will become much more important hereon after. The main aim of this study is the development of numerical tools to enhance tire and wear modeling for future tire design. Laboratory Abrasion & Skid Tester (LAT100) and Linear Friction Tester (LFT) experiments are conducted. The use of wear parameters calibrated in the LFT experiments on an LAT100 FEM model yields high and inaccurate results. Such inaccuracies are assumed to be caused by the larger sliding distance of the sample in an LFT experiment compared to an LAT100 one. Hence, a numerical calibration on the LAT100 model can be one way to determine the wear parameters. Numerical calibration indicated the use of a numerical model (e.g. FEM, meta model) as the target function for optimization. For LAT100, the advantage of this method is the accurate calculation of variables that cannot be extracted experimentally such as the contact surface area and the slip velocities.

An ALE adaptive meshing approach is used for wear simulation. This approach allows for an independent flow of material within the mesh. Nonetheless, the node to element connectivity remains unchanged. Due to this constraint, wear amounts are limited by the element volume at the contact zone. If that volume is exceeded, failure of the simulation occurs. To extend the capabilities of this wear modeling approach beyond the failure limits, the development of meta models is needed. Although more advanced meta models could be used for the prediction of wear such as neural networks (NNs) or support vector regressions (SVRs) a more simplified approach is adopted in here. For this numerical calibration, using a series of surface functions for

DOI: 10.1201/9781003516880-30

different LAT100 test conditions as meta model is enough to achieve the desired objectives of the study. A coupling is performed between the meta and finite element modeling (FEM) under a hybrid approach. Finally, a numerical calibration is performed using an optimization algorithm and the hybrid model for the identification of wear parameters.

2 EXPERIMENTAL SETUP

2.1 *Laboratory Abrasion & Skid Tester (LAT100)*

The Laboratory Abrasion & Skid Tester 100 (LAT100) is a device to measure the abrasion resistance of rubber compounds using a rotating rubber wheel. In a typical LAT100 experiment, a sample is subjected to a pre-set load, slip angle, and speed as it runs on a dry disc for a pre-determined distance. The weight of the sample is measured before and afterwards. In this study, the influence of the rotational speed, slip angle, and driving distance of the wheel is investigated. The same compound and abrasive surface (LAT disc grit 60) are used across all tests. The energy dissipated per sliding distance W is calculated as

$$W = F \sin \alpha, \quad (1)$$

where F is the side force and, α is the slip angle (Heinz & Grosch 2007) Moreover, the wheel is in a free rolling condition in its tangential direction since the wheel rotates freely with the motion of the grinding disk. Therefore, the projection of the side force on the tangential direction is the only energetic contribution in Equation (1). The LAT100 experiments are conducted by CEAT Tyre Ltd., India. A total of 10 different testing conditions are performed, which are presented in Table 1. The overall representation of the LAT100 experiment is shown in Figure 1 while the wear mass loss at different test conditions is depicted in Figure 2.

Table 1. LAT100 test conditions.

Test condition	Speed [km/h]	Slip angle [◦]	Distance [m]
C1	2.5	5.5	500
C2	2.5	5.5	1400
C3	2.5	9	600
C4	2.5	13	200
C5	12	5.5	1500
C6	12	9	600
C7	12	16	140
C8	25	5.5	2500
C9	25	9	600
C10	25	16	250

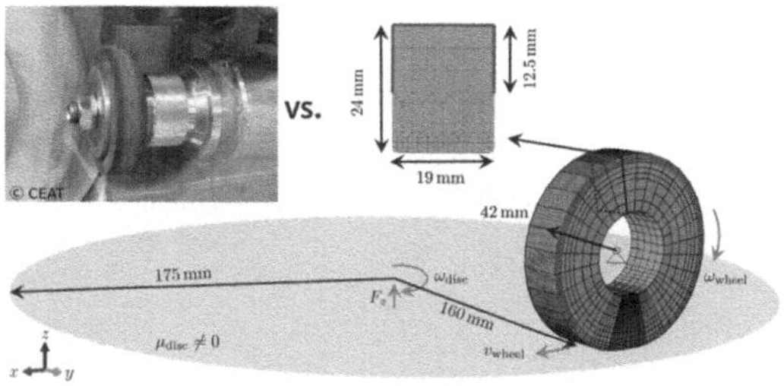

Figure 1. LAT100 grinding disk and sample geometry.

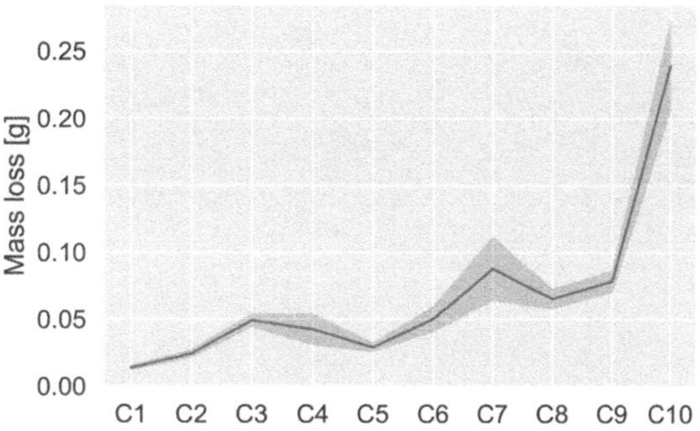

Figure 2. LAT100 wear mass results for different test conditions (filling around lines represents one standard deviation Â± the average - experimental results provided by CEAT).

2.2 *Linear Friction Tester (LFT)*

LFT tests are conducted to identify the friction behavior of rubber compounds at different pressures and speeds. The tests are performed using the High-speed Linear Tester (HiLiTe) by the Institute of Dynamics and Vibration Research (IDS) at Leibniz Universit¨at Hannover, Germany. Using a pre-set sliding velocity and normal force, a rectangular sample block is slided on a track. The block is first accelerated on an aluminum sheet until the required speed is reached. During sliding on the abrasive surface, the normal force and the tangential friction force are monitored. The coefficient of friction μ is calculated by dividing the tangential force by the normal force at the steady state condition. The mass of the block is measured before and after the test for wear calculation. The results of μ and mass loss from the HiLiTe are shown in Figure 3 and 4. The friction coefficient μ is characterized by a fast increase at speeds between 1mm/s and 110mm/s and a minor decrease afterward with an overall higher value for lower pressures. On the other hand, the mass loss increases with increasing pressure and speed.

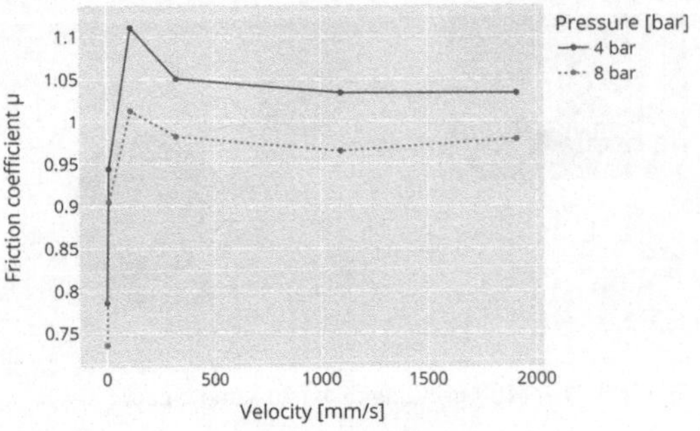

Figure 3. LFT friction coefficient at different vertical loads and speeds.

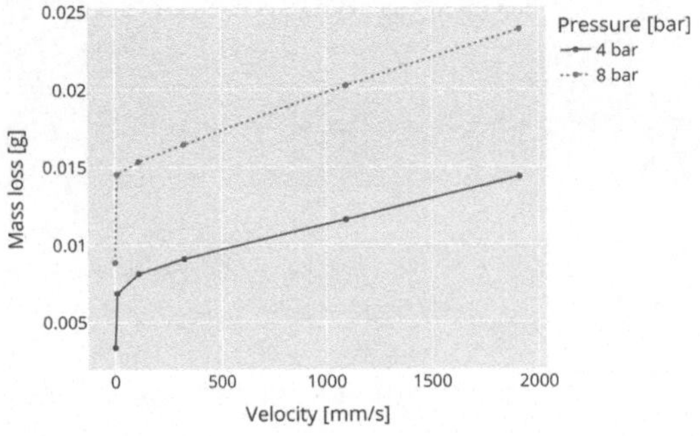

Figure 4. LFT wear mass at different loads and speeds.

3 WEAR MODELLING

3.1 *LAT100 FEM discretization*

The grinding disk surface is modeled as a rigid analytical surface. The GROSCHwheel is modeled with 8-node hybrid linear brick elements with constant pressure and an augmented LAGRANGIAN node to surface contact formulation. Transport velocity is applied to the wheel and the grinding disk to simulate steady state rolling. For wear simulations, the ALE adaptive meshing (also referred to as adaptive meshing) is used (ABAQUS 2014). Wear is simulated by the movement of the mesh independently of the material, which means that the material points are not fixed to the mesh. The topology of the mesh and node to element connectivity remains unchanged. This adaptive meshing faces two additional challenges. First, the algorithm is unable to handle more wear than the overall volume of contact elements. Elements become too small and element volume becomes negative causing the simulation to abort. On the other hand, inaccuracies advecting the deformation gradient are created (ABAQUS 2014) when using hyperelastic materials due to their high nonlinearity. Therefore, a linear elastic constitutive law is used.

3.2 *Material model*

Since a linear elastic material model is used for the rubber, no viscoelastic behavior is directly modeled. Instead, dynamic mechanical analysis (DMA) test results are used to derive the YOUNG's modulus. An assumption is taken so that the YOUNG's modulus of the material at a specific test condition is equal to the complex modulus derived from the DMA experiment at the frequency corresponding to the rotational frequency of the GROSCH wheel at that condition. The YOUNG's modulus E_{f_i} at a frequency f_i is derived as

$$E_{f_i} = \left\| E^*_{f_i} \right\| = \sqrt{E'^2_{f_i} + E''^2_{f_i}}, \tag{2}$$

where $E^*_{f_i}$, $E'^2_{f_i}$ and $E''^2_{f_i}$ are the complex, storage and loss modulus at a frequency f_i from the master curve, respectively.

3.3 *HUEMER friction model*

The HUEMER friction model is sz on experimental investigations and a pressure-dependent friction law (Huemer, Liu, Eberhardsteiner, & Mang 2000). This model takes into account the effects of pressure and sliding velocity to estimate the friction coefficient,

$$\mu(p, v) = \frac{\alpha |p|^{n-1} + \beta}{a + \frac{b}{|v|^{1/m}} + \frac{c}{|v|^{2/m}}}, \tag{3}$$

where α, β and n are parameters related to contact pressure and, a, b, c and m are constants related to the sliding velocity. Using a least square algorithm, calibration of the parameters is performed iteratively. The algorithm aims to minimize the difference between the experimental data points $\mu_{\text{exp},i,j}$ and the numerically calculated point $\mu_{i,j}$,

$$S = \sum_{i,j}^{k,l} \left(\mu_{i,j} - \mu_{\text{exp},i,j} \right)^2 \Rightarrow \min, \tag{4}$$

where k is the number of pressure points, and l is the number velocity points (Huemer, Liu, Eberhardsteiner, & Mang 2001). For unconstrained problems throughout this study, a LEVENBERG-MARQUARDT least square algorithm, implemented by SciPy based on the formulation of (Mor´e 1978), is used. Throughout the study, several nonlinear target functions are used (e.g. Equation (4)). For this reason, the mean average percentage error (MAPE) is used instead of the usual coefficient of determination (R^2) to evaluate the calibration results. MAPE is a percentage-based evaluation metric for regression problems and is calculated as

$$\text{MAPE} = \frac{1}{n} \sum_{i}^{n} \frac{|y_i - f_i|}{|y_i|}, \tag{5}$$

where n is the number of data points, y_i is the experimental value, and f_i is the predicted result. The parameter calibration for the HUEMER friction model is conducted based on the LFT test results. The MAPE of the calibration is 2.69%.

3.4 *Wear model*

In this study, a combination of ARCHARD's wear model (Achard 1953) and SCHALLAMACH's abrasion law (Schallamach & Grosch 1981) presented in (Hartung, Garcia, Berger, Hindemith, Wangenheim, & Kaliske 2021) is applied. The model used in this work is based on recent publications (Li, Li, &Wang 2020, Nguyen, Zheng, Schmerwitz, & Wriggers 2018),

$$\frac{\dot{V}_s}{A_s} = k_w \left(\frac{\dot{\mathcal{D}}_s}{A_s}\right)^{a_w}, \qquad (6)$$

where $\dot{V}_w$ is the wear rate, A_s is the streamline area, and k_w and a_w are wear constants to be identified. The frictional energy dissipation rate

$$\dot{\mathcal{D}}_s = \int_a ||\tau_T|| \, ||v_T|| \, \mathrm{d}a, \qquad (7)$$

is calculated as the integral of the shear stresses τ_T and the sliding velocity v_T over the contact surface.

4 EXTENSION OF FEM WEAR MODELING USING META MODELS

4.1 *Failure surface identification*

In this section, an attempt is made to find the failure surface that separates failed and successful simulations under varying wear parameters. Investigating the failure surface properties allows for a reduction in computational time. The failure or success of a simulation is predicted before the simulation. A failure surface identification algorithm is developed. First, a_w is initialized as 0 such that $\dot{V}_w = k_w$ and a starting point of $k_w = 0.005\ mm^2/N$ is assumed. If the simulation fails, k_w is decreased by 10%. Whereas if the simulation is successful k_w is increased by 10%. The procedure is looped until a change between failure or success region is detected. Afterward, a_w is set to 0.05, and a similar procedure of increasing or decreasing k_w by 10% is repeated. Each time a change is detected, a_w is increased by 15%. The algorithm stops after detecting 50 region changes. Finally, points that did not occur immediately before or after a change are removed and the remaining data is then fitted by

$$k_w = \gamma_1 \exp(\gamma_2 {a_w}^{\gamma_3}), \qquad (8)$$

where γ_1, γ_2 and γ_3 are constants. In Figure 5, the fitted failure surface for test condition C6 is shown.

As pointed out in Figure 6, the fitting is producing inaccurate results, especially in areas with low k_w. Fortunately, such values of k_w are too low and not considered in practice. For this reason, the fitting is only evaluated within a range of 10^{-10} to $1.46 \cdot 10^{-4} mm^2/N$ for k_w, and $5 \cdot 10^{-2}$ to 3 for a_w.

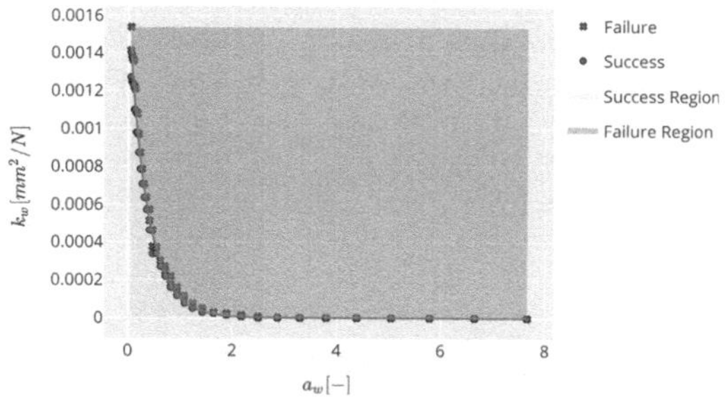

Figure 5. Failure surface for test condition C6 in linear scale.

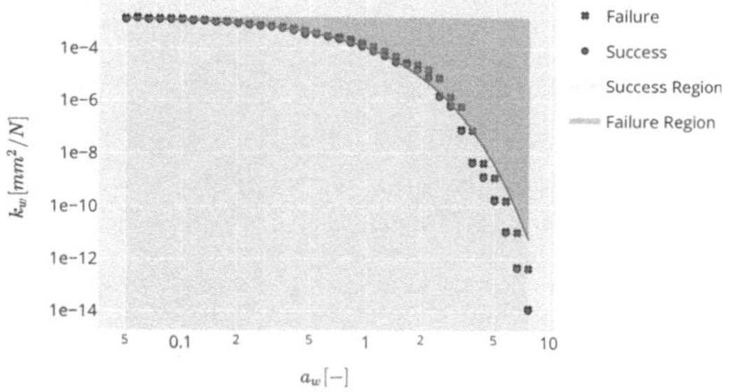

Figure 6. Failure surface for test condition C6 in logarithmic scale.

Table 2. Fitting results for surface failure algorithm.

Test	γ_1 $[mm^2/N]$	γ_2 [-]	γ_3 [-]	MAPE [%]
C1	0.0360	-5.1669	0.3430	14.90 %
C2	0.0001	-0.5753	1.8177	15.26 %
C3	0.0001	-0.5065	1.8784	18.46 %
C4	0.0001	-0.0772	3.5451	20.8 %
C5	0.0003	-1.3283	1.9822	39.28 %
C6	0.0029	-3.2791	0.7210	59.05 %
C7	0.0005	-0.9351	1.8519	39.68 %
C8	0.0002	-1.4570	2.3826	50.42 %
C9	0.0003	-0.7830	2.9526	49.61 %
C10	0.0005	-1.2315	1.9464	34.70 %

4.2 *Fitting of meta models*

Three-dimensional surfaces, in which k_w and a_w are inputs and wear mass is the output, are generated as meta models. Within the limits discussed previously, a 25 x 25 grid mesh is created in which points are evenly distributed in the logarithmic space. Wear mass is evaluated by performing FEM simulations at each point of the grid and for each test condition.

The fitted failure surface is used to pre-evaluate each point on the grid and save on computational time. After the evaluation, successful points are used to fit the following surface function

$$m = \zeta_1 k_w \zeta_2^{\zeta_3 a_w + \zeta_4}, \tag{9}$$

where m is the wear mass in grams and ζ_1, ζ_2, and ζ_3 are constants which have to be identified. Figure 7 shows the meta model and FEM simulation results for test conditions C3. The results for all test conditions are presented in Table 3.

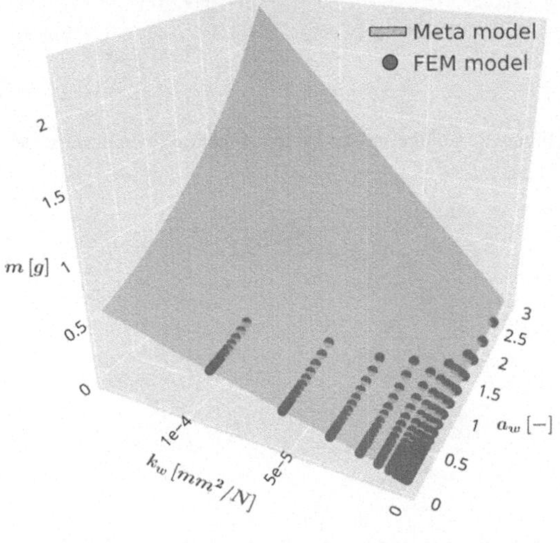

Figure 7. Meta model for test condition C3.

Table 3. Fitting results for the meta models.

Test	ζ_1 $[gN/mm^2]$	ζ_2 [-]	ζ_3 [-]	ζ_3 [-]	MAPE [%]
C1	2.7373	5.3330	-0.0973	4.2326	5.53 %
C2	3.8043	3.6030	-0.1291	6.0836	6.06 %
C3	3.2306	6.5770	0.2540	3.7536	6.10 %
C4	3.3023	3.8542	0.6788	4.3823	6.25 %
C5	1.6967	4.5983	0.9858	4.6286	4.72 %
C6	4.0063	4.5151	1.4530	3.4517	6.22 %
C7	2.7384	3.8573	2.1576	2.9725	18.26 %
C8	3.0884	4.3625	1.5332	4.2275	4.61 %
C9	2.9588	3.3681	2.4602	3.8801	7.59 %
C10	3.0341	4.2058	2.5482	2.5870	11.84 %

5 NUMERICAL CALIBRATION OF WEAR PARAMETERS

The use of wear parameters from the previously conducted LFT experiment yields high and inaccurate amounts of wear. Hence, other approaches are needed for the calibration of the wear parameters for the LAT100 experiment. Using the LAT100 test results, wear parameter calibration is conducted through experimental calibration, the hybrid model, and the standalone meta model. In (Hartung, Garcia, Berger, Hindemith, Wangenheim, & Kaliske 2021), the wear parameters are fitted using an LFT experiment in which the guiding velocity is the same as the slip velocity, which is not the case in a LAT100 experiment. Instead, a FEM simulation can be used for the calculation of the slip velocities. Moreover, a numerical calibration can also be one way to identify the wear parameters. For the calibration using the hybrid and the standalone meta models, k_w and a_w are input parameters and the wear mass m is the output. For the hybrid model, the calibration is done in a constrained way according to the previously set limits as this produces the best MAPE. A Trust Region Reflective (TRF) constrained minimization algorithm is used for the calibration implemented using SciPy (Branch, Coleman, & Li 1999). The hybrid model acts independently of the optimization algorithm. As shown in Figure 8, the optimization algorithm passes different k_w and a_w values to the hybrid model which sends back a predicted wear mass for each test condition. If a set of k_w and a_w values is outside the success region, the meta model returns the wear mass prediction and vice versa. This way, the minimization algorithm is not aware of the changes inside the hybrid model, and coupling between the meta and FEM model works seamlessly. For the experimental and meta model calibration, the unconstrained algorithm is used since the optimum solution is found outside the limits. The calibration results are shown in Table 4, as well as in Figures 9, and 10. The results show that experimental calibration yields the lowest MAPE of 56%. On the other hand, the results of the hybrid model have a slightly lower MAPE (58%). Finally, the standalone meta model performed the worse with a MAPE of 65%.

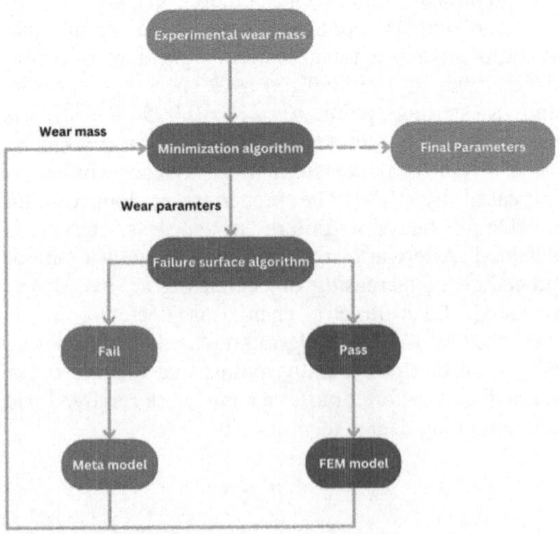

Figure 8. Numerical calibration algorithm.

Table 4. Calibration results for different approaches.

Model	k_w $[mm^2/N]$	a_w [-]	MAPE [%]
Experimental	3.1338e-07	2.6469	56 %
Hybrid	1.1917e-07	2.7640	58 %
Meta	2.5475e-08	3.2089	65 %

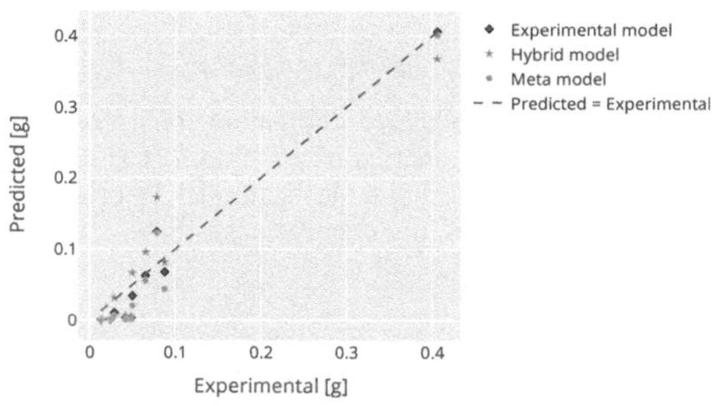

Figure 9. Comparison of measured and predicted mass loss in linear scale (experimental results provided by CEAT).

6 CONCLUSIONS

Under certain boundary conditions, the ALE adaptive meshing algorithm may decrease some element volumes to zero causing the simulation to fail. For this reason, a failure surface algorithm is suggested to identify the surface at which the simulation fails. This algorithm assists the development of the meta models to extend FEM models and overcome the limitation of adaptive remeshing. The use of the meta model allowed for a numerical calibration of wear parameters under a hybrid meta-FEM approach. This approach is able to achieve a close accuracy to that of experimental calibration. While experimental calibration yields the lowest MAPE of 56%. On the other hand, the hybrid model achieves a slightly lower MAPE (58%). Finally, the standalone meta model performed the worse with an MAPE of 65%.

ACKNOWLEDGMENTS

This project has received funding from the European Union's Horizon 2020 research and innovation program under grant agreement No 958174. This project is co-financed from tax revenues on the basis of the budget adopted by the Saxon State Parliament. The authors would also like to gratefully acknowledge CEAT Tyre Ltd. for their assistance in conducting the LAT100 experiments and funding of the LFT experiments described in this paper.

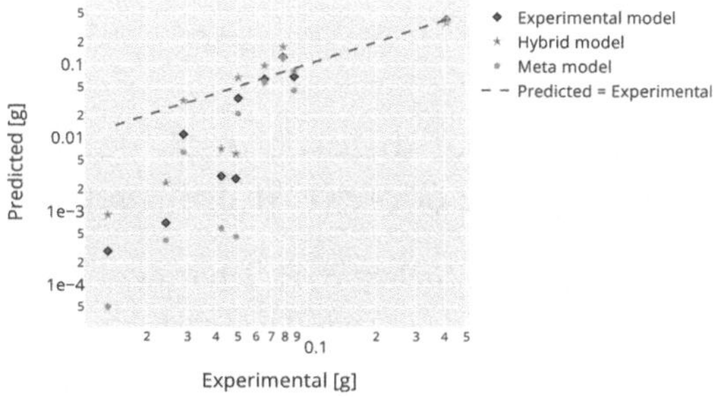

Figure 10. Comparison of measured and predicted mass loss in logarithmic scale (experimental results provided by CEAT).

REFERENCES

ABAQUS (2014). 2014. *Version 6.14 User's Manual*. Providence, RI, USA: Dassault Syst`emes Simulia Corp.

Achard, J. (1953). Contact and rubbing of flat surface. *Journal of Applied Physics* 24, 981–988.

Branch, M. A., T. F. Coleman, & Y. Li (1999). A subspace, interior, and conjugate gradient method for large-scale boundconstrained minimization problems. *SIAM Journal on Scientific*
Computing 21, 1–23.

Hartung, F., M. A. Garcia, T. Berger, M. Hindemith, M. Wangenheim, & M. Kaliske (2021). Experimental and numerical investigation of tire tread wear on block level. *Lubricants* 9, 113.

Heinz, M. & K. Grosch (2007, 09). A laboratory method to comprehensively evaluate abrasion, traction and rolling resistance of tire tread compounds. Volume 80, pp. 580–607.

Huemer, T.,W. Liu, J. Eberhardsteiner, & H. Mang (2001). A 3d finite element formulation describing the frictional behaviour of rubber on ice and concrete surfaces. *Engineering Computations* 18, 417–437.

Huemer, T., W. Liu, J. Eberhardsteiner, & H. A. Mang (2000). Entwicklung eines Wirklichkeitsnahen Reibungsmodells f¨ur Gummi zum Zwecke Dreidimensionaler Traktionssimulationen. Technical report, Universit¨at Wien.

Kole, P. J., A. J. L¨ohr, F. G. Van Belleghem, & A. M.Ragas (2017). Wear and tear of tyres: a stealthy source of microplastics in the environment. *International Journal of Environmental Research and Public Health* 14, 1265.

Li, Z., Z. Li, & Y.Wang (2020). An integrated approach for friction and wear simulation of tire tread rubber. part i: friction test, characterization, and modeling. *Tire Science and Technology* 48, 123–145.

Mor´e, J. J. (1978). The levenberg-marquardt algorithm: Implementation and theory. In G. A. Watson (Ed.), *Numerical Analysis*, Berlin, Heidelberg, pp. 105–116. Springer Heidelberg.

Nguyen, V., D. Zheng, F. Schmerwitz, & P. Wriggers (2018). An advanced abrasion model for tire wear. Wear 396, 75–85.

Schallamach, A. & K. Grosch (1981). Tire traction and wear. *Mechanics of Pneumatic Tires* 5, 365–474.

Constitutive Models for Rubber XIII – Dal (Ed.)
© 2025 The Author(s), ISBN 978-1-032-85138-9

Comparison of different methods for implementing consistent tangent moduli in the context of nonlinear material models

D. Pornhagen, S. Siebert & P. Schneider
Deutsches Institut für Kautschuktechnologie e.V., Hannover, Germany

ABSTRACT: In this contribution we discuss and compare different approaches to implement consistent tangent moduli via an analytic form of differentiation, finite differences and automatic differentiation. The advantages and disadvantages of these methods are compared using an isotropic hyperelastic model in principal stretches. Hereby a special focus is set on the numerical accuracy with coalescing eigenvalues.

1 INTRODUCTION

The implementation of a constitutive material model within an implicit Finite Element Method ("FEM") code requires, in addition to the definition of a stress tensor as a function of the deformation gradient, the calculation of the consistent tangent modulus, often referred to as Jacobian matrix. The tangent modulus is needed for the stepwise solution of the nonlinear equilibrium equation by means of Newton iteration. Thus, the exact and consistent formulation influences the speed of convergence. For the calculation of the stress and the tangent modulus, the first and second order derivatives of the strain-energy density function are necessary for constitutive models.

The most obvious determination of the derivative is the analytical derivative, which can be done manually or automated. Especially for complex material models, this method can lead to fairly complex expressions which do not necessarily result in numerically fast and efficient code. Furthermore, there are known numerical instabilities associated with the same and similar principal stretches (Connolly et al. 2019).

Another approach is the determination of the derivative via finite differences (Miehe 1996). This approach is easy to implement and requires little revisions when the underlying material model changes. However, the computational effort is high and numerical errors can lead to slow convergence.

An exact calculation of the derivative can be carried out using automatic differentiation. This approach takes advantage of the fact that most algorithms consist of a sequence of basic arithmetic operations and basic functions. The partial derivatives of the composition function of such algorithms can be calculated quickly and accurately, which means that automatic differentiation is already considered a helpful tool in the development phase of constitutive modeling (Rothe and Hartmann 2015).

2 MODELLING

2.1 *Kinematics*

When considering a body $\mathcal{B}$ under finite strain, the function $\vec{\phi}$ describes a mapping between material points $\vec{X} \in \mathcal{B}$ from the reference configuration and spatial points $\vec{x} \in b$ from the current configuration. For the description of the deformation, it is useful to define the deformation gradient $\mathbf{F} := \partial\vec{\phi}(\vec{x})/\partial\vec{X}$. The resulting volume change can be described via $J := \det(\mathbf{F})$.

In the deformation gradient, the pure stretches are superimposed by rigid body rotations. To describe the material behavior, it is useful to describe the pure stretch of the body using

$$\mathbf{C} = \mathbf{F}^T \cdot \mathbf{F} \tag{1}$$

the right Cauchy-Green deformation tensor. The symmetric tensor can be split by the spectral decomposition

$$\mathbf{C} = \sum_{i=1}^{3} \lambda_i^2 \left(\vec{N}_i \otimes \vec{N}_i\right) \tag{2}$$

into products of squared principal stretches λ_i^2 and dyadic products of their eigenvectors $\vec{N}_i$. An analytical solution to the eigenvalue problem can be found using the characteristic polynomials of these symmetrical second-order tensors. However, numerical inaccuracy occurs for equal or coincident eigenvalues (Hartmann 2019). In this work, a robust analytical method from (Harari and Albocher 2022) was used, which first finds the unique, non-coalescing eigenvalue and then determines the remaining eigenvalues using deflation.

DOI: 10.1201/9781003516880-31

2.2 *Material model*

In the context of hyperelasticity, the existence of a strain energy density function Ψ is postulated. For an efficient description of the material behavior of rubber, it is advisable to split the strain energy function additively into an isochoric $\bar{\Psi}$ and a volumetric $\tilde{\Psi}$ part due to the quasi-incompressibility. Plagge et al. (2020) proposes a simplified form of the extended tube model ("SET") to describe the isochoric part

$$\bar{\Psi} = \frac{G}{2} \frac{\bar{I} - 3}{1 - \frac{\bar{I}-3}{n}}, \tag{3}$$

with the material parameters G and n. With this model, the strain energy can be calculated with a small number of parameters and using only the first invariant $\bar{I}_1$ of the isochoric right Cauchy-Green deformation tensor $\overline{\mathbf{C}} := J^{-\frac{2}{3}}$ without determining the principal stretches. This makes the model well suited for engineering tasks.

For the volumetric part

$$\tilde{\psi} = \frac{\kappa}{2}(J-1)^2 \tag{4}$$

offers a simple and convex form and is often used for modeling rubber-like materials, with κ beeing the bulk modul.

3 DIFFERENTIATION METHODS

3.1 *Analytical derivative*

Analogous to the strain energy, the stress can be split additively

$$\mathbf{S} = \overline{\mathbf{S}} + \tilde{\mathbf{S}} \tag{5}$$

into the isochoric stress $\overline{\mathbf{S}} = 2\partial\bar{\Psi}/\partial\mathbf{C}$ and volumetric stress $\tilde{\mathbf{S}} = 2\partial\tilde{\Psi}/\partial\mathbf{C}$. This can be done in the same way for the tangent modulus via

$$\mathbb{C} = \overline{\mathbb{C}} + \tilde{\mathbb{C}} \tag{6}$$

with the isochoric $\overline{\mathbb{C}} = 2\partial\overline{\mathbf{S}}/\partial\mathbf{C}$ and volumetric $\overline{\mathbb{C}} = 2\partial\overline{\mathbf{S}}/\partial\mathbf{C}$ part. In the following, the partial derivatives are determined analaytically, first for the isochoric and then for the volumetric part.

For rubber-like material the isochoric contribution of the strain energy $\bar{\Psi}(\bar{\lambda}_i)$ is often described as a function of the isochoric stretches $\bar{\lambda}_i = J^{-\frac{1}{3}}\lambda_i$, which are the eigenvalues of the isochoric deformation gradient $\overline{\mathbf{F}} := J^{-\frac{1}{3}}\mathbf{F}$.

Using the chain rule, the stress can be calculated by

$$\overline{\mathbf{S}} = 2\frac{\partial\bar{\Psi}}{\partial\mathbf{C}} = \sum_{a=1}^{3} \beta_a \lambda_a^{-2} \vec{N}_a \otimes \vec{N}_a \tag{7}$$

with the stress coefficient β_a

$$\beta_a = \bar{\lambda}_a \frac{\partial\bar{\Psi}}{\partial\bar{\lambda}_a} - \frac{1}{3}\sum_{b=1}^{3} \bar{\lambda}_b \frac{\partial\bar{\Psi}}{\partial\bar{\lambda}_b}. \tag{8}$$

The summation symbol is intended to make it clear that the index b is repeated, whereas the index a is not.

By deriving Equation (7) and reapplying the chain rule again, the material tangent can be determined as follows

$$\begin{aligned}\overline{\mathbb{C}} = &\sum_{a,b=1}^{3} \left(\frac{\gamma_{ab}}{(\lambda_a)^2(\lambda_b)^2} - 2\delta_{ab}\beta_a(\lambda_a)^{-4} \right) \\ &\left(\vec{N}_a \otimes \vec{N}_a \otimes \vec{N}_b \otimes \vec{N}_b \right) \\ &+ \sum_{a,b=1;a\neq b}^{3} \frac{\beta_b(\lambda_b)^{-2} - \beta_a(\lambda_a)^{-2}}{(\lambda_b)^2 - (\lambda_a)^2} \\ &\left[\vec{N}_a \otimes \vec{N}_b \otimes \left(\vec{N}_a \otimes \vec{N}_b + \vec{N}_b \otimes \vec{N}_a \right) \right]\end{aligned} \tag{9}$$

with the tangent coefficient γ_{ab}

$$\begin{aligned}\gamma_{ab} = &\left(\frac{\partial^2\bar{\Psi}}{\partial\bar{\lambda}_a\partial\bar{\lambda}_b}\bar{\lambda}_a\bar{\lambda}_b + \frac{\partial\bar{\Psi}}{\partial\bar{\lambda}_a}\delta_{ab}\bar{\lambda}_b \right) \\ &+ \frac{1}{9}\sum_{c,d=1}^{3} \left(\frac{\partial^2\bar{\Psi}}{\partial\bar{\lambda}_c\partial\bar{\lambda}_d}\bar{\lambda}_c\bar{\lambda}_d + \frac{\partial\bar{\Psi}}{\partial\bar{\lambda}_c}\delta_{cd}\bar{\lambda}_d \right) \\ &- \frac{1}{3}\sum_{c=1}^{3} \left(\frac{\partial^2\bar{\Psi}}{\partial\bar{\lambda}_a\partial\bar{\lambda}_c}\bar{\lambda}_a\bar{\lambda}_c + \frac{\partial\bar{\Psi}}{\partial\bar{\lambda}_c}\delta_{ac}\bar{\lambda}_c \right) \\ &- \frac{1}{3}\sum_{c=1}^{3} \left(\frac{\partial^2\bar{\Psi}}{\partial\bar{\lambda}_b\partial\bar{\lambda}_c}\bar{\lambda}_b\bar{\lambda}_c + \frac{\partial\bar{\Psi}}{\partial\bar{\lambda}_b}\delta_{bc}\bar{\lambda}_c \right).\end{aligned} \tag{10}$$

The second term in Equation (9) leads to numerical problems if the principal stretches are equal or nearly equal. Thereby, not only the difference of the squared stretches tends to zero, but also the difference between the associated principal stresses. In the case of equal principal stretches, the term in (9) can be calculated by

$$\begin{aligned}&\lim_{\lambda_a \to \lambda_b} \frac{\beta_b(\lambda_b)^{-2} - \beta_a(\lambda_a)^{-2}}{(\lambda_b)^2 - (\lambda_a)^2} = \\ &\lambda_b^{-4}\left(\frac{1}{2}\gamma_{bb} - \beta_b \right) - \frac{1}{2}\lambda_a^{-2}\lambda_b^{-2}\gamma_{ab}\end{aligned} \tag{11}$$

using L'Hopital's rule (Chadwick and Ogden 1971, Holzapfel 2000). A more detailed examination of the numerical error for equal or nearly equal aspect ratios can be found in (Connolly et al. 2019). Therein, a threshold value is discussed for coinciding eigenvalues with $|\lambda_a - \lambda_b| < 10^{-6}$ when the equation (11) is better suited.

A more detailed derivation of the isochoric stress and tangent modulus can be found in (Simo and Taylor 1991)

In a much simpler form and without the discussed numerical problems, the volumetric component of the stress

$$\tilde{\mathbf{S}} = J\frac{\partial \tilde{\Psi}}{\partial J}\mathbf{C}^{-1} \tag{12}$$

and the material tangent

$$\tilde{\mathbb{C}} = 4\left[\frac{\partial \tilde{\Psi}}{\partial J}\frac{\partial^2 J}{\partial \mathbf{C}\partial \mathbf{C}} + \frac{\partial^2 \tilde{\Psi}}{\partial J^2}\left(\frac{\partial J}{\partial \mathbf{C}} \otimes \frac{\partial J}{\partial \mathbf{C}}\right)\right] \tag{13}$$

can be calculated.

3.2 *Finite difference ("FD")*

Analogous to the approach in (Miehe 1996), we use the unit vectors $\vec{e}_1$, $\vec{e}_2$ and $\vec{e}_3$ to consider a particular perturbation

$$\hat{\varepsilon}_{ij} = \frac{\varepsilon}{2}\left(\vec{e}_i \otimes \vec{e}_j + \vec{e}_j \otimes \vec{e}_i\right) \tag{14}$$

in the entry ij with the magnitude ε. With this perturbation, a central difference quotient can be used to approximate the stress via

$$S_{ij} \approx \frac{\Psi(\mathbf{C} + \hat{\varepsilon}_{ij}) - \Psi(\mathbf{C} - \hat{\varepsilon}_{ij})}{\varepsilon}. \tag{15}$$

In order to obtain the second-order derivative especially for the mixed derivatives, it is useful to develop the equation (15) by a second central difference quotient and to determine the material tangent

$$\begin{aligned}\mathbb{C}_{ijkl} \approx &\big[\Psi(\mathbf{C} + \hat{\varepsilon}_{ij} + \hat{\varepsilon}_{kl}) - \Psi(\mathbf{C} - \hat{\varepsilon}_{ij} + \hat{\varepsilon}_{kl})\big] \\ &-\Psi(\mathbf{C} + \hat{\varepsilon}_{ij} - \hat{\varepsilon}_{kl}) - \Psi(\mathbf{C} - \hat{\varepsilon}_{ij} - \hat{\varepsilon}_{kl})\big]\varepsilon^{-2}\end{aligned} \tag{16}$$

by a good approximation.

For $\varepsilon \to 0$ the two formulations converge to the exact values. As the perturbation decreases, the numerical cancellation error increases. Accordingly, the magnitude of the perturbation must be meticulously selected in order to minimize the error.

Based on a Taylor series, difference quotients of higher order can be derived identically to the equations (15) and (16). As the order increases, the error decreases, but the function evaluations also increase.

3.3 *Automatic differentiation ("AD")*

In this contribution, automatic differentiation is carried out using the Juliana package "ForwordDiff" (Revels et al. 2016). Within this package, the partial derivation is calculated using dual numbers (Clifford 1871). Analogous to the complex numbers, the dual numbers have two base elements with a non-real unit ι. The non-real unit fulfills the condition $\iota^2 = 0$. In a Taylor series expansion

$$f(a + b\iota) = f(a) + \frac{\partial f}{\partial a}b\iota, \tag{17}$$

this property ensures that terms of higher order vanish. This shows that the derivative of the function $f(a)$ corresponds to the "dual" part of $f(a + b\iota)$ multiplied by b. By overloading all required elementary functions, the derivation of any generic algorithm can be determined by evaluating the dual part of every used elementary function. In order to calculate derivatives of higher orders, further non-real units can be added in the sense of the hyper-dual numbers. A more detailed explanation can be found in (Fike and Alonso 2011).

4 NUMERICAL VALIDATION

4.1 *Method*

The previously presented methods for determining the derivative of the isochoric part are now to be examined with regard to their accuracy. For the analysis, the SET model (3) is used with typical material parameters for filled rubber of $G = 1\,\text{MPa}$ and $n = 50$. The model depends on the first invariant $\bar{I}$, for which the analytic implementation of the tangent modulus $\overline{\mathbb{C}}$ is described in the literature (Holzapfel 2000) and can be done without numerical difficulties. With the help of the relation

$$\bar{I} = \text{tr}(\overline{C}) = \bar{\lambda}_1^2 + \bar{\lambda}_2^2 + \bar{\lambda}_3^2 \tag{18}$$

the model can also be describe via the principal stretches. Analogous to the investigation in (Connolly et al. 2019), the invariant implementation of $\overline{\mathbb{C}}^{\bar{I}}$ is used in order to have an exact reference. The reference is used to compare it with different differentiation methods based on the principle stretches $\overline{\mathbb{C}}^{\bar{\lambda}}$. Using Voigt notation, a relative error measure

$$E_{\mathbb{C}} = \sum_{I,J=1}^{6} \left|\overline{\mathbb{C}}_{IJ}^{\bar{\lambda}} - \overline{\mathbb{C}}_{IJ}^{\bar{I}}\right| / \sum_{I,J=1}^{6} \left|\overline{\mathbb{C}}_{IJ}^{\bar{I}}\right|, \quad (19)$$

is introduced to compare all 36 independent entries of the tangent modulus.

This error is subsequently determined using two particular deformations. The first deformation describes a plane strain ("ps") tension with the maximum elongation λ, leading to three unique eigenvalues. The second deformation describes a uniaxial ("ua") tension with the maximum elongation λ and two identical eigenvalues. The uniaxial tension is superimposed with a small perturbation ε to investigate the behavior with coincident eigenvalues. The associated deformation gradients are described by

$$\boldsymbol{F}_{\text{ps}} = \begin{bmatrix} \lambda & 0 & 0 \\ 0 & 1 & 0 \\ 0 & 0 & \frac{1}{\lambda} \end{bmatrix}; \boldsymbol{F}_{\text{ua}} = \begin{bmatrix} \lambda & 0 & 0 \\ 0 & \frac{1}{\sqrt{\lambda}} + \varepsilon & 0 \\ 0 & 0 & \frac{1}{\sqrt{\lambda}} \end{bmatrix}. \quad (20)$$

4.2 *Results*

The relative error $E_{\mathbb{C}}$ under planar tension is shown in Figure 1 with changing stretches. In addition to a central difference quotient with order two $\mathcal{O}2$, an order of four $\mathcal{O}4$ and six $\mathcal{O}6$ were also examined. The number of required function evaluations and the used perturbance are shown in table 4.2. As expected, the relative error decreases with increasing order from an mean average of $E_{\mathbb{C}} = 9.1 \cdot 10^{-7}$ to $E_{\mathbb{C}} = 2.1 \cdot 10^{-8}$ and then $E_{\mathbb{C}} = 1.8 \cdot 10^{-9}$ respectively.

In addition, the relative error was determined using the automatic and analytic differentiation. For these two methods, the error is in the range of the machine accuracy $E_{\mathbb{C}} = 1.8 \cdot 10^{-9}$.

Table 1. Number of function evaluations for S and $\mathbb{C}$ for central difference quotient of different error order without consideration of symmetries.

	function evaluations		
accuracy	S	$\mathbb{C}$	ε
$\mathcal{O}2$	12	144	$1.4 \cdot 10^{-5}$
$\mathcal{O}4$	24	576	$1.2 \cdot 10^{-5}$
$\mathcal{O}6$	36	1296	$5.0 \cdot 10^{-4}$

The relative error for the uniaxial tension load case is shown in Figure 2. The magnitude of the perturbation ε was varied with regard to exactly two identical eigenvalues. In addition, the arithmetic mean was determined from various extension states between $\lambda = 1$ and $\lambda = 3$. When using the finite difference methods, the error shows no significant difference to the behavior under plane tension. This indicates that no numerical problems occur with equal or coincident principal stretches. However, a robust method for determining the eigenvalues and vectors is essential. The error is similarly consistent when using automatic differentiation. As under ps loading, the error remains within the machine accuracy range.

Using the analytical method, the error at $\varepsilon < 4 \cdot 10^{-8}$ is within the range of machine accuracy. However, the error then increases up to the a numerical threshold of equal eigenvalues to $E_{\mathbb{C}} = 3.6 \cdot 10^{-12}$. The increase is due to the fact that l'Hospital's rule is applied even though the eigenvalues are not exactly equal. Afterwards the error then decreases because the numerical error decreases due to the similar eigenvalues from Equation (9).

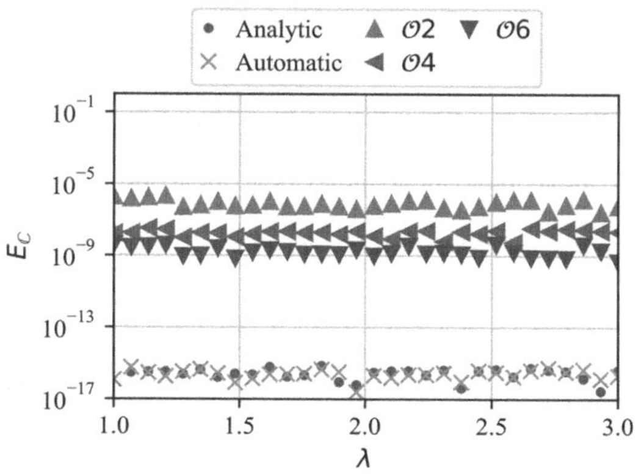

Figure 1. Relative error of the elasticity tensor from the different methods under plane strain tension with increasing deformation.

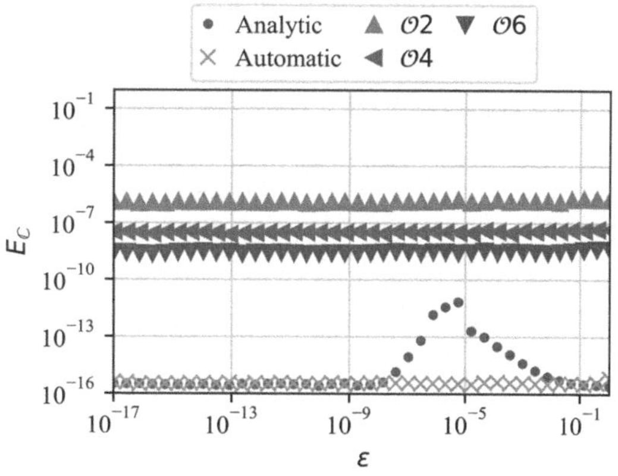

Figure 2. Relative error of the elasticity tensor from the different methods with varying perturbation of the uniaxial deformations averaged over several stretches.

5 FINITE ELEMENT METHOD INVESTIGATION

5.1 *Method*

When choosing the differentiation method for engineering applications, the influence on the speed of convergence is important in addition to a simple implementation and execution speed. This means that an exact calculation of the tangent modulus is not necessary, but only a sufficiently accurate one that does not influence the speed of convergence. The rate of Convergence q is suitable for describing the speed of convergence, which can be defined via

$$R_{n+1} \leq d \cdot R_n^q \quad (21)$$

using the residual of the last R_n and the current R_{n+1} iteration as well as a constant d. Newton-Raphson method. With a quadratic convergence rate the number of correct decimal places double with each iteration. The order of convergence can be determined

$$q \approx \frac{\log(R_{n+1}/R_n)}{\log(R_n/R_{n-1})} \quad (22)$$

approximately (Senning 2007).

To compare the convergence rate for different differentiation methods, two finite element benchmark simulations are carried out. One simulation describes a displacement-controlled uniaxial stretching of a unit cube with equidistant steps of $\Delta\lambda = 0.1$ per step to a total of $\lambda_{\max} = 4$. In the second simulation, the unit cube is sheared in two directions simultaneously. The displacement on one surface is set to zero in all spatial directions, while an increasing force is applied on the opposite side. The simulation was implemented in Ferrite (Carlsson et al. 2023), which is a general open FE toolbox.

The material model is again the SET with the material parameters and the implementations from the last section. For the volumetric part equation (4) is used with $\kappa = 2000\,\text{MPa}$.

In order to minimize the influence of the quasi-incompressibility on the convergence, a modified 2-field formulation is used (Brink and Stein 1996). Here, a second field p is introduced, which can be interpreted as a perturbed Lagrange multiplier. A quadratic Ansatz is used for the displacements and a linear one for the new field parameter.

5.2 *Results*

The mesh used in the uniaxial tensile testis shown in Figure 3, which consists of 3072 tetrahedral elements. Next to it is the deformed block with the magnitude of the displacements.

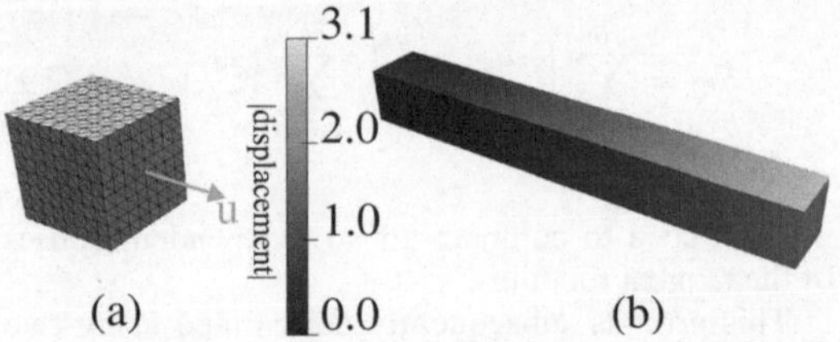

Figure 3. FE simulation of the uniaxial tensile test: (a) undeformed mesh with displacement boundary condition (b) magnitude of displacements of the deformed body with $\lambda = 4$.

The arithmetic mean of the convergence rate of all iterations per deformation step is shown in Figure 4. Within a deformation step, the estimate of the convergence rate varies greatly. Nevertheless, the criterion provides an opportunity to discuss the perfomance of the differentiation methods qualitatively.

Basically, the invariant, automatic and analytical methods provide almost identical results. For lower elongations ($\lambda < 1.7$), the theoretical value of two is approximately achieved. The value then drops to $q \approx 1.7$. The methods using the FD show poorer results. Not only is the value of the convergence rate lower, but the value scatters strongly over the course of the deformation. This results in the $\mathcal{O}2$ method requiring a total of two more iterations in order to calculate the entire deformation. However, the resulting time gain is negligible compared to the total of 118 iterations for all other methods.

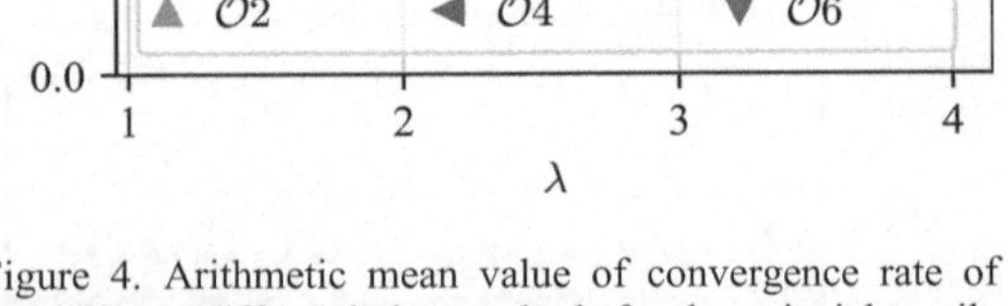

Figure 4. Arithmetic mean value of convergence rate of the different differentiation methods for the uniaxial tensile test.

The undeformed mesh for the shearing of the cube is shown together with the magnitude of displacements after the last step in Figure 5.

The result of the averaged convergence rates is shown in Figure 6. In general, it can be shown that the values for the second load case are higher than in the first one and also higher than the theoretical value. On the one hand, this is due to the fact that equation (22) is only an approximation. On the other hand, the load

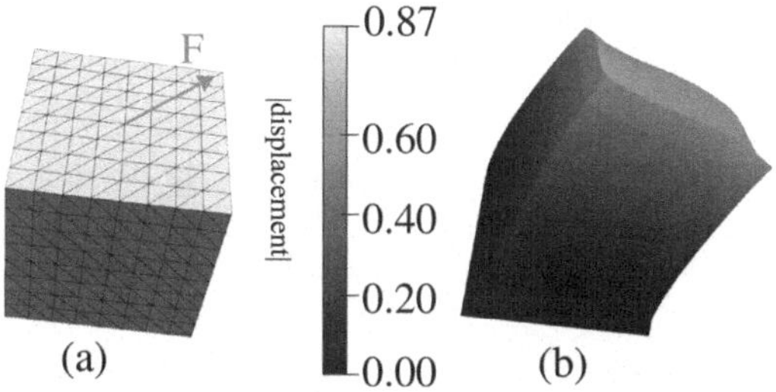

Figure 5. FE simulation of the of the sheared cube: (a) undeformed mesh with displacement boundary condition (b) magnitude of displacements at the last step.

increment in shear is significantly lower than in uniaxial tension. Furthermore, it can be seen that the FD methods of order $\mathcal{O}4$ and $\mathcal{O}6$ do not differ significantly from the exact methods. Only the method of order $\mathcal{O}2$ has a low convergence rate. Nevertheless, this does not lead to more iterations in this case.

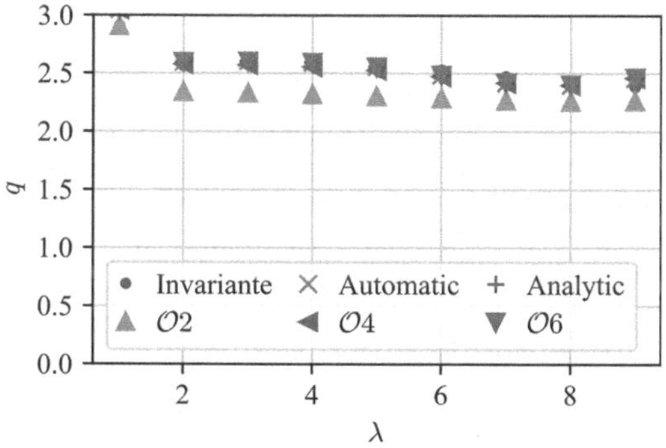

Figure 6. Arithmetic mean value of convergence rate of the different differentiation methods for sheared cube.

6 CONCLUSION

In this contribution, different methods for the calculation of the tangent modulus for constitutive material models in principal stretches were compared. Their accuracy was investigated, whereby the case of two equal or nearly equal principal stretches was considered. It was shown that the error for FD and AD is not influenced by coinciding principal stretches. Furthermore, it was shown that the error for the AD was even smaller than for the analytical method.

Furthermore, FE simulations were carried out to test the practical use of the methods. This showed that both FD and AD are good methods for achieving good convergence, particularly in the development phase of material models. The automatic differentiation lead to almost identical results as the analytical methods.

REFERENCES

Brink, U.&E. Stein (1996). On some mixed finite element methods for incompressible and nearly incompressible finite elasticity. *Comput. Mech. 19*, 105–119.

Carlsson, K., F. Ekre, & Contributors. (2023). Ferrite.jl (v0.3.14).

Chadwick, P. & R. W. Ogden (1971). A theorem of tensor calculus and its application to isotropic elasticity. *Arch. Ration. Mech. Anal. 44*, 54–68.

Clifford, M. (1871). Preliminary sketch of biquaternions. *Proc. Lond. Math. Soc. s1-4*, 381–395.

Connolly, S., D. Mackenzie, & Y. Gorash (2019). Isotropic hyperelasticity in principal stretches: Explicit elasticity tensors and numerical implementation. *Comput. Mech. 64*, 1273–1288.

Fike, J. & J. Alonso (2011). *The Development of Hyper-Dual Numbers for Exact Second-Derivative Calculations*. American Institute of Aeronautics and Astronautics.

Harari, I. & U. Albocher (2022). Computation of eigenvalues of a real, symmetric 3 × 3 matrix with particular reference to the pernicious case of two nearly equal eigenvalues. *Int. J. Numer. Methods Eng. 124*, 1089–1110.

Hartmann, S. (2019). Computational aspects of the symmetric eigenvalue problem of second order tensors. *Tech. Mech.- Eur. J. Eng. Mech. 23*, 283–294.

Holzapfel, G. A. (2000). *Nonlinear solid mechanics: a continuum approach for engineering*. Chichester, England: John

Wiley & Sons. Miehe, C. (1996). Numerical computation of algorithmic (consistent) tangent moduli in large-strain computational inelasticity. *Comput. Meth. Appl. Mech. Eng. 134*, 223–240.

Plagge, J., A. Ricker, N. Kröger, P. Wriggers, & M. M. Kl¨uppel, (2020). Efficient modeling of filled rubber assuming stressinduced microscopic restructurization. *Int. J. Eng. Sci. 151*, 103291.

Constitutive Models for Rubber XIII – Dal (Ed.)
© 2025 The Author(s), ISBN 978-1-032-85138-9

A design method for developing rubber anti-vibration mounts for high frequency applications in EV suspension systems

A.K. Serbest, S. Koçak, B. Zanbak & S. Gazi
Yamas Yaşar Makine Kalıp Oto Yedek Parça San. Tic. A.Ş., Bursa, Türkiye

M. Yazıcı
Prof. Dr., Department of Automotive Engineering, Uludağ University

ABSTRACT: Rapid development of industries such as aerospace, defense and automotive has made vibration control and isolation a critical issue. Electrical Vehicles (EV), in automotive industry, brings additional engineering problems for anti-vibration applications in High Frequency Range (50-2500Hz). Isolating the vibration on EV engine requires additional design and material development activities comparing with Internal Combustion Engines (ICE).

For long time viscoelastic materials such as Rubber are widely in use for vibration isolation applications. In this study, high-frequency behavior of viscoelastic materials was analyzed virtually in Finite Element Analysis by using DMA and RPA test results as material inputs. Frequency dependent responses of the rubber mounts has been examined. The obtained storage and loss modulus data in frequency domain were used to simulate high frequency behaviors and modify the frequency band of isolators that resonates at high frequencies. Parametrical geometry modification has been iterated until required performance obtained.

To overcome the difficulties for obtaining Storage and Loss modules at high frequencies, additional studies were conducted by using machine learning methods to increase the accuracy of FEA results. Proposed approach provides promising results to get quick responses depending on design modifications. This makes the selection of optimum viscoelastic materials quicker and provides a design method for high frequency applications.

1 INTRODUCTION

Control of noise and vibration in electric vehicles (EVs) is becoming increasingly important every day, particularly when evaluating driving performance and cabin comfort. With the need to preserve driving performance data in internal combustion engine (ICE) vehicles while ensuring no change in cabin comfort, there arises a necessity to develop engineering approaches. While vibration sources within vehicles are typically classified as road, air and engine, in EVs, motor vibration has been moved into the noise spectrum. Consequently, noises that were previously unnoticed by drivers became audible or perceptible in EVs. To mitigate this effect, which negatively impacts cabin comfort, updates to technical specifications by automotive OEM companies aim to expand the frequency band to be monitored from 50-400 Hz to the range of 50-2500 Hz.

As a result of studies on this subject, (Elahinia, March, 2010) focused on semi-active hydraulic engine mounts for controlling noise and vibration levels in Hybrid Electric Vehicles (HEVs) and Plug-in Hybrid Electric Vehicles (PHEVs). Hydraulic engine mounts are vibration damping elements that exhibit different damping characteristics in high amplitude low frequency or low amplitude high frequency loads. Dampers produced using multiple sub-components are preferred in luxury class vehicles due to cost and production complexity. Additionally, since vibrations from motors in EVs are consistently defined as low amplitude high frequency, the proposed method involves additional engineering work. In the study by (Tosun, 2019), vibration originating from road and air sources was addressed, examining technologies for reducing cabin noise using dampers made from MCU material. As demand increases, there has been an observed increase in the number of studies on known techniques, and the need has arisen to develop approaches related to testing products using the finite element method (FEM) to test products. In their study, (Danko, 2021) examined the dynamic modeling of dampers to be used in EVs using the finite element method. Analysis conducted in the range of 0-2500 Hz for rubber dampers was published using FEM. As studies progressed, it was observed that as the natural

*Corresponding author: alikamil.serbest@yamas.com.tr

DOI: 10.1201/9781003516880-32

frequencies of dampers remained within a region of 0-2500 Hz, resonance phenomena began to occur during operation. This negatively affects damping performance. A study conducted at Ohio State University (Xianpai, 2015) mathematically modeled the damping characteristics of damping elements for EVs. This mathematical model was used to aim for resonance frequencies to be moved outside the operating band. In a study conducted at the University of Zagreb (Kovarik, 2015), the dynamic damping characteristics of motor damping elements were mathematically modeled according to frequency. Similar to the Ohio study, recommendations were made to move the resonance frequencies of dampers outside the operating band. In addition to mathematical models, there is a need for physical testing methods to verify conceptual designs. In her study, (Uz, 2015) mathematically modeled the operation of systems to test dampers. The study showed that when the structural resonance frequencies of the systems used for testing are outside the measurement band, the measurement results deviate from accuracy. Therefore, the design of test systems has been identified as another challenge for the development of measurement systems for conceptual designs. This challenge will not only apply to test systems but also to the design of fixtures to be used for connecting dampers to the test system. A study conducted at the Frauenhofer Institute (Troge, 2014) described a test system suitable for conducting measurements at high frequencies, which compensates for structural resonances based on modal exciters with internal sensors. The described system consists of seismic mass and electromagnetic exciters working in concert. Instantaneous data collected from force sensors and accelerometers are used to calculate system resonance, which is then subtracted from measurement values. The separated acceleration data is then fed back into the system for correction, ensuring stable measurements. The acquisition costs of the system and measurement discrepancies resulting from user-induced differences in system usage indicate the need for simpler systems. Studies related to the design of damping products in (Patent No. US2021/0222753, 2021) claim that innovative designs are capable of moving the resonance frequencies of dampers outside the operating band.

In light of these information, this study was conducted in this article with the idea that the acquisition of material models for damping designs in the 0-2500 Hz operating band using finite element analysis, parametric and iterative solutions for moving the results of designs outside the operating band, and the need for a test system for the simple and rapid testing of designed products.

2 METHOD AND MATERIAL

In line with the study objectives, the aim was defined as determining the finite element method (FEM) approach required for the development of a damping element operating in the high-frequency band and developing a test approach that allows the analysis results to be tested on a prototype. The study was initiated with the technical specification of the rubber bushing that dampens and supports the electric motor used in EV vehicles, which was selected as a pilot. The design inputs, including geometric constraints and static resistance requirements, are as shown in Table 1. The dynamic stiffness values dependent on the static stiffness values are also shown in Figure 2. The process of developing a chemical recipe dependent on the rubber recipe was not evaluated within the scope of the study.

Within a design cycle where static and dynamic resistance were targeted, CAE activities were completed. Material data for the rubber raw material required for finite element analysis were collected from test samples produced using geometries defined in standards representing hyperelastic material models for static resistance. The viscoelastic material data necessary for dynamic stiffness analyses were obtained from tests conducted on Rubber Process Analyzer (RPA) and Dynamic Mechanical Analyzer (DMA) test devices. Using the material data obtained from both static and dynamic analyses, designs were finalized. Prototype molds of the frozen design were manufactured.

Accelerometers positioned on a modal shaker, free-moving masses, and specially designed test fixtures for the product were utilized in the setup of the test apparatus to be used in testing the prototypes. Samples produced from prototype molds were tested in the test setup and made comparable with FEM results.

Concurrently, machine learning studies were conducted to determine the viscoelastic properties of rubber material within the 0-2500 Hz band. Tests of multiple products were conducted in the apparatus mentioned in (Troge, 2014). Test results and design parameters of the product geometry were used together with a machine learning approach, and

Table 1. Static stiffness targets.

Item	Target
Stiffness [N/mm]	Y: 800
	X: 1300
	Z: 200
Max. Disp [mm]	Y: +/- 7
	X: +/- 6,5
	Z: **
Linear Disp [mm]	Y: +/- 5
	X: +/- 4,5
	Z: 3
Rubber	42-48ShA
Bolt	M12, 10.9 89kN Preload

efforts were made to calculate the viscoelastic material parameters of the rubber material.

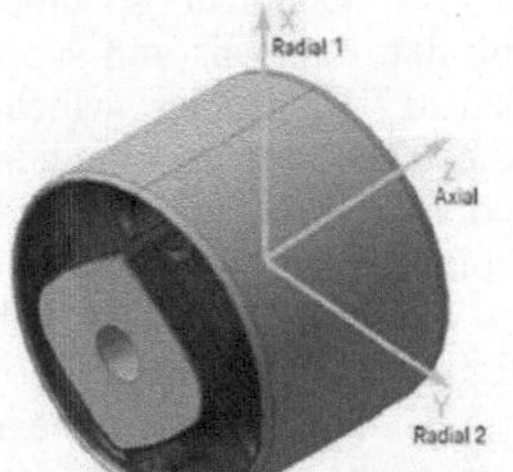

Figure 1. Coordinate system for rubber mount.

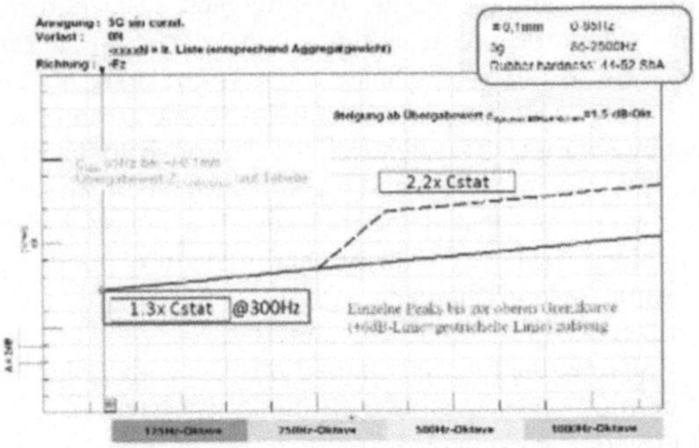

Figure 2. Dynamic stiffness targets.

2.1 *Geometric design and static stiffness studies*

CAD model created considering the assembly area boundary conditions for the static stiffness targets specified in Table 1 Static Stiffness Targets. The model consists of a profile made from an aluminum extrusion with a special section in the section, a tube made from an aluminum extrusion in the outer section and a rubber bonding these two parts. As assembly boundary, a slot with a width of 50mm and a diameter of Ø70mm was determined and as movement limits, ±7mm in the radial direction Y and ±6.5mm in the X direction were set. The widths of the rubber arms and the rubber hardness were determined as design parameters and iterative FEA studies were conducted. Mooney Rivlin 2 coefficient model was used as the material model for static stiffness analyses. The C10 and C01 coefficients required for the model to work were obtained from tests using rubber models with geometries compatible with the test apparatus shown in Figure 4.

Figure 3. CAD ve FEA models for static analysis.

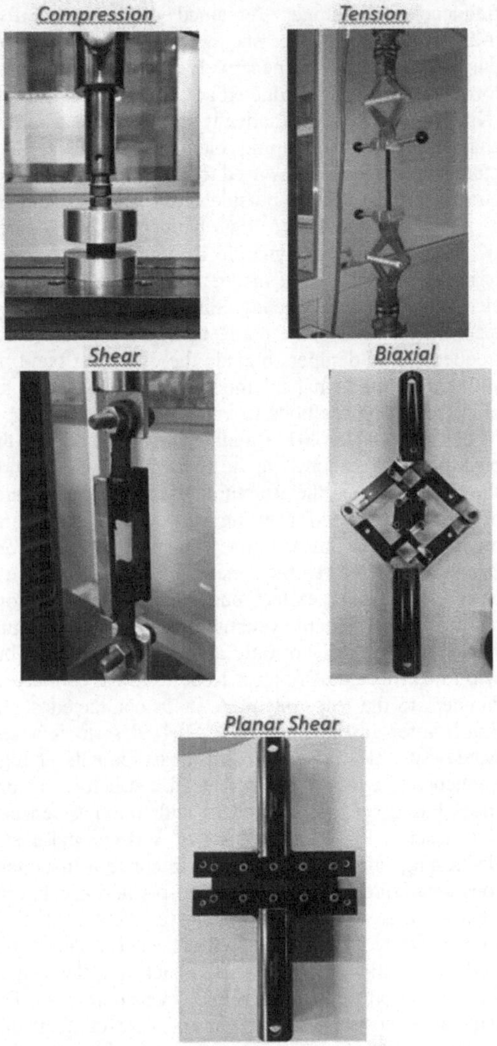

Figure 4. Static test rigs.

2.2 *Test rig design for high frequency tests*

As mentioned in the introduction, both the resonance frequencies of the test system and the test fixtures used must be outside the measurement band for the damper to be tested at high frequency. However, instead of reducing the mobility of the modal shaker with the weight of the fixture, the aim was to shift the measurement capability to the test side by designing the lightest possible fixture. It was aimed to use the minimum number of connecting elements in the connections of the fixture to the test system, considering that the connecting elements would have different mode shapes at high frequencies. The body of the test fixture was machined as a single piece of aluminum material. Fixtures used in the high-frequency test device Figure 5 were used for tests.

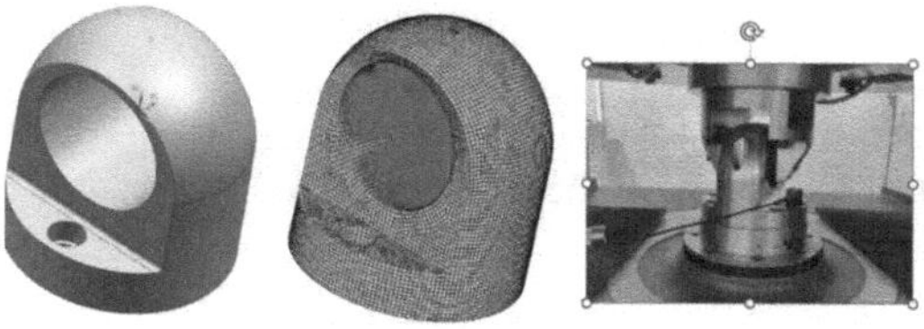

Figure 5. High frequency test apparatus.

In the scope of the study, a fixed body attached to the modal shaker table, a test sample mounted inside this body and freely oscillating masses mounted on top of the test sample were used. The frequency values obtained as a result of the modal analysis of the fixed body were ensured to be outside the pre-amplitude band. The freely oscillating masses represent multiple measurement parameters for verifying the test method. Similarly, for the verification of the test method, only a sample consisting of rubber and the rubber damping element developed within the scope of the study were used.

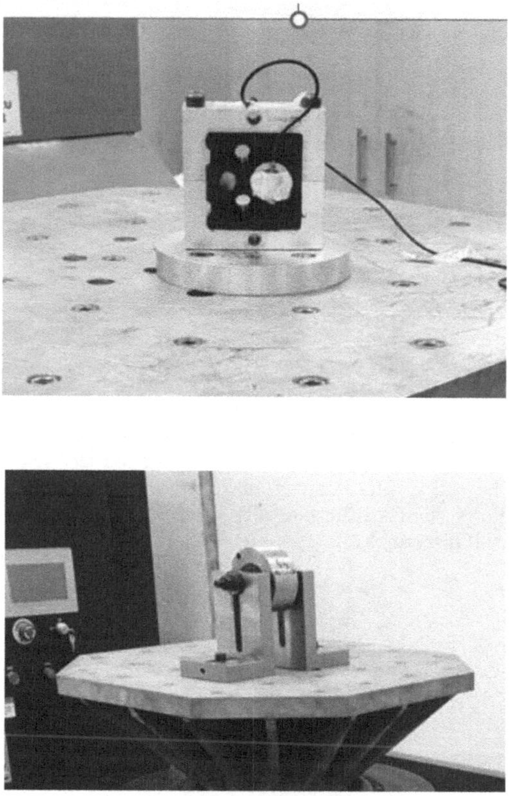

Figure 6. Test Rigs for developed test method.

2.3 *RPA tests for material properties*

In order to obtain the loss and storage modulus parameters of rubber materials, a testing device capable of operating within the frequency band is required. Due to the design challenges of creating a testing device capable of testing within the high-frequency band, the decision was made to utilize RPA (Rubber Process Analyzer) and DMA (Dynamic Mechanical Analyzer) devices. The RPA 2000 Test device can be programmed to perform frequency sweep or strain sweep tests on materials in the 0-35 Hz band using uncured rubber materials. During the test, crosslinking occurs in the elastomer material, allowing the rheological properties and frequency-dependent parameters of the material to be obtained. Additionally, the DMA test device can conduct material tests in the 0-400 Hz band. For this purpose, vulcanized samples designed to fit the geometry of the test device should be used. Within the scope of the study, the results from RPA and DMA were compared. Due to the similarity of the results, the decision was made to proceed with RPA within the scope of the study. The RPA device used is shown in Figure 7. Finite element analysis was performed on a simple disk model and two selected models from suspension bushes using data collected from the materials. By comparing the analysis results with the test results, the accuracy of the dynamic analysis results in the 1-35 Hz band was demonstrated.

Figure 7. RPA 2000 Test rig and Test Sample.

2.4 *Machine learning for material parameters*

Machine learning approach was used to obtain the loss and storage modulus values at high frequencies. Prototype vibration damping bushes with known geometry were tested in the test device mentioned in (Ref.7) 1-2500 Hz band. The test results were obtained using prototypes of bushes produced from rubber of different hardness but with the same geometry. The test results were introduced to the Odyssee software to obtain the loss and storage modulus values required for dynamic analyses. It is known that rubber compounds with different hardness will have different material parameters. FEA models of the prototypes were created. A loop was established between FEA and Odyssee. This loop aimed to calculate the loss and storage moduli on the optimization surface until the FEA test result matched the test results with 80% accuracy. The comparison between the results obtained from RPA and the material data obtained through machine learning concluded that the RPA device could be used in the design phase of the project.

Figure 8. High frequency test bench.

2.5 *Dynamic stiffness FEA analysis*

Using the material parameters obtained in the frequency band of 1-2500 Hz, the dynamic analyses of the product (Figure 3) were completed after completing the static analyses. Parallel to this, prototypes produced were tested in the developed test setup shown in Figure 6. The objective under this heading is to validate both the material data and the developed test setup, which are the ultimate goals of the study. Figure 9 shows the work flow of the study.

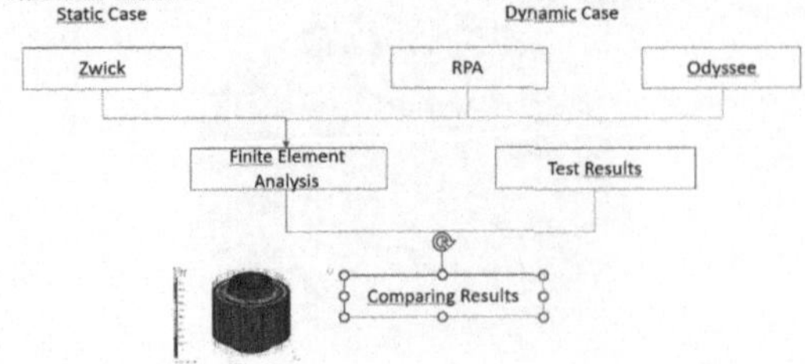

Figure 9. Workflow of study.

3 RESULTS

3.1 *Static and dynamic stiffness FEA analysis for developed anti-vibration mount*

The static test results for the developed bushing showed an 85% correlation with the test results. Plot 1 shows static stiffness results in radial 1, radial 2 and axial directions.

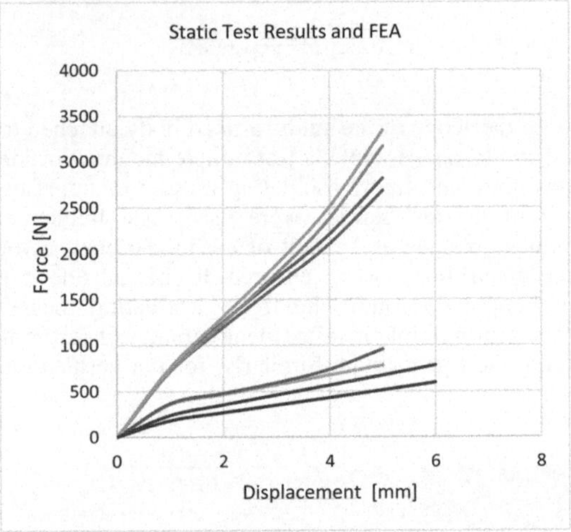

Plot 1. Static stiffness of developed mount (radial1, Radial2 and Axial Dir.)

Figure 10. CAD ve FEA models for static analysis.

The dynamic stiffness tests for the developed bushing were conducted in the developed test setup. The test results showed an 80% correlation. Plot 2 shows static stiffness results in radial 1, radial 2 and axial directions.

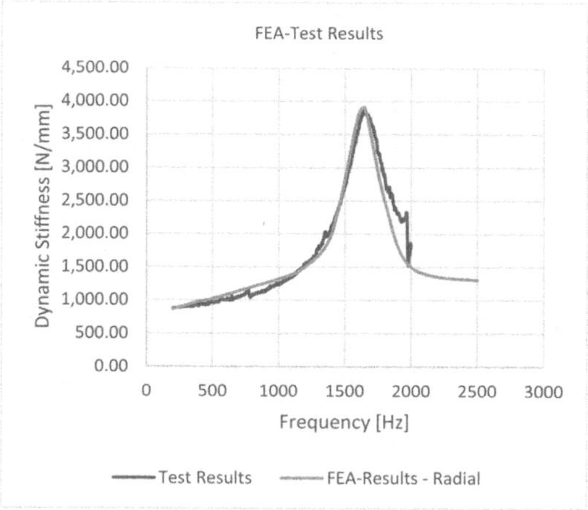

Plot 2. FEA vs Test Results for developed mount in radial direction.

3.2 *RPA correlation*

The results from RPA started with a disk sample and were subsequently validated on two rubber-metal vibration mounts. The results could be calculated with over 90% accuracy in both radial and axial directions.

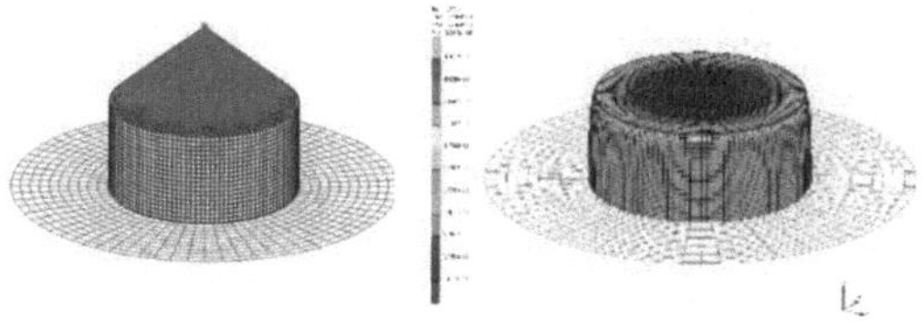

Figure 11. Rubber Disc FEA models.

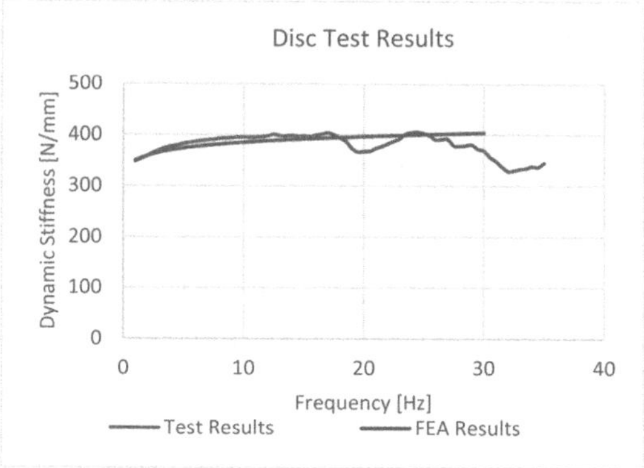

Plot 3. Rubber Disc FEA and Dynamic stiffness results.

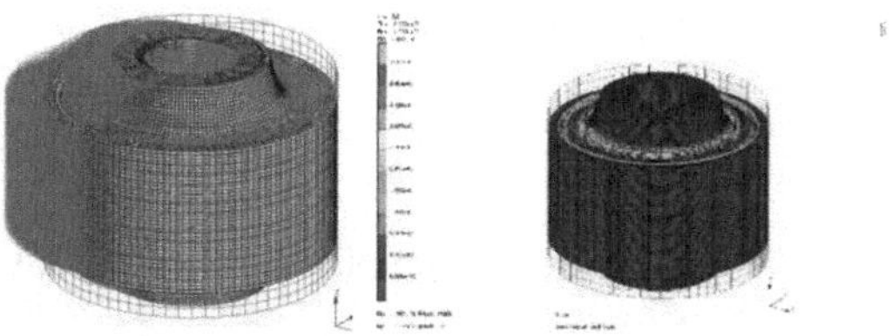

Figure 12. Sample Rubber Bush for dynamic correlation.

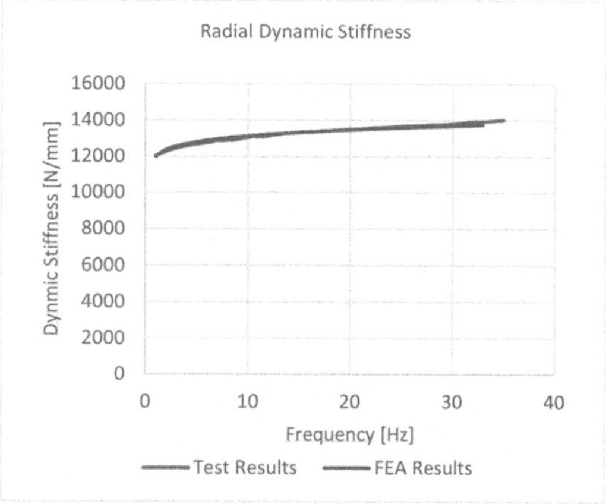

Plot 4. Rubber bush dynamic stiffness vs FEA.

3.3 *High frequency test bench results*

With support of machine learning approach, dynamic stiffness of reference mount can be simulated within 1-2500 Hz frequency band. Plot 5 shows the correlation between FEA and Test Results.

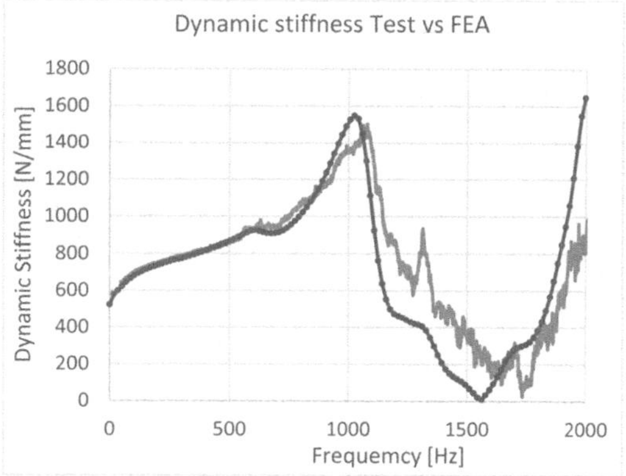

Plot 5. Dynamic stiffness vs FEA results for reference mount.

4 CONCLUSION

As a result of a detailed study, although results with all the features of an expensive test system could not be obtained, an FEA approach with over 80% accuracy, suitable for use in the design phase, has been developed. The test setup based on the modal shaker has been the greatest support in the development of this approach. Although there were challenges in processing and evaluating the data collected during the use of the test setup, such as extracting acceleration, displacement, and force data vectorially, meaningful conclusions have been reached. The collection of viscoelastic properties of recipe-based rubber compounds from the high-frequency band remains an area open for further development. Both the capabilities of test samples and test devices are intriguing for new studies. As a result of the study, hyperplastic and viscoelastic material data in the range of 45-70 ShA rubber hardness for use in the 1-2500 Hz band have been recorded in the database. With the developed setup and FEA approach, the development of motor mounts, gearbox mounts, or suspension bushes for EV vehicles has become possible.

Throughout the continuation of the study, with the support of enthusiastic team members, the research has expanded at every stage. We are indebted for their contributions to YAMAS Inc., to Prof. Dr. Murat Yazıcı from Bursa Uludağ University Automotive Department and to all team members.

REFERENCES

1. Elahinia, M. 2010. Reducing Noise and Vibration of Hydraulic Hybrid and Plug-In Hybrid Electric Vehicles.
2. Danko J. 2021. Dynamic Properties Modeling Analysis of the Rubber-Metal Elements for Electric Drive.
3. Tosun. M. 2019. Elektrikli Araçlarda Müşteri Beklentilerini Karşılamaya Yönelik Geliştirilen NVH Teknoloji Trendleri.
4. Xianpai, Z. 2015. Analysis of Motor Vibration Isolation System with Focus on Mount Resonances for Application to Electric Vehicles.
5. Kovarik, B. 2016. Modeling of the Nonlinear Frequency Dependent Engine Mount Dynamic Stiffness.
6. Uz, C. 2015. Design Of a Test Setup for The Characterization of Vibration Isolators.
7. Troge. J. 2014. Acoustical simulation and evaluation of load-dependent hydraulic engine mounts using two-port description.
8. Abe. J. 2021. US2021/0222753 Cylindrical antivibration device for motor mount.
9. Geisen. J. 2022 US2022/0042567 Bush.

(Elahinia, March, 2010)
(Danko, 2021)
(Tosun, 2019)
(Xianpai, 2015)
(Kovarik, 2015)
(Uz, 2015)
(Troge, 2014)
(Patent No. US2021/0222753, 2021)
(Patent No. US2022/0042567, 2022)

Fatigue and fracture

Constitutive Models for Rubber XIII – Dal (Ed.)

A stochastic phase-field approach for the failure of rubberlike materials

Kemal Açıkgöz, Bülent Efe Tanış & Hüsnü Dal
Department of Mechanical Engineering, Middle East Technical University, Ankara, Türkiye
Computational Micromechanics Laboratory, CAD/CAM Robotics Center, Middle East Technical University, Ankara, Türkiye

ABSTRACT: Although in the classical phase-field approach the assumption is a homogenized distribution of the critical energy release rate over the material, leading to catastrophic failure and hence brittle fracture; the experimental evidence regarding the fracture behaviour of rubber-like materials implies ductile-like progressive failure. To characterize and model the functioning of rubber-like materials, the phenomena is extremely vital as fatigue failure crack growth is holds substantial importance in engineering design processes. In this work, the critical entropic energy distribution is taken as a spatially varying material parameter to represent the stochasticity. Conducting a series of fracture experiments using die-cut v-shaped double edge notched specimens of unfilled styrene-butadiene rubber (SBR), stochastic properties of the rubbers were able to be shown over load-displacement curves. Results demonstrated ductile-like fracture in contrast with laser-cut specimens where the data obtained indicated the exhibition of a pronounced tendency towards sharp brittle fracture. Analyzing the results of the work, It is observed that ductile-like fracture propagates intermittently and follows intricate irregular paths even on homogeneous specimens. This complex behaviour challenges the conventional notions of fracture mechanics in rubber-like materials and provides significant insight for engineering design and materials science, leading the research to pave the way for more reliable engineering applications.

1 INTRODUCTION

Design and reliability of rubber components heavily depends on the prediction of failure in such materials. In this context, predictive theories that account for crack initiation, crack propagation under multiaxial deformation states is of crucial importance. There is a complex interplay between void initiation, void coalescence and formation of macroscopic crack, see Dal et al. (2018). Particularly, description of a failure surface that accounts for volumetric, and multiaxial shear deformations and incorporation into gradient-type damage or the so-called *phase-field* approach to diffusive crack is still an open issue, see Açıkgöz & Dal (2022). In this work, we confine ourselves to the quasi-static and monotonically increasing deformations where the rate-dependent and stable fatigue crack growth mechanisms are excluded. The rupture of rubberlike materials dates back to successive experimental works of Rivlin & Thomas (1953), Thomas (1955), Greensmith & Thomas (1955) and Greensmith (1956) where tearing energy for rubbers was defined as a critical energy release rate in the sense of Griffith (1920). Lake & Thomas (1967) proposed a simple *molecular-theoretical model*, that relates critical surface or tearing energy to the molecular structure of rubber. A theory that incorporates bond rupture through atomic potentials has been proposed by Dal & Kaliske (2009). Such an ansatz lack the mesh objectivity leading to mesh dependent localization during crack propagation, see also Behnke et al. (2013).

The classical theory of brittle fracture which was laid out by Griffith (1920) and Irwin (1962) has several shortcomings on determining the crack initiation, curvilinear crack paths, crack kinking and branching angles. The phase field approach aims to circumvent these shortcomings by providing a diffusive crack topology, which smears out the sharp crack surfaces, with the help of an auxilliary field variable. This is achieved by variational methods based on energy minimization proposed by Francfort & Marigo (1998). For detailed account of the subject we refer to Bourdin et al. (2000). Therein, regularized setting is obtained by Γ-convergence theory of Mumford & Shah (1989). For the application of the crack phase field approach to rubberlike materials, we refer to Schanzel et al. (2013a, 2013b). Moreover, a finite viscoelastic rheological model based on Reese & Govindjee (1998) is successfully incorporated in the phase-field modeling by Yin & Kaliske (2020) for rate dependent effects for both the mechanical and fracture responses. A theory that combines a generalized energy based failure that captures multiaxiality i.e. dilatational and isochoric deformations was proposed in our former contribution (Açıkgöz and Dal 2022).

DOI: 10.1201/9781003516880-33

The paper is organized as follows: In Sections 2 and 3 we briefly layed out the fundamentals of the multi-physics problem of phase-field by introducing field variables and summarizing the extended eight-chain model. In Section 4, a stochastic extension of this particular phase-field approach is introduced. In Section 5 the results of incorporating a stochastic approach is discussed along with experimental and simulation results. Finally, concluding remarks close the paper.

2 FUNDAMENTALS OF THE MULTIPHYSICS PROBLEM

In this section we will briefly cover the extended eight-chain model and associate its terms with the history fields of the phase-field model. For further information on the hyperelastic constitutive models, we refer to Dal et al. (2021).

2.1 *Stored energy functional for the hyperelastic solid*

In this part, we will briefly outline the extended-eight chain model for the hyperelastic response of rubber-like materials, see Dal et al. (2020). The compressible eight-chain model can be written as

$$\Psi(J,\lambda_r) = \tilde{\psi}_v(J) + \tilde{\psi}_f(\lambda_r) \tag{1}$$

where $\tilde{\psi}_v(J)$ and $\tilde{\psi}_f(\lambda_r)$ are purely volumetric (energetic) and entropic components of the free energy function. The entropic response of the eight-chain model depends only on the I_1 term, where

$$\lambda_n = \sqrt{\frac{\lambda_1^2+\lambda_2^2+\lambda_3^2}{3}} = \sqrt{I_1/3} \tag{2}$$

$\mathcal{L}$ is defined as the mean-square root average of the principal stretches λ_i in the sense of Arruda & Boyce (1993). The free-chain part of the stored energy function for the eight-chain model incorporates the inverse Langevin function[1] $\mathcal{L}^{-1}(\lambda_r)$ in its definition along with $\sqrt{N}$ as the chain extensibility limit characterized by the number of segments N, that is

$$\tilde{\psi}_f(\lambda_r) = \mu N\left(\lambda_r\mathcal{L}^{-1}(\lambda_r) + \ln\frac{\mathcal{L}^{-1}(\lambda_r)}{\sinh\mathcal{L}^{-1}(\lambda_r)}\right), \tag{3}$$

where $\lambda_r = \lambda_n/\sqrt{N}$ is the relative average network stretch such that $\lambda_n \in [0,N)$, $\lambda_r \in [0,1)$, and μ is the shear modulus. A volumetric response can be obtained for compressible behavior with an ansatz

$$\tilde{\psi}_v(J) = \frac{\chi}{4}\left[(\ln J)^2 + (J-1)^2\right] - \frac{\mu}{3}\frac{3N-1}{N-1}\ln J, \tag{4}$$

where $\chi = \kappa - \frac{2}{3}\mu$ is the first Lamé constant, κ is the bulk modulus, and $J = \sqrt{I_3(C)} = \lambda_1\lambda_2\lambda_3$ is the volumetric stretch. The extended eight-chain model (Dal, Gültekin, and Açıkgöz 2020) micromechanically motivates an entropic I_2 term for the tube-like constraint of the free-chains which relies on the definition,

$$I_2 = \nu_1^2 + \nu_2^2 + \nu_3^2 \tag{5}$$

where $\nu_i = J/\lambda_i$ are *areal stretches*. Similar to (2), a network averaged areal stretch can be defined

$$\nu_n = \sqrt[3]{\frac{\nu_1^2+\nu_2^2+\nu_3^2}{3}} = \sqrt[3]{\frac{I_2}{3}}. \tag{6}$$

The constraint part of the extended eight-chain model reads

$$\tilde{\psi}_c(\nu_n) = \mu_c(\nu_n - 1) \tag{7}$$

where μ_c is the associated shear modulus. Finally, the complete free energy function expression for the extended eight-chain model

$$\psi(J,\lambda_r,\nu_n) = \tilde{\psi}_v(J) + \tilde{\psi}_f(\lambda_r) + \tilde{\psi}_c(\nu_n), \tag{8}$$

can be written with modification on the volumetric part

$$\tilde{\psi}_v(J) = \begin{array}{l} \frac{\chi}{4}\left[(\ln J)^2 + (J-1)^2\right] \\ -\frac{\mu}{3}\frac{3N-1}{N-1}\ln J - \frac{4}{9}\mu_c\ln J \end{array} \tag{9}$$

where the last term in (4) and the last two terms in (9) are added in order to normalize the stresses at undeformed configuration. The extended eight-chain model has the added benefit of performing better for (equi)biaxial deformation fits, with the additional (3rd) parameter μ_c.

3 BALANCE EQUATIONS OF THE MULTI-FIELD PROBLEM

The energy storage functional E for the specific choice of free energy function (8) can be written as

$$E(\varphi,d) = \int_B \Psi(J,\lambda_r,\nu_n,d)dV \tag{10}$$

with

$$\Psi := g_v(d)\tilde{\psi}_v(J) + g_f(d)\tilde{\psi}_f(\lambda_r) + g_c(d)\tilde{\psi}_c(\nu_n) \tag{11}$$

where $g_v(d)$, $g_f(d)$, and $g_c(d)$ are monotonically decreasing functions which govern the degradation of the separate terms of (8). The terms $\tilde{\psi}_v(J)$, $\tilde{\psi}_f(\lambda_r)$, and $\tilde{\psi}_c(\nu_n)$ represent the intact reference energies of

1 For the computation of the inverse Langevin function, we refer to Basdemir & Dal (2022) and references therein.

the volumetric, free-chain, and constraint parts respectively. The rate of energy storage reads

$$\varepsilon(\dot{\varphi}, \dot{d}; \varphi, d) := \frac{d}{dt}E = \int_B [\boldsymbol{P} : \dot{\boldsymbol{F}} - f\dot{d}]\, dV \qquad (12)$$

where, we introduce, per definition, the first Piola-Kirchhoff stress expression

$$\begin{aligned} \boldsymbol{P} &:= \partial_F \psi(\boldsymbol{F}, d) \\ &= g_v(d)\boldsymbol{P}_v + g_f(d)\boldsymbol{P}_f + g_c(d)\boldsymbol{P}_c \end{aligned} \qquad (13)$$

and the energetic force

$$\begin{aligned} f := \; & -\partial_d \Psi(J, \lambda_r, \nu_n, d) = -g_v'(d)\,\tilde{\psi}_v(J) \\ & -g_f'(d)\,\tilde{\psi}_f(\lambda_r) - g_c'(d)\,\tilde{\psi}_c(\nu_n) \geq 0 \end{aligned} \qquad (14)$$

which is work conjugate to $\dot{d}$. The external mechanical loading is defined by the external load power functional

$$\mathcal{P}(\dot{\varphi}) = \int_{\mathcal{B}} \tilde{\gamma} \cdot \dot{\varphi}\, dV + \int_{\partial \mathcal{B}_t} \tilde{t} \cdot \dot{\varphi}\, dV, \qquad (15)$$

where $\tilde{\gamma}$ and $\tilde{t}$ are the volume specific body force and the prescribed surface tractions, respectively. A viscous dissipation functional, i.e.

$$\mathcal{D}_\eta(\dot{d}, \beta; d) = \int_{\mathcal{B}} [\beta \dot{d} - \frac{1}{2\eta}\langle t_c(\beta; d, \nabla d)\rangle^2] \mathrm{d}V, \qquad (16)$$

where the viscosity η determines the viscous overforce governing the evolution of $\dot{d}$. In (16) the positive values for the threshold function t_c are always filtered out owing to the ramp function $\langle x \rangle = (x + |x|)/2$. Accordingly, the corresponding viscous rate-type potential reads

$$\Pi_\eta = \varepsilon + \mathcal{D}_\eta - \mathcal{P}. \qquad (17)$$

In the rate-independent setting, the particular choice of the free energy (11) results in the the crack driving force (14) which takes the following form

$$\begin{aligned} f &= g_c \delta_d \gamma_l := \frac{g_c}{l}[d - l^2 \Delta d] \\ &= -g'_v(d)\Psi_v(J) - g_f'(d)\Psi_f(\lambda_n) - g_c'(d)\Psi_c(\nu_n) > 0\,. \end{aligned}$$

Let the following functions be defined

$$h_i(d) = -g_i'(d), \quad \text{where} \quad i = \{v, f, c\}. \qquad (18)$$

Then the crack driving force (14) can be recast as

$$f = h_v(d)\Psi_v(J) + h_f(d)\Psi_f(\lambda_n) + h_c(d)\Psi_c(\nu_n)\,.$$

Replacing $\psi_{cr} = g_c/l$, we obtain the equation

$$[d - l^2 \Delta d] = h_v(d)\bar{\mathcal{H}}_v + h_f(d)\bar{\mathcal{H}}_f + h_c(d)\bar{\mathcal{H}}_c$$

for the evolution of the phase field in the case of loading *and* unloading. Note that this equation equips the crack topology by the local *crack source* on the right hand side. We introduce the *local history field* of a maximum positive reference energy of the polymer network

$$\bar{\mathcal{H}}_i := \max_{s \in [0,t]} \langle \frac{\psi_i}{\psi^i_{cr}} - 1 \rangle \qquad (19)$$

obtained in a typical multiaxial loading scenario, where $i = \{v, f, c\}$ for volumetric, free-chain, and constraint part responses. The variational formulation requires $\psi^v_{cr} = \psi^f_{cr} = \psi^c_{cr} = g_c/l$. Herein, distinct local evolution is foreseen for the energetic and entropic response of the material. In order to (i) enforce irreversiblity, e.g. to prevent crack healing, and (ii) to describe a distinct elastic surface, the specific form (22) suppresses damage evolution below the critical energy threshold ψ_{cr} and uses the maximum value of local crack source throughout the loading history. In this respect, the Macaulay brackets filter out the positive values for $(\psi_i/\psi^i_{cr} - 1)$ for $i = \{v, f, c\}$ and keeps the solid intact until the failure surface is reached. In the above expression, free-chain and topological constraint parts of entropic response are considered separately. This scenario is the most general case for the crack initiation and propagation. Such an ansatz allows distinct cavitation and shear-type rupture modes. A more specific form can be obtained if the degradation functions for the free-chain and constraint response are identical $g_f(d) = g_c(d)$ such that

$$[d - l^2 \Delta d] = h_v(d)\bar{\mathcal{H}}_v + h_f(d)\bar{\mathcal{H}}_e \qquad (20)$$

and a specific choice of local crack driving forces in the sense

$$\bar{\mathcal{H}}_e := \max_{s \in [0,t]} \langle \frac{\psi_f(\lambda_n, s) + \alpha_c \psi_c(\nu_n, s)}{\psi^e_{cr}} - 1 \rangle\,, \qquad (24)$$

where

$$\alpha_c = \frac{\psi^f_{cr}}{\psi^c_{cr}} \quad \text{and} \quad \psi^e_{cr} = \psi^f_{cr}. \qquad (25)$$

With this notion at hand, the proposed fracture phase field model may be reduced to the compact two-field coupled problem

$$\begin{aligned} 1:&\quad \mathrm{Div}\, \boldsymbol{P} + \rho_0 \bar{\gamma} = 0, \\ 2:&\quad d - l^2 \Delta d = h_v(d)\bar{\mathcal{H}}_v + h_f(d)\bar{\mathcal{H}}_e, \end{aligned} \qquad (26)$$

which determines the current deformation and phase fields φ and d in terms of the definitions (13), (23), and (24).

4 STOCHASTICITY ON CRITICAL ENERGIES

In this framework, material failure initiates upon reaching the critical entropic energy ψ^e_{cr}. In our in-house

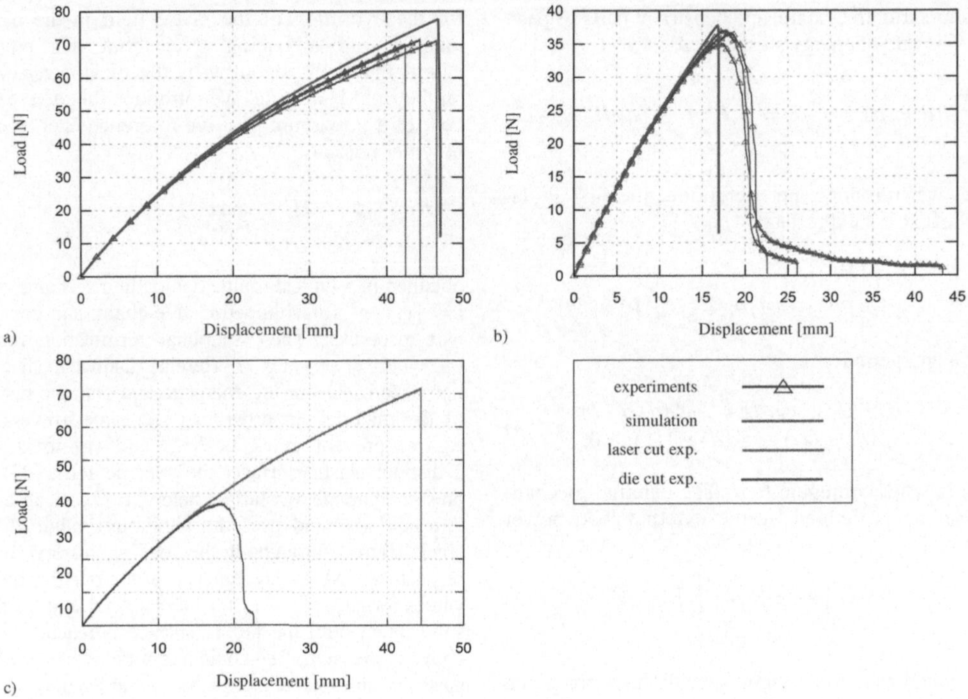

Figure 1. Experiments and simulation results for a) laser cut specimens, and b) die-cut specimens, and c) comparison of average behaviors of laser-cut and die-cut experiments.

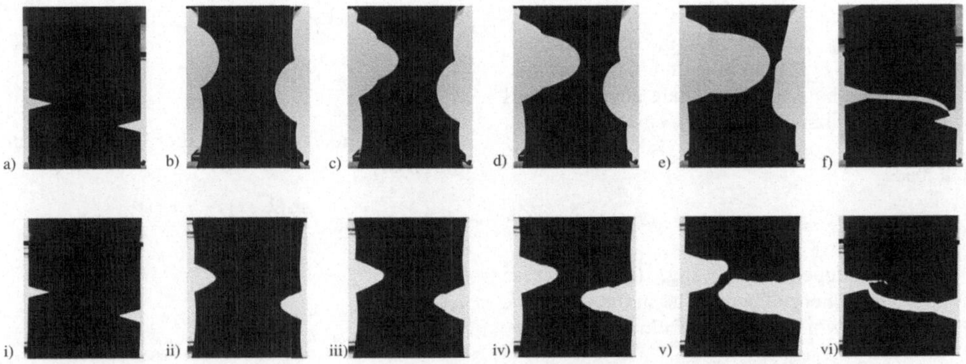

Figure 2. Experiment snapshots for laser-cut (a-f) and die-cut (i-vi) asymmetric V-notched specimens.

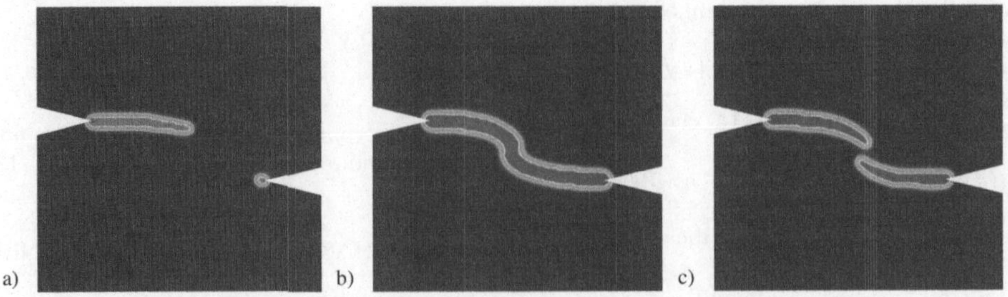

Figure 3. Simulation results for same stochastic parameters μ, and σ^2.

experiments, for a homogeneous die-cut specimen, the fracture pattern was jagged and deviated from the expected paths. This deviation is attributed to varying critical energy release rates throughout the material's geometry. When a single ψ^e_{cr} value is used uniformly across the geometry, the result is a brittle and catastrophic failure, analogous to a dam breach, where the crack propagates until complete rupture. Conversely, when ψ^e_{cr} is treated as a spatially variable material parameter field, with different values at each material point, a progressive failure mechanism is observed. This occurs because the propagating crack may halt upon encountering a region with a higher ψ^e_{cr} until this higher value is also exceeded. This variable ψ^e_{cr} field results in a jagged crack path, diverging from the theoretical homogeneous behavior observed with a uniform ψ^e_{cr} parameter.

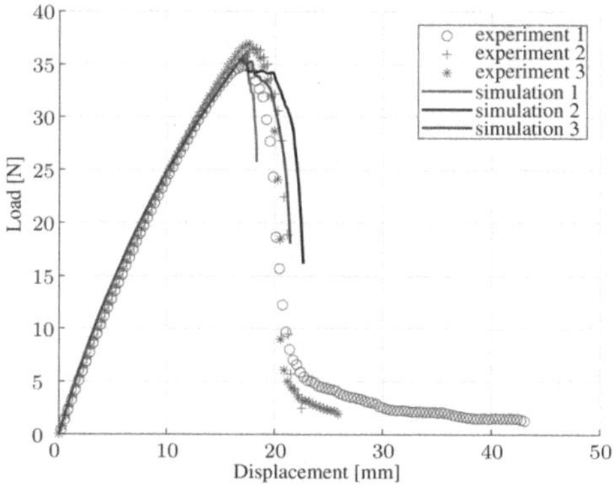

Figure 4. Load displacement curves for asymmetric experiments and stochastic phase-field simulations.

In this study, the ψ^e_{cr} material parameter field is generated by assigning normally distributed random ψ^e_{cr} values to each element node. The ψ^e_{cr} field is defined on the reference configuration and remains constant throughout deformation. Each material point (Gauss point) is subjected to this ψ^e_{cr} field via standard finite element interpolation. Specifically, we utilize

$$\psi^e_{cr}(\boldsymbol{X}) \sim \mathcal{N}(\mu, \sigma^2), \tag{27}$$

$$f(\psi^e_{cr}) = \frac{1}{\sqrt{2\pi\sigma^2}} \exp\left(\frac{-(\psi^e_{cr} - \mu)^2}{2\sigma^2}\right) \tag{28}$$

where (24) represents samplings for each $\boldsymbol{X}$ from normal distribution (25) for ψ^e_{cr}, here, $\mu = \bar{\psi}^e_{cr}$ is the mean, and σ^2 is the variance. This model allows for the incorporation and adjustment of ductile-like behavior by controlling the standard deviation σ. The load-displacement experiments can be utilized to tune the μ and σ parameters. In addition, cyclic fracture tests in pure-shear experiments form can be carried out to fine tune the σ value. In this work, we considered asymmetrically cut V-notched specimens to show the effect of σ on crack paths and load-displacement curves.

5 RESULTS AND DISCUSSION

The extended eight-chain model parameters are obtained first, using a series of uniaxial, equibiaxial, and volumetric compression tests. These are provided in Table 1. To illustrate both brittle and ductile fracture behavior observed in experiments, asymmetrically cut V-notched unfilled SBR specimens were employed. The fracture toughness of these specimens significantly increases when cut using a laser. The laser process cauterizes the specimen edges, thereby causing an increase in the energy required to initiate fracture. A notable distinction in fracture toughness is observed between laser-cut and die-cut specimens (cf. Figure 1). Due to the increase in the fracture toughness, laser-cut specimens can withstand higher displacements and result in a catastrophic failure. In both Figures 1a) and b), the simulation results for the framework presented in this paper without stochasticity is also provided,i.e., the ψ^e_{cr} field is homogeneous on the entire geometry. Without the stochasticity, the phase-field simulations can exhibit only brittle failure behavior, which is akin to brittle failure exhibited by laser-cut specimens. The laser-cut and die-cut specimens provide a contrasting view between brittle and ductile like behavior and provide a pronounced and obvious examples for both cases. To further illustrate the key aspects of this work, consider the snapshots of the fracture for laser and die-cut specimens given in Figure 2(a-f). The crack path for laser-cut specimen displays a sharp path, and the corresponding load-displacement curve in Figure 1a) exhibits a brittle failure. On the other hand, the die-cut specimen snapshots in Figure 2(i-vi) displays a jagged and irregular crack path, which is also manifested on load-displacement curves in Figure 1b) as a ductile-like behavior.

The simulation results for homogeneous and stochastic critical entropic energies are given in Figure 4 for die-cut specimens. The load-displacements after the initiation of the fracture follows a ductile-like profile, which can be adjusted using the σ^2 value on the ψ^e_{cr} field. On the other hand, the mean failure initiation point is adjusted using the $\mu = \bar{\psi}^e_{cr}$ parameter. In the Figure 3, the final achieved steps of the simulation snapshots are provided. Since the fracture phenomenon is highly nonlinear, and the solution strategy is quasi-static, obtaining the completed fracture simulations is strenuous. That's the reason one can observe incomplete fracture patterns in Figure 3 and also in corresponding load-displacement curves in Figure 4.

The exact implementation details of this framework involves obtaining the average critical entropic energies, as well as utilizing artificial stiffness of $k = 10^{-2}$ in the sense of Miehe et al. (2010). A single $l = 1$ [mm] length scale parameter is utilized for all simulations, which corresponds to the size of 7 elements around the crack path.

Table 1. The material parameters for the unfilled SBR obtained using a sereis of uniaxial, equibiaxial, and volumetric compression tests.

Material parameters for the ground state				
	$\mu[MPa]$	$N[-]$	$\mu_c[MPa]$	$X[MPa]$
Unfilled SBR	0.34	90	0.91	462

This framework provides a way to simulate progressive failure. Therefore, we can see ductile-like failure during crack evolution, cf. Figures 4 and 1. This paves the way of modeling fatigue failure in rubber-like materials. A progressive failure mechanism, therefore, is introduced in the phase-field method that can also be coupled with a shrinking failure surface idea based on a damage mechanism to potentially allow physically realistic fatigue life-time predictions.

6 CONCLUSION

In this study, we introduce an energy-based failure criterion derived from the distinct distortional and dilatational terms of the extended eight-chain model. Subsequently, we develop a novel phase-field approach incorporating these terms, enabling prediction of failure under both entropic and energetic deformations. To address entropic deformations specifically, the energy-based criterion is integrated with a stochastic method, yielding realistic fracture simulations that can accurately depict crack propagation paths. The load-displacement curves exhibit ductile-like behavior consistent with observations from our in-house experiments on die-cut asymmetric V-notched specimens.

ACKNOWLEDGEMENT

The financial support provided by TUBİTAK under grant number 315M140 and METU Scientific Research Project Coordination Unit under grant DOSAP-B-302-2019-10052 are gratefully acknowledged.

REFERENCES

Açıkgöz, K. and H. Dal (2022). A generalized phase-field approach for the failure of rubberlike materials. *Constitutive Models for Rubber XII*, 312–320.

Arruda, E. M. and M. C. Boyce (1993). A three–dimensional constitutive model for the large stretch behavior of rubber elastic materials. *Journal of the Mechanics and Physics of Solids 41*, 389–412.

Başdemir, S. and H. Dal (2022). A one-pass predictor-corrector algorithm for the inverse langevin function. *Mathematics and Mechanics of Solids*, doi:10.1177/10812865221102549.

Behnke, R., H. Dal, G. Geißler, B. Näser, C. Netzker, and M. Kaliske (2013). *Macroscopical Modeling and Numerical Simulation for the Characterization of Crack and Durability Properties of Particle-Reinforced Elastomers*, pp. 167–226. Berlin, Heidelberg: Springer Berlin Heidelberg.

Bourdin, B., G. A. Francfort, and J. J. Marigo (2000). Numerical experiments in revisited brittle fracture. *Journal of the Mechanics and Physics of Solids 48*, 797–826.

Dal, H., K. Açıkgöz, and Y. Badienia (2021). On the Performance of Isotropic Hyperelastic Constitutive Models for Rubber-Like Materials: A State of the Art Review. *Applied Mechanics Reviews 73*(2), 020802.

Dal, H., B. Cansz, and C. Miehe (2018). A three-scale compressible microsphere model for hyperelastic materials. *International Journal for Numerical Methods in Engineering 116*(6), 412–433.

Dal, H., O. Gültekin, and K. Açıkgöz (2020). An extended eight-chain model for hyperelastic and finite viscoelastic response of rubberlike materials: Theory, experiments and numerical aspects. *Journal of the Mechanics and Physics of Solids 145*, 104159.

Dal, H. and M. Kaliske (2009). A micro-continuum-mechanical material model for failure of rubber-like materials: Application to ageing-induced fracturing. *Journal of the Mechanics and Physics of Solids 57*(8), 1340–1356.

Francfort, G. A. and J. J. Marigo (1998). Revisiting brittle fracture as an energy minimization problem. *Journal of the Mechanics and Physics of Solids 46*, 1319–1342.

Greensmith, H. (1956). Rupture of rubber. iv. tear properties of vulcanizates containing carbon black. *Journal of Polymer Science 21*, 175–187.

Greensmith, H. and A. Thomas (1955). Rupture of rubber. iii. determination of tear properties. *Journal of Polymer Science 18*, 189–200.

Griffith, A. A. (1920). The phenomena of rupture and flow in solids. *Philosophical Transactions of the Royal Society London A 221*, 163–198.

Irwin, G. R. (1962). Crack-extension force for a part-through crack in a plate. *Journal of Applied Mechanics 29*, 651–654.

Lake, G. and A. Thomas (1967). The strength of highly elastic material. *Proceedings of the Royal Society of London. Series A, Mathematical and Physical Sciences 300*, 108–119.

Miehe, C., M. Hofacker, and F. Welschinger (2010). A phase field model for rate-independent crack propagation: Robust algorithmic implementation based on operator splits. *Computer Methods in Applied Mechanics and Engineering 199*(45), 2765–2778.

Mumford, D. and J. Shah (1989). Optimal approximations by piecewise smooth functions and associated variational problems. *Communications on Pure and Applied Mathematics 42*, 577–685.

Reese, S. and S. Govindjee (1998). A theory of finite viscoelasticity and numerical aspects. *International Journal of Solids and Structures 35*(26), 3455–3482.

Rivlin, R. and A. Thomas (1953). Rupture of rubber. i. characteristic energy for tearing. *Journal of Polymer Science X, No. 3*, 291–318.

Schänzel, L., H. Dal, and C. Miehe (2013). On micromechanically-based approaches to failure in polymers. *PAMM 13*(1), 557–560.

Schanzel, L., H. Dal, and C. Miehe (2013). Phase field modeling of fracture in rubbery polymers. *Constitutive Models for Rubber VIII*, 335–341.

Thomas, A. (1955). Rupture of rubber. ii. the strain concentration at an incision. *Journal of Polymer Science 18*, 177–188.

Yin, B. and M. Kaliske (2020, Feb). Fracture simulation of viscoelastic polymers by the phase-field method. *Computational Mechanics 65*(2), 293–309.

Constitutive Models for Rubber XIII – Dal (Ed.)
© 2025 The Author(s), ISBN 978-1-032-85138-9

Evaluating the cavitation of a penny shaped flaw under hydrostatic tension through fracture mechanics

R.J. Windslow & T.W. Hohenberger
SLB, Houston, Texas, USA

J.J.C. Busfield
School of Engineering and Material Science, Queen Mary University of London, London, UK

ABSTRACT: Fracture of elastomers is a well-studied topic with an established analytical solution. The solution can be applied to a range of flaws, both surface and embedded, to evaluate the fracture potential under a far field strain. However, these equations do not actually cover all potential fracture cases. Elastomeric behavior is split into deviatoric and volumetric; however, typical fracture analysis evaluates deviatoric fracture potential only, not volumetric. A special case of fracture under volumetric conditions is cavitation. Here an embedded flaw undergoes far field hydrostatic tension causing unstable growth to the point of rupture. Cavitation is typically evaluated in terms of instability mechanics rather than fracture mechanics, as prior fracture mechanics approaches are typically restricted to small strains, limiting their ability to predict cavitation to the point of failure. Here, using Finite Element Analysis, the tearing energy of a flaw undergoing cavitation is studied and used to develop an analytical solution capable of modelling cavitation using a fracture mechanics approach that is valid for high strains.

1 INTRODUCTION

1.1 *Fracture mechanics of elastomers*

The inherent fracture resistance of an elastomer is typically evaluated by studying how a surface level edge crack propagates through the material. For example, using the trouser tear test piece, the planar edge crack growth test or related geometries that were proposed by Rivlin & Thomas (1953). Whilst characterisation tests evaluate crack growth behavior in terms of surface flaws, in the majority of engineering applications failure initiates from internal flaws that propagate to sufficient sizes such that they compromise the integrity of the part.

Given that an elastomer's tearing energy relationship to crack growth rate is an inherent material property, tear strength relationships developed from edge cracked test specimens do translate to internal flaws. However, the relationship between crack length (or radius) and tearing energy does vary significantly between internal flaws and surface flaws. This variance is captured through the proportionality parameter, k.

In general, the tearing energy, T, is found to form a linear relation to the crack's length/radius, c, with the far field cracking energy density, w_c, and the proportionality parameter, k.

$$T = kw_c c \qquad (1)$$

Historically, the proportionality parameter was only well defined for uniaxial deformations (Gough & Muhr 2005, Yeoh 2006). However, recent work by Windslow et al. (2020) derived a biaxiality dependent proportionality parameter, allowing a single equation to evaluate the tearing energy through multiple deformation modes including planar and biaxial tension.

Through these equations a crack's potential to fracture under far field load is relatively well described. However, these equations do not actually cover all potential fracture cases. The issue is that these tearing energy equations relate to fracture under deviatoric loading conditions, not volumetric.

Elastomeric behavior is split into deviatoric and volumetric; however, typical fracture analysis evaluates deviatoric fracture potential only. It has been shown that volumetric compression leads to improved fracture resistance, as the hydrostatic pressure prevents voids from opening (Lake et al. 1992). The opposite is true for volumetric tension; under this condition, the hydrostatic pressure causes the flaw to expand rapidly leading to a high-risk fracture process. Under these conditions the crack is known to propagate to failure; however, the underlying mechanics are very different relative to the deviatoric modes.

Expansion of an internal void to fracture under volumetric tension is referred to as cavitation. It

DOI: 10.1201/9781003516880-34

plays a critical role in many processes from rapid gas decompression damage in elastomer seals to debonding failure in elastomer joints.

1.2 *Cavitation type fracture*

Cavitation in elastomers was first studied by Gent & Lindley (1959) through what is referred to as the poker chip test. A thin circular disk of elastomer was bonded to two metal surfaces which were pulled apart to induce a hydrostatic tension within the disk. This led to a network of cavities arising within the elastomer ultimately leading to failure.

By treating the problem as a single spherical void in an infinitely large Neo-Hookean system, they determined that the pressure acting on the void's wall was a function of the elastomer's shear modulus, G, and the principal stretch in the void's wall, λ, Equation 2. Later research confirmed the validity of this equation through FEA (Busfield et al. 1996).

$$P = \frac{G}{2}\left(5 - \frac{4}{\lambda} - \frac{1}{\lambda^4}\right) \qquad (2)$$

By realizing that the void must rupture if stretch tended towards infinity, they contended that the critical pressure, P_C, at which a void cavitated was a function of the material's shear modulus:

$$P_C = \frac{5G}{2} \qquad (3)$$

Another key point from this equation is the undeformed state, λ=1. This suggests that the void wall pressure must be greater than 1.5x the elastomer's stiffness before the void can expand. This then defines an initiation pressure, P_0, below which the elastomer can resist cavitation.

$$P_0 = \frac{3G}{2} \qquad (4)$$

Critically, this approach evaluates the failure point not through the tearing energy, but rather when a critical pressure is reached. It essentially treats cavitation as an instability problem rather than a fracture process. That is not to say the equation is irrelevant to fracture analysis; rather it provides critical information on resisting cavitation. Because of this, elastomers used in the Oil and Gas industry typically have very high Young's moduli, around 15MPa.

Plotting Equation 2 in Figure 1, it becomes clear as to why this is an instability problem. Initially the elastomer behaves in a stiff manner, requiring a large change in pressure to expand the void. As the void grows there is less resistance to expansion until ultimately a small change in pressure causes a great deal of expansion and possibly damage. This is therefore an instability problem.

In their initial model, Gent & Lindley (1959) assumed the voids were spherical. Gent & Wang (1991) later revisited this proposing that they were better modelled as flat penny shaped cracks. In the same work they found that the critical pressure hypothesis only held true for voids with an initial radius between 0.5μm and 1mm. Smaller voids required greater energy to rupture due to surface tension effects and had higher strains in their walls, two conditions that are not amenable to a Neo-Hookean model.

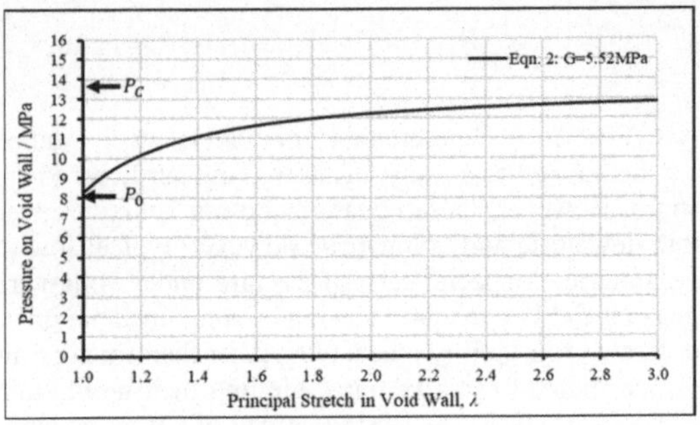

Figure 1. Elastic instability of an expanding spherical void in a finite Neo-Hookean body.

As described above, there are benefits to evaluating this problem through instability mechanics. However, this is best suited to single cycle, instant failure scenarios. It cannot account for how time or loading cycles may allow damage to accumulate leading to gradual failure. In reality incremental damage from time or loading cycles plays a major role, particularly in engineering applications. For example, a de-gassing seal will be bled over a period of time. Evaluating how incremental damage causes a crack to propagate to the point of failure requires a fracture mechanics approach to be implemented rather than an instability approach.

Stevenson & Morgan (1995) and Busfield et al. (1996) attempted to treat cavitation as a fracture process rather than an instability by applying fracture mechanics to the cavitation of a penny shaped flaw. Both theorems were based on Sneddon (1946) in which a penny shaped flaw of radius, c, was modelled at the center of an infinitely large sheet. In this case, the elastic energy local to the flaw, U, is related to the material's Young's modulus, E, the Poisson's ratio, v, the pressure acting on the flaw's wall, P, and the crack's radius.

$$U = \frac{8(1-\nu^2)}{3E}P^2c^3 \qquad (5)$$

In the case of Busfield et al., their adaption started by assuming the elastomer was incompressible, such that $3G = E$. By then defining that each face of the disk-shaped crack had an area, $a = \pi c^2$, the tearing energy, T, could be represented as:

$$T = \frac{\Delta U}{\Delta a} = \frac{P^2 c}{\pi G} = \frac{2}{\pi} wc \tag{6}$$

Note, this equation takes a similar form to Equation 1, where the pressure over shear modulus term defines the strain energy density, w, for a Neo-Hookean material, and $2/\pi$ represents the proportionality parameter. Nevertheless, both these approaches only hold true to relatively low strains. The issue here is that as the void opens, the shape transitions from penny shaped flaw to spherical flaw. Taking a disk-shaped flaw in a unit cube, initially only the pressure normal to the penny shaped flaw is actively causing the flaw to expand; pressure in the other planes act to close the flaw. As the penny shaped flaw expands it transitions through ellipsoid forms until it becomes spherical. Through this transition period the pressure in the other planes gradually begin to positively influence the void's expansion. As such, the accuracy of these initial approaches is limited to the small strain region.

In more recent times, researchers have looked to expand the cavitation studies to higher strains (Lopez-Pamies 2011a, 2011b). However, the functions derived are too complex for practical use. In this paper, a practical large strain solution will be developed for cavitation using Finite Element Analysis (FEA) to study the transition of the flaw and using localised fracture mechanics to quantify the tearing energy. From this the relationship between transition shape to tearing energy will be used to form a single equation for evaluating fracture during cavitation.

2 FINITE ELEMENT STUDY

2.1 *Finite element study*

To evaluate cavitation through fracture mechanics, an FEA model was built in Abaqus. The model consisted of a 2D axisymmetric body with a centralized internal flaw or radius, c, modelled using a seam. The body was modelled sufficiently large relative to the flaw to meet far field conditions. During the analysis an external pressure was applied to the external faces, ramping linearly through the step time, generating hydrostatic tension. In turn, the hydrostatic tension caused the internal flaw to expand, Figure 2.

The tearing energy at the crack was evaluated using a J-Integral approach with 12 contour paths. This allowed the tearing energy to be tracked as a function of applied pressure. Note, the hydrostatic tension was applied at the far field, not the void wall, so hence it should not void the flaw surface pressure assumption required by the J-Integral approach.

The analysis was run with 5 crack lengths of varying size to better capture the size dependence. The

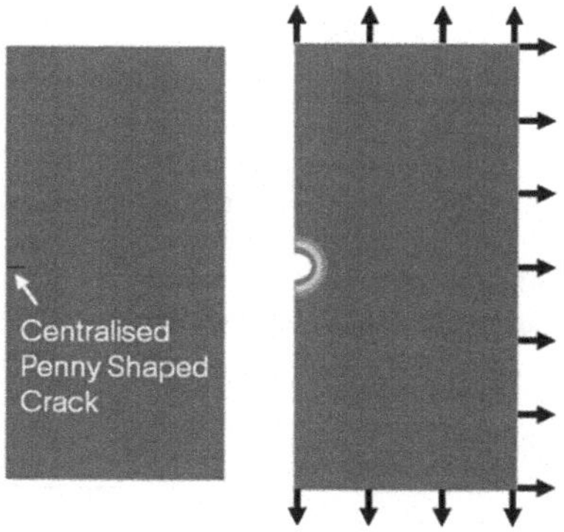

Figure 2. Axisymmetric Neo-Hookean body with an internal penny shaped flaw subject to external hydrostatic tension.

elastomer body was modelled using an incompressible Neo-Hookean material model. This kept the shear modulus constant, with the shear modulus known to influence cavitation, Equation 2. To study the effect of shear stiffness the analysis was run using 3 shear moduli; 5.517MPa, 8.276MPa and 11.034MPa.

3 RESULTS AND ANALYSIS

3.1 *Tearing energy results*

For each stiffness, the tearing energy was plotted versus pressure for the varying crack lengths. Given the tearing energy varies linearly with crack length, Equation 1, the authors have previously suggested normalising the tearing energy against the crack radius (Windslow et al. 2020). This is referred to as the proportioned cracking energy density (PCED), Equation 7. Through this approach, the tearing energy for each shear modulus case can be defined through a single profile, with the PCED accounting for the crack radius dependence, Figure 3.

$$\text{PCED} = \frac{T}{C} = kw_c \tag{7}$$

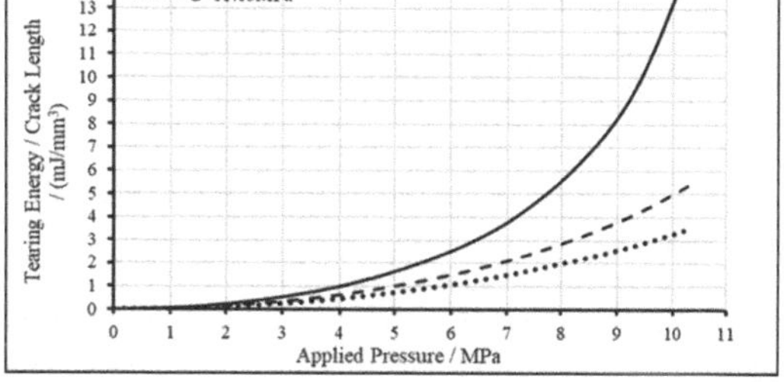

Figure 3. Effect of shear modulus and pressure on tearing energy viewed in terms of PCED, Equation 7.

3.2 *Small stain comparison*

To demonstrate the small strain limitation of prior solutions, in Figure 4 the Busfield equation, Equation 6, is compared against the FEA results. As shown, the equation is effective at relatively low pressures relative to the material's stiffness, but quickly under-estimates the tearing energy as the pressure increases. This is because the equation is limited to penny shaped flaws, so cannot account for the flaw's transition to spherical form under hydrostatic tension.

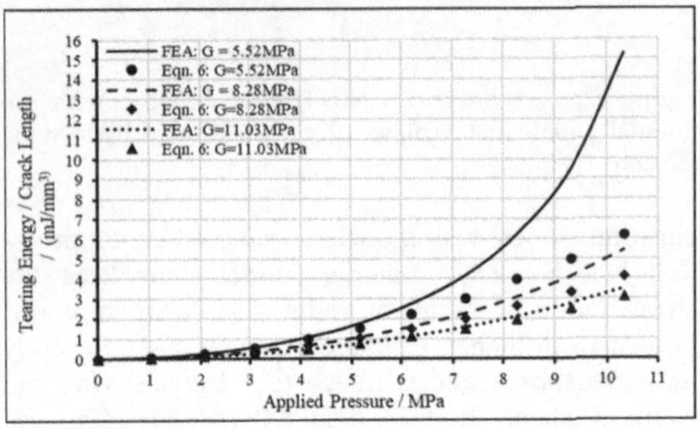

Figure 4. Comparison of the small strain solution, Equation 6, against the FEA predicted tearing energy.

3.3 *Shape factor of flaw*

Given this finding a critical aspect of cavitation is the geometry transition. This must be quantified effectively within any solution. Figure 5 illustrates the microscale perspective of this transition.

Consider a 3D elastomer body that contains a penny shaped flaw of initial radius, c_0, and an extremely thin depth, h_0. As the body undergoes hydrostatic tension, P, in the far field region, the hydrostatic tension causes the flaw to expand. The flaw's expansion is biased towards the large circular face of the penny shaped flaw. This causes the flaw to gradually transition from its initial penny shape form to a full sphere through a series of ellipsoids of major radii, c_t, and minor radii, h_t. The expansion causes tearing energy to be generated at the tip of the flaw as it expands, leading to rupture.

By following the deformation of the modelled crack, the rate at which the flaw transitions from penny shaped to spherical with applied pressure can be tracked, Figure 6. Given that both penny shape and spherical geometries are special cases of ellipsoids, the current geometric state of the flaw can be defined in terms of a shape function that represent the ratio of the minor radius to the major radius. This is also shown in the plot. Interestingly, despite the difference in growth rate between the major and minor radii, the shape factor in the low-mid strain region has a relatively linear relationship with pressure. There is a clear gradual transition from penny to spherical form.

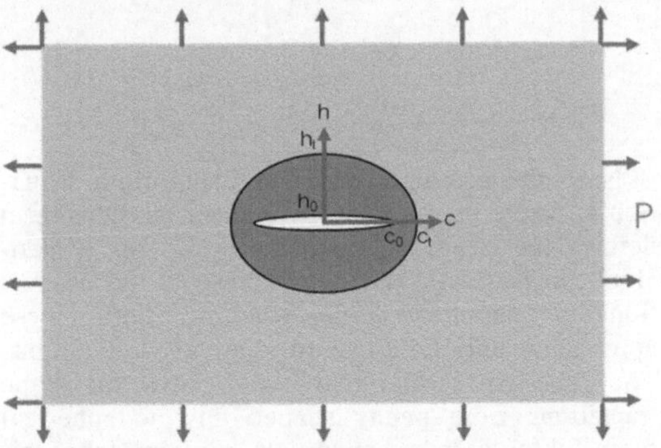

Figure 5. An internal penny shaped flaw under external pressure transitioning into spherical form through a series of ellipsoids.

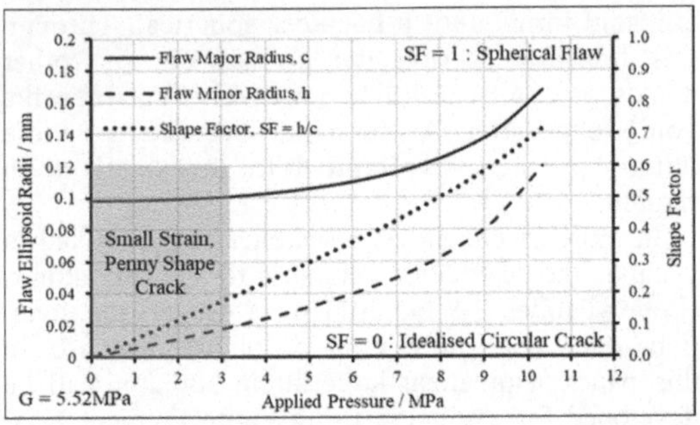

Figure 6. Transformation of the penny shaped flaw through a series of ellipsoid forms under external pressure, G=5.52MPa.

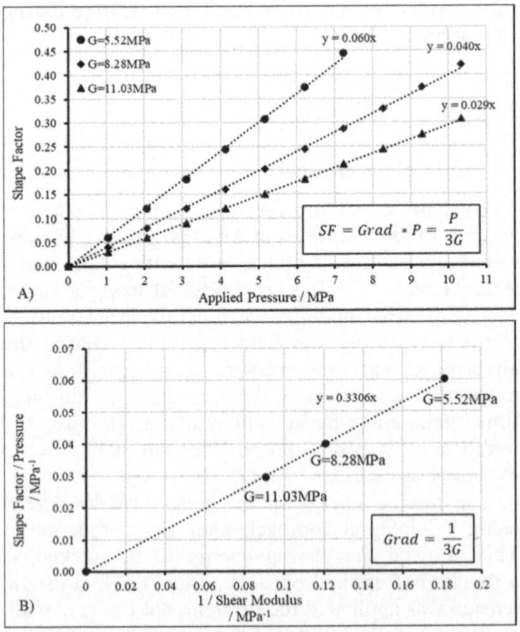

Figure 7. a) Linear relationship between shape factor and pressure for the different moduli. b) Relationship between the shape factor rate and the inverse of the shear modulus.

In Figure 7a the shape factor is plotted for the three shear moduli. As expected from Equation 2, the flaw's transition rate, as defined through the shape factor, is higher for softer materials. There is less resistance to the flaw's expansion.

The shape factor should be unitless, so given there is a linear relationship between shape factor and pressure, the gradient of each line in Figure 7a probably relates to the inverse of the shear modulus for that line. Plotting these gradients against the inverse of the shear moduli, Figure 7b, gives the following relationship:

$$SF \approx \frac{P}{3G} \tag{8}$$

It is worth noting that for the incompressible state $3G=E$. As such the equation is essentially showing the linear transformation of the crack with pressure.

3.4 *Cavitation dependence on shape factor*

Now that the shape factor is defined, it can be used to scale the small strain solution. As a first estimate, Mueller & Knauss (1971) proposed a scaling relationship for fracture under expanding conditions such as swelling tests, Equation 9. Here T_s is the swollen tearing energy and λ_s is the linear stretch from swelling. A squared relationship is required given the change in length is being related to change in area of fracture surface.

$$T_s = \frac{T}{\lambda_s^{\,2}} \tag{9}$$

Given the shape factor is essentially a stretch measurement, it follows that its correction may follow a similar relationship. Furthermore, given the correction is from penny shaped to spherical flaw, a reducing correction is required to meet the two known boundary conditions. Firstly, when $SF=0$ the solution should match the small strain equation, Equation 6. However, when $SF=1$ the tearing energy tends towards infinite which captures the instability of cavitation. This leads to the following equation which meets these key requirements:

$$\frac{T}{c_0} = \frac{\frac{P^2}{\pi G}}{(1 - SF^2)} = \frac{\frac{P^2}{\pi G}}{\left(1 - \left(\frac{P}{3G}\right)^2\right)} \tag{10}$$

Figure 8a compares Equation 10 against the FEA predicted profiles. As shown, the equation expands Busfield's small strain solution to medium strains. However, the solution still falls short at mid-high strains; it is not working effectively as the spherical cavitation is approached.

A deeper inspection suggests this relates to the instability zone. Based on instability mechanics the critical cavitation pressure tended towards, $2.5G$, Equation 3. Based on the shape factor function, the instability point when $SF \rightarrow 1$ is given by $3G$. This suitable difference explains why the shape factor form of the equation, Equation 10, is underestimating the tearing energy in the high strain zones; it is resisting the instability.

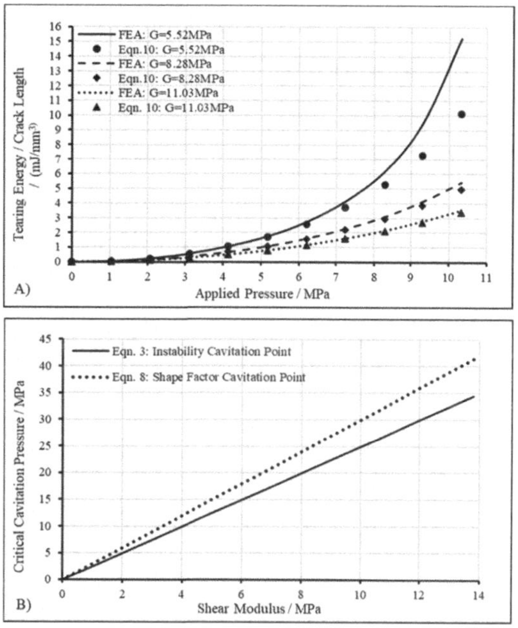

Figure 8. a) Comparing the shape factor correction, Equation 10, against the FEA predicted PCED. b) Comparison of the cavitational pressures using instability mechanics and shape factor.

3.5 *Correction for instability*

Given the above, to correct for the instability point, it is proposed to replace the shape factor function, Equation 8, in Equation 10, with the instability mechanics definition, Equation 3. This results in Equation 11. In this equation, the onset of instability should follow that of the instability mechanics theorem.

$$\frac{T}{c_0} = \frac{\frac{P^2}{\pi G}}{\left(1 - \left(\frac{2P}{5G}\right)^2\right)} \tag{11}$$

The updated equation is compared against the FEA solutions in Figure 9 showing excellent correlation. Importantly, the equation maps the profile of the softest material through the instability point very well.

Converting Equation 11 from PCED to tearing energy form gives a final equation for the tearing energy during cavitation. This allows incremental damage due to hydrostatic tension to be quantified.

$$T = \frac{\frac{P^2}{\pi G}}{\left(1 - \left(\frac{2P}{5G}\right)^2\right)} c_0 \tag{12}$$

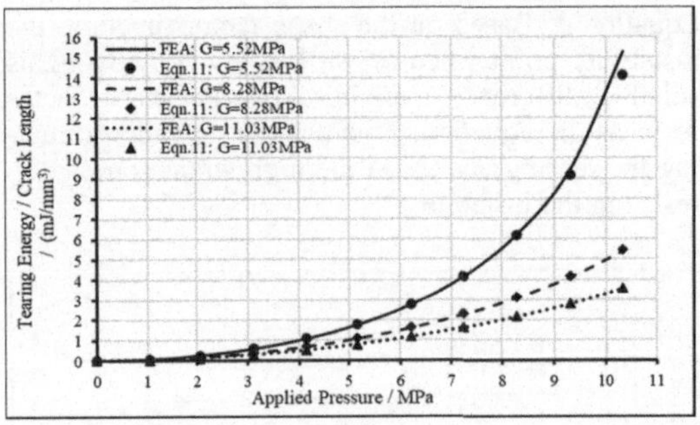

Figure 9. Comparing the instability correction, Equation 11, against the FEA predicted PCED demonstrates its effective fit across the pressure range and stiffnesses.

4 DISCUSSION

The equation, Equation 11, highlights the importance of the shear modulus on the cavitational process; it is the key term that resists the void's expansion. Interestingly, the equation does not provide an onset pressure as seen from the instability approach, Equation 4. This is because the instability approach treats the void as spherical throughout the entire deformation, whereas the proposed equation begins from penny shaped. Penny shaped flaws have sharp tips which act as stress concentrators, hence generate tearing energy at relatively low deformations.

The equation opens improved routes for evaluating cavitation processes. The instability equation provides a critical cavitation pressure, Equation 3, but in practice the flaw will have ruptured before this is reached if the initial flaw shape is not spherical. Through the updated fracture mechanics approach, the critical pressure can be better defined should the tearing energy value for instant failure be known. The equation can also be combined with static fatigue curves to evaluate time dependent cavitation which is critical for evaluating the bleed down rates of high-pressure elastomeric seals.

5 CONCLUSION

A large strain fracture mechanics solution is proposed for modelling cavitation of internal flaws due to external hydrostatic tension. The equation is found to be effective throughout the pressure range, mapping well through the critical cavitation zone. The work demonstrates how the flaw's geometry transitions, defining this through a shape function. The shape function is then found to correlate well with the instability definition from Gent and Lindley (1959). By correcting the small strain solution through the shape transition, a unified equation is developed that models cavitation accurately to the critical failure point. This was validated through FEA. Once again, the shear modulus of the elastomer body is found to critically resist cavitation.

REFERENCES

Busfield, J.J.C., Davies, C.K.L., & Thomas, A.G. 1996. Aspects of fracture in rubber components. *Progress in Rubber and Plastics Technology Sciences*, 12(3): 191–207.

Gent, A.N. & Lindley, P.B. 1959. Internal Rupture of Bonded Rubber Cylinders in Tension. *Proc. R. Soc. Lond. A*, 249(1257): 195–205.

Gent, A.N. & Wang, C. 1991. Fracture mechanics and cavitation in rubber-like solids. *Journal of Materials Science*, 26(12): 3392–3395.

Gough, J. & Muhr, A.H. 2005. Energy Release Rates for Small Cracks in Rubber Components. In P.E. Austrell & L. Kari (eds), *Constitutive Models for Rubber IV: Proceedings of the Fourth European Conference on Constitutive Models for Rubber, vol 4*: 51–57. London: Taylor & Francis Group.

Lake, G.J., Lawrence, C.C., & Thomas, A.G. 1992. Tear behaviour of rubber and the effect of pressure. In *Rubbercon 1992 Conference Proceedings*: 75–89. London.

Lopez-Pamies, O., Idiart, M.I., & Nakamura, T. 2011a. Cavitation in elastomeric solids: I – A defect-growth theory. *Journal of the Mechanics and Physics of Solids*, 59: 1464–1487.

Lopez-Pamies, O., Nakamura, T., & Idiart, M.I. 2011b. Cavitation in elastomeric solids: II - Onset-of-cavitation surfaces for Neo-Hookean materials. *Journal of the Mechanics and Physics of Solids*, 59: 1488–1505.

Mueller, H.K. & Knauss, W.G. 1971. The Fracture Energy and Some Mechanical Properties of a Polyurethane Elastomer. *Transactions Of The Society Of Rheology*, 233: 217–233.

Rivlin, R.S. & Thomas, A.G. 1953. Rupture of Rubber I. Characteristic energy for tearing. *J. Polym. Sci* 10: 291–318.

Sneddon, I.N. 1946. The Distribution of Stress in the Neighbourhood of a Crack in an Elastic Solid. *Proc. R. Soc. Lond. A*, 187: 229–260.

Stevenson, A. & Morgan, G. 1995. Fracture of Elastomers by Gas Decompression. *Rubber Chemistry and Technology*, 68(2): 197–211.

Windslow, R.J., Hohenberger, T.W. & Busfield, J.J.C. 2020. Determination of the Loading Mode Dependence of the Proportionality Parameter for the Tearing Energy of Embedded Flaws in Elastomers Under Multiaxial Deformations. In G. Heinrich, R. Kipscholl & R. Stoček (eds), *Fatigue Crack Growth in Rubber Materials. Advances in Polymer Science, vol 286*: 445–465. Cham: Springer.

Yeoh, O.H. 2006. Strain Energy Release Rates for Some Classical Rubber Test Pieces by Finite Element Analysis. In V.A. Coveney (eds), *Elastomers and Components: Service Life Prediction – Progress and Challenges*: 75–89. London: Woodhead Publishing.

Constitutive Models for Rubber XIII – Dal (Ed.)
© 2025 The Author(s), ISBN 978-1-032-85138-9

Is J-integral at crack onset an intrinsic property of filled elastomers? Experimental testing and finite element modelling for fracture toughness evaluation

I. Denora, A. Bianucci & C. Marano
Dipartimento di Chimica, Materiali e Ingegneria Chimica "Giulio Natta", Politecnico di Milano, Milano, Italy

A. Kaiser
ARLANXEO Deutschland GmbH

ABSTRACT: The fracture behaviour of two filled NBR based compounds, differing in filler type (carbon black or silica) and content (75 and 60 phr respectively) has been evaluated performing fracture tests under different loading conditions. Single edge notched tension (SENT) tests, tests in pure shear (PS) deformation conditions on single edge notched specimens and equibiaxial (EB) tensile tests on centre-notched square shaped specimens have been carried out at room temperature at a nominal deformation rate of 1 min^{-1}. Different approaches were adopted for the material's fracture toughness evaluation. From both SENT and PS tests, for which analytical expressions for the tearing energy are available, the material's tearing energy was evaluated. From PS tests, J-integral at fracture onset was also evaluated, since also this energy-based parameter can be determined analytically in this case. Further, once defined the materials' constitutive equations, J-integral has been numerically evaluated at fracture onset in uniaxial tensile, PS and EB tests. A comparison between the different parameters have been performed for the two materials in relation to the evaluation method adopted.

1 INTRODUCTION

Elastomers are widely used to produce industrial products, such as seals, tires, and hoses. In order to increase their mechanical response, they are reinforced with nanometric filler particles, such as carbon black (CB) or silica (Si) (Bokobza, 2017; Donnet, 1998; Fröhlich et al., 2005). In real applications the integrity of these components should be assured for safety reasons, so it is important to evaluate the fracture behaviour of the adopted materials. Through a fracture mechanics approach, in the past some methods to evaluate the fracture toughness of rubbers have been developed. Energy-based parameters, such as the J-integral (Rice, 1968) or the tearing energy (Rivlin & Thomas, 1953) have been proposed as expression of material fracture toughness. By definition, an intrinsic material property should have the same value irrespective of the loading conditions. This was verified in unfilled rubbers just for tearing energy but it is necessary to verify if it is still valid for the commonly used reinforced ones. The research work here reported has been performed within the Polymers4Hydrogen COMET Project. The work has the aim to assess the suitability of the energy-based parameters proposed in literature to be used to evaluate fracture toughness also for filled rubbers, by comparing the values of these parameters obtained analytically or evaluated numerically for tests performed in different loading conditions.

2 MATERIALS AND METHODS

2.1 *Materials and specimens' geometry*

Two filled NBRs has been supplied by Arlanxeo Germany GmbH: one material contains 75 phr of CB while the other one 60 phr of Si. The two compounds will be referred to respectively as NBR/CB and NBR/Si.

Sulphur-based curing of the compounds was performed at 170°C for 5 minutes. Dumbbell specimens with a gauge length of 25 mm and tight stripes with gauge length of 70 mm were die-cut from compression moulded rectangular sheets about 1.5 mm thick and were used for uniaxial tensile tests. PS specimens (wide stripes with dimensions 20x100x1.5 mm) and square shaped specimens (with dimensions 128x128x1.5 mm) were produced respectively for tests in pure shear deformation conditions and equibiaxial tension.

DOI: 10.1201/9781003516880-35

For fracture tests, a notch was introduced with a razor blade at mid height on one side of the pure shear specimens and of the tight stripes (which are now called "single edge notched tension", SENT, specimens) and in the center of the square shaped specimens, with the notch plane parallel to one of loading directions.

A paint-based speckle pattern has been applied on specimens by using an airbrush to measure strains through Digital Image Correlation (DIC) analysis.

2.2 *Mechanical testing*

Tensile tests in uniaxial loading conditions and pure shear deformation conditions were carried out with an Intron 5967 dynamometer while tensile tests in equibiaxial loading conditions were performed using a home-made biaxial dynamometer. All tests were performed at 23°C in a displacement control mode at a nominal strain rate of 1 min^{-1}. Tests were video-recorded in order to measure deformations by DIC analysis. Engineering stresses and engineering strains are used to plot the mechanical response of the materials. For fracture tests, the fracture onset was identified by analyzing the video recordings, as the instant when the deformed notch changed its shape from blunt to sharp.

2.3 *Preliminary characterization and material constitutive equation definition*

A preliminary mechanical characterization of the materials was performed, carrying out cycling tests in uniaxial tension on dumbbell specimens, up to different maximum strains to evaluate the strain dependency of materials' dissipated strain energy density. This quantity was calculated as the difference between the total and the stored strain energy densities, which can be obtained as the stress-strain curve integral evaluated on the loading and unloading stress-strain curves respectively.

Moreover, monotonic tensile tests were conducted up to failure in both uniaxial and equibiaxial tension and in pure shear deformation conditions. A MATLAB code was used to optimize the simultaneous fitting of the nominal stress-strain curves according to a third Yeoh model and the values of the model constant terms were obtained for the studied materials. According to this model, the strain energy density (W) is expressed as

$$W = C_1(I_1 - 3) + C_2(I_1 - 3)^2 + C_3(I_1 - 3)^3 \quad (1)$$

where I_1 is the first stretch invariant and C_1, C_2 and C_3 are the constants. In particular, C_1 is half the shear modulus G and the conditions $C_1>0$ and $C_1+C_2>0$ have to be respected.

2.4 *Fracture mechanics parameters*

Quasi-static fracture tests have been performed on notched specimens in pure shear deformation and uniaxial and equibiaxial tensile conditions.

For pure shear configuration, J-integral (Kim & Joe, 1989; Rice, 1968) and tearing energy – T (Rivlin & Thomas, 1953) have been evaluated at crack onset using respectively the following analytical formulations:

$$J_c = \eta \frac{U_c}{t(w-a)} \quad (2)$$

$$T_{c\_PS} = u_{0c} h \quad (3)$$

where, for equation 2, U_c is the input energy at crack onset, t is the specimen thickness, w is the specimen width, a is the initial crack length, η is a dimensionless factor equal to 1, and, for equation 3, h is the fracture PS specimen's gauge length in its undeformed state and u_{0c} is the material strain energy density evaluated in pure shear deformation conditions, at the same strain at which crack onset occurs in the fracture PS test. For the latter quantity, the results from the cyclic loading-unloading test described in Paragraph 2.3 were used.

Using the materials' constitutive equations, a numerical evaluation of the value of J-integral at crack onset was also obtained, by finite element simulation, calculating as the line integral on a path surrounding the crack tip:

$$J_c = \int_{\Gamma} \left(W dy - V_i \frac{dl_i}{dx} ds \right) \quad (4)$$

where Γ is the contour, W is the strain energy density, V_i and l_i are the traction vector and the displacement vector at a point on a path respectively, ds is the length increment along the path and x and y are coordinates of a Cartesian reference system, considering the notch parallel to x-axis. The whole pure shear specimen has been modeled and the specimen elongation at crack onset (obtained from experimental tests) was used as a boundary condition. A mesh containing eight node (quadratic quadrilateral) plane stress elements with fully integration was used.

For SENT tests, tearing energy was calculated analytically as follows:

$$T_{c\_SENT} = \frac{2\pi}{\sqrt{\lambda_c}} \frac{U_c a}{twh} \quad (5)$$

where U_c is the input energy at crack onset, a is the initial crack length, twh is the unstrained specimen volume and λ_c is the draw ratio (or stretch) at crack onset. As for the J-integral, it was evaluated only through finite element simulations in a similar way as for pure shear tests.

For equibiaxial fracture tests there is no analytical formulation to evaluate the fracture toughness, so only the J-integral parameter was evaluated, through the finite element modeling. In this case, the specimen's central part with the whole notch was modeled and the displacement field measured by DIC during experimental tests was set as boundary conditions (Caimmi et al., 2015). Also in this case, a mesh containing eight node (quadratic quadrilateral) plane stress elements with fully integration was used.

3 RESULTS

3.1 *Preliminary characterization*

The results of tests performed in uniaxial tension are reported in Figure 1 for the two materials. Figure 2 reports the percentage of the strain energy density which is dissipated during a loading-unloading cycle as a function of the applied strain. Even though a material comparison is not one of the goal of this paper, it can be observed that: (i) the NBR/Si compound is more deformable and has a higher strength at break than the NBR/CB compound; (ii) for both the materials, the dissipated energy density fraction increases with the applied deformation, almost in the whole explored deformation range; (iii) for the same deformation increment, the strain energy density fraction increases more for NBR/CB respect to NBR/Si.

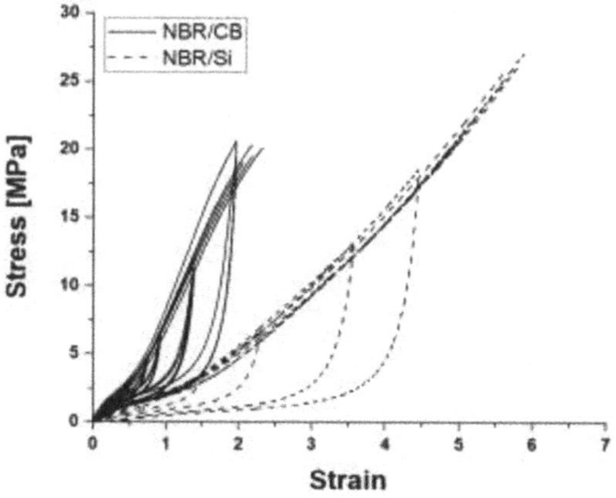

Figure 1. Stress-strain curves of monotonic and cycling tests in uniaxial tension, performed on NBR/CB and NBR/Si.

3.2 *Materials constitutive equation definition*

The experimental data obtained from tests in uniaxial and equibiaxial tension conditions for NBR/CB and NBR/Si and, for the former material, also in pure deformation conditions are reported in Figure 3 and

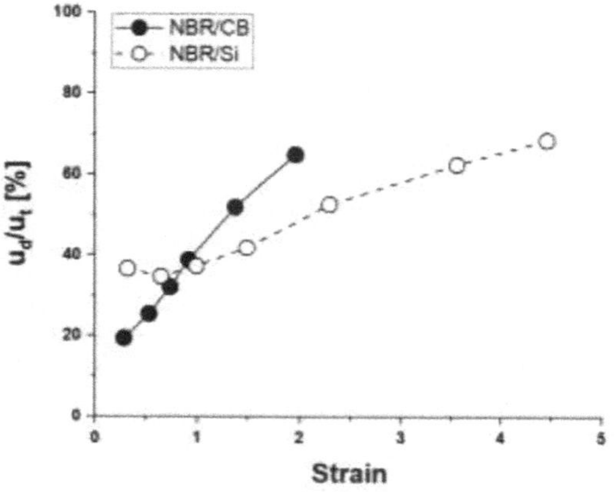

Figure 2. Strain dependency of the dissipated strain energy density (u_d) evaluated as percentage of the total strain energy density (u_t) for NBR/CB (full symbols) and NBR/Si (open symbols).

Figure 4 respectively together with the FEM predictions. The third order Yeoh model constants obtained for the two materials and are reported in Table 1.

Table 1. Third order Yeoh model material constants.

Material	G	C2	C3
NBR/CB	2.6910	0.2502	-0.0116
NBR/Si	1.0200	0.0279	-0.0002

It can be observed that, despite some deviations at the lower strains, the numerical models well predict the experimental results.

3.3 *Fracture toughness*

Figures 5 and 6 report for NBR/CB and NBR/Si respectively, the values of: (i) the J-integral evaluated analytically, according to equation 2, at crack onset in fracture tests performed in pure shear deformation conditions ($J_{AN\ PS}$); (ii) the tearing energy evaluated analytically from fracture tests in pure shear deformation conditions (T_{PS}) and from SENT tests (T_{SENT}) following equation 3 and equation 5 respectively; (iii) the J-integral calculated numerically according to equation 4 in the different loading conditions (J_{FEM_SENT}, J_{FEM_PS}, J_{FEM_EB}). In Figure 7, the strain measured through DIC analysis on the notch plane at crack initiation from pure shear fracture tests is reported as a function of the distance from the crack tip for the two materials: these data were used for T_{PS} evaluation, as detailed in Paragraph 2.4.

For a general comparison between the two materials, it can be observed that NBR/Si has a higher fracture toughness compared to NBR/CB, a result which is consistent with the higher dissipative response of the material in the range of the strains experienced at crack initiation. Further, for NBR/Si the crack onset occurs at considerably higher strains with respect to NBR/CB.

For the NBR/CB compound, the value of J-integral analytically obtained for the pure shear fracture test (J_{AN_PS}) is similar to the J-integral value obtained from FEM for the same test (J_{FEM_PS}). These values are also close to the J-integral value numerically obtained for the SENT test (J_{FEM_SENT}), but significantly lower than the J-integral integral value numerically obtained for the fracture test performed in equibiaxial tension conditions (J_{FEM_EB}). When comparing the tearing energy values, a lower value was obtained for the SENT tests ($T_{PS}>T_{SENT}$). This result is not consistent with the considerations reported in (Kim & Joe, 1989) on unfilled rubbers: since in the unconstrained sides of the pure shear specimen the strain energy density is lower than in it is in pure shear deformed zone, the strain energy density used in equation 3 should be underestimated, and T_{SENT} is expected to be higher than T_{PS}. Nevertheless, considering the data dispersion, it can be stated that for the NBR/CB compound there is a good agreement between the fracture toughness related results differently obtained.

As regards NBR/Si, for which only SENT and PS fracture tests have been carried out, not only the values of $J_{AN\ PS}$ and of J_{FEM_PS} were found to be comparable, but also for the tearing energy evaluated in the PS fracture test, T_{PS}, a similar value of J-integral was obtained. On the contrary, from the SENT fracture tests, the evaluated T_{SENT} resulted to be lower than the calculated J_{FEM_SENT} and both these two parameters resulted to be significantly higher than those determined from the PS fracture tests. Further analyses should be conducted to better understand the results, that could be related to the fact that the identification of the fracture onset as the instant when the notch changes its shape from blunted to sharp, was not easy when the low stiffness NBR/Si was tested in SENT tests. It would also need to be verified whether the used SENT specimen geometry can be adopted independently of the elastic modulus of the material.

4 CONCLUSIONS

The value of the J-integral at fracture onset for two filled rubber compounds, a carbon black filled NBR and a silica filled NBR, have been evaluated following different methods: (i) by using the analytical formulation of J-integral valid for pure shear deformation conditions, (ii) by a numerical evaluation of J-integral simulating SENT fracture tests and fracture tests in PS deformation conditions and in equibiaxial tension.

Going beyond what it is stated in the title of this work, a comparison between J-integral at fracture onset and the tearing energy, evaluated using

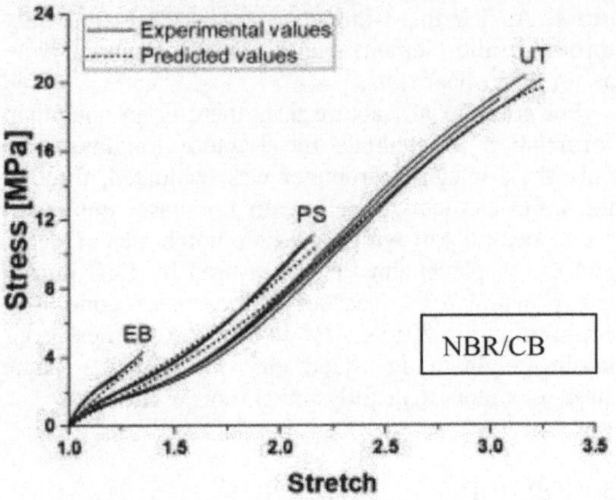

Figure 3. Experimental and predicted data of the stress-stretch curves for NBR/CB in uniaxial (UT) and equibiaxial (EB) tension and in pure shear deformation (PS) conditions.

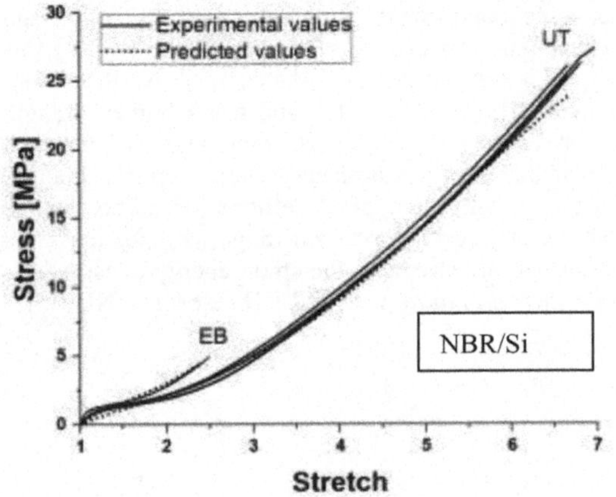

Figure 4. Experimental and predicted stress-stretch curves for NBR/Si in uniaxial (UT) and equibiaxial (EB) tension.

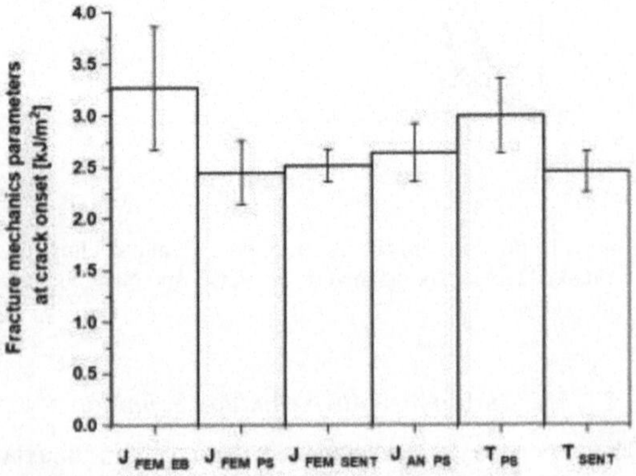

Figure 5. Fracture toughness obtained with different methods for NBR/CB.

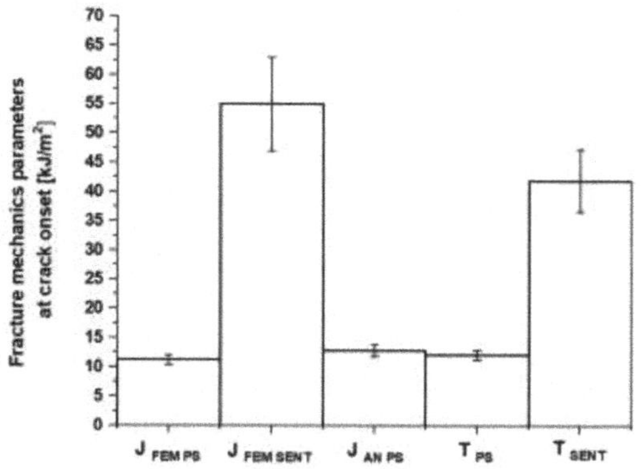

Figure 6. Fracture toughness obtained with different methods for NBR/Si.

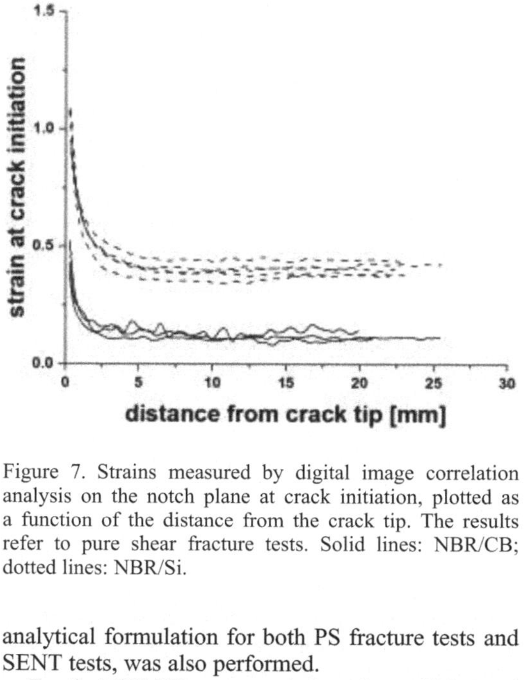

Figure 7. Strains measured by digital image correlation analysis on the notch plane at crack initiation, plotted as a function of the distance from the crack tip. The results refer to pure shear fracture tests. Solid lines: NBR/CB; dotted lines: NBR/Si.

analytical formulation for both PS fracture tests and SENT tests, was also performed.

For the NBR/CB compound, the values of J-integral at fracture onset measured by the different methods are comparable and are close to those obtained for the tearing energy.

Differently, for what concerns the NBR/Si compound, J-integral at fracture onset and tearing energy are similar if compared within the same test configuration. Larger values of J-integral and tearing energy were obtained from the SENT tests. Further analyses are required to justify the results obtained.

ACKNOWLEDGEMENTS

The research work was performed within the COMET-project "Polymers4Hydrogen" (project-no.: 21647053) at Politecnico di Milano within the framework of the COMET-program of the Federal Ministry for Climate, Action, Environment, Energy, Mobility, Innovation and Technology and the Federal Ministry for Digital and Economic Affairs, Austria with contributions by Polymer Competence Center Leoben (PCCL), Montanuniversität Leoben (Department Polymer Engineering and Science, Chair of Chemistry of Polymeric Materials, Chair of Materials Science and Testing of Polymers), Technical University of Munich, Tampere University of Technology, Bundesanstalt für Materialforschung und -prüfung (BAM) and Arlanxeo Deutschland GmbH, ContiTech Rubber Industrial Kft., Peak Technology GmbH, SKF Sealing Solutions Austria GmbH, Faurecia.

REFERENCES

Bokobza, L. (2017). Mechanical and Electrical Properties of Elastomer Nanocomposites Based on Different Carbon Nanomaterials. *C*, *3*(4), 10. https://doi.org/10.3390/C3020010

Caimmi, F., Calabrò, R., Briatico-Vangosa, F., Marano, C., & Rink, M. (2015). J-Integral from Full Field Kinematic Data for Natural Rubber Compounds. *Strain*, *51*(5), 343–356. https://doi.org/10.1111/STR.12145

Donnet, J. B. (1998). Black and White Fillers and Tire Compound. *Rubber Chemistry and Technology*, *71*(3), 323–341. https://doi.org/10.5254/1.3538488

Fröhlich, J., Niedermeier, W., & Luginsland, H. D. (2005). The effect of filler–filler and filler–elastomer interaction on rubber reinforcement. *Composites Part A: Applied Science and Manufacturing*, *36*(4), 449–460. https://doi.org/10.1016/J.COMPOSITESA.2004.10.004

Kim, B. H., & Joe, C. R. (1989). Single specimen test method for determining fracture energy (Jc) of highly deformable materials. *Engineering Fracture Mechanics*, *32*(1), 155–161. https://doi.org/10.1016/0013-7944(89)90213-0

Rice, J. (1968). A Path Integral and the Approximate Analysis of Strain Concentration by Notches and Cracks. *Journal of Applied Mechanics*, *35*, 379–386. https://doi.org/10.1115/1.3601206

Rivlin, R. S., & Thomas, A. G. (1953). Rupture of rubber. I. Characteristic energy for tearing. *Journal of Polymer Science*, *10*(3), 291–318. https://doi.org/10.1002/pol.1953.120100303

Constitutive Models for Rubber XIII – Dal (Ed.)
© 2025 The Author(s), ISBN 978-1-032-85138-9

Influence of strain induced crystallisation on the dynamic crack propagation resistivity of NR vulcanisates

N. Hanne, C. Egelkamp & J. Meier
Deutsches Institut für Kautschuktechnologie e.V., Hannover, Germany

P. Junker
Institut für Kontinuumsmechanik (IKM), Leibniz Universität Hannover, Germany

ABSTRACT: Ongoing demands on the improvement of service life performance of highly stressed elastomer parts require transferable and efficient laboratory evaluation procedures to estimate the suitability of the materials for safety-relevant usage. For analysis of the mechanical fracture behaviour often dynamic fatigue crack growth (FCG) tests on notched specimens are performed. Recently, company COESFELD MATERIAL TEST and the DIK developed an enhanced evaluation method which combines the automatic Tear- Fatigue-Analysis with Digital Image Correlation (DIC) and Post-Processing to get spatially resolved strain and stress information in the vicinity of the crack tip. While the strain information obtained by DIC are highly accurate, the stress information depends on the assumed material model, which is applied within a mapping procedure to locally transfer local strain to local stress. Since elastomers generally display a complex material behaviour involving e.g., hyper elasticity, relaxation, energy dissipation and in some cases, strain induced crystallisation (SIC), it is necessary to develop constitutive models in respect for these phenomena. Accordingly, an energetical model considering crystallisation is presented which can be applied for the interpretation of FCG experiment results, especially with respect to the significance in the prediction of local strain fields in the crack tip vicinity.

1 INTRODUCTION

1.1 *Phenomena of strain induced crystallisation*

Strain induced crystallisation (SIC) is a phenomenon occuring especially in vulcanisates of Natural Rubber (NR) but also in some synthetic rubbers like Polyisoprene Rubber (IR). It's occurance is assumed to be the one of the causes for the good mechanical properties in particular for its good resistivity against fatigue crack growth under cyclic loading (Le Cam and Toussaint 2010). When exposed to large strains some of the amorphous aligned polymer chains rearrange to a formation of small crystallites with orientation in direction of the strain. This effect was first found by Katz (1925) using wide-angle-X-ray-spectrographie (WAXS). Since then WAXS is the method of choice for evaluating the kinetics or fraction of cristallinity in NR. It has been the subject of various investigations on different scales. Many works have focused on microstructural effects to describe the phenomenon, for example (Plagge and Klüppel 2018), who also described the influence of temperature on crystallization behaviour. One approach to a microstructural description for the prediction of macroscopic material behavior is given by Mistry & Govindjee (2014), while Aygün & Klinge (2020) describe a macroscopic continuum mechanics approach, which is also based on an energetic consideration, different to the one presented here by the derivation of the evolution of crystallinity from the dissipation potential. The approach presented here is borrowed from the modelling of shape memory alloys from Junker (2014).

While SIC is qualitatively assumed to strenghten the resistivity against FCG it makes the estimation of FCG-resistance more complex. Because of the strain exagerration in the crack tip crystallites appear in the crack tip forcing the crack to bypass the areas with high amount of crystallinity.

1.2 *Measuring crack growth and digital image correlation*

While this Work aims to evaluate the influence of SIC to the FCG-behaviour of NR materials undergoing cyclic loadings it is necessary to get relatable spatial resolved strain data in the vicinity of the crack tip. In a previous project with the company COESFELD MATERIAL TEST a fully automated Tear-Fatigue-Analyser (TFA) was enhanced to generate fullfield strain images by using the method of digital image correlation (DIC).

DOI: 10.1201/9781003516880-36

The basic idea of DIC is to compare images with different deformation shapes and to track the change in the position of surface points. To identify surface points, the sample must be prepared with a speckle pattern. The DIC calculation is performed with Ncorr, a subset-based DIC toolbox for MATLAB (Blaber et al. 2015). In subset-based DIC calculations, the investigated image region of interest (ROI) is divided into a number of subsets within the the ROI. The deformation in a subset is assumed to be homogeneous. For tracking, the algorithm identifies the corresponding subsets in a reference image and in a current image. Therefore, the position of each subset can be determined for different time steps.

From the 2D-DIC-Analysis one gets displacement values u and v for horizontal and vertical directions respectivly. These can be computed into a displacement gradient **H** by a plane-fit over a certain area to eliminate noisy values from the DIC-meaurement. From **H** the deformation gradient can be obtained by $\mathbf{F} = \mathbf{H} + \mathbf{1}$.

$$\mathbf{H} = \begin{pmatrix} \frac{\partial u}{\partial x} & \frac{\partial u}{\partial y} & 0 \\ \frac{\partial v}{\partial x} & \frac{\partial v}{\partial y} & 0 \\ 0 & 0 & H_{33} \end{pmatrix} \quad (1)$$

$$\mathbf{F} = \mathbf{H} + \mathbf{1} = \begin{pmatrix} \frac{\partial u}{\partial x} + 1 & \frac{\partial u}{\partial y} & 0 \\ \frac{\partial v}{\partial x} & \frac{\partial v}{\partial y} + 1 & 0 \\ 0 & 0 & F_{33} \end{pmatrix} \quad (2)$$

While the value for H_{33} is not defined the material is assumed to be incompressible and the load is assumed as plane stress. Therefore the value for F^*_{33} can be obtained by:

$$\det(\mathbf{F}) = 1 \Rightarrow F_{33} = (F_{11} \cdot F_{22} - F_{12} \cdot F_{21})^{-1} \quad (3)$$

2 CONSTITUTIVE MODELLING

This section briefly describes the structure of the constitutive model, starting with the thermodynamic basis. The Yeoh model is used to describe the material behavior, as it provides a simple basis for building a hyperelastic model to begin with. Finally, the evolution equations are derived and the numerical implementation of these is briefly described.

2.1 *Thermodynamic framework*

The used model for calculating and mapping the crystallisation onto the strainfield generated by DIC is based on the principle of the minimum of the dissipation potential stated as eq. (4), where Ψ denotes the Helmholtz free energy and $\mathcal{D}$ the dissipation function.

$$\mathcal{L} = \dot{\Psi} + \mathcal{D} \rightarrow \text{stat.}(\dot{\xi}) \quad (4)$$

Generally the Helmholtz energy depends on a set of external and a set of internal variables. Here, the external variable is represented by the externally applied strain **C**, and the volume fraction of the crystallinity ξ is introduced for the internal variable. Accordingly, the rate of the Helmholtz energy can be written as:

$$\dot{\Psi} = \frac{\partial \Psi}{\partial \mathbf{C}} : \dot{\mathbf{C}} + \frac{\partial \Psi}{\partial \xi} : \dot{\xi} \quad (5)$$

The stationary condition reads:

$$\frac{\partial \mathcal{L}}{\partial \dot{\xi}} = \frac{\partial \Psi}{\partial \xi} + \frac{\partial \mathcal{D}}{\partial \dot{\xi}} = 0 \quad (6)$$

wich is the famous Biot equation, wich can be recast as Equation (7) with the thermodynamic driving force $P := -\partial\Psi/\partial\xi$

$$\frac{\partial \mathcal{D}}{\partial \dot{\xi}} = P \quad (7)$$

2.2 *Assumption for the energy density*

We model the Helmholtz free energy of natural rubber with a phenomenological Yeoh-prototype energy with the three stiffness parameters c_i as shown in Equation (8).

$$\Psi = \sum_{i=1}^{3} c_i \cdot (\bar{I}_1 - 3)^i \quad (8)$$

Here we model the effective Yeoh-parameters via a harmonic mean, i.e. Equation (9), in contrast other approaches like (Aygün and Klinge 2020) formulate the mechanic energy densities as sum of the hyperelastic and crystalline parts.

$$c_i(\xi) = \left[\frac{\xi}{\xi_{max}} \cdot c_{i,x}^{-1} + \left(1 - \frac{\xi}{\xi_{max}}\right) \cdot c_{i,0}^{-1}\right]^{-1} \quad (9)$$

Consequently, we find $c_i = c_{i,0}$ for non-crystallised materrial and $c_i = c_{i,x}$ for a material with maximum crystallisation, where, $c_{i,x}$ and $c_{i,0}$ represent the constants for the crystalline and non-cristalline phases.

In Figure 1 an examplarily comparison of energy density for non-crystallising (Equation 8) and crystallising material model is shown. It can be seen that increasing the crystallinity ξ reduces the energy density $\Psi(\bar{I}_1, \xi)$.

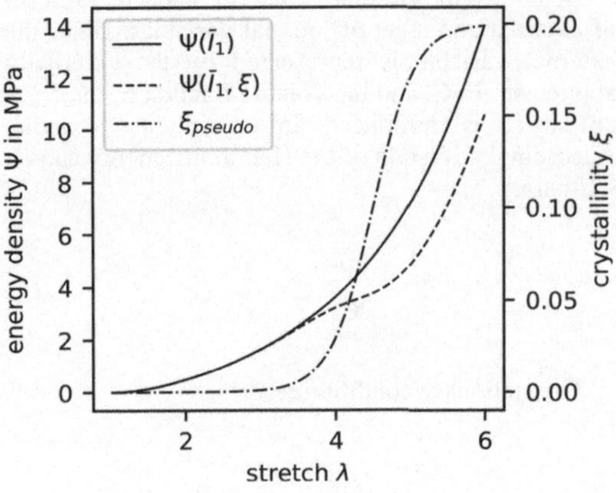

Figure 1. Strain dependency of Yeoh-type energy density with and without crystallinity ξ.

2.3 *Thermodynamic driving force*

Like described previously in Section 2.1 the thermodynamic driving force is defined by the derivation of the energy density with respect to the crystallinity variable ξ.

$$P = -\frac{\partial \Psi}{\partial \xi} =$$

$$-\sum_{i=1}^{3} \frac{c_{i,x} \cdot c_{i,0}\,(c_{i,x} - c_{i,0})}{(c_{i,x} \cdot \xi_{max} - c_{i,x} \cdot \xi + c_{i,0} \cdot \xi)^2} \cdot (\bar{I}_1 - 3)^i \tag{10}$$

2.4 *Dissipation potential*

Principle dissipation

While crystallinity is developed due to the externally applied force, energy is dissipated depending on the rate of crystallisation. According to the setup model in Section 2.1 the evolution of the volume fraction of the crystallinity ξ can be derived from the principle dissipation potential $\mathcal{D}_p(\dot{\xi})$. We made use of a dissipation function wich is homogeneous of order one and two, i.e., Equation (11), representing a ratedependend dissipation.

$$\mathcal{D}_p := \frac{1}{2}\eta \cdot \dot{\xi}^2 + r \cdot |\dot{\xi}| \tag{11}$$

Here η represents a viscosity or a kind of a reaction inertia and r is an ernergetic threshold value for the onset of crystallisation. The influence of this viscosity on the dependence of the dissipation potential $\mathcal{D}_p$ on the rate of crystallinity is shown qualitatively in Figure 2 the rate of crystallisation can now be determined from Equation (7), specified as Equation (12). However, our approach yields the famous Reuβ energy bound for small deformations ($\bar{I}_1 \approx 3$), c.f., Junker (2014).

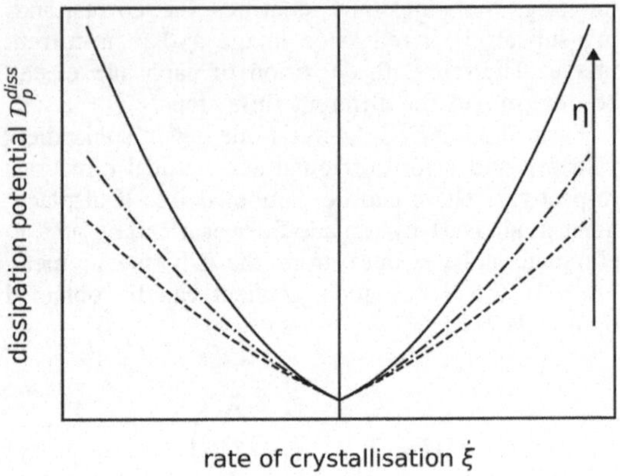

Figure 2. Dissipation potential $\mathcal{D}^{diss}$ with respect to rate of crystallisation for different values of η.

$$P = \eta \cdot \dot{\xi} + r \cdot \mathrm{sgn}(\dot{\xi}) \tag{12}$$

from wich follows Equation (13)

$$\dot{\xi} = \frac{1}{\eta} \cdot [P - r]_+ \cdot \mathrm{sgn}(\dot{\xi}) \tag{13}$$

follows. Here the expression $[P - r]_+$ ensures that the it does not return any negative values in the event that $P < r$. In the Following it is denoted by $\phi_+ := [P - r]_+$ for which holds:

$$\phi_+(x) := <x> = \begin{cases} x & \text{for } x>0 \\ 0 & \text{else} \end{cases} \tag{14}$$

Adapted dissipation for limited criystallinity

To account for the constraint $\xi \in [0, \xi_{max}]$ with some maximum degree of crystallinity ξ_{max}. We introduce a suitable mapping by Equation (15)

$$\xi(\alpha) := \frac{1}{1 + e^{-(\alpha+s)\cdot a}} \cdot \xi_{max} \tag{15}$$

such that

$$\dot{\xi} = \frac{\partial \xi}{\partial \alpha} \cdot \dot{\alpha} \quad (16)$$

Consequently, we update the crystallinity state via:

$$\dot{\alpha} = \left[\frac{\partial \xi}{\partial \alpha}\right]^{-1} \cdot \dot{\xi} \quad (17)$$

Here ξ_{max}, the upper limit of the crystallinity is determined by empirical measurements. The parameter a is a gradient parameter that controls how steep the function $\xi(\alpha)$ rises while the parameter s shifts the function horizontally. Both are also determined empirically. The function from Equation (15) is shown in Figure 3 qualitatively for different values of a and with $s = 0$.

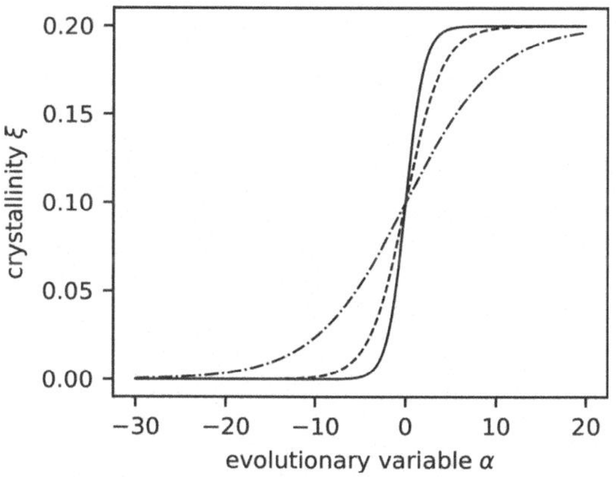

Figure 3. Switching function $\xi(\alpha)$ for different values of the gradient parameter a and s = 0.

3 NUMERICAL RESULTS

Main part of the calculation is the evolution of α. The value for α_n at the current time increment n is first calculated as seen in Equation (18).

$$\alpha_n = \alpha_{n-1} + \frac{\Delta t}{\eta} \cdot \phi_+(\bar{I}_{1,n}, \xi_{n-1}) \frac{P(\bar{I}_{1,n}, \xi_{n-1})}{|P(\bar{I}_{1,n}, \xi_{n-1})|} \quad (18)$$

Afterwards the current crystallinity can be determined from Equation (15) and subsequently the energydensity from Equation (19)

$$\Psi_n = \Psi(\bar{I}_{1,n}, \xi(\alpha_n)) \quad (19)$$

The resulting stress is calculated from the current state of the energy density and stretch state:

$$\sigma_n = \sigma(\Psi_n, \bar{I}_{1,n}) \quad (20)$$

An example calculation of the crystallisation development undergoing uniaxial tension with parameters as shown in Table 1 is presented briefly in the following.

The elastic parameters $c_{i,0}$ used for the Yeoh-based model are based on those of an unfilled natural rubber fitted in a strain range without any crystallinity, while the parameters $c_{i,x}$ are based on the same material in a strain range with a known high value of crystallinity. The other parameter values, especially the one for Equations (15) and (18), were selected so that the calculated degree of crystallinity corresponds to an experimentally determined value found for an unfilled NR (Rault, Marchal, Judeinstein, & Albouy 2006).

Figure 4 shows the result of the exemplary calculation of strain crystallisation. It is clear to see that the degree of strain crystallisation jumps from value zero to a positive finite value in the first time step, which is not to be expected. This indicates a non vanishing equilibrium value in the non deformed material state within this model. Eventually in Figure 5 the stress curves for crystallising and non crystallising material is shown. Here it has to be mentioned that the curve for the non-crystallising material is in fact an extrapolated curve from fitting paramaters estimated at lower strains.

Table 1. Parameters for exemplaric calculation.

Parameter	Value
strainsteps	1000
max. Strain ε	600 %
Yeoh-Parameter in MPa	for Eq. ?? and 10
c_1	$2.4035 \cdot 10^{-1}$
c_2	$1.0 \cdot 10^{-8}$
c_3	$1.6074 \cdot 10^{-4}$
$c_{x,1}$	$2.6439 \cdot 10^{-1}$
$c_{x,2}$	$0.9 \cdot 10^{-8}$
$c_{x,3}$	$1.9289 \cdot 10^{-4}$
η	6.98 (Equation 18)
r	0.12
Parameter for est. ξ	Equation (15)
a	1
s	0
ξ_{max}	0.2

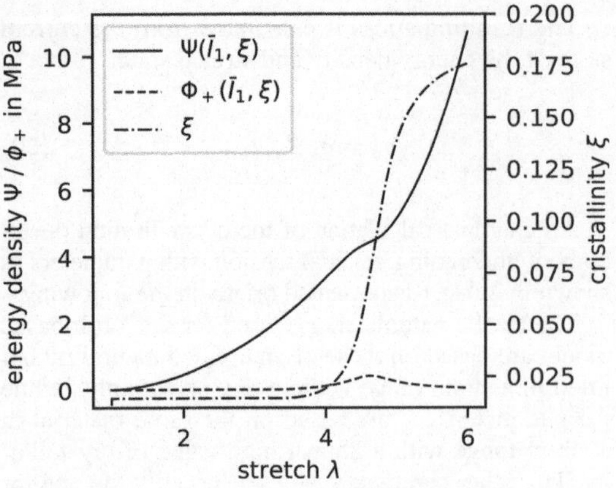

Figure 4. Energy density $\Psi(\bar{I}_1, \xi)$, reaction function $\phi_+(\bar{I}_1, \xi)$ and volume fraction of crystallinity from example calculation with uniaxial load.

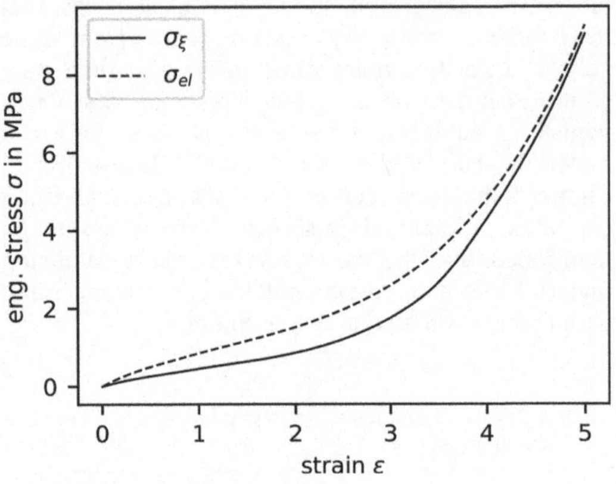

Figure 5. Stress calculated from energy density with crystallisation an without.

4 CONCLUSIONS AND OUTLOOK

In this work, an energy-based model was suggested to calculate the formation of crystalline phase components as a result of mechanical strain. Exemplarily calculations for the strain case of uniaxial tension using parameters from empirical measurements of natural rubber type vulcanisates were carried out. As can be clearly seen in Figure 4, the present model shows a basic value of strain crystallisation without any significant strain energy being present. This could be remedied by expanding the previously purely mechanical model to include a caloric contribution. In addition, the model already contains a formalism to calculate the process of SIC melting.

Hence, the model seems promising to further extension to take into account melting effects, since strain crystallisation and melting experimentally display a pronounced hysteresis behavior.

Already in the present state, this concept can be used to map the SIC onto DIC based strainfields in the vicinity of a growing crack. This will be the next steps of the authors.

REFERENCES

Aygün, S. & S. Klinge (2020, July). Continuum mechanical modeling of strain-induced crystallization in polymers. *International Journal of Solids and Structures* 196-197, 129–139.

Blaber, J., B. Adair, & A. Antoniou (2015, July). Ncorr: Opensource 2D digital image correlation matlab software. *Experimental Mechanics 55*(6), 1105–1122.

Junker, P. (2014, October). An accurate, fast and stable material model for shape memory alloys. *Smart Materials and Structures 23*(11), 115010.

Katz, J. R. (1925, May). Was sind die Ursachen der eigentümlichen Dehnbarkeit des Kautschuks?: Ueber die Aenderung des Röntgenspektrums des Kautschuks bei der Dehnung. *Kolloid-Zeitschrift 36*(5), 300–307.

Le Cam, J.-B. & E. Toussaint (2010, May). The Mechanism of Fatigue Crack Growth in Rubbers under Severe Loading: The Effect of Stress-Induced Crystallization. *Macromolecules 43*(10), 4708–4714.

Mistry, S. J. & S. Govindjee (2014, January). A micromechanically based continuum model for strain-induced crystallization in natural rubber. *International Journal of Solids and Structures 51*(2), 530–539.

Plagge, J. & M. Klüppel (2018, May). A Theory Relating Crystal Size, Mechanical Response, and Degree of Crystallization in Strained Natural Rubber. *Macromolecules 51*(10), 3711–3721.

Rault, J., J. Marchal, P. Judeinstein, & P. A. Albouy (2006, November). Stress-Induced Crystallization and Reinforcement in Filled Natural Rubbers: 2 H NMR Study. *Macromolecules 39*(24), 8356–8368.

Constitutive Models for Rubber XIII – Dal (Ed.)
© 2025 The Author(s), ISBN 978-1-032-85138-9

Analysis of the durability damage scenarios of air spring sleeves with axial reinforcements based on computer tomography and digital image processing

R. Klauke & P. Charrier
Vibracoustic, – ZI Carquefou France, Advanced Engineering, ESM

N. Serrurier
ENSTA Paris – Institut des Sciences de la Mécanique et Applications Industrielles (IMSIA)

M. Le Saux
ENSTA Bretagne, UMR CNRS, IRDL, Brest, France

M. Skoglund
Helmut Schmidt Universität Hamburg, Hamburg, Germany

ABSTRACT: Air springs are flexible components used in various applications such as vehicles, machinery, and industrial equipment. They are made of a multilayer composite material based on rubber and reinforcing textile cords. The service life and reliability of air springs depend heavily on their fatigue behaviour, which is influenced by various factors such as temperature, pressure, load and environmental conditions. To better understand and model the fatigue behaviour of air spring sleeves, it is necessary to identify and quantify the damage mechanisms and processes inside the individual layers. In this work, a new tool is presented that is able to distinguish the components of an air spring sleeve using volumetric CT-data. The tool is based on advanced greyscale detection technology to recognize and segment the different layers and materials in the data of a scanned sleeve. It allows to determine the geometry, thickness, and volume fractions of each layer and to detect potential defects such as cracks, delaminations, or holes. The tool is applied to various damage scenarios for the fatigue behaviour of air springs. The focus is put on fatigued axial sleeves subjected to cyclic loading. Various CT scan data sets of axial sleeves are analysed and compared to their initial state after being exposed to different load cases. It is shown how the tool helps to track the damage evolution in each layer and how to identify the critical areas. This paper provides the first results of the damage scenarios for fatigued axial sleeves. The presented tool will allow an in-depth analysis of the fatigue behaviour of axial air spring composite materials in future and can also be applied to other types of air spring sleeves or similar structures. It can provide an important contribution to improving the service life prediction and design of air springs.

1 INTRODUCTION

The latest achievements in computer architecture and the development of cost-efficient computer tomographs enable the use of 3D-radioscopy for product development and for industrial R&D. State of the art postprocessing tools offer a wide range of features but are comparatively expensive and often difficult to automate. Therefore, the development of a highly specialized single purpose tool based on open source is ideal for certain products or product groups, assuming that they consist mostly of similar components. This is the case for axial air spring sleeves.

1.1 *Components of an axial air spring sleeve*

Axial air spring sleeves are made out of a rubber-cord-composite. It usually comes in the shape of a thin-walled hollow cylinder. The walls of the sleeve can be separated into three layers, as depicted in Figure 1. The three layers of an air spring are as follows:

- The inner layer (layer 1) is a thin rubber layer that seals the air inside the bellows. It must be airtight and flexible to withstand the pressure and movement of the air. Usually, chloroprene or another airtight rubber material is used.
- The middle layer (layer 2) is a reinforced rubber layer that provides the mechanical strength and

DOI: 10.1201/9781003516880-37

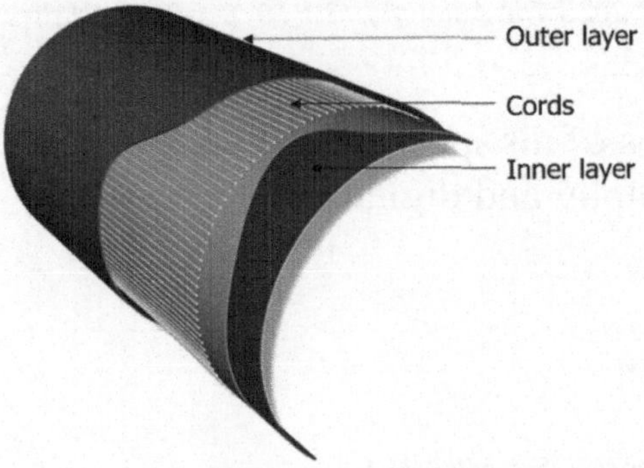

Figure 1. The three layers of an axial air spring sleeve.

stability of the bellows. It contains cords of textile or metal that are embedded in the rubber matrix. The cords can be arranged axially or crosswise, depending on the expected load directions.

- The outer layer (layer 3) is another rubber layer that protects the bellows from external influences such as abrasion, UV radiation, ozone, chemicals or temperature. It must be resistant and elastic to cope with the various environmental conditions. Here too, chloroprene or a similar rubber material is often used (see Schmidt 2023 and Serrurier 2023).

2 DEVELOPMENT OF A POSTPROCESSING TOOL FOR AXIAL SLEEVES

2.1 *Features of the tool*

This report documents the development of a Python script presented in Serrurier for the automatic evaluation of computed tomography data of air spring bellows. The script enables the detection and quantification of damages or defects inside axial air spring bellows, which can affect the performance and safety of the system. The algorithm works in several steps. First, a background subtraction is performed to separate the data of the air spring sleeve from the surroundings. Next, a random-walk algorithm (see Grady and Canny) is applied to segment the air spring sleeve into its individual components: the cord-layer and the two rubber matrix layers. The segmented components are then labeled with markers, which can also be displayed in color. Once the cord- and the matrix-layers are identified, various features are calculated, which can indicate changes in the structure or the material. These can include material inclusions, cracks in the rubber matrix, or voids at the transition of the cord to the rubber matrix. The algorithm provides a set of metrics, which assess the basic size of the reinforcements of the air spring bellows to enhance the crack detection. Additionally, the algorithm can generate visualizations of the segmented components and the detected flaws, cracks and inclusions, which can help the user to inspect the condition of the air spring bellows more easily.

2.2 *Automated virtual segmentation of strip samples*

To develop and verify the algorithm, strip specimens were initially used for validation. The samples were cut out of air spring bellows like is shown in Figure 2. The strip specimens have a width of about 1 cm and a length of about 30 cm. They contain several cords embedded parallel in a rubber matrix. The strip specimens are stretched uniaxially to represent the load a material point is objected to, when passing the rolling fold. That corresponds to the load case for a force-deflection scenario of an air spring. Due to this load, cracks can form in the rubber matrix or at the interface between cord and matrix over time, which can impair the strength and tightness of the air spring.

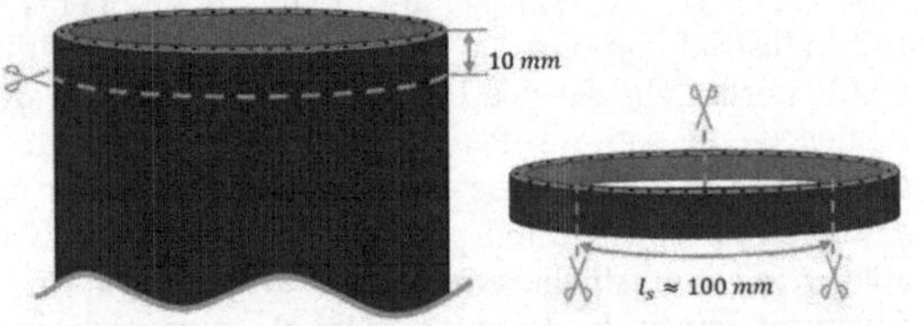

Figure 2. How to cut strip samples from an axial air spring sleeve.

The effectiveness of the algorithm can be demonstrated with the strip specimens, as detected cracks can be verified by cutting and analyzing the strips. For this purpose, the strips are cut lengthwise after the CT scan and the individual layers are examined under a microscope. In this way, the cracks that the algorithm has detected can be visually inspected, quantified and measured. This method allows to evaluate and adjust the accuracy and reliability of the algorithm if necessary.

In order to distinguish the different materials or structures in the CT data of the air spring sleeves, a histogram is created that shows the values of the different gray shades of the voxels. The histogram has different peaks that stand for background, matrix, cords, and cracks. Figure 3 shows an example of such a histogram with four areas of interest: void, fiber, rubber matrix and the separation value to distinguish the rubber from the fiber.

Usually, the histogram shows areas without useful information on the left and right side. To refine the segmentation, the histogram is cropped in order to remove superfluous areas at the beginning and end. These areas usually only have noise or artifacts that can interfere with segmentation. By removing these areas, the range of the gray scales is reduced, and the

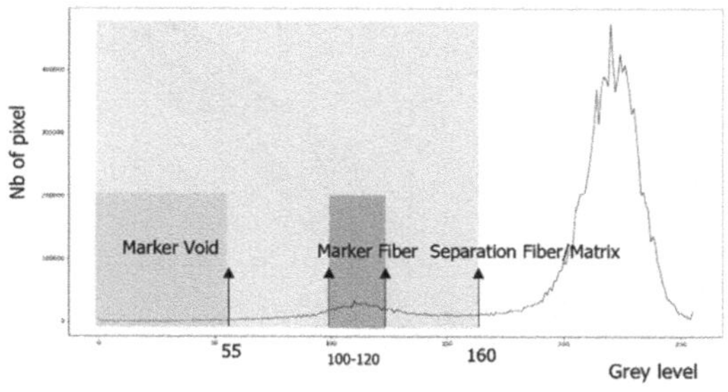

Figure 3. Histogram showing the numbers of the gray values of the voxels and four peaks.

individual contrasts of the voxels are clearer. This allows for a more precise classification of the voxels into the appropriate segments.

2.3 *Cracks detected in a strip sample cut from a sleeve after service life*

To improve the image quality and facilitate the crack detection, the script was developed based on the random walk algorithm, which was found to be most reliable in the segmentation process. During the final segmentation step, the script classifies the voxels into two classes: either they belong to a crack or not. The algorithm uses the information about the gray values of the voxels and their spatial proximity to other voxels to calculate the probability that a voxel belongs to a crack. The voxels that have a high probability of belonging to a crack are marked as flaw, while the other voxels keep their segmented marker. To achieve this, the algorithm first assigns each voxel a classification. Voids inside the matrix, without any cords in proximity are defined as cracks immediately. Crack in the interface between cord and matrix are more difficult to detect, as voids can also occur inside the cords. Here the basic information of the cord's average cross-section drives the detection of interface cracks. All voids in the proximity of cords but are outside a bounding circle are interface cracks. In Figure 4, the result of the algorithm for a spot in the air spring bellows is shown, where several cracks are present. In the image cracks and voids inside cords are not yet distinguished. The cracks run along the cord structure and can occur both within and between the cords. Some cracks are very small and barely visible, others are much larger and wider. The algorithm is able to detect and mark most of the cracks, even if they consist of only a few voxels. It is noteworthy that the algorithm also detects a crack in the matrix that is not connected to a cord structure. This shows that the algorithm works independently of the orientation or position of the cracks.

The cracks are an important indicator of the material degradation and the service life of the air spring bellows. Therefore, it is of great interest to

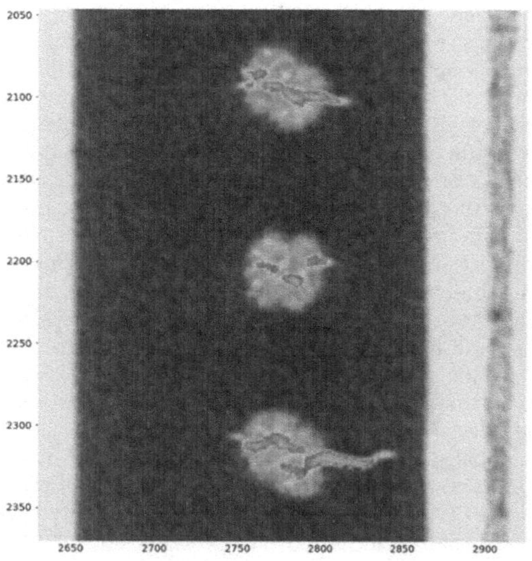

Figure 4. Segmentation of rubber, cords and cracks.

analyze the cracks quantitatively and track their development over time. The detection and marking of the cracks are the first step for such an analysis.

However, the data in the cord cross-sections are still very noisy, which makes the detection of the cracks difficult. In addition, the resolution of the CT data is not high enough to capture the microscopic processes in the cords that lead to the cracks. Therefore, it is currently not possible to make an accurate statement about what happens in the cords and how the cracks form and propagate there.

2.4 *Data boundary conditions*

In order to apply the algorithm for detecting and marking the cracks, the data from the computer tomograph must meet certain requirements, which are described in the following.

On the one hand, the resolution of the CT images plays an important role for the quality of the results. The resolution must be high enough to make the cracks and their shape visible, but not too high to perform the evaluation in a reasonable time. A too high resolution leads to an increase in the file size of the images and the storage space required for processing. This limits the number of images that can be processed at the same time and slows down the algorithm. An example: with a computer that has 128 GB of RAM memory, the algorithm is able to process about 150 images in the resolution of 5000 x 5000 pixels at once. If the resolution is doubled, only a quarter of this number can be processed. Computers with less RAM are therefore inefficient for the algorithm and may not provide any results.

On the other hand, the histogram of the gray values must be narrow to allow a clear separation between the different components of the air springs. Metal parts in the CT scan cause artifacts, which appear as bright points or stripes in the image. These artifacts spread the histogram and reduce the contrast between the gray levels. As a result, many details are lost and many smaller cracks or defects are not detected. For this reason, constructive measures have been taken to scan the air springs under slight pressure without any metal parts. For this purpose, plastic parts from a liquid 3D printer were used, which are stable enough for the pressure build-up and have the necessary strength for the construction. The plastic parts also have the advantage that their density is at the same scale like rubber and therefore do not cause any artifacts.

Finally, it should be mentioned that the image stacks were subjected to a series of additional preparation steps beforehand to improve the data quality and speed up the evaluation. This includes sorting out unnecessary images, cropping to interesting areas and a rough adjustment of the gray values. These steps were performed manually with the help of another script that allows a visual inspection of the image stacks.

With these requirements, the algorithm for segmenting and marking the cracks was applied. The duration of the evaluation with a conventional CAD workstation is about half an hour. Here, the segmentation and the reconstruction of the cracks have the largest share of the processing time. The results of the algorithm are presented in the next section.

2.5 *Results for an air spring sleeve*

Non-destructive testing of air spring sleeves is an important step in ensuring their quality and performance. In order to detect the possible defects in the cuffs, an optical scanning method with a special image processing algorithm was used. Not only individual strip test specimens, but also complete air springs were examined. Analysis of the image data revealed that there are two basic categories of defects in the air spring boots: defects in the matrix that are independent of the cords, and defects that occur at the cord-matrix interface. The defects at the interface are dominant both quantitatively and qualitatively, as they greatly reduce the mechanical properties and tightness of the sleeve and can thus lead to premature failure. An example of a crack that runs from the interface to the outside is given in Figure 5.

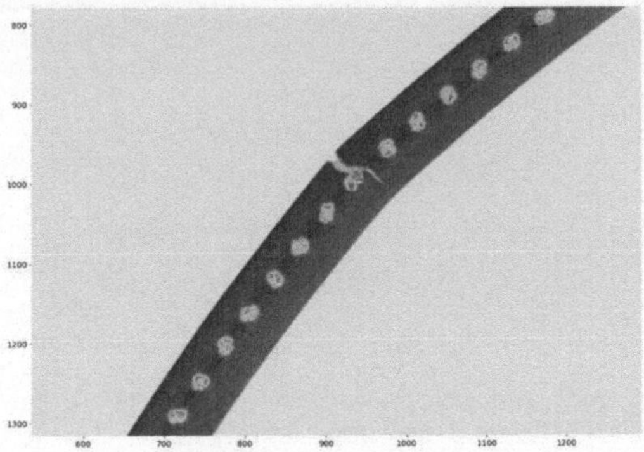

Figure 5. Axial sleeve with detected crack.

3 CONCLUSIONS

The document presents a non-destructive testing method for air spring sleeves using an tomographic scanning technique and a special image processing algorithm. The method can detect defects in both single strip specimens and complete air springs. The defects are classified into two main types: defects in the matrix that are independent of the cords, and defects at the interface between cord and matrix. The latter type is more prevalent and detrimental, as it reduces the mechanical and sealing properties of the sleeve and leads to premature failure. For the investigation of the crack evolution during service life the method will be used in the context of interrupted fatigue tests of axial air spring sleeves, where the composite will be analyzed after enduring a certain portion of their fatigue life.

REFERENCES

Serrurier, N. 2023. Development of an automatic image processing method based on tomography to decompose the different elements and defects of an air spring sleeve. FISE23 - MAMS, *ENSTA Bretagne*.

Schmidt, P. 2023. Untersuchung der Schädigungsmechanismen und Ableitung eines einfachen Schädigungsmodells für Luftfedern anhand von Bauteilmessungen und Messungen an Streifenprüfkörpern. Masterthesis, *Technische Universität Hamburg, Institut für Strukturmechanik im Leichtbau*. Hamburg.

Grady, L. 2006. Random Walks for Image Segmentation. *IEEE TRANSACTIONS ON PATTERN ANALYSIS AND MACHINE INTELLIGENCE, Vol. 28, No. 11*.

Canny, J. 1986. A Computational Approach to Edge Detection, *IEEE Trans. Pattern Analysis and Machine Intelligence*. 8:679–714, 1986 DOI:10.1109/TPAMI.1986.4767851.

Constitutive Models for Rubber XIII – Dal (Ed.)
© 2025 The Author(s), ISBN 978-1-032-85138-9

A rubber damage criterion with three principal components for antivibration fatigue design

R.K. Luo
Department of Engineering and Technology, Trelleborg IAVS, Leicester, UK

ABSTRACT: Durability is an essential requirement to ensure that rubber products operate safely in a dynamic environment during the required service life. An accurate prediction of the fatigue life of a product depends on the reliability of the criterion used in the design stage. It is hypothesised that the crack is initiated due to the effective tensile stress σ_t under the fatigue load. To be applicable to multimode loading and multiaxial conditions, the magnitude of σ_t is defined as a function of its three principal components. This criterion is developed from the concept of effective stress with the ability to account for both positive and negative R ratios. Three published experimental data sets were used to validate this methodology, including both crystalizing rubber and noncrystalizing rubber. This potential novel approach could be used for antivibration design stage and failure analyses.

1 INTRODUCTION

Durability is an essential requirement to ensure that rubber products operate safely in a dynamic environment during the required service life. An accurate prediction of the fatigue life of a product depends on the reliability of the criterion used in the design stage. Many damage criteria have been proposed and applied to different rubber specimens and products, such as strain energy density (Liang et al. (2021) and Boulenouar et al. (2016)), CED (Cracking Energy Density) criterion (Mars (2001), Mars et al. (2005) & Xu et al. (2021)), strain-based criteria ((Zarringhalami & Fatemi (2013), Belkhiria et al. (2020), Luo (2022ab), (2023)) and stress-based criteria (Luo (2004), Verron (2005), Saintier (2006), Luo (2022)). The above damage criteria were also applied to industrial cases (Luo & Wu (2006), Wu et al. (2006), Samad (2011), Luo (2021)). When R ratios (the ratio between the minimum variable and the maximum variable) are involved, generating Haigh diagrams is a direct approach (Cadwell (1939), Cam et al. (2013), Champy et al. (2021)). However, enough data needs to be collected to provide suitable applications. Mars and Fatemi (Mars and Fatemi (2003)) and Wang et al. (Wang et al. (2014)) realised this issue and proposed fitting functions with adjustable constants to simulate fatigue damage under different R ratios.

In this study, we propose an effective stress criterion to calculate fatigue damage under positive and negative R ratios using an analytical function without any adjustable parameters. The paper is organised as follows. First, we briefly describe a history of the development of this criterion, followed by a validation procedure that involves three published experimental data with different R ratios. The findings are summarised in the conclusion section.

2 DEVELOPMENT HISTORY OF THE EFFECTIVE TENSILE STRESS

In 2004, the effective stress criterion was proposed based on the zero R ratios (the ratio between the minimum stress and the maximum stress) (Luo et al. (2004)). Its form is written as follows:

$$\sigma_f = \sqrt{\sigma_{f1}{}^2 + A\sigma_{f2}{}^2 + B\sigma_{f3}{}^2} \qquad \sigma_{f1} > 0, \\ \sigma_{f1} \geq \sigma_{f2} \geq \sigma_{f3}, \quad -1 < A \leq 1, \ -1 < B \leq 1 \tag{1}$$

$\sigma_{fi}(i = 1, 2, 3)$ are the three principal components of σ_f. A and B are weighting parameters. Several assumptions were made in equation (1):

a. All three principal stresses should be considered instead of the maximum tensile stress alone.
b. A (B) are weighting parameters, their values depend on $\sigma_{f2}(\sigma_{f3})$.
c. The damage generated from $\sigma_{f2}(\sigma_{f3})$ cannot be greater than the damage generated from σ_{f1}.
d. The geometric meaning can be illustrated in Figure 1 (a).

DOI: 10.1201/9781003516880-38

Equation (1) describes an ellipsoidal failure envelope. Under this definition, any point on the ellipsoidal surface gives the same fatigue damage caused by repeated cyclic loading.

In fact, A (B) cannot be specified accurately. For safety, we treat A(B) as an integer 1. Therefore, equation (1) can be expressed as

$$\sigma_f = \sqrt{\sigma_{f1}^2 + \sigma_{f2}^2 + \sigma_{f3}^2} \; \sigma_{fi} > 0, \; i = 1,2,3 \quad (2)$$

The geometric meaning of Equation (2) is illustrated in Figure 1(b), which

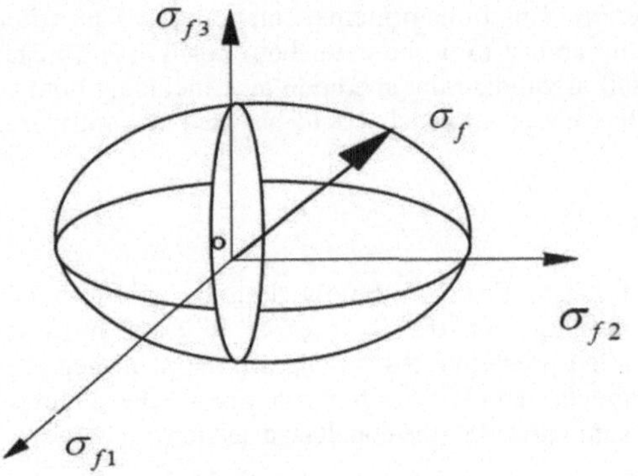

(a) Geometric illustration of Equation (1).

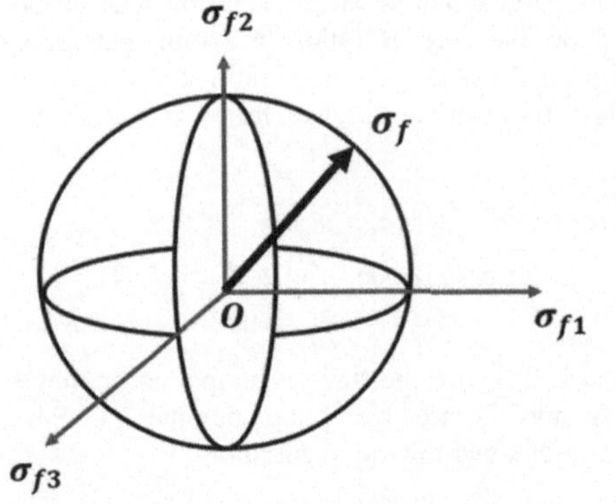

(b) Geometrical illustration of Equation (2).

Figure 1. The illustration of equations, $\sigma_{fi} > 0$.

In 2021, the effective stress criterion was further developed to compare the stress range and the current stress simultaneously and to take into account a part of the compressive stress provided that the compression portion of σ_{fi} is not greater than the tension portion of σ_{fi}, see Luo (2021). This methodology is usually applicable to a preloading procedure, as the R ratio is not always equal to zero. In addition, the rubber material has a known S-N curve.

Assuming σ_{fi} contains both tension portion σ_{fit} and compression portion σ_{fic}, then

$$\sigma_{fi} = 2\sigma_{fit}, \qquad \text{if } \sigma_{fit} < |\sigma_{fic}| \quad (3)$$

$$\sigma_{fi} = \sigma_{fit} + |\sigma_{fic}|, \text{ if } \sigma_{fit} \geq |\sigma_{fic}| \quad (4)$$

In 2022, fatigue loading cases with both positive and negative R ratios (Wang et al. (2014)) started to challenge this criterion, which has led to a new criterion: the effective tensile stress criterion; see Luo (2022c).

When a cyclic load runs between a base state with stress $\boldsymbol{\sigma}_I$ and a current state with stress σ_{II}, and fatigue damage caused can be assessed by comparing a stress range $\Delta\boldsymbol{\sigma} = \boldsymbol{\sigma}_{II} - \boldsymbol{\sigma}_I$ and the current stress $\boldsymbol{\sigma}_{II}$, which leads to effective tensile stress σ_t:

$$\sigma_t = \sqrt{\sigma_{t1}^2 + \sigma_{t2}^2 + \sigma_{t3}^2} \; \sigma_{t1} > 0, \quad \sigma_{t1} \geq \sigma_{t2} \geq \sigma_{t3} \quad (5)$$

$\sigma_{ti}(i = 1,2,3)$ are the three principal components of σ_t. To obtain σ_{ti}, let the three principal components of the stress increment $\Delta\boldsymbol{\sigma}$ are $\Delta\sigma_i$ $(i = 1,2,3)$. Let the projections of the current stress $\boldsymbol{\sigma}_{II}$ on $\Delta\sigma_i$ are $\sigma_{II\Delta}i$ $(i = 1,2,3)$, we have

$$\sigma_{ti} = 0, \qquad \text{if } \sigma_{II\Delta i} \leq 0 \text{ or } \Delta\sigma_i \leq 0 \quad (6)$$

$$\sigma_{ti} = \Delta\sigma_i, \text{ if } \sigma_{II\Delta i} \geq \Delta\sigma_i \quad (7)$$

$$\sigma_{ti} = \sigma_{II\Delta i}, \text{if } \sigma_{II\Delta i} < \Delta\sigma_i \quad (8)$$

Assuming that the three principal components are equal to 1 MPa, the value of σ_t is 1.73 MPa. If the first two principal stresses are 1 MPa and the minimum one is 0, the value of σ_t is 1.41MPa. Therefore, considering all the principal values would give the maximum damage value rather than just using one principal value alone.

To determine the direction of the crack plane, we need to obtain the direction of σ_t. Assume the projections of three components as

$$\sigma_{t1} = (\sigma_{t1l}, \sigma_{t1m}, \sigma_{t1n}) \\ \sigma_{t2} = (\sigma_{t2l}, \sigma_{t2m}, \sigma_{t2n}) \\ \sigma_{t3} = (\sigma_{t3l}, \sigma_{t3m}, \sigma_{t3n}) \quad (9)$$

and $\sigma_{ti} > 0\,(i = 1,2,3)$,

we have the direction cosine $\boldsymbol{d}$ of σ_t as

$$\begin{aligned} \boldsymbol{d} &= (d_l,\ d_m, d_n) \\ d_l &= (\sigma_{t1l} + \sigma_{t2l} + \sigma_{t3l})/\sigma_t \\ d_m &= (\sigma_{t1m} + \sigma_{t2m} + \sigma_{t3m})/\sigma_t \\ d_n &= (\sigma_{t1n} + \sigma_{t2n} + \sigma_{t3n})/\sigma_t \end{aligned} \tag{10}$$

Their relationship is illustrated in Figure 2.

We will use the published experimental data from Wang et al. (2014) to verify this damage criterion.

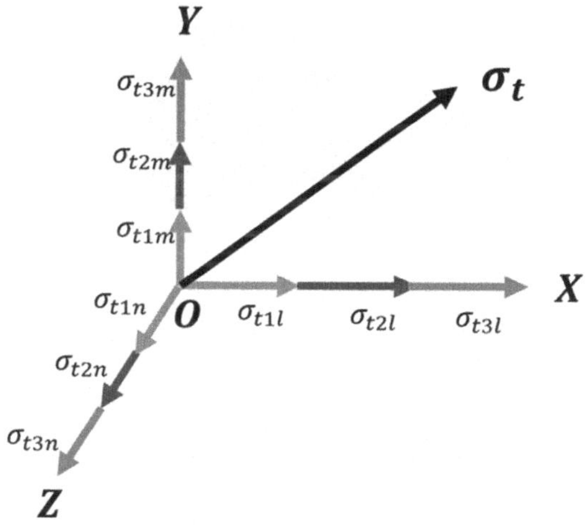

Figure 2. Illustration of the effective tensile stress and its three principal components.

3 VALIDATIONS

3.1 *Constitutive model*

Rubber materials are usually modelled using the hyperelastic approach based on the strain energy density; see Ogden (1984), Luo (2020), and Abaqus user manual (2020). The main equations for a hyperelastic model are given below.

$$W = W_I(\bar{I}) + W_J(J) \tag{11}$$

where W is the strain energy density. W_I is related to the deviatoric mechanical behaviour of the material and W_J describes the volumetric behaviour of the material.

$\bar{I}$ is a set of the invariants of the left Cauchy-Green deformation tensor $\boldsymbol{b}$ that can be expressed using the space-dependent displacement vector $\boldsymbol{u} = [\boldsymbol{u_1}(\boldsymbol{X_1}, \boldsymbol{X_2}, \boldsymbol{X_3})\boldsymbol{u_2}(\boldsymbol{X_1}, \boldsymbol{X_2}, \boldsymbol{X_3})\boldsymbol{u_3}(\boldsymbol{X_2}, \boldsymbol{X_2}, \boldsymbol{X_3})]^{\boldsymbol{T}}$, its gradient $\boldsymbol{H}$ and the deformation gradient tensor $\boldsymbol{F=I+H}$

$$\boldsymbol{F} = \left\{ \begin{matrix} 1 + \frac{\partial u_1}{\partial X_1} & \frac{\partial u_1}{\partial X_2} & \frac{\partial u_1}{\partial X_3} \\ \frac{\partial u_2}{\partial X_1} & 1 + \frac{\partial u_2}{\partial X_2} & \frac{\partial u_2}{\partial X_3} \\ \frac{\partial u_3}{\partial X_1} & \frac{\partial u_3}{\partial X_2} & 1 + \frac{\partial u_3}{\partial X_3} \end{matrix} \right\} \tag{12}$$

$$\boldsymbol{b} = \boldsymbol{F} \cdot \boldsymbol{F^T} \ \left(\ b_{ij} = F_{ik}\ F_{jk}\ \right) \tag{13}$$

The first part of $\bar{I}$ read as

$$\bar{I}_1 = \frac{I_1}{J^{2/3}} \tag{14}$$

$$I_1 = \text{trace}(\boldsymbol{b}) \tag{15}$$

$J$is the Jacobian of the deformation gradient:

$$J = \sqrt{\det(\boldsymbol{b})} = \det(\boldsymbol{F}) \tag{16}$$

The second part of $\bar{I}$ is

$$\bar{I}_2 = \frac{I_2}{J^{4/3}} \tag{17}$$

$$I_2 = \frac{1}{2}(I_1^2 - \boldsymbol{b} : \boldsymbol{b}) \tag{18}$$

The stress-strain law can be obtained by differentiating the strain energy density W using standard methods, and the results are given below.

$$\begin{aligned} \sigma_{ij} = \tfrac{2}{J}\Big[\tfrac{1}{J^{\frac{2}{3}}}\big(\tfrac{\partial W}{\partial \bar{I}_1} + \bar{I}_1\ \tfrac{\partial W}{\partial \bar{I}_2}\big)b_{ij} - \big(\bar{I}_1\ \tfrac{\partial W}{\partial \bar{I}_1} + \\ 2\,\bar{I}_2\ \tfrac{\partial W}{\partial \bar{I}_2}\big)\tfrac{\delta_{ij}}{3} - \tfrac{1}{J^{\frac{4}{3}}}\ \tfrac{\partial W}{\partial \bar{I}_2} b_{ik}b_{kj}\Big] + \tfrac{\partial W}{\partial J}\ \delta_{ij}. \end{aligned} \tag{19}$$

Where σ_{ij} are the components of the symmetric Cauchy stress tensor (true stresses) and δ_{ij} is the Kronecker delta.

The polynomial model has been used frequently in industry to simulate the response of rubber structures. The formula of this material model can be written as follows.

$$W = \sum\nolimits_{i+j=1}^{N} C_{ij}\,(\bar{I}_1 - 3)^i(\bar{I}_2 - 3)^j \tag{20}$$

where C_{ij} is constant.

This model is used for simulation in section 3.2 (N=1).

3.2 *Cylindrical dumbbell specimens*

We take experimental data from Wang et al. (2014) to validate the proposed criterion. Figure 3 (a) shows the dumbbell sample for fatigue tests with 30 cases (7 cases for R≈0, 9 cases for R > 0 and 14 cases for R<0). The total height of the sample was 80 mm with a larger radius of 23.5 mm at both ends. The specimens were made from filled natural rubber. The obtained S-N curve (Wöhler curve) achieves a correlation coefficient of 0.86 with a scatter factor of 1.3, plotted in Figure 3(b).

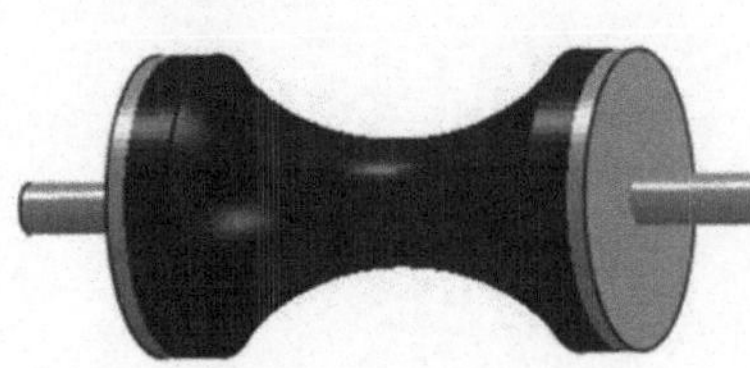

(a) The cylindrical dumbbell specimen.

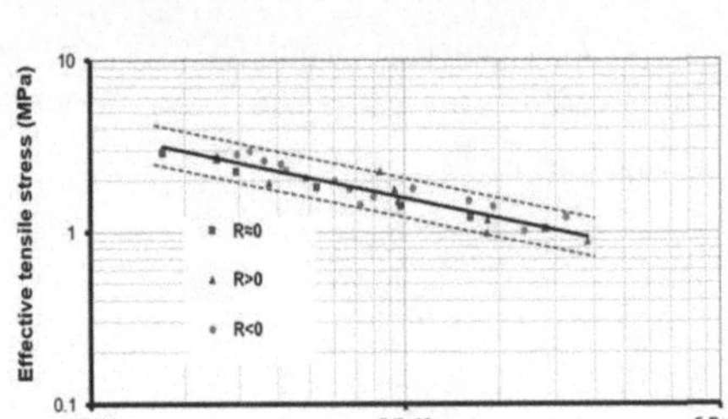

(b) The S-N curve of all fatigue cases with a scatter factor of 1.3, taken from Luo (2022a).

Figure 3. The cylindrical dumbbell sample and the results.

3.3 *AE2 specimens*

The AE2 specimen was 30mm high and its middle section had a radius of 2mm, as shown in Figure 4 (a). The samples were made from noncrystalizing rubber, i.e. filled SBR (Styrene-Butadiene Rubber). A second-order Ogden form was used to simulate the rubber response under the loadings Ayoub et al. (2012).

128 fatigue cases were calculated and produced very good result (scatter band of 1.35 was achieved), see Luo (2023b).

Figure 4. AE2 specimen.

3.4 *AE42 specimens*

In reference Ayoub (2014), AE42 samples, made from the same material as those in AE2 samples, were used to perform 40 fatigue tests in three loading modes (R = 0): tension, torsion, and tension + torsion. For the tension loads, there were 17 cases; for the torsion loads, there were 10 cases; and for the combined tension-torsion cases, there were 13 cases. The sample was 30 mm in height and 20 mm in diameter at the middle section, as shown in Figure 5. the σ_t response has achieved a scatter band of 1.3 with $R^2 = 0.95$, see Luo (2023b).

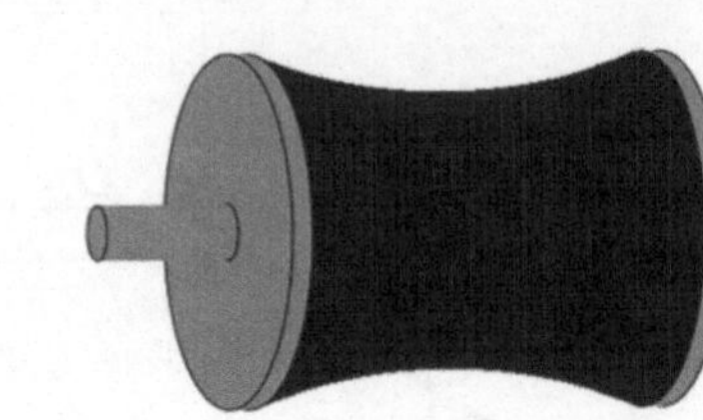

Figure 5. AE42 specimen.

3.5 *Effect of geometry*

The AE2 and AE42 specimens were made from the same rubber material. The only difference is in the geometry profile. To test the applicability of the damage criterion, we can unify the responses of AE2 and AE42 to generate a single S-N curve for both samples. Figure 6 shows the unified stress response in all 168 fatigue cases, which has achieved a scatter band of 1.35 with $R^2 = 0.91$.

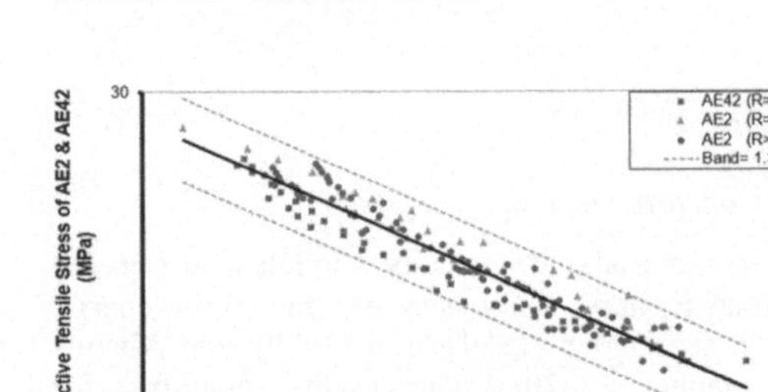

Figure 6. σ_t response of both AE2 and AE42 specimens in 168 fatigue cases, taken from Luo (2023b).

4 CONCLUSIONS

In this investigation, the proposed criterion has been applied to both crystalizing and noncrystalizing rubbers using cylindrical dumbbell specimens with both positive and negative R ratios in 30 cases, AE2 specimens, and AE42 specimens with positive R ratios in 168 cases. The main findings are summarised below.

- The proposed fatigue criterion is a potentially novel approach for predicting fatigue life.
- The maximum values of the damage criteria were located at crack locations observed in the experiments.
- The proposed criteria unified all multiaxial experimental data in different loading modes and R ratios in this study. Strong correlations between the prediction and the observed failure cycles with a range of $R^2 \geq 0.86$ in all predictions.

This potential novel approach could be used for antivibration design stage and failure analyses.

REFERENCES

Ayoub, G., Naït-Abdelaziz, M., Zaïri, F., Gloaguen, J.M., Charrier, P., 2012. Fatigue life prediction of rubber-like materials under multiaxial loading using a continuum damage mechanics approach: Effects of two-blocks loading and R ratio, Mechanics of Materials, 52, 87–102.

Ayoub, G., Naït-Abdelaziz, M., Zaïri, F., 2014. Multiaxial fatigue life predictors for rubbers: Application of recent developments to a carbon-filled SBR, International Journal of Fatigue, 66, 168–176.

Belkhiria S, Hamdi A, Fathallah R. 2020. Strain-based criterion for uniaxial fatigue life prediction for an SBR rubber: comparative study and development. Proc Inst Mech Eng Part L: J Mater: Des Appl;234(7):897–909.

Boulenouar A, Benseddiq N, Merzoug M, Benamara N, Mazari M. 2016. A strain energy density theory for mixed mode crack propagation in rubber-like materials. J Theor Appl; 54(4):1417–31.

Cadwell S, Merrill R, Sloman C, Yost, F. 1939. The dynamic fatigue life of rubber, SAE technical paper 390023.

Cam JL, Huneau B, Verron B. 2013. Fatigue damage in carbon black filled natural rubber under uni- and multiaxial loading conditions, International Journal of Fatigue, Volume 52, 82–94.

Champy C, Le Saux V, Marco Y, Glanowski T, Charrier P, Hervouet W. 2021. Fatigue of crystallizable rubber: generation of a Haigh diagram over a wide range of positive load ratios. Int J Fatigue; 150:106313.

Dassault Systems. 2020, Abaqus user manual, Dassault Systems, USA.

Liang C, Gao Z, Hong S, Wang G, Kwaku Asafo-Duho BM, Ren J, et al. 2021. A fatigue evaluation method for radial tire based on strain energy density gradient. Adv Mater Sci Eng 2021:1–12.

Luo RK, Wu WX, Cook PW, Mortel WJ. 2004. An approach to evaluate the service life of rubber springs used in rail vehicle suspensions. Proceedings of the Institution of Mechanical Engineers, Part F: Journal of Rail and Rapid Transit; 218(2):173–177.

Luo RK, Wu WX. 2006. Fatigue failure analysis of anti-vibration rubber spring. Eng Fail Anal 13(1):110–6.

Luo RK. 2020. Simulation methods for rubber antivibration systems. Singapore: World Scientific Publishing Co. Pte. Ltd.

Luo R.K. 2021. Effective stress criterion for rubber multiaxial fatigue under both proportional and non-proportional loadings, Engineering Failure Analysis, Volume 121:105172.

Luo RK. 2022a. Shear criterion and experimental verification for antivibration fatigue design. Exp Mech 62, 537–547.

Luo RK. 2022b. Effective strain criterion under multimode and multiaxial loadings – a rubber S-N curve with the scatter-band factor of 1.6 from 90 fatigue cases. eXPRESS Polym Lett.16(2):130–41.

Luo, R. 2022c. Stress-Based S–N Base Function with Shear Modulus for Fatigue Assessment on Industrial Isolators. Exp Mech. 62, 1059–1066.

Luo RK. 2023a. Fatigue damage criteria for antivibration systems with filled natural rubber under negative and positive R ratios, Mechanics of Materials, Volume 179, 104588.

Luo, R. K. 2023b. Multimode Fatigue Criteria for Filled Non-crystalizing Rubber under Positive R Ratios. Journal of Applied and Computational Mechanics, 9(4): 1049–1059.

Mars W.V. 2001. Multiaxial fatigue crack initiation in rubber, Tire Sci. Technol. 29, 171–185.

Mars WV, Fatemi A. 2003. A phenomenological model for the effect of R ratio on fatigue of strain crystallizing rubbers. Rubber Chem Technol, 76(5):1241–58.

Mars W.V, Kingston JGR, Muhr A, Martin S, Wong KW. 2005. Fatigue life analysis of an exhaust mount. In: Austrell P, Kari L., editors, Proceedings of the 4th European conference on the constitutive models for rubber. London: Taylor and Francis. 23–9.

Ogden RW. 1984. Non-linear elastic deformations. Eng Anal 1(2):119.

Saintier N, Cailletaud G, Piques R. Multiaxial fatigue life prediction for a natural rubber. Int J Fatigue 2006; 28(5-6):530–9.

Samad MSA, Ali A, Sidhu RS. D. 2011. Durability of automotive jounce bumper. Mater Des; 32(2):1001–5.

Verron E. 2005. Prediction of fatigue crack initiation in rubber with the help of configurational mechanics. In: Austrell P, Kari L, editors, Proceedings of the 4th European conference on the constitutive models for rubber. Taylor and Francis, London, 3–8.

Wang X-L, Shangguan W-B, Rakheja S, Li W-C, Yu B. 2014. A method to develop a unified fatigue life prediction model for filled natural rubbers under uniaxial loads. Fatigue Fract Engng Mater Struct, 37(1):50–61.

Xu X, Zhou X, Liu Y. 2021.Fatigue life prediction of rubber-sleeved stud shear connectors under shear load based on finite element simulation. Eng Struct 227:111449-1–111449-13.

Zarrin-ghalami T, Fatemi A. 2013. Fatigue life predictions of rubber components: applications to an automobile cradle mount. Proc Inst Mech Eng Part D: J Automobile Eng. 227(5):691–703.

Constitutive Models for Rubber XIII – Dal (Ed.)
© 2025 The Author(s), ISBN 978-1-032-85138-9

Fatigue design of automotive parts: Application of machine learning methods to design parts according to a stiffness loss criterion

F. Porhiel, P. Charrier & C. Champy
Vibracoustic Nantes, Engineering and Simulations Methods (ESM) Department, Carquefou, France

Y. Marco
ENSTA Bretagne, UMR CNRS, IRDL, Brest, France

F. Szmytka
ENSTA Paris, IMSIA - UMR EDF/CNRS/ENSTA, UME, Palaiseau, France

ABSTRACT: The main objective of this study is to challenge the use of machine learning techniques to model the stiffness loss occurring during durability tests after crack initiation without explicitly simulating the crack, on automotive parts. Extensive studies have been conducted at Vibracoustic to investigate the durability of carbon-filled natural rubber, under various test conditions. This database includes notably relaxing uniaxial tension tests on multiple samples geometries, and a part of them will be our focus in this paper. By using a sub-sample of the total available database, the authors aim to first explain the different steps taken towards the creation of a FAIR data-model. Our first findings confirm that some techniques from the Machine Learning field can be proficiently used in this case study. The next step will be to use an enlarged database with more loading conditions and geometries including real parts.

1 GENERAL INSTRUCTIONS

Durability in carbon-filled natural rubber is a well-known topic, with relatively accurate lifetime predictions for automotive parts. Actual standard calculation approach only focuses on hotspots predictions and small cracks appearance, which is not entirely coherent with cars manufacturers specifications for an end-of-life functional criterion:

1. No visible cracks in case of technical inspection;
2. Maximal functional static stiffness loss of 20%.

This second condition is not actually explicitly considered in the Vibracoustic virtual design loop. Currently, the FEA outputs are the hotspots for crack appearance. Their severities are ranked thanks to a damage evaluation. However, the identified most damaging hotspot might not be leading the functional stiffness loss due to its crack growth rate. To consider this issue, the company does not want to conduct extensive crack propagation simulations during design loops. Vibracoustic then decided to explore the usage of Machine Learning (ML) techniques for stiffness loss predictions. Therefore, the industrial goal motivating the PhD study, and consequently this paper, is to enhance the partner company's capabilities to assess the severity of hotspots, based on inputs obtained with no extensive FEA simulations.

2 METHODS

2.1 *Material and samples*

The studied elastomer is based on natural rubber. The specimens used in this paper were manufactured under equal molding settings and from a single material batch to reduce the sources of scattering. The compound is the same as the one used in (Warneboldt *et al.*, 2022) and (Le Gal *et al.*, 2023). The samples chosen in this study are part of the large panel of samples/parts available in Vibracoustic. They are molded using injection press and can be tested for durability in tension, torsion, compression or even under various loadings combinations. Loadings can either be relaxing or not.

2.2 *Experimental setup*

All the Hourglass-shaped samples have been tested on test benches devoted to research studies in Vibracoustic, similar to the one used in (Ostoja-Kuczynski, 2005). Those research benches can be

DOI: 10.1201/9781003516880-39

used for durability tests, in tension, torsion, compression and every combination of those 3 loadings. The benches are driven either in displacement/angle or in force/torque. The ambient temperature can be monitored and enforced thanks to an oven, and the skin temperature at the failure location area is measured thanks to a pyrometer.

2.3 *Post-treatment method*

A common durability measure is the number of cycles until a crack is initiated (Ni). To do so, a post-treatment method was proposed by (Ostoja-Kuczynski *et al.*, 2003). It is based on the fact that during most cycles of the fatigue test, the microscopic cracks in stages 0, 1, 2 and 3 (from decohesion to coalescence), do not influence the macroscopic stiffness (Masquelier, 2014). The initial stiffness evolution is therefore linked to the viscoelasticity of elastomers and can be modelled by the following equation: $K = a.\log(N) + K_0$, with K the macroscopic stiffness, and N the number of cycles.

The deviation from this equation allows the detection of cracks large enough (coalesced at phase 3) to affect the stiffness evolution. Using the same principle, a stiffness loss of X% can be derived, by establishing a threshold at $(100 - X)\%$ of the semi-Loglinear regression fitted on the viscoelastic part of the curve. This gives the number of cycles after a X% stiffness loss, see Figure 1. The following notation will be considered throughout the entire paper.

N_i corresponds to the number of cycles to reach the initiation of a large enough crack to have visible impact on the static stiffness (Ostoja-Kuczynski *et al.*, 2003). N_x corresponds to the number of cycles to reach a stiffness loss of X%, compared to the one at mid-life, obtained for $N = N_i/2$. Some parameters are used in this post-treatment, such as the percentage of points to remove, to obtain a linear regression of the viscoelastic phase. All those parameters are stored in our data-model for traceability concerns.

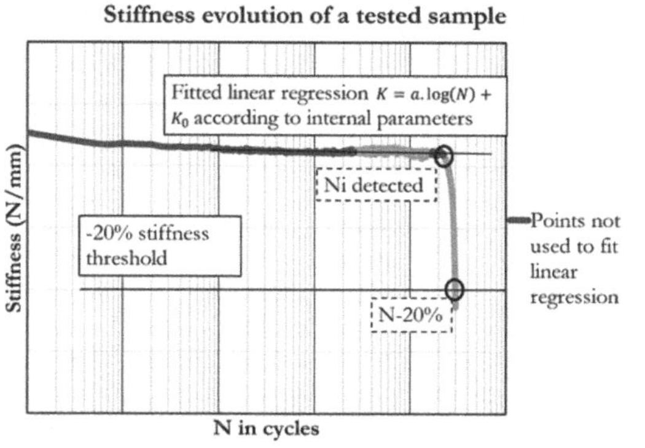

Figure 1. Example of a classical stiffness evolution curve and the posttreatment method used.

2.4 *Test campaign*

In this paper, reduced Wöhler curves were measured, under uniaxial relaxing tension, see Figure 2. The lifetimes are comprised between 10^4 and 10^6 cycles. For each sample type, four load levels are tested, and for each level, five samples are used. This procedure allows us to account for fatigue natural scattering.

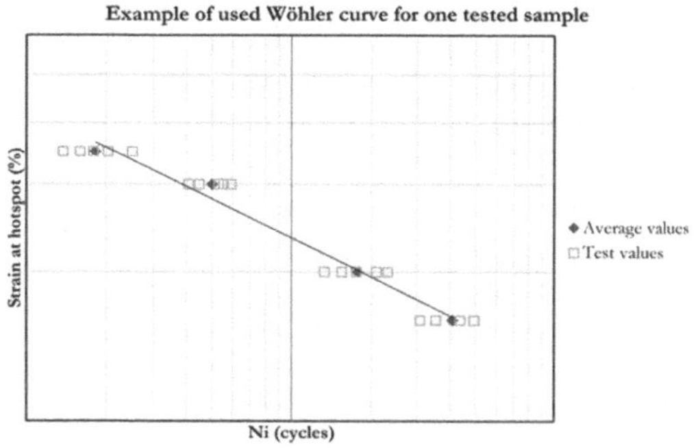

Figure 2. Illustration of the used Wöhler curves for one sample.

2.5 *Finite Element Analyses (FEA)*

To study the local mechanical fields of the different samples during the fatigue tests, FEA calculations were set-up, using Abaqus software. While no cracks are here considered in the meshes, other methods including cracks could be studied in the future for this study. Samples are all meshed with axisymmetric elements of four nodes (Abaqus reference: CAX4H), and average element size of 0.5mm. A Mooney-Rivlin constitutive model is used, its parameters are fitted on uniaxial compression, tension, and shear tests, according to Vibracoustic internal procedure. The desired boundary conditions are applied to master nodes, same as in (Felipe Mouallem De Assis, 2022). Figure 3 shows that, with the implemented model, good correlations are obtained between FEA and measurements at mid-life during the fatigue test for the load-displacement curves.

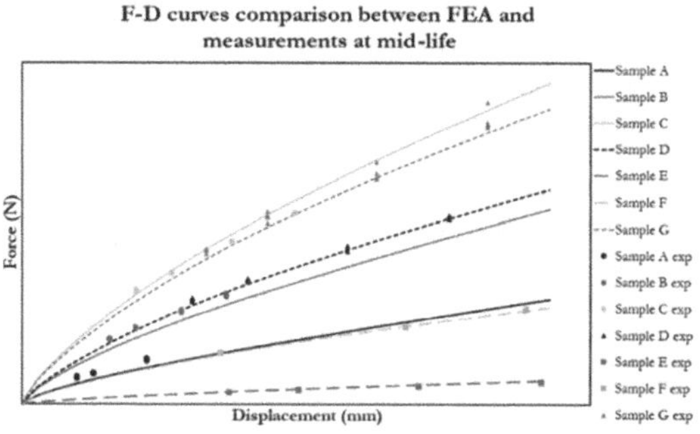

Figure 3. Load-displacement curves, correlations between measurements and FEA simulations.

The hotspot, where the crack initiates, is captured thanks to an internal Vibracoustic fatigue criterion. At this element's centroid, all evaluated mechanical measures are stored and added to the data-model: strain and stress tensors, elastic energy, etc.

To perform ML using these values, a structured data-model was first needed and therefore established.

3 DATA-MODELING AND DATA PREPARATION FOR MACHINE LEARNING (ML)

3.1 *Content and formatting of a data-model*

ML approaches are dependent on the quality of the used data-model. Consequently, the authors will insist on a good quality data-model definition.

3.1.1 *Definition of a good data-model*

A data-model refers to data structuration, as an ensemble of facts, figures, etc. It rules the way it must be stored, accessed, or retrieved. Three main types of data-models exist, see (Mayr & Thalheim, 2021):

- Conceptual data-model: for a global overview;
- Logical data-model: for a more precise overview, where data attributes, data types and relationships between data are specified;
- Physical data-model: for a detailed overview with technical information such as how data is stored.

A commonly accepted definition to assess the quality of a data-model lies in the key-word FAIR, used to describe good practices for data management:

- Findable: easy to find with good documentation;
- Accessible: define access conditions for all users;
- Interoperable: used format is standard;
- Reusable: detailed instructions for easy usage;

3.1.2 *Definition of a database management system*

The database management system must also be defined, often as either SQL (Structured Query Language) or NoSQL ("not only SQL"). SQL systems have a fixed structure, allowing for fast data queries but storing new unseen data will be complex. On the other hand, NoSQL systems have a more flexible architecture, but can take longer to answer queries. Data-modeling also needs correct identifiers: to split data in multiple sub-tables instead of one. Therefore, unique, and easily understandable identifiers are used to link data from place A (e.g. material information), with others stored in place B (e.g. FEA meshes).

3.1.3 *How to ensure good quality data*

To realize relevant analysis and reuse legacy data, a good quality of data is needed and ensured using a time-consuming pre-processing step. It accounts for better trust in future analysis. Data pre-processing regroups analysis and cleaning steps. In this study, used data were generated during the past 20 years. An exploration step was first conducted, to well understand data generation, formatting, missing values, etc.

For the cleaning step, each variable/feature should be a column (e.g. maximum enforced displacement), each observation a row (e.g. each conducted fatigue test) and each cell a single value (e.g. for 3rd tested sample, the value of maximum enforced displacement). Also, one does not want to mix up different value types, such as floats or strings. Standardized writing convention prevents the model from getting polluted by different ways of writing the same thing.

In the present study, for now we are using SQL type for our data-model. It is composed of around 9.900 tests and is divided into three tables as can be seen in Table 1, linked together with unique identifiers.

Table 1. Overview of some information from used data-model.

Test table	Material table	FEA table
Signal description;	Compound specifications;	Mechanical values at hotspot;
Post treated data N_i, $N_{X\%}$, $Etc.$;	Molding information;	Simulation parameters;
Etc.	Etc.	Etc.

From this database, only a reduced database of 140 tests has been extracted and selected for this paper. This selection is statistically representative and yet reduced enough to allow for fast treatment. All the samples were tested in uniaxial relaxing tension.

3.2 *Preparation of the data-model to apply ML techniques*

ML techniques are mainly based on statistical tools, with pre-requisites to apply (Telmen, 2023).

3.2.1 *Imputation*

The data used must be "continuous", meaning no blanks/null values (not machine understandable). While all observations with missing values could be removed, physical hidden phenomenon could easily be discarded. Another approach is then to impute blanks with average values, predictions based on other features, etc. In this study, the skin temperature at the hotspot was often missing, due to a sensor issue. The average temperature from all the samples of the same geometry, and load level was then imputed.

3.2.2 *Categorical encoding*
For statistics, character chains are also not usable. A commonly used method to exploit categorical features, other than labelling, is to convert the string values to scalars. In this paper, the sample type (A to G) could be converted to a number between 1 and 7. This would give predominant weight to sample G. The preferred method is then to apply one-hot encoding, as detailed in (Jason Brownlee, 2020). Each label (here sample A, sample B, etc.) is considered as a feature with value of 1 or 0. Still, it increases the number of features in the ML models. The number of features to keep, based on their utility, will be discussed later on.

3.2.3 *Scaling*
Scaling prevents having one input mainly driving the final output. The "MinMax" algorithm as defined in (Pedregosa *et al.*, 2011), scales all values between 0 and 1. See Figure 4, where the authors compare its dependency on a known outlier. As shown in Figure 4, this algorithm seems highly dependent on outliers.

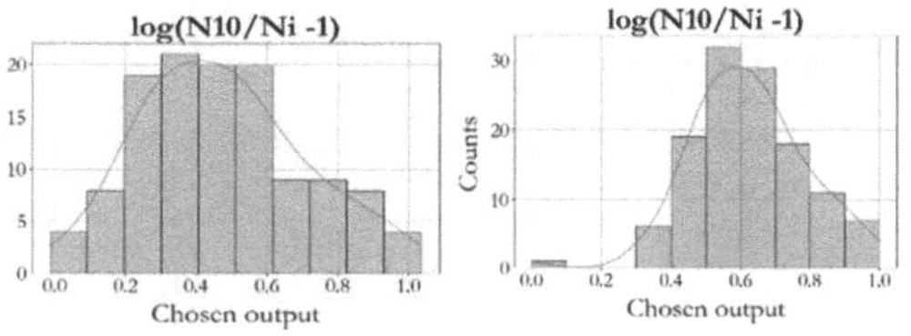

Figure 4. Distribution of scaled chosen output, using "Min-Max" algorithm, with removed (left) or kept (right) known outliers.

On the contrary, the "Quantile" algorithm defined in the same publication, maps each feature to a gaussian distribution, centering values on 0 with a standard deviation of 1. It is very robust to outliers, as seen in Figure 5, as it renders them indistinguishable from the data population. This might however lead to a loss of information for infrequent data, as their values will be scaled to the distribution boundaries.

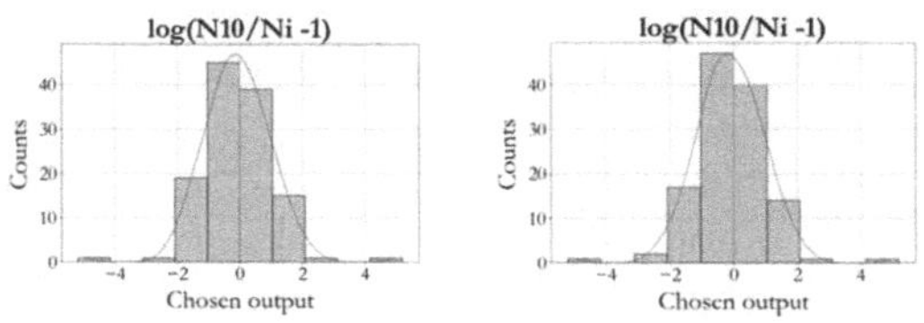

Figure 5. Distribution of scaled chosen output, using "Quantile" algorithm, with removed (left) or kept (right) known outliers.

For the future shown results, the "Quantile" scaling algorithm is chosen, to reduce the dependency to outliers. This choice is first motivated by an initial limited knowledge about the presence of outliers. Still, as it is good practice to identify outliers, it will be the focus of the next section. Indeed, they give informative insight on the tested data, and account for a better understanding of data generation history (e.g. how exactly outliers occurred).

3.2.4 *Identify outliers*
Identifying outliers is a key pre-processing step for ML as they can highly impact statistical analyses. Outliers can be valid results; in this case, they must be carefully considered. Others can be erroneous and not representative. In this study, outliers could be linked to a default in a sample (e.g. inhomogeneous dispersion of carbon black during the mixing process could result in a bigger particle than usual). Choosing to remove them is a choice to thoroughly justify.

To identify outliers, different statistic methods exist: For example, the standard deviation or the inter-quartile range (IQR). Depending on the threshold applied (ex: 3σ for standard deviation), outliers are identified. In this study, the outlier identified (by IQR range) was removed from the database as it seemed to be a molding default, see Figure 6.

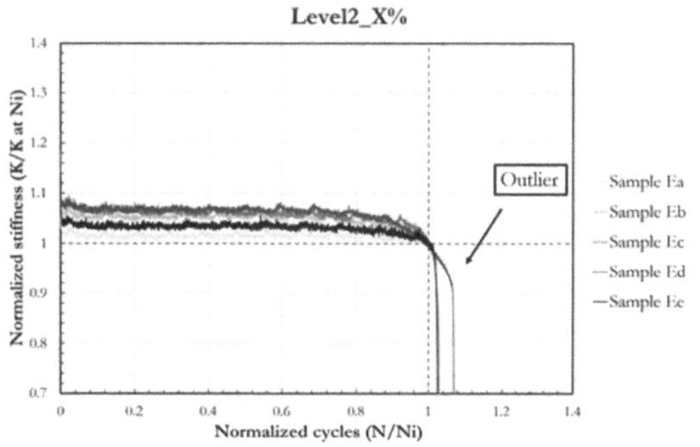

Figure 6. Evolution of normalized stiffness over normalized cycles for sample E, and a given load level.

3.3 *Application of ML to elastomers*

ML algorithms are a pertinent tool to leverage information contained in big amount of data, if it is coupled with physical analyses. The authors offer to show an illustration of how classical ML models can be applied to perform a study on crack propagation in elastomers.

3.3.1 *ML models*
ML models are algorithms that learn patterns from a given database, to then make predictions or evaluate tendencies without human interaction. Varying the ML type used can lead to an increasing complexity of the output prediction (for linear regressions, two coefficients, for deep neural networks thousands of coefficients are adjusted). There exist four main types of ML models (Brownlee, 2019).

- Supervised learning: labeled data are used in the training of the algorithm. It consists, in this study, in having a relative propagation rate linked to a stiffness loss, for each test data and train an algorithm with the extended database (from the material, test and FEA tables). The algorithm then learns which inputs drives the rate of the crack growth.
- Unsupervised learning: unlabeled data are given to the algorithm, that tries to identify patterns.
- Semi-supervised learning: first a small quantity of labeled data are given to the model to build a general trend, and then large amounts of unlabeled data are added. One of the main applications of such approach is speech or image recognition.
- Reinforcement learning: it is based on trials and errors to optimize step by step the results.

In this study, for now mostly supervised learning was used. For such an approach, two main kinds of algorithms can be used: regression and classification. Simply said, regression models will try to predict a scalar output, e.g. a crack propagation rate value. On the other hand, classification models will try to predict a class, e.g. is the crack propagating fast or slow. Models can often perform both regression and classification tasks. Some of them were identified as candidates for crack growth rate evaluation for this study, and are listed below:

- Decision trees;
- Support Vector Machine (SVM);
- K-Nearest Neighbors (KNN);
- Ensemble algorithms.

Following a previous internal study, the decision was made to use a powerful ensemble predictor, called gradient-boosting regressor (GBR), which can be easily implemented from the python library Scikit-learn (*Scikit-learn user guide*). GBR is constructed by combining several weak learners, such as decision trees. The model is built sequentially, with each new learner trained to correct the errors made by the previous ones, by fitting a new tree to the gradient of the loss function (Natekin & Knoll, 2013).

3.3.2 *Choice of the output*

The motivation of this study is to be able to predict a crack propagation rate, based on the stiffness loss caused by a macroscopic crack growth, after its initiation. Therefore, the selected output must mirror this expectation. Following a previous study (Felipe Mouallem De Assis, 2022), the output $\text{Log}(N_{10}/N_i - 1)$ was selected, as it is naturally shaped as a gaussian distribution, as seen in Figure 7.

3.3.3 *Choice of the inputs*

Concerning the choice of inputs, multiple approaches are available. One of them could be to feed the algorithm with all the available features. This is not really a proper way, especially in the present study since there are existing correlations between them. This can lead to over-learning, meaning the model has become too precise on training data, and therefore cannot predict at all unseen data. Another approach is to perform a preliminary physical analysis, to determine which inputs are of interest, by answering the following questions:

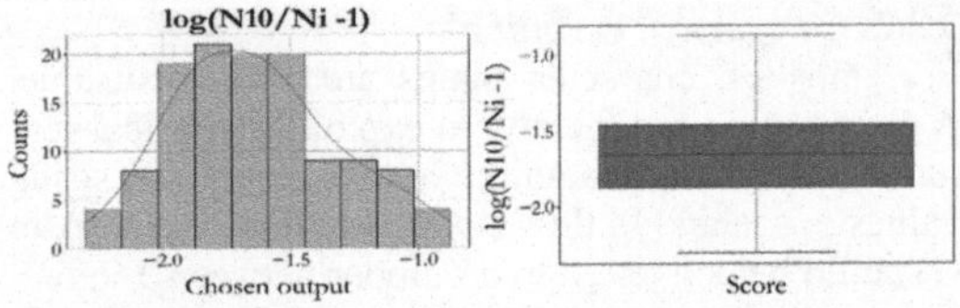

Figure 7. Distribution of the chosen output (without outliers) before scaling.

- Does their variation influence the model?
- Are they correlated with other inputs?

At first, for this study, a broad range of features was selected to ensure objectivity in the future final choice of inputs. This includes experimental (frequency, ambient temperature, material data, etc.), geometrical (volume, length, etc.) and FEA data (strain tensor, stress tensor, etc.).

As the test are uniaxial relaxing tension tests, using one unique batch, at room temperature and no pre-aging, some inputs could be removed (e.g. material data). A second more detailed analysis was then conducted to remove correlated data. For this purpose, a Principal Component Analysis (PCA) algorithm, coded in scikit-learn, was used. This is a dimensionality reduction algorithm: its task is to reduce the number of variables while keeping as much as possible the initial information. This is done by finding inputs combinations explaining as best as possible the total variance, as detailed in (Jolliffe & Cadima, 2016).

Some simple conclusions can therefore be derived from this analysis: e.g. the PCA algorithm identified that λ_I, σ_I, I_1, I_2 are correlated. This was expected as, from mechanical reasoning:

- σ_I corresponds to the stress value in the maximum principal direction, here the tension one;
- λ_I corresponds to the strain value in the maximum principal direction ($\lambda_I = \varepsilon_I + 1$);
- I_1 and I_2 are the first and second invariants of the Green-Lagrange tensor.

As the tests are all simple tensions on dumbbell specimens with the same compound, λ_I and σ_I are directly linked (see Figure 8). Similarly, the biaxiality ratio staying close to -0.5 ($B = \ln(\lambda_I)/\ln(\lambda_{II})$) and the volume being almost constant ($\lambda_I \cdot \lambda_{II} \cdot \lambda_{III} = 1$), the invariants I_1 and I_2 are only depending on λ_I. This verification is necessary to assess the level of trust in this PCA method. It offers new perspectives, for future studies on more complex test, to help reduce the number of features given to the model.

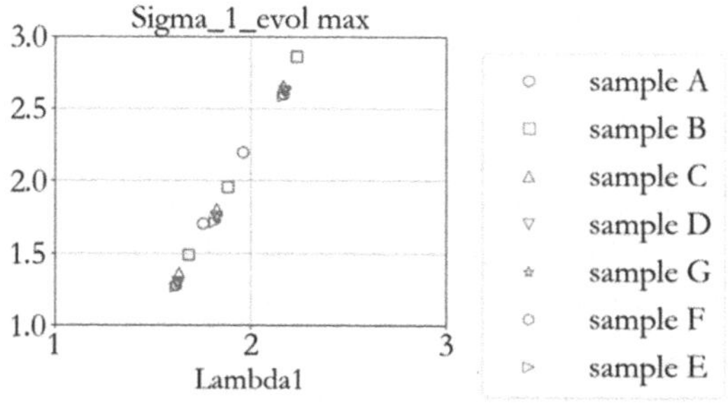

Figure 8. Evolution of σ_{I} as a function of λ_1.

The final chosen inputs will not be completely described here for confidentiality purposes.

3.3.4 *Choice of machine learning algorithm*

As already stated in 3.3.1, the gradient-boosting regressor (GBR) was selected in the present study. To evaluate the predictor's performance, a common approach is to split the database in two distinct sets: a training and a testing one. The authors chose to keep, according to various authors in the literature, 80% of the data for the training (112 tests) and 20% for the testing (28 tests). This is still up to debate in the scientific community. The split was chosen to ensure that all different samples are represented in same proportions in the training and testing set.

First, the algorithm is applied on the training set (including inputs and an output being $\log(N_{10}/N_i - 1)$) and then predictions are made on the testing set. The mean squared error is estimated between predictions and outputs to evaluate the performance of the model. Other error models exist, they are not detailed here.

Hyperparameters are intrinsic parameters of the predictor, that enables better predictions. But those parameters require tuning, which is one of the main difficulties. For the GBR, which is a combination of weak learners such as decision trees, it would be for example the depth of the used trees, the total number of trees, etc. Common algorithms exist in the literature to estimate hyperparameters value, such as the one used in this study, GridSearchCV (Fabian Pedregosa, et al., 2011). The training set is split in new training-testing folds (K-folds method), helping to not overfit the data. For each fold, the set of hyperparameters is optimized, to obtain the best training and testing R^2 regression score (the scoring method can be adapted).

3.4 *Predictions from the chosen model*

As shown in Figure 9, the model can predict unseen data quite well with the identified inputs of interest. That result is here valid for uniaxial relaxing tension tests. Since the model has only been trained on specific tests, it is still an open question on how it would perform for other loading conditions.

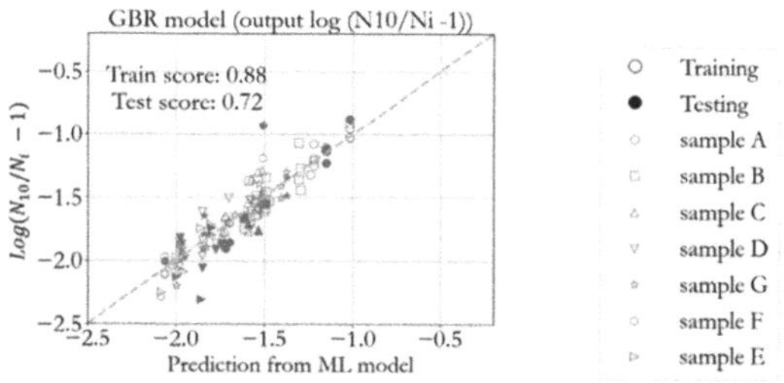

Figure 9. Application of GBR to the selected sub-database.

3.5 *Applicability to other test data*

Other tests were then post-treated: their loading condition was changed, from uniaxial relaxing tension to relaxing torsion. Other parameters such as material, molding, temperature, etc. remain unchanged.

Figure 10 shows that the previously identified model is quite bad at predicting the crack propagation rate for these new points. This highlights physical phenomenon occurring during the relaxing torsion fatigue tests, that can't be captured by using only relaxing tension ones. Indeed, with the given dataset for training, the model couldn't comprehend for example the importance of the strain maximal principal direction, as it was always constant in relaxing tension.

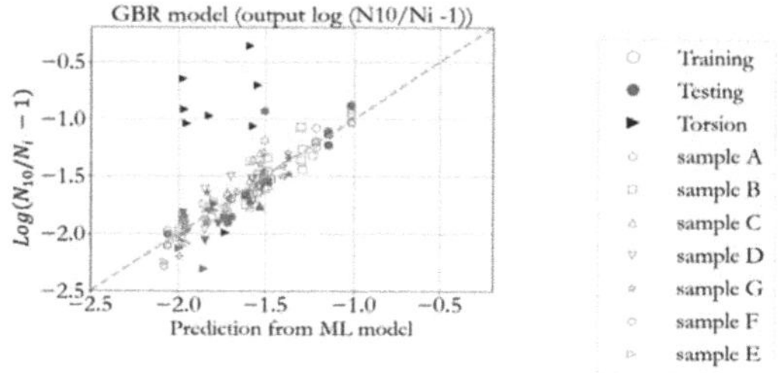

Figure 10. Application of the established GBR model to new data points in relaxing torsion fatigue tests.

4 SYNTHESIS AND CONCLUSIONS

The presented study was conducted as a first approach to use ML for a crack propagation rate estimation based on stiffness loss, to discuss on how to use these kinds of algorithms as comprehension tools. One main finding here is that a database constructed only of uniaxial relaxing tension tests cannot be used in the context of relaxing torsion. This is an expected result, that demonstrates the limits of ML models and shows the importance of having a database representative of different test conditions. The main outlook for this study is to enlarge the used database, to challenge ML potential, and finally, to use it to highlight better the physics behind the tests.

ACKNOWLEDGEMENT

The authors would like to thank ANRT for its financial support (grant number 2022/1746).

REFERENCES

Brownlee, J. (2019) *Master Machine Learning Algorithms: Discover How They Work and Implement Them From Scratch.*

Fabian Pedregosa, et al. (2011) Scikit-learn: Machine Learning in Python.

Felipe Mouallem De Assis (2022) Investigation of Machine Learning Algorithms Applied to the Development of a Fatigue Criterion for Crack Propagation in Elastomers.

Jason Brownlee (2020) Data preparation for machine learning.

Jolliffe, I.T. & Cadima, J. (2016) Principal component analysis: a review and recent developments in *Philosophical transactions.*

Le Gal, et al. (2023) Influence of curing temperature and pressure on the fatigue properties of carbon-black filled natural rubber: An analysis based on the sulfur network. *Constitutive models for rubber XII: Proceedings of the 12th European Conference on Constitutive Models for Rubber.*

Masquelier, I. (2014) *Influence de la formulation sur les propriétés en fatigue d'élastomères industriels.*

Mayr, H.C. & Thalheim, B. (2021) The triptych of conceptual modeling. *Software and Systems Modeling.*

Natekin, A. & Knoll, A. (2013) Gradient boosting machines, a tutorial. *Frontiers in neurorobotics.*

Ostoja-Kuczynski, E. (2005) *Comportement en fatigue des élastomères: application aux structures antivibratoires pour l'automobile.*

Ostoja-Kuczynski, et al. (2003) Crack initiation in filled natural rubber: experimental database and macroscopic observations. In: *Constitutive models for rubber III: Proceedings of the Third European Conference on Constitutive Models for Rubber*

Telmen, B. (2023) *Elaboration de stratégie des réparations additive de structure soumise à des chargements cycliques. Application a l'acier 316L pour composant nucléaire.*

Warneboldt, I., et al. (2022) Experimental investigations about complex non-relaxing fatigue loads for carbon-black filled natural rubber. *International Journal of Fatigue.*

Constitutive Models for Rubber XIII – Dal (Ed.)
© 2025 The Author(s), ISBN 978-1-032-85138-9

Cavity formation and stress distribution at notch tips of rubber vulcanizates

T. Tada & M. Naito
Sumitomo Rubber Industries Ltd, Kobe, Japan

M. Ishikawa
Emertus Yamagata University, Yonezawa, Japan

K. Schneider & E. Euchler
Leibniz Institute of Polymer Research Dresden, Dresden, Germany

G. Heinrich
TUD Dresden University of Technology, Dresden, Germany

ABSTRACT: Crack initiation of rubber vulcanizate is investigated under constrained tensile deformation in terms of cavitation and stress deformation. Thin disc-shaped specimens, with various shape factors, which is a ratio of radius to thickness were used to vary the degree of constrained condition. Here, the degree of constraint becomes higher for higher shape factor. In general, higher constrained conditions result in higher expansion stress to initiate cavitation. It has already been elucidated that cavity size distribution is strongly influenced by the shape factor of specimens by using tensile tests equipped with dilatometer or X-ray microtomography. The size of cavities became smaller for higher shape factor due to higher strain constraints and larger contribution of volumetric deformation as compared to shear deformation. Stress-strain relation of rubber specimens also depended on the shape factor and yielding point can be found at the onset strain of cavitation. In the present study, Finite Element Analysis is employed on notched rubber specimen to investigate the effect of strain constraints on the stress distribution at crack tips. It was found that stress concentration at the crack tip was mitigated by reducing the Poisson's ratio. The mechanism of cavity formation in the rubber matrix will be also discussed based on the FEA results.

1 INTRODUCTION

It is assumed that the fracture of polymeric materials begins when cracks propagate due to stress concentration caused by strain constraints around defects in the matrix of a rubber vulcanizate. In fracture mechanics, the strain energy release rate, or J-integral, is used as the resistance to crack propagation[1-4]. Some study results have also been reported regarding other polymers[5,6]. On the other hand, to accelerate crack propagation, the magnitude of stress concentration must exceed the cohesive strength of the polymer material[7], but there are only a few cases in which this detail has been investigated for polymers. Using a notch as a model of a rubber defect, its deformation and stress state can be analyzed using the Finite Element Analysis (FEA). The mechanism of improving the strength of rubber by reducing the stress concentration caused by strain constraint by adjusting the bulk modulus to a low value is examined.

2 FINITE ELEMENT ANALYSIS AT THE CRACK TIP

Figure 1 shows the double-edge notched specimen model of rubber vulcanizates. Right-hand coordinate system is employed, in which thickness, tensile and width directions are defined as x, y and z axile, respectively. The width of the strip, z, is large enough to fix the strain at the midst of width to be z=0. It is expected that with this model we will be able to investigate stress concentrations, while satisfying the plane strain at the crack tip. Edge notches were applied to the model symmetrically as shown in Figure 1. The notch tip radius was set to 0.1 mm and the width of the model sample was set to 40 mm so that plane strain is always kept during the deformation. The total thickness and ligament length were 20 mm and 14 mm, respectively. Ogden's constitutive relationship expressed by Equation 1 was used as the constitutive expression formulating the relationship between stress and strain in 3D deformation

DOI: 10.1201/9781003516880-40

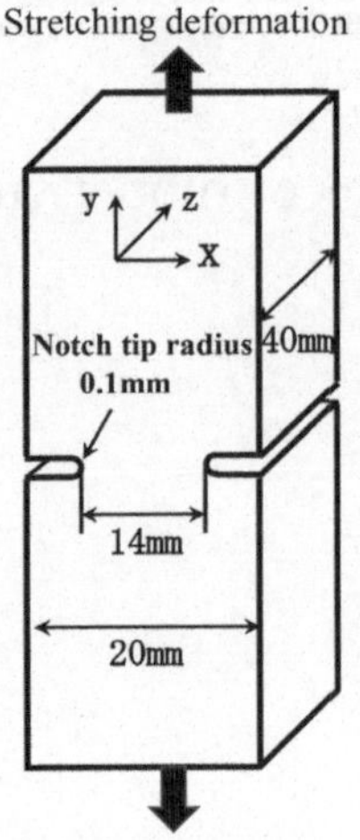

Figure 1. Deformation under plane strain using a thick sheet with notches.

of rubber. The material parameters, n = 4, a were estimated by curve fitting with the experimental results. The estimation results are shown in Table 1. FEA of notched samples was performed using Abaqus.

$$W = \sum_{n=1}^{n=4} \frac{\mu_n}{\alpha_n} \left(\lambda_1^{\alpha_n} + \lambda_2^{\alpha_n} + \lambda_3^{\alpha_n} - 3\right) \qquad (1)$$

Table 1. Ogden material parameter values.

μ	μ_1 -0.8763	μ_2 0.7782	μ_3 0.7526	μ_4 0.0002
α	α_1 -1.1734	α_2 1.5914	α_3 -1.173	α_4 5.7832

2.1 *Evaluation of stress and strain concentration*

Figure 2 shows the effect of Poisson's ratio on stress (a) and distance from the notch tip (b) as a function of the distance from the notch tip. The Poisson's ratio was varied between 0.4 and 0.499. Tensile stress, σ_y, and strain, ε_y, show their maximum value at the notch tip and decrease rapidly to the global value as the distance from the notch tip increases. The magnitude of the stress in the y direction depends on both the stress σ_x and the stress σ that suppress the Poisson contraction in the direction perpendicular to it. When the value of the elastic modulus (σ_y /ε_y) in the y direction is small at a certain position away from the notch tip, the stress that constrains strain is small, and the deformation there is dominated by shear deformation. On the other hand, the higher elastic modulus, the larger the strain constraint acts, and the deformation there is dominated by volumetric deformation.

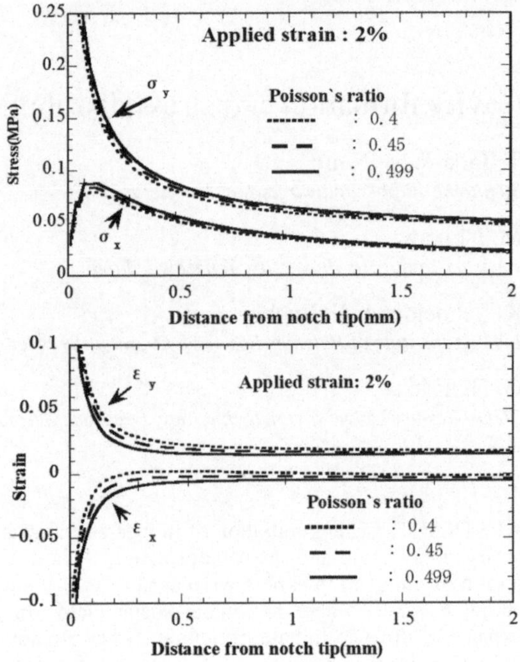

Figure 2. Effect of bulk modulus on stress and strain distribution depending on the distance from thr tip of the notch.

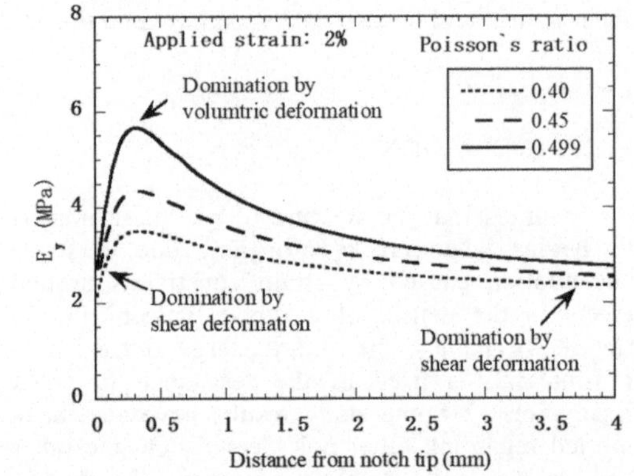

Figure 3. Variations in elastic modulus(Ey) in the y direction depending on the distance from the tip of the notch.

Figure 3 shows the elastic modulus (E_y) defined as (σ_y /ε_y) in Figure 2. The modulus in y axis (E_y) provides information about the contribution of bulk modulus on the strain constraint as a function of the distance from the notch tip. The low E_y near the notch tip suggests that shear deformation is dominant in this region because shear modulus is much smaller than the volumetric modulus. E_y increases with increasing the distance from the notch tip up to 0.5 mm and reach a maximum value, reflecting the volumetric deformation dominant in the region. With increasing distance from the notch tip, the deformation mode is again controlled by shear deformation in line with reducing E_y and the effect of strain constraints disappear. It is clearly shown that volume deformation has a large contribution to the stress concentration near the notch tip.

2.2 *Change in shape of notch tip*

Obviously, the shape of notch tip varies under different loading conditions. Figure 4 shows the effect of Poisson's ration on the contour of crack tip at a global tensile strain of 2 %. The range of Poisson's ratio is corresponding to Figure 2. It can be seen that the original notch tip with a radius of 0.1 mm is enlarged by the positive stress in the x and y axles, as same as the expansion stress of a balloon, resulting in a large increase in the strain around the notch. As shown in Figure 4, the notch tip radius was not strongly dependent on Poisson's ratio or in other words, bulk modulus. However, the position of notch tip moved inside the specimen as the Poisson's ratio approaches to 0.5, resulting in decreasing notch tip radius, being ascribed to increasing resistance against volumetric deformation. Consequently, it is found that tensile strain at notch tip becomes smaller and stress concentration is more enhanced if bulk modulus becomes larger. It is also found that if bulk modulus or Poisson's ratio decreases, the difference of modulus along the x-axis (E_x) around notch tip becomes smaller and the distribution of modulus becomes more homogeneous.

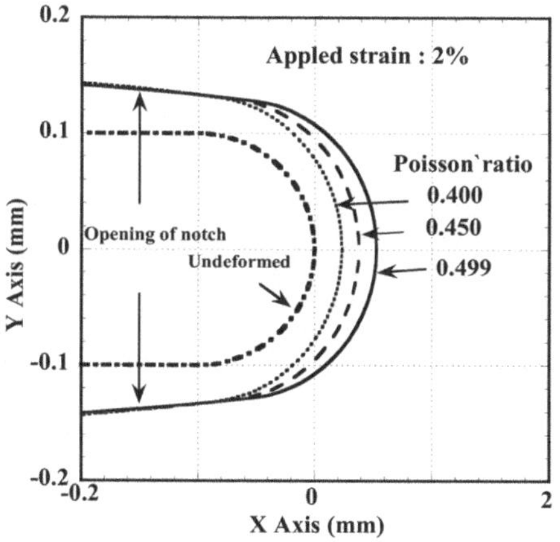

Figure 4. Change in shape of notch tip due to bulk modulus.

3 DISCUSSION

3.1 *Stress concentration due to strain constraint*

In general, the deformation of a solid body is composed of incompressible shear deformation and volumetric deformation along with volume change. Since the resistance of volumetric deformation is greater than that of shear deformation, large stress is generated because bulk modulus (K) is much higher than shear modulus (G) – for rubbers, it´s in the order of three decades. When a homogeneous solid body is subject to uniaxial extension, the strain perpendicular to the tensile direction contracts due to the Poisson effect in order to preserve incompressibility. If a void exists in a uniform solid, the modulus of the solid body is greatly different around the void, and the stress field around the voids becomes three-dimensional as same as the case of stress field around the notch tip. As shown in Figures 2 and 3, the position at which maximum value of E_y was observed roughly corresponds to the position of the maximum σ_x in Figure 2 indicating that constraint stress is generated perpendicular to the tensile direction to constraint expansion strain against Poisson contraction. The Poisson contraction is higher as the Poisson's ratio is closer to 0.5, the bulk elastic modulus becomes higher, and the resistance to volumetric deformation also becomes larger. Pure elongation shown in Figure 5 is known as the deformation with the strongest strain constraint[8-14]. Pure elongation is categorized as a deformation that prohibits contraction of cross-section and completely compete with Poisson contraction which drives contraction of specimen perpendicular to tensile direction. The stress in each direction caused by the strain ε_y in the y direction of the elastic body can be expressed as follows using the shear G and K.

$$\sigma_{yp=}\ \left(K+\frac{4}{3}G\right)\varepsilon_y \tag{2}$$

$$\sigma_{xP}\left(\text{or } \sigma_{zp}\right) = \left(K-\frac{2}{3}G\right)\varepsilon_y \tag{3}$$

Since the tensile modulus of pure elongation in Equations 2 and 3 include the bulk elastic modulus

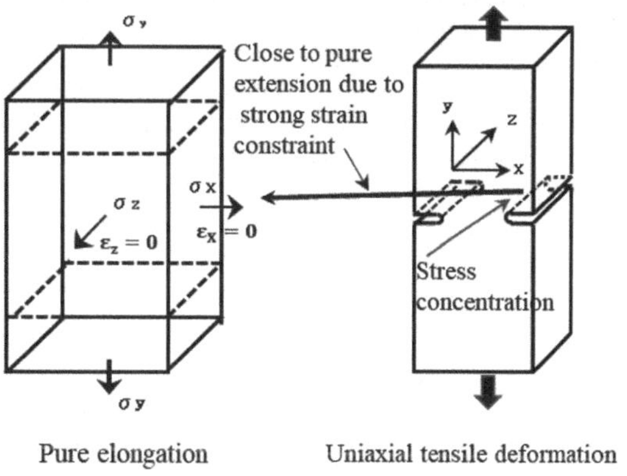

Figure 5. Stress concentration at the tip of a plan strain notch.

(K), it can be seen that it has a larger value compared to the tensile modulus of uniaxial elongation[13,14] because K is much higher than G, as described later. The stress-strain field of notch tip would be considered satisfying the plane strain as shown in Figure 5. The strain in both the z and x directions near the notch tip are constrained by the notch surface because it is not possible to contract in x and z directions enough to preserve incompressibility while ε_y becomes very high as shown in Figure 2. If the deformation is close to pure extensional strain constraint, high stresses are generated there. In other words, stress concentration occurs.

Since the elastic modulus of pure extension in Equation 4 includes the bulk elastic modulus (K), the bulk elastic modulus (K) of the rubber is involved in the magnitude of the stress generated at the tip of the notch. It would be reasonable to define the ratio (SC) of the elastic modulus of pure elongation σ_{yp} to the elastic modulus of rubber (E) when there is no notch as a defect and no stress concentration is involved, as an index to describe the magnitude of stress concentration.

$$SC = (K + 4G/3)/E \qquad (4)$$

The proposed measure of stress concentration uses the elastic modulus of pure extension. Since it is difficult to completely reduce the strain due to Poisson contraction to zero in the x and z directions at the tip of the notch, the actual magnitude of stress concentration is estimated to be much lower than this guideline. But it would be useful as a guideline of the magnitude of stress concentration.

3.2 *Fracture of unfilled rubber in uniaxial extension*

There are several reports that suggest that there is a large distribution in crosslink density in the matrix of rubber vulcanizates. Jaravel impregnated rubber with hydrogen gas at high pressure until equilibrium condition and observed an inhomogeneous structure that was presumed to be voids formed by the subsequent pressure reduction[15]. The locations of void formation is macroscopically non-uniformly distributed. Furthermore, macroscopically nonuniform dispersion is observed in the formation of nanovoids observed in the stretching test of Pancake specimens[16]. Although the formation mechanism of nanovoids with non-uniform distribution at these formation positions does not seem to be clear, the movement of molecular chains may be affected by the formation and progression of crosslinks. If this happens, likely the crosslinking density will be distributed depending on the position as crosslinking progresses. The distribution of crosslink density can remain the distribution of local elastic modulus. If there exists a distribution in the elastic modulus, strain constraint would occur around the region with high crosslinking density and high elastic modulus, and the deformation there will approach pure extension, making it easier to form nanovoids as shown in Figure 6. Formation of nanovoids eliminates the interaction between adjacent molecular chains between crosslinks, so the applied stress in the nanovoid region decreases, and expansion stress is generated at the tip of the nanovoid extension region due to strain constraint similar to a notch. As the load increases, the elongation strain of the crosslinked molecular chains in the region where the nanovoids are formed increases, and eventually the crosslinks break and a crack is formed. As a result, even greater strain is constrained at the tip of the crack formed from the nanovoid region, resulting in high stress concentration as shown in Figure 7. The formation of voids at the tip of rubber notches was investigated by H. Zhang[17]. Stress concentration guideline (SC or Ey) is estimated to be about 180 to 200, using literature values and assuming that the shear modulus of unfilled rubber vulcanizates is ca.1 MPa and, thus, the bulk modulus is 500 to 600 MPa. This is an extremely large value. With high bulk modulus and strong strain constraints, the more easily nanovoids are formed.

In general, the strain at break of unfilled rubber vulcanizates is as large as 1.5, but no volumetric expansion is observed until the beginning of rupture, indicating that the deformation of unfilled rubber is nearly constant in volume. It is natural that high stress concentration due to strain constraint contributes to failure. Unlike thermoplastic polymers, which are capable of plastic deformation, in the case of rubber vulcanizates, up to the criterion of crack initiation, most of the energy applied during deformation should be stored as elastic strain energy. Therefore, the rate of change in the elastic strain energy released by crack propagation tends to be greater than the rate of change in the energy consumed during crack propagation. Although the deformation before fracture is large, it is presumed that it leads unfilled rubber to brittle fracture under uniaxial tension.

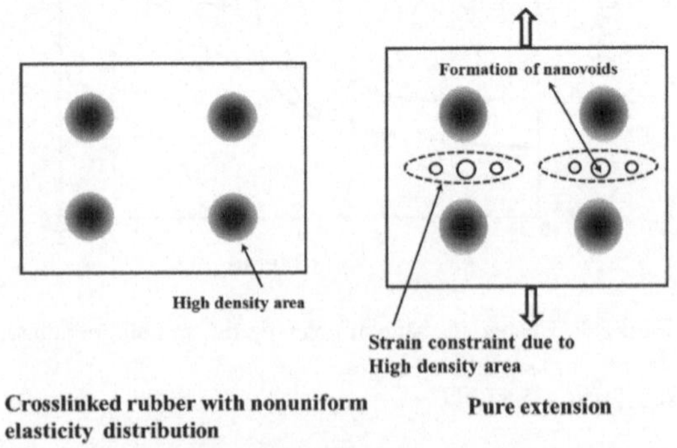

Figure 6. A model for the formation of nanovoids due to strain restraint due to the dispersion of regions with different crosslink densities.

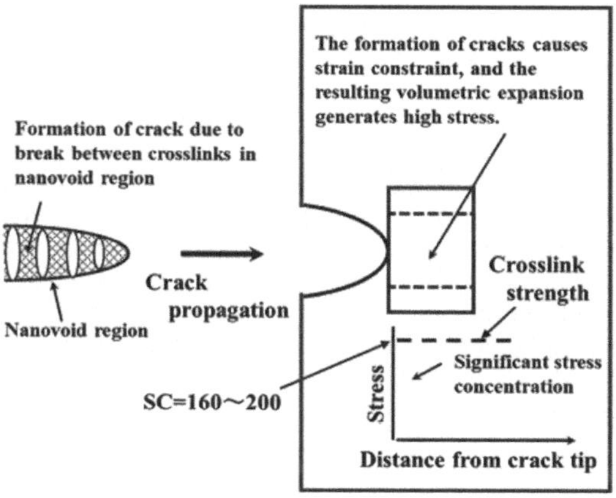

Figure 7. Strain restraint and generation of expansion stress due to cracks formed by cutting crosslinks in unfilled rubber.

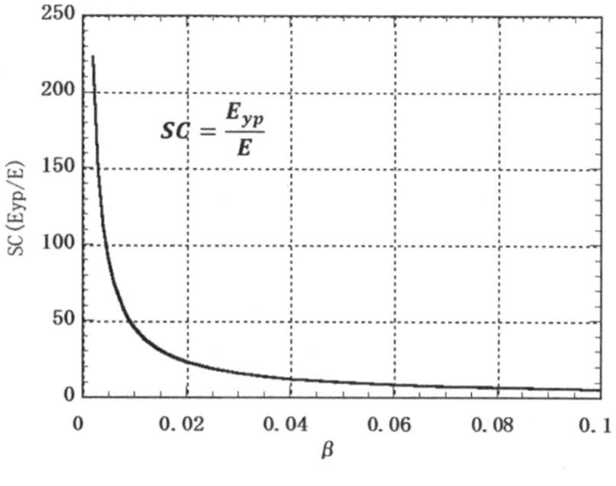

Figure 8. Variation of rough estimate of stress concentration (SC) calculated from Equation 4 using Equation 5 due to content of void.

3.3 *Improvement of rubber durability*

Equation 4 show that stress concentration due to strain constraint can be alleviated by adjusting the bulk modulus of the solid to a small value. It would not be easy to adjust the bulk modulus at a low level by adjusting the crosslinking or molecular structure of rubber[18] while Poisson's ratio of rubber decreases by adding filler particles, such as carbon black.

The molecular chains between crosslinks are moving due to micro-Brownian motion. If the interaction between adjacent molecular chains is small, nanovoids can be easily formed when expansion stress is applied throughout a deformation mode like the pure extension. When nanovoids are formed, contribution of shear deformation becomes possible around the inner walls of the nanovoids. Since the shear modulus for unfilled rubber vulcanizates is, in general, an order of 1 MPa and the bulk modulus of 500-600 MPa, it can be inferred that when nanovoids expand under the control of shear deformation, the resistance is greatly reduced as compared to that under the control of volume deformation of continuum. The change of the bulk modulus of an elastic material containing voids (K (β)) due to the content of voids (β) can be expressed relatively easily by the following equation using mechanics of materials [19-21].

$$K(\beta) = \frac{4G}{3}\left(\frac{1}{\beta} - 1\right) \tag{5}$$

Figure 8 shows the change in guideline of stress concentration (SC) with void content calculated from Equation 4 using the bulk modulus calculated from Equation 5.

It can be seen that even a small number of voids reduces the bulk modulus and the stress concentration is rapidly relaxed. This behavior was confirmed by the FEA results in Figure 3 where the bulk modulus decreases as a decrease of Poisson's ratio and the increase of tensile modulus (Ey) was suppressed due to strain constraint at the tip of the notch. As a result, the stress distribution at the notch tip is averaged, and the stress concentration is alleviated.

There is an experimental approach to generate nanovoids in rubber vulcanizate by applying a load in uniaxial stretching. It is to incorporate fine CB particles into a rubber matrix. CB particles dispersed in rubber form clusters, which are uniformly dispersed. The uniaxial stretching causes the filled rubber of the CB to extend, and the width of the sample simultaneously decreases due to Poisson contraction. At this time, the distance between CB grains naturally decreases with the elongation. Above a certain limit of elongation, the CB particles aligned perpendicular to the elongation direction approach and collide. In subsequent further stretching, Poisson contraction in the vicinity of the CB grain becomes difficult due to its high stiffness of CB cluster. As a result, expansion stress is generated between the grains of CB due to the strain constraint, which leads to the formation of nanovoids on rubber area. This has been confirmed by direct evaluation of the sample dimensions in uniaxial stretching[14], or by evaluation of small-angle X-ray scattering[23]. At higher loads, the crosslinks in the rubber will break and voids will form, followed by the formation of cracks. At this time, the difference from the rupture of unfilled rubber, as shown in Figure 9, is that when cracks are formed in filled rubber, a large number of nanovoids are already generated around the cracks. Therefore, the bulk modulus of unfilled rubber containing voids should be greatly relaxed as compared to unfilled rubber. The degree of stress concentration due to strain restraint at defects in CB-filled rubber

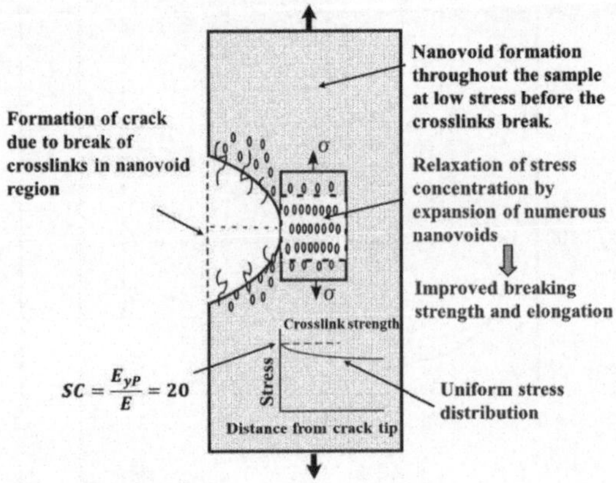

Figure 9. Improvement of rubber strength using relaxation of stress concentration by uniform nanovoids.

can be estimated from the results of the previously reported extension tests using Pancake specimens.

The SC of rubber filled with CB is estimated to be around 20. Since that of unfilled rubber is about 160 to 200, it can be seen that the degree of stress concentration in CB-filled rubber is alleviated by more than one order of magnitude compared to unfilled rubber, and as a result, the breaking strength is greatly improved. In this example, a blend of hard particulates leads to the formation of numerous nanovoids at low load, greatly relaxing the strain constraint and improving stiffness and strength. The proposed guideline of stress concentration adequately describes this behavior.

A mechanism that changes the manifestation of bulk modulus from a large resistance to density reduction to expansion due to shear deformation around voids and relaxes the bulk modulus has already been reported for improving the toughness of thermoplastic polymers[24]. It is well known that highly rigid thermoplastic polymers begin to break at the crazes, which are composed of voids and oriented molecular chains. A.S. Argon has pointed out that the unstably expansion of voids leads to the formation of craze and subsequent brittleness destruction[25]. Such brittle fracture can be greatly improved by forming an elastomer blended with a large number of uniform voids under a small load.

The improvement of strength of rubber by greatly relaxing the strain constraint by forming voids is basically the same as the mechanism of improving toughness by giving priority to forming voids by adding an elastomer with low strength to the brittle polymer material.

4 CONCLUSION

The distribution of stress and strain at the notch tip of rubber near plane strain was analyzed by the finite element method, and the effect of bulk modulus on stress concentration due to strain constraint was clarified. Based on the results, it was proposed the ratio of the elastic modulus of pure extension to the elastic modulus of uniaxial extension of the solid as a guideline of the magnitude of stress concentration due to strain constraint derived from the structure of the solid. It was evaluated the guideline of stress concentration for pure rubber and CB-filled rubber and showed that the extremely large strength improvement of CB-filled rubber can be clearly explained by the relaxation of this guideline of stress concentration.

REFERENCES

1) A.A. Griffith, Phil.Trans. Roy.Soc., Vol.221(1920), p163.
2) E. Orowan, Reports on Progress in Physics, Vol 12 (1949),p152.
3) G.R. Iwin, Fracturing of Metals, ASM (1948), p.152.
4) J.R. Rice, J. Appl. Mech., Vol.35(1968), p379.
5) Y. Fukabori, J. Soc. Rub. Sci. and Tec, Japan,Vol50 (1977), p409.
6) M. Rink, L. Andena,and C. Marano, Engineering Fracture Mechanics, Vol 127 (2014) p.46.
7) Timoshenko S.P. and Goodier J.N. Theory of Elasticity, 3rd ed., (1970).
8) A.N. Gent J. Mater. Sci. 26. 3392(1991).
9) P.A. Kakavas, Polymer Engineering And Science, FEBRUARY 37, 2 (1997).
10) A. Dorfmann, K.N.G. Fuller, R.W. Ogden, Inter national Journal of Solids and Structures 39 1845–186 (2002).
11) J.M. Ball, philosophical Transaction of Royal Society. London A306 557(1982).
12) S. Biwa, International Journal of Non-Linear. Mechanics 41 1084 (2006).
13) M. Drass, J. Schneider, S. Kolling, Materials and Design 138, 71(2018).
14) N. Ait Hocine, A. Hamdi, M. Nait Abdelaziz, P. Heuillet, F. Zairi, International Journal of Solids and Structures 48 1248 (2011).
15) J. Jaravel, S. Castagnet, JC. Grandidier, G. Benoît Polymer Testing 30 (2011) 811–818.
16) E. Eucher, K. Schneider, G. Heinrich, T. Tada and M. Ishikawa. KGK 09 32(2018).
17) H. Zhang, Arthur K.Scholz,Jordan de Cre voisier, Daniel Berghezan, Theyencheri Nara yanan, Edward J. Kramer, Costantino Creton1, Journal of Polymer Science, PART B: Polymer Physics 53, 422–429 (2015).
18) L. Courtney Sherman, C. Robert Zeigler, E. Nikhil Verghese, J. Maurice Marks, Polymer 49, 1164(2008).
19) E. Euchler, R. Bernhardt, K. Schneider, G. Heinrich, S. Wießner and T. Tada. Polymer 187, 122086(2020).
20) M.Drass, J. Muth, C. Louter, J. Schneider, Glass Struct. Eng. 4,433–448 (2019).
21) M. Drass, P. A. Du Bois, J. Schneider, S. Kolling, Glass Struct. Eng. 5, 41–65 (2020).
22) M. Drass, N. Bartels, J. Schneider, D. Klein Glass Struct. Eng. 5,67–82 (2020).
23) H. Zhang, A. K. Scholz, F. Vion-Loisel, Y. Merckel, M. Brieu, H. Brown, S. Roux,E.J. Kramer, C. Creton, Macromolecules, 46,900(2013).
24) M. Ishikawa and I. Chiba, Polymer 31,1232 (1990).
25) A.S.Arogn, Phil.Mag., 36, 1195(1977).

Constitutive Models for Rubber XIII – Dal (Ed.)
© 2025 The Author(s), ISBN 978-1-032-85138-9

A comparative case study for fatigue lifetime predictors on EPDM rubber

Bülent Efe Tanış, Selçuk Başdemir & Hüsnü Dal
Mechanical Engineering Department, Middle East Technical University, Ankara, Türkiye

Halil Bilal
Bayrak Lastik Sanayi ve Ticaret A.Ş., Bursa, Türkiye

ABSTRACT: Prediction of lifetime for rubber-like materials subjected to uniaxial and multiaxial loads is an essential requirement in design of rubber components. The current state-of-art predictions rely on a range of stretch, stress, and energy metrics; each with its own formal definition. Strain-based methodologies include maximum principal stretches, Green-Lagrange strain, and the left Cauchy-Green tensor, while stress-based techniques encompass minimum stress and maximum principal stress, Cauchy stress, and Eshelby stress using configurational mechanics. To exemplify energy-based approaches, critical plane approach of Mars, utilizing cracking energy density, has also gained prominence as an influential method in this context. Despite the extensive literature available on these methods, a comprehensive comparative study on EPDM rubber has been conspicuously lacking. This study addresses this gap by conducting a review and comparison of several fatigue predictors through a numerical case study. The comparison is based on obtained lifetime of the benchmark geometry, a two-pipe bushing under multiaxial loading conditions, directly. The outcomes of this case study are expected to provide insights for lifetime prediction of rubber components.

1 INTRODUCTION

Rubber-like materials consist of relatively long polymeric chains having a high degree of flexibility and mobility. Due to their molecular structure, rubber-like materials have many unique properties unlike metals and are commonly used in many engineering applications that need high elastic deformations, high damping properties, low thermal conductivity and low electrical conductivity. They can be used in applications where sealing, shock resistance or vibration dampening are fundamental needs of the system or structure. Since rubber-like materials have many unique properties, unique formulations of rubbers are also formulated and manufactured according to specific needs. Some of those formulations can be listed as natural rubber, nitrile-butadiene subber, styrene-butaediene rubber and ethylene-propylene-diene and EPDM.

The mechanical behavior of rubber-like materials is highly complex due to their large deformability, near-incompressibility, strain crystallization, strain-softening, and sensitivity to both high temperatures and the filler content. These characteristics make it quite challenging to predict the mechanical and fatigue properties of such materials. The basic mechanical response of rubber is often considered hyperelastic, see Dal et al. (2020, 2021) and references therein. Nonetheless, the fatigue life of rubber components is affected by numerous factors, including loading history, environmental conditions, rubber formulation, and dissipative elements within the constitutive response of the rubber, such as viscoelasticity and strain-induced crystallization (Mars and Fatemi 2004, Dal and Kaliske 2009, Behnke et al. 2013, Tee et al. 2018). It has been demonstrated that viscoelasticity significantly influences crack growth under cyclic loading (Näser et al. 2009). Additionally, strain-induced crystallization is known to enhance fatigue life under non-relaxing conditions ($R>0$) (Li et al. 2015). Fatigue behavior under repeated loading in rubber can generally be divided into two stages: the time required for an initial crack to form from an existing micro-defect or flaw and the stable crack propagation leading to complete failure (Mars and Fatemi 2002, Verron and Andriyana 2008a). Crack nucleation techniques often employ maximum principal stretches, principal stresses, strain energy density, or cracking energy density (Mars 2002a) as fatigue predictors, whereas crack growth methods are primarily based on the tearing energy concept t (Rivlin and Thomas 1953). On a microscopic scale, the research by Le Cam et al. (2004) indicated that fatigue cracks originate from cavitation, which results from the decohesion between zinc oxides and the rubber matrix in carbon-black-filled natural rubber. Le Cam et al. (2013) further established that fatigue cracks initiate

DOI: 10.1201/9781003516880-41

from microstructural defects with diameters not exceeding 400 μm, and that macroscopic crack initiation corresponds to microscopic crack growth, validating recent energy-based approaches used to predict fatigue crack initiation in rubbers. A multiscale approach for the mechanical response of elastomers including cavities was treated in Dal et al. (2018).

EPDM rubber is widely used due to its excellent resistance to tearing, abrasion, and ozone degradation. Along with its numerous advantages, it is also one of the cost-effective polymers in the industry. Since rubber components are subjected to cyclic dynamic loads during service, investigating fatigue in rubber-like materials becomes essential. Over the past few decades, several predictors have been introduced to estimate the fatigue lifetime of rubber-like materials. These predictors can be classified into two main approaches: crack nucleation and crack propagation based approaches. Crack nucleation based predictors mainly use alternative stress and/or strain measurements to predict the fatigue lifetime. On the other hand, crack propagation based predictors use energy methods and/or configurational mechanics as estimators. In this paper, well accepted fatigue lifetime predictors are compared by multiaxially loaded benchmark geometry.

2 EXISTING FATIGUE LIFETIME PREDICTORS

Since Wöhler's fatigue lifetime prediction method for linear elastic materials, in order to characterize fatigue lifetime of rubber, various approaches are utilized which pivot around crack nucleation and crack propagation. Crack nucleation approach is focused on predicting the lifetime of rubbers by addressing the material's initial stage without any macroscopic crack formed, whereas in crack propagation approach, behaviour concentrated on is the advance of pre-existing macroscopic cracks. Crack nucleation approach for rubber dates back to works of Cadwell (Cadwell, Merrill, Sloman, & Yost 1940). Crack propagation based predictors try to estimate fatigue lifetime of a rubber by introducing a pre-existing crack instead of internal defect. There have also been hybrid models such as critical plane analysis using cracking energy density (Mars 2002b) that both uses crack nucleation and crack propagation experiments to determine the fatigue behaviour of rubber. (Bilal et al. 2022)

2.1 *Crack nucleation based predictors*

Contrary to crack propagation based methods, crack propagation based predictors assume no pre-existing cracks. In this approach, crack is nucleated to a predetermined flaw size in order to imply fatigue failure. The occurrence of crack originates from unification of internal defects.

2.1.1 *Strain-based predictors*

Due to operational loads on structures advantaging rubber materials being mainly displacement based, early works of rubber fatigue-lifetime prediction branches out from strain based predictions. The very first approach for this predictor is from the works of Cadwell et al. (1940). As a result of the mentioned research, it is established that although the maximum stretch level of a loading type is important for the fatigue lifetime of rubber, minimum stretch of the loading also comprised a crucial part of components' lifetime. Works dating back to Ro (1989) also investigated the effect of minimum strain, mean strain and maximum strain in their work for SBR. It was shown that the fatigue life depends on the applied strain amplitude rather than mean strain. However, strain as a fatigue lifetime predictor failed to unify uniaxial and multiaxial data in the work.

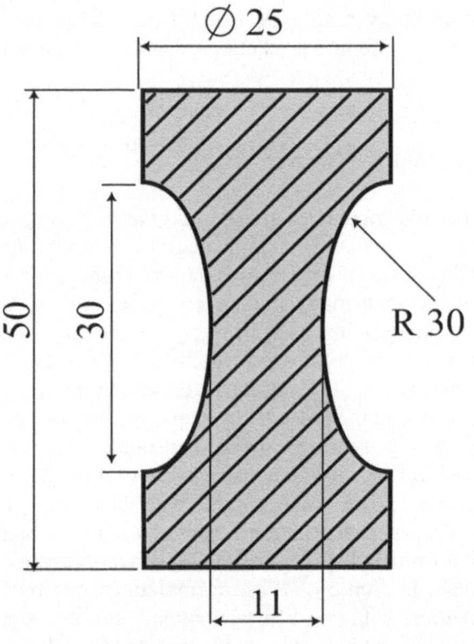

Figure 1. Dumbbell specimen.

2.1.2 *Stress-based predictors*

Through the literature, stress based lifetime prediction for rubber varies in terms of forms of the stress and their configuration. Approaches have been developed around several key stress descriptors, including maximum principal engineering stress, maximum Cauchy stress, mean stress of loading cycles and von Mises' stress by the works of many Andre et al. (1999), Saintier et al. (2006), (Abraham, Alshuth, & Jerrams 2005). Moreover, for EPDM rubber, maximum engineering stress is analyzed in literature i.e. Abraham et al. (2001). In the aforementioned study, increase in maximum stress resulted in increase in fatigue lifetime in the presence of an increasing preload. Therefore the work states that the maximum engineering stress criterion on its own does not seem to capture the service life behaviour accurately. In their work, Verron and Andriyana proposed a new predictor using Eshelby stress tensor Verron and Andriyana (2008b). Contrary to Abraham, Verron and Andriyana found out that stress-based predictors can be used to predict the fatigue lifetime of rubber coupled with the stress history. However, it is stated that further investigation is required due to the non-unique configuration of deformation gradient to elastic and inelastic parts.

2.2 *Crack propagation based predictors*

Crack propagation approach assumes a pre-existing crack on material and characterizes the fatigue behaviour of rubber from advance of the crack. The approach is first introduced in 1965 (Lake & Lindley 1965). The work analyzed the correlation of rubber fatigue lifetime with tearing energy on natural rubber.

2.2.1 *Energy-based predictors*

Although theorizing a predictor was not the scope of the aforementioned work, they showed that tearing energy and fatigue-lifetime are correlated with concurrent increase and decrease in the presence of external effects such as inclusions of antioxidants. Another energy based approach for lifetime prediction is cracking energy density proposed by Mars(Mars 2002b). This new method defined an energy parameter, cracking energy density, which represents stored elastic energy that is available to crack on a material plane. However, due to the nature of methodology and implementation of crack nucleation test to find required prediction parameters, the predictor cannot be strictly stated as a propagation based predictor.

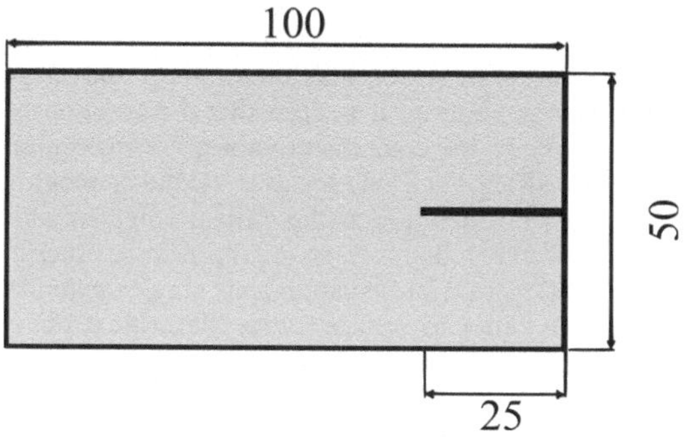

Figure 2. Planar tension specimen.

3 EXPERIMENT METHODOLOGY

Experiments of the work revolved around two main pivotal axes: crack propagation tests and fatigue-lifetime estimation tests, aiming to obtain relevant information to relate specific loading conditions with fatigue lifetime of rubber. For the work, EPDM formulated rubber with 66-shore hardness is used due to its robustness and versatility in various applications varying from defence systems to automotive industry. The rubber is provided by Bayrak Lastik Sanayi ve Ticaret A.Ş. as the partner of the project.

3.1 *Fatigue experiments on EPDM rubber with 66 shore hardness*

When conducting the fatigue experiments, dumbbell specimens with of 11 mm gauge diameter and 30 mm gauge length are prepared. Specimens are produced using vertical injection moulding. We used MTS 831 fatigue testing device for fatigue lifetime experiments. Rubbers were subjected to strain-controlled loading in a fully-relaxed sinusoidal load type. Experiments were conducted at 5 Hz frequency.

3.2 *Crack propagation experiment setup*

To proceed with the crack propagation experiments, the research employed planar tension geometry as depicted in Figure 2 with 10 mm gauge length and 25 mm notch length. Preparation of specimens involved the production of 2 mm rubber sheets with compression press. Subsequent to the preparations of rubber sheets, sheets were laser-cut into 50 mm x 100 mm pieces with precision laser-cutting CNC machine around its outer boundaries. Laser-cut rubber was notched vertically with a sharp-knife by 25 mm. Afterwards, the specimens are mounted at 20 mm each side, resulting in a gauge length of 10 mm, 1/10 of its horizontal span, in order to facilitate a planar tension test using the specified geometry.

For the experiments, we used Zwick-Roell universal testing device, coupled with monitoring the crack propagation by SONY ZV-E10 4K video camera. An LED light is placed behind the specimen and ISO value of the 4K video camera was adjusted accordingly so that images we obtained outlined the edge crack. Subsequent to obtaining the images, resultant frames are processed and converted to black and white. Processed images, yielding calculated pixel measurements, were then converted into milimeters for physical quantitative analysis. Data analysis was done by using our in-house software, written in MATLAB®. The approach for calculating crack propagation "da" adopted measuring crack propagation in each cycle as Euclidian instead of horizontal projection. In order to obtain crack propagation distance, frames at 20% strain at each cycle are recorded so that the crack tip is barely visible and no loss occurs due to crack closure. This also helped obtaining the crack tip in Lagrangian framework instead of Eulerian.

To ascertain appropriate loading conditions, we conducted preliminary uniaxial tests on planar tension geometry for the purpose of identifying the failure point of rubber under given geometry.

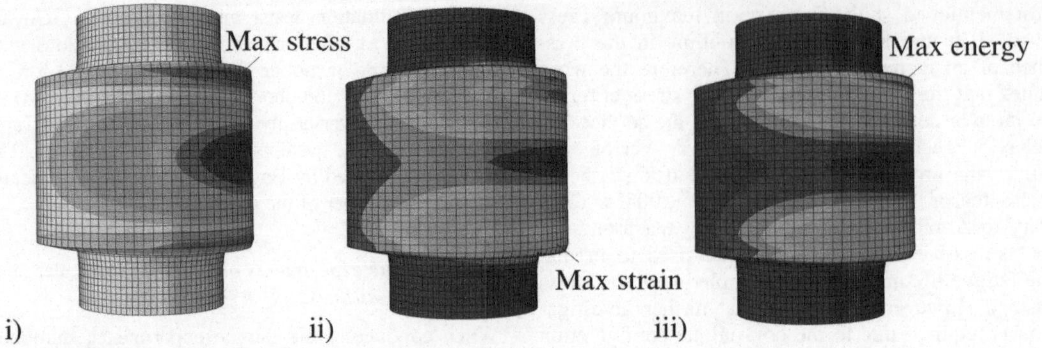

Figure 3. Critical region for i) maximum stress ii) maximum strain iii) maximum strain energy.

Afterwards, we proceeded with the propagation test in fully-relaxed tensile mode with strain-controlled ramp-loading. Loading pattern was configured in a sawtooth waveform, dictated by the capabilities of testing machine. In order to minimize viscoelastic effects, 20 mm/min strain rate is specified without unduly prolonging the time span of experimental procedure. The experimental protocol incrementally raised the strain level by 1.2% following the completion of each cycle.

4 FINITE ELEMENT ANALYSIS OF EXPERIMENTS

Cyclic fatigue lifetime experiments of EPDM rubber is modeled on finite-element-analysis software, ABAQUS. The main motivation in modelling the experiments was to obtain relevant stress and strain values to correlate with the experimental results. Before the modelling process, EPDM rubber is characterized by uniaxial, equibiaxial and cyclic tests in order to obtain Ogden parameters of the material. Material parameters are presented as:

Table 1. Ogden parameters for the EPDM rubber.

Parameter	Unit	1	2	3
α	[-]	−2.548	0.148	3.858
μ	[MPa]	0.099	0.640	0.364
r	[-]	1.268		
m	[MPa]	1.189		
β	[-]	0.004		

5 RESULTS AND DISCUSSION

For fatigue lifetime experiments of EPDM rubber, conducted tests yielded maximum of 1264700 cycles for 0.72 nominal strain and minimum of 1550 cycles for 1.50 nominal strain. The data is then used to find relevant fatigue fitting parameters. For predicting the fatigue lifetime, an exponential fit is carried out:

$$N = \alpha e^{\beta \psi} \quad (1)$$

where α and β are fit parameters and ψ is the fatigue-predictor parameter. The data is then log fitted and prediction equations are found.

For $\psi = \varepsilon_{nominal}$,

$$N = 3,188,492,678e^{-10.744\psi} \quad (2)$$

For $\psi = \sigma_{cauchy,max}$,

$$N = 94,792,964e^{-1.590\psi} \quad (3)$$

For $\psi =$ Strain energy (U),

$$N = 26,135,699.8e^{-4.880\psi} \quad (4)$$

For maximum stress, maximum strain and maximum energy criteria, it is seen that the location of critical point in the component varies. For maximum stress and strain, the same location on the element is taken into consideration as the critical point. In addition to given results for three main predictor criteria, results of the crack propagation tests are presented in 5 and 3 in order to proceed with the critical plane theory. Repetitive validation tests using SBR, NR and NR+SR specimens are performed for the obtained propagation results. Although those rubber formulations concluded in crack propagation in horizontal direction, EPDM rubber with 66 shore hardness and 90 phr filler showed crack propagation with a sharp angular direction. When collecting propagation data, Eulerian distance of the crack propagation is taken into account directly. Consecutive to

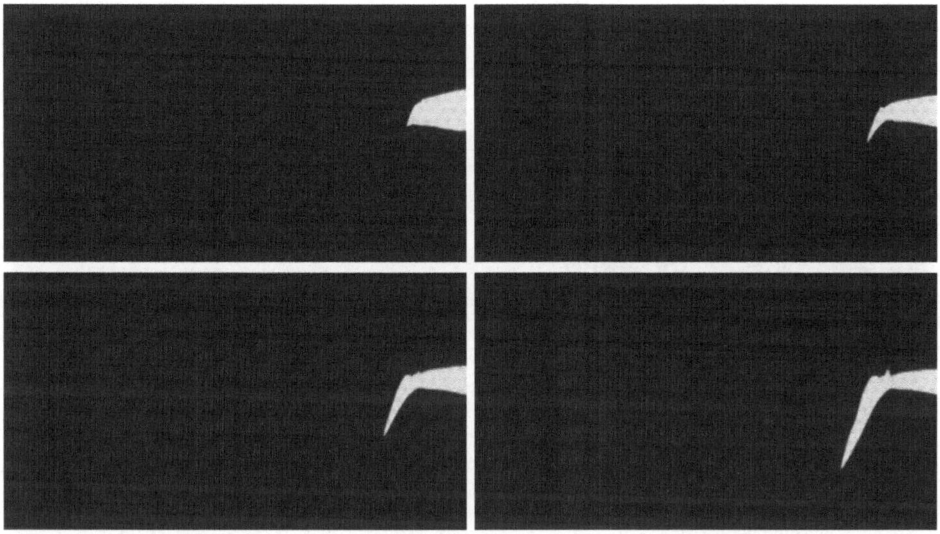

Figure 4. Crack propagation of 90 phr 66 shore EPDM rubber.

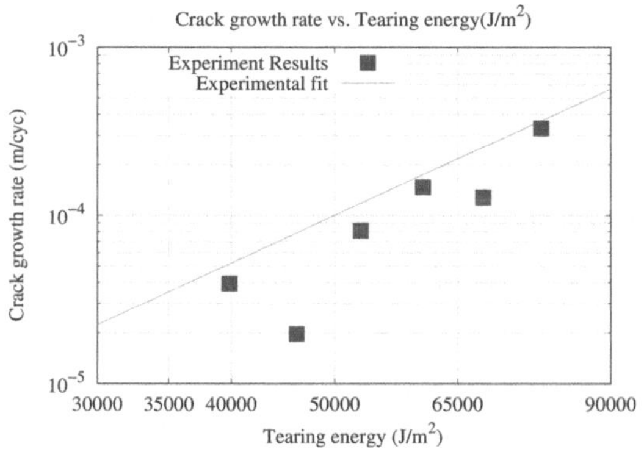

Figure 5. Crack growth rate vs. Tearing energy.

obtaining the crack propagation results, to find cracking energy density and fatigue lifetime, crack growth rate r_c, tearing energy T_c and power-law parameter F must be found. In addition, to obtain crack precursor c_0, fatigue life equation shown in equation 5 must be fitted as well. Hence, a multi-objective fitting between two equations, fatigue lifetime and crack propagation rate equations, is carried out. Fit shown in Figure 5 yielded fit performance of $R^2 = 0.6944$ with $r_c = 0.399$, $T_c = 80000\,\frac{N}{m}$ and $F = 2.9394$ parameter set.

$$N = \int_{c_0}^{c_f} \frac{dc}{r_c\left(\frac{2kWc}{T_c}\right)^F} \tag{5}$$

Crack precursor size is found by equation 5 as $0.1\,mm$ and constant k is found as $k = 3.2839$. With the simultaneous fit, high cycle fatigue is found as $N_H = 798259$ cycles whereas low-cycle fatigue is found as $N_L = 25360$ cycles. Sequential to setting up the prediction parameters for EPDM rubber, a two-pipe bushing was modeled for applying the predictors. Loading mode for the bushing was set in radial direction, and hence yielded multiaxial loads on the part. First portion of the loading is taken as a swaging process. Then, the component is loaded radially, creating both tensile and compressive portions on the part multiaxially. Radial loading condition of bushing is defined as fully relaxed between $0\,N$ and $4000\,N$. Geometry used has the following geometrical properties: $d_i = 23$ mm, $d_o = 37$ mm, $t = 1.24$ mm and $L = 47.5$ mm. Analysis of the geometry yielded 2.517 [MPa], 1.348, 1.244 [MPa] for maximum Cauchy stress $\sigma_{cauchy,max}$, maximum nominal strain E_{max} and maximum strain energy density, U, as respectively. Results are utilized in fitting equations to predict lifetime of two piped bushing. Afterwards, fatigue lifetime calculations using critical plane analysis with cracking energy density as predictor were carried out. Analysis is done with 0.1 step size increments on ABAQUS. After exporting deformation gradients from software, r_c, T_c, F, c_0 and deformation gradient F are used to obtain fatigue lifetime for EPDM rubber using critical plane analysis. An in-house software was developed for the analysis. Obtained prediction equation parameters and respective lifetimes are presented in Table 2.

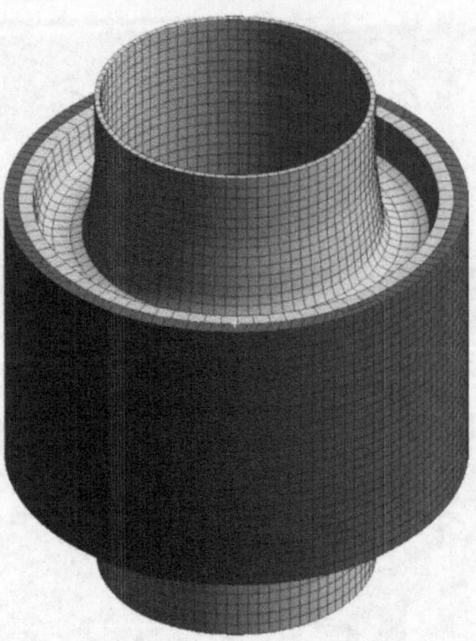

Figure 6. Two-pipe bushing geometry.

Table 2. Fatigue lifetime results for the EPDM rubber.

Parameter	Max	N
σ_{cauchy}	2.517 [MPa]	1.73×10^6
E	1.348	1.636×10^3
U	1.244 [MPa]	6.036×10^4
Critical plane		165

6 CONCLUSION

In this work, fatigue lifetime of a two-piped rubber bushing under multiaxial loading is analyzed with several approaches to compare prominent predictors in the literature. It can be observed that existing crack precursors do not have unified results with each other. All the approaches examined have yielded lifetime results at different levels, with different order of magnitudes. It shows that the prediction of fatigue lifetime for rubber-like materials is still a field full of uncertainties.

The stress-based predictor provided the highest life prediction, while the strain-based and energy-based predictors predicted a shorter life. Besides, critical plane analysis gives the shortest lifetime. This indicates that in addition to loading conditions, the precursor size and its orientation are also important factors for lifetime estimation in rubber-like materials. Furthermore, it is noted that with multiaxial operating condition based loading, aforementioned fatigue lifetime predictors do not yield unified results. These facts demonstrate how worth it is still to study on the fatigue lifetime prediction of rubber-like materials.

ACKNOWLEDGEMENT

The financial support provided by the support of METU Scientific Research Project Coordination Unit under grant GAP-302-2021-10747 is gratefully acknowledged.

REFERENCES

Abraham, F., T. Alshuth, & S. Jerrams (2001, September). Dependence on mean stress and stress amplitude of fatigue life of EPDM elastomers. *Plastics, Rubber and Composites 30*(9), 421–425.

Abraham, F., T. Alshuth, & S. Jerrams (2005, May). The effect of minimum stress and stress amplitude on the fatigue life of non strain crystallising elastomers. *Materials & Design 26*(3), 239–245.

Andre, N., G. Cailletaud, & R. Piques (1999). Haigh diagram for fatigue crack initiation prediction of natural rubber components. *Kautschuk Gummi Kunststoffe*.

Behnke, R., H. Dal, G. Geißler, B. Näser, C. Netzker, & M. Kaliske (2013). *Macroscopical Modeling and Numerical Simulation for the Characterization of Crack and Durability Properties of Particle-Reinforced Elastomers*, pp. 167–226. Berlin, Heidelberg: Springer Berlin Heidelberg.

Bilal, H., F. Bilbay, M. Şahin, I. Şahin, Y. Vardar & H. Dal (2022). Effect of temperature and filler ratio on fatigue life of rubber. *Constitutive Models for Rubber XII*, 365–370.

Cadwell, S. M., R. A. Merrill, C. M. Sloman, & F. L. Yost (1940, January). Dynamic Fatigue Life of Rubber. *Industrial & Engineering Chemistry Analytical Edition 12*(1), 19–23.

Publisher: American Chemical Society.

Dal, H., K. Açıkgöz, & Y. Badienia (2021). On the performance of isotropic hyperelastic constitutive models for rubber-like materials: A state of the art review. *Applied Mechanics Reviews 73 73*, 020802.

Dal, H., B. Cansız, & C. Miehe (2018). A three-scale compressible microsphere model for hyperelastic materials. *International Journal for Numerical Methods in Engineering 116(6)*, 412–433.

Dal, H., O. Gültekin, & K. Açıkgöz (2020). An extended eight-chain model for hyperelastic and finite viscoelastic response of rubberlike materials: Theory, experiments and numerical aspects. *Journal of the Mechanics and Physics of Solids 145*, 104159.

Dal, H. & M. Kaliske (2009). Bergström-boyce model for non- linear finite rubber viscoelasticity: Theoretical aspects and algorithmic treatment for fe method. *Computational Mechanics 44*, 809–823.

Lake, G. J. & P. B. Lindley (1965). The mechanical fatigue limit for rubber. *Journal of Applied Polymer Science 9*(4), 1233–1251.

Le Cam, J.-B., B. Huneau, & E. Verron (2013). Fatigue damage in carbon black filled natural rubber under uni- and multiaxial loading conditions. *International Journal of Fatigue 52*, 82–94.

Le Cam, J.-B., B. Huneau, E. Verron, & L. Gornet (2004). Mechanism of fatigue crack growth in carbon black filled natural rubber. *Macromolecules 37*(13), 5011–5017.

Li, F., J. Liu, W. Mars, T. W. Chan, Y. Lu, H. Yang, & L. Zhang (2015). Crack precursor size for natural rubber inferred from relaxing and non-relaxing fatigue experiments. *International Journal of Fatigue 80*, 50–57.

Mars, W. & A. Fatemi (2002). A literature survey on fatigue analysis approaches for rubber. *International Journal of Fatigue 24*(9), 949–961.

Mars, W. & A. Fatemi (2004). Factors that affect the fatigue life of rubber: a literature survey. *Rubber Chemistry and Technology 77*(3), 391–412.

Mars, W. V. (2002a). Cracking Energy Density as a Predictor of Fatigue Life under Multiaxial Conditions. *Rubber Chemistry and Technology 75*(1), 1–17.

Mars, W. V. (2002b, March). Cracking Energy Density as a Predictor of Fatigue Life under Multiaxial Conditions. *Rubber Chemistry and Technology 75*(1), 1–17.

Näser, B., M. Kaliske, H. Dal, & C. Netzker (2009). Fracture mechanical behaviour of visco-elastic materials: application to the so-called dwell-effect. *ZAMM - Journal of Applied Mathematics and Mechanics/Zeitschrift fAr Angewandte Mathematik und Mechanik 89*(8), 677–686.

Rivlin, R. S. & A. G. Thomas (1953). Rupture of rubber. i. characteristic energy for tearing. *Journal of Polymer Science 10*(3), 291–318.

Ro, H. S. (1989, January). Modeling and interpretation of fatigue failure initiation in rubber related to pneumatic tires. *Theses and Dissertations Available from ProQuest*, 1–166.

Saintier, N., G. Cailletaud, & R. Piques (2006, May). Multiaxial fatigue life prediction for a natural rubber. *International Journal of Fatigue 28*(5), 530–539.

Tee, Y. L., M. S. Loo, & A. Andriyana (2018). Recent advances on fatigue of rubber after the literature survey by mars and fatemi in 2002 and 2004. *International Journal of Fatigue 110*, 115–12.

Verron, E. & A. Andriyana (2008a). Definition of a new predictor for multiaxial fatigue crack nucleation in rubber. *Journal of the Mechanics and Physics of Solids 56*(2), 417–443.

Verron, E. & A. Andriyana (2008b, February). Definition of a new predictor for multiaxial fatigue crack nucleation in rubber. *Journal of the Mechanics and Physics of Solids 56(2)*, 417–443.

Constitutive Models for Rubber XIII – Dal (Ed.)
© 2025 The Author(s), ISBN 978-1-032-85138-9

Rupture mode transition of strain induced crystallizing rubber

K. Tsunoda
Sustainable and Advanced Materials Division, Bridgestone Corporation, Tokyo, Japan

K. Urayama
Department of Material Chemistry, Graduate School of Engineering, Kyoto University, Kyoto, Japan

ABSTRACT: Strain induced crystallizing rubber shows an abrupt drop of elastic stored energy density at break (W_b) at a critical initial crack length (c_o^*) in tensile edge crack test. It is elucidated that the abrupt change of W_b at c_o^* reflects the rupture mode transition between cyclic crack growth process at $c_o<c_o^*$ as a result of SIC strengthening effect at the advancing crack tip, and catastrophic fracture process at $c_o>c_o^*$ without a significant SIC effect. At $c_o<c_o^*$ where energy release rate (G) is enhanced by SIC effect, postponing catastrophic fracture takes place and G can be explained by $G = 1/2 \cdot (c_o/B)^{1/2}$. Proposed theory suggests that the maximum G in $c_o<c_o^*$ region can be regarded as the characteristic energy release rate enhanced by the SIC effect (G_{SIC}). Edge crack test method reveals significant differences in G_{SIC} and threshold temperature between natural rubber and synthetic polyisoprene rubber.

1 INTRODUCTION

Natural rubber (NR) is the most widely utilized biomaterial polymer in the rubber industry. The most significant benefit of NR is superior physical properties, especially high strength originated from strain induced crystallization (SIC). In addition, as there is increasing interest in environmental issues and momentum toward the realization of a sustainable society, the importance of utilizing natural resources is highlighted in recent years (Kohjiya *et al.* 2021). Deep understanding of NR performance and its relationship with micro/macro structure and/or impurities is one of the utmost concern research fields as it will lead to maximize NR performance and new insight for methodologies to apply another bio-source poly isoprene, like Guayule, and propose molecular design direction for recycled isoprene monomer. Rubber is visco-elastic material, and hence its strength is strongly associated with viscoelastic energy dissipation. Additionally, as for SIC rubber, like NR, its strength is governed by SIC in addition to viscoelastic energy dissipation. Here, it is worth to note that SIC originated strength has a threshold condition for each of temperature and strain rate as SIC is kinetic event (Thomas *et al.* 1970). The methodology of the quantification of the SIC polymer's two strength components and the threshold conditions of SIC effect remains to be established, despite their importance.

NR shows an abrupt drop of elastic stored energy density to break (W_b) when varying initial crack length (c_o) in tensile testing with single side pre-notched tensile specimen (which we call edge crack test) (Thomas *et al.* 1967, Bell *et al.* 1982, Hammed *et al.* 1983, Hammed *et al.* 1999, Hammed *et al.* 2002, Rong *et al.* 2016). The corresponding schematic is shown in Figure 1a. We revisit this unique rupture mode transition, and the establishment of the methodology to quantify each of SIC originated and viscoelastic energy dissipation originated strengths, and the threshold conditions for SIC effect (Tsunoda *et al.* 2023).

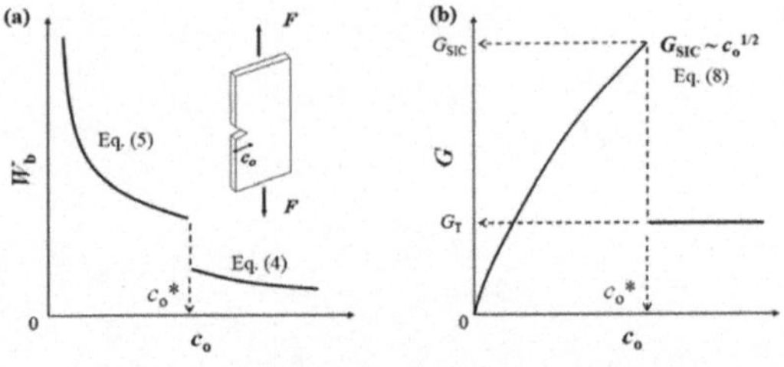

Figure 1. Schematic illustration of the relationship between elastic stored energy density at break (W_b) and initial crack length (c_o) in single side pre-notched tensile test (a) and determination of two strength component G_{SIC} and G_T (b).

2 THEORY

2.1 *Fracture mechanics of rubber*

Crack growth in rubber has been successfully described utilizing a fracture mechanics approach

DOI: 10.1201/9781003516880-42

utilizing energy release rate, G (Rivlin *et al.* 1953). The magnitude of G is defined as

$$G = -1/t \cdot (\partial U/\partial c)_l \qquad (1)$$

where c is the length of a crack in a body of thickness t and U is the total elastic stored energy. Hence, G is the required energy to drive a crack at given rate and temperature. The magnitude of G is not a material constant due to irreversible energy dissipation process, and hence the magnitude depends on crack growth rate, extension rate and ambient temperature (Andrews *et al.* 1974). In case of non-strain crystallizing rubber, like Styrene Butadiene Rubber (SBR), G increases with increasing extension rate and decreasing temperature. This reflects the dominant influence of the visco-elastic energy dissipation. On the other hand, in case of strain crystallizing rubber, like NR and synthetic poly-isoprene rubber (IR), the influences of extension rate and temperature becomes less pronounced. This is expected to result from strain induced crystallization (SIC), whose effect on crack growth outweighs the visco-elastic effect. For single side pre-notched tensile specimen, schematically shown in Figure 1(a), G is given by

$$G = 2kWc \qquad (2)$$

where c is the initial crack length, W is the elastic stored energy density in the bulk of the specimen and k is a slightly varying function of the extension ratio, λ, which is given approximately as (Greensmith *et al.* 1963)

$$k = \pi/\sqrt{\lambda} \qquad (3)$$

2.2 *Abrupt change in fracture toughness, W_b, at a critical crack length, c_o^**

Thomas and Bell reported unique W_b transition with varying c_o in NR vulcanizates utilizing single side pre-notched tensile specimen (Thomas *et al.* 1967, Bell *et al.*1982, Hammed *et al.* 1983, Hammed *et al.* 1999, Hammed *et al.* 2002, Rong *et al.* 2016). W_b decreases with increasing c_o and an abrupt fall of W_b occurs at critical crack length, c_o^*. This unique abrupt drop can be observed specifically in SIC rubber, like NR, but not observed in non-strain crystallizing rubber, like SBR. Thomas proposed a quantitative theory to explain this characteristic behavior of W_b in SIC rubber. If sufficient SIC does not take place at ahead of advancing crack tip, then failure occurs in a manner dictated by the catastrophic tear manner. The corresponding W_b can be expressed by

$$W_b = G_T/2kc \qquad (4)$$

where G_T is the critical energy release rate. If sufficient SIC takes place at ahead of the advancing crack tip, crack grows in cyclic crack growth manner until the failure occurs in catastrophic tear manner. In this condition, W_b can be represented by

$$W_b = 1/\left(4kc_o{}^{1/2}B^{1/2}\right) \qquad (5)$$

where B is the cyclic crack growth characteristic for each material and can be determined from cyclic crack growth measurements under large strain.

2.3 *Extension of existing theory*

As for single side pre-notched tensile specimen, G is given by equation (2). Here, for cyclic crack growth, G also can be written as

$$G = 2kW(c_o + \Delta c) \qquad (6)$$

From Equations (5) and (6), then W_b is now given as

$$G = 1/2 \cdot (c_o/B)^{1/2}(1 + \Delta c/c_o) \qquad (7)$$

Then, if $\Delta c \ll c_o$, we yield

$$G \approx 1/2 \cdot (c_o/B)^{1/2} \qquad (8)$$

The increment Δc, which is characterized by the interval between adjacent striation lines in the fracture surface, is significantly smaller than c_o. Equation (8) (i.e., $c_o = 4BG^2$) expects that G increases in proportion to $c_o{}^{1/2}$ when $c_o < c_o^*$, reflecting the feature of the cyclic crack growth manner (Tsunoda *et al.* 2023).

When c_o exceeds c_o^*, G is expected to drop abruptly to G_T in equation (4) because the fracture occurs via catastrophic crack growth, where no significant SIC near the crack tip takes place. The c_o dependence of G in this scenario is shown in Figure 1b. The maximum value of G at c_o^* can be regarded as the characteristic energy release rate enhanced by the SIC effect (G_{SIC}). In the case of non-SIC rubber, G in cyclic crack growth which is a subcritical process is significantly smaller than G_T. In the edge-crack tensile tests of SIC rubber, G in cyclic crack growth mode can be larger than G_T, because the SIC near the crack tip pronouncedly enhances G.

3 EXPERIMENT

3.1 *Materials*

Compound formulation used in this study is shown in Table-1. All compounds are mixed with 0.5L capacity internal mixer. Mixing condition were controlled chamber temperature of 120 ºC for 180 seconds at master batch mixing and 80 ºC for 60 seconds at final mixing. Vulcanized rubber sheet of 150 × 75 × 2 mm was prepared with hot press. Cure temperature was 160 ºC and cure time were 8 minutes for NR-0A and NR-0B, 14 minutes for IR-0 and 34 minutes for SBR-0. Curing agents were selected for adjusting average molar volume between cross-links to be around 7000 to 7500 g/mol.

Table 1. Compound formulation.

	NR-0A	NR-0B	IR-0	SBR-0
NR (RSS #3)	100.0	100.0		
IR2200			100.0	
SBR1500				100.0
Stearic Acid	2.0	2.0	2.0	2.0
TMDQ	0.3	0.3	0.3	0.3
6PPD	1.0	1.0	1.0	1.0
Sulfur	1.4	1.5	1.5	1.5
DPG	0.3			1.0
CBS	1.8	1.5	1.5	
MBTS				0.6
TBBS				0.6
Zinc Oxide	3.0	3.0	3.0	3.0

3.2 *Tensile measurement*

Tensile strip specimen (50 × 5 × 2 mm) was used. Initial crack is inserted into the center of single side of tensile strip specimen with a specially designed cutting apparatus. Initial crack length was varied from 0.25 to 3.5 mm in 0.25 mm interval. Tensile measurements are carried out utilizing commercially available tensile testing machine (SHIMAZU AGX-Plus 1kN) in controlled temperature chamber at a strain rate of 0.5/s. For the investigation on the effect of temperature, ambient temperature is varied from 25 to 90 ºC.

3.3 *Fracture surface observation*

Fracture surface observation after edge crack test is carried out utilizing a digital microscope (Hi-rox Co., RH-2000) equipped with a variable lighting adopter. Precise length of c_o is also measured with this microscope after edge crack test.

3.4 *Wide angle X-ray scattering measurement*

Wide angle X-ray scattering (WAXS) observations were carried out using synchrotron radiation at beam line BL03XU of SPring-8 (JASRI). The wavelength of X-ray was 0.1 nm, the camera length was 130 mm. A CMOS camera (Hamamatsu Photonics K.K. Orca Flash-4.0) with an image intensifier was used as a detector. Sequential WAXS images were recorded every one second after sample extension started. The X-ray scattering intensity of observed 2D WAXS image at λ, $I_x^{OBS}(q, \lambda)$, was obtained by integrating for $\beta = \pm 10°$, where β is azimuth angle, after correcting effects of the dark noise of detector, air scattering and sample thickness. As no reflection was observed at $\lambda \leq 2$, the X-ray scattering intensity from crystal ($I_x(q, \lambda)$) was obtained by

$$I_x(q,\lambda) \equiv I_x^{OBS}(q,\lambda) - I_x^{OBS}(q,\lambda \leq 2) \quad (9)$$

We focused on the peak intensity of 200 reflection $I_x(200)$.

4 RESULTS AND DISCUSSIONS

4.1 *Validation of theoretical model*

Figures 2a and 2b display the W_b and the G values calculated using Equation (2) as a function of c_o for SBR-0 at 25 ºC, respectively. W_b decreases with increasing c_o monotonically, and the resultant G values are constant. This result demonstrates that G_T at given temperature and strain rate for non-SIC rubber is independent of c_o.

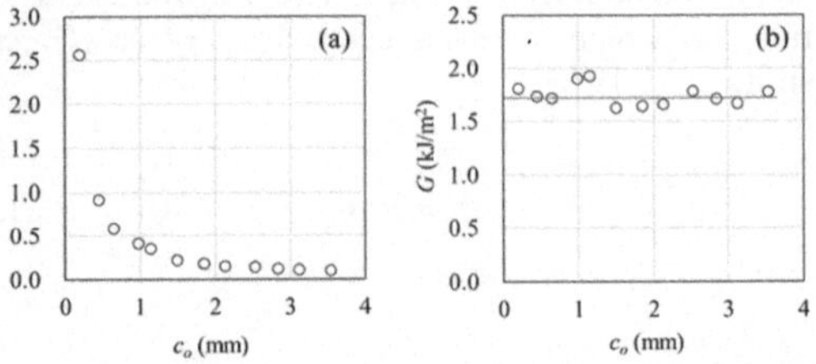

Figure 2. (a)Strain energy density at break (W_b) as a function of initial crack length (c_o) for SBR-0 at 25ºC and strain rate of 0.5/s, (b)c_o dependence of energy release rate (G) calculated from the data (a) and equation (2).

Figures 3a and 3b illustrate the corresponding plots for NR-0A. W_b shows an abrupt fall at a critical initial crack length c_o^*. Resultant G increases with increasing c_o at $c_o < c_o^*$, then fall on to a constant value of G_T. Solid line in the region of $c_o < c_o^*$ in Figure-2(b) is calculated by Equation (5) using B as a fitting parameter (fitted value of $B = 3.65 \times 10^{-13}$ N^{-2} m^3). The imposed dashed line is computed by Equation (5) utilizing B value of 4.06×10^{-13} N^{-2} m^3, which is separately evaluated in cyclic crack growth test (Figure 4). The good agreement of the experimental data and the calculated curves at $c_o < c_o^*$ validates the proposed theory

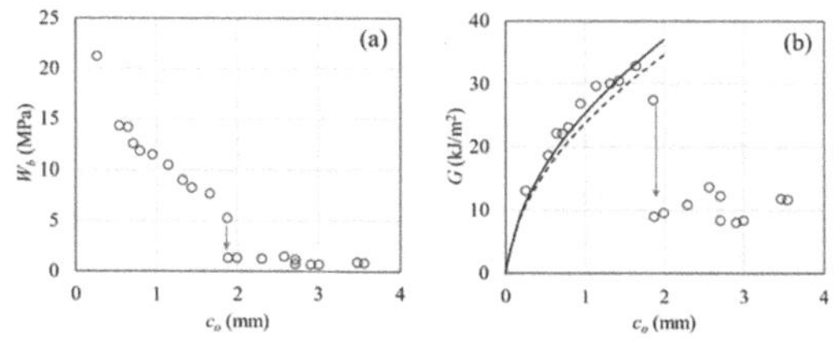

Figure 3. (a) Strain energy density at break (W_b) as a function of initial crack length (c_o for NR-0A at 25°C and strain rate of 0.5/s, (b)c_o dependence of energy release rate (G) calculated from the data (a) and equation (2). The solid line represents the fitted curve of equation (8) to the data using B=3.65×10^{-13} N^{-2} m^3 which is obtained as fitting parameter. The dashed line depicts the curve of equation (8) with the B value (4.06×10^{-13} N^{-2} m^3) obtained using the cyclic loading test for the same specimen.

which regards the rupture mode at $c_o < c_o^*$ as cyclic crack growth process. The theory is further validated by the fracture surface observation (Figure 5). The fracture surface at $c_o < c_o^*$ has specific striations specific to cyclic crack growth. In contrast, in the fracture surface at $c_o > c_o^*$, no specific striation is observed, indicating the crack growth process governed by catastrophic failure.

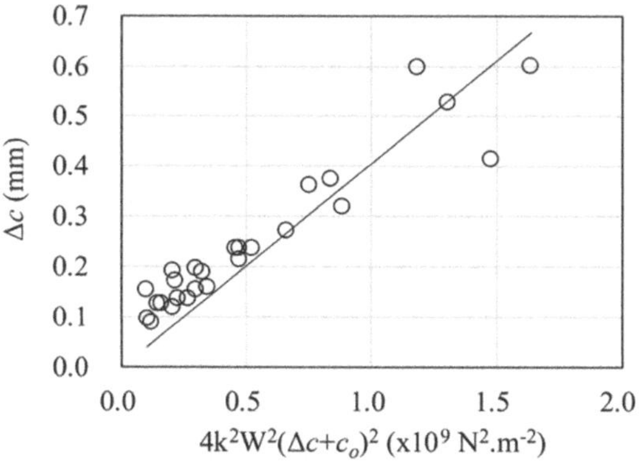

Figure 4. Incremental crack length (Dc) as a function of $4k^2 W^2(c_o$+D$c)^2$ for NR-0A in the cyclic loading test. The line indicates the result of the linear regression. The slope corresponds to the B value of 4.06×10-13 N^{-2}m^3.

4.2 *Effect of temperature*

Figures 6a and 6b display W_b and G as a function of c_o for NR-0B at various ambient temperatures (T), respectively. W_b decreases with increasing c_o and show an abrupt fall at critical c_o^* at T < 80°C. Beyond 90°C the abrupt change is no longer detected. Additionally, the c_o^* value decreases with increasing T. The maximum G value at c_o<c_o^*, which can be regarded as G_{SIC}, decreases with increasing T. This result indicates that SIC is suppressed at elevated temperatures. G values in c_o>c_o^* region shows constant value (G_T) for each temperature. G_T value decreases with increasing temperature as G_T is merely governed

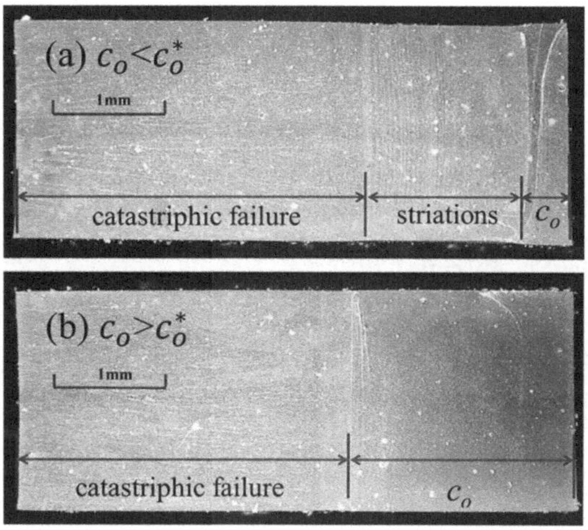

Figure 5. Fracture surfaces for NR-0A (a)c_o<c_o^*, (b)c_o>c_o^*.

by viscoelastic energy dissipation at given temperature, which decreases with increasing temperature.

The corresponding plots for IR-0 are shown in Figure 7. IR-0 exhibits qualitative same trend as

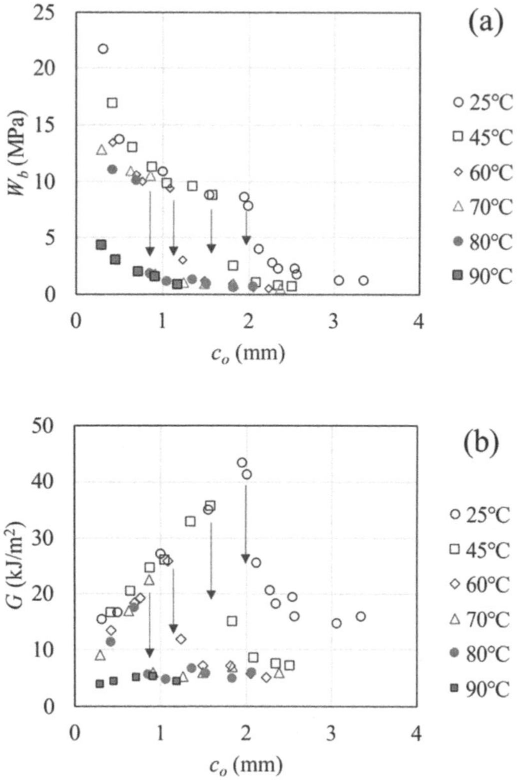

Figure 6. (a) W_b as a function of c_o over the wide range of temperature for NR-0B, (b)G calculated with equation (2) as a function of c_o over the wide range of temperature for NR-0B.

NR-0B. However, maximum temperature showing finite G_{SIC} was 60°C for IR-0, which is 20°C lower than that of NR-0B.

Figure 8 summarizes G_{SIC} and G_T as a function of T for NR-0B and IR-0. No appreciable difference in G_T is observed between NR-0B and IR-0. In contrast, NR-0B shows significantly larger values of the magnitude of G_{SIC} at each T and the threshold temperature for SIC effect than IR-0.

Effect of T on SIC behavior is investigated by WAXS analysis with unnotched tensile specimens. Figure 9a displays $I_x(200)$ as a function of nominal tensile strain at various T for NR-0B. At 27 °C, crystallization occurred at 230% of strain, and crystallinity continuously increases with increasing strain. With elevating T, SIC onset strain increases. The occurrence of SIC is appreciable at $T<90$°C, while the crystal diffraction is no longer observed at 110 °C. The corresponding plots for IR-0 are shown in Figure 9(b). IR-0 exhibits qualitatively similar behavior, but the vanishing temperature for SIC effect is 20 °C lower than that for NR-0B. This gap is similar to the threshold temperature gap observed in edge crack testing (Figure 8). This consistency validates the interpretation for the abrupt change of W_b in this study.

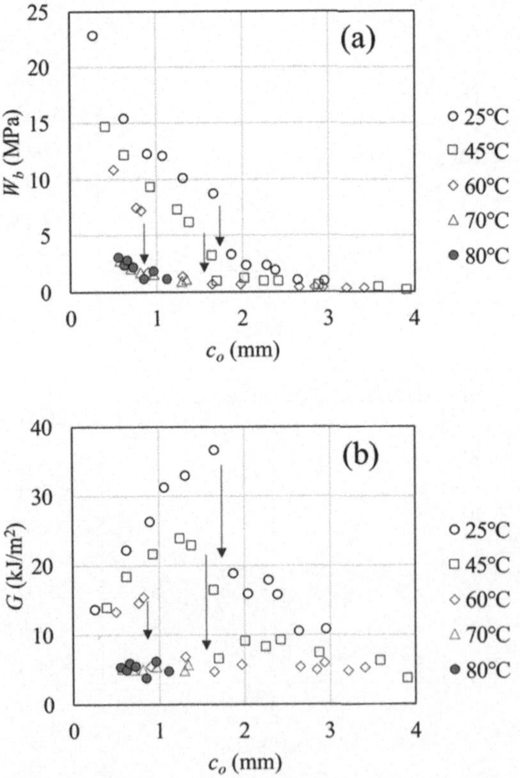

Figure 7. (a)W_b as a function of c_o over the wide range of temperature for IR-0, (b)G calculated with equation (2) as a function of c_o over the wide range of temperature for IR-0.

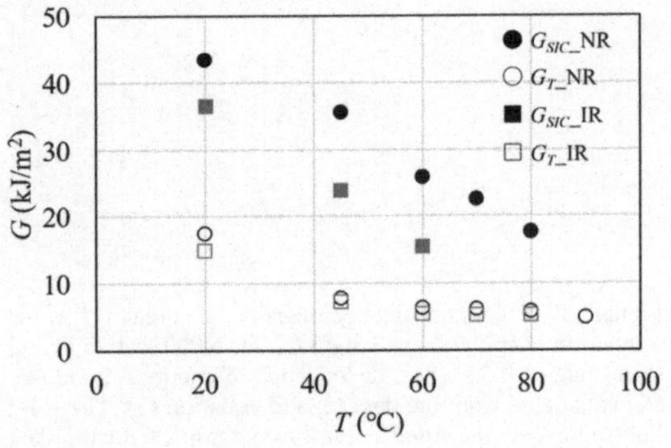

Figure 8. (a)Temperature dependence of G_{SIC} and G_T for NR-0B and IR-0.

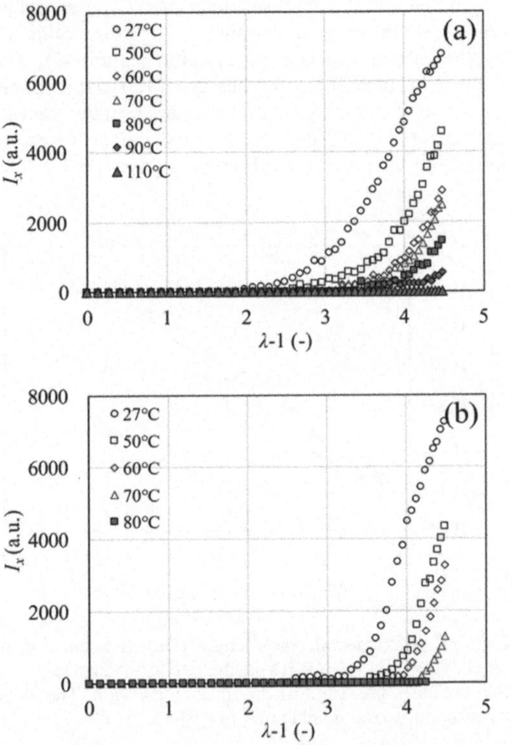

Figure 9. Scattering intensity of the $I_x(200)$ diffraction of the SIC crystal as a function of the imposed tensile strain for (a) NR-0B and (b) IR-0 at wide range of temperatures.

5 CONCLUSION

A discontinuous variation in W_b at a critical crack length c_o^*, which is specifically observed in SIC rubber, results from a change in the rupture mechanism between a cyclic crack growth process where SIC effect is pronounced at the advancing crack tip, and a catastrophic crack growth process without appreciable SIC effect. The theory proposed in this study facilitates the separate evaluations of the two strength components, G_{SIC} and G_T, and the threshold conditions for the onset of pronounced SIC effect around a crack tip, by the edge crack tensile tests. This methodology revealed the superiority of NR relative to synthetic IR in both G_{SIC} and threshold temperature.

ACKNOWLEDGEMENT

This work was partly supported by JST, CREST (grant number JPMJCR2091), Japan.

REFERENCES

Andrews, E.H. (1974) A generalized theory of fracture mechanics, J. Mater. Sci., 9, 887–894.

Bell, C.L.M., Stinson, D. & Thomas, A.G. (1982) Measurement of Tensile Strength of Natural Rubber Vulcanizates at Elevated Temperature, Rubber Chem. Technol., 55(1), 66–75.

Greensmith, W.H. (1963) Rupture of rubber III, Determination of tear properties, J. Applied Polym. Sci., Vol.7, 993–1002.

Hamed, G.R. (1983) Effect of Crosslink Density on the Critical Flaw Size of a Simple Elastomer, Rubber Chem. Technol., 56(1), 244–251.

Hamed, G.R. & Park, B. H. (1999) The Mechanism of Carbon Black Reinforcement of SBR and NR Vulcanizates, Rubber Chem. Technol., 72(5), 946–959.

Hamed, G.R. & Rattanasom, N. (2002) Effect of Crosslink Density on Cut Growth in Black-Filled Natural Rubber Vulcanizates Rubber Chem. Technol., 75(5), 935–942.

Kohjiya, S. & Ikeda, Y. (2021), Chemistry, Manufacture and Applications of Natural Rubber, Woodhead Publishing, 2nd edn.

Rivlin, R.S. & Thomas, A.G. (1953), Rupture of Rubber. I. Characteristic Energy for Tearing, J. Polym. Sci., 10(3), 291–318.

Rong, G., Hamed, G.R. & Jiang, J. (2016) Comparison of the Strength of Normal and Edge-Cut Tensile Specimens of Styrene- Butadiene Rubber and Natural Rubber with Similar Crosslink Density, Rubber Chem. Technol., 89 (4), 631–639.

Thomas, A.G. (1967) Fracture of Rubber, The Physical Basis of Yield and Fracture Conference Proceedings, Oxford, 134–143.

Thomas, A.G. & Whittle, J.M. (1970) Tensile Rupture of Rubber, Rubber Chem. Technol., 43(2), 222–228.

Tsunoda, K., Kitamura, Y., Urayama, K. (2023) Transition of rupture mode of strain crystallizing elastomers in tensile edge-crack tests, Soft Matter, 19, 1966–1976.

Aging

Constitutive Models for Rubber XIII – Dal (Ed.)
© 2025 The Author(s), ISBN 978-1-032-85138-9

Novel approach to modelling chemical stress relaxation using finite element analysis

A.M. Duncan & K. Akutagawa
School of Engineering and Materials Science, Queen Mary University of London, London, UK

J.L. Ramier
SLB Cambridge Research, Cambridge, UK

J.J.C. Busfield
School of Engineering and Materials Science, Queen Mary University of London, London, UK

ABSTRACT: A novel finite element analysis (FEA) approach for modelling the effects of ageing rubber under strain is presented. Chemical stress relaxation results in inhomogeneous, anisotropic material behavior together with permanent set in parts post-ageing. Existing models struggle to accurately capture all these various aspects, and those models that have been proposed are often too complex for straightforward implementation in FEA models. Commercial software such as ABAQUS lacks built-in capabilities for modelling chemical stress relaxation. This paper introduces a new method based on the dual-network theory to model chemical stress relaxation in ABAQUS. The method represents complex aged rubber parts as the superposition of different rubber components with geometries either modelled on the as moulded geometry or on the geometry in which the ageing was done. This approach simplifies the modelling process while accurately capturing the effects of chemical stress relaxation without requiring user subroutines. The paper details this innovative method and demonstrates its application on complex parts. Validation experiments conducted on aged nitrile butadiene rubber sheets aged at constant strain show close agreement between simulations and experimental data, confirming the approach's effectiveness.

1 INTRODUCTION

1.1 *Chemical stress relaxation*

When subjected to a constant strain at elevated temperature, rubber undergoes a phenomenon known as chemical stress relaxation. Stress relaxation is characterized by a drop in stress overtime in the direction of the applied strain, and a permanent set upon release of that stress (Curro et al., 1975), (Tobolsky et al. 1944). Tobolsky and co-workers (Tobolsky et al., 1944), (Andrews et al., 1946), (Tobolsky & Callinan, 1960), (Tobolsky et al., 1972) proposed that chemical stress relaxation was the consequence of chemical reactions taking place in the rubber molecules that break and reform crosslinks in the polymer network. The rates of these chain scissioning and crosslinking reactions have been shown to increase with oxygen and heat (Celina, 2013), as does the chemical stress relaxation rate (Tobolsky & Thomas 1960). Both reactions proceed concurrently but at differing rates. The dual-network theory suggests that the results of both chemical reactions can be treated as the breakdown of the original network and the simultaneous creation of a new network in the deformed configuration. These networks are treated as separate, acting in parallel with no interactions. The original network is in equilibrium with the original unstrained stretch, while the new network is in equilibrium with the stretch at which it was formed. Both networks contribute to the stiffness of the material. Figure 1 helps to illustrate this theory.

In Figure 1, the sample that has length L_u is initially at rest. It is composed only of the original, as molded polymer network, represented with solid blue lines. The sample is then stretched to length L_s and held at an elevated temperature. After ageing, the original network has been diminished, while a new polymer network represented with hollow red lines has been formed. Following the test, the sample retains both networks, which, at any length L_x will have different extension ratios. The stress at length L_x can be calculated using the dual-network theory.

DOI: 10.1201/9781003516880-43

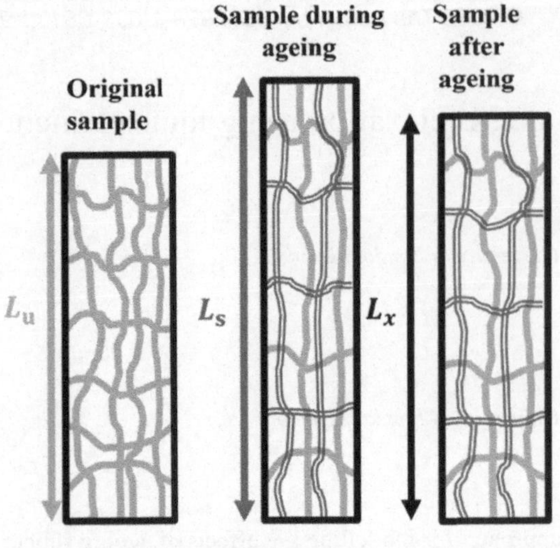

Figure 1. Diagram illustrating breakdown of original network (solid lines) and formation of new network (hollow lines) (Adapted from Duncan et al. 2023).

1.2 *Dual-network theory*

A simple model based on the dual-network theory for a pure elastomer, with no reinforcement, is presented. It is assumed that after ageing there are two rubber networks acting in parallel that are independent of each other. Each is in equilibrium with the length at which it was formed. Each network's contribution to the stress is proportional to its network density. Using these assumptions, Equation 1 is derived:

$$\sigma_T = \frac{\nu_u}{\nu_{u0}} f_1(\lambda_u) + \frac{\nu_s}{\nu_{u0}} f_1(\lambda_s) \tag{1}$$

where:

$$\lambda_u = \frac{L_x}{L_u} \qquad \lambda_s = \frac{L_x}{L_s}$$

σ_T is the total stress at length L_x. ν_{u0}, ν_u and ν_s are the network density of the original network before ageing, the original network remining after ageing and the new network formed during the ageing, respectively. λ_u and λ_s are the extension ratios of the original and new network, respectively. L_u, L_s and L_x are the length of the sample before ageing, during ageing and at the stress σ_T, respectively. And $f_1(\lambda_u)$ is the stress-extension function of an unaged sample.

In Equation 1 the first term represents the contribution of the original network and the second term the contribution of the new network. Both terms are multiplied by some function of the extension ratio, but critically the extension ratio is defined differently for each term. Since the original network is in equilibrium with the sample before testing, its extension ratio is the ratio of the test length and the original length of the sample, whereas the second term defines the extension ratio as the ratio of the test length divided by the length the sample was aged at.

A commonly used function for Equation 1 is derived from the neo-Hookean strain energy function (Rivlin 1948), which in the case of an incompressible material in simple tension can be shown to simplify to Equation 2; see Treloar (1975):

$$f_1(\lambda) = k_0(\lambda - \lambda^{-2}) \tag{2}$$

where λ is the extension ratio in the direction of principle stretch, and k_0 is a constant. Substituting Equation 2 into Equation 1 results in Equation 3:

$$\sigma_T = \frac{\nu_u}{\nu_{u0}} k_0 \left[\lambda_u - (\lambda_u)^{-2}\right] + \frac{\nu_s}{\nu_{u0}} k_0 \left[\lambda_s - (\lambda_s)^{-2}\right] \tag{3}$$

Tobolsky's (1944) dual-network theory was developed for simple ageing scenarios on unfilled elastomers, it is only capable of accounting for chemical ageing and is unable to account for other factors that cause differences in stiffness between unaged and aged samples; for example, a loss of additives such as plasticiser. An updated improved dual-network theory was developed by Duncan et al (2023) to address these drawbacks. This dual-network theory can be summarized using Equation 4:

$$\sigma_T = \left(1 + \frac{\nu_s}{\nu_u}\right)^{-1} f_2(\lambda_u) + \left(1 + \frac{\nu_u}{\nu_s}\right)^{-1} f_2(\lambda_s) \tag{4}$$

where $f_2(\lambda_u)$ is the stress-extension function of a sample aged under no strain.

Just as with Equation 1, the first and second term represent the contribution of the original and new network respectively. However, by using $f_2(\lambda_u)$, derived for an aged sample, instead of $f_1(\lambda_u)$, factors that cause differences in stiffness between unaged and aged samples are properly accounted for. Tobolsky's (1944) dual-network theory uses the Neo-Hookean strain energy function, which works well for unfilled rubber. The assumptions in both Tobolsky's (1994) original dual-network theory and the improved dual-network theory presented by Duncan et al. (2023) allow for any appropriate strain energy function to be used to derive $f_1(\lambda_u)$ or $f_2(\lambda_u)$ (Dinari et al., 2021), (Duncan et al., 2022), (Duncan et al., 2023). In this work the Yeoh strain energy function is used (Yeoh, 1990). In simple tension using the Yeoh strain energy function for $f_2(\lambda_u)$ results in equation 5:

$$f_2(\lambda) = 2(\lambda - \lambda^{-2})\left[C_{10} + 2C_{20}(I_1 - 3) + 3C_{30}(I_1 - 3)^2\right] \tag{5}$$

where I_1 is the first invariant and C_{10}, C_{20} and C_{30} are material constants.

1.3 *Modelling the dual-network theory in FEA*

There are many challenges associated with modelling chemical stress relaxation in finite element models. After ageing at set strains rubber materials exhibit both permanent set and anisotropy. Modelling either of these requires the use of user-defined materials subroutines which require the creation of their own material models. Complex parts will also often have different strain levels throughout. This results in different material properties at different locations in the part. Modelling this adds another layer of complexity to the required user subroutines. Existing models struggle to accurately capture all these various aspects, and those models that have been proposed are often too complex for straightforward implementation in finite element analysis (FEA) models.

Achenbach (2000) shows the effectiveness of using the dual-network theory in FEA but does not detail how the dual-network theory was implemented. Shyu (2015) shows an innovative way of applying the dual-network theory by adapting the Bergstrom-Boyce model, an existing viscoelastic model that is already built into ABAQUS. This method can closely approximate the dual-network theory; however, it has not seen widespread adoption. This may be because while the idea is simple, in practice adapting the Bergstrom-Boyce model in ABAQUS, in the ways set out in the paper, requires complex codes and the use of user subroutines.

2 MODELLING APPROACH

This section details the modelling approach used to examine chemical stress relaxation. Figure 2 helps to illustrate the approach. In short, this approach avoids modelling one complex part, with permanent set, anisotropy, and varying material properties throughout the part, but instead it models two simple parts with uniform isotropic material properties and then ties the two parts together. Each part represents one of the two networks present after ageing.

1. First the unaged part is modelled and loaded to the desired deformed state, at which the simulated ageing takes place.
2. The model then needs to output and store the position of all external surfaces/edges of the unaged part, after deformation.
3. Next a new model is made with two parts, one part with the same geometry as the unaged part before loading, and one with the same geometry as the unaged part after the loading was applied, Figure 7 shows an example of this on a D-shaped seal.
4. The two parts are then given different material properties. These material properties correspond to the first and second term in Equation 4 for the original and new network part respectively. This means that the test data used to fit the stress-strain model should be that of a sample aged under no strain ($f_2(\lambda_u)$) multiplied by $(1+(\nu_s/\nu_u))^{-1}$ for the original network part and $(1+(\nu_u/\nu_s))^{-1}$ for the new network part. The details of how to calculate the variables in Equation 4 are covered previously in Duncan et al. (2023).
5. All external surfaces/edges of the original network part and new network part should be made coincident. This should be done so that each point of the original network is matched to where that point deformed to on the new network part. This is best done as its own loading step using multiple point constraints that gradually bring the distance between these points to zero.
6. Finally, the same boundary conditions and surface interactions used for the unaged part in the first model are applied to the original network only—there is no need to apply these boundary conditions and surface interaction to the new network part as they are still tied together.

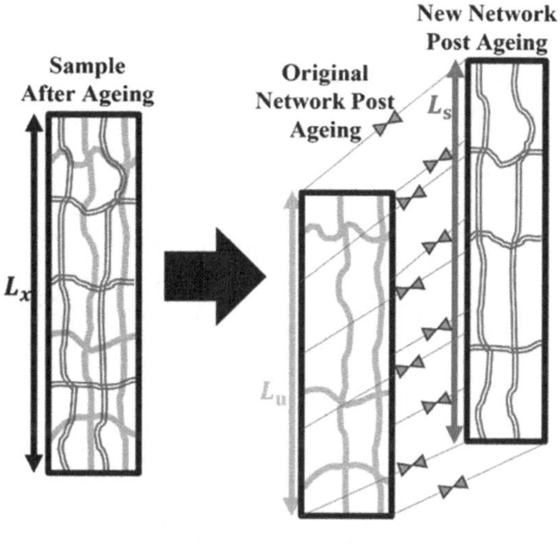

Figure 2. Diagram illustrating modelling approach. Figure shows original network (solid line) and new network (hollow lines) separately with both networks joined by connections.

All strains (from the undeformed state) and surface interactions such as contact forces can be examined directly from the original network part. All other internal measurements such as stress or heat buildup must be examined separately for the original and new network part. If the total stress or the heat buildup are needed, then the values from both parts will need to be summed in a post-processing step.

3 TEST METHOD

3.1 *Material*

All tests were carried out on the same batch of carbon-black-filled nitrile butadiene rubber (NBR) compound. The most important details of the formulation are given in Table 1. The compound was mixed in an internal mixer and then re-milled immediately prior to molding in a two-roll mill and cured into 2-mm thick sheets using a hot press at 125 °C for 30 minutes.

Table 1. Formulation of NBR compound used.

Component	Details	Filler Loading/pphr
Elastomer	NBR (33% ACN content)	100
Carbon black	N550	60
Crosslink	Sulphur (efficient vulcanization system)	

The NBR compounds also contain the typical small amounts of plasticizer, activators, accelerators, and antioxidants that would be widely adopted in commercial practice.

3.2 *Sample preparation*

NBR was aged under no strain, under a pure shear strain, and under simple tension. All samples were aged at 125 °C for 24 hours in an air oven. For simple tension a sheet of rubber 40 mm by 195 mm by 2 mm was gripped in a stretching frame 165 mm apart and stretched to 181.5 mm (10% strain), then aged; see Figure 3. For pure shear, a sheet of rubber 195 mm by 70 mm by 2 mm was gripped in a stretching frame 40 mm apart and stretched to 44 mm (10% strain).

After ageing, rubber sheets were left to relax overnight before samples were cut out. Dog-bone samples matching dimensions from ASTM No. D416-16 (2021), with a cross-sectional area of 2 mm by 2 mm, were cut out of the sheets in the direction of the applied strain (0°) and perpendicular to the applied strain (90°); see Figure 3. In all samples 0° was lined up with the milling direction of the rubber sheet.

All cut-outs were taken close to the center of the rubber sheets to ensure the effects of the clamps were negligible and the area samples were cut from was in either homogenous pure shear or simple tension.

3.3 *Stress-strain testing*

Samples were tested on a screw-driven Instron 5967, using a video extensometer to track strain. All tests were done with a crosshead displacement of 1mm/s. Samples were tested from 0% to 100% strain, at least three repeats for each sample type were done to check for good repeatability.

3.4 *FEA simulation setup*

The FEA simulation presented in this paper was run on ABAQUS CAE 2023. All parts use the built-in ABAQUS Yeoh function for hyperplastic materials. The Yeoh function parameters were fitted in ABQUS using

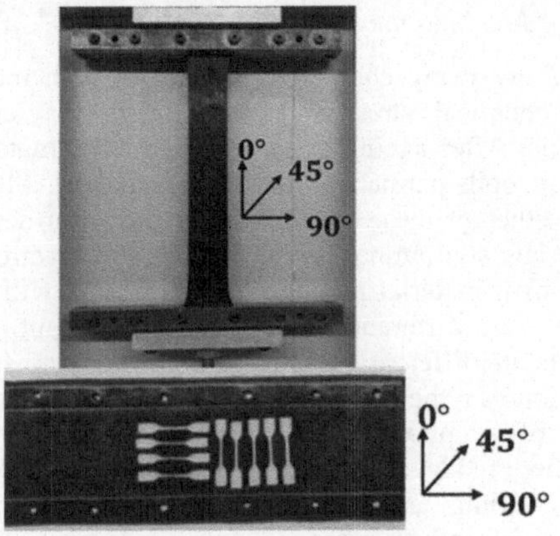

Figure 3. Image of rubber sheet being aged under simple tension (top). Image of dog-bone cut-outs taken from a rubber sheet after being aged in pure shear (bottom) (image taken from Duncan et al. 2022).

test data. Models are meshed using first-order quadrilateral plane stress elements, CPS4R, with enhanced hourglassing control and distortion control. The simulations were set up following the steps detailed in section 2. The network density ratio (ν_s/ν_u) was derived using instructions from Duncan et al. 2023, in all the simulations it was set to a value of 0.978.

4 RESULTS

Figure 4 compares results for an unaged NBR sample from a tensile test and from FEA simulations. This figure gives an idea of how close of a fit can be expected from the FEA model before ageing. A good ageing model should have similarly good fits. The values measured for C_{10}, C_{20} and C_{30} for the unaged part were 1.2056, -0.0799 and 0.1024 respectively. For the aged parts C_{10}, C_{20} and C_{30} were set to 1.307, -0.1567 and 0.0049 respectively.

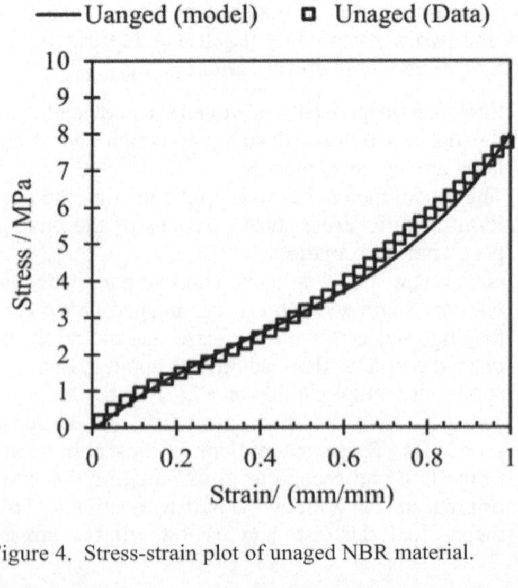

Figure 4. Stress-strain plot of unaged NBR material.

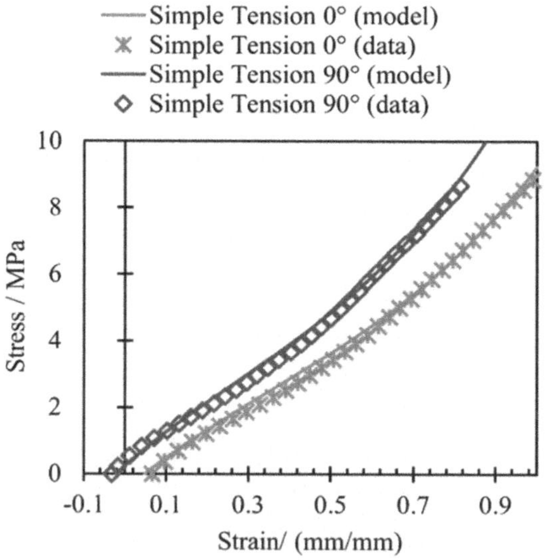

Figure 5. Stress-strain plot of NBR material aged at 10% strain in simple tension for 24 hours at 125°C. Samples cut in the direction of strain (0°) and perpendicular to the strain (90°).

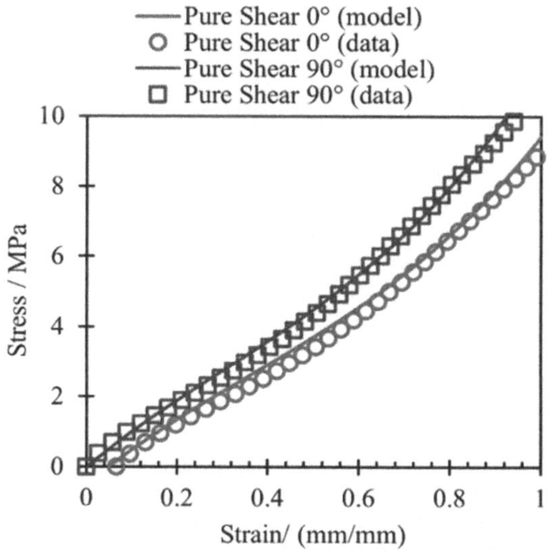

Figure 6. Stress-strain plot of NBR material aged at 10% strain in pure shear for 24 hours at 125°C. Samples cut in the direction of strain (0°) and perpendicular to the strain (90°).

Figures 5 and 6 show the stress-strain plots from tensile samples cut out of the NBR sheet aged in simple tension and in pure shear, respectively, and compares the collected data to FEA simulations using the dual-network theory. Both figures show a close fit between FEA simulations and the collected data. In both cases the FEA simulations were able to accurately capture permanent set, changes in stiffness, and the anisotropy caused by ageing at set strain.

5 EXAMPLE

The results in Section 4 show that the approach used in this paper can accurately capture the effects of chemical ageing in simple scenarios with uniform strain across the sample. However, as stated in the Introduction, one of the challenges in modelling chemical stress relaxation is that most real-world ageing scenarios are not at a homogenous strain. This results in different material properties at different locations in a part post-ageing.

This section shows an example of a more complex scenario. A D-shaped seal is compressed by a rigid body and the part undergoes a simulated ageing of 24 hours at 125°C, the same ageing conditions as were tested in Section 4. The simulation was set up as described in Section 2. This example used the built-in ABAQUS Yeoh function for hyperplastic materials and used test data to fit the parameters. Models were meshed using first-order quadrilateral plane stress elements, CPS4R, with enhanced hour-glassing control and distortion control.

Figure 7 shows how this example was developed, with the original geometry of the part in light blue, and the deformed geometry in red. The figure also highlights some of the connections between the two parts. For Figures 8, 9 and 10 the new network part is hidden, and only the original network is shown.

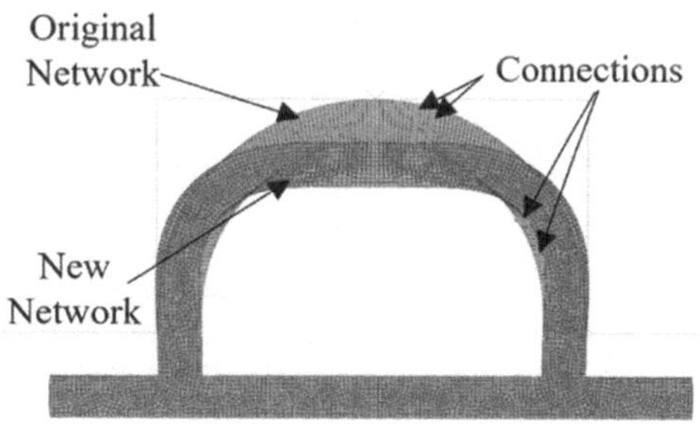

Figure 7. D-shaped seal example, showing both part instances of the original (light blue) and new network (red) and highlighting some connections between the two instances.

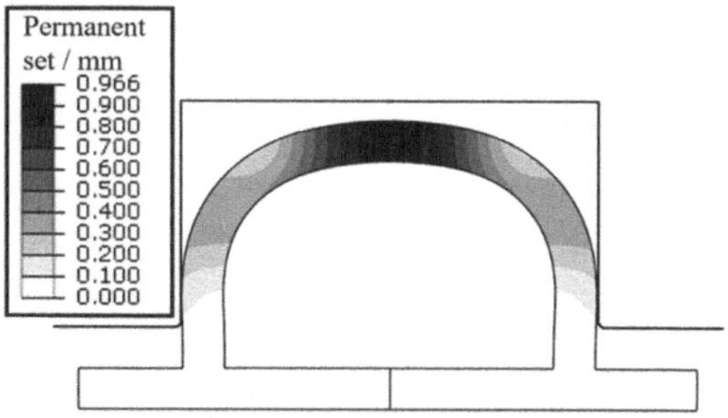

Figure 8. Simulated permanent set of D-shaped seal after 24 hours of simulated ageing at 125°C in the compressed state.

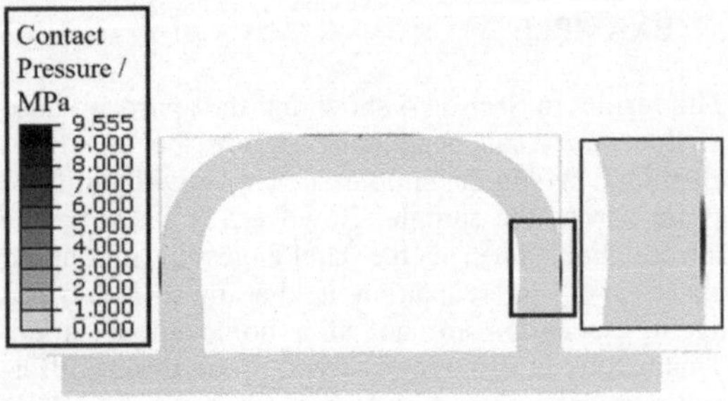

Figure 9. Simulated contact pressure of D-shaped seal after 24 hours of simulated ageing at 125°C in the compressed state.

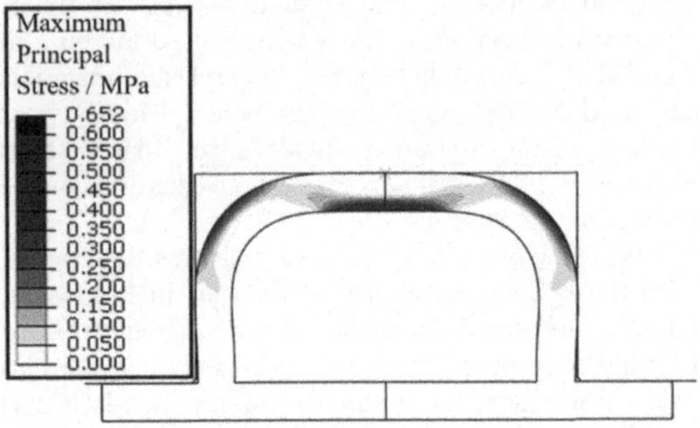

Figure 10. Simulated maximum principal stress in the original network of a D-shaped seal after 24 hours of simulated ageing at 125°C in the compressed state.

Figure 8 shows the aged part unloaded with no external forces acting on it. The displacement map shows the permanent set caused by ageing. Figure 9 shows the contact pressure between the old network part and the rigid body surface. Because the new network part has no surface interactions, this shows the total contact pressure. In this example, the maximum contact pressure dropped from 12.35 MPa to 9.56 MPa as a result of ageing. Figure 10 shows the maximum principal stress in the original network part only. Since the new network part also experiences stress, this does not show the total stress in the part; therefore, to calculate the total stress in the part, the stress from the two parts would have to be summed in a post-processing step.

6 DISCUSSION

The ability to examine the stresses in each network individually could allow for the future study into how ageing under strain affects different aspects of rubber behavior. For example, if fracture in aged parts is dependent on the total stress/tearing energy in a part or if different stresses/tearing energies in the two networks play a role. Or, to evaluate the extent to which the differences in the stress between the new and original network affect cyclic stress relaxation in parts after ageing.

In many real-world examples, parts are not under constant strain but rather a cyclic strain. Bjork & Stenberg (1990) showed that the dual-network theory could be adapted for cyclic applications. The method involves splitting the new network into multiple networks, each with its own equilibrium length corresponding to a length at different timesteps in the cyclic loading. Similarly, the method in this paper could be adapted to cyclic applications by replacing the single new network part with multiple new network parts.

7 CONCLUSION

A new method, based on the principles of the dual-network theory, for modelling chemical stress relaxation using finite element analysis (FEA) is shown. The approach used simplifies the modelling process by representing aged rubber components as multiple simple parts, eliminating the need for user subroutines in ABAQUS. The model is examined using test data measured from sheets of nitrile butadiene rubber (NBR) aged at set homogenous strains and this shows a very close agreement with the experimental data. The method is also shown to be able to be applied to more complex geometries such as an extruded seal profile.

ACKNOWLEDGEMENTS

The authors would like to thank SLB Cambridge Research (SCR), for providing the materials for this study. Thank you to SCR, Queen Mary University of London and the Engineering and Physics Science Research Council (EPSRC) for funding this work.

REFERENCES

American Society for Testing and Materials (2021). Standard test method for vulcanized rubber and thermoplastic elastomers – Tension ASTM No. D412-16(2021)

Achenbach, M. 2000. Service life of seals–numerical simulation in sealing technology enhances prognoses. *Computational Materials Science* 19(1-4): 213–222.

Andrews, R. D., Tobolsky, A. V., & Hanson, E. E. 1946. The theory of permanent set at elevated temperatures in natural and synthetic rubber vulcanizates. *Journal of Applied Physics* 17(5): 352–361.

Björk, F., & Stenberg, B. 1990. Stress relaxation of a nitrile rubber surrounded by an oil that increases the network density. *Polymer* 31(9): 1649–1657.

Celina, M. C. 2013. Review of polymer oxidation and its relationship with materials performance and lifetime prediction. *Polymer Degradation and Stability* 98(12): 2419–2429.

Curro, J. G. & Salazar, E. A. 1975. Physical and chemical stress relaxation of elastomers. *J Appl Polym Sci* 19(9): 2571–2581.

Dinari, A., Zaïri, F., Chaabane, M., Ismail, J., & Benameur, T. 2021. Thermo-oxidative stress relaxation in carbon-filled SBR. *Plastics, Rubber and Composites* 50(9): 425–440.

Duncan, A. M., Akutagawa, K., Ramier, J. L., & Busfield, J. J. C. 2022. Anisotropy of nitrile butadiene rubber induced by thermal ageing at fixed strain. In *Constitutive Models for Rubber XII*, Milan: CRC Press, 434–439.

Duncan, A.M., Akutagawa, K., Ramier, J.L. and Busfield, J.J. 2023. Improved dual-network model for aging of rubber composites under set strains. *Macromolecules* 56(17): 6704–6711.

Rivlin, R. S. 1948. Large elastic deformations of isotropic materials IV. Further developments of the general theory. *Philosophical Transactions of the Royal Society of London. Series A, Mathematical and Physical Sciences* 241(835): 379–397.

Shyu, G-D. 2015. On the modeling of chemical stress relaxation of elastomers. *Fall 188th Technical Meeting of the Rubber Division, ACS*, 13-15 October, Cleveland, OH.

Tobolsky, A. V., Prettyman, I. B., & Dillon, J. H. 1944. Stress relaxation of natural and synthetic rubber stocks. *Rubber Chemistry and Technology* 17(3): 551–575.

Tobolsky, A. V. & Callinan, Thomas D. 1960. Properties and structure of polymers. *Journal of Electrochemical Society* 107.10: 243C.

Tobolsky, A. V, Takahashi, Y., & Naganuma, S. 1972. Effect of additional cross-linking of continuous chemical stress relaxation of cis-polybutadiene. *Polymer Journal* 3(1): 60–66.

Treloar, L. G. 1975. The physics of rubber elasticity.

Yeoh, O.H. 1990. Characterization of elastic properties of carbon-black-filled rubber vulcanizates. *Rubber Chemistry and Technology* 63(5): 792–805.

Constitutive Models for Rubber XIII – Dal (Ed.)
© 2025 The Author(s), ISBN 978-1-032-85138-9

Time homogenization in modelling of rubber damage and ageing

J. Heczko
NTIS - New Technologies for the Information Society, Faculty of Applied Sciences, University of West Bohemia, Pilsen, Czech Republic

ABSTRACT: The contribution is focused on cumulation of the effects of damage and ageing of elastomers on their stiffness properties. An earlier model is enhanced by introducing permanent set to an approximation of hyperelastic parameters as functions of time and loading type. The nature of the model, however, does not offer any straightforward way of extrapolation in case of different loading modes or service conditions. Therefore, a complex material model that takes into account different mechanisms of ageing and damage is considered. In order to enable numerical simulations of high-cycle loading, the method of asymptotic series is applied resulting in a time-homogenized version of the model. In addition to savings in computational time, the model is formally similar to the original approximation.

1 INTRODUCTION

Considering a part in service conditions, various effects influence its mechanical properties, e.g. mechanical loading, temperature, radiation or chemical agents. In order to study such effects, different angles of view may be combined and different levels of complexity of the description might be achieved. Obviously, simple description is preferable for an efficient simulation of real world parts. This has already been attempted in (Heczko & Kottner 2015), where the changes in mechanical properties were modelled using a closed formula for the parameters of the Ogden hyperelastic model as a function of time. Similar formulae have also been used by other researchers (Li, Zhao, Ma, & Liu 2022). The most severe drawback of the approach (Heczko & Kottner 2015) was the neglecting of permanent set.

In the present paper, the original approximation is revisited. The introduction of permanent set together with different strain energy density function leads to a significant improvement in the quality of the model, while maintaining its simplicity. The second part of the article deals with a constitutive model of rubber ageing and damage with the aim of simplifying it to a form comparable to the original approximation.

2 MATERIAL MODEL

The material model in consideration was previously used by the author (Heczko, Kottner, & Kossa 2021, Heczko 2022) and is based on the generalized Yeoh strain energy density (Hohenberger, Windslow, Pugno, & Busfield 2019) and the dynamic network model of the ageing effects (Naumann & Ihlemann 2015).

The total deformation gradient $\mathbf{F}$ is split into a volumetric, $\hat{\mathbf{F}}$, and isochoric part, $\bar{\mathbf{F}}$,

$$\mathbf{F} = \bar{\mathbf{F}}\hat{\mathbf{F}} \tag{1}$$

and the isochoric part is further split into permanent and elastic deformation

$$\bar{\mathbf{F}} = \mathbf{F}_2 \mathbf{F}_1 \,, \tag{2}$$

$$\bar{\mathbf{C}} = \mathbf{F}^{\mathrm{T}}\mathbf{F} = \mathbf{F}_1^{\mathrm{T}}\mathbf{F}_2^{\mathrm{T}}\mathbf{F}_2\mathbf{F}_1 = \mathbf{F}_1^{\mathrm{T}}\mathbf{C}_2\mathbf{F}_1 \,, \tag{3}$$

where $\mathbf{C}_2$ corresponds to the elastic part of deformation and the strain energy density (SED) should therefore depend on it. An equivalent approach is, however, to express SED in terms of an internal variable $\mathbf{C}_1^{-1}$ that corresponds to permanent set. The effect of fatigue damage is expressed by means of an inner variable D; $D = 0$ denotes undamaged material while $D = 1$ corresponds to a totally damaged material with no residual strength.

Neglecting the spatial non-homogeneity due to diffusion-limited oxidation (DLO), the following simplified model is obtained:

$$W = (1 - D) \sum_i K_i \nu \left(\mu \left(\mathbf{C}_1^{-1} \cdot\cdot \bar{\mathbf{C}} - 3\right)\right)^{p_i} \,, \tag{4}$$

DOI: 10.1201/9781003516880-44

$$\dot{\nu} = k_S \nu , \quad \nu(0) = 1 , \tag{5}$$

$$\dot{\mu} = k_R \mu , \quad \mu(0) = 1 , \tag{6}$$

$$\frac{d\mathbf{C}_1^{-1}}{dt} = k_R \left(\bar{\mathbf{C}}^{-1} - \mathbf{C}_1^{-1} \right) , \quad \mathbf{C}_1^{-1}(0) = \mathbf{I} , \tag{7}$$

$$\dot{D} = \left(-\frac{1}{A} \frac{\partial W}{\partial D} \right)^a , \tag{8}$$

where $k_S < 0$ and $k_R > 0$ are material parameters that may depend on temperature, oxygen concentration or other environmental factors. The resulting 2nd Piola-Kirchhoff stress is evaluated as

$$\mathbf{S} = 2(1-D) \sum_i K_i \mu \nu p_i \left(\mu \left(\mathbf{C}_1^{-1} \cdot\cdot \bar{\mathbf{C}} - 3 \right) \right)^{p_i - 1} I_3^{-1/3} \left(\mathbf{C}_1^{-1} - \tfrac{1}{3} \left(\mathbf{C}_1^{-1} \cdot\cdot \bar{\mathbf{C}} \right) \bar{\mathbf{C}}^{-1} \right) . \tag{9}$$

3 CLOSED FORMULA MODEL

The complex model of ageing is now reformulated as a hyperelastic model with permanent set whose parameters are time-dependent. This way, the time-dependent parameters may be pre-computed or estimated using approximate formulae. A numerical simulation at a particular time then does not require a full integration in time.

By examination of (9), the following values of parameters and stress are obtained at time t:

$$K_i'(t) = K_i (1-D) \mu(t)^{p_i} \nu(t) , \tag{10}$$

$$p_i'(t) = p_i = \text{const.} , \tag{11}$$

$$\mathbf{S} = 2 \sum_i K'_i p'_i \left(\mathbf{C}_1^{-1} \cdot\cdot \bar{\mathbf{C}} - 3 \right)^{p'_i - 1} \cdot = \cdot I_3^{-1/3} \left(\mathbf{C}_1^{-1} - \tfrac{1}{3} \left(\mathbf{C}_1^{-1} \cdot\cdot \bar{\mathbf{C}} \right) \bar{\mathbf{C}}^{-1} \right) , \tag{12}$$

i.e. the only principal difference from a hyperelastic model is incorporating permanent set in terms of $\mathbf{C}_1^{-1}$. Note that no particular form of $\mathbf{C}_1^{-1}(t)$ is assumed.

A similar result would be obtained if one used a suitable approximation instead of (10) and (11). An even greater amount of freedom and computational efficiency can be gained by this approach, but at the potential cost of obtaining a non-physical model. Another potential caveat would be that such approximation would usually not capture any time-history of loading or coefficients. Based on the nature of the ageing process, as expressed by (5)-(7), a suitable choice of such more permissive approximation would involve exponential functions, e.g. the form used in (Heczko & Kottner 2015):

$$K_i''(t) = K_{i0} \left(1 - \sum_{j=1}^{N_{Ki}} A_{Kij} \left(1 - \exp\left(B_{Kij} t \right) \right) \right) , \tag{13}$$

$$p_i''(t) = p_{i0} \left(1 - \sum_{j=1}^{N_{pi}} A_{pij} \left(1 - \exp\left(B_{pij} t \right) \right) \right) , \tag{14}$$

$$\mathbf{C}_1^{-1''}(t) = \left(\mathbf{I} - \sum_{j=1}^{N_{C1}} \mathbf{A}_{C1j} \left(1 - \exp\left(B_{C1j} t \right) \right) \right) . \tag{15}$$

For the sake of stability of the resulting model, the B-parameters should be negative and the sum of the A-parameters should not exceed 1:

$$B_{\cdot j} < 0 , \quad \forall j , \quad \sum_j A_{\cdot j} \leq 1 . \tag{16}$$

4 TIME HOMOGENIZATION

As stated above, the closed-form approximation, despite possibly quite general, does not guarantee any extrapolation quality. For this reason, a constitutive relation that treats properly different mechanisms might be preferable, however, this usually comes at a huge computational cost. In this section, the method of asymptotic series is applied to the ageing model in order to reduce the necessary effort of its integration in time. This approach has already been applied in case of fatigue life estimations (Oskay & Fish 2004, Heczko & Kottner 2018), for further explanation see e.g. (Aubry & Puel 2014).

Starting with the simplified model (4)-(8) (i.e. neglecting DLO), the evolution equations are mutually independent – with the exception of damage – and therefore the homogenization procedure can be applied to each equation (5)-(7) separately and (8) being solved afterwards.

In case of evolution of the ageing parameters μ and ν, the only time-varying parameters are the coefficients k_R and k_S. The reason for their possible time-dependency are e.g. periodic temperature or oxygen concentration changes. In the more general case of DLO, the ageing parameters would also be influenced by deformation. Regarding permanent set, expressed through $\mathbf{C}_1^{-1}$, the time-dependent driving parameters are both the coefficient k_R and the isochoric part of total deformation, $\bar{\mathbf{C}}$.

Starting with the first case, only μ will be shown since the equation for ν is formally identical. Two time scales are considered, the slow time t and fast time τ with a very small ratio $\xi = \frac{t}{\tau}$. The ageing parameter μ, assumed being a function of these two time variables, is expressed using the asymptotic expansion

$$\mu(t,\tau) = \mu_0(t,\tau) + \xi\mu_1(t,\tau) + O(\xi^2)\,. \tag{17}$$

The total time derivative is, by chain rule,

$$\frac{d\mu}{dt} = \dot{\mu} + \xi^{-1}\mu'\,, \tag{18}$$

where the partial derivatives are

$$\dot{\mu} = \frac{\partial \mu}{\partial t} = \dot{\mu}_0 + \xi\dot{\mu}_1 + O(\xi^2)\,, \tag{19}$$

$$\mu' = \frac{\partial \mu}{\partial \tau} = \mu'_0 + \xi\mu'_1 + O(\xi^2)\,. \tag{20}$$

Moreover, the coefficient k_R is assumed to consist of a slow and fast component:

$$k_R(t,\tau) = k_{R\,\mathrm{slow}}(t) + k_{S\,\mathrm{fast}}(\tau)\,. \tag{21}$$

Substituting (17)-(21) into (6)

$$\begin{aligned}\frac{d\mu}{dt} &= \frac{1}{\xi}\mu'_0 + \dot{\mu}_0 + \mu'_1 + \xi\dot{\mu}_1 + O(\xi^2) = \\ &= k_R(t,\tau)\left(\mu_0 + \xi\mu_1 + O(\xi^2)\right)\,.\end{aligned} \tag{22}$$

Collecting the terms with the same power of ξ:

$$\xi^{-1} : \mu'_0 = 0\,, \tag{23}$$

$$\xi^0 : \dot{\mu}_0 + \mu'_1 = (k_{R\,\mathrm{slow}}(t) + k_{S\,\mathrm{fast}}(\tau))\mu_0\,. \tag{24}$$

The first equation states that μ_0 is independent of τ. Applying the integral average operator[1] to the other equation, a homogenized evolution equation is obtained:

$$\langle \dot{\mu}_0 + \mu'_1 \rangle = \dot{\mu}_0 = k_{R\,\mathrm{slow}}\mu_0 + \langle k_{R\,\mathrm{fast}}\mu_0 \rangle\,. \tag{25}$$

In case of (7), the same process is:

$$\mathbf{C}_1^{-1}(t,\tau) = (\mathbf{C}_1^{-1})_0(t,\tau) + \xi(\mathbf{C}_1^{-1})_1(t,\tau)\ldots\,, \tag{26}$$

$$\bar{\mathbf{C}}^{-1}(t,\tau) = (\bar{\mathbf{C}}^{-1})_0(t,\tau) + \xi(\bar{\mathbf{C}}^{-1})_1(t,\tau)\ldots\,, \tag{27}$$

$$\frac{d\mathbf{C}_1^{-1}}{dt} = \dot{\mathbf{C}}_1^{-1} + \xi^{-1}\mathbf{C}_1^{-1'}\,, \tag{28}$$

$$\frac{d\mathbf{C}_1^{-1}}{dt} = \frac{1}{\xi}(\mathbf{C}_1^{-1})_0' + (\dot{\mathbf{C}}_1^{-1})_0 + (\mathbf{C}_1^{-1})_1' + \xi(\dot{\mathbf{C}}_1^{-1})_1 + O(\xi^2)\,. \tag{29}$$

Again, collecting the terms with equal powers of ξ:

$$\xi^{-1} : (\mathbf{C}_1^{-1})_0{}' = 0\,, \tag{30}$$

$$\xi^0 : \;(\dot{\mathbf{C}}_1^{-1})_0 + (\mathbf{C}_1^{-1})_1{}' = k_R\left(\left(\bar{\mathbf{C}}^{-1}\right)_0 - (\mathbf{C}_1^{-1})_0\right)\,. \tag{31}$$

Thus, $(\mathbf{C}_1^{-1})_0$ does not depend on τ and from the second equation:

$$\begin{aligned}\langle(\dot{\mathbf{C}}_1^{-1})_0 + (\mathbf{C}_1^{-1})_1{}'\rangle &= (\dot{\mathbf{C}}_1^{-1})_0 = \\ &= \left\langle k_R\left(\bar{\mathbf{C}}^{-1}\right)_0\right\rangle - \langle k_R\rangle(\mathbf{C}_1^{-1})_0\,.\end{aligned} \tag{32}$$

A point worth noting is that using the averaged relations not only permits larger steps in time, but also reduces the complexity of the problems being solved by decoupling the equations. The only way the averaged (macroscopic) evolution equations depend on deformation is through their coefficients, which are obtained by solving a single period of loading. But this micro-problem is only a time-dependent nonlinear elasticity with no evolution of the internal variables.

A numerical example was solved to test the above described formulas. The state of uniaxial stress and homogeneous ageing was assumed, with prescribed deformation

$$\lambda(t) = 1 + \lambda_{\mathrm{amp}}\sin^2\left(\frac{2\pi t}{T}\right)\,, \tag{33}$$

with time period $T = 1\,\mathrm{s}$ and magnitude $\lambda_{\mathrm{amp}} = 0.1$. The ageing coefficients were chosen constant in time here

$$k_R = 1.2\cdot 10^{-3}\,\mathrm{s}^{-1}\,,\quad k_S = -1.0\cdot 10^{-3}\,\mathrm{s}^{-1}\,. \tag{34}$$

Time dependent coefficients have been tested on other occasions. The other parameters of the model were:

$$K_1 = 1.0\,\mathrm{Pa}\,, p_1 = 1.0\,, \tag{35}$$

1 Assuming time period of the fast component T, $\langle\alpha\rangle = \frac{1}{T}\int_0^T \alpha(t,\tau)d\tau$. For a function α, periodic in τ, $\langle\alpha'\rangle = 0$.

$$A = 5.0\,\text{Pa}\,, a = 3.0\,. \tag{36}$$

The resulting time-histories of ageing variables are shown in Figure 1. Very good agreement between full integration and the homogenized model can be seen in all ageing variables.

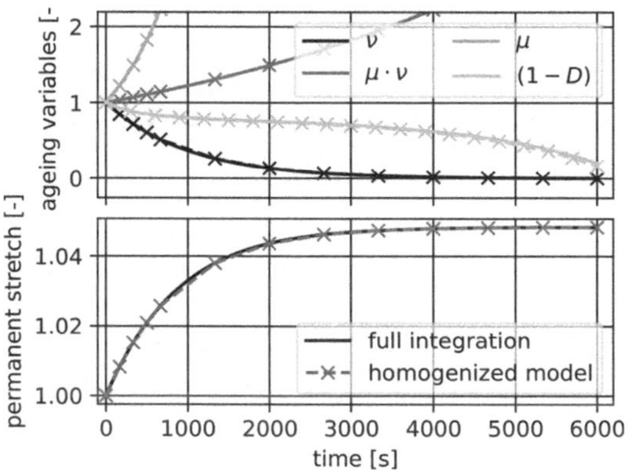

Figure 1. Comparison of full integration and homogenized model.

Table 1 lists computation times for different numbers of cycles. The full integration scales proportionally to cycle count. The time discretization in case of the homogenized model was fixed (8 points, first 5 with smaller time-step), therefore its computational cost remains low. In a real-world problem, the numbers of loading cycles would probably be even higher, thereby further increasing the advantage of using the homogenized model.

Table 1. Computation times for different numbers of loading cycles.

# of cycles	Full	Homogenized
	[s]	[s]
1200	3.31	0.588
$12 \cdot 10^3$	32.4	0.714
$12 \cdot 10^4$	328	1.17

4.1 *Analogy to the exponential form*

Assuming constant operating conditions in time – i.e. $\left\langle k_{\mathrm{R}}\left(\bar{\mathbf{C}}^{-1}\right)_0\right\rangle = \text{const.}$ and $\langle k_{\mathrm{R}}\rangle = \text{const.}$ – (32) can be solved as

$$\mathbf{C}_1^{-1}(t) = \frac{\left\langle k_{\mathrm{R}}\left(\bar{\mathbf{C}}^{-1}\right)_0\right\rangle}{\langle k_{\mathrm{R}}\rangle} + \\ + \left(\mathbf{C}_1^{-1}(0) - \frac{\left\langle k_{\mathrm{R}}\left(\bar{\mathbf{C}}^{-1}\right)_0\right\rangle}{\langle k_{\mathrm{R}}\rangle}\right) \exp(-\langle k_{\mathrm{R}}\rangle t)\,. \tag{37}$$

Comparing the structure of this solution to that of (15), their equivalence can be seen with $N_{C1} = 1$ and

$$\mathbf{A}_{C1} = \mathbf{C}_1^{-1}(0) - \frac{\left\langle k_{\mathrm{R}}\left(\bar{\mathbf{C}}^{-1}\right)_0\right\rangle}{\langle k_{\mathrm{R}}\rangle}\,, \tag{38}$$

$$B_{C1} = -\langle k_{\mathrm{R}}\rangle\,. \tag{39}$$

Similarly, the time history of the ageing variable μ might be expressed as

$$\mu(t) = \mu(0)\exp(\langle k_{\mathrm{R}}\rangle t)\,, \tag{40}$$

which results in the following set of parameters:

$$A_{\mu 1} = 1\,, \quad B_{\mu 1} = \langle k_{\mathrm{R}}\rangle\,. \tag{41}$$

5 COMPARISON TO EXPERIMENTAL DATA

5.1 *Time domain*

The data of (Heczko & Kottner 2015) are presented here and the simplified model of Sec. 3 is calibrated to them. Cylindrical specimens were loaded in compression. The loading history consisted of repeating two blocks:

- Four quasi-static cycles between 0 and 25% nominal strain.
- 5000 cycles of harmonic displacement superposed to a static value. Different values of amplitude and mean value of displacement were tested – the loadcases/tests are denoted K01, K02, and K03.

The goal here is to fit the quasi-static stress-strain curves by the simplified model (12). Only the approximation (13)-(15) is presented here.

Figure 2 shows the calibrated parameters as a function of time. A generalized Yeoh model with two terms was chosen over three terms, since the latter led to non-unique calibration results, while the two-term model resulted in well-posed calibration problems and in the expected continuous time-dependencies of its parameters. Figure 3 shows a comparison of stress-strain curves at different stages of the fatigue process. Parameters of (13)-(15) are shown in Tab. 2 The key observation here is that the exponents p_1 and p_2 are not constant, which is in contrast with (11).

5.2 *Multiaxial loading*

In this section, the assumption $W = W(I_1)$ is tested using stress-strain curves obtained in different deformation modes. Each deformation mode corresponds to a trajectory in the I_1-I_2 plane, therefore if I_1

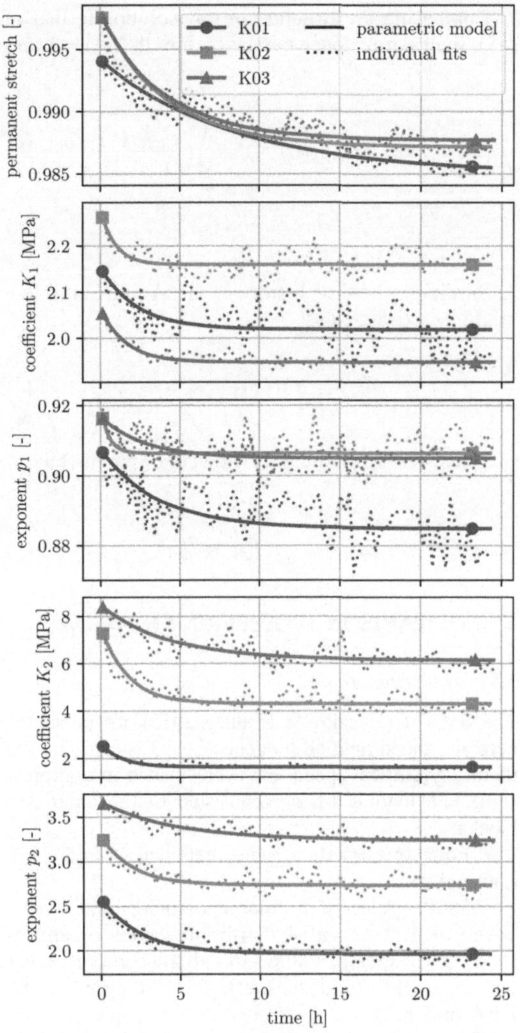

Figure 2. Parameters of the simplified model (13)-(15) calibrated to high-cycle fatigue data.

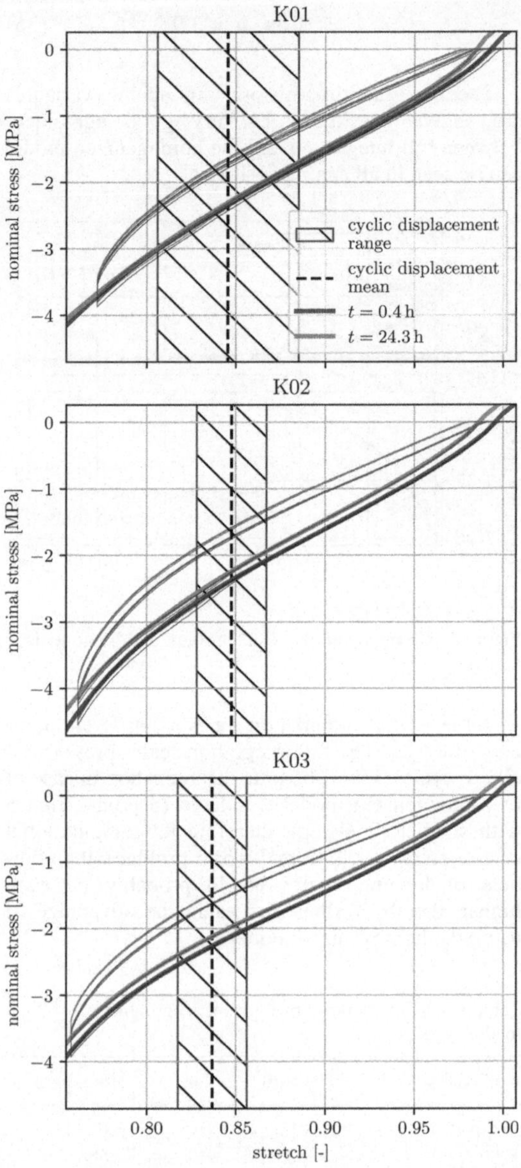

Figure 3. Comparison of stress-strain curves – simplified model and experimental data.

is fixed and the values of W are compared at different values of I_2 one should be able to decide whether $W = W(I_1)$ or if a more general function should be chosen.

Figure 4 shows such comparison for a single rubber compound in its original state and in the state after approximately one year in service. For the new material, the results are somewhat inconclusive, especially due to large data spread and perhaps some systematic error in shear experiments (these are the middle column in each group of symbols). The tension and compression data (left and right column of symbols in each group), however still suggest a slightly decreasing tendency. Contrary to that, the used material clearly shows increasing tendency of W with respect to I_2. As a result, a more general form of W might be necessary if multiaxial loading were of interest.

6 CONCLUSIONS

The exponential approximation is capable of exact description of stress-strain relations as a function of time under constant loading conditions. The

Table 2. Calibrated values of the approximation (13)-(15).

	K01	K02	K03
A_{K1}	$6.20 \cdot 10^{-2}$	$5.16 \cdot 10^{-2}$	$5.57 \cdot 10^{-2}$
B_{K1}	$-1.18 \cdot 10^{-4}$	$-2.95 \cdot 10^{-4}$	$-1.70 \cdot 10^{-4}$
A_{p1}	$2.50 \cdot 10^{-2}$	$1.39 \cdot 10^{-2}$	$1.28 \cdot 10^{-2}$
B_{p1}	$-7.61 \cdot 10^{-5}$	$-5.36 \cdot 10^{-4}$	$-9.89 \cdot 10^{-5}$
A_{K2}	$3.71 \cdot 10^{-1}$	$4.28 \cdot 10^{-1}$	$2.75 \cdot 10^{-1}$
B_{K2}	$-2.10 \cdot 10^{-4}$	$-1.66 \cdot 10^{-4}$	$-6.38 \cdot 10^{-5}$
A_{p2}	$2.39 \cdot 10^{-1}$	$1.64 \cdot 10^{-1}$	$1.16 \cdot 10^{-1}$
B_{p2}	$-1.12 \cdot 10^{-4}$	$-1.40 \cdot 10^{-4}$	$-5.89 \cdot 10^{-5}$
$[\mathbf{A}_{C1}]_{11}$	$-1.87 \cdot 10^{-2}$	$-2.19 \cdot 10^{-2}$	$-2.07 \cdot 10^{-2}$
B_{C1}	$-3.28 \cdot 10^{-5}$	$-6.72 \cdot 10^{-5}$	$-6.87 \cdot 10^{-5}$

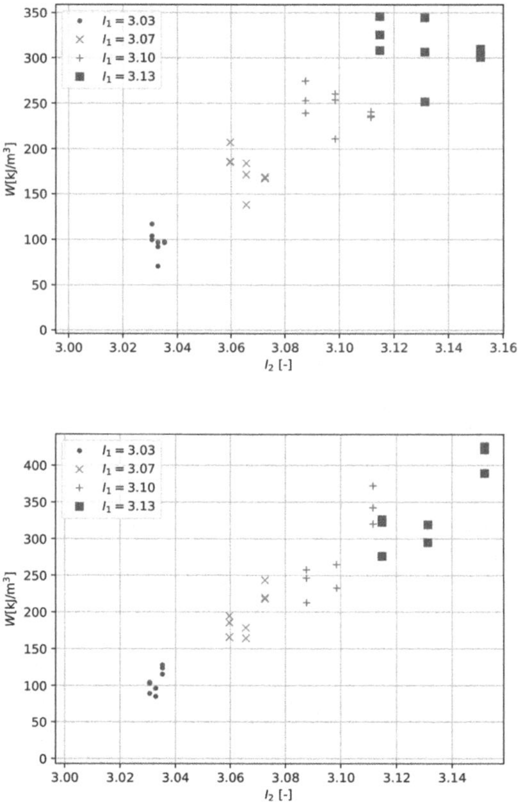

Figure 4. Strain energy density as a function of strain invariants I_1 and I_2 evaluated from experimental data. Top: new material, bottom: the same material used for one year in service. Uniaxial tension, compression and simple shear data were used – different values of I_2 correspond to different deformation modes while I_1 is fixed for each group of symbols.

homogenized evolution equations for the ageing and damage variables enable efficient numerical simulations of combined ageing and fatigue under various service conditions. The efficiency is achieved both by allowing for large time steps and by decoupling the governing equations. Despite only 1D problems have been shown, the approach is general and can be easily modified to include diffusion effects. The comparison of the homogenized model to experimental data revealed some of its possible modifications, such as exponents of the strain energy density function being time-independent or its possible enhancement by I_2-dependence. These will be addressed in future work.

REFERENCES

Aubry, D. & G. Puel (2014). Two-timescale homogenization method for the modeling of material fatigue. Volume 10. Institute of Physics Publishing.

Heczko, J. (2022). Numerical modelling of oxidative ageing in deformed state. *Constitutive Models for Rubber XII – Proceedings of the 12th European Conference on Constitutive Models for Rubber, 2022*, 452–456.

Heczko, J. & R. Kottner (2015). Modelling of changes inmechanical properties of rubber under cyclic loading. *Constitutive Models for Rubber IX - Proceedings of the 9th European Conference on Constitutive Models for Rubbers, ECCMR*, 411–415.

Heczko, J. & R. Kottner (2018). Modeling of material damage using finite elements and time homogenization in case of finite strain. *Applied Mathematics and Computation 319*, 264–273.

Heczko, J., R. Kottner, & A. Kossa (2021). Rubber ageing at elevated temperature – model calibration. *European Journal of Mechanics - A/Solids 89*(May), 104320.

Hohenberger, T. W., R. J. Windslow, N. M. Pugno, & J. J. Busfield (2019). A strain-energy function to model low and high strain non-linearities in highly filled elastomers. *Constitutive Models for Rubber XI - Proceedings of the 11th European Conference on Constitutive Models for Rubber, 2019*, 273–278.

Li, Y., G. Zhao, Y. Ma, & R. Liu (2022). Effect of alternation of aging and seawater erosion on properties of rubber material used in lead rubber bearing. *Journal of Renewable Materials 10*(6), 1641–1658.

Naumann, C. & J. Ihlemann (2015). A dynamic network model to simulate chemical aging processes in elastomers. *Constitutive Models for Rubber IX - Proceedings of the 9th European Conference on Constitutive Models for Rubbers, ECCMR*, 39–46.

Oskay, C. & J. Fish (2004, 9). Fatigue life prediction using 2-scale temporal asymptotic homogenization. *International Journal for Numerical Methods in Engineering 61*, 329–359.

Constitutive Models for Rubber XIII – Dal (Ed.)
© 2025 The Author(s), ISBN 978-1-032-85138-9

Energy flow in a filled rubber isolator at physical room temperature ageing

L. Kari & M. Vizcaíno-Vergara
KTH Royal Institute of Technology, Stockholm, Sweden

M. Vizcaíno-Vergara & J.J.C. Busfield
Queen Mary University London, London, UK

ABSTRACT: This study explores the physical ageing of rubber, traditionally associated with temperature changes, with a focus on room temperature effects, particularly in carbon black-filled rubber. Contrary to previous beliefs, experiments reveal that it takes up to four days for the shear modulus to equilibrate at room temperature. The findings have significant implications for the mechanical behavior of systems containing such rubber components during temperature variations. The paper concentrates on modeling torsional energy flow in filled rubber isolator subjected to a temperature shift at room temperature. The constitutive model incorporates a novel approach considering contributions from free volume and configurational changes to represent the impact of physical ageing on the shear modulus. The resulting torsional energy flow exhibits shifts in levels, peak frequencies and trough frequencies across the audible frequency range, persisting even after temperature stabilization. This highlights the crucial role of physical ageing in the design of vibration isolation systems.

1 INTRODUCTION

Rubber undergoes physical ageing, a natural progression wherein its molecular structure, initially disturbed from thermodynamic equilibrium, gradually returns to its equilibrium state (Cangialosi et al. 2013). This process involves various physical transformations such as changes in volume, entrapment and release of polymer chain ends and reorganization of polymer chains. Appropriate thermodynamic parameters for describing these processes encompass free volume and enthalpy (Odegard & Bandyopadhyay 2011, Cangialosi et al. 2013). During temperature changes, free volume initially disrupted from thermodynamic equilibrium naturally reverts to its equilibrium state across various temperature ranges, including glassy, transition and rubbery phases. In the rubbery phase, this reversion is rapid due to short relaxation times of molecular processes, leading to swift equilibrium attainment. Conversely, in transition and glassy phases, the process is notably slower due to longer relaxation times, resulting in gradual equilibrium achievement (Greiner, R. & Schwarzl 1984, Odegard & Bandyopadhyay 2011, Cangialosi et al. 2013, Kari 2017a,b). Physical ageing time denotes the duration required to attain thermodynamic equilibrium and is contingent upon factors such as temperature, thermal history, encompassing other thermodynamic records and the molecular structure (Odegard and Bandyopadhyay 2011, Cangialosi et al. 2013).

Recent studies have revealed that significant physical ageing can occur considerably above the glass transition region (Tunnicliffe, L.B. & Busfield 2016, Vizcaíno-Vergara et al. 2019, 2022, 2023, 2024).

This study utilizes a recently developed physical ageing model (Vizcaíno-Vergara et al. 2024), which incorporates constitutive fractional derivative order evolution alongside free volume evolution, to simulate torsional energy flow in a filled rubber isolator system experiencing a temperature shift at room temperature.

2 METHOD

2.1 *Torsional vibration isolator*

Imagine the torsional vibration isolator depicted in Figure 1, characterized by its length L, radius R and density ρ_0 at temperature T_0. Crafted from filled rubber, it is stimulated at one end by angular displacement

$$u_\theta = \varphi e^{i2\pi f}, \tag{1}$$

while remaining fixed at the opposite end ($z = L$). The torsional driving point and torsional transfer stiffness read

DOI: 10.1201/9781003516880-45

$$k_{\mathrm{D}} = \frac{M_\theta(z=0)}{u_\theta(z=0)} = kR^4\mu \frac{\pi}{2\tan(kL)} \tag{2}$$

and

$$k_{\mathrm{T}} = \frac{M_\theta(z=L)}{u_\theta(z=0)} = kR^4\mu \frac{\pi}{2\sin(kL)}, \tag{3}$$

respectively, where the torsional wave number

$$k = \frac{2\pi f}{c}, \tag{4}$$

the wave velocity

$$c = \sqrt{\frac{\mu}{\rho}}, \tag{5}$$

M_θ, i and μ are the torsional moment, imaginary unit and the shear modulus, respectively.

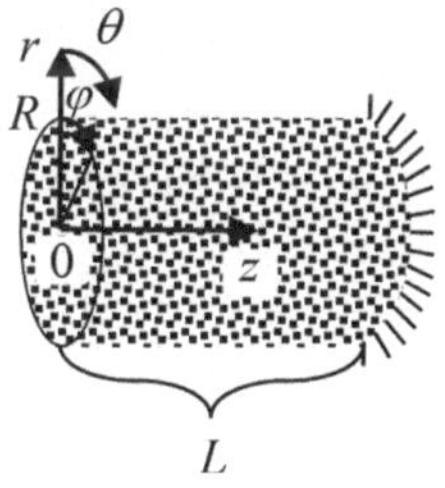

Figure 1. Torsional vibration isolator.

2.2 *Physical ageing*

The fractional differential evolution equation from Vizcaíno-Vergara et al. (2024), at a physical ageing time t_{ph} for fractional free volume $f_{\mathrm{T}} = (v_T - v_{\mathrm{cry}T})/v_{\infty T}$ and at a temperature T, reads

$$\tau_T^{\alpha_{t\mathrm{ph}}} D^{\alpha_{t\mathrm{ph}}}[f_T] = \mathrm{sgn}[f_{\infty T} - f_T]\,|f_{\infty T} - f_T|^{\gamma}, \tag{6}$$

where v_T, $v_{\mathrm{cry}T}$, $v_{\infty T}$ and $f_{\infty \mathrm{T}}$ are the specific volume, specific volume at crystalline state, equilibrium specific volume and the equilibrium fractional free volume, respectively at temperature T; τ_T is the relaxation time, $0 < \alpha_{t\mathrm{ph}} < 1$ is the Caputo fractional derivative order, which depends on the physical time t_{ph}, $\gamma > 1$ is a time stretching parameter and sgn is the sign function.

It is possible to show (Kari 2017a, Vizcaíno-Vergara et al. 2024) that the shear modulus at temperature T reads

$$\mu_T = \frac{T_0+\Delta T}{T_0}\frac{(1-\beta\Delta T)\mu_{\infty 0}}{1+f_T(t_{\mathrm{ph}})-\Delta\beta(A_2+\Delta T)} \times \left\{ 1 + \frac{\Delta[\mathrm{i}2\pi f\mu_\vartheta]^{\alpha_{t\mathrm{ph}}}\, 10^{\alpha_{t\mathrm{ph}} A_1 \left[\frac{A_2\Delta\beta}{f_T(t_{\mathrm{ph}})} - 1\right]}}{\Delta + [\mathrm{i}2\pi f\mu_\vartheta]^{\alpha_{t\mathrm{ph}}}\, 10^{\alpha_{t\mathrm{ph}} A_1 \left[\frac{A_2\Delta\beta}{f_T(t_{\mathrm{ph}})} - 1\right]}} \right\}, \tag{7}$$

during the physical ageing, after a temperature shift of $\Delta T = T - T_0$, where $\mu_{\infty 0}$ is the equilibrium shear modulus at temperature T_0, Δ is a stress relaxation intensity, f is the frequency, β and β_{cry} are the thermal volume expansion coefficients of rubber at the rubbery and crystalline state, respectively, $\Delta\beta = \beta - \beta_{\mathrm{cry}}$; μ_ν is a material parameter and, finally, A_1 and A_2 are the constants in the William–Landel–Ferry (WLF) shift function. The evolution of shear modulus with physical ageing time (7) incorporates the evolution of fractional free volume and of the fractional derivative order. The latter facilitates ongoing physical ageing despite the stabilization of fractional free volume, considering potential configurational changes.

2.3 *Energy flow*

Consider the torsional vibration source in Figure 2, conceptualized as a rigid cylindrical steel mass with length L_{s}, radius R_{s} and density ρ_{s}, possessing a torsional moment inertia $J_{\mathrm{s}} = \pi\, L_{\mathrm{s}}\, \rho_{\mathrm{s}}\, R_{\mathrm{s}}^4/2$. It is subjected to a torsional moment M_{s}, inducing an angular displacement φ_{s}. The time averaged energy flow into the vibration isolation system $P_{\mathrm{in}} = \mathrm{Re}[M_{\mathrm{s}}\, \mathrm{conj}(2\pi\, \mathrm{i}\, f\, \varphi_{\mathrm{s}})]/2$ and out of the system $P_{\mathrm{out}} = \mathrm{Re}[(k_{\mathrm{T}}\, \varphi_{\mathrm{s}} - k_{\mathrm{D}}\, \varphi(z = L))\, \mathrm{conj}(2\pi\, \mathrm{i}\, f\, \varphi_{\mathrm{s}}\, (z = L))]/2$, respectively, where Re and conj denote the real part and conjugate, respectively. The semi-infinite cylinder is made of steel with density ρ_{s} and shear modulus μ_{s}.

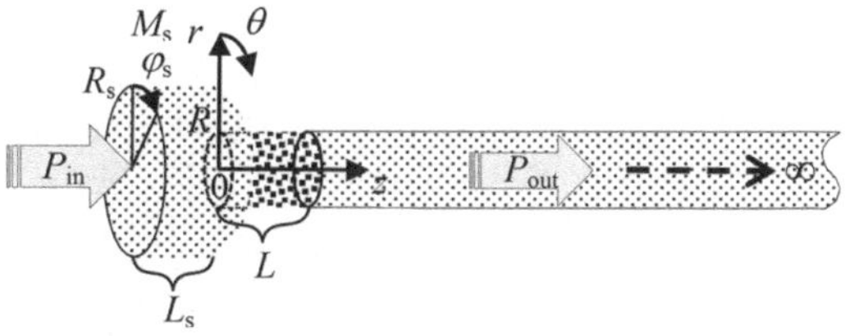

Figure 2. Torsional energy flow from a torsional source via a filled vibration isolator into a semi-infinite cylinder made of steel.

3 RESULTS AND DISCUSSIONS

3.1 *Temperature change and model parameters*

The vibration isolator is assumed to undergo a temperature shift from T_0 = 313K (+40°C) to T = 283K (+10°C); that is, a shift through the room temperature range.

This study utilizes a recently developed physical ageing model (Vizcaíno-Vergara et al. 2024), which incorporates constitutive fractional derivative order evolution alongside free volume evolution. The evolution of the former is supposed to read

$$\alpha_{t\text{ph}} = \alpha_\infty + \{h[t_{\text{ph}} - t_{\text{start}}] - h[t_{\text{ph}} - t_{\max}]\}\{k_{\text{up}}\log_{10}(t_{\text{ph}}) + c_{\text{up}}\} + \{h[t_{\text{ph}} - t_{\max}] - h[t_{\text{ph}} - t_{\text{stop}}]\}\{k_{\text{down}}\log_{10}(t_{\text{ph}}) + c_{\text{down}}\}, \tag{8}$$

where α_∞ is the equilibrium fractional derivative order; t_{start} and t_{stop} are the starting and stopping times for the change of the fractional derivative order, respectively, where α is reaching its maximum $\alpha_\infty + \Delta\alpha$ at $t_{\max}$: $t_{\text{start}} < t_{\max} < t_{\text{stop}}$ and $\Delta\alpha > 0$. The coefficients

$$k_{\text{up}} = \frac{\Delta\alpha}{\log_{10}(t_{\max}) - \log_{10}(t_{\text{start}})} \tag{9}$$

and

$$k_{\text{down}} = \frac{\Delta\alpha}{\log_{10}(t_{\max}) - \log_{10}(t_{\text{stop}})}, \tag{10}$$

while the parameters

$$c_{\text{up}} = \alpha_\infty + \Delta\alpha - k_{\text{up}}\log_{10}(t_{\max}) \tag{11}$$

and

$$c_{\text{down}} = \alpha_\infty + \Delta\alpha - k_{\text{down}}\log_{10}(t_{\max}). \tag{12}$$

The evolution satisfies the requirements $\alpha_{t\text{ph}} = \alpha_\infty$ as $t_{\text{ph}} \to 0$ and ∞. Additionally, it exhibits a max(α) > α_∞ consistent with the measurement outcomes (Vizcaíno-Vergara et al. 2024).

The model parameters and their values are given in Table 1. The main rubber parameters are from Kari et al (2001) with values shifted to the reference temperature T_0 = 313K. However, the rubber treated in present paper contains reinforcing filler while that in Kari et al (2001) is unfilled. The reader is referred to Kari et al. (2001) for specifics on the measurements, optimization procedures and the rubber material utilized.

3.2 *Physical ageing evolution*

The resulting evolution of the fractional free volume is shown in Figure 3 versus a wide spectrum of physical ageing times, ranging from 10^{-13} s, comparable to the period time of atom-to-atom oscillations, to 10^5 s, slightly exceeding a day. The solid curve ($f_{\gamma=2}$) is derived from the values specified in Table 1, whereas the dashed curve ($f_{\gamma=1}$) assumes a specific value γ = 1 while keeping the other values the same. Similarly, the dash-dotted curve ($f_{\Delta\alpha=0}$) assumes a constant fractional derivative order at its specific equilibrium value with $\Delta\alpha$ = 0 while keeping the other values the same. For clarity, the fractional derivative order evolution, applying the values specified in Table 1, is shown as a solid curve while the equilibrium fractional derivative order is shown as a diamond-dashed line in the same Figure 3. The former curve is displaying a triangle shape within the considered logarithmic physical ageing time range while the latter curve is constant.

Table 1. Model parameters and their values.

Parameter	Value	Parameter	Value
T_0	313 K	t_{start}	100 fs
T	283 K	$t_{\max}$	100 s
R	10.0 mm	t_{stop}	100 ks
L	30.0 mm	$\mu_{\infty 0}$	858 kN/m^2
ρ_0	974 kg/m^3	μ_v	4.47 µs/m^2
R_s	50.0 mm	△	276
L_s	50.0 mm	γ	2.00
ρ_s	7800 kg/m^3	B	660 µK^{-1}
μ_s	78.5 GN/m^2	$\Delta\beta$	440 µK^{-1}
α_∞	0.657	A_1	5.41
$\Delta\alpha$	0.100	A_2	167 K

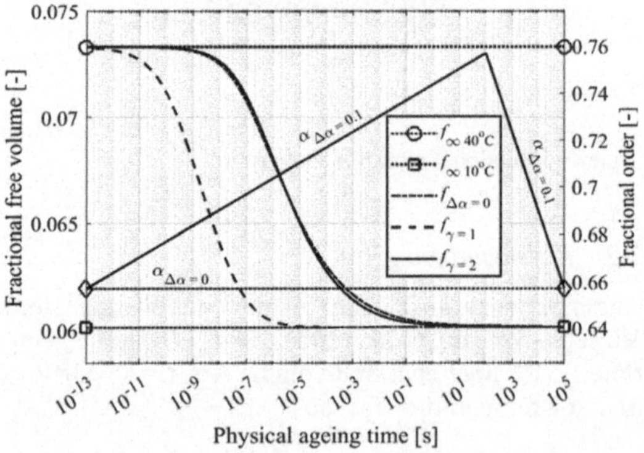

Figure 3. Evolution of the fractional free volume (left *y*-axis) and fractional order derivative (right *y*-axis) versus physical ageing time (*x*-axis).

Clearly, the fractional free volume, disrupted from thermodynamic equilibrium by a sudden temperature change, naturally moves towards restoring its equilibrium state of fractional free volume. At the physical ageing time 10^{-13} s, the primary fractional free volume ($f_{\gamma=2}$) remains in practice unchanged from its equilibrium at 40°C, despite the temperature being adjusted to 10°C. At the physical ageing time 10^{-10} s, this fractional free volume begins to undergo a more pronounced change, experiencing a sharp decline at 10^{-6}s, plateauing and stabilizing from 10^2 s. It should be noted that even after the fractional free volume has stabilized at its equilibrium, the evolution of fractional derivative orders persists for decades.

Essentially, the fractional free volume corresponding to $\gamma = 1$ is shifted to substantially lower physical times as compared to the primary curve; it begins to undergo the pronounced change already at 10^{-13} s and is stabilizing already from 10^{-4} s.

Finally, the fractional free volume corresponding to $\Delta\alpha = 0$ shows a slightly earlier pronounced change, a slightly later plateauing and stabilizing while displaying a slightly less sharp decline at 10^{-6} s.

3.3 *Shear modulus evolution*

The evolutions of both the magnitude of shear modulus $|\mu|$ and the loss factor Im μ/Reμ at $T_{\text{ph}} = 0^+$, 10^{-7}, 10^{-1} and 10^5 s while using the parameter values specified in Table 1, where Im is imaginary part, are depicted in Figure 4 across the entire audible frequency range, spanning from 20 to 20 000 Hz.

Evidently, the shear modulus exhibits a significant dependency on both the frequency and physical ageing time, even after the fractional free volume has attained equilibrium. The primary explanation for the latter is that the evolution of fractional derivative order continues for decades even after the fractional free volume has reached its equilibrium.

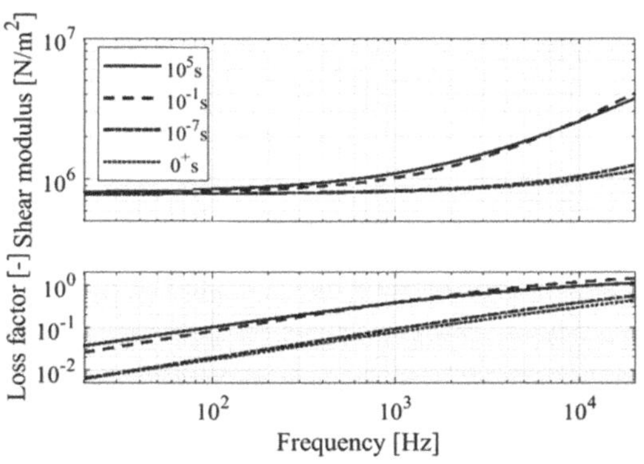

Figure 4. Evolutions of the magnitude and loss factor of the shear modulus versus frequency.

3.4 *Torsional energy flow evolution*

The evolutions of the energy flow transmissibility into the vibration isolation system $T_{\text{in}} = P_{\text{in}} / P_{\text{no isolator}}$ and out of the system $T_{\text{out}} = P_{\text{out}} / P_{\text{no isolator}}$ at $t_{\text{ph}} = 0^+$, 10^{-7}, 10^{-1} and 10^5 s are depicted in Figure 4 across the entire audible frequency range, while using the parameter values specified in Table 1, where $P_{\text{no isolator}}$ is the time averaged energy flow into (and out of) the system without vibration isolator and $T_{\text{no isolator}} \equiv 1$.

It is evident that the energy flow transmissibilities demonstrate a notable dependence on both frequency and physical ageing time, even after the fractional free volume has reached equilibrium. Specifically, there is a noticeable difference in the energy flow transmissibity between the physical ageing times of 10^{-1} and 10^5 s. These differences correlate with variations in both the magnitude and loss factor of the shear modulus at those specific physical ageing times, as illustrated in Figure 4. Moreover, the energy flow transmissibilities exhibit peaks and troughs, attributed to internal resonances occurring within the vibration isolator. At the longer physical ageing times (10^{-1} and 10^5 s), the peaks and troughs diminish due to the heightened loss factor, as depicted in Figure 4.

Specifically, the discernible peaks below 3000 Hz in Figure 5 are shifted towards higher frequencies for the physical ageing time of 10^5 s compared to the time of 10^{-1} s. This shift is attributed to an increased shear modulus magnitude, as illustrated in Figure 4. Finally, there exists a significant distinction between the energy flow transmissibilty into the vibration isolator system and that flowing out of it, primarily attributed to the conversion of vibration energy into heat within the vibration isolator.

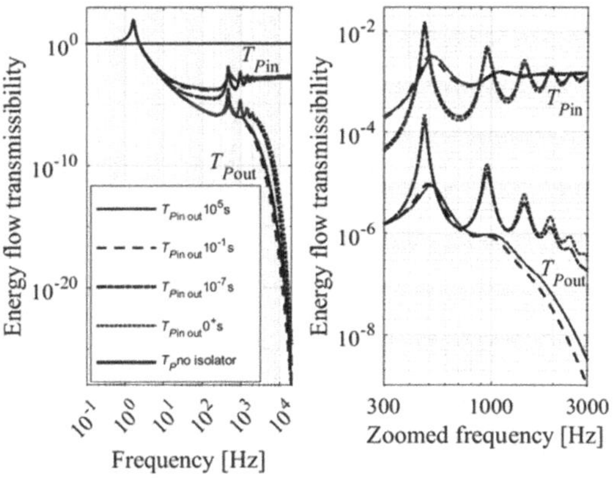

Figure 5. Evolutions of the torsional energy flow transmissibility into and out of the vibration isolation system versus frequency.

REFERENCES

Cangialosi, D., Boucher, V.M., Alegría, A., Colmenero, J. 2013. Physical aging in polymers and polymer nanocomposites: recent results and open questions. *Soft Matter* 9: 8619–8630.

Greiner, R. & Schwarzl, F.R. 1984. Thermal contraction and volume relaxation of amorphous polymers. *Rheologica Acta* 23: 378–395.

Kari, L. 2017a. Dynamic stiffness of chemically and physically ageing rubber vibration isolators in the audible frequency range. Part 1: constitutive equations. *Continuum Mechanics and Thermodynamics* 29: 1027–1046.

Kari, L. 2017b. Dynamic stiffness of chemically and physically ageing rubber vibration isolators in the audible frequency range. Part 2–waveguide solution. *Continuum Mechanics and Thermodynamics* 29: 1047–1049.

Kari, L., Eriksson, P. & Stenberg, B. 2001. Dynamic stiffness of natural rubber cylinders in the audible frequency range using wave guides. *Kautschuk Gummi Kunststoffe* 54: 106–113.

Odegard, G.M. & Bandyopadhyay, A. 2011. Physical aging of epoxy polymers and their composites. *Journal of Polymer Science. Part B: Polymer Physics* 49: 1695–1716.

Tunnicliffe, L.B. & Busfield, J.J.C. 2016. Reinforcement of rubber and filler network dynamics at small strains. In Stöckelhuber, K., Das, A. & Klüppel, M. (eds), *Designing of elastomer nanocomposites: from theory to applications. Advances in Polymer Science* 275. Cham: Springer.

Vizcaíno-Vergara, M. Kari, L. & Busfield, J.J.C. 2019. Free volume evolution equation for physical ageing of filler reinforced rubber. *Constitutive Models for Rubber XI*: 548–553. Boca Raton: CRC Press.

Vizcaíno-Vergara, M. L., Tunnicliffe, L.B., Busfield, J.J.C. & Kari, L. 2022. Viscoelastic characterisation of carbon black reinforced rubber using the torsion pendulum: Guidelines and temperature sweep results. *Polymer Testing* 107692.

Vizcaíno-Vergara, M. L., Kari, L., Tunnicliffe, L.B. & Busfield, J.J.C. 2023. Evolution of the viscoelastic properties of filler reinforced rubber under physical aging at room temperature. *Polymers* 15: 1806.

Vizcaíno-Vergara, M. Kari, L. & Busfield, J.J.C. 2024. Extended constitutive model for physical ageing of filler reinforced rubber. *To be submitted to an international scientific journal.*

Constitutive Models for Rubber XIII – Dal (Ed.)
 ISBN 978-1-032-85138-9

Evolution of crack depth in natural rubber due to ozone loading

C. Treib & M. Johlitz
Institute for Mechanics, Department of Aerospace Engineering, University of the Bundeswehr, Munich, Germany

ABSTRACT: Natural rubber degrades due to environmental influences: thermal ageing, oxidation, or other chemical change results in hardening or softening effects in the surface or bulk of components. Besides oxygen, ozone is a powerful aggressor attacking the double bonds present in the rubber molecule. Even at low concentrations to be found at ground level, cracking is inevitable despite using chemical or physical antiozonants. Most common are diamines and wax which both show limited benefits depending on the application temperature, friction and simultaneous mechanical strain. Thus, this contribution focuses on carbon-black filled natural rubber without antiozonants. Size and depth of cracks limit the lifetime of components. Therefore, predicting the maximal crack depth is crucial for components configuration and replacement schedule. An artificial ageing procedure is developed to simulate and accelerate the natural degradation process. Temperature is used to separate ozone ageing from thermo-oxidative degradation. Ageing parameters range up to 75 pphm ozone concentration for maximum 111 hours, and static and cyclic deformation during the ageing process are compared. The resulting crack depth is analysed in the cross-section, revealed by punching samples after the ageing process. Subsequently, the quantitative results are used to fit a proposed evolution equation to predict the crack depth after specific ageing states. It regards the varying parameters: ageing time, ozone concentration, deformation type and level. The overall fit concludes the cracks do not stagnate but their growth is slowing down with ageing time.

1 INTRODUCTION

With above 11 million metric tons of natural rubber (NR) consumed in 2022 (Malaysian Rubber Board 2023) the global consumption is rising. The natural resource is applied mostly in tyres besides isolators for bridges or below constructions exposed to earthquakes, bearings, sealings and hoses. Plenty sectors like construction, combustion or aviation industry rely on the rubber supply. As not yet sufficient alternatives of hevea brasiliensis are market-ready, the understanding of ageing processes is necessary for efficient replacement at predicted lifetimes.

Ozone causes cracks even at very low concentrations. Lewis (1986) states, cracks evolve within some days at only 1–3 pphm (parts per hundred million) ozone concentration in strained NR, a concentration which is observed in the environment by Lake & Mente (1992). Rodrigues, Santos, Feitosa, Ricardo, & Heatley (2004) confirm cracking due to comparable concentrations of ozone connected to the split of the double bond in backbone of NR. The electrophilic attack of the double bond by ozone and following reaction steps are analysed by Criegee (1975), called the *Criegee mechanism*, which is widely accepted (Christian Geletneky 1998, Salisbury et al. 2001, Cataldo 2019). Besides the low ozone level, strain is necessary for crack initiation and propagation, stronger than shallow crazing (Lewis 1986). Depending on the surface smoothness Braden & Gent (1960) state, lacking a threshold stress, no cracks evolve. However, the threshold of 3–5 % strain is met easily in any application. In addition, cyclic strains are common for NR parts and cause cracking even in NR compounds protected by antiozonants (Cataldo 2019, Huntink et al. 2004). The evolution with ageing time for thermo-oxidative ageing is often approximated by the Arrhenius' law describing a linear relation between the logarithmic ageing time and the material characteristic (Herzig 2020, Ehrenstein & Pongratz 2007). A different approach is developed by Dakin (1948) assuming a monomolecular reaction rate and connecting it to a measureable physical deterioration. The latter leads to an exponential function for the evolution of the reaction product with the reaction time stopping at total consumption depending on the initial concentration. The maximal crack depth approximated by the exponential formulation is limited to the one measured at the longest ageing time conducted, discovered in a former study (Treib, Johlitz, & Lion 2024). Therefore, an evolution equation exceeding the highest measured crack depth is necessary.

DOI: 10.1201/9781003516880-46

Regarding the evolution of cracks with increasing ozone concentration, several authors (Braden and Gent 1960, Spreckels et al. 2010) state a constant crack growth proportional to ozone concentration, using up to 50 pphm.

In most use cases for NR, cyclic deformation is present for which in combination with ozone loading both antiozonants fail to completely prevent cracking, as stated by Cataldo (2019) and Treib & Johlitz (2022). Consequently, in this contribution the crack evolution of unprotected rubber is analysed. For the crack depth seems not to approach a final maximum even without antiozonants applied, the evolution shall be reanalyzed for accelerated ageing in the lab. To accelerate the ageing process, the ozone concentration is increased compared to the ground-level concentration most applications face during lifetime. Simultaneously to the ozone concentration mechanical load is applied. Static strain is used of 20, 30 and 40 %, a cyclic amplitude of 10 % strain on top of 30 % static strain and a complete deloading in the cycle of 20 % strain amplitude on top of 20 % static strain. After the accelerated ageing process, the crack depth is analysed under an optical microscope. The thus obtained evolution of crack depth over time and over ozone concentration is fitted to generate a crack evolution equation. Extrapolating to lower ozone values, the crack depth is approximated for realistic load cases and the time necessary for equal crack depth in accelerated tests can be predicted.

2 EXPERIMENTAL

2.1 *Material*

The material used is a conventionally manufactured NR (CV 60) from hevea brasiliensis. It is sulphur vulcanized with zink oxide and stearic acid using the accelerator TBBS[1]. Besides rubber the main component is carbon black, reinforcing the material. The mixing ratio is given in Table 1.

Table 1. Natural rubber compound/phr[a].

NR CV 60	100
Carbon Black N330	40
Zinc Oxide 3	3
Stearic Acid	1
Accelerator TBBS	1
Sulphur	2

[a] parts per hundred rubber

2.2 *Ground-level ozone*

Most NR components are situated on ground level. Ground-level ozone concentrations are measured by governmental environmental stations as by Bundesumweltamt (2019). These meteorological stations publish 1 to 8-hour averages using a slow chemical measurement method. The unit used is µg/m^3 referring to barometric pressure at 20 °C. It can be converted in the volume fraction unit of pphm used in laboratory ozone analysers due to its pressure independence.

In general, fluctuations are strong between day and night as between winter and summer. Maximal values reached 300 µg/m^3 resp. 15 pphm in summer afternoons whereas the yearly average ranges from 40–60 µg/m^3 resp. 2–3 pphm. The average is slightly trending to higher values stated by Bundesumweltamt (2019). The influence of urban or rural area is measurable as well as the difference to e.g. the marine boundary layer, where Salisbury, Rickard, Monks, Allan,Bauguitte, Penkett, Carslaw, Lewis, Creasey, Heard, Jacobs, & Lee (2001) states a background level at nighttime of 3–4 pphm to be measured at the Mace Head Atmospheric Research Station in Ireland. Besides the location, air pollution with precursors to ozone creation can locally change the resulting ozone concentration stated by De, White, & Naskar (2001)and Kruza, Lewis, Morrison, & Carslaw (2017). Especially combustion and volatile organic compounds increase ozone concentrations in areas with smog and heat waves measured up to 16 pphm in China, 20 pphm in France, 40 pphm in southern (Kruza, Lewis, Morrison, & Carslaw 2017). Despite changes in ozone concentration, the average level will be used in Sec. 3.3 to relate laboratory tests to realistic outdoor loading.

2.3 *Artificial ageing*

Artificial ageing needs to be speed up to shorten testing times compared to components lifetime of several years. Regarding thermo-oxidative ageing mostly temperature is used to gain fast results and extrapolate them with laws like Arrhenius' (Neuhaus 2019). However, to analyse ozone ageing isolated from thermo or thermo-oxidative ageing the temperature is chosen not to exceed 40 °C. Thus, no strong acceleration of these ageing processes is ensured, as ageing times needed for temperatures below 60 °C span several weeks (Broudin et al. 2018).

The ozone concentration in a climate simulator is feasible with an accuracy of ± 5 pphm. Thus low concentrations of few pphm as on ground level are not reasonable for artificial ageing. With 25 pphm the lowest ozone concentration chosen as ageing parameter is already causing an acceleration in degradation. Using the ozone concentration as ageing

1 *N*-tert-Butyl-2-benzothiazolesulfenamide

accelerator is therefore inevitable but practical to not interfere with accelerated oxidative ageing due to the limit of 40 °C. Hence, ozone concentrations applied are 25 pphm, 36 pphm, 52 pphm, and 75 pphm regarding the accuracy of about 5 pphm in the laboratory device, *SIM7500-TH* by *Anseros Klaus Nonnenmacher GmbH.*

Ageing time is crucial to the ozone induced degradation process since cracks grow with ageing time under strain, stated by Braden & Gent (1962), Lewis (1986) and Lake (1970). The onset of cracking is stated after 3 hours at 50 pphm by Kamaruddin & Muhr (2018) for strained samples. For low concentrations of 1 – 3 pphm Lewis (1986) states ageing times of a few days for cracks to appear under tensile strain.

Braden & Gent (1962) state a threshold strain for cracks to appear about 3–5%. Mechanical strain is necessary for cracks to appear, otherwise only a surface layer is affected as Treib, Lion, & Johlitz (2022) conclude. Therefore, different strain levels of 20, 30 and 40 % are applied and a cyclic amplitude of 10 % strain on top of 30 % static strain as well as a cyclic amplitude of 20 % strain on top of 20 % static strain. The latter is chosen for Lindley (1973) states an facilitated tearing process for relaxation to the initial state during cyclic loading. The applied ageing parameters for artificial, accelerated ageing are listed in Table 2.

Table 2. Ageing parameters.

Temperature/°C	40			
Humidity/rH	50			
Ozone concentration/pphm	25	36	52	75
Ageing time/h	4	12	37	111
Deformation static/%	20	30	40	
Deformation cyclic/%	30±10	20±20		

2.4 *Crack detection*

After artificial ageing cracks are apparent on the surface. However, to reveal the crack depth in the cross-section, the samples are cut by punching stripes. They are only cut from the center part of the aged bar to ensure 1-dimensional ageing. For NR is somewhat tacky, the sample stripes are slightly strained to view open cracks under an optical microscope. The microscope *VHX-2000* by *KEYENCE* is used. Within the software of the microscope a calibrated length measurement is used to measure the crack depth of the samples, see Figure 1. To estimate the accuracy of the measurement 100 readings of one crack are taken, resulting in a standard deviation of 12.4 μm.

3 SIMULATION OF CRACK EVOLUTION

To fit an equation to the crack data over ageing time or over ozone concentration the data is normalized to the maximum at each concentration resp. time.

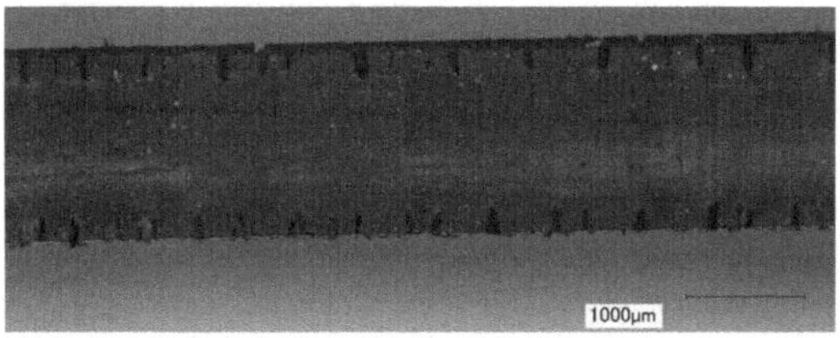

Figure 1. Opened cracks in the cross-section of a sample aged at 30(±10) % strain at 75 pphm for 37 h.

Even without normalization the difference between the static strain levels in not correlating to the strain levels, thus all static strains are considered together.

3.1 *Linear regression for crack evolution depending on ozone concentration*

The linear approach with ozone concentration holds up to the applied 52 pphm regarding ageing times longer than 12 h. Shorter ageing times include to much deviation due to the experimental process and are excluded from the linear regression on the normalized data depicted in Figure 2 for static strain during ageing, in Figure 3 for cyclic load of 30(±10) % strain and 20(±20) % strain.

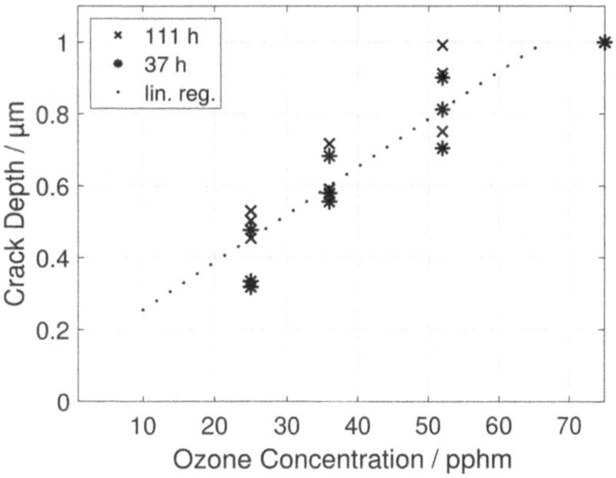

Figure 2. Normalized crack data and linear regression for 37 and 111 hours of static strained samples.

For the applied concentration of 75 pphm the crack evolution changes its rate for both static and cyclic loading. In conclusion due to raising the ozone concentration above 52 pphm, a change of the ageing process might take place. Accelerated lab tests should not exceed 50 pphm to ensure a linear extrapolation to lower ozone concentrations as well as exclude short ageing times.

3.2 *Formulation for crack evolution depending on ageing time*

Dakin (1948) assumes a monomolecular reaction to gain an approach for the time evolution of the deterioration of an insulation material. The proposed

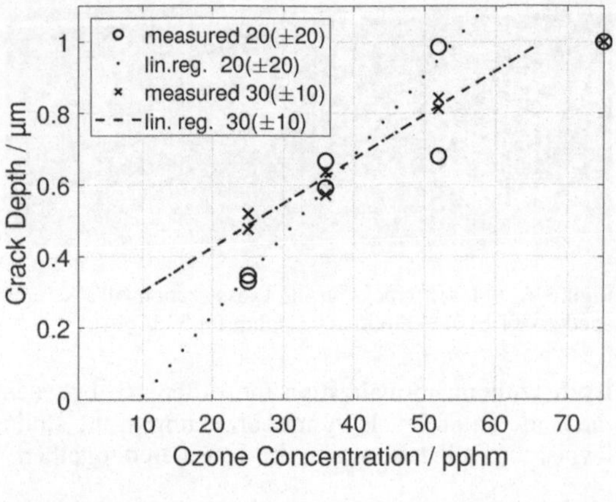

Figure 3. Normalized crack data and linear regression for samples with 30(±10) % and 20(±20) % strain during ageing.

kinetic approach leads to an exponential decrease of reactant concentration $c_a(t)$ with ageing time t, alike Equation (1):

$$c_a(t) = c_{ao} \cdot \exp\ (-k \cdot t), \tag{1}$$

depending on the reaction velocity k and the initial concentration of the reactant c_{ao}.

Though the requirement of a monomolecular reaction does not hold for ozonolysis, the resulting exponential reaction has been used as a first approach to model the growth of cracks in NR in Treib, Johlitz, & Lion (2024). The available number of double bonds is correlated to the concentration c_a that is reduced by the cleavages due to ozone. The counterpart of scissions is then correlated to the crack depth r_t:

$$r_t(t) = r_{max} \cdot (1 - \exp\ (-k \cdot t)), \tag{2}$$

with the maximum crack depth r_{max} to be reached correlating to the initial concentration of double bonds available.

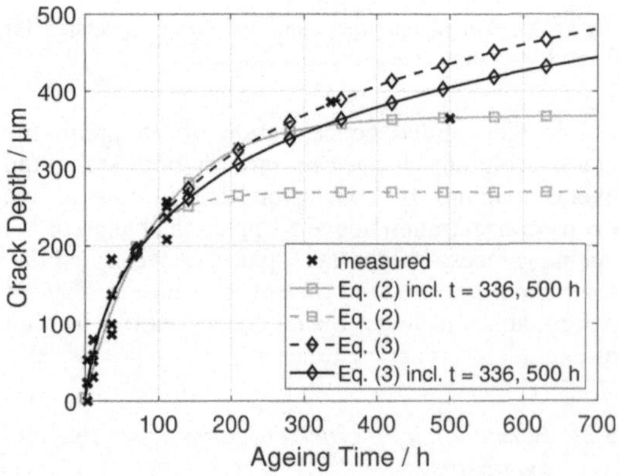

Figure 4. Crack data for 25 pphm with static strain and simulation of Equations (2) and (3) up to 111 h ageing time and including the crack depth after 336 and 500 h.

New data on crack depth after longer ozone ageing times at 25 pphm disprove the assumption of a stagnation in crack depth within 111 h ageing time assumed in Treib, Johlitz, & Lion (2024). A diffusion limit as in unstrained ozone-ageing might lead to a deceleration of crack growth velocity but not a complete stagnation of crack development. Thus, an evolution equation capable of depicting a crack depth rising beyond the crack depth of the longest ageing time conducted is necessary. Based on the assumption of a linear logarithmic correlation of time and ageing process by Ehrenstein & Pongratz (2007), a modified logarithmic equation is developed:

$$r_t(t) = D \cdot \ln\ (1 + k \cdot t)), \tag{3}$$

with a factor D on the overall crack depth and parameter k affecting the evolution with ageing time. The result is in good correlation with the measured data for 336 h and 500 h at 25 pphm even excluding these measurements, see Figure 4. In addition, the simulation is conservative in predicting deeper cracks when excluding longer ageing data which is not the case for the exponential function. Consequently, Equation (3) is chosen for the higher ozone concentrations to avoid underestimation of crack depths:

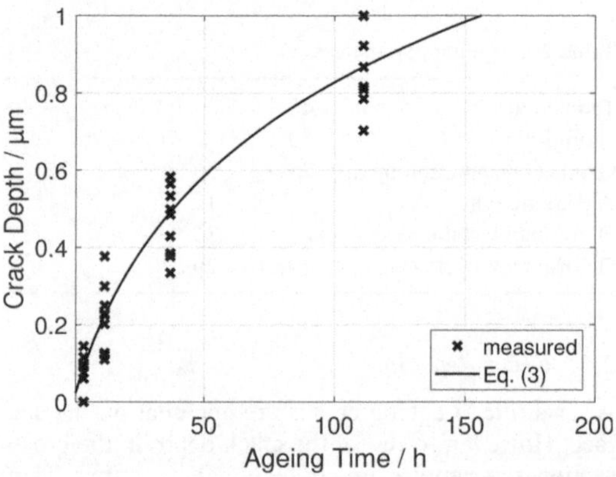

Figure 5. Normalized crack data for 25, 36, and 52 pphm with static strain and simulation of Equation (3) up to 111 h ageing time.

After normalizing the crack data of each concentration to its maximum crack depth, Equation (3) is used for simulation, see Figure 5. The concentration of 75 pphm is excluded for its nonlinear behaviour with ozone concentration. The simulation parameters are listed in Table 3. Likewise for cyclic loading during the ageing process, Equation (3) is capable of approximating the normalized crack data, see Figure 6. Comparing the different mechanical strains during ageing, its influence on the crack evolution besides the final cracking depth is apparent.

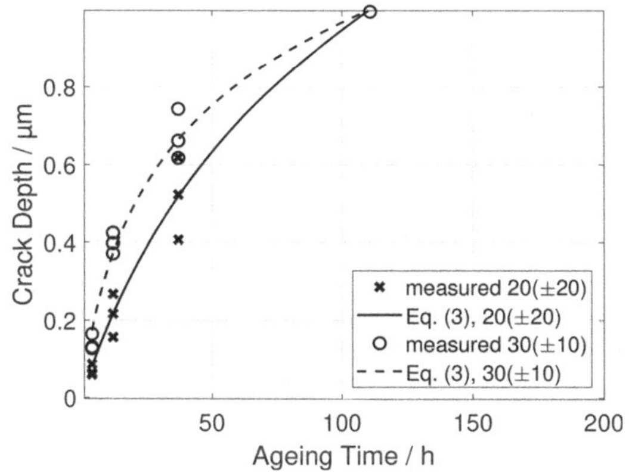

Figure 6. Normalized crack data for 25, 36, and 52 pphm with cyclic strain of 20(±20) % and simulation of Equation (3) up to 111 h ageing time.

Table 3. Simulation parameters for Equation 3 including ageing times up to 111 h for normalized crack depth at 25, 36, and 52 pphm.

Fig.	Mechanic Strain/%	D/µm	$k/\frac{1}{h}$
5	20, 30, 40	0.443	0.0548
6	30(±10)	0.335	0.1694
6	20(±20)	0.649	0.0331

3.3 *Combined evolution equation*

The logarithmic evolution over ageing time in Equation 3 is multiplied by the ozone concentration to add the linear dependency described in Sec. 3.1 resulting in:

$$r_t(t, c_{ozon}) = D \cdot c_{ozon} \cdot \ln\ (1 + k \cdot t), \qquad (4)$$

with crack depth r_t in µm, ageing time t in h, ozone concentration c_{ozon} in pphm, factor D in µm/pphm and velocity k in 1/h.

The evolution of the crack depth and the obtained simulation parameters to the experimental data are plotted in Figures 7, 8, and 9 and given in Table 4.

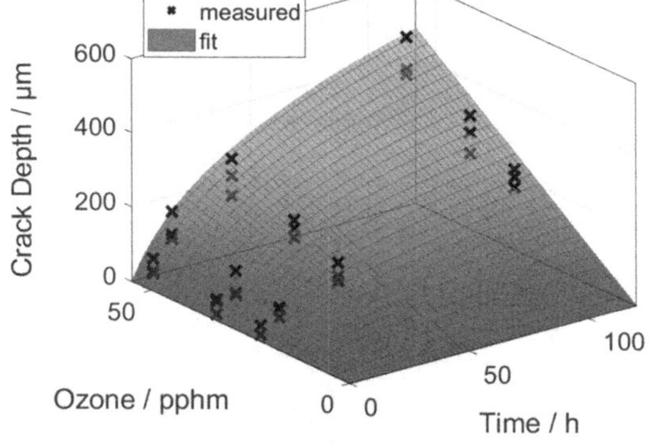

Figure 7. Crack data for 25, 36, and 52 pphm at static strain and simulation of Equation (4).

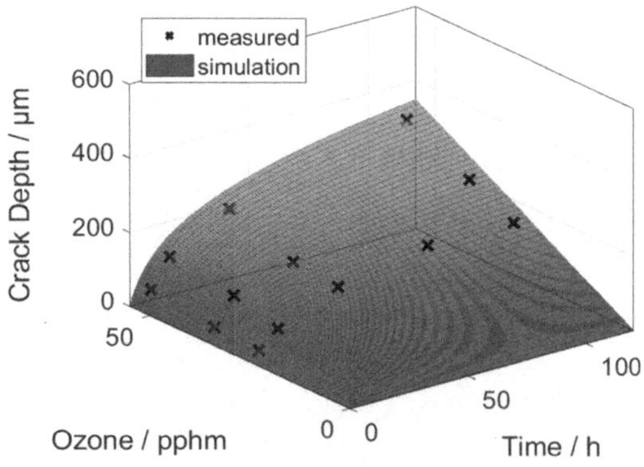

Figure 8. Crack data for 25, 36, and 52 pphm at cyclic strain of 30(±10) % and simulation of Equation (4).

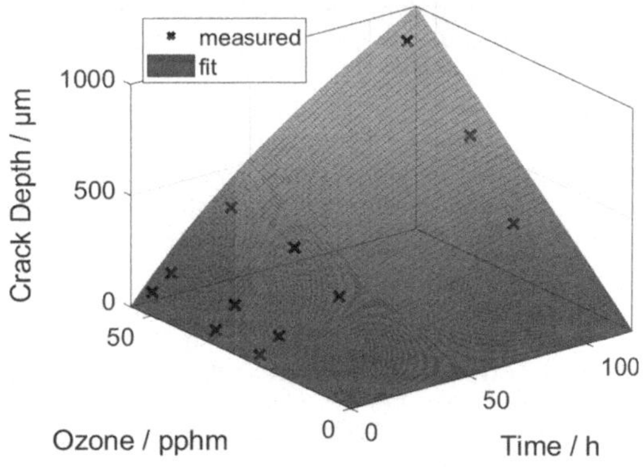

Figure 9. Crack data for 25, 36, and 52 pphm at cyclic strain of 20(±20) % and simulation of Equation (4).

As for the simulation over ageing time the parameters show high concordance, regardless weather crack depths after longer ageing times at 25 pphm are considered. Contrary, the amplitude of cyclic strain has a strong impact on the crack evolution.

Using the simulation result an approximation of crack depth for lower ozone concentrations is possible for each type of mechanical strain tested. For example 20 years of exposure at 3 pphm under static strain would lead to 112 µm crack depth but for cyclic strain of 30(±10) % only 61 µm is expected, whereas the cyclic strain of 20(±20) % including complete deloading in the cycle leads to deepest cracks of 479 µm.

With the relation between the accelerated lab test to real loading, a necessary testing time to simulate specific conditions accelerated can be drawn:

$$t_{lab} = \frac{\exp(\frac{c_{oz,real}}{c_{oz,lab}} \cdot \ln(1 + k \cdot t_{real})) - 1}{k} \qquad (5)$$

The exemplary condition of 3 pphm for 20 years under static strain would thus take 24.7 h at 30 pphm in the artificial ageing procedure.

Table 4. Simulation parameters for Equation (4).

Fig.	Strain/%	$D/\frac{\mu m}{pphm}$	$k/\frac{1}{h}$
7	20, 30, 40	4.0265	0.0621
including additional ageing times	20, 30, 40	3.8978	0.0685
8	30(±10)	1.9235	0.2163
9	20(±20)	21.01	0.0114

4 CONCLUSION AND OUTLOOK

Opposing former results, the crack depth does not completely stagnate with ageing time but the growth rate slows down tremendously for unprotected NR even under cyclic loading. Over time the developed modified linear logarithmic approach approximates all strain states tested during artificial ageing satisfactory considering the scatter in experimental data. Regarding the evolution of crack depth depending on ozone concentration, a maximum concentration of 52 pphm is recommended to obtain a linear relation. This limits the acceleration of the artificial ageing by increased ozone concentration as conducted in this study. Thus, a combination of the evolution over ageing time and ozone concentration is used to simulate the experimental crack depth values. A prediction equation for the crack progress is obtained and consequently applied to extrapolate the accelerated ageing tests to realistic ground-level ozone concentrations.

Additional parameters, including the mechanical strain applied in the evolution equation based on more experimental data is a favourable future goal. In addition, long term testing at ground-level conditions is necessary to validate the extrapolated cracking depth at lower ozone concentrations and confirm the linear relation in the extrapolated ranges. Since the ground-level ozone is not constant, ageing at different ozone concentrations applied consecutively is a next step to relate the accelerated ageing procedure to real loading. Finally, including antiozonants the evolution equation must be re-evaluated for the migration and consumption and blocking effects change the reaction process.

REFERENCES

Braden, M. & A. Gent (1960). The attack of ozone on stretched rubber vulcanizates. II. Conditions for cut growth. *J. Appl. Polym. Sci. 3*(7), 100–106.

Braden, M. & A. Gent (1962). The attack of ozone on stretched rubber vulcanizates. III. Actions of antiozonants. *J. Appl. Polym. Sci. 6*(22), 449–455.

Broudin, M., Y. Marco, V. LeSaux, P. Charrier, W. Hervouet, & P. Y. LeGac (2018). Investigation of thermooxidative ageing effects on the fatigue design of automotive anti-vibration parts. *MATEC Web of Conferences*.

Bundesumweltamt (2019, July). Aktuelle Luftdaten. www.umweltbundesamt.de/daten/luft/luftdaten/karten, Accessed 16 Nov 2021.

Cataldo, F. (2019). Protection mechanism of rubbers from ozone attack. Ozone: *Sci Eng 41*(4), 358–368.

Christian Geletneky, S. B. (1998). The mechanism of ozonolysis revisited by ^{17}o-nmr spectroscopy. *European Journal of Organic Chemistry*.

Criegee, R. (1975). Mechanism of ozonolysis. *Angew. Chem. Int. Edition 14 14*(11), 745–752.

Dakin, T. W. (1948). Electrical insulation deterioration treated as a chemical rate phenomenon. *Transactions of the American Institute of Electrical Engineers 67*(1), 113–122.

De, S. K., J. R. White, & K. Naskar (2001). *Rubber technologist's handbook, Volume 2*. Rapra Technology Limited.

Ehrenstein, G. & S. Pongratz (2007). Beständigkeit von Kunststoffen. Carl Hanser Verlag.

Herzig, A. (2020). *Thermo-oxidative ageing of elastomers. A contribution to the experimental investigation and modelling*. Ph. D. thesis, Universität der Bundeswehr München.

Huntink, N. M., R. N. Datta, & J. W. M. Noordermeer (2004). Addressing durability of rubber compounds. *Rubber Chem Technol 77*(3), 476–511.

Kamaruddin, S. & A. H. Muhr (2018). Investigation of ozone cracking on natural rubber. *J Rubber Res 21*(2), 73–93.

Kruza, M., A. C. Lewis, G. C. Morrison, & N. Carslaw (2017). Impact of surface ozone interactions on indoor air chemistry: *A modeling study. Indoor Air 27*(5), 1001–1011.

Lake, G. J. (1970). Ozone cracking and protection of rubber. *Rubber Chem Technol 43*(5), 1230–1254.

Lake, G. J. & P. G. Mente (1992). Ozone Cracking and Protection of Elastomers at High and Low Temperatures. *J. nat. Rubb. Res. 7*(1), 1–13.

Lewis, P. M. (1986). Effect of ozone on rubbers: countermeasures and unsolved problems. *Polym. Degrad. Stab. 15*(1), 33–66.

Lindley, P. B. (1973). Relation between hysteresis and the dynamic crack growth resistance of natural rubber. *Journal of Fracture 9*(4), 449–462.

Malaysian Rubber Board (2023, July). Consumption of natural and synthetic rubber worldwide from 1990 to 2022 (in million metric tons). In Statista. from https://www.statista.com/statistics/275399/worldconsumption-of-natural-and-synthetic-caoutchouc/, Accessed February 20, 2024.

Neuhaus, C. (2019). *Methodenentwicklung für die Lebensdaueruntersuchung thermomechanisch beanspruchter Elastomerlager*. Ph. D. thesis, Universität der Bundeswehr München, p. 26 ff.

Rodrigues, F. H. A., E. F. Santos, J. P. A. Feitosa, N. M. P. S. Ricardo, & F. Heatley (2004). Ozonation of unstretched natural rubber film from Hevea brasiliensis studied by ozone consumption and 13C MNR. *Polymer International*, 733–739.

Salisbury, G., A. R. Rickard, P.S. Monks, B. J. Allan, S. Bauguitte, S. A. Penkett, N. Carslaw, A. C. Lewis, D. J. Creasey, D. E. Heard, P. J. Jacobs, & J. D. Lee (2001). Production of peroxy radicals at night via reactions of ozone and the nitrate radical in the marine boundary layer. J. *Geophys. Res. 106*(D12), 12669–12687.

Spreckels, J., U. Weltin, M. Flamm, & T. Steinweger (2010). Consideration of environmental influences on fatigue tests of elastomer components. *Constitutive Models for Rubber VI: Proceedings of the 6th European Conference on Constitutive Models for Rubber.*
Treib, C. & M. Johlitz (2022). Ozone ageing of polydimethylsiloxane coated natural rubber. In *Consitutive Models for Rubber XII*, pp. 408–414. CNC Press Taylor and Francis Group.
Treib, C., M. Johlitz, & A. Lion (2024). *Lectures Notes on Advanced Structured Materials 2*, Chapter Crack Evaluation of Ozone-Aged Elastomers. Springer, Cham.
Treib, C., A. Lion, & M. Johlitz (2022). FT-IR Microscopic Analysis of the Chemical Change in the Molecular Structure after Ozone Ageing of Natural Rubber. In *Lectures Notes on Advanced Structured Materials*, pp. 241–252. Springer.

Industrial applications

Constitutive Models for Rubber XIII – Dal (Ed.)
© 2025 The Author(s), ISBN 978-1-032-85138-9

Finite element analysis of elastomeric bearings under cyclic shear loading

B. Biçer & S. Özüpek
Boğaziçi University, Istanbul, Türkiye

ABSTRACT: The study is concerned with the analysis of high damping rubber (HDR) isolation bearings under seismic excitations. The emphasis of the study is on the formulation of a constitutive model for HDR accounting for large strains, cyclic loading hysteresis and rate dependence while being computationally efficient. The calibration of the constitutive model and its validity are done using experimental data available in the literature. The calibrated constitutive model is used in the analysis of an industrial rubber bearing subjected to compression and shear loadings.The sensitivity of the response to various parameters is investigated.

The agreement of the predictions with the limited test data on HDR bearing is encouraging. Bearing shear stiffness and damping ratio are calculated within ranges available in the literature. The effect of loading rate on the hysteresis amount follows the trend observed in the experiments.

1 INTRODUCTION

Laminated elastomeric bearings are used in seismic isolation industry to support buildings and bridges while providing flexibility and energy dissipation against earthquake excitations. High damping rubber bearings (HDRB) are widely used as seismic isolators due to their high damping capability.

When a structure is subjected to seismic movements, HDR generally shows large deformations, hysteresis, time dependence and softening. Each of these effects have implication on stress response of the elastomer.

Large deformation is a geometric nonlinearity that needs to be taken into account for accurate representation of the elastomer behavior. Hysteresis is a material nonlinearity observed in cyclic response of the elastomer. It is a measure of the dissipated energy, hence its representation in HDR constitutive model is particularly important. Rate dependence of rubber response may be implemented using viscoelastic material model. However, characterization of elastomers as a viscoelastic material requires extensive testing which is not widely available for HDRB. Softening of rubber is observed as the stress reduction between the first and the successive loading cycles. This nonlinearity is called Mullin's effect and and often becomes negligible after a few cycles.

The following literature review summarizes various modeling approaches to represent the important aspects of the HDR mechanical behaviour.

In the study by Amin et al. (2006), rate dependent behavior of HDR under cyclic compression and shear was investigated. The loading part was found to vary significantly with the strain rate while the unloading part had weak dependence on the rate. A finite strain viscoelastic model was proposed. The model represented the monotonic response quite well, while cyclic test simulations were less successfull. The model has the advantage of having few parameters to be calibrated. A computational strategy for this model was developed in Amin et al. (2015) and implemented in a finite element code. The implementation was validated for monotonic loading. Two-dimensional and three-dimensional finite element analysis of an industrial HDRB were conducted to investigate the effect of shape factor and shear strain rate on the model response.

In the study by Alotaibi et al. (2019), two-dimensional finite element analysis of an elastomeric layered bearing subjected to cyclic loading with increasing shear amplitudes was conducted. Behavior of the elastomer was modelled using a nonlinear visco-elastic model. The results extracted from the predicted hystheresis loop for each loading cycle indicated that while the effective damping increased with increasing shear amplitudes, the effective stiffness remained the same.

In the study by Bhuiyan et al. (2015),HDR was tested in cyclic shear, multi-step relaxation and simple relaxation. The hystheresis loops in cyclic tests increased with increasing strain rates. An elasto-viscoplastic material model for HDR was developed and calibrated. Although the model was shown to represent the measured hystheresis, it was not validated for test data not used in the calibration.

In the study by Kalfas et al. (2017) hystheresis behavior of rubber was modeled using Ogden hyper-elastic function and hystheretic model with parameters from Bergstrom and Boyce (1998).A numerical study using the model showed that an increase of compression load magnitude reduced the shear

DOI: 10.1201/9781003516880-47

stiffness and increased the damping capacity of the laminated bearing. The model was not validated with test data.

In the study by Nguyen et al. (2015), a one-dimensional mechanical model was proposed to represent rate dependent hystheresis behaviour of HDRB. The model aims to represent the overall response of the bearing specimen rather than the constitutive behaviour of the HDR. Model parameters were determined from multi-step relaxation and sinusoidal cyclic test data performed on HDRB specimen at various temperatures. The model was validated for cyclic shear loading at moderate rates.

As most of the references point out, hysteresis behaviour of rubber still needs extensive research and requires high accuracy model outputs under combined shear and compression loading. The present work focuses on representing the mechanical response of HDR with a computationally efficient finite strain formulation that accounts for rate dependence and predicts hystheresis in cyclic loading. Details of the constitutive formulation are described in Section 2. Finite element analysis of a full-size bearing under cyclic loading is presented in Section 3. The bearing response is validated for monotonic loading and numerical experiments are performed to show the rate effect on the HDRB response. Results and conclusions are presented in Section 4. All analyses were completed in commercial general purpose finite element software ABAQUS Simulia Inc. (2014).

2 CONSTITUTIVE MODELING OF THE ELASTOMER

- HDR is represented with a rate-dependent model valid for large deformations (Bergstrom and Boyce 1998). The model is micromechanism-inspired and combines the response of an equilibrium network A with the time dependent response of network B. The stress response of network A under the deformation gradient **F** is characterized by an isotropic hyperelastic function. In this study Yeoh energy formulation is used. For network B, **F** is decomposed into elastic, $\mathbf{F}^e_B$, and inelastic parts, $\mathbf{F}^{cr}_B$. These are related to the deviatoric Cauchy stress $\mathbf{S}_B$ through an evolution equation for $\mathbf{F}^{cr}_B$

.
$$\mathbf{F}^e_B.\mathbf{F}^{cr}_B.\mathbf{F}^{cr-1}_B.\mathbf{F}^{e-1}_B = \dot{\in}^{cr}\frac{\mathbf{S_B}}{\sigma_B}$$

where $\sigma_B = \sqrt{\frac{3}{2}\mathbf{S}_B : \mathbf{S}_B}$ is the effective creep stress, and $\dot{\in}^{cr}$ is the effective creep strain rate driven by

$$\dot{\in}^{cr} = A[\lambda^{cr}_B - 1 + E]^C(\sigma_B)^m$$

where $\lambda^{cr}_B = \sqrt{\frac{1}{3}\mathbf{I} : (\mathbf{F}^{cr^T}_B.\mathbf{F}^{cr}_B)}$ is the chain stretch.

Parameters to be determined are a scaling constant A, creep strain exponent C, effective stress exponent m, parameter E that helps regularize the creep strain rate in the vicinity of the undeformed state, and a stress-scaling factor, S, that defines the ratio of the stress carried by network B to that carried by network A under identical elastic stretching.

Uniaxial compression monotonic test data at 0.001/s rate is used to calculate the Yeoh parameters. Unixial compression cyclic test data at 0.001/s rate is used to calculate the hysteresis model parameters. Since only uniaxial compression data was used in the model calibration the agreement between the model and test data for shear loading was limited to 100% shear strain. The validation of the calibrated model for cyclic uniaxial compression and simple shear loadings at rates not used in the calibration are shown in Figures 1 and 2.

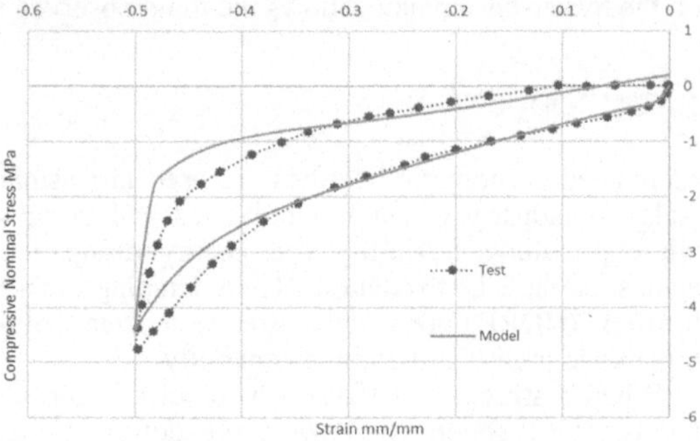

Figure 1. Uniaxial compression at 0.024/s. (Amin et al. (2015)).

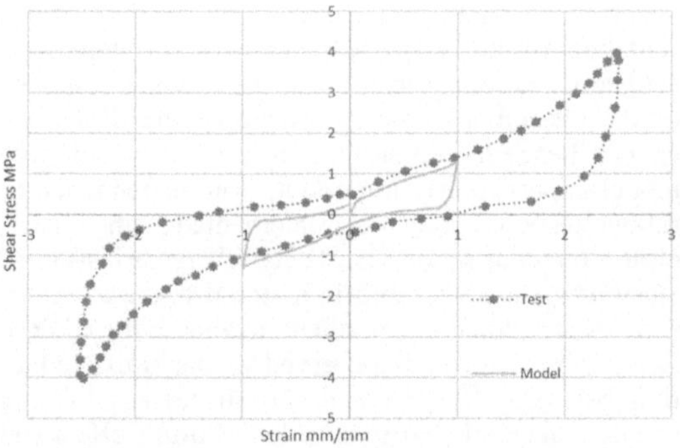

Figure 2. Simple shear at 0.4/s (Amin et al. (2006)).

3 FINITE ELEMENT ANALYSIS OF AN INDUSTRIAL BEARING

An industrial rectangular laminated bearing was selected from the literature (Amin et al. 2015). The geometry details are given in Table 1.

The elastomer was represented with the constitutive model described in Section 2, and the steel was

Table 1. Geometry data of HDRB (Amin et al. 2015).

Cross section	240 mm x 240 mm
Number of rubber layers	6
Number of steel layers	5
Thickness of each rubber layer	5 mm
Thickness of each steel layer	2.3 mm

represented with an elastoplastic material model with kinematic hardening. Plane strain analysis was carried out and both the rubber and the steel were meshed with linear elements. The bearing was bounded to the top and bottom rigid surfaces. For sides of the bearing contact with friction coefficient $\mu = 0.3$ was defined. The analysis loads were selected to correspond to average compression stress of 6 MPa applied at 0.001/s strain rate, and cyclic shear applied at strain rates 0.05/s and 1.5/s. Deformed conuration of the bearing is shown in Figure 3. Limited validation of the model was possible since only monotonic test data was available (Amin et al. 2015). Close correlation between the model prediction and experiment, as shown in Figure 4, validates the developed finite element model. Therefore the model may be used for various numerical experiments to understand the effect of material, geometry and loading parameters on the response.

In the following a study conducted to determine the effect of the elastomer compressibility and strain rate, on the shear stiffness, dissipated energy and damping ratio is presented. In particular, the elastomer was modeled as an incompressile material and with compressibility $K_0/\mu_0 = 100$, where K_0, μ_0 are the initial bulk and shear moduli, respectively.

Shear stiffness was calculated as the shear stress over the shear strain at maximum displacement. Dissipated energy was calculated as the area enclosed by the reaction force-displacement hysteresis curve. The damping ratio was calculated as (Quaglini et al. 2016).

$$\zeta = \frac{2}{\pi} * \frac{E_D}{K_{eff} * (u_{max} - u_{min})^2}$$

where E_D represents the dissipated energy, K_{eff} is the slope of shear force-displacement curve at maximum strain and u are displacements.

4 RESULTS AND CONCLUSIONS

Analysis results at slow and fast shear strain rates for an incompressible and compressible material models are shown in Figures 5 and 6, Tables 2 and 3. The following conclusions can be made:

- Analyzed bearing has shear stiffness in the range of 1.10 to 1.47. These values are comparable with Murota et al. (2021) which reports range of 0.68 to 1.56 for the stiffness.
- Analyzed bearing has damping ratio between 0.2 to 0.23. This is in agreement with Troy et al. (2001) where the range is given as 0.10 to 0.28.
- Strain rate dependence is dominant in the rubber. Faster loading shows larger shear stiffness, dissipated energy and damping ratio compared to slower loading.
- Lower rubber compressibility results in slight increase in shear stiffness, dissipated energy and damping ratio. It is noted that the results not discussed in this paper showed that the compressibility effect on the compression load is very significant.
- Overall, shear strain rate has a considerable effect on the calculated values, for all compressibility levels. This conclusion is in agreement with Yun et al. (2019).

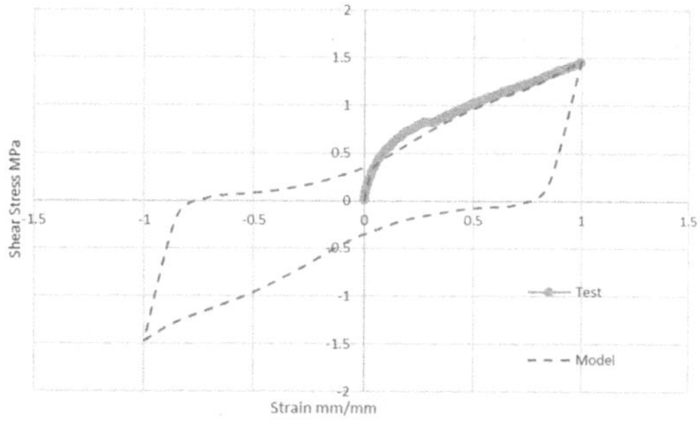

Figure 4. Monotonic shear test (Amin et al. 2015) and cyclic shear prediction at of 1.5/s.

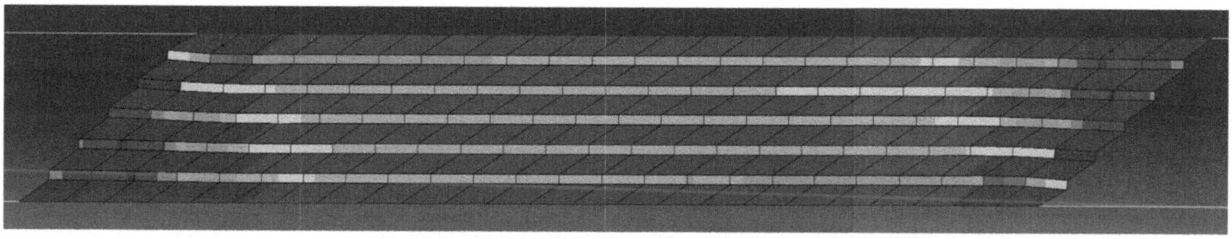

Figure 3. Deformed conuration of the HDRB.

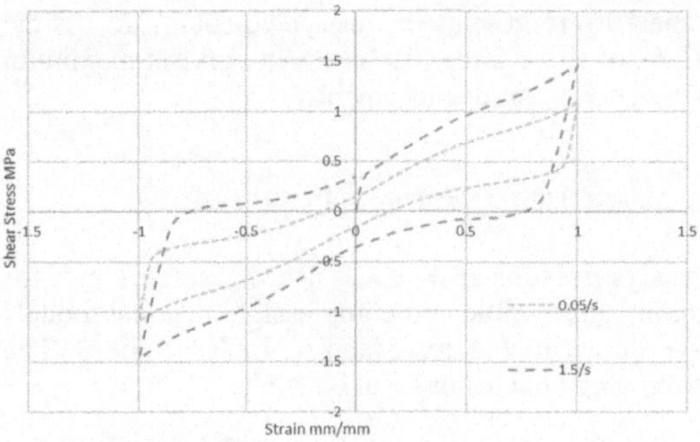

Figure 5. Cyclic shear of compressible material model at different rates.

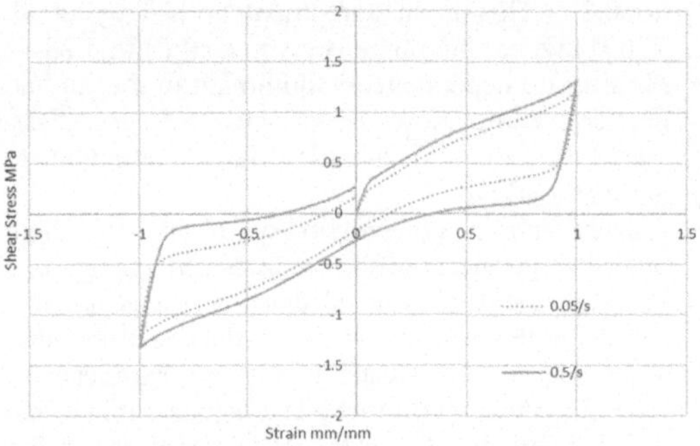

Figure 6. Cyclic shear of compressible material model at different rates.

Constitutive model completed in this work can further be used to determine structural responses, damping capabilities, fatigue life and possible reinforcement requirements of the isolation systems. Possible alternative designs can be further analyzed without the need of full-scale tests.

Calibration of time dependent hysteresis parameters are done by trial-error process. This calibration can be further improved through optimization.

In the study, rate dependence is combined with a hyperelastic material model. Further improvement to constitutive model can be done by adding Mullin's effect. Although it is often observed that after 5 to 10 cycles, Mullin's effect perishes, contrary test results exist. Modeling of the cyclic softening in multiple cycles may be needed in such cases.

Table 2. Shear strain rate effect for incompressible material.

Shear Strain	Dissipated	Shear stiffness	Damping
Rate (1/s)	Energy (Nmm)	(MPa)	Ratio
0.05	2.34E+06	1.097	0.197
1.5	3.58E+06	1.469	0.225

Table 3. Shear strain rate effect for compressible material.

Shear Strain	Dissipated	Shear stiffness	Damping
Rate (1/s)	Energy (Nmm)	(MPa)	Ratio
0.05	2.54E+06	1.205	0.214
0.5	2.95E+06	1.317	0.206

REFERENCES

Alotaibi, E., N. Nassif, R. AlSodi, S. Ayman, & I. H. Fattouh (2019, April). Assessment of hyper-visco-elastic seismic isolator behaviour using finite element. In D. Brown and S. Green (Eds.), *Proceedings of the 4th World Congress on Civil, Structural, and Environmental Engineering Conference, Rome*, Italy. ACM.

Amin, A. F. M. S., T. Hossain, & Y. Okui (2015). Nonlinear viscosity law in finite-element analysis of high damping rubber bearings and expansion joints. *Journal of Engineering Mechanics 141*(6). 1610–1657.

Amin, A. F. M. S., A. Lion, S. Sekita, & Y. Okui (2006). Nonlinear dependence of viscosity in modelling the rate-dependent response of natural and high damping rubbers in compression and shear: Experimental identification and numerical verification. *International Journal of Plasticity 22*(9), 1610–1657.

Bergstrom, J. S. & M. C. Boyce (1998). Constitutive modeling of the large strain time-dependent behavior of elastomers. *Journal of the Mechanics and Physics of Solids 46*(5), 931–954.

Bhuiyan, A. R., H. Mitamura, T. Imai, & Y. Okui (2015). Arheology model of high damping rubber bearings for seismic analysis. *International Journal of Solids and Structures 46*(7), 1778–1792.

Kalfas, K. N., S. A. Mitoulis, & K. Katakalos (2017). Numerical study on the response of steel-laminated elastomeric bearings subjected to variable axial loads and development of local tensile stresses. *Engineering Structures* 134, 346–357.

Murota, N., S. Suzuki, T. Mori, K. Wakishima, B. Sadan, C. Tuzun, F. Sutcu, & M. Erdik. (2021). Performance of high-damping rubber bearings for seismic isolation of residential buildings in turkey. *Soil Dynamics and Earthquake Engineering 143*(2), 0–0.

Nguyen, D. A., J.Dang, Y.Okui, A.F.M.S.Amin, S.Okada, & T.Imai. (2015). An improved rheology model for the description of the rate dependent cyclic behavior of high damping rubber bearings. *Soil Dynamics and Earthquake Engineering 77*(10), 416–431.

Quaglini, V., i. P. Dubin, & V.G. (2016). Experimental assessment of high damping rubber under combined compression and shear. *Journal of Engineering Materials and Technology 138*, 011002–1.

Simulia Inc. (2014). *Abaqus 6.14 Documentation*. Simulia Inc.

Troy, M., W. Andrew, & T. Andrew (2001). Cyclic behaviour of high-damping rubber bearings. *In Proceeedings of Fifth World Congress on Joints, Bearings and Seismic Systems for Concrete Structures, Rome, Italy.*

Yun, C., C. Chao, M. Qianqian, J. Huanjun, & W. Zhiwei (2019). Study on mechanical properties of high damping viscoelastic dampers. *Advances in Structural Engineering 22*(14), 2925–2936.

Constitutive Models for Rubber XIII – Dal (Ed.)
© 2025 The Author(s), ISBN 978-1-032-85138-9

Hyperelastic-viscoplastic and crack growth behavior of thermoplastic vulcanizates (TPVs)

Aizeti Burgoa & Kepa Zulueta
Leartiker S. Coop, Markina-Xemein, Bizkaia, Spain

Will Mars, Tom Ebbott & Mark Bauman
Endurica LLC, Findlay, Ohio, USA

ABSTRACT: In the field of recyclable elastomeric materials, thermoplastic vulcanizates (TPVs) are attracting a growing interest in transport-related industries, such as automotive and railway, due to their light weight, easy and cost-effective processing and design flexibility. The microstructure of TPVs is characterized by crosslinked rubber particles dispersed in a continuous thermoplastic matrix. This unique microstructure generates a highly nonlinear mechanical behavior that depends on stress, strain history, temperature, and frequency, characterized by remarkable inelastic effects and viscoplastic behavior. Therefore, in order to ensure a competitive and confident application of TPV materials, the complexity of their mechanical properties must be taken into account. In this work, a TPV material with potential applications in the automotive industry is characterized for hyperelastic, viscoplastic and crack growth characteristics. Mechanical properties are measured and analyzed in the flow and transverse directions. This study continues the effort towards advanced modeling and simulation workflows looking for enhanced thermomechanical and fatigue life predictions in TPV-based components.

1 INTRODUCTION

Elastomers are widely used in the automotive industry. For example, elastomeric anti-vibration components are made of thermoset rubbers. However, new drivers related to requirements of the new sustainable mobility industry are pushing towards the substitution of thermoset rubbers by thermoplastic elastomers (TPEs). Among TPEs, thermoplastic vulcanizates (TPVs) represent an interesting material choice for anti-vibration products due to their light weight, recyclability, ease of processing by means of injection molding and design flexibility.

Previous work on fatigue and fracture has demonstrated the challenge of using conventional methods for crack growth characterization in TPVs due in part to the large amounts of inelastic deformation (Buroga et al. 2020, 2022, and Mars et al. 2017, 2022).

In this work, significant anisotropy in stiffness and fatigue properties is demonstrated. These phenomena are due to the flow during injection molding and are interpreted in terms of the orientation of EPDM particles together with PP crystals and chain alignment observed in microstructural characterization tests.

Although anisotropy introduces complexity, it also opens potential new opportunities to develop TPV-based anti-vibration components with superior elasticity and customizable static and dynamic mechanical properties.

Even though the stress-strain deformation behavior of TPVs has been the focus of several experimental and theoretical studies, the assessment of the plastic deformation and notable time-and temperature-dependent behavior in the fatigue performance of TPV materials remains elusive. This work represents a small advancement in the overall understanding of TPV mechanical behavior with the goal to aid more widespread usage.

2 EXPERIMENTAL

2.1 *Materials*

A commercial PP/EPDM TPV from Celanese Corporation has been selected for the study. The trade name and main properties of the TPV materials are presented in Table 1.

2.2 *Injection molding of samples*

TPV plaques of 180x50x1 mm^3 and 180x50x4 mm^3 were fabricated by injection moulding in an Insert v200 (ENGEL GmbH, Schwertberg, Austria)

DOI: 10.1201/9781003516880-48

Table 1. Trade name and main properties of selected TPV material. (Tensile properties are transverse to flow).

Material trade name	Hardness, Shore A	Density	Tensile strength at break	Tensile stress at 100% strain	Elongation at break
Santoprene™ 101-73	78	0.97	7.98MPa	3.44MPa	478%

machine following the material manufacturer's processing recommendation. As shown in Figure 1, ISO 37 type 3 dumbbell-shaped specimens were die cut following transverse (90° orientation) and longitudinal (0° orientation) directions with respect to the flow. Planar tension (pure shear) specimens were also cut out in the same orientations.

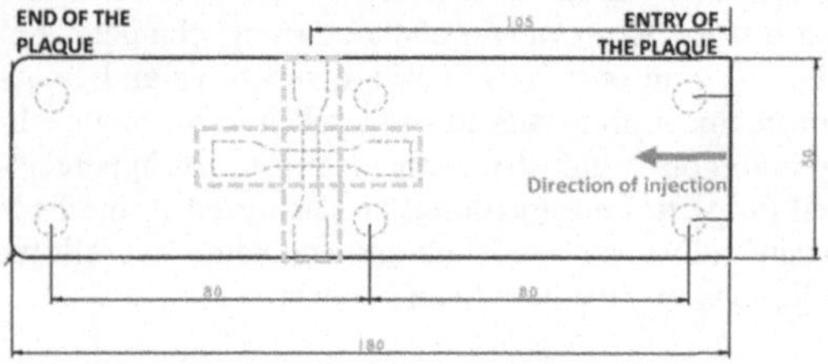

Figure 1. Location of the studied samples in the injected TPV plaques.

2.3 *Quasi -static mechanical characterization*

The quasi-static mechanical properties of the TPV samples were measured by performing cyclic simple tension, planar tension and biaxial tests at room temperature. Typical test configurations are shown in Figure 2.

Figure 2. Example hyperelastic test specimens, left-to-right, simple tension, planar tension (pure shear), and biaxial tension. (Courtesy of Axel Products).

Simple tension tests were conducted in two directions to investigate the degree of anisotropy due to the injection molding. Due to the large amount of anisotropy, the equi-biaxial tests need further investigation and are not reported here.

2.4 *Dynamic mechanical analysis*

The variation of the viscoelastic properties with temperature, frequency, static prestrain and dynamic strains was analyzed on a Metravib DMA+300 in shear and tension modes.

2.5 *Fracture and crack growth characterization*

Fatigue crack growth characterization was performed using planar tension (pure shear) crack growth specimens. Due to the injected TPV plaque size, mini-pure shear specimens, 50 mm long by 5 mm high were used with thicknesses of 1 mm and 4 mm. The crack length and crack growth were measured using digital camera images.

3 RESULTS AND DISCUSSION

3.1 *Cyclic stress-strain behavior*

The longitudinal and transverse cyclic stress-strain response is displayed in Figure 3 showing the pronounced anisotropy with a 2x difference in stiffness due to orientation. The TPV shows a pronounced loss of hysteresis when moving from the first to the fifth stabilized cycle, suggesting a significant Mullins softening effect. Another important phenomenon that appears when subjecting the TPV material to cyclic loading conditions is the permanent deformation, which is a consequence of the viscoplastic nature, and represents the unrecoverable deformations when a strain threshold is reached.

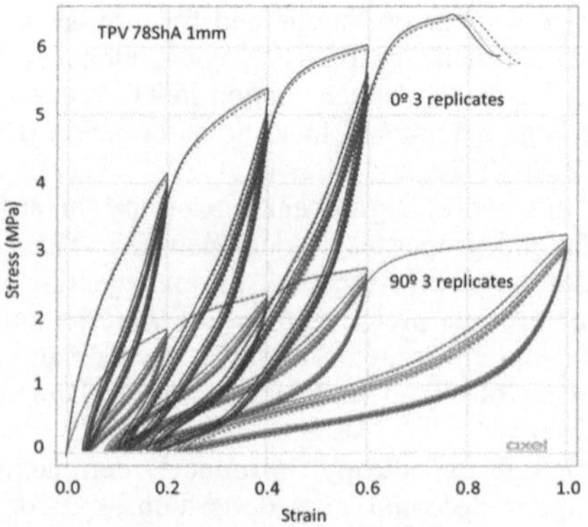

Figure 3. Simple tension for the subject TPV in two orientations illustrating the strong anisotropy.

Cyclic deformation and microstructural changes of TPVs are rarely correlated by models in the literature. Burgoa, et al. (2020) reported that the plastic deformation of TPVs is driven by the yielding, buckling, and bending of the thin thermoplastic matrix ligaments and the elastic recovery ability of the rubber droplets, which is in accordance with the microstructural model known as the Soliman model (Figure 4).

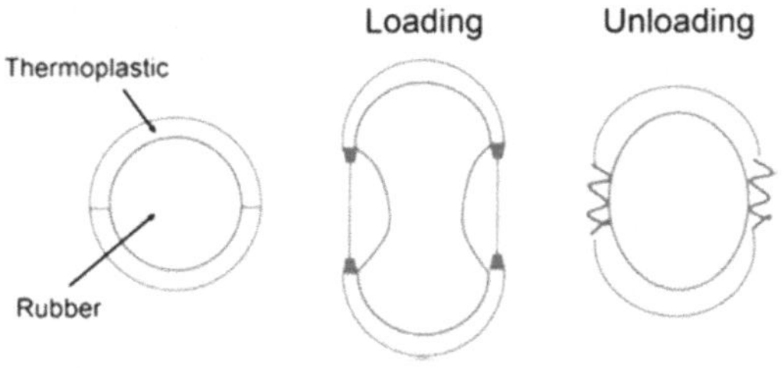

Figure 4. Schematic representation of the deformation mechanism of TPV materials based on Soliman model.

At low strains both the PP matrix phase and the EPDM rubber droplets may deform elastically and so the permanent deformation is quite minimal at the initial strain levels. With increasing strain level, the semicrystalline PP matrix will start to yield in the regions where matrix ligaments are thinnest. Upon unloading, the elastic forces of the stretched interconnected rubber network are able to pull back the plastically deformed thin ligaments by either bending or buckling. The loss of stiffness during the reloading step may be attributed to the damage generated by the bending or buckling of the thin plastic ligaments during the unloading. The thicker thermoplastic ligaments of the matrix will interconnect the rubber droplets, forming an elastic interconnected network structure of rubber droplets and slightly deformed thicker zones of the PP matrix. When the strain level is increased the thin matrix ligaments will continue to yield in order to conform to the elastic deformation of the rubber droplets.

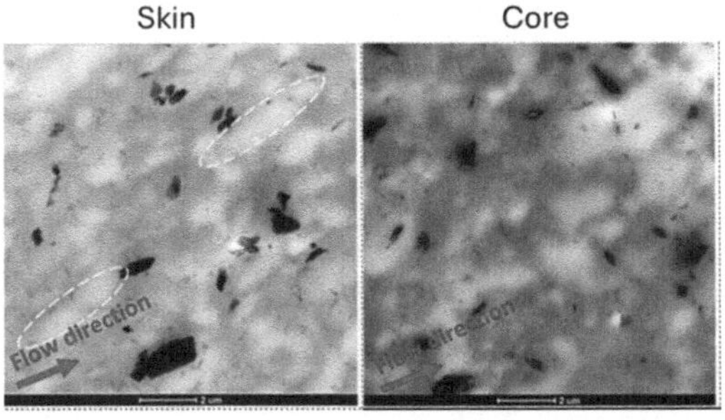

Figure 5. Flow induced anisotropic morphology in a ShoreA87 TPV 3.2mm thickness plaque. The dashed lines show the elongated EPDM phase. (Burgoa et al. 2023).

Moreover, there is a pronounced flow induced anisotropic effect, as a lower permanent deformation is observed in sampled cut in the transverse direction. From the microstructural point of view, the orientation of EPDM particles together with the PP crystals and chain alignment observed in the previous works (by TEM and AFM images) appear to be key features that lead to the differences in permanent deformations between the specimens cut in the transverse and longitudinal directions with respect to the flow direction (Burgoa et al. (2023)). Figure 5 shows one example of the flow-induced anisotropic morphology in a ShoreA87 TPV similar to the material in this study.

The skin morphology, near the surface of the 3.2 mm thickness shows much more anisotropy compared to the core which was taken near the center of the thickness direction.

3.2 *Dynamic mechanical behavior*

Figures 6-8 present the dependence of shear storage modulus *G'*, loss modulus *G''* and tanδ on the dynamic amplitude and frequency for the 1 mm thick samples cut longitudinal and transverse to the flow direction. The Payne softening effect and dynamic rigidification when increasing the frequency is readily observed. Both the storage and loss moduli of the TPV decrease significantly in the transverse direction, but there is no significant effect of orientation on the tanδ damping.

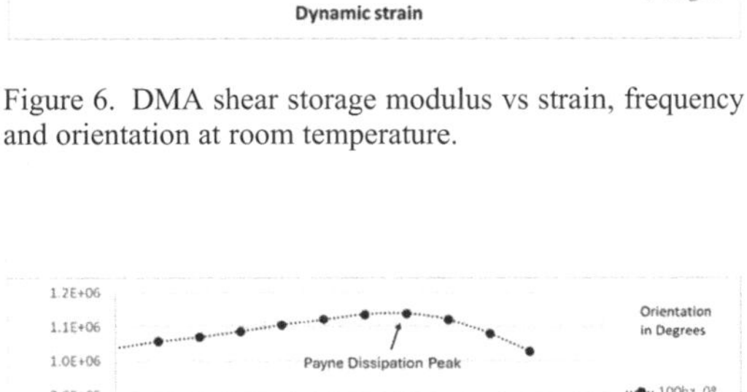

Figure 6. DMA shear storage modulus vs strain, frequency and orientation at room temperature.

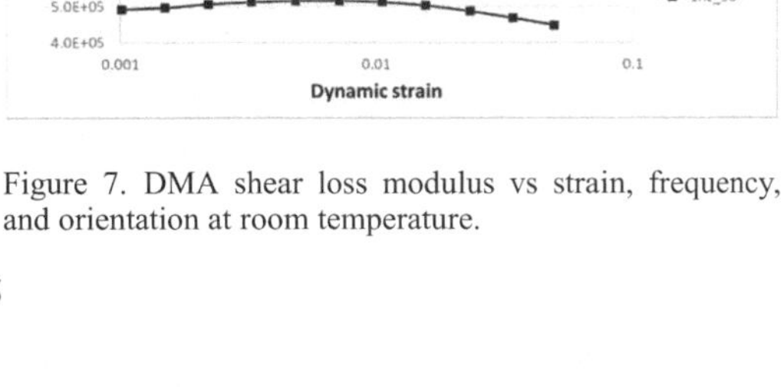

Figure 7. DMA shear loss modulus vs strain, frequency, and orientation at room temperature.

A static pre-deformation superimposed with a small harmonic oscillation can be found in many elastomeric applications where the elastomeric segment is used to attenuate the vibrations of industrial structures. Therefore, another interesting case to consider is the dependence of TPV systems on the static pre-deformation and oscillation amplitude. Frequency sweep DMA experiments were conducted in tension with pre-strains ranging from 1% to 15% and dynamic strain amplitudes ranging from 0.05% to 5% for the 1 mm and 4 mm samples cut longitudinal and transverse to the flow direction. Some of the results are shown in Figures 9 and 10.

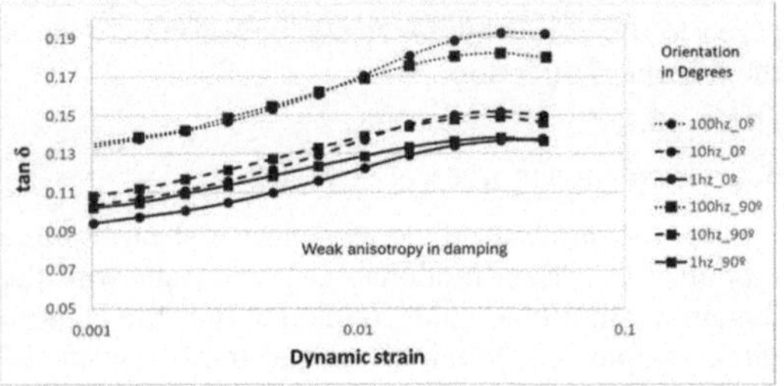

Figure 8. DMA shear tanδ vs strain, frequency, and orientation at room temperature.

Figure 9 displays the dependence of the tensile storage modulus on frequency and the effect of pre-strain. For the moderate pre-strains investigated, the stiffness decreases with pre-strain as expected.

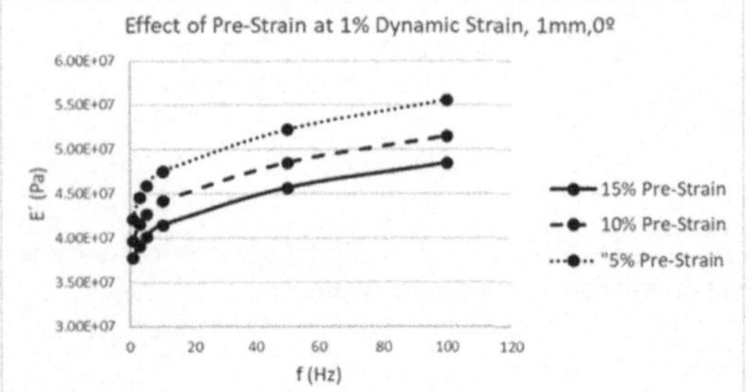

Figure 9. DMA tensile storage modulus vs frequency for different pre-strains.

Figure 10 shows the impact of plaque thickness and flow orientation. The degree of anisotropy in tension for the 1mm plaques is greater than that in shear. The drop in tension modulus from 0° orientation to 90° in Figure 10 is a factor of 2 (consistent with the quasi-static results in Figure 3), while a similar analysis of Figure 6 shows only about a 20% reduction in shear storage modulus.

The lack of significant anisotropy for the 4mm specimens in Figure 10 is also interesting. This is partially explained by the morphology shown in Figure 5. The 4mm plaques will have a higher percentage of material that is like the "core" toward the mid-thickness that shows very little anisotropy in the microstructure.

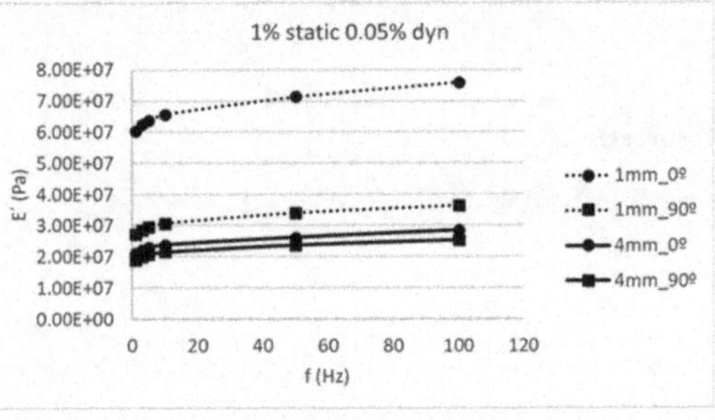

Figure 10. DMA frequency seeps in tension showing the effect of plaque thickness on anisotropy.

3.3 *Fracture and crack growth behavior*

For tear strength (T_c) and crack growth measurement, planar tension (pure shear) specimens were used with dimensions 50x5x1 mm^3 with a 15 mm long initial crack. Intrinsic strength (T_0) tests were also conducted on a similar geometry but with the addition of a razor blade cutting to advance the crack front. Crack growth tests were also performed on 4mm thick specimens, but the cracks tunneled in such a manner that the crack tip was obscured from the image recording cameras, so the results are not reported here.

Figures 11 and 12 show the T_c tests with 3 replicates for 23° C, and 80° C, respectively. The stiffness drops by about 25%, while the tear strength drops by a factor of three – 6.7 kJ/m^2 to 2.2 kJ/m^2 – when the temperature is raised from 23° to 80° C.

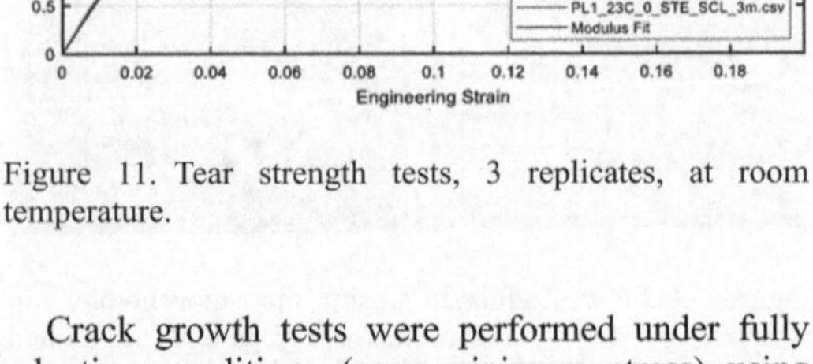

Figure 11. Tear strength tests, 3 replicates, at room temperature.

Crack growth tests were performed under fully relaxting conditions (zero minimum stress) using a ramped maximum strain from 3-15% with a planar

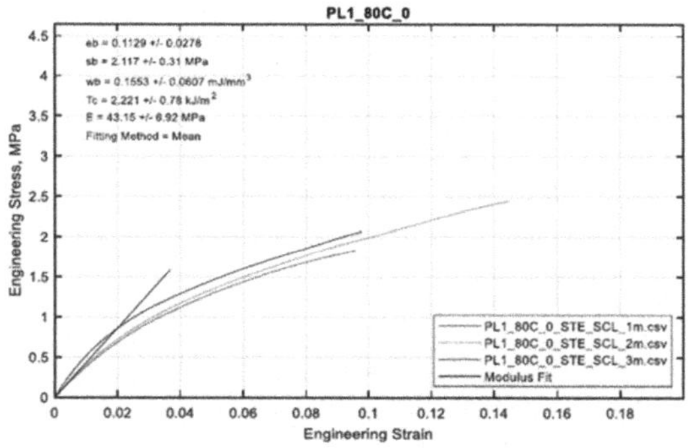

Figure 12. Tear strength tests, 3 replicates, at 80° C.

tension (pure shear) specimen. Crack contour images were taken periodically throughout the tests and used to measure the crack length. Results for the 90° orientation, where the crack growth is parallel to the flow direction, are presented first.

Figure 13 shows the progressive crack contours for three replicates. Images are plotted in increments of 5040 cycles. The similarity and consistency among the three replicates and the smooth self-similar crack growth is remarkable.

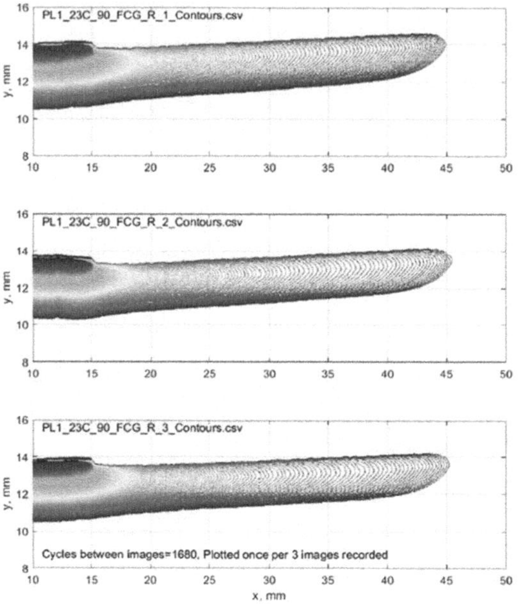

Figure 13. Planar tension crack growth contours. 90° orientation. Three replicates. Cycles between images = 5040.

Figure 14 shows the same plots for the 0° orientation, where the crack growth is perpendicular to the flow direction. The crack growth is sometimes non-existent and other times slow and jagged, like a natural rubber under non-relaxing conditions.

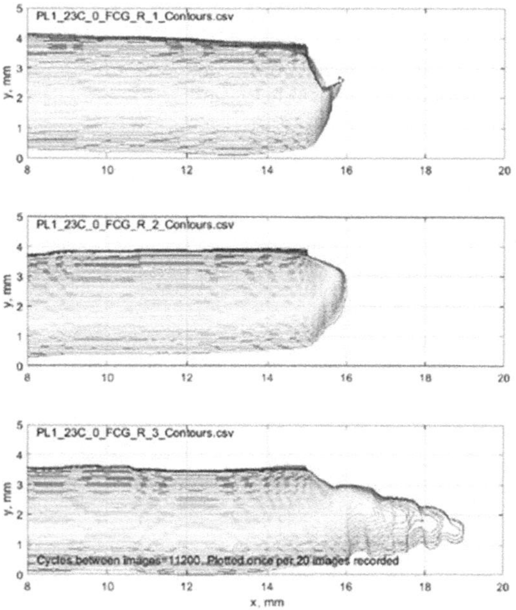

Figure 14. Planar tension crack growth contours. 0° orientation. Three replicates. Cycles between images = 224,000.

The crack growth data are reduced in order to determine a Thomas power-law relationship for crack growth rate vs. tearing energy (energy release rate).

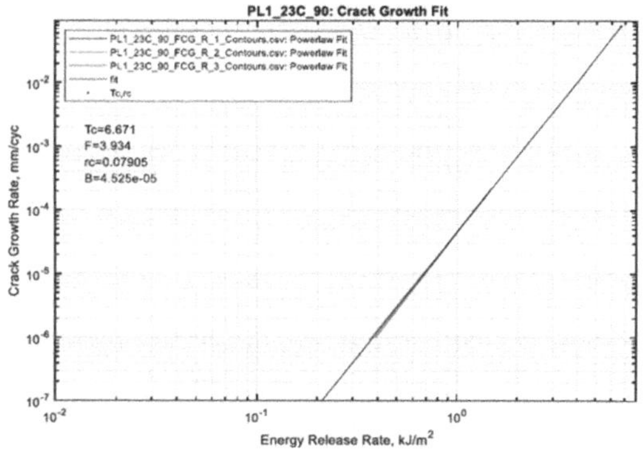

Figure 15. Crack growth rate vs. Energy Release Rate (Tearing Energy). Thomas power-law plots 90° orientation. Three replicates.

The Thomas power law for the 90° orientation is shown in Figure 15. The results for the three replicates are nearly identical—a result not usually found in fatigue testing. The results for the 0° orientation are shown in Figure 16, and the difference among the replicates is evident. The coefficients and power exponents for the Thomas Law are shown in Table 2. There is a very significant difference in crack growth behavior, and the resulting rates of crack propagation are different by 2-3 orders of magnitude. Critical tear strength values are available only for the 0° orientation. Further experiments will be conducted to determine T_c in the 90° orientation.

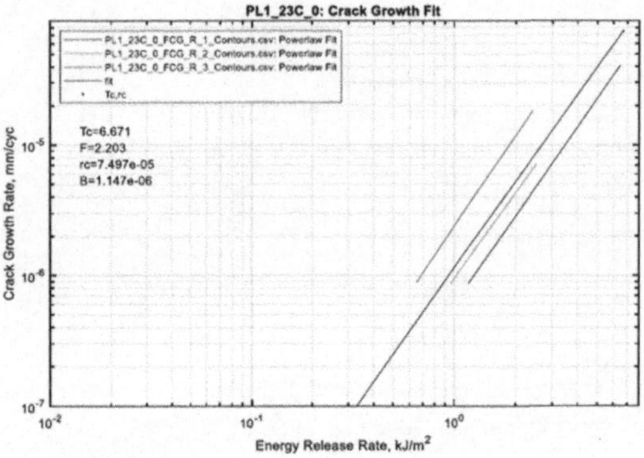

Figure 16. Crack growth rate vs. Energy Release Rate (Tearing Energy). Thomas power-law plots 0° orientation. Three replicates.

Table 2. Thomas power-law parameters.

Orientation	Coefficient (d*c*/d*N* at *T*=*T*c)	Power Exponent
0°	7.5E-5 mm/cyc	2.20
90°	7.9E-2 mm/cyc	3.93

Intrinsic strength measurements were also carried out in the manner outlined in Lake & Yeoh (1978). At this writing only the 90° results are available with a measured intrinsic strength, T_0 value of 44.9 J/m^2.

4 CONCLUSIONS

Thermoplastic vulcanisates offer an interesting alternative to conventional carbon-black filled elastomers for many automotive and anti-vibration components. Some advantages include recyclability and manufacturing simplicity. However, behaviors such as significant anelasticity (visco-plasticity) and flow-induced anisotropy need to be characterized and understood in order to enable the design and use of components with these materials.

This work demonstrates the significant flow-induced anisotropic effects on elastic, viscoelastic, and crack growth behavior. Stiffness differences due to orientation were measured to be around a factor of 2 in quasi-static tension, and the 0° orientation was about 25% higher than the 90° orientation in DMA shear. Anisotropic differences are expected to be less in shear compared to tension, since the stiffer regions are not directly extended when testing in shear.

The difference in crack growth behavior due to orientation was very significant, both in behavior and in the resulting crack growth rates. The crack growth in the 90° orientation, when the crack growth was parallel to the flow direction, was faster and very regular and repeatable, indicative of a weaker material. The crack growth in the 0° orientation, when the crack growth is perpendicular to the flow direction was slow and jagged showing promise as a stronger material similar to a carbon-black filled natural rubber compound.

REFERENCES

Burgoa, A., Arriaga, A., Zulueta, K., Acuña, EM., Laza, JM., Hernández, R., & Vilas, JL., 2020. Experimental investigation of the nonlinear quasi-static and dynamic mechanical behavior of novel PA6/XHNBR thermoplastic vulcanizates: Linking mechanical nonlinearities to microstructural features, *Materials Today Communications*, **25**, 101395.

Burgoa, A., Arriaga, A., Badiola, JH, Ibarretxe, J., Iturrondobeitia, M., Martinez-Amesti, A., & Vilas, JL., 2022. On the injection molding of thick-walled thermoplastic vulcanizates: linking static and dynamic mechanical properties with morphology, *Polymer International*, **72**, 508–519.

Lake, G. & Yeoh, OH. 1978. Measurement of rubber cutting resistance in the absence of friction, *Intl J. Fracture*, **14**, 509.

Mars, W. V. & Ellul, M. D. 2017. Fatigue characterization of a thermoplastic elastomer. *Rubber Chemistry and Technology*, 90(2), 367–380.

Mars, W. V., Isasi, M., Barinagarrementeria, X., & Ebbott, T. G. 2022. Material plane identification and tracking for small cracks in presence of large inelasticity. In *Constitutive Models for Rubber XII*, pp. 305–311, CRC Press.

Constitutive Models for Rubber XIII – Dal (Ed.)
© 2025 The Author(s), ISBN 978-1-032-85138-9

Finite element modelling for fatigue in polymer heart valves

R.S. Patel & G.D. Moggridge
Department of Chemical Engineering and Biotechnology, University of Cambridge, Cambridge, UK

R.J. Windslow & T.W. Hohenberger
SLB, Houston, Texas, USA

J.J.C. Busfield
School of Engineering and Material Science, Queen Mary University of London, London, UK

ABSTRACT: The traditional route to develop a new polymer heart valve (PHV) involves rigorous validation by experimental approaches which can take many years to complete. For this reason, there are currently no PHVs on the market for clinical use. Computational modelling offers an alternative route to study heart valve behaviour and fatigue. This work uses finite element analysis (FEA) on a polymer heart valve to compare the fatigue performance of different block copolymers: a styrene-based triblock copolymer, poly(styrene-*block*-ethylene/butylene-*block*-styrene) (SEBS) and two types of polyurethane-based multiblock copolymers, polyisobutylene-based polyurethane (PIB-PU) and Elast-Eon, a polydimethylsiloxane-based polyether urethane (PDMS-PEU). The FEA method is based on cracking energy density, calculated using an approach that requires biaxiality of the stretch ratios to be determined. The materials were tested to obtain inputs for the PHV model: (1) cyclic stress-strain test data were fitted to a Neo-Hookean relationship to obtain shear moduli and compressibility parameters for the model, (2) fatigue crack growth experiments were carried out to measure A and F materials characteristics for the power law regime 3 of the tearing energy and crack growth rate relationship and (3) accelerated fatigue tests of rigid post PHV prototypes enabled measurement of the number of cycles to failure (N_f) in samples to validate the model experimentally. The model predicted the durability of SEBS20, PIB-PU and Elast-Eon rigid post PHVs to last 12.5 million, 139.4 million, and 1919.8 million cycles. The corresponding experimental results to validate the PHV model for each material were 19.9 million, 122.6 million, and 128.6 million for a specified inherent flaw size (c_0) of 0.025 mm. Therefore, the FEA results obtained are in a suitable range for fatigue life prediction, with the exception of Elast-Eon whose fatigue test data was obtained separately from literature and may not be appropriate. Overall, this approach provides an effective framework for fatigue life prediction which can be used to study the suitability of current and new material candidates for PHVs in timescales of months rather than years. This has positive implications for the rapid development of new PHVs with potential for clinical application in the future.

1 INTRODUCTION

Polymer heart valves (PHVs) are an essential breakthrough that have the potential to improve quality of life for those in need of heart valve replacements. Given the application, in-situ testing is unfeasible, so design and validation is limited to long form physical testing and/or computational modelling approaches. Physical testing, whilst helpful, cannot fully capture the intricacies of human form, and long timeframes of multiple years to obtain results restrict efficient progress. Hence, high-fidelity FEA models are critical to supplement validation of these parts.

Research on modelling PHVs typically splits into two categories: finite element analysis (FEA) studies and advanced fluid-structure-interaction (FSI) analysis (Bavo et al. 2016, Zakerzadeh et al. 2017, Luraghi et al. 2017). An important consideration that is less studied in the literature is the fatigue life. This is critical for material selection and design optimisation. Whilst there are some studies evaluating this aspect, most use a Wöhler type approach which is not aligned with the larger body of research on elastomer fatigue. Thus, the approach presented here is rooted in elastomer-based fracture mechanics and evaluates fatigue within a more conventionally accepted framework.

The concept of a PHV is an entirely polymer-based prosthesis designed to overcome the challenges faced with current clinical prostheses. This includes mechanical heart valves made from rigid synthetic materials and biological tissue-based heart valves which come with the compromises of limited

DOI: 10.1201/9781003516880-49

biocompatibility and durability, respectively. PHVs have the potential to resolve these issues, but they are not yet adopted for clinical use.

Prior to clinical use, a PHV must be fully validated through a portfolio of physical and computational testing governed by the regulatory standard, ISO 5840. This defines the computational requirements, in particular on simulating intended use conditions and a basic fatigue life prediction model (BSIB 2021).

A critical aspect of the fatigue approach is to compare FEA predicted fatigue life against physical test data. Given this regulatory document is relatively broad it does not differentiate between heart valve forms; hence the fatigue prediction approach is not specifically tailored towards PHVs.

The present work centres on evaluating the fatigue life of the J6-Poli-Valve design PHV (Stasiak et al. 2020) and aims to develop an approach based on computational modelling to compare the fatigue life of candidate materials more effectively than through the ISO standard. Physical testing is used to calibrate an elastomer's fatigue properties, which is then used within an FEA model to predict the fatigue life. The FEA model uses a cracking energy density (CED) framework to predict the local tearing energy and hence fracture potential. From this, fatigue life predictions for the PHV are made. This study was carried out for a range of PHV materials.

To validate the model's predictions, fatigue tests on rigid post PHVs were run experimentally. Rigid post PHVs were used given their propensity to fail earlier than a more realistic constraint system (Patel 2023). The accelerated failure of rigid post prototypes is obtained by testing the integrity of the PHV leaflets which enables durability results to be obtained in a timeframe of months rather than years.

2 METHOD

2.1 *Materials and sample preparation*

The PHV materials studied are detailed in Table 1, representing a range of block copolymers. The table includes their hard and soft segments and the proportions by weight. These include a commercial grade styrene-based triblock copolymer, SEBS, a poly(styrene-*block*-ethylene/butylene-*block*-styrene) which was used previously in the J6-Poli-Valve, and two polyurethane-based multiblock copolymers - a lab-synthesised polyisobutylene-based polyurethane (PIB-PU) and Elast-Eon, a commercial siloxane-based polyurethane (PDMS-PEU). Elast-Eon is a reference biostable block copolymer used in other implant applications.

It is known that the hard segment content and distribution strongly influences mechanical and fatigue properties of these polymers. Phase separation of hard and soft segment blocks in SEBS20 can cause their agglomeration into separate domains with long-range order. These styrene domains are cylindrical, and therefore produce melt flow induced anisotropy. As PIB-PU and Elast-Eon are multiblock copolymers, they form a disordered domain structure with no long-range order, meaning they are isotropic. Considering this, all samples for material characterisation were solvent cast (SC) from the polymer with tetrahydrofuran (THF) solvent. This ensured isotropic properties in all three materials.

Table 1. Block copolymers studied for the PHV model.

Material	Hard Segment (H)	Soft Segment (S)	H:S Ratio
SEBS20	Styrene	Ethylene/butylene	20:80
PIB-PU	Polyurethane	Polyisobutylene	30:70
Elast-Eon	Polyether urethane	Polydimethylsiloxane	40:60

2.2 *Cyclic mechanical test*

An Instron ElectroPuls E10000 fitted with a 250 N dynamic load cell was used to conduct cyclic stress-strain tests on ISO-37 Type 4 dumbbell specimens. All tests were carried out at a frequency of 1 Hz under displacement control. Preconditioning was carried out for 100 cycles at a strain 10% greater than the test strain and then cycled at 100% strain to 1000 cycles to ensure the behaviour was stable. Stress-strain data for the 1000th cycle was used to calibrate the hyperelastic functions used in the FEA models.

2.3 *Crack growth experiment*

Cyclic crack growth tests were run to calibrate the fatigue profiles for the material selection. This was carried out over two stages on pure shear specimens. Firstly, cyclic planar tensile tests were run to calibrate the strain energy density, SED or w, followed by crack growth tests. Rectangular samples, 30 mm x 130 mm cut from SC sheet samples were prepared for SEBS20 and PIB-PU. Due to limitations in material supply, Elast-Eon could not be tested in the same trial. However, crack growth in Elast-Eon has previously been reported by Chaffin et al. (2013) so their results are referenced. After gripping the samples, the height of the test region measured 10 mm ensuring that width to height ratio with surface crack was maintained with at least a 10:1 ratio (Mars & Fatemi 2002).

For the initial mechanical characterisation, single, uncut samples were loaded onto a Stable Microsystems Texture Analyser fitted with a 300N load cell set to a cycling frequency of 1 Hz. Samples were cycled at progressively increasing and decreasing strains from 30-110% in 10% intervals, each for 100 cycles. The stress-strain (σ-ε) data for each cycle was post processed using a MATLAB script. This converted the force/displacement profile into an equivalent stress/strain curve, then used numerical integration to calculate the SED, Equation 1. The planar mechanical test acted as a preconditioning stage, stretching the sample to elevated strains, and reducing the influence of cyclic stress softening on the fatigue crack growth measurements.

$$w = \int_0^{\varepsilon} \sigma \, \mathrm{d}\varepsilon \tag{1}$$

For the crack growth test, a 5 mm cut was introduced to the edge of the sample using a scalpel. The samples were then cycled through a range of strains, increasing in 10% intervals. Generally, samples were tested between 30% and 110% strain, with a few tested at higher levels. Typically, the samples were tested for 12,000 cycles at the lower strain range; however, the number of cycles were reduced as the strain amplitude increased.

The sample was cycled at each strain until there was measurable crack growth. Using correlated digital imagery, the crack's length was monitored throughout the test, allowing the crack growth rate to be evaluated in each strain stage. Using Equation 2 the tearing energy, T, was calculated in terms of the peak SED per cycle and the sample's undeformed height, h. Once known, the relationship between tearing energy and crack growth rate was plotted and used to calibrate cyclic fatigue parameters, A and F, Equation 3. This applies up to a critical tearing energy, T_c, normalised against the transition tearing energy, T_t.

$$T = wh \tag{2}$$

$$\frac{\mathrm{d}a}{\mathrm{d}N} = A\left(\frac{T}{T_t}\right)^F \qquad T_t \leq T \leq T_c \tag{3}$$

2.4 *Durability testing*

DuraPulse™ heart valve durability testers by ElectroForce were used to measure (N_f) of PHV prototypes. Using the rapid-failure rigid post design, the durability of PHVs made from each of the three materials were tested. The PHV prototypes were compression moulded and loaded into a specially designed rigid post holder and cycled open and closed as shown in Figure 1. Testing PHVs with the stent structure held fixed in place alters the deformation such that the stent remains relatively rigid when the leaflets cycle. The method is necessary to ensure timely sample failure within months, as test lengths on normal PHVs would last multiple years.

The PHV prototypes were cycled at an accelerated frequency of 30 Hz, as compared to the more typical 1 Hz in the case of a native heart valve. The test temperature was 37°C. The setup ensured a 13.3 kPa (100 mmHg) pressure difference across the closed PHV for at least 5% of each cycle which is sustained for > 95% of the total number of cycles. This was done in accordance with ISO 5840:2021.1 standards.

An increase in pressure and motor output readings from their steady-state values was used to evaluate failure in PHV prototypes. Cracks visible in the samples were observed after testing was stopped. It is important to note that the rapid-failure rigid post testing configuration primarily examines the integrity of the PHV leaflets, so comes with the compromise of not being fully reflective of the free post loading conditions for a normal PHV prototype. However, it is effective for comparing a range of experimental variables, in particular, the durability performance of these materials. It is also functional as a benchmark for calibrating the fatigue modelling approach.

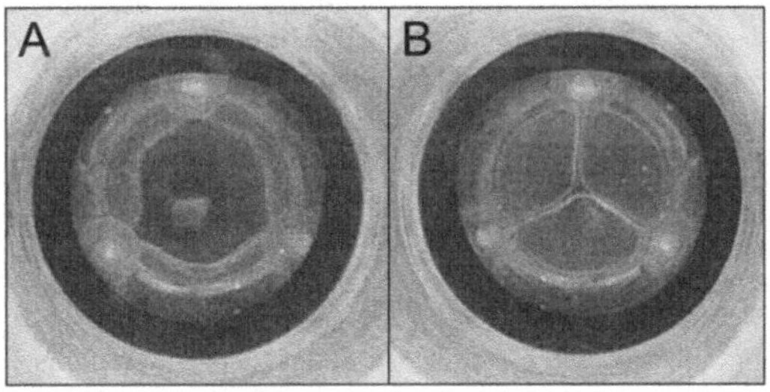

Figure 1. Compression moulded PHV prototype held in a rigid post holder for rapid-failure accelerated durability testing: A) while opening; B) upon closing.

2.5 *Finite element analysis*

Using a customised VUMAT code, the fatigue life of the components was evaluated. An FEA model was built of the PHV and simulated through a single loading cycle consisting of the valve opening and closing due to applied surface pressures. The VUMAT code expanded a typical Neo-Hookean hyperelastic model to include a fracture mechanics framework. Throughout the simulation, the VUMAT calculated the current biaxiality and cracking energy density, CED or w_c, in each element from which the tearing energy could be determined (Windslow et al. 2020). Once the tearing energy was known, the number of cycles to failure could be predicted using the cyclic fatigue parameters. The VUMAT stored the lowest fatigue life achieved for each element to allow the overall life of the PHV to be predicted using FEA.

The basis of this method is the relationship between CED and SED with biaxiality. Based on physical testing and initial FEA, PHVs are relatively small strain applications with strains typically below 20% for these materials. As such, the small strain CED ratio equation provided in Mars & Fatemi (2006) and updated in Zine et al. (2006) is applicable, Equation 4, where ν is Poisson's ratio and θ is crack angle relative to the stretch term λ_1. This allows the ratio of CED to SED to be evaluated as a function of biaxiality, with the biaxiality defined using Equation 5.

$$\frac{w_c}{w} = \frac{(Bv+1)\cos^2\theta + (B^2 + Bv)\sin^2\theta}{B^2 + 2Bv + 1} \quad (4)$$

$$B = \frac{\ln\lambda_2}{\ln\lambda_1} \quad (5)$$

Once the CED is known, it can be related to the tearing energy. Windslow et al. (2020) derived a biaxiality dependent proportionality parameter, k, suitable for complex multi-axial deformations beyond uniaxial tension, Equation 6, where m_m, b_m, m_b, and b_b are FEA-derived fitting parameters (see original paper for their determination). Through this, an equation was derived for the tearing energy of internal penny shaped flaws under complex deformation as a function of the biaxiality and CED, Equation 7. Note, however, the biaxiality was defined in terms of the principal stretch components, Equation 8, to remove the need for critical plane analysis. The VUMAT implemented this equation, Equation 8, within the fracture mechanics framework, with the fitting parameters calibrated to an appropriate Neo-Hookean model.

$$k = \frac{3}{2\pi}\left(\frac{(m_m B + b_m)}{\sqrt{\lambda_{max}}}(m_b B + b_b)\right) \quad (6)$$

$$T = \frac{3}{2\pi}\left(\frac{(m_m B + b_m)}{\sqrt{\lambda_{max}}}(m_b B + b_b)\right)w_c c \quad (7)$$

$$B = \frac{\ln\lambda_{min}}{\ln\lambda_{max}} \quad (8)$$

It should be noted that Equation 4 needs to be defined through the biaxiality definition in Equation 5, before being related to the load mode equivalent biaxialities defined in Equation 8. After this, a relationship between the CED to SED ratio and the biaxiality from Equation 8 is formed and fitted numerically. Figure 2 demonstrates the fitted relationship of CED/SED to biaxiality. In this case, the fitting function is only applied on the range (-2,-1) because the CED/SED ratio is constant between (-1,-0.5) (Windslow et al. 2020).

The VUMAT works as follows: In the first instance the code works as a standard Neo-Hookean hyperelastic model. Once the stress in each element has been determined, the code enters the fracture mechanics framework. Here, the stretch in the element is used to determine the principal components from which the biaxiality is determined using Equation 8. Given the SED is already calculated per the

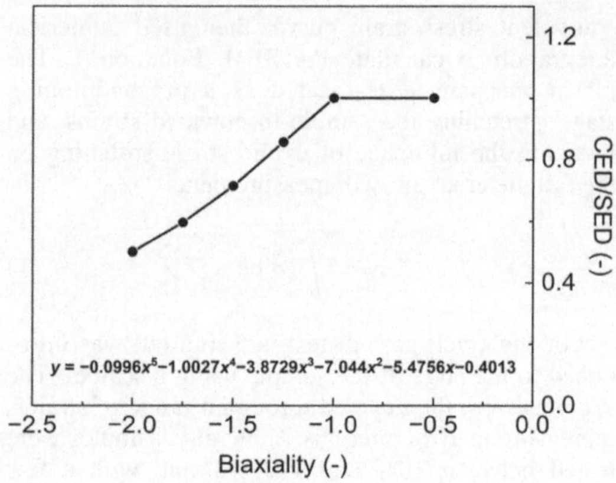

Figure 2. CED/SED small strain relationship to biaxiality.

Neo-Hookean section, once the biaxiality is known, the CED can be determined using the small strain solution, Equation 4. The proportionality parameter is then calculated for the current biaxiality through Equation 6 with the fitting parameters defined for a Neo-Hookean energy function. Once the proportionality parameter and CED are defined, the fatigue life can be estimated using the familiar fatigue approach, Equation 9, where A and F are the scalar and power terms of the material's fatigue profile and calibrated through laboratory data according to Equation 3, where c_0 is the inherent flaw size. These parameters are input material parameters to the subroutine. The VUMAT used Equation 9 to determine the current fatigue life based on the state in the current step, and then compared it to the prior minimum value to determine the lowest fatigue life for the element.

$$N_f \approx \frac{1}{F-1}\frac{1}{A(kw_c)^F}\frac{1}{C_0^{F-1}} \quad (9)$$

3 RESULTS AND DISCUSSION

3.1 *Cyclic mechanical test*

Cyclic stress-strain testing was required to obtain an input signal for tensile test data for the model which accounted for stress softening in the materials. Figure 3 shows the 1000th cycle stress-strain curves for SEBS20, PIB-PU and Elast-Eon, fitted to a Neo-Hookean relationship in Abaqus with Equation 10. The SEBS20 is the most flexible PHV, whereas the PU-based PHVs are stiffer due to their higher hard segment contents. Table 2 gives material model fitting parameters.

$$w = \frac{G}{2} * (\bar{I}_1 - 3) + \frac{1}{D_1}(J-1)^2 \quad (10)$$

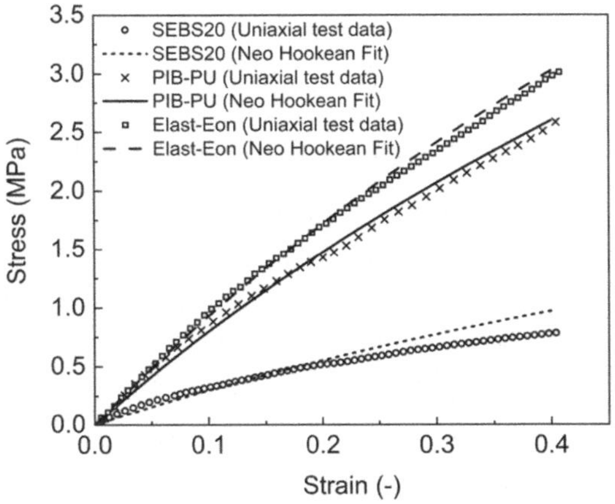

Figure 3. Neo Hookean fit of materials to uniaxial cyclic stress-strain test data of the 1000th cycle, used to model stress softened material behaviour in the PHV application.

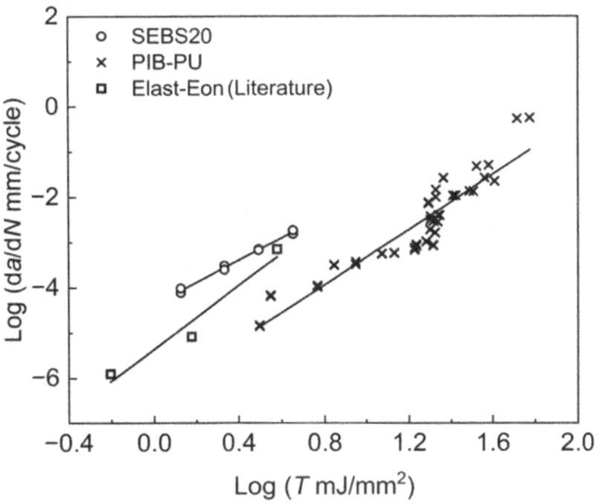

Figure 4. Crack growth experiment results for tearing energy against crack growth rate for PIB-PU and SEBS20 presented alongside literature data on Elast-Eon by Chaffin et al. (2013).

Table 2. Coefficients for Neo Hookean fit of stress-strain data.

Material	G/2	D_1
SEBS20	0.55	0.0036641
PIB-PU	1.46	0.0013719
Elast-Eon	1.70	0.0011770

3.2 *Crack growth experiment*

Figure 4 presents the crack growth behaviour of SEBS20, PIB-PU and Elast-Eon, the latter of which was taken from the literature, Chaffin et al. (2013). Higher tearing energy for a given crack growth rate corresponds to greater fatigue resistance. As such the order of increasing fatigue performance here are SEBS20, Elast-Eon and PIB-PU. Regime 3 crack growth is selected for the characterisation of fatigue parameters (Mars & Fatemi 2002). The linear relationships for each material have been fitted across the full datasets covering regime 3, since it is most reflective of the crack growth mode in the application. The fatigue parameters, A and F, which are the y-intercept and gradient of these lines on the plot, respectively, have been calculated for each material. Table 3 gives the results which were used directly in the CED approach for the FEA model. Considering that Equation 3 applies here, for a T_c of 1, the units of A are reduced to mm/cycle as it is normalised to the tearing energy unit.

It should be noted that there are doubts regarding the fatigue data input for Elast-Eon. Considering it has the highest hard segment content, it is expected to have the greatest fatigue resistance. Furthermore, given it was the stiffest and toughest material (see Figure 3), it should have the greatest fatigue strength. This is not shown in the data. There are questions regarding the test set-up used in the prior work, given the planar sample did not maintain the 10:1 width to height ratio once the edge crack was introduced. This may mean the fatigue data is incorrectly evaluated as planar crack growth. Future work will re-visit the fatigue strength of this material.

Table 3. Crack growth fatigue parameters of three materials studied for the PHV model.

Material	A (mm/cycle)	F (-)
SEBS20	4.25×10^{-5}	2.44
PIB-PU	5.98×10^{-6}	2.43
Elast-Eon	4.35×10^{-6}	3.52

3.3 *Durability testing*

The rapid-failure rigid post experiment was used to test the durability of samples made from SEBS20 (n = 3), PIB-PU (n = 2) and Elast-Eon (n = 4) to provide an experimentally determined N_f to which the computational model outputs can be compared. The average durability of rigid post PHV prototypes of 0.36 mm leaflet thickness made from SEBS20, PIB-PU and Elast-Eon were 19.9 million, 122.6 million and 128.6 million, respectively. Generally, failure occurred either at the leaflet edge or the commissure where the leaflets connect to the posts. Often, crack growth proceeded along the stent-leaflet boundary.

3.4 *Finite element analysis*

The simulated PHVs were opened and closed with pressures correlating to percentage opening and closing values determined during physical testing. The pressures applied in the FEA model were scaled until the same level of opening or closing was achieved as seen in video footage of the PHV leaflets. Results for PHV opening and closing were obtained individually for each material, based on the deformation observed during experimentation. Table 4 presents the results for closing pressures which primarily dictates the PHV durability.

Table 4. Percentage of closing and corresponding pressure for 0.36 mm leaflet thickness PHV prototypes.

Material	Closing (%)	Applied pressure (kPa)
SEBS20	5.19	51.5
PIB-PU	8.10	51.0
Elast-Eon	6.62	100.0

The model outputs a contour plot of the minimum fatigue life across the PHV body. The predicted failure locations were found to correlate with the experimentally observed failure modes discussed above, suggesting failure along the leaflet edge close to the stent boundary. The cracking energy density approach was used to compare tested PHV durability to that determined by FEA. Results are summarised in Table 5. The fatigue life predictions worked well for SEBS20 and PIB-PU, with fatigue life reasonably predicted for sensible inherent flaw sizes. Given the limited crack growth test data from literature for Elast-Eon which had the longest tested fatigue life, and the fact that this compares samples from different batches, the variability seen can be expected. With this, crack growth testing under the same conditions applied for SEB20 and PIB-PU is required. Though overall, these results point towards a highly effective framework for fatigue life prediction for PHVs.

Table 5. Durability determined by FEA model of rigid post PHVs made from each material, for a sensible range of c_0 values.

Material	N_f (Million)			
		c_0 (mm)		Tested
	0.050	0.025	0.010	result
SEBS20	4.6	12.5	46.9	19.9
PIB-PU	51.9	139.4	514.3	122.6
Elast-Eon	334.9	1919.8	19304.6	128.6

4 CONCLUSIONS

An FEA model for a PHV has been developed to accurately simulate the nature of valve opening and closing in rigid post accelerated durability testing. To determine input data for the model: (1) Neo-Hookean fitting of stabilised stress-strain data was used to identify shear moduli and compressibility parameters, (2) crack growth tests enabled A and F for each material to be determined and (3) the opening and closing pressures for SEBS20, PIB-PU and Elast-Eon PHVs with 0.36 mm leaflet thickness were modelled based on experimentally measured percentages followed by determining the pressure required to achieve this state in the FEA model.

The durability of PHVs has been calculated based on the relationship between CED and biaxiality through the implementation of a subroutine. Validation of the model results by comparison to experimental durability measurements has shed light on the promising framework this work provides for fatigue life prediction for PHVs. The results obtained by the model for SEBS20 and PIB-PU are within an acceptable range of approximation when compared to experimental results. It has also been highlighted that application of consistently acquired crack growth test data is important for obtaining reliable fatigue life predictions from the model.

Future work on the FEA model will look towards simulating the normal PHV under free post conditions to explore the PHV durability in service. To ensure accurate predictions for Elast-Eon, crack growth tests will be conducted by the same method applied to SEBS20 and PIB-PU to obtain comparable results.

ACKNOWLEDGEMENTS

The authors thank T. Deodhar, Y. M. Tseng, C. Tantisuwanno, A. Joy, A. N. Pathan, Y. O. Yepes and J. Eagan for the synthesis and preparation of PIB-PU sheet specimens. R. Windslow and T. Hohenberger thank SLB for permission to participate in this research by illustrating how to apply the FEA technique for fatigue life prediction.

REFERENCES

Bavo, A. M., Rocatello, G., Iannaccone, F., Degroote, J., Vierendeels, J. & Segers, P. 2016 Fluid-Structure Interaction Simulation of Prosthetic Aortic Valves: Comparison between Immersed Boundary and Arbitrary Lagrangian-Eulerian Techniques for the Mesh Representation, *PLoS One*, 11(4), 1–17.

Zakerzadeh, R., Hsu, M. C., & Sacks, M. S. 2017. Computational methods for the aortic heart valve and its replacements, *Expert Rev Med Devices*, 14(11), 849–866.

Luraghi, G., Wu, W., De Gaetano, F., Rodriguez Matas, J. F., Moggridge, G. D., Serrani, M., Stasiak, J., Costantino, M. L. & Migliavacca, F. 2017. Evaluation of an aortic valve prosthesis: Fluid-structure interaction or structural simulation? *J Biomech*, 58, 45–51.

B.S.I.B. STANDARD, 2021. BS EN ISO 5840 - 1 : 2021 BSI Standards Publication Cardiovascular implants — Cardiac valve prostheses, Part 1: General Requirements.

Stasiak, J. R., Serrani, M., Biral, E., Taylor, J. V., Zaman, A. G., Jones, S., Ness, T., De Gaetano, F., Costantino, M. L., Bruno, V. D., Suleiman, S., Ascione, R. & Moggridge, G. D. 2020. Design, development, testing at ISO standards and in vivo feasibility

study of a novel polymeric heart valve prosthesis, *Biomaterials Science*, 8(16), 4467–4480.
Patel, R. S. 2023. Material selection and manufacture for a polymer heart valve application, PhD Thesis. University of Cambridge.
Chaffin, K. A., Wilson, C. L., Himes, A. K., Dawson, J. W., Haddad, T. D., Buckalew, A. J., Miller, J. P., Untereker, D. F. & Simha, N. K. 2013. Abrasion and fatigue resistance of PDMS containing multiblock polyurethanes after accelerated water exposure at elevated temperature, *Biomaterials*, 34(33), 8030–8041.
Mars, W. V. & Fatemi, A. 2002. A literature survey on fatigue analysis approaches for rubber, *Int J Fatigue*, 24(9), 949–961.
Mars, W. V. & Fatemi, A. 2006. Multiaxial stress effects on fatigue behavior of filled natural rubber, *Int J Fatigue*, 28(5–6), 521–529.
Zine, A., Benseddiq, M., Nait-Abdelaziz, M., Hocine, N. A. & Bouami, D. 2006. Prediction of rubber fatigue life under multiaxial loading, *Fracture of Nano and Engineering Materials and Structures – Proceedings of the 16th European Conference of Fracture*, 769–770.
Windslow, R. J., Hohenberger, T. W. & Busfield, J. J. C. 2020. Determination of the Loading Mode Dependence of the Proportionality Parameter for the Tearing Energy of Embedded Flaws in Elastomers Under Multiaxial Deformations, *Advances in Polymer Science*, 286, 445–465.

Constitutive Models for Rubber XIII – Dal (Ed.)

Service life estimation of axial air spring sleeves

M. Skoglund & F. Mantwill
Helmut-Schmidt-University/University of the Federal Armed Forces Hamburg, Professorship for Machine Elements and Computer Aided Product Design, Hamburg

R. Klauke
Vibracoustic SE & Co. KG, Advanced Engineering (ESM), Hamburg

ABSTRACT: The service life of air spring sleeves is an important factor for the reliability and safety of vehicles with air spring/ damper systems (Schmidt 2023). This paper presents first experimental investigations of the service life of such sleeves with axially oriented reinforcing cords, using a specially developed test rig. In addition to investigating various rolling geometries under cyclic axial displacement, the test rig also allows for superimposed cardanic excitations, variations of the ambient temperature and of the pressure inside the air spring. The service life of air spring sleeves is described using the Wöhler concept (Wöhler 1870), which is applied to the experimentally determined service life data. In order to validate the Wöhler concept, simulation methods (FEM) are used. The air spring sleeves under investigation are made out of a composite material, which can be modeled in different degrees of complexity (Abaqus 2023, MSC 2020, Donner 2017). In this work, various modeling approaches are compared, focusing on the correct prediction of a potential failure location. The actual quantification of the lifetime in the hotspot is of secondary importance in the context of this paper. Linear damage accumulation is used as the basis for lifetime prediction. The damage totals are compared and correlated with the experimental data. The results show to what extent the Wöhler concept provides good qualitative correlation with the experiments. Assessing the complexity of the presented simulation methods, an efficient basis for the service life analysis of air spring sleeves is given.

1 INTRODUCTION

Throughout the past years air springs became an established part of modern mobility. With a spring rate based on air compression they allow for adjustment to a wide range of loading and driving conditions. Increased demand leads to technological advances and major improvements on the state of the art for air spring sleeves. A robust durability prediction is crucial in this competitive market environment. While models and predictions for solid rubber components are widely advanced, air spring sleeves continuously pose a challenge being composites with thin cross-sections under a complex load scenario.

1.1 *Axial air spring sleeves*

The sleeve is one of the main components of a rolling lobe air spring. It enables for vertical deflection as well as for compliance to superimposed loads induced by the chassis. Axial sleeves mostly have one axial layer of reinforcements embedded in a chloroprene rubber matrix as shown in Figure 1a). Their advantage lies in a low harshness and a relatively high tolerance against torsional loads compared to cross-ply sleeves. In the pressurized air spring, the sleeve forms the so-called rolling fold in the unconfined gap between the rolling piston and the outer support as shown in Figure 1a).

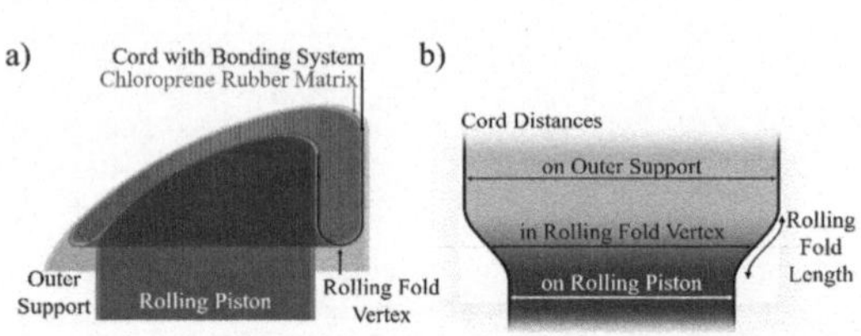

Figure 1. a) Schematic cut through an air spring with axial sleeve; b) Representation of the distance between two neighboring reinforcements within the rolling lobe.

Upon pure axial deflection, the sleeve rolls through the rolling fold. With the given rotational symmetry, the strains on the material induced by the diameter change along the rolling fold are schematically depicted in Figure 1b) as distances between two neighboring cords.

Throughout their service life, sleeves are subjected to high loads as well as to a wide range of

DOI: 10.1201/9781003516880-50

loading frequencies and temperatures. A robust assessment of these influences on service life has a direct impact on the final design. Therefore, it plays an important role in the product development process. Accordingly, complementing a simple test setup by a robust service life estimation contributes significantly to quick development loops.

2 EXPERIMENTAL INVESTIGATIONS

2.1 *Concept of the double rolling fold test rig*

Failures in the field occur in the rolled-through area of the sleeve in the form of a crack in the rubber matrix close to the reinforcements. In order to reproduce these failures in an experimental environment, a so-called double rolling fold test rig has been developed and is in use for detailed investigations on the service life evolution. The test setup is depicted in Figure 2.

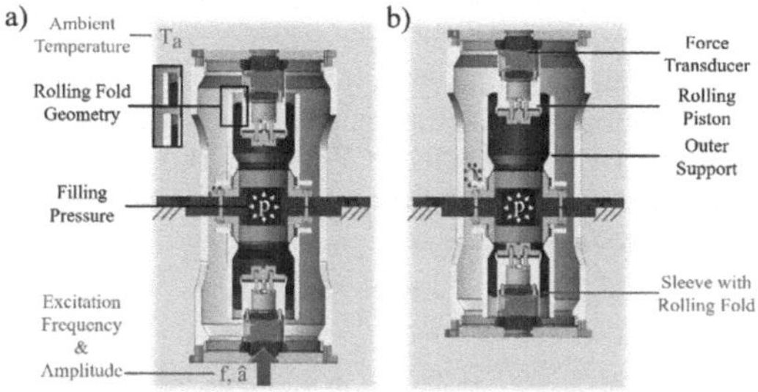

Figure 2. Schematic of the double rolling fold test rig a) In undeflected state with test parameters and b) In undeflected state with components.

It allows for adaption to geometric boundary conditions, namely rolling piston and outer support diameters as well as for a wide range of axial deflection amplitudes and loading frequencies. Furthermore, filling pressure and ambient temperature can be adjusted.

The double rolling lobe concept is first introduced in (Wode 1995). Two rolling lobe air springs share a common air volume which results in a constant pressure throughout the whole deflection cycle. In addition to previous applications of the test method (Wode 1995, Förster 2012), the test rig design at hand allows for individual replacement of each sleeve within a test nest. Furthermore, the rolling force is measured independently by a force transducer in the rolling piston (Lips et al. 2023). In addition, a variety of load cases from pure axial deflection to superimposed torsional and cardanic loads can be applied. Only axial loads are investigated in the context of this paper, as they already show a majority of the occurring failure scenarios and form the basis for a comparative analysis with superimposed load cases.

2.2 *Test program*

With the given constellation of sleeve and test rig, the following parameters are varied in order to investigate their influence on the service life of the sleeve composite:

a) Filling pressure
b) Rolling fold width
c) Excitation frequency
d) Ambient temperature

Based on those parameters, the test program depicted in Table 1 has been carried out. Only one parameter at a time is varied in order to assess relevant tendencies. The full variation range for each parameter is also given in Table 1.

Table 1. Test program for basic durability investigations on the double rolling fold test rig.

	Pressure	Freq.	Fold Width	Amp.	Temp.
Config.	p [bar]	f [Hz]	b_{RF} [mm]	$\hat{a}$ [mm]	T [°C]
Initial	2	4	20	51.4	23
Variation a)	**10**	4	20	51.4	23
Variation b)	2	4	**15**	**43.6**	23
Variation c)	2	**2**	20	51.4	23
Variation d)	2	4	20	51.4	**65**
Range	2…16	2…6	10…20	35.7… 51.4	23… 90

The initial configuration for the measurement series comprises an internal pressure of 2 bar, an ambient temperature of 23°C (room temperature) and an excitation frequency of 4 Hz for a rolling fold width of 20 mm. All tests in this program are conducted with a strain ratio R=0.

For the air spring design, the pressure and the effective area defined by the geometric boundary conditions for the rolling fold (see Lips et al. 2023) are major design variables. On component level, the pressure is determining the preload on the reinforcements and the surrounding rubber membrane. While field operating pressures for axial sleeves are higher, a low pressure for the initial configuration was chosen to allow for comparability to simplified component tests as conducted in (Schmidt 2023, Nguyen 2023). On component level, a narrower rolling fold results in a reduction in strain amplitude (see Figure 1b)). In the air spring, a decrease in rolling fold width requires higher pressures and an increased internal air volume to realize the same reaction force and characteristics.

It is important to note that different rolling fold widths result in different displacements required for the rolling lobe vertex to be subjected to the minimum and maximum strains (Lips et al. 2023).

Therefore, test amplitudes are normalized to the amplitude required for a reference sleeve length of 20 mm to be subjected to the full strain range. Regarding fatigue life this ensures quantitative comparability and exceeds the threshold amplitude defined in (Rambacher 2014).

Besides the design variables, applied loads and environmental conditions are largely influencing the service life. While the strain amplitude is varied with the rolling fold width, the loading frequency is altered in a separate configuration. This is to cover possible self-heating effects induced by higher frequencies. These add on the environmental temperature, posing a great challenge for automotive rubber parts. In order to assess the influence of the ambient temperature on durability, a temperature variation is carried out.

2.3 *Test results*

As the durability tests are conducted on component basis and under operational conditions, the classic Wöhler approach (see Wöhler 1870) is adapted: Instead of homogeneous deformations with defined stress-strain-states, part Wöhler curves, considering the above-mentioned parameters are shown and compared. A hot spot location is not considered, as the part Wöhler curves are of quantitative nature only. Regardless, for the campaign at hand, failures exclusively occurred in the reference region in the form of a cord-parallel crack resulting in pressure loss. Figure 3 shows the results for the pressure variation.

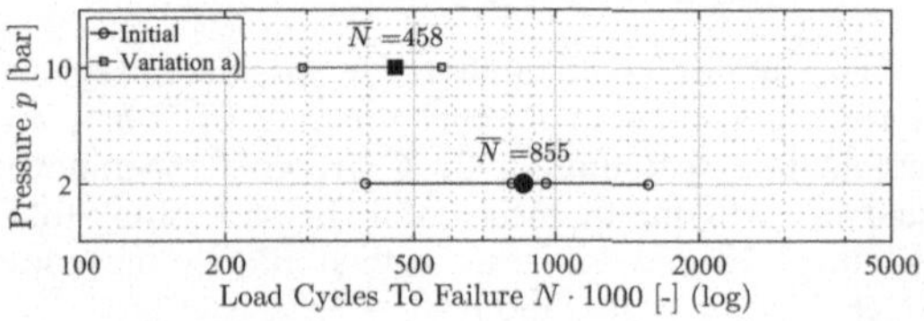

Figure 3. Fatigue life on the double rolling fold test rig for the Initial Configuration and for a variation of the filling pressure (Variation a)).

While the ordinate axis indicates the varied parameter, durability results are displayed on a logarithmic scale on the abscissa. Assuming a log-normal distribution of the results, the respective logarithmic mean value $\bar{N}$ is computed using Equation 1 and expressing a 50% failure probability (see DIN 2016).

$$\bar{N} = N_{50\%} = 10^{\left(\frac{1}{n}\sum_{i=1}^{n}\log N_i\right)} \tag{1}$$

where n is the number of samples per parameter and N are the endured load cycles for sample i.

It can be seen that an increase of pressure by a factor of five results in mean fatigue life being halved for the study at hand.

The sensitivity of the service life to the width of the rolling fold is given in Figure 4.

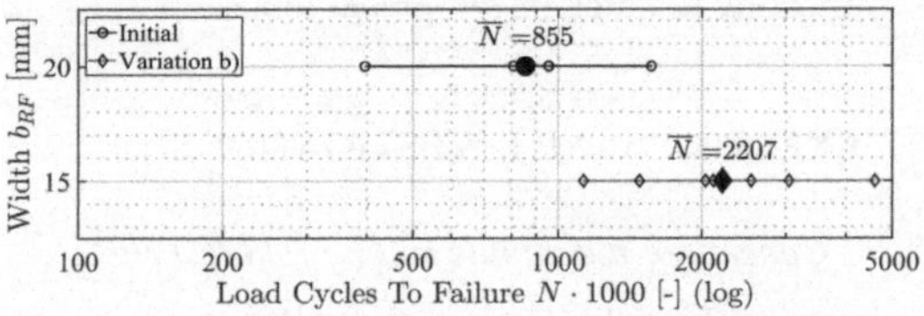

Figure 4. Fatigue life on the double rolling fold test rig for the Initial Configuration and for a variation of the rolling fold width (Variation b)).

As can be seen, a rolling fold width reduction by 5 mm results in a significantly increased fatigue life.

Given the thin cross-section of the sleeves in a relatively large environment and without mentionable thermodynamic effects through air compression, frequency-induced self-heating is assumed to be negligible within the range defined in Table 1. Accordingly, the results in Figure 5 show a minor frequency effect. For future tests, this allows for reduced testing time.

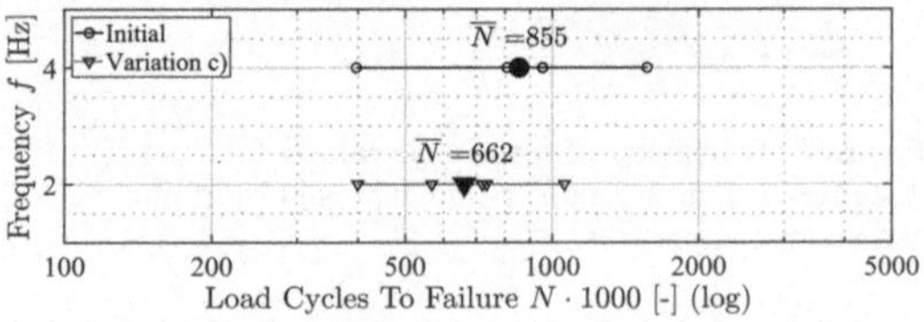

Figure 5. Fatigue life on the double rolling fold test rig for the Initial Configuration and for a variation of the excitation frequency (Variation c)).

Looking at aging under increased temperatures though, the testing frequency directly affects the exposition time. Despite these effects being investigated in future studies, a temperature variation has already been carried out for this paper. The results are shown in Figure 6.

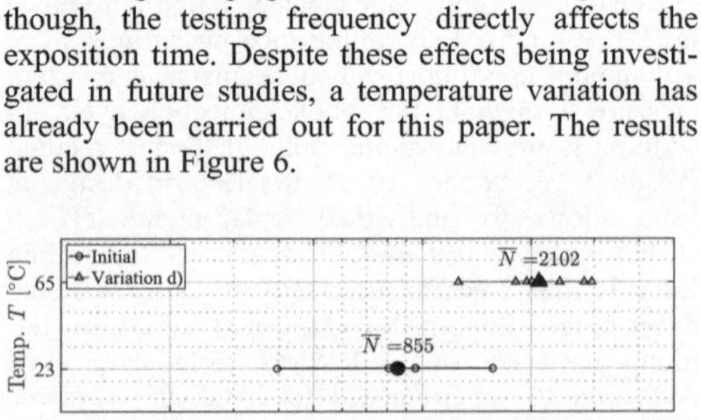

Figure 6. Fatigue life on the double rolling fold test rig for the Initial Configuration and for a variation of the ambient temperature (Variation d)).

It can be observed that temperature seemingly increases the lifetime on part level. While this seems contradictory to common literature about rubber durability (see Mars & Fatemi 2004), possible reasons for this inconsistency are discussed in Section 4.

In order to summarize the results for a first sensitivity analysis, a normalization to the respective parameter range in Table 1 is applied according to Equation 2.

$$S_{norm,\,P} = \left(\frac{\bar{N}_P - \bar{N}_{initial}}{\bar{N}_{initial}}\right) \cdot \left(\frac{V_P - V_{initial}}{R_P}\right)^{-1} \quad (2)$$

where $S_{norm,\,P}$ is the normalized sensitivity to the parameter P. $\bar{N}_P$ and $\bar{N}_{initial}$ are the mean load cycles for the respective configuration and the initial configuration (see Equation 1). V_P and $V_{initial}$ are the parameter values for the configuration at hand and the initial configuration. R_P is the parameter range as given in Table 1. The results are shown in Figure 7.

It gives the sensitivity to parameter changes for the mean values as well as for respective minimal and maximal endured load cycles. A sensitivity of *0* would indicate no influence of a parameter change.

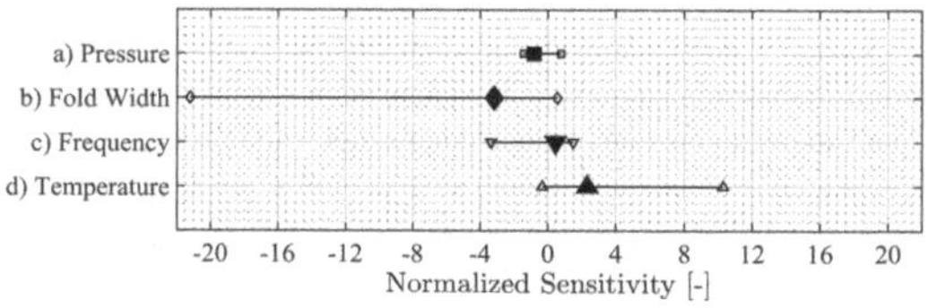

Figure 7. Normalized Sensitivity of fatigue life to changes of pressure, rolling fold width, excitation frequency and temperature.

A negative sensitivity value corresponds to an inversely proportional relation between altered parameter and fatigue life. It is to be noted, that this representation is only valid for the parameter changes presented in this paper, neglecting possible durability optimums or non-linearities. However, some general assumptions can be made: A pressure increase corresponds to a relatively slight decrease in mean fatigue life. A frequency change seems to be of minor effect. Rolling fold width and temperature however, significantly influence fatigue life with the aging related effect of temperature being especially controversial.

3 SIMULATIONS

3.1 *Simulation methods*

For the durability prediction of axial sleeves, a suitable FEA-formulation is needed. On the one hand, all important aspects need to be part of the model. On the other hand, less details lead to a more efficient and stable computation. A good compromise is the basis for a meaningful prediction. State of the art simulations for composites are based on rebar elements. Due to the selected load case with axial loads only, and the rotational symmetry of the sleeves, it is sufficient to simulate only a sector. For the modelling of the composite, three approaches are used for a comparative analysis.

1. Reinforcements based on rebar-elements
2. Modelling the reinforcements of the composite as stiff thin anisotropic plastic rods
3. Consideration of the complex yarn geometry and the interaction with the rubber matrix (see Donner 2017).

In practice, the rebar element formulation is mainly used to derive basic air spring characteristics like force-deflection curves, resulting stiffnesses or to check kinematics on a macroscopic scale. Due to its nature, rebar elements cannot highlight dedicated hotspots at the interface between cord and matrix. Therefore, they are not suited for a precise service life estimation but rather for basic A-B comparisons.

A more complex approach includes a model of the cords in the form of plastic rods as presented in (Hohmann 2013). Although the cord stiffness is based on an anisotropic material formulation, the stress strain state at the hotspot cannot be used for a strain based durability approach, as these simplified cords have a non-negligible stiffness against compressive loadcases, which is in contrast to real cords. Still, a qualitative A-to-B comparison, also in terms of hotspot locations, is valid. The benefit of this formulation is that a higher order material model can be used for the rubber component, taking important mechanical properties like Payne- and Mullins-effect into account. The used parameters are fitted to the air spring sleeve material and ensure the smoothing of strain hot spots due to the Mullins effect and thus prevent a too conservative durability prediction.

The most complex approach used in the context of this paper is based on (Donner 2017, Khattabi 2023). Using this approach allows for a detailed analysis of the stress-strain state in the hot spot. Therefore, it is perfectly suited for a durability estimation.

3.2 *Simulation results*

Using the boundary conditions defined in chapter 2, and a basic global durability approach, method 2 yields the result shown in Figure 8.

A hot spot is indicated at a plausible location, the underlying structural values however are not matching reality. Due to the lacking precision this could easily lead to misjudgments of the service life. Therefore, method 2 bears a high risk for misinterpretations. In comparison to that, method 3 is highly detailed as depicted in Figure 9.

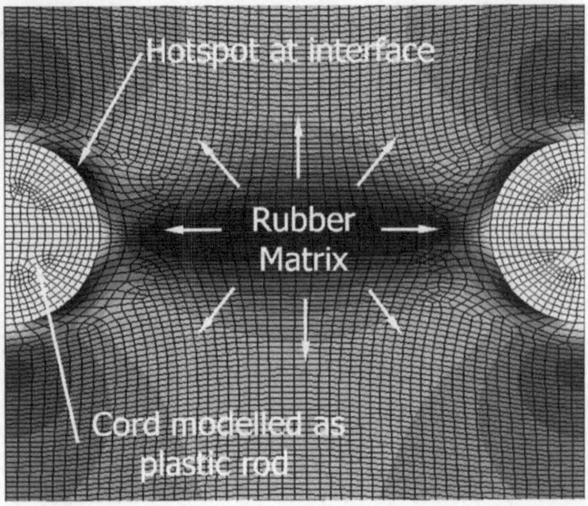

Figure 8. Method 2 with Hot Spot.

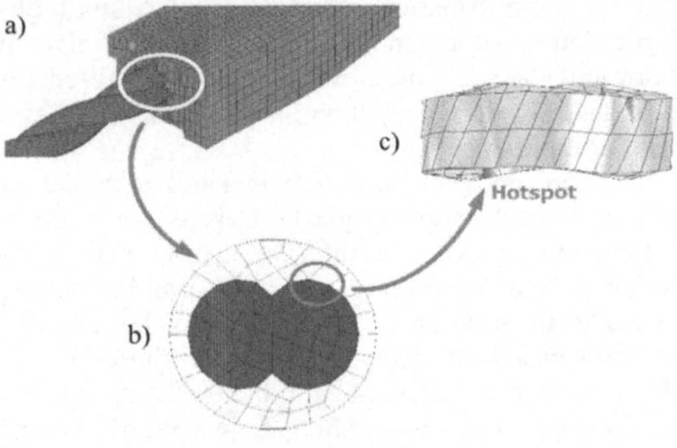

Figure 9. Method 3 with a) detailed yarn embedded in matrix. Strain hotspot location on b) cross-section and c) cord surface.

Despite high computation times needed for this approach, the information stored in the hot spots can be plausibly used for a local durability prediction concept.

4 DURABILITY CONCEPT

4.1 *Damage criterion and accumulation*

For the investigations carried out in the context of this paper, linear damage accumulation (Miner 1945) and maximum principal strains are used as damage criteria. With these assumptions at hand, a simple comparison of the measured fatigue life and the simulation results for given boundary conditions can be done.

The comparison is carried out for the initial configuration using simulation results for the rolling fold and transferring them on a hotspot Wöhler curve obtained by (Schmidt 2023). Knowing that pressure, R-ratio and other quantities play a role, this basic approach is used to show the general feasibility.

4.2 *Validation against measurements*

Figure 10 shows the Wöhler curve from (Schmidt 2023) and the corresponding part-Wöhler curve carried out in the context of this paper with the Initial Configuration and Variation b).

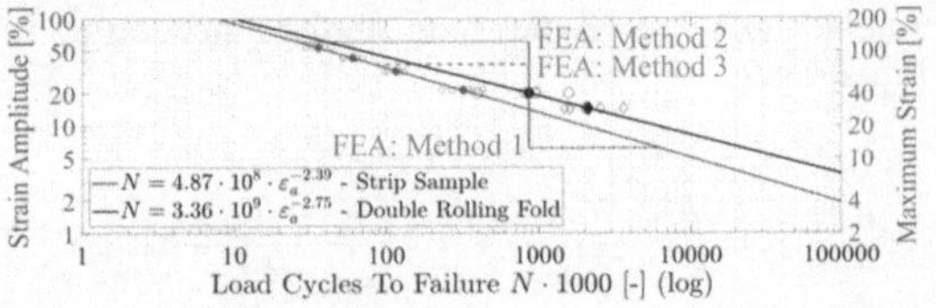

Figure 10. Wöhler curve from strip sample and part Wöhler curve from double rolling fold test rig with simulation results for R=0 and room temperature.

All measurements were taken with an R-ratio of R=0 and under room temperature (23°C). The left ordinate gives the strain amplitude. The right ordinate shows the corresponding maximum strains.

In (Schmidt 2023), linear damage accumulation is proven valid for the strip sample. Accordingly, simulation results should be applicable. However, the part and the hotspot Wöhler curves show a distinct offset as well as a deviation in slope.

Table 2 gives the measurement results for the Initial Configuration on the Double Rolling Fold Test Rig and the respective result for the strip sample. Using these, predictions of the service life for the three FEA methods are computed by projecting the maximum principle strains on the hotspot Wöhler curve (see Figure 10).

Table 2. Service Life from durability tests for the Initial Configuration against predictions based on simulation approaches.

	Max. Global Strain	Hotspot Strain	Load Cycles
	ε_{max} [%]	ε_{HS} [%]	$\bar{N} \cdot 1000$ [-]
Double Rolling Fold	42.1	-	**855**
Strip Sample	42.1	...-	**396**
Simulation Method 1	42.1	12.4	*6204*
Simulation Method 2	42.1	121.7	*27*
Simulation Method 3	42.1	75.1	*84*

As expected, the rebar approach (method 1) yields a relatively low maximum strain which results in a massive overestimation of service life. Method 2 yields a contrary result based on very high local strains at the interface between the plastic rod and the matrix. Method 3 seems to deliver a reasonable result, the derived prediction however is far below the actually achieved fatigue life.

This may be attributed to different damage scenarios of strip sample and part as indicated by the offset between the Wöhler curves. Another reason could be the parameter set fitted to the sleeve material, resulting in mismatched simulated strains. In order to test these assumptions, simulations for different rolling fold widths will be carried out as part of future investigations.

4.3 *Temperature effects*

The simulations in this paper were carried out for material parameters under room temperature. Transferability to higher temperatures would require changes in the material model, especially with regards to aging. Aging processes under raised test temperatures lead to a permanent set in the sleeve, thus diminishing the effective strain amplitude over time. This setting behavior is discussed in (Lips et al. 2023), where it could be shown, that the effect is reversible under cyclic loading.

In Figure 11 however, a permanent deformation within the area of the initial rolling fold *(Marker 4)*, the area under loading *(Marker 5)* and for permanently strained areas on the outer support *(Markers 1 to 3)* is observed for samples cycled under raised temperature.

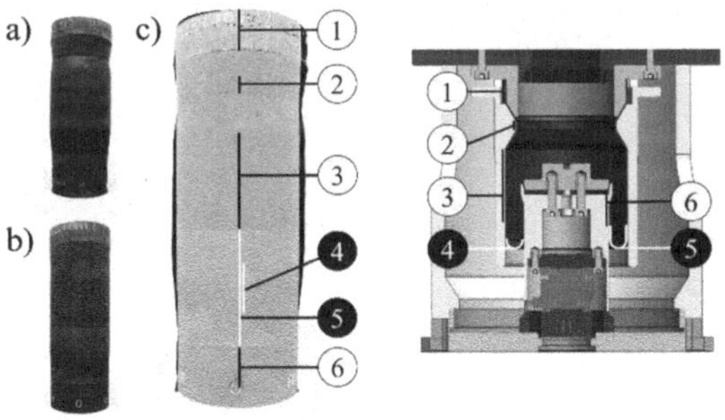

Figure 11. Axial sleeves after durability test under different ambient temperatures. a) 65°C b) 23°C and c) Diameter comparison with regards to positions in the test rig.

Subsequently, the applied load is in reality a collective with continuously decreasing strain amplitudes. This results in an improved service life throughout the experiments (see Figure 6). Those improvements have to be considered with care, as excessive aging leads to destruction of the rubber's structure and therefore to a decrease of life. In addition, these observations made under laboratory conditions do not necessarily reflect aging effects in the field, where an improvement of lifetime due to aging has not been observed yet.

5 SUMMARY & OUTLOOK

This paper presents the first investigations towards a durability prediction of axial air spring sleeves, using simplified experiments on a double rolling fold test rig and FE simulations. By varying certain boundary conditions, load cases and environmental conditions, a sensitivity to the investigated parameters is derived. It can be shown, for the variations at hand, that loading frequency and filling pressure have a minor influence on fatigue life. Variations of the rolling fold width and ambient temperature however, largely affect the endurable load cycles. A first assumption regarding the improved performance under raised ambient temperature is given. A validation for this hypothesis will be carried out in the future, as well as an expansion of the test program with different R ratios.

For R = 0 and one configuration, FE simulations with different levels of complexity were carried out. It is shown, that the hot spots for the chosen damage criterion of maximum principal strain fit the failure spots in the measurements.

Using a linear damage accumulation and Wöhler results obtained on simplified strip specimens, no qualitative correlation can be shown. Possible reasons for this mismatch may lie in different damage scenarios or in the material model.

Future investigations will involve Computer Tomography analysis of axial sleeves after failure and in the course of interrupted fatigue tests in order to assess the damage mechanism and damage evolution. This information can be used to improve simulations and model-based predictions. Subsequently, it allows for developing and assessing even more simplified and robust test specimens to further reduce testing efforts and to speed up development loops.

REFERENCES

Schmidt, P. 2023. Untersuchung der Schädigungsmechanismen und Ableitung eines einfachen Schädigungsmodells für Luftfedern anhand von Bauteilmessungen und Messungen an Streifenprüfkörpern, Master Thesis, Hamburg University of Technology.

Wöhler, A. 1870. Über die Festigkeitsversuche mit Eisen and Stahl. *Zeitschrift für Bauwesen Vol. 20*: 73-106

ABAQUS Theory Manual. 2023, Dassault Systèmes.

MSC. 2020. MARC 2020 Volume A: Theory and User Information. MSC. Software Corporation.

Donner, H. 2017. FEM-basierte Modellierung stark anisotroper Hybridcord-Elastomer-Verbunde, Dissertation, TU Chemnitz.

Wode, S. 1995. Experimentelle Untersuchungen zum Ermüdungsverhalten von Rollbalg-Luftfedern, Dissertation, Leibnitz Universität Hannover.

Förster, K. 2012. Ein Beitrag zur Untersuchung der Lebensdauerabschätzung und des Walkverhaltens von Luftfederbälgen mittels statistischer Methoden, Dissertation, Universität der Bundeswehr Hamburg. Aachen: Shaker Verlag.

Lips, M.; Breitenbach, S.; Schmidt, P.; Mantwill, F. 2023. Walkkräfte an Axialbälgen von Rollbalg-Luftfedern. *Tagungsband Ilmenauer Federntag*. Ilmenau: Verlag ISLE.

Nguyen, T.T. 2023. Analysis of the fatigue strength of elastomer components in air springs to validate a dynamic test on alternative samples, Master Thesis, Hamburg University of Applied Sciences.
Rambacher, C. 2014. Vereinfachte Lebensdauerersatzprüfungen für Luftfedern, Dissertation, Universität der Bundeswehr Hamburg. Aachen: Shaker Verlag.
Deutsches Institut für Normung e. V. (DIN) 2016. Schwingfestigkeitsversuch – Durchführung und Auswertung von zyklischen Versuchen mit konstanter Lastamplitude für metallische Werkstoffproben und Bauteile, DIN 50100.
Miner, M. A. 1945. Cumulative damage in fatigue. *Journal of applied mechanics 12 (3)*: 159–164.
Hohmann, F. 2013. Vergleich von Simulationsansätzen in der FEM von Luftfedern mit Axialbalg hinsichtlich Fadenhaftung und Submodelltechnik, Master Thesis, Hamburg University of Applied Sciences
Khattabi, O. 2023. Internal project report. *Modeling, simulation and design of air springs with axial sleeves*. Technical University of Chemnitz, Institut für Festkörpermechanik. Unpublished.
Mars, W.V., Fatemi, A. 2004. Factors that Affect the Fatigue Life of Rubber: A Literature Survey. *Rubber Chemistry and Technology 77 (3)*: 391–412.
Freund, M. 2013. Verallgemeinerung eindimensionaler Materi-almodelle für die Finite-Elemente-Methode, Dissertation, Düsseldorf: VDI-Verlag.

Smart elastomer materials applications and modelling

Constitutive Models for Rubber XIII – Dal (Ed.)
© 2025 The Author(s), ISBN 978-1-032-85138-9

Viscoelastic effects within the sound reflection of elastomer attenuation layers

L. Schasse & J. Meier
Deutsches Institut für Kautschuktechnologie e.V., Hannover, Germany

E. Schmidtke
Wehrtechnische Dienststelle für Schiffe und Marinewaffen, Maritime Technologie und Forschung (WTD 71), Eckernförde, Germany

ABSTRACT: Elastomer components typically exhibit a pronounced temperature and frequency-dependent mechanical behaviour, which is usually tested exclusively by dynamic shear tests. These material properties are also reflected in the acoustic structure-borne sound properties, which, however, are additionally influenced by the compression modulus, which, according to recent studies, also shows viscoelastic contributions. In this study, the influence of this material behaviour on the attenuation of a structure-borne sound wave passing through an elastomer layer is considered numerically. The simulations are carried out using a model coupling acoustics and structure mechanics built with representative unit cells and periodic boundary conditions. The simulation results clearly show the effect of an appropriate viscoelastic material modelling on the predicted overall reflection damping of unstructured and structured attenuation layers.

1 INTRODUCTION

Modelling of the dynamic-mechanical characterization of elastomers is widely done within the frame of linear viscoelasticity together with the assumption of the thermo-rheological simplicity (Elias 1996). The mechanically accessible frequency range is limited to several hundred Hertz but acoustic transmission in the ultrasonic regime has been shown useful to acquire direct material response up to one Mega-hertz (Kroll 2006, Maslak 2015).

In previous studies, the authors discussed homogenized material properties of massive and of porous material in the sense of shear modulus and longitudinal modulus, both with respect to frequency.

Additionally, it has been shown recently that the structure-borne sound properties of elastomers depend significantly on the shear-related damping properties, but are also influenced by the compression modulus, which is also found to display viscoelastic contribution (Schasse 2022).

In the following, the technical relevance of the viscoelastic behavior of the compression modulus will be discussed by means of simulation for the case of the sound reflection of elastomer attenuation layers.

Here, the influence of the material-specific viscoelasticity for structured and non-structured layers in an underwater structure is simulatively analyzed for elastomer-based acoustic damping layers. For this purpose, the reflection passage of a structure-borne sound wave through an elastomer layer is analyzed using an acousto-mechanically coupled model, which is constructed from a representative unit cell and periodic boundary conditions.

2 MOTIVATION

Generally, for the modelling of acoustic relevant damping modes only shear related viscoelasticity is assumed, e.g. in constrained layer damping treatments for used for vibration damping with reasonable predictions for the functionality (Gröhlich 2021).

One might ask for a molecular scaled reasoning, why the compression modulus should be considered as viscoelastic. One possible consideration on this is the following.

Technical polymers are mostly of organic type, i.e. macromolecules with carbon backbone.

If one considers other carbon-based materials, coal or a diamond, their reaction force on deviatoric shear or on volumetric compression is of an energy-elastic nature and results from the local response of covalent bonds on changes of length or angle. The corresponding moduli for e.g. semi-anthracite are of the order 3 GPa and 8 GPa for the shear modulus and the compression modulus, respectively.

DOI: 10.1201/9781003516880-51

In contrast to this, it is well known that the shear modulus of polymers (and of elastomers, too) is of enthropy elastic nature and strongly correlates to the free volume available for the coordinative change of conformation or for reptation (Doolittle 1951). The outcome is a shear modulus that is three decades lower than that of coal. As the compression modulus is also lower by a less pronounced factor of about one decade, it can be assumed that also with respect to compression of polymers or elastomers the free volume is of importance for the local material response.

Recent studies have confirmed such a frequency dependent viscoelastic behavior of the compression modulus (Schasse 2022).

From a technical point of view, it is worth to estimate the relevance of this findings. It can be argued, that due to the large Poisson ratio of typically about 0.499 the viscoelastic nature of the compressions modulus only becomes apparent under conditions where the energy density is forced to contain a substantial volumetric contribution. This can be assumed, if the energy transport velocity reaches the structure born sound velocity, since in this case viscoelasticity alone is not sufficient to describe the material behavior. Instead, the compression modulus and the density are known to both contribute to the structure borne sound velocity.

Here, the case of the attenuation of transmitted structure-born sound within a rubbery layer is discussed for a broad frequency range.

And the question put-forth is, whether the frequency dependency of the compression modulus is relevant for the absorption properties of an elastomer attenuation layer.

3 EXPERIMENTAL PROCEDURE

3.1 *Materials*

The material in investigation is based on a styrene-butadiene rubber (SBR) Sprintan SLR 4602 with a glass transition outside the frequency ranges of both measurement ranges described below, the mechanical (< 100 Hz) and the acoustic (> 100 kHz) technique. The full compound contains the polymer, 10 phr processing oil, but no filler reinforcement. The curing system is a semi-efficient combination of sulfur and TBBS. Vulcanization is done at 160 °C with T_{90} time including an extra minute for each millimeter of sample thickness.

3.2 *Dynamic mechanical thermal analysis*

The dynamic mechanical thermal analysis (DMTA) was carried out for stripe-shaped specimens with a cross-section of 2 mm x 10 mm and a clamp distance of 30 mm.

In shear mode at amplitudes of 0.5 % a frequency sweep from 0.1 to 10 Hz (10 steps per decade) was performed, sequentially repeated at temperatures from −50 °C to +40 °C (steps of 5 °C).

3.3 *Ultrasonic transmission*

The measurement setup of the ultrasonic (US) transmission technique is described in detail in Schasse 2022. An ultrasonic transmitter is used to send acoustic pulses towards coin-shaped rubber specimens, using multiple excitation frequencies between 250 and 1000 kHz.

From the acoustic signals recorded at the receiver transducer opposite to the transmitter material properties such as the acoustic phase velocity c_{ph} and the attenuation coefficient α were calculated. The complex speed of sound c^* with respect to angular frequency ω is:

$$c^* = \left(\frac{1}{c_{\text{ph}}} + \frac{i\alpha}{2\omega}\right)^{-1} \tag{1}$$

By temperature variation from −30 to +60 °C, the basis for the creation of acoustic master curves is achieved.

To compare acoustic and mechanical measurements the longitudinal modulus M^* can be calculated from the complex speed of sound considering the material density ρ:

$$M^* = (c^*)^2 \cdot \rho \tag{2}$$

3.4 *Creation of master curves*

For a thermo-rheological simple material, the dependency of the dynamic moduli on temperature and frequency is coupled by a shift factor a_T, following the WLF equation at temperatures below the glass transition temperature (Williams 1955):

$$\log(a_T) = -\frac{C_1(T - T_{\text{ref}})}{C_2 + (T - T_{\text{ref}})} \tag{3}$$

With a given reference temperature T_{ref}. The associated angular frequency ω_{m} of the master curve is described as follows (Rubinstein & Colby 2003):

$$\omega_{\text{m}} = a_T(\text{T}) \cdot \omega \tag{4}$$

This was done for both the dynamic shear modulus G^* and the complex longitudinal modulus M^* acquired by ultrasonic transmission. In the latter case the master curve creation was done with respect to the reciprocal speed of sound $1/c^*$ (compare Equation 1) followed by a calculation of the modulus M^* according to Equation 2.

Reference temperature for the master curves was set to 20 °C.

4 MATERIAL MODELLING

4.1 *Linear viscoelasticity of shear modulus*

Dynamic properties of a non-porous (massive) material can be described by a Prony series for the linear viscoelasticity of the shear stress response.

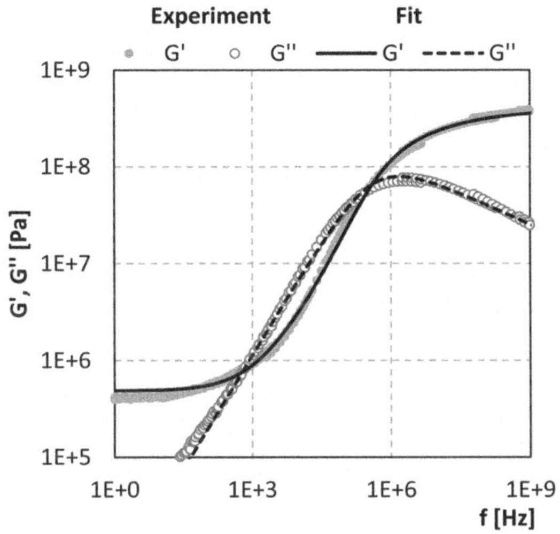

Figure 1. Frequency dependency of dynamic storage (G′) and loss (G′′) shear moduli; mastercurves from experiment and fitting according to Havriliak-Negami function; master temperature 20 °C.

Alternatively, a frequency dependence of Havriliak-Negami type can be used:

$$G^* = G_\infty + \frac{G_0 - G_\infty}{(1 + (i\omega\tau)^\alpha)^\beta} \tag{5}$$

This generalization of the Debye relaxation process reduces for $\beta = 1$ to the Cole-Cole equation, and for $\alpha = 1$ to the Cole-Davidson equation.

The shear moduli master curves G' (elastic) and G'' (viscous) of the material and a curve fitting according to Equation (5) is shown in Figure 1. The fit parameters are $G_\infty = 0.486$ MPa, $G_0 = 428$ MPa, $\tau = 0.445$ μs, $\alpha = 0.781$, $\beta = 0.278$.

4.2 *Linear viscoelasticity of compression modulus*

From ultrasonic transmission results a master curve of the longitudinal modulus together with the damping coefficient of the material.

The dependency of the longitudinal modulus on shear and on compression components can be considered in two ways.

A first assumption for the viscoelastic behavior of the longitudinal modulus considers a constant compression modulus K_{stat} independent of frequency:

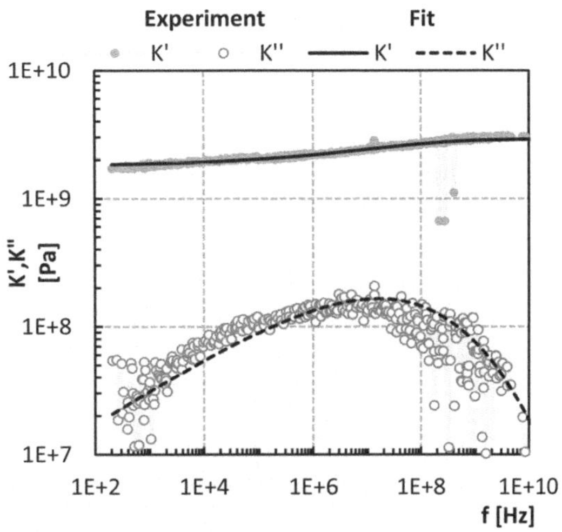

Figure 2. Frequency dependency of dynamic storage (K′) and loss (K′′) compression moduli; mastercurves from experiment and fitting according to Havriliak-Negami function; master temperature 20 °C.

$$M^*{}_{K,stat} = K_{stat} + \frac{4}{3} G^* \tag{6}$$

This reflects a "shear-only" viscoelasticity of the longitudinal modulus.

A second assumption considers a longitudinal modulus M^* with frequency dependency for both, shear modulus and compression modulus

$$M^* = K^* + \frac{4}{3} G^* \tag{7}$$

With the complex compression modulus

$$K^* = K' + i\, K'' \tag{8}$$

The master curves of the viscoelastic compression moduli K' (elastic) and K'' (viscous) of the material together with a curve fitting according to Equation (5) is shown in Figure 2. The identified parameters are $K_\infty = 1794$ MPa, $K_0 = 2890$ MPa, $\tau = 0.01044$ ns, $\alpha = 0.2717$, $\beta = 6.584$.

5 SIMULATION

5.1 *Reference geometry*

To evaluate the relevance of viscoelasticity in technical applications, an elastomer absorption layer is chosen a reference case. Such layers are used in massive form or with cavities arranged in a regular structure.

Here, a structured layer version is considered and the geometry is based the geometries used by Ye 2018.

5.2 *Simulation setup*

The full geometry of the unit cell for the structured case is shown in Figure 3.

A substrate of steel is covered by an elastomer layer of 50 mm thickness. This layer incorporates air filled cavities which are aligned to each other forming a regular grit structure.

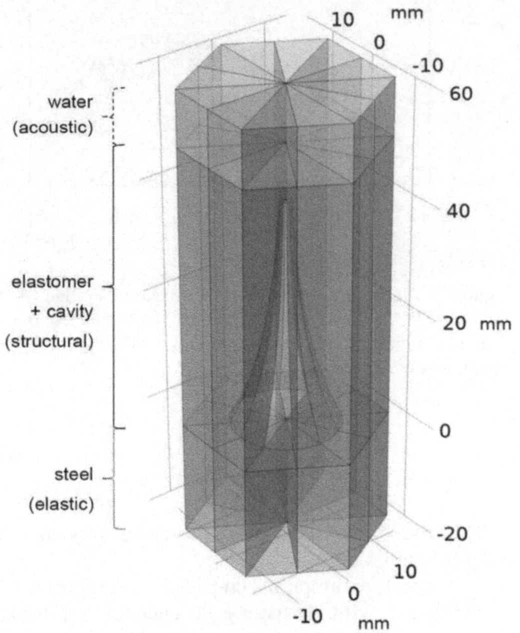

Figure 3. Geometry and material of unit cell for simulation of structured damping layer.

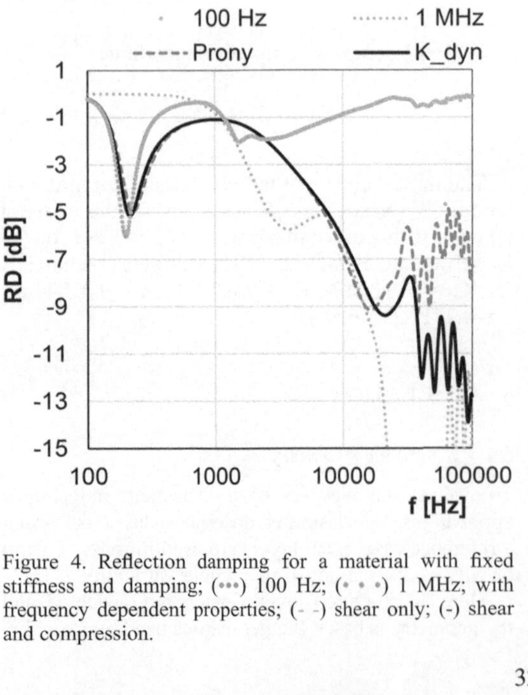

Figure 4. Reflection damping for a material with fixed stiffness and damping; (•••) 100 Hz; (• • •) 1 MHz; with frequency dependent properties; (- -) shear only; (-) shear and compression.

Adjacent to the elastomer layer is sea water.

The dynamic model considers a planar incident wave approaching perpendicularly from the water towards the elastomer.

To reflect the intended material properties of the elastomer the energy transport within the materials was done in the following way: acoustic in the water, structural dynamic in the elastomer, elastic in the steel.

The simulations were done with Comsol Multiphysics.

6 RESULTS AND DISCUSSION

6.1 *Dynamic behavior in shear and compression*

When comparing the viscoelastic behaviour of the shear and compression moduli, the shear response displays a typical glass transition with a step-like stiffness increase of three decades, whereas the compression response shows only a stiffness increase by a factor two.

This indicates that the main underlying relaxation contributing to the compression modulus are short range chain configuration changes of beta relaxation type. These are associated with much shorter relaxation times in comparison to the larger ones dominating the shear response which also originate from chain reptation.

6.2 *Simulation of an attenuation layer*

By comparison of the intensity of the incoming wave with the outgoing one the effective reflection damping (RD) can be evaluated. This is shown in dependence of frequency for four material concepts in Figure 4.

Two material concepts consider no frequency dependency of the material properties. The fixed shear modulus and attenuation coefficient were evaluated at a reference frequency (cases *100 Hz* and *1 MHz*). In both cases, the compression modulus is 1717 MPa.

The curve labeled as *Prony* reflects a viscoelastic shear modulus behavior as shown in Figure 1 but fitted according to a Prony relaxation series. Again a compression modulus of 1717 MPa is considered.

It can be seen that the first reflection damping minimum is at the same frequency 200 Hz as in the case *100 Hz* and can be identified as the first mass oscillation mode. But this minimum is shifted to 3 kHz in the case *1 MHz* since this material parameters are too stiff in this frequency range. At higher frequencies above 10 kHz, the *Prony* cases displays an RD behavior with values around -7 dB, located in-between of the fixed-modulus cases.

Finally, the material case with viscoelastic shear and viscoelastic compression modulus results in an RD characteristic labelled as *K-dyn*. This material displays similar behavior as the *Prony* case up to 6 kHz, possibly due to the dominating shear contribution to energy transport in this regime. But at higher frequencies the deviations become more apparent, presumably due to the higher compressive part of the energy transport which can be damped only by the material of the *K-dyn* case.

Obviously, the difference in the viscoelastic material consideration has a pronounced effect in the higher kilo-Hertz regime amounting to additional -6 dB for frequencies up to 100 kHz.

7 CONCLUSION

It was demonstrated, that the consideration of visco-elasticity for the attenuation of sound absorbing layers can be taken into account either within the shear modulus only or in both, shear and compression moduli.

The latter case requires experimentally evaluation of viscoelastic compression behavior which can be derived from master curves from ultrasonic transmission measurement. This is possible due to a recently introduced new method, by extending the frequency range of a temperature dependent ultrasonic transmission measurement.

The correlation of the shear viscoelasticity with the compression viscoelasticity can be understood in principle with reference to the free volume concept. This improved acoustic measurement allows to obtain deeper insights into polymer dynamics especially regarding the dynamic compressibility of elastomers.

A simulative study on the influence of a viscoelastic treatment of the compression modulus on the reflection damping of a sound absorption layer shows significantly different behavior with respect to a material concept with constant compression modulus, most pronounced at frequencies above 10 kHz.

In summary, it can be said that the structure-borne sound properties of elastomers appear to be of an inertia-viscoelastic nature. That means the energy transport can only happen predominantly in shear if the material inertia does not interfere. In acoustic terms this means that the viscoelasticity of the compression modulus comes apparent when the specific wavelength is of the order of or smaller than the geometry of the stimulated elastomer.

Further research about the viscoelastic nature of the compression behavior of rubber can focus on the expected relaxation function which correlates to the viscoelastic spectrum.

Additionally, studies on the compression visco-elasticity might aim to investigate the relevance for the dynamic response of rubber mounts.

REFERENCES

Eckert-Kastner, S., Alshuth, T., & Schuster, R. H. (2004). *Untersuchung der dynamischen Hochfrequenzeigenschaften von Elastomeren mit Hilfe eines Ultraschallspektrometers.* KGK Kautschuk Gummi Kunststoffe, 57, 423–429.

Elias, H.G. (1996). *Polymere: Von Monomeren und Makromolekülen zu Werkstoffen.* Heidelberg: Hüthig & Wepf Verlag.

Gröhlich, M., Lang, A., Böswald, M., Meier, J. (2021). *Viscoelastic damping design – thermal impact on a constrained layer.* Materials & Design, pp. 1-17, 207-109885 2021.

Kroll, J., Alshuth, T., & Schuster, R. H. (2006). *Dynamische Moduli von Elastomeren in Ul-traschall und Dynamisch-Mechanischer Analyse.* KGK Kautschuk Gummi Kunststoffe, 59, 184–189.

Maslak, A., Lang, A., Meier, J., & Alshuth, T. (2015). *High frequency investigation of elastomer properties with glass spheres using ultrasonic spectrometer.* In Constitutive Models for Rubber IX (S. 461–466). London: B. Marvalova, I. Petrikova.

Meier, J., Schasse, L. &, Schmidtke, E. (2024). *Simulationen zur Viskoelastizität in Elastomerdämpfungsschichten.* In P. Leistner (Hrsg.), Fortschritte der Akustik - DAGA 2024 (in print). Stuttgart: Deutsche Gesellschaft für Akustik e.V. (DEGA).

Rubinstein, M., & Colby, R. H. (2003). Polymer physics (Bd. 23). Oxford University Press New York.

Schasse, L., Schmidtke, E., & Meier, J. (2022). Charakterisierung der Dynamischen Eigenschaften von Elastomeren mittels Ultraschalltransmission. In P. Leistner (Hrsg.), Fortschritte der Akustik - DAGA 2022 (S. 1398–1400). Stuttgart: Deutsche Gesellschaft für Akustik e.V. (DEGA).

Schasse, L., Schmidtke, E., & Meier, J. (2022). *Charakterisierung der Dynamischen Eigenschaften von Elastomeren mittels Ultraschalltransmission.* In P. Leistner (Hrsg.), Fortschritte der Akustik - DAGA 2022 (S. 1398–1400). Stuttgart: Deutsche Gesellschaft für Akustik e.V. (DEGA).

Williams, M. L., Landel, R. F., & Ferry, J. D. (1955). The Temperature Dependence of Relaxation Mechanisms in Amorphous Polymers and Other Glass-forming Liquids. Journal of the American Chemical Society, 77, 3701–3707. doi:10.1021/ja01619a008.

Ye, C., Liu, X., Xin, F. & Lu, T.J. (2028). Influence of hole shape on sound absorption of underwater anechoic layers, Journal of Sound and Vibration, 426, 54–74.

Recyclable elastomer systems design and modelling

Constitutive Models for Rubber XIII – Dal (Ed.)
© 2025 The Author(s), ISBN 978-1-032-85138-9

Rubber wastes as enhancers of the strain induced crystallization and elastocaloric properties of natural rubber

N. Candau & M.L. Maspoch
Departament de Ciència i Enginyeria de Materials (CEM), Universitat Politècnica de Catalunya BarcelonaTech (UPC), Barcelona, Spain

ABSTRACT: Waste rubber from the pneumatic tyre (GTR) are blended with a natural rubber (NR) matrix, and the blend is vulcanized by dicumyl peroxide. The GTR, spatially dispersed in the NR matrix, are found to show a nucleating ability on strain induced crystallization in the NR matrix. This results in an improvement of the elastocaloric effect generated by the SIC (exothermic) during loading and the melting of strain induced crystals during unloading (endothermic). This improvement is attributed to a strain localization in the NR matrix due to week deformability of the GTR particles. In addition to the amount of GTR, the increase of crosslink density of the blends also results in a significant improvement of the eC effect. This is explained by the role of crosslinks as nucleating sites for SIC, mostly at the origin of the eC effect. This investigation would be of interest for the design of elastocaloric waste-based rubber composites for eco-friendly heating/cooling applications.

1 INTRODUCTION

Sustainable energy technology and materials recycling are two fundamental topics that aim to contributing to reduce the environmental impact of materials and energy production. Caloric materials possess promising properties that can contribute to achieve this aim.

Contrarily to most of the caloric materials, natural rubber (NR) is soft, cheap and biodegradable, simultaneously solving some of the major engineering and sustainability issues of artificial heating and/or cooling. It has been reported that the rubber heating induced by uniaxial deformation, so-called elastocaloric effect (eC) is ascribed to the strain induced crystallization (SIC) as shown by X-rays (Dart, Anthony, & Guth 2002). Such strain induced crystallization in rubber has been predicted latter by Flory in 1947 (Flory 1947). Strain induced crystallization can be found in natural but also synthetic elastomers such as isoprene rubber (IR) (Trabelsi, Albouy, & Rault 2004), chloroprene rubber (CR) (Le Gac, Albouy, & Petermann 2018). Nonetheless, SIC in NR is known to be more effective as compared to other synthetic elastomers (Candau et al. 2016) that is commonly accepted to be partly due to the high proportion of NR chains with cis 1,4 configuration (>98%) that facilitates their alignment upon deformation.

Nowadays, the eC effect in NR is still a subject of active investigation. Most of the experimental studies were aimed to optimize the eC effect: stretching level (Xie, Sebald, & Guyomar 2015), strain rate and environmental temperature (Xie, Sebald, & Guyomar 2017), etc. One practical advantage of eC properties in natural rubber is the possibility to modulate the strain induced crystallization kinetics with strain. At a deformation of 600%, its lowest crystallization time was found around few decades of milliseconds. As a consequence, rubber can be heated rapidly up to a decade of degrees after the completion of its crystallization process. Another remarkable advantage of natural rubber is the weak mechanical energy required to activate its eC effect. It has a lower stress level as compared to the current standard vapor-compressors (two order of magnitude lower than in shape memory alloys), which proves its technical feasibility and facilitates technology transformation. Hence, the development of sustainable energies, such as heat pump or cooling machines based on eC effect of natural rubber has received a considerable interest from the academic community.

However, some works remain to do to make rubber an efficient eco-friendly eC material: *(i)* the partial replacement of the natural rubber matrix by waste rubber particles and *(ii)* a physical description of the eC effect in these natural/wastes rubber blends to fine tune the cooling properties. While natural rubber is more abundant that rare earth metals, it is also prone to a risk of supply shortages. To overcome this issue, natural rubber can be partially replaced by wastes. It is a challenge to reintroduce waste rubber into the production cycle in the rubber industry as – in comparison to the fresh rubber – the added value is limited, especially for waste contents above 50 wt.%. Nonetheless, it has been shown that

DOI: 10.1201/9781003516880-52

the incorporation of crystallizable wastes, such as waste Chloroprene rubber (CR) (Salleh, Ahmad, and Ismail 2016), waste Latex (Mathew et al. 2001), or Ground Tire Rubber (GTR) (Candau, Oguz, et al. 2021) into NR enhances mechanical reinforcement of the resulting composites. In spite of these evidences of crystallization, and hence potential caloric properties in these above-mentioned natural/wastes rubber blends, their development as candidates for heating/cooling systems has received far less attention. As examples, a characterization of barocaloric effects (Bom et al. 2020) and eC effect (Candau, Vives, et al. 2021) in natural/wastes rubber blends had been performed.

Despite numerous investigations on the mechanical reinforcement in virgin/waste rubber composites, there is a lake of experimental observation of the occurrence of SIC and eC in these materials. To address this issue, we focus in this paper on the mechanical reinforcement and strain induced crystallization in natural rubber/ground tire rubber (NR/GTR) composites, using in-situ X-ray scattering. In addition, the eC effect has been investigated by in-situ by infrared thermography. For the physical understanding of the eC properties, we used a thermodynamic frame able to predict both the heat variations and the mechanical behavior arising from thermoelastic effects. Finally, heterogeneities of the vulcanized NR/GTR blends were investigated.

2 MATERIALS AND EXPERIMENTAL METHODS

2.1 *Materials*

The natural rubber (NR) is an SMR (Standard Malaysian Rubber) CV60 (Mooney Viscosity ML 1+4, 100 °C: 55-60), supplied by the company Akrochem (USA), with 0,15% of hydroxylamine added to the latex stage to prevent the raw rubber stiffening while storing. Ground tire rubber (GTR) was supplied by the company J. Allcock & Sons Ltd (United Kingdom) using the transformation of tire buffing into finer rubber crumbs via a controlled cryo-grinding. GTR is composed of 55 wt.% of rubber (85 wt.% of natural rubber, NR, and 15 wt.% of Styrene Butadiene Rubber, SBR) and 45 wt.% of carbon black (CB). The GTR crumbs are free of contaminants such as textile, metal and road dirt. The GTR particles were subsequently sieved using a vibratory sieve shaker (Analysette 3, Germany) with a mesh 120's (size <125 μm). Before processing, the GTR particles were dried overnight in a vacuum oven (Vaciotem-TV, J.P. SELECTA®, Spain) to prevent humidity absorption, over silica gel at 70 °C to remove any moisture. The natural rubber was masticated inside the chamber of an internal mixer (Brabender Plastic-Corder W50EHT, Brabender GmbH & Co., Germany) at a temperature of 80 °C, for 5 minutes and a rotation speed of 40 rpm. After 5 minutes of mastication, the GTR was added. After 5 more minutes the dicumyl peroxide (DCP) was added (1.5 wt.% of the NR) as vulcanizing agent and mixed for 5 minutes. The obtained masterbatch containing NR, GTR and DCP was vulcanized according to the estimate of the optimal time at 170 °C under 4 MPa.

2.2 *Tensile tests with in-situ Infra-X-Rays measurements*

A miniature tensile test machine was used to record 2D-WAXS patterns from specimens subjected to tensile strains. Tensile test-bars with a 0.5 × 4 mm2 rectangular cross-section and 5 mm gauge length were prepared. The experiments were carried out owing to a Xeuss 2.0 apparatus (Xenocs) and the two-dimensional X-Rays scattering patterns are recorded by a Pilatus 200k hybrid pixel detector (Dectris) with a specimen-detector distance of 140 mm. The experiments are performed in transmission mode with a beam size of 500 μm × 500 μm. The exposure time is equal to 30 seconds. The intensity Id due to air scattering and direct beam were measured in absence of any sample. They were removed from the total measured intensity scattered in the presence of the material Each scattering pattern was integrated azimuthally. The deconvolution of the curve I=f(2θ) enabled the extraction of the intensity at the peak top and the width at half height of each crystalline peak and the intensity at the peak top of the amorphous phase. The crystallinity index CI was then determined as follows CI=Ic/(Ia+Ic) where Ic and Ia are the intensity of the crystalline and amorphous peaks respectively.

2.3 *Tensile tests with in-situ Infra-red thermography measurements*

Incremental uniaxial tensile tests have been performed up to various strains up to failure, at the speeds of 14 $mm.s^{-1}$. The specimens have a gauge length of 15 mm, so that the nominal strain rate was estimated equal to 5.6 × 101 min^{-1}. The tension testing machine is an electromechanical Zwick 2.5 kN equipped with a load sensor of 100 N. During the tests, the sample temperature was recorded by means of an infrared camera which is a Cedip Silver (Flir). The infrared sensor is 320 × 240 pixels, and the spatial resolution is 620 μm per pixel. The camera was turned on at least 3 hours before the tests such that its internal temperature stabilizes. The acquisition frequency is 25 Hz. The (Non-Uniformity Correction of the Captor (CNUC) was used with a homemade calibration. The corresponding polynomial, unique for all pixels, is T (DL) = −7.8827.101 + 2.3105.10 −2 × DL − 1.5942.10−06 × DL2 + 4.2811.10−11 × DL3 where DL stands for digital level. Because of the large deformations imposed on the samples, a motion compensation technique was used in order to obtain the temperature at the sample center during the duration of the tests. This was achieved by computing a marker tracking algorithm on the infrared images. A specific target, with a high emissivity contrast, was fixed to the moving clamp. At the initial state of the tests, the coordinates of the edges of the clamps were manually set using the contrast between the clamps and the background. The distance between the target and the edge of the moving clamp was assumed constant. The horizontal coordinate of the center of the sample was also assumed to be constant. At each time (or image) the vertical coordinate of the center of the sample was computed to correspond to the mid-distance between the clamps.

2.4 *High resolution Infra-red thermography measurements*

Room temperature uniaxial tensile tests were performed on a universal testing machine Zwick/Roell (Z005) equipped with a 5 kN force sensor. The crosshead speed during the loading and unloading phases was chosen equal to 3000 mm/min corresponding to a nominal strain rate of 2×10^2 min^{-1} according to the specimen dimensions (15 mm length). The rubber specimens were stretched up to a deformation of 500% followed by a relaxation step in the deformed state for one minute. After the relaxation step, the specimens were unloaded with the same nominal strain rate of 2×10^2 min^{-1} down to the relaxed state (zero force) followed by another relaxation for one more minute. The temperature field on the front face of the samples was recorded using an Infra-red camera (InfraTech ImageIR® 8800) equipped with a MCT detector with a temperature resolution at 30 °C higher than 0.035 K. The distance between the IR camera and the specimen was chosen sufficiently close (60 mm) to ensure a spatial resolution sufficiently high allowing to record images with fine pixel size (15 μm). The size of the observation zone is 64 × 64 pixels2 (1 × 1 mm^2). Image data was synchronized with analogical data of the tensile test machine, namely time, force and displacement between the grips. The acquisition frequency of the IR images was chosen equal to 100 Hz in order to capture fast temperature changes that may occur within the different phases of the tests. These frequencies are high enough to ensure a sufficiently low strain increment between two images during the loading and/or unloading (maximum of 8% of deformation) or sufficiently fast acquisition time during relaxation steps (10 ms). The IRBIS 3.1 professional software (InfraTech ImageIR® 8800) was used to extract punctual temperature values along the specimen longitudinal axis. The central part of the specimen where the temperature data was extracted shows a rather homogeneous temperature field at the mm scale.

3 RESULTS AND DISCUSSION

3.1 *Strain induced crystallization*

We investigated SIC in the NR/GTR materials by in situ WAXS to obtain quantitative data on the crystal structure. The 2D scattering patterns obtained at different specimen deformation show the transformation from an initially isotropic amorphous state to an anisotropic semi-crystalline state as evidenced by the appearance of the crystalline plane reflections (002), (200), (120) and (210) typical from monoclinic crystal structure with the c axis parallel to chains axis and oriented in the tensile direction. The WAXS crystallinity index, reveals that SIC appears at a deformation around 300% for the vulcanized NR (Figure 1), and the maximum crystallinity index measured at a deformation around 570% is around 8 vol.%. Such behavior is consistent with what is observed for peroxide cured NR with a similar network chains density. Interestingly, SIC is found to appear at a lower strain in the NR/GTR composites. This demonstrates a nucleation ability of the GTR particles on SIC in NR, associated with a faster increase of crystalline phase in the NR/GTR10 and NR/GTR20 with strain. It results, in the final stages of the tensile test, in an optimum crystallinity index for these two materials.

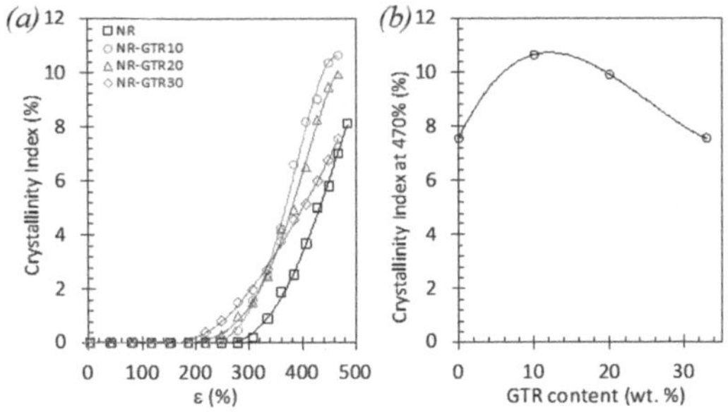

Figure 1. (a) WAXS crystallinity index versus the strain for the different NR/GTR composites and (b) the crystallinity index at the strain of 470% as a function of the GTR wt. fraction in the NR/GTR composites.

As evidenced by a more limited ability of the NR/GTR33 to crystallize at highest strain, an exceedingly high amount of GTR may tend to limit the crystallization ability of NR. While the decreasing GTR inter particle distance may initially contribute to a strain localization and by inference favors the occurrence of SIC, a further development of SIC at larger strain is slowed down, that can arise from a decreasing nucleation and/or crystal growth rates.

3.2 *Elastocaloric effect*

Fast loading is associated with a self-heating on the rubber specimen surfaces (Figure 2) that may find several origins: thermoelastic effects and/or crystallization, as has been shown in the case of vulcanized natural rubber. To confirm the respective contribution of thermoelastic effects and crystallization to heating, we estimate the temperature rise induced by the thermoelastic effect in the frame of the gaussian approximation (dotted line in Figure 2). The temperature increase estimated from gaussian approximation is found to follow the experimental data up to a strain corresponding to the strain upturn attesting for the occurrence of SIC. This suggests the description of thermoelastic effects assuming gaussian chains to nicely describe both mechanical and temperature experimental data in this strain range. Self-heating due to thermoelastic effects is found to increase while increasing GTR content that directly derives from the strain amplification effect. The sudden increase of the temperature with strain at a critical strain around 300% for NR and slightly below for the NR/GTR blends can be ascribed to a transition from dominant thermoelastic (below the critical strain) to a dominant crystallization effect (above the critical strain).

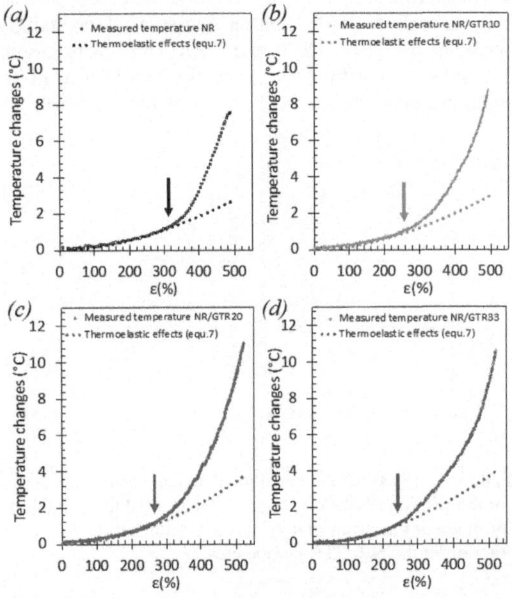

Figure 2. (a, b, c, d) temperature at the specimen surface measured on the center of the specimen for the same materials (solid lines) during uniaxial single loading performed at room temperature and at the nominal strain rate of 2 × $102\ min^{-1}$ for the series of rubber: NR, NR/GTR10, NR/GTR20 and NR/GTR33. The dotted lines represent the fitting using the theoretical temperature rise due to thermoelastic effects within the gaussian approximation and the arrows indicate the strain at which a deviation between experimental data and the thermoelastic contribution.

One may note however that other dissipative mechanisms contributing to self-heating may occur. This is the case for carbon black or silica filled rubber that can undergo viscoelastic effect such as molecular sliding at filler/rubber interface, or damage arising from debonding/cavitation in the vicinity of the fillers. Dissipative mechanisms at the interface between rubber and carbon black particles in the GTR particles are expected to be negligible in our materials due to the reasonable content of carbon black in our materials as compared to industrial filled elastomers. However, distinct dissipative mechanisms such as molecular friction or chains breakage at NR/GTR interface can occur. Nonetheless, incremental loading on our materials have been shown to exhibit low permanent strain, with a maximum of 35% in NR/GTR33, that is found three times lower than in conventional industrial filled deformed in close loading conditions. This suggests that such dissipative mechanisms are not dominant in our materials and by inference they expect to not have strong impact on the heating response.

3.3 *Heterogeneous elastocaloric effect*

To access a more local information on elastocaloric properties related to the distribution of waste rubber particles, an infra-red camera was placed closer to the rubber specimen to get higher resolution. The temperature at the specimen surface was tracked during the following test: fast loading (a-b), relaxation in the deformed state (b-c), fast unloading (d-e) and maintain in the relaxed state (e-f), as seen on Figure 3a-b. A micron scale observation of the elastocaloric effect in NR reveals a slight anisotropy with filament-like features oriented along the stretching axis. This is consistent with thermal features associated with entropic elasticity and latent heat arising from thermoelastic effect and SIC respectively and that occur due to the highly orientation of the rubber chains along the stretching axis. Contrarily, the NR/GTR blends show distinct elastocaloric features. Temperature heterogeneities are induced by the presence of waste particles, which is clearly visualized by increasing the image contrast (Figure 3c-d). It seems that the surrounding areas of the particles show slightly higher heating during loading that may arise from stress concentration in their vicinity, near the interface with NR matrix. Conversely, during cooling, the same type of localized areas, of the size of the waste particles, are found colder as compared to the matrix. This likely arises from localized strain, inducing localized thermoelastic effect and SIC during loading and hence subsequent higher cooling due to release of these to effects.

The relative temperature fluctuation (in %) during the tests had also been measured as $\sigma/|T|$ with T the average temperature specimen and σ the absolute temperature fluctuation around T. This relative temperature fluctuation is difficult to interpret during the tests, as the area under analysis is moving. Nonetheless, just before loading and right after unloading, states (a) and (e) respectively, can be compared, as the same specimen area is analyzed. State (a) serves as a reference indicating a pre-existing relative temperature fluctuation prior the tensile test, that is equal to 0.8% (corresponding to an absolute temperature fluctuation of ± 0.27 °C) and 0.6% (corresponding to an absolute temperature fluctuation of ±0.20 °C) in case of NR and NR/GTR respectively likely arising from differences in the emissivity due to irregularities on the specimen surface, inclination in the specimen in the area studied or due to the presence of wastes containing CB fillers. One may note that the surface emissivity of carbon black filled rubber is close to a black body one and is fixed at 0.94 or above, and shown to not depend on the carbon black content nor on the applied deformation. Right after the end of the unloading, relative temperature fluctuation is found equal to 0.8% (corresponding to an absolute temperature fluctuation of ± 0.27 °C), in the case of NR which is the exact same value as before loading. Nonetheless it is found equal to 1.1% (corresponding to an absolute temperature fluctuation of ± 0.37 °C) in the case of NR/GTR blend, an increment of ± 0.17 °C as compared to fluctuation before the tensile test. This is likely due to remaining temperature heterogeneities in the surrounding region of waste particles, likely due to the discharge of the local stress associated with the release of entropic elasticity and melting of SIC crystals.

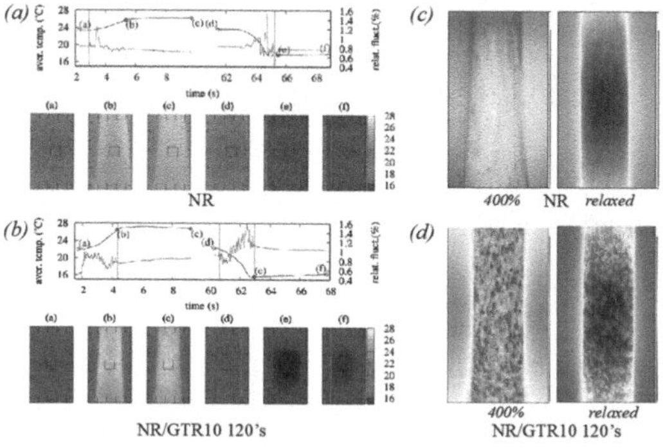

Figure 3. Average temperature and relative fluctuation of temperature at the specimen surface during fast loading (a-b), relaxation in the deformed state (b-c), fast unloading (d-e) and relaxation in the undeformed state (e-f) for NR (a) and NR/GTR10 (b). Images recorded during loading phase at around 400% of deformation (top figure) and at the end of unloading (bottom figure) with high contrast, for NR (c) and NR/GTR10 (d) respectively.

4 CONCLUSION

The mechanical reinforcement and strain induced crystallization in natural rubber/ground tire rubber (NR/GTR) composites were examined. The large strain mechanical reinforcement in NR/GTR has been found to mainly relate to (i) the reinforcing effect of the waste GTR particles owing their high carbon black content, and (ii) their nucleating effect for SIC in the NR matrix inducing additional reinforcement through the formation of oriented strain-induced crystallites. In addition, the elastocaloric effect has been investigated by in-situ by infrared thermography, revealing that highest heat exchanges were observed for natural/wastes rubber blend using 20 wt.% of wastes that were ascribed to (i) an improved thermoelastic effect and (ii) an improved ability of the natural rubber matrix to crystallize, both resulting from a strain amplification in the rubber phase in presence of undeformable carbon black aggregates in the waste particles. Finally, the heterogeneities of SIC and eC effect the vulcanized NR/GTR blends were investigated. The presence of waste rubber particles was found to modify the temperature field during both heating and cooling, from a ligament type feature to a sea island type feature showing temperature fluctuations that we state to be ascribed to strain localization in the regions near the reinforcing waste particles.

ACKNOWLEDGEMENTS

We acknowledge the Spanish Ministry of Science and Innovation (Project TED2021-129952A-C33, and the group eb-POLICOM/Polímers i Compòsits Ecològics i Biodegradables, a research group of the Generalitat de Catalunya (Grant 2021 SGR 01042).

REFERENCES

Bom Nicolau Molina, Érik Oda Usuda, Mariana da Silva Gigliotti, Denílson José Marcolino de Aguiar, William Imamura, Lucas Soares Paixão, and Alexandre Magnus Gomes Carvalho. 2020. Waste Tire Rubber-Based Refrigerants for Solid-State Cooling Devices. *Chinese J. Polym. Sci2020, Vol. 38, Pages 769-775* 38 (7): 769–75. https://doi.org/10.1007/s10118-020-2385-y.

Candau Nicolas, Laurent Chazeau, Jean-Marc Chenal, Catherine Gauthier, and Etienne Munch. 2016. A Comparison of the Abilities of Natural Rubber (NR) and Synthetic Polyisoprene Cis-1,4 Rubber (IR) to Crystallize under Strain at High Strain Rates. *Physical Chemistry Chemical Physics* 18 (5): 3472–81. https://doi.org/10.1039/C5CP06383C.

Candau Nicolas, Oguzhan Oguz, Carlos Eloy Federico, Gregory Stoclet, Jean-François Tahon, and Maria Lluïsa Maspoch. 2021. Strain Induced Crystallization in Vulcanized Natural Rubber Containing Ground Tire Rubber Particles with Reinforcement and Nucleation Abilities. *Polymer Testing* 101. 107313. https://doi.org/10.1016/j.polymertesting.2021.107313.

Candau Nicolas, Eduard Vives, Ana Inés Fernández, and Maria Lluïsa Maspoch. 2021. Elastocaloric Effect in Vulcanized Natural Rubber and Natural/Wastes Rubber Blends. *Polymer* 236: 124309. https://doi.org/10.1016/j.polymer.2021.124309.

Dart S. L., R. L. Anthony, and Eugene Guth. 2002. Rise of Temperature on Fast Stretching of Synthetics and

Natural Rubbers. American Chemical Society. https://doi.org/10.1021/ie50395a020.
Flory Paul J. 1947. Thermodynamics of Crystallization in High Polymers. I. Crystallization Induced by Stretching. *The Journal of Chemical Physics* 15 (6): 397–408. https://doi.org/10.1063/1.1746537.
Le Gac Pierre-Yves, Pierre-Antoine Albouy, and Denis Petermann. 2018. Strain-Induced Crystallization in an Unfilled Polychloroprene Rubber: Kinetics and Mechanical Cycling. *Polymer* 142 (April): 209–17. https://doi.org/10.1016/j.polymer.2018.03.034.
Mathew G, R. P Singh, N. R Nair, and S Thomas. 2001. Recycling of Natural Rubber Latex Waste and Its Interaction in Epoxidised Natural Rubber. *Polymer* 42 (5): 2137–65. https://doi.org/10.1016/S0032-3861(00)00492-4.
Salleh S. Z., M. Z. Ahmad, and H. Ismail. 2016. Properties of Natural Rubber/Recycled Chloroprene Rubber Blend: Effects of Blend Ratio and Matrix. *Procedia Chemistry*, 5th International Conference on Recent Advances in Materials, Minerals and Environment (RAMM) & 2nd International Postgraduate Conference on Materials, Mineral and Polymer (MAMIP), 19 (January): 346–50. https://doi.org/10.1016/j.proche.2016.03.022.
Trabelsi S., P.-A. Albouy, and J. Rault. 2004. Stress-Induced Crystallization Properties of Natural and Synthetic CIS-Polyisoprene. *Rubber Chemistry and Technology* 77 (2): 303–16. https://doi.org/10.5254/1.3547825.
Xie Zhongjian, Gael Sebald, and Daniel Guyomar. 2015. Elastocaloric Effect Dependence on Pre-Elongation in Natural Rubber. *Applied Physics Letters* 107 (8): 081905. https://doi.org/10.1063/1.4929395.
Xie Zhongjian, Gael Sebald, and Daniel Guyomar. 2017. Temperature Dependence of the Elastocaloric Effect in Natural Rubber. *Physics Letters A* 381 (25): 2112–16. https://doi.org/10.1016/j.physleta.2017.02.014.

Constitutive Models for Rubber XIII – Dal (Ed.)

The effect of sunflower oil on the tread of passenger tyres

F. Doğan & C. Özlüer
PETLAS Tyre Corporation, Kırşehir, Turkey

ABSTRACT: In this study, the usage of sunflower oil was investigated in comparison with petroleum based oil in passenger tyre tread compounds. Four identical recipes were prepared and quantities of the oils varied from 0 to 15 phr. Advantages and disadvantages of oil types were compared in terms of cure characteristics, mechanical and dynamic properties of rubber compounds. The study showed that cure characteristics in terms of torque difference and scorch time remained similar as the amount of sunflower oil was increased. Moreover, tensile and elongation at break tended to increase as increasing loading level of the sunflower oil. Also, tear strength was improved as sunflower oil was added. In addition, ice grip and rolling resistance properties were better than those of compound which is used with petroleum based oil.

The results reveal that the sunflower oil can be considered as good alternative to petroleum based oil as processing agent for tire tread compounds.

1 INTRODUCTION

Rubber compounds include rubber, reinforcing fillers, process oils, process aids, stabilizers and vulcanization agents. Thus, the properties of rubber compounds are mainly dependent on the amount and dispersion of those ingredients.

Fillers are used for the reinforcement of rubbers. However, their processability can be difficult especially in higher amounts. To solve that problem, oils, plasticizers and process aids are used in the recipe.

Oils can enhance the dispersion of the compounds via reducing the viscosity of fillers and provide better compatibility by integrating to the compound structure.

There are many different petroleum-based oils in terms of chemical compositions like paraffinic, naphthenic and aromatic.

As K. Sahakaro et al. mentioned in their work the usage of distillate aromatic extract (DAE) oils which are one of the petroleum-based oil is very common in tire sector (Sahakaro & Beraheng, 2011). Nevertheless, it is widely acknowledged that there are high amounts of polycyclic aromatic hydrocarbons (PAHs) in DAE oils (Boonrasri, Sae-Oui, Reungsang, & Rachtanapun, 2021). As A. Tarantini et al. remarked that numerous PAHs have been found to be carcinogenic or to have harmful effects on human organisms (Tarantini et al., 2011). It is advised to replace DAE oil with low PAH petroleum-based oils, such as naphthenic oil (NTO) and treated distillate aromatic extract (TDAE) oil. However, as J. Niemczyk pointed out that there can be still carcinogens in petroleum-based clean oils (Niemczyk, 2012).

Regarding to those risks, there have been many studies to develop bio-based oils like soybean oil, palm oil, corn oil, etc. Researchers have studied a wide range of natural oils to create more compatible and better performance characteristics of rubber compounds in tyre industry while providing sustainable and non-toxic products.

Sunflower oil is one of the most widely used bio-based oils. Since it is an ester of oleic acid and it has the double bond in its structure as shown in Figure 1, it is integrated into the structure of the compound by vulcanization and causes the Tg of the polymer to decrease. This means that it can improve tyre performance on snow and ice-covered roads and simultaneously reduce rolling resistance.

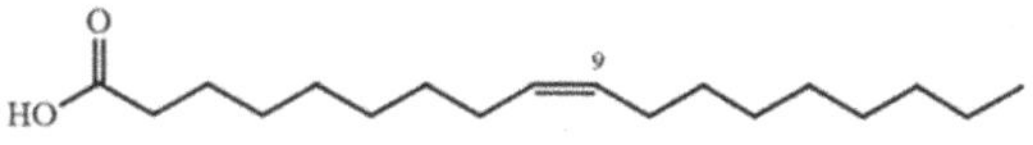

Figure 1. Structure of sunflower oil.

Moreover, due to its chemical structure, it can penetrate between polymer chains and increase the space between the chains, which allows the chains to move or rotate freely. Thus, it has the advantages of increasing the flexibility and durability of rubber compounds.

In this study, the usage of sunflower oil in replacement of petroleum based oil is compared.

2 MATERIALS AND METHODS

2.1 *Materials*

Solution Styrene Butadiene Rubber (Kumho SOL 5270H) used in this study was purchased from Korea Kumho Petrochemicals Co. Ltd. (Seul, Korea).

DOI: 10.1201/9781003516880-53

Highly cis-polybutadiene rubber with Neodymium Catalyst (BR1245-ND) was taken from Sibur Holding (Voronezh, Russia).

Bis-(triethoxysilylpropyl)disulfide (TESPD) and precipitated silica (Ultrasil 7000 GR, CTAB external surface 145-175 m^2/g, BET total surface 155-195 m^2/g) were supplied by Evonik Industries (Antwerpen, Belgium) and Egesil (Adapazarı, Turkey), respectively. Zinc oxide and stearic acid were obtained from Metal Oksit Chemistry Ind. Co. (İstanbul, Turkey) and Acid Chem Inc. (IOI Group) (Penang, Malesia), respectively. 2,2,4-Trimethyl-1,2-Dihydroquinoline (TMQ) (Vulkanox HS/LG) was purchased from Lanxess (Brunnsbuttel, Germany). N-(1,3-Dimethylbutyl)-N'-Phenyl-P-Phenylenediamine (6PPD) (Santoflex 6PPD-Pastilles) was taken from Flexysy Chemical Belgium NV (Antwerpen, Belgium). Soluble sulphur was supplied by Altınkükürt Chemical Materials Ind. Trade Co. Ltd. (Kırıkkale, Turkey). N,N-Diphenylguanidine (DPG Powder) was obtained from RDC (Milano, Italy). N-Cyclohexyl-2-Benzo Thiazylesulfenamide (CBS) was purchased from Shandong Sunshine Chemical (Quingdao, China). TDAE oil was taken from Nynas (Nynashamn, Sweden) and sunflower oil was supplied (Turkey).

2.2 *Methods*

For all compounds, the recipes were prepared based on the formulations in Table 1 using a laboratory mixer (108-367-62 Buzuluk, Komarov, Czech Republic). In Table 1, the units were given as "part-per-hundred-rubber (phr)" unit. The laboratory mixer has two-roll mill with 2.5 liters capacity.

Table 1. Compound recipes.

Raw Materials	Trial-1 phr	Trial-2 phr	Trial-3 phr	Trial-4 phr
SBR	70	70	70	70
CBR	30	30	30	30
Silica	90	90	90	90
SI75	7.2	7.2	7.2	7.2
TDAE Oil	15	10	5	0
Sunflower Oil	0	5	10	15
Stearic Acid	2.5	2.5	2.5	2.5
Zinc Oxide	1.5	1.5	1.5	1.5
TMQ	1	1	1	1
6PPD	2	2	2	2
Sulphur	1.5	1.5	1.5	1.5
DPG	1.5	1.5	1.5	1.5
CBS	2	2	2	2

The mixing procedure was carried out with a three-step mixing process as illustrates in Table 2. In the first step, rotor speed and temperature were set 50 rpm and 90°C, respectively. The fill factor was at 0.75. Rubbers (SBR and CBR) were masticated for 30 s prior to the addition of silane and most of silica. After mixing for 30-35 s, oil and stearic acid were added. After about 3 minutes of mixing, masterbatch was decanted at 140°C. In the second step, the rotor speed and temperature were set at 50 rpm and 70°C, respectively. The masterbatch was mixed with the remaining silica, zinc oxide, TMQ and 6PPD for about 3 min and then the obtained masterbatch was decanted at 140°C. In the final step, the masterbatch from the second step was mixed with sulphur, DPG and CBS for about 1 minute and then the masterbatch was decanted at 110°C.

Table 2. Mixing process.

Mixing Process
Step I
Rubbers
0.70 of silica + silane
Oil + stearic acid
Dump temperature: 140°C
Step II
Masterbatch
0.30 of silica + zinc oxide + TMQ + 6PPD
Dump temperature: 140°C
Final Step
Masterbatch
Sulphur + DPG + CBS
Dump temperature: 110°C

After that, it sheeted on a laboratory two-roll mill (14201P2 Buzuluk, Komarov, Czech Republic) and kept at 23±2°C for at least 1 hour prior to preparing test samples.

Mooney viscosity of the compounds was measured by a Mooney viscometer (Mooney MV 2000 Alpha Technology, USA) at 100°C in accordance with ASTM D1646.

Cure characteristics of the rubber compounds were determined via a moving die rheometer (MDR 2000, Alpha Technology, USA) at 170°C in line with ASTM D5289. Scorch time and cure time are represented by t_5 and t_{90}, respectively. Torque difference (max. torque - min. torque) was calculated and used to simulate the crosslink density.

Measurement of storage shear modulus (G') as a function of strain of the rubber compounds was carried out using a rubber process analyzer (Premier RPA, Alpha Technology, USA) at frequency and temperature of 10 Hz and 60°C, respectively.

The rubber compounds were vulcanized by compression molding at 170°C in a hydraulic press (1645 Saspol, Vigevano, Italy) according to their optimum cure time (t_{90}) obtained from the moving die rheometer (MDR). The vulcanized samples were stored at room temperature for at least 8 h before testing.

Tensile properties were measured using a universal testing machine (model 3366; Instron Corporation, USA) equipped with pneumatic clamp and extensometer, in accordance with ISO37. The ring-shaped specimens were cut from the rubber vulcanized sheets having thickness of approximately 4 mm. The specimens were tested using a crosshead speed of 500 mm/min.

Damping properties were measured as a function of temperature by using a dynamic mechanical analyzer; DMA (DMA-Q800 TA Instruments, USA). Tan values at 50°C and 10 °C which are directly related to rolling resistance and wet grip index of tires are reported. The test was performed in tension mode at a constant frequency of 5 Hz. The temperature was scanned from –110 to 80°C with a heating rate of 1°C/min.

Hardness was measured in accordance to ASTM D2240 by using Shore A durometer hardness tester (Type: BS 61, Bareiss, Germany).

A Laboratory Abrasion Tester 100 (LAT 100, VMI, The Netherlands) was used to estimate abrasion resistance, ice grip, wet grip and rolling resistance of the tire treads in conditions which were 30° C, -10°C, 5°C and 60°C, respectively.

Cylindrically-shaped test pieces were tested on Schob type Rebound Resilience (Model 5109, Zwick Roell) in accordance with ASTM D7121 to evaluate heat-build up parameter of compounds.

3 RESULTS AND DISCUSSION

3.1 *Effects of oils on cure characteristics*

Petroleum-based oil and sunflower oil were compared in terms of cure characteristics via MDR. As shown in Table 3, maximum torque and torque difference remained similar as the amount of sunflower oil was increased. In addition to that, scorch time (t_5) and cure time (t_{90}) were not significantly affected with the type of oil. This means that both TDAE oil and sunflower oil have similar cure characteristics.

Table 3. Cure characteristics of compounds.

Properties	Trial-1	Trial-2	Trial-3	Trial-4
ML (dNm)	6.10	5.90	4.99	4.85
MH (dNm)	27.48	27.61	27.90	27.50
MH-ML (delta)	21.38	21.71	22.91	22.65
t_5 (min)	17.08	18.00	19.11	20.00
t_{30} (min:mm)	2.12	2.12	2.34	2.56
t_{60} (min:mm)	3.05	3.02	3.24	3.36
t_{90} (min:mm)	5.71	5.67	5.43	5.75

On the other hand, mooney viscosity decreased as illustrated in Figure 2. It is an expected result because of the plasticization effect of sunflower oil.

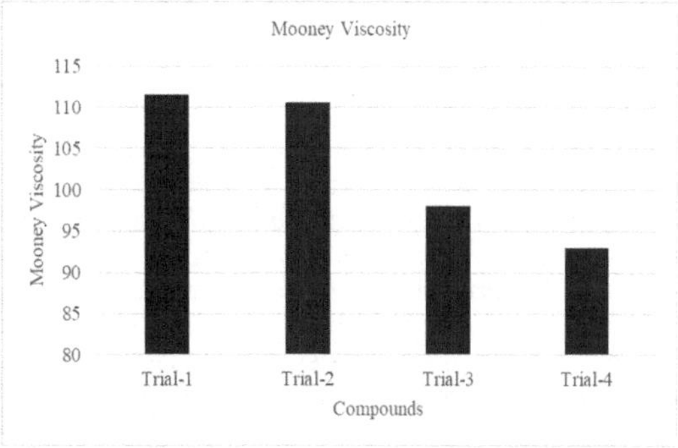

Figure 2. Comparison of mooney viscosity.

3.2 *Effects of oils on mechanical properties*

The mechanical properties of petroleum-based oil and sunflower oil were investigated via ring-shaped samples in Tensile Testing Machine. As shown in Table 4, 100% modulus and 200% modulus remained similar. Besides, since elongation at break increased with the addition of sunflower oil, Trial-3 and Trial-4 reached 300% modulus. Moreover, tensile increased in accordance with elongation at break. In addition, hardness was not significantly affected.

Table 4. Comparison of mechanical properties.

Properties	Trial-1	Trial-2	Trial-3	Trial-4
100 % Modulus (MPa)	3.00	3.00	2.90	3.00
200 % Modulus (MPa)	6.90	6.80	6.40	6.75
300 % Modulus (MPa)	-	-	11.70	12.50
Tensile (MPa)	10.70	11.10	13.80	15.80
Elongation at break (%)	271	279	339	363
Hardness (Shore A)	80	78	78	78

On the other hand, tear strength was improved as the sunflower oil was added as illustrated in Figure 3. However, there was not much significant difference in the compounds with sunflower oil as seen in Trial-2, Trial-3 and Trial-4.

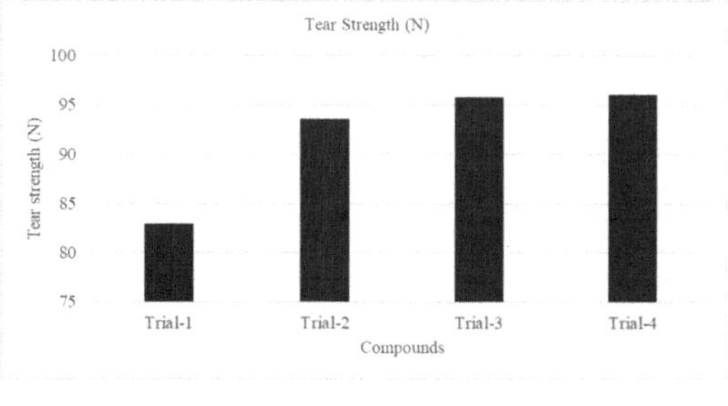

Figure 3. Comparison of tear strength.

3.3 *Effects of oils on dynamic properties*

The dynamic properties of petroleum-based oil and sunflower oil were evaluated via DMA (dynamic mechanical analyzer). Here, -20°C was used to simulate ice traction, 10°C was used for wet traction, 30°C was used for dry traction and 50°C was used for rolling resistance parameter. As shown in Figure 4, although dry grip and wet grip were better in the recipes with the dominant amount of petroleum-based oil, ice grip and rolling resistance of the compound were enhanced with the addition of sunflower oil.

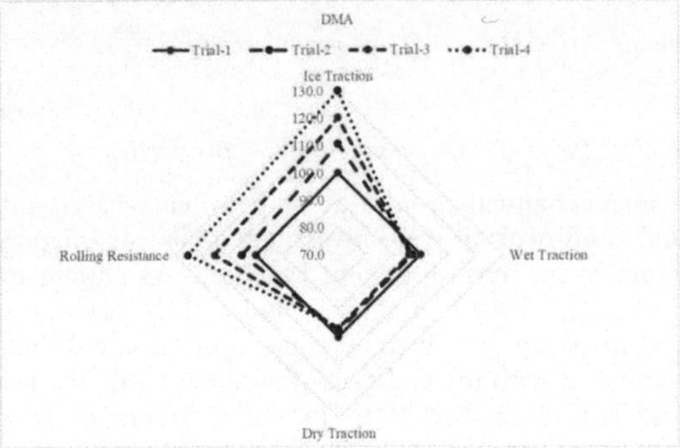

Figure 4. Comparison of DMA results.

The improvement of rolling resistance and payne effect could be seen also in RPA results as shown in Table 5.

Table 5. Comparison of RPA results.

Properties	Trial-1	Trial-2	Trial-3	Trial-4
Tand at 60°C (9 %)	0.205	0.195	0.188	0.180
G'at 60°C (9 %) (MPa)	3.98	3.94	3.88	3.80

Furthermore, heat-build up improvement could be seen in resilience test as illustrated in Figure 5.

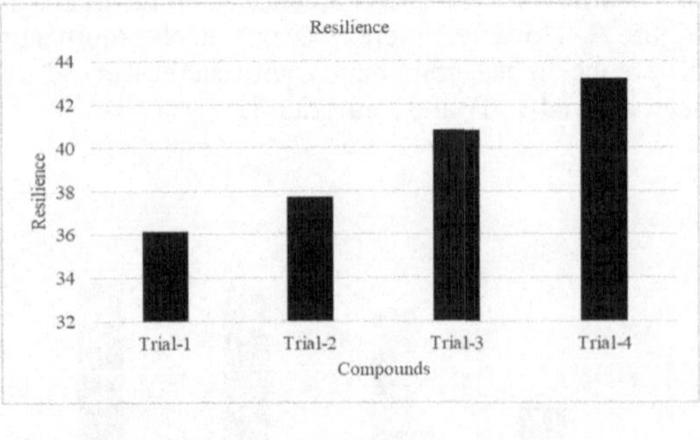

Figure 5. Comparison of resilience.

In LAT100 results, the enhancement of ice grip and rolling resistance could be seen as shown in Figure 6. On the other hand, wet grip was a little bit lower since ice grip was improved while abrasion was not significantly affected.

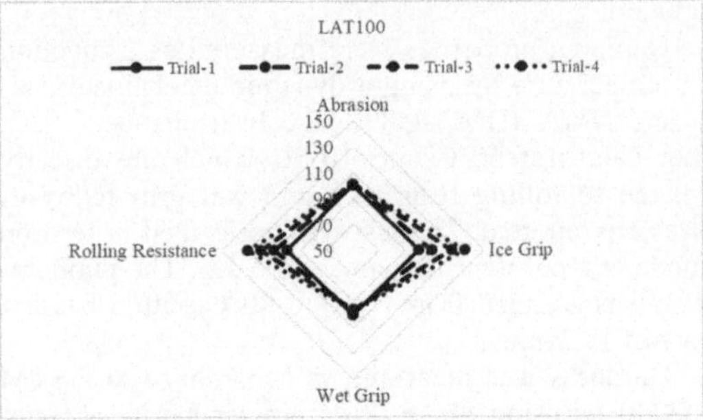

Figure 6. Comparison of LAT100 results.

4 CONCLUSIONS

Sunflower oil has been tested in passenger tyre tread compound recipes in place of TDAE oil and has been investigated in the scope of cure characteristics, mechanical and dynamic properties. At the end of the study, cure characteristics has remained almost identical. However, mechanical properties in terms of tensile and elongation at break are enhanced. Moreover, a significant improvement is noted especially in ice grip and rolling resistance values. In addition to that heat build-up based on resilience is enhanced with sunflower oil.

As a result, it is found that sunflower oil is a better alternative to petroleum-based oil in terms of both sustainability and environmental concerns and tyre performance characteristics.

REFERENCES

Boonrasri, S., Sae-Oui, P., Reungsang, A., & Rachtanapun, P. (2021). New vegetable oils with different fatty acids on natural rubber composite properties. *Polymers*, *13*(7), 1108.

Niemczyk, J. (2012). Some 'clean'-oils may contain carcinogens-Orgkhim. *European Rubber Journal*, *194*, 12.

Sahakaro, K., & Beraheng, A. (2011). Epoxidized natural oils as the alternative safe process oils in rubber compounds. *Rubber Chemistry and Technology*, *84*(2), 200–214.

Tarantini, A., Maître, A., Lefèbvre, E., Marques, M., Rajhi, A., & Douki, T. (2011). Polycyclic aromatic hydrocarbons in binary mixtures modulate the efficiency of benzo[a]pyrene to form DNA adducts in human cells. *Toxicology*, *279*(1-3), 36–44.

Constitutive Models for Rubber XIII – Dal (Ed.)
© 2025 The Author(s), ISBN 978-1-032-85138-9

Author Index